FUNDAMENTALS OF

Pharmacology for Veterinary Technicians

SECOND EDITION

Join us on the web at

Agriscience.delmar.com

FUNDAMENTALS OF

Pharmacology for Veterinary Technicians

SECOND EDITION

Janet Amundson Romich, DVM, MS

DELMAR
CENGAGE Learning™

Australia • Canada • Mexico • Singapore • Spain • United Kingdom • United States

DELMAR
CENGAGE Learning

Fundamentals of Pharmacology for Veterinary Technicians, Second Edition
Janet Amundson Romich, DVM, MS

Vice President, Career and Professional Editorial: Dave Garza

Director of Learning Solutions: Matthew Kane

Acquisitions Editor: Benjamin Penner

Managing Editor: Marah Bellegarde

Senior Product Manager: Darcy M. Scelsi

Editorial Assistant: Scott Royael

Vice President, Career and Professional Marketing: Jennifer Baker

Marketing Manager: Erin Brennan

Marketing Coordinator: Jonathan Sheehan

Production Director: Carolyn Miller

Production Manager: Andrew Crouth

Senior Content Project Manager: Elizabeth Hough

Senior Art Director: David Arsenault

Technology Project Manager: Patricia Allen

Cover Photography Credits: Cover Image (background) copyright Catalin Stefan under license from iStockphoto.com. Image (vitamin inject) copyright Dennis Guyitt under license from iStockphoto.com. Image (heartworm treatment) copyright Dennis Guyitt under license from iStockphoto.com. Image (veterinary administers medication to a bull) copyright Jim Parkin Under license from Shutterstock.com. Image (vaccinating a sheep) copyright Chris Turner. Under license from Shutterstock.com. Image (cat doctor) copyright Yellowj. Under license from Shutterstock.com. Image (dog eye) copyright Kudrashka-a. Under license from Shutterstock.com.

For product information and technology assistance, contact us at **Cengage Learning Customer & Sales Support, 1-800-354-9706**

For permission to use material from this text or product, submit all requests online at **www.cengage.com/permissions.** Further permissions questions can be e-mailed to **permissionrequest@cengage.com**

Library of Congress Control Number: 2010920623

ISBN-13: 978-1-4354-2600-9
ISBN-10: 1-4354-2600-2

Delmar
5 Maxwell Drive
Clifton Park, NY 12065-2919
USA

Cengage Learning is a leading provider of customized learning solutions with office locations around the globe, including Singapore, the United Kingdom, Australia, Mexico, Brazil, and Japan. Locate your local office at: **international.cengage.com/region**

Cengage Learning products are represented in Canada by Nelson Education, Ltd.

To learn more about Delmar, visit **www.cengage.com/delmar**

Purchase any of our products at your local college store or at our preferred online store **www.CengageBrain.com**

Notice to the Reader

Publisher does not warrant or guarantee any of the products described herein or perform any independent analysis in connection with any of the product information contained herein. Publisher does not assume, and expressly disclaims, any obligation to obtain and include information other than that provided to it by the manufacturer. The reader is expressly warned to consider and adopt all safety precautions that might be indicated by the activities described herein and to avoid all potential hazards. By following the instructions contained herein, the reader willingly assumes all risks in connection with such instructions. The publisher makes no representations or warranties of any kind, including but not limited to, the warranties of fitness for particular purpose or merchantability, nor are any such representations implied with respect to the material set forth herein, and the publisher takes no responsibility with respect to such material. The publisher shall not be liable for any special, consequential, or exemplary damages resulting, in whole or part, from the readers' use of, or reliance upon, this material.

Printed in the United States of America
8 9 10 11 12   21 20 19 18 17

CONTENTS

CHAPTER 5

VETERINARY DRUG USE, PRESCRIBING, ACQUISITION, AND PHARMACY MANAGEMENT 90

CHAPTER 6

SYSTEMS OF MEASUREMENT IN VETERINARY PHARMACOLOGY 129

CHAPTER 7

DRUGS AFFECTING THE NERVOUS SYSTEM 177

CHAPTER 12

URINARY SYSTEM DRUGS   343

CHAPTER 13

DRUGS AFFECTING
MUSCLE FUNCTION   361

CHAPTER 14

ANTIMICROBIALS   371

CHAPTER 15

ANTIPARASITICS   415

PREFACE

It is an exciting time to be in the veterinary profession! There are limitless opportunities to be part of a profession that continues to expand its field of knowledge. One of the fields that is ever expanding is the discipline of pharmacology. One reason for the increased interest in veterinary pharmacology is the expansion of drug choices available for animal use. New categories of drugs and new uses for existing drugs continue to be presented in the literature on a regular basis. Clients' willingness to treat their animals and manage clinical disease has also increased the use of pharmacological agents in the veterinary field.

Another cause of interest in veterinary pharmacology is that we have entered the age of consumer advertising in the veterinary profession. Ready or not, we are being dragged into the brave new world of pharmacology-savvy clients armed with their Internet research on the latest medications available to treat their pets. Newstand magazines contain full-color ads for drugs such as Frontline®, Cosequin®, and Rimadyl®. Television ads featuring friendly veterinarians discussing flea and tick control with not-so-compliant companion animal owners are now frequently seen by the viewing public. Veterinary pharmaceuticals are now being marketed and sold directly to the consumer. How are we, as veterinary professionals, preparing ourselves to respond to this change in the marketplace—a change that will require us to stay current in the knowledge of drugs and their applications?

The key ingredient in professional preparedness is a good understanding of the fundamentals of pharmacology. This text is designed to give veterinary professionals the solid foundation in pharmacology that the professional habit of staying current in emerging trends in pharmacology is built upon. Building a solid foundation of understanding in pharmacology requires a textbook in an easy-to-read format that provides practical applications of new information and review of concepts, calculations, and critical thinking skills. These applications during a course of study using this text will give student technicians ample opportunity to develop their confidence and proficiency in the area of pharmacology. Developing the confidence in applying pharmacological agents to specific medical uses and acquiring the language needed to explain these uses in both professional and lay terms are considered the benchmark for veterinary professionals.

This textbook and the accompanying learning materials make the process of understanding, applying, and staying current of the changes in pharmacology as straightforward as possible. This textbook presents brief but thorough explanations and reviews of basic terminology, anatomy and physiology, and disease processes needed to understand how drug therapy is utilized. Introductory case studies in each chapter presented as "Setting the Scene," practice calculations throughout the text and on StudyWARE™, and interesting facts presented as "Clinical Ques" provide the learner with many ways to examine the material presented in pharmacology. Summary sections and chapter reviews provide the learner guidance in understanding the key points of each pharmacological topic. An introduction to how drugs are developed, made, and delivered to the veterinary community provides the learner with key insights into the world of pharmacology.

ABOUT THE AUTHOR

Dr. Romich received her Bachelor of Science degree in Animal Science from the University of Wisconsin–River Falls, and her Doctor of Veterinary Medicine and Master of Science degree from the University of Wisconsin–Madison. She worked as a pharmacy technician in both human and veterinary settings while attending veterinary school. Her master's thesis was based on FDA research for a veterinary pharmaceutical company. Currently Dr. Romich teaches at Madison Area Technical College in Madison, Wisconsin, where she has taught and continues to teach a variety of science-based courses. Dr. Romich was honored with the Distinguished Teacher Award in 2004 for her use of technology in the classroom, advisory and professional activities, publication list, and fund-raising efforts. She is also a member of an IACUC member for a hospital research facility. Dr. Romich authored the textbook *An Illustrated Guide to Veterinary Medical Terminology, 3rd edition, with* StudyWARE™*. Understanding Zoonotic Diseases*, and co-authored *Delmar's Veterinary Technician Dictionary*. Dr. Romich remains active in veterinary practice through her relief practice, where she works in both small animal and mixed animal practices.

REVIEWERS OF THE SECOND EDITION

Tony Beane, DVM
SUNY Canton
Canton, New York

Laura Earle, DVM
Brevard Community College
Cocoa, Florida

Bruce Ferguson, DVM, MS
Holistic Veterinary Care
Karawara, Western Australia

Doug Kneuven, DVM
Beaver Animal Clinic
Beaver, Pennsylvania

Bonnie Loghry, RVT
Yuba College
Marysville, Louisiana

Scott F. Long, RPh, PhD
Southwestern Oklahoma State
    University
Weatherford, Oklahoma

Shawn Messonnier, DVM
Pet Care Naturally™
Plano, Texas

Paul Porterfield, DVM
Central Carolina Community
    College
Sanford, North Carolina

Elizabeth Thomovsky, DVM
University of Wisconsin-Madison
Madison, Wisconsin

Sarah Wagner, DVM, PhD
North Dakota State University
Fargo, North Dakota

First Edition

Kenneth Brooks, DVM, Diplomate
  ABVP
Lodi Veterinary Hospital
Lodi, Wisconsin

Grace A. Cooper, CVT
Wanderer Kennels
Madison, Wisconsin

Robert Garrison, DVM, MS, Diplomate
  ABMM
Wisconsin State Laboratory of
  Hygiene
Madison, Wisconsin

Kelly Gilligan, DVM
Four Paws Veterinary Hospital
Prairie du Sac, Wisconsin

Linda Kratochwill, DVM
Crow-Goebel Veterinary Clinic
Scanlon, Minnesota

James Meronek, DVM
USDA-APHIS Veterinary Services
Madison, Wisconsin

Thomas Colville, DVM
North Dakota State University
Fargo, North Dakota

Richard Flora, DVM
Newberry College
Newberry, South Carolina

James Kennedy, DVM
Colorado State University
Rocky Ford, Colorado

Paul Porterfield, DVM
Central Carolina Comunity
  College
Sanford, North Carolina

NOTICE TO THE READER

Every attempt has been made to assure that the information included in this book is correct; however, errors may occur, and it is suggested that the reader refer to original references or the approved labeling information of the product for additional information.

NEW TO THIS EDITION

Critical Thinking Questions have been added to the end of each chapter to aid the reader in using problem-solving skills and applying concepts. Expanded artwork was added throughout each chapter to help clarify difficult concepts.

Chapter 1

- Improved definitions relevant to pharmacy terminology
- Summary of the chronology of drug regulation in the United States
- Discussion of the Food Animal Residue Avoidance Databank (FARAD)

Chapter 3

- Greater emphasis on safety of drug administration and accountability
- Discussion of the six rights of medication administration

- Added discussion of additional routes of drug administration such as intramammary injections, vaginal route, and transdermal

Chapter 4

- Enhanced discussions of diffusion
- Additional discussion of patient factors affecting drug absorption
- Enhanced discussion of tolerance to drugs
- Additional discussion on measuring drug action

Chapter 5

- Added discussion of compounding
- Enhanced discussion of veterinary/client/patient relationship
- Added discussion of electronic and paperless record keeping
- Added discussion of pharmacy economics

Chapter 6

- More discussion and examples of conversions between units in measurement systems and between measurement systems
- More practice problems at the end of the chapter

Chapter 7

- Added discussions on the following: benzodiazepines, lorazepam, potassium bromide, butyrophenones, dexmedetomidine, naloxone, propofol, and sevoflurane
- Discussion of add-on treatments for refractory seizures: levetiracetam, zonisamide, gabapentin, and felbamate
- Expansion of the discussion of pain and classification of pain
- Expansion of the discussion of opioids

Chapter 8

- Expansion of the discussion of blood vessels and blood
- Added discussion of cardiovascular diseases: congestive heart failure, cardiac arrhythmias, and alterations in blood pressure
- Added discussion of benzimidazole-pyridazinones
- Added discussion of calcium channel blockers
- Added discussion of clopidogrel bisulfate
- Expanded discussion of erythropoietin
- Added discussion of cyanocobalamine and folic acid

Chapter 9

- Added discussion of respiratory conditions
- Added discussion of drugs used to treat asthma and COPD

Chapter 10

- Enhancements to content on diabetes mellitus and insulin
- Enhancements to the discussion of hyperadrenocorticism

Chapter 11

- Added discussion of gastrointestinal disorders
- Enhanced descriptions of anticholinergic drugs
- Expanded discussion on causes of vomiting
- Add discussion of neurokinin (NK_1) receptor antagonists
- Added discussion on appetite-altering drugs

Chapter 12

- Added discussion of urinary system disorders

Chapter 13

- Added discussion of muscular disorders

Chapter 14

- Added discussion of pathogenic microorganisms and disease
- Discussed FDA approvals, withdrawal times, and withholding times for drugs
- Added discussion of carbapenems, monobactams, aminocoumarins, and diterpines
- Added discussion of controlling growth in microorganisms

Chapter 15

- Added discussion related to parasites and diseases in animals
- Added discussion of depsipeptides including emodepside
- Added discussion of equine protozoal myeloencephalitis

Chapter 16

- Added discussion related to giving aspirin to pets
- Added discussion of the following drugs: selective COX-2 inhibitors, diclofenac sodium, piroxicam, and gold salts
- Enhanced discussion of pain

Chapter 18

- Included discussion of ophthalmic drugs other than the topical route
- Use of prostaglandins and alpha-adrenergic agonists to treat glaucoma
- Expanded discussion on otitis interna

Chapter 19

- Added discussion of administration of intraperitoneal fluids
- Added discussion of fluid overload
- Added discussion of adverse reactions to many of the drugs covered in the chapter
- Added discussion of administration through a special administration chamber
- Expanded content on emergency drugs

Chapter 20

- Added discussion of neoplasms
- Added discussion of pulse dosing
- Updated and expanded discussion of OSHA regulations of antineoplastic agents
- Added discussion of enzyme inhibitors

Chapter 21

- Added discussion of adjuvants
- Added discussion of polyneucleotides

Chapter 22

- Added discussion of anticonvulsants
- Added discussion of pheromones

Chapter 23

- Added discussion of hawthorn berry, ginger, and milk thistle

Appendices

- Appendix E: The Dos and Don'ts of Drug Administration
- Appendix K: Euthanasia Procedure
- Appendix J: Mathematics Review

StudyWARE™

How to Use StudyWARE™ **to Accompany** *Fundamentals of Pharmacology for Veterinary Technicians, Second Edition*

The StudyWARE™ software helps you learn terms and concepts in *Fundamentals of Pharmacology for Veterinary Technicians, Second Edition.* As you study each chapter in the text, be sure to explore the activities in the corresponding chapter in the software. Use StudyWARE™ as your own private tutor to help you learn the material in your *Fundamentals of Pharmacology for Veterinary Technicians, Second Edition,* textbook.

Getting started is easy. Install the software by inserting the CD-ROM into your computer's CD-ROM drive and following the on-screen instructions. When you open the software, enter your first and last name so the software can store your quiz results. Then choose a chapter from the menu to take a quiz or explore one of the activities.

Menus

You can access the menus from wherever you are in the program. The menus include Quizzes and other Activities.

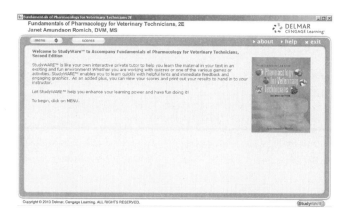

Quizzes. Quizzes include multiple choice, fill in the blank, and matching questions. You can take the quizzes in both practice mode and quiz mode. Use practice mode to improve your mastery of the material. You have multiple tries to get the answers correct. Instant feedback tells you whether you're right or

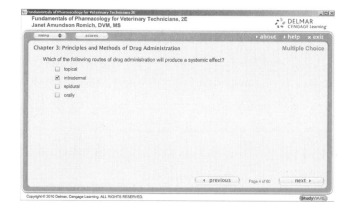

wrong and helps you learn quickly by explaining why an answer was correct or incorrect. Use quiz mode when you are ready to test yourself and keep a record of your scores. In quiz mode, you have one try to get the answers right, but you can take each quiz as many times as you want.

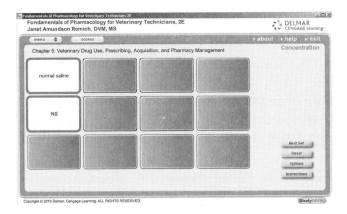

Scores. You can view your last scores for each quiz and print your results to hand in to your instructor.

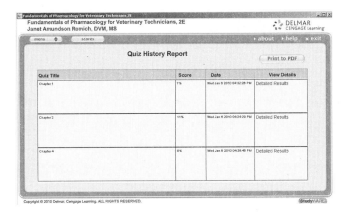

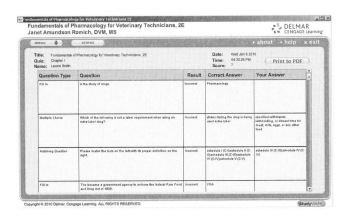

Activities. Activities include image hangman, concentration, and championship game. Have fun while increasing your knowledge!

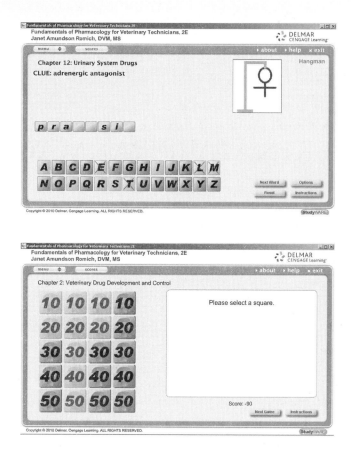

Review. Content related to review of math skills has also been provided. Additional quizzes can be printed and taken to refresh and continue to hone math skills and basic pharmacological calculations problems.

CHAPTER 1

A Brief History of Veterinary Pharmacology

OBJECTIVES

Upon completion of this chapter, the reader should be able to:

- describe the history of pharmaceuticals and their development.
- define key terms used in pharmacology.
- define the FDA's role in drug approval and drug monitoring.
- compare and contrast the use of prescription drugs, over-the-counter drugs, extra-label drugs, and controlled substances.
- describe a veterinarian/client/patient relationship.
- explain the differences between C-I, C-II, C-III, C-IV, and C-V drugs.

KEY TERMS

biologics
controlled substance
drug
Drug Enforcement Administration (DEA)
extra-label drug
Food and Drug Administration (FDA)
kinetics

over-the-counter (OTC) drugs
pharmacodynamics
pharmacokinetics
pharmacology
pharmacotherapeutics
pharmacotherapy
prescription drugs
veterinarian/client/patient relationship

Setting the Scene

The owner of a sick dog comes into a veterinary clinic seeking treatment for the animal. The owner is depending on the veterinarian to prescribe drugs that will effectively treat the dog's illness, yet be safe for the dog and manufactured properly. How can this owner be sure that the drugs prescribed for his dog are safe and effective? How can the veterinary technician convince this owner that this is true? Can an explanation of how drugs are manufactured and how their safe use in animals is monitored be given to the owner of the sick dog? What agencies regulate drugs? Are all drugs regulated in the same way?

A HISTORY OF VETERINARY PHARMACOLOGY

Veterinary medicine has existed since ancient times, as evidenced by archeological ruins in India of a hospital for horses and elephants that operated in 5000 B.C. However, *veterinary pharmacology*—the study and use of drugs in animal health care—is a much younger specialty. Its origins are traced to the early 1700s, when multiple epizootics (an epidemic within an animal population) in western Europe wiped out most of its cattle population. The first outbreak, which occurred in 1709, was of a disease called rinderpest, a highly fatal and contagious viral disease causing fever, anorexia, and ocular discharge in cattle. The second outbreak occurred in 1712, resulting from anthrax, a highly fatal bacterial disease causing rapid death in cattle and potentially causing disease in humans. The third outbreak occurred in 1755 from foot-and-mouth disease, a highly contagious viral disease causing fever and blisters in cloven-hooved animals. Europeans realized, in the wake of these catastrophes, how little anyone knew about animal diseases. In France, this concern stimulated the establishment of five veterinary colleges in the 1760s. Austria, Germany, the Netherlands, England, and Scotland followed France's lead in the late 1790s. By the 1850s, veterinary colleges were being organized in America, with a veterinary college opening in Philadelphia in 1852 and another one opening in Boston in 1854. Since then, the number of American veterinary colleges has increased to twenty-eight in the United States and five in Canada.

What did the establishment of veterinary colleges have to do with the birth of veterinary pharmacology? These veterinary colleges were founded as adjuncts to schools of medicine, the curricula of which included *materia medica*, the study of the physical and chemical characteristics of materials used as medicines. Originally, these materials were natural plant components. As scientists extracted and synthesized more sophisticated drugs from these plant components, *materia medica* gave way to a new field called pharmacology and veterinary *materia medica* became veterinary pharmacology. Even today, many drugs come from plants, bacteria, and animal sources (Table 1-1). Many anticancer drugs have been discovered in plants, most antibiotics have been discovered in soil bacteria, and many hormonal drugs are processed from animal tissue.

Table 1-1   Drug Sources

DRUG SOURCE	EXAMPLE
Minerals	sulfur, iron, electrolytes
Botanical (from plants)	digitalis, opioids
Animal	insulin, thyroid hormone, lanolin
Synthetic (manmade or engineered)	aspirin, steroids, procaine
Biological (from molds or bacteria)	antibiotics, ergot

These natural substances may now be made semi-synthetically, with substances chemically modified from a natural source. The vast majority of drugs currently in use are made by synthetic means, through chemical reactions in a laboratory. These agents are synthesized after determination of how the chemical structure of a compound relates to its pharmacological properties. Because synthetic drugs are made in the laboratory, they tend to have greater purity than those that are naturally derived.

PHARMACEUTICAL TERMINOLOGY

Clinical Que

A **drug** is a substance used to treat, prevent, or diagnose disease in animals.

Understanding pharmacological terms is fundamental to understanding pharmacology. The root *pharmaco-* is Greek for "drug" or "medicine"; therefore, the term **pharmacology** is the study of drugs (-logy is the suffix that means "study of"). *Pharmacotherapeutics*, *pharmacokinetics*, and *pharmacodynamics* are three terms that are used to explain the mechanics of veterinary pharmacology.

Pharmacotherapy is the treatment of disease with medicines. The use of medicine to treat disease has been prevalent for thousands of years. Hippocrates revolutionized medicine in ancient Greece by using medicines to heal illness. The use of antibiotics to treat a skin infection in dogs is an example of pharmacotherapy.

Pharmacotherapeutics is the field of medicine that studies drug use in the treatment of disease. Veterinary pharmacotherapeutics involves investigating how a sick animal responds to drugs.

Pharmacokinetics describes the "motion of drugs" and is used to indicate the study of the absorption, distribution, biotransformation (or metabolism), and excretion of drugs (**kinetics** is the medical term for the scientific study of motion). Understanding pharmacokinetics helps determine the amount of a drug that is in the body based on how quickly the medicine gets to the desired site and how long that medicine stays in the animal's body. Veterinary pharmacokinetics involves understanding how the motion of drugs in the body effects or brings about an animal's response to them. Clinically, pharmacokinetic considerations include the onset of drug action, drug half-life, timing of the peak effect of the drug, duration of the drug's effects, metabolism of the drug, and the site of excretion. Chapter 4 will take a closer look at the principles of pharmacokinetics.

Pharmacodynamics is the study of the mechanisms of action of a drug and involves understanding the interactions between the chemical components of living systems and the drugs that enter those systems. All living organisms function by a series of complicated, continual chemical reactions, and when a new chemical enters the system, multiple changes in cell function may occur. Veterinary pharmacodynamics involves the study of a healthy animal's response to drugs to determine their effects on the physiological and biochemical systems of the body. Understanding the interaction between a drug and its receptor is an example of pharmacodynamics. Clinically, pharmacodynamics is used to avoid drug interactions and side effects, which may result when drugs are administered to an animal.

Clinical Que

Although the FDA regulates the development and sale of drugs in the United States, local laws further regulate the distribution and administration of drugs. In most cases, the strictest law is the one that prevails in legal proceedings.

Clinical Que

It is important to note that FDA regulations do not cover certain medically significant compounds known as **biologics** (therapeutic agents derived from living organisms, such as vaccines, antibodies, and toxoids). Biologics are governed by regulations of the United States Department of Agriculture (USDA) and are brought to market under an entirely different system from that by which drugs are brought to market.

REGULATION OF DRUG PRODUCTS

The **Food and Drug Administration (FDA)**, became a government agency to enforce the federal Pure Food and Drugs Act of 1906. Before 1906, drug manufacturers had no obligation to establish the safety, purity, or effectiveness of their drugs; therefore, many questionable or even harmful products were legally sold. The Pure Food and Drug Act of 1906 established standards for drug strength and purity, and guidelines for drug labeling.

During its first three decades of operation the FDA was a small agency with limited influence and authority. The drugs available in that era were primarily from botanical (plant) and biological (molds or bacteria) sources. Public concern focused on three botanical drugs: ergot, a derivative of rye fungus used to induce labor and treat migraines; quinine, a derivative of tree bark used to treat malaria; and digitalis, a derivative of the foxglove plant used to treat cardiac failure. All three drugs perform well if correctly dosed, but all are quite toxic if overdosed. At that time, dosing and overdosing were frequently a matter of trial and error. The FDA had little power to determine and enforce correct dosage information, and in 1933, the FDA introduced a bill to revise and strengthen the 1906 Pure Food and Drug Act. A five-year legislative battle ensued, until a drug disaster killing over 100 people occurred. In 1937, the drug Elixir of sulfanilamide was distributed in a vehicle of diethylene glycol, which was never tested on people. It turns out that diethylene glycol is toxic to people. To help regulate drug dosing, Congress passed the federal Food, Drug, and Cosmetic Act of 1938, which required that a drug be adequately tested to demonstrate its safety when used as its label directs. The 1938 law greatly expanded the power and responsibilities of the FDA, and this expansion continued in the postwar years, as chemicals became widely used for drugs, cosmetics, food additives, and pesticides. In 1972, the Act was amended to include many more protections such as over-the-counter drug review to enhance the safety, effectiveness, and labeling of drugs sold without a prescription.

The Center for Veterinary Medicine

The FDA is headed by a commissioner and organized into a number of different centers, each performing a specific function. The FDA's Center for Veterinary Medicine (CVM) ensures that approved veterinary medicines will not harm animals, or at least that the harm a drug produces will be outweighed by its benefit. The FDA-CVM prohibits the sale and use of a drug that would cause animals to suffer serious health problems. For example, a dog with tachycardia (rapid heartbeat) is given a pill that returns the heartbeat to a normal rate within two hours. When the dog eats a bowl of food, the heart medication causes him to vomit, making him feel poorly again, but for a different reason. The drug that corrected the dog's tachycardia made him vomit and feel very uncomfortable. The FDA-CVM would determine whether the discomfort caused by the medication was acceptable in comparison to the benefit of lowering the dog's heart rate. In this case, if the vomiting subsides and causes minimal discomfort, the drug's ability to slow the heartbeat far outweighs the unpleasant but tolerable side effect of vomiting. The FDA would approve this drug and advise medical professionals of the side effects of the drug. The FDA thus strives to protect consumers, health professionals, and animals by maximizing the benefit of drugs while minimizing their dangers.

The FDA's power to protect veterinarians and animals is a result of the 1968 amendments to the Act concerning "New Animal Drugs." These amendments require a drug manufacturer to demonstrate that its drug is safe for animals and does what the label claims. However, the 1968 amendments also affect veterinarians in a completely different way: Drug manufacturers must now provide both a reliable analytical method to detect drug residues in food derived from animals and an acceptable drug withdrawal period following a food-producing animal's last dose. These provisions seek to assure consumers that dairy, poultry, and meat products are drug-free. Since the 1968 amendments, veterinarians must take care to instruct livestock owners about the laws governing any drugs administered to their animals.

Figure 1-1 provides a timeline for drug regulation in the United States.

Clinical Que

An idiosyncratic reaction (an abnormal response to a drug that is peculiar to an individual animal) does not cause the FDA to disapprove a drug. FDA approval of a particular drug is based on the drug's effects and side effects in many test animals.

1906	The original Pure Food and Drug Act is passed by Congress on June 30 and signed by President Theodore Roosevelt. It prohibits interstate commerce in misbranded and adulterated foods and drugs. The Meat Inspection Act is passed the same day (this Act was passed due to concern over insanitary conditions in meatpacking plants and the use of poisonous preservatives and dyes in foods).
1914	The Harrison Narcotic Act imposes upper limits on the amount of opium, opium-derived products, and cocaine allowed in products available to the public; requires prescriptions for products exceeding the allowable limit of narcotics; and mandates increased record-keeping for physicians and pharmacists that dispense narcotics.
1933	FDA recommends a complete revision of the obsolete 1906 Pure Food and Drug Act. The first bill is introduced into the Senate, launching a five-year legislative battle.
1937	Elixir of sulfanilamide, containing the poisonous solvent diethylene glycol, kills 107 persons, many of whom are children, dramatizing the need to establish drug safety before marketing and to enact the pending food and drug law.
1938	The federal Food, Drug, and Cosmetic Act of 1938 is passed by Congress, containing new provisions such as authorizing factory inspections and requiring new drugs to be shown safe before marketing. This act started a new system of drug regulation.
1941	FDA drastically revises manufacturing and quality control standards due to nearly 300 deaths and injuries as a result of distribution of sulfathiazole tablets tainted with the sedative phenobarbital. The incident prompts the beginning of good manufacturing practices (GMPs).
1951	Durham-Humphrey Amendment defines the kinds of drugs that cannot be used safely without medical supervision and restricts their sale to prescription by a licensed practitioner.
1965	Drug Abuse Control Amendments are enacted to deal with problems caused by abuse of depressants, stimulants, and hallucinogens.
1968	Animal Drug Amendments place all regulation of new animal drugs under one section of the Food, Drug, and Cosmetic Act (Section 512) making approval of animal drugs and medicated feeds more efficient.
1970	The Comprehensive Drug Abuse Prevention and Control Act replaces previous laws and categorizes drugs based on abuse and addiction potential.
1970	Environmental Protection Agency established; takes over FDA program for setting pesticide tolerances.
1972	Over-the-Counter Drug Review is initiated to enhance the safety, effectiveness and appropriate labeling of drugs sold without prescription.
1972	Regulation of biologics (including serums, vaccines, and blood products) is transferred from National Institutes of Health to FDA.

Figure 1-1 Chronology of Drug Regulation in the United States. (Adapted from Cener for Drug Evaluation and Research at http://www.fda.gov/cder/about/history/time1.html) *(Continued)*

1983	Orphan Drug Act passed, enabling FDA to promote research and marketing of drugs needed for treating rare diseases.
1984	Drug Price Competition and Patent Term Restoration Act expedites the availability of less costly generic drugs by permitting FDA to approve applications to market generic versions of brand-name drugs without repeating the research done to prove them safe and effective.
1988	Generic Animal Drug and Patent Term Restoration Act extends to veterinary products benefits given to human drugs under the 1984 Drug Price Competition and Patent Term Restoration Act. Companies can produce and sell generic versions of animal drugs approved after October 1962 without duplicating research done to prove them safe and effective. The act also authorizes extension of animal drug patents.
1994	Dietary Supplement Health and Education Act establishes specific labeling requirements, provides a regulatory framework, and authorizes FDA to mandate good manufacturing practice regulations for dietary supplements. This act defines "dietary supplements" and "dietary ingredients" and classifies them as food.
1994	Animal Medicinal Drug Use Clarification Act (AMDUCA) allows veterinarians to prescribe extra-label use of veterinary drugs for animals under specific circumstances. In addition, the legislation allows licensed veterinarians to prescribe human drugs for use in animals under certain conditions.
1996	Animal Drug Availability Act adds flexibility to animal drug approval process, providing for flexible labeling and more direct communication between drug sponsors and FDA.
1997	Food and Drug Administration Modernization Act reauthorizes the Prescription Drug User Fee Act of 1992 and mandates the most wide-ranging reforms in agency practices since 1938. Provisions include measures to accelerate review of devices, advertising unapproved uses of approved drugs and devices, health claims for foods in agreement with published data by a reputable public health source, and development of good guidance practices for agency decision-making.
2001	Minor Use and Minor Species Health Act is similar to the human Orphan Drug Act of 1983. It is intended to provide FDA-authorized drugs for those less common species and indications (provides labeled drugs for needy minor species and provides major species (cats, dogs, horses, cattle, swine, turkey, chickens) with needed therapeutics for uncommon indications called minor uses). The Minor Use and Minor Species Animal Health Act encourages the development of treatments for species that would otherwise attract little interest in the development of veterinary therapies.
2003	The Animal Drug User Fee Act (ADUFA) permits FDA to collect subsidies for the review of certain animal drug applications from sponsors, analogous to laws passed for the evaluation of other products FDA regulates, ensuring the safety and effectiveness of drugs for animals and the safety of animals used as foodstuffs.
2008	The Animal Generic Drug User Fee Act (AGDUFA) permits FDA to study ways to improve the timeliness and predictability of the animal generic drug review process. The goals are to shorten the time to review and act on submissions by increasing staff and resources.

Delmar/Cengage Learning

Figure 1-1 Chronology of Drug Regulation in the United States. (Adapted from Cener for Drug Evaluation and Research at http://www.fda.gov/cder/about/history/time1.html) *(Continued)*

CATEGORIES OF DRUG PRODUCTS

Some drugs are available without a prescription for treating a variety of conditions. These drugs are referred to as **over-the-counter (OTC) drugs** and may be purchased by the client without a prescription. Some of these drugs were approved as prescription drugs but later were found to be very safe and useful in patients without the need of a prescription. Some of these drugs were not rigorously screened and tested by the current drug evaluation protocols because they were "grandfathered" into use by their longtime application. In 1972, the Over-the-Counter Drug Review was initiated to enhance the safety, effectiveness, and labeling of OTC drugs. Many OTC drugs do not have a significant potential for toxicity when "taken as directed" nor do they require special administration; however, there are several problems related to OTC drug use. These problems include the drugs' potential to mask the signs of underlying disease, making diagnosis more difficult; to interact with prescription drug therapy; to cause serious complications if they are not taken as directed. Aspirin is a common OTC human drug used in veterinary medicine. Aspirin must be given less frequently in both cats and dogs than it is given in humans. Acetaminophen is another OTC human drug used in veterinary medicine. Acetaminophen at any dose is contraindicated for use in felines because of its profound toxicity.

Prescription drugs, which the FDA regulates, are limited to use under the supervision of a veterinarian or physician because of their potential danger, toxicity concerns, administration difficulty, or other considerations. Every drug has the potential to cause harm (if given for the wrong reason, to the wrong animal, or in the wrong amount); therefore, they must be regulated to ensure safe use. Prescription drugs for animals can be obtained only through a veterinarian via a prescription. Before a prescription drug can be prescribed for an animal, a **veterinarian/client/patient relationship** must exist. Animals need to be seen and examined by a veterinarian, who assumes responsibility for making clinical assessments based on sufficient knowledge about the health of the animals and their need for treatment and follow-up care. Prescription drugs must be labeled with the statement or legend: "Caution: Federal law restricts the use of this drug to use by or on the order of a licensed veterinarian." Figure 1-2 shows a prescription drug label.

Sometimes veterinarians utilize drugs that are not FDA approved in the treatment of animals for a particular disease or condition in a particular species. Many times this lack of FDA approval is because the drug was tested for human use and the cost to conduct experiments in animals for animal-use approval would not be financially rewarding for the drug manufacturer. Veterinarian discretion allows veterinarians to use drugs in a manner not indicated by the labeling. This is termed **extra-label drug** use and is defined as the use of a drug

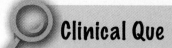

Clinical Que

The veterinarian/client/patient relationship exists when the veterinarian has assumed the responsibility for making clinical judgments, the client has agreed to follow the veterinarian's instruction, the veterinarian has sufficient knowledge of the animal to make the diagnosis, and the veterinarian is available for follow-up evaluation in the event of an adverse reaction or failure of treatment. The veterinarian can determine that currently available FDA-approved drugs are not effective in the animal to be treated, thus making the use of a non-FDA-approved drug necessary.

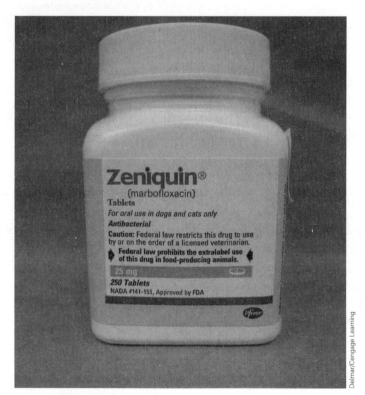

Figure 1-2 Example of a prescription drug label

Clinical Que

Drugs are deemed extra-label when they are used in a different species; for a different reason (medical condition); or at a different dosage, frequency, or route of administration; or when a different withdrawal time is used.

in a manner not specifically described on the FDA-approved label. Extra-label use of certain approved animal drugs, and approved human drugs for animals under certain conditions, is allowed under the Animal Medicinal Drug Use Clarification Act of 1994 (AMDUCA). The key constraints of AMDUCA are that any extra-label use must be by or on the order of a veterinarian within the context of a veterinarian/client/patient relationship. The requirements for records and labels when using extra-label drugs are found in Table 1-2.

This extra-label use must also not result in drug residues in food-producing animals. Drug residues can be monitored and prevented by proper identification and tracking of food-producing animals and determining an extended period for drug withdrawal before marketing milk, meat, or eggs from treated animals. The Code of Federal Regulations lists drugs that are subject to prohibitions on extra-label use. The drugs on this list cannot be used in animals.

The Food Animal Residue Avoidance Databank (FARAD) is a computer-based system designed to provide livestock producers and veterinary professionals with information on how to avoid drug, pesticide, and environmental contaminant residue problems. FARAD contains current label information including withdrawal times of all drugs approved for use in food-producing animals in the United States; official tolerances for drug and pesticides in tissues, eggs, and milk; and scientific articles with data on residues, pharmacokinetics, and the fate of chemicals in food animals. Drugs prohibited for use in livestock are also provided through FARAD. Information about FARAD may be obtained from their web site at http://www.farad.org/index.html or via telephone at 1-888-USFARAD.

Table 1-2   Labeling and Recording Requirements for Use of Extra-Label Drugs

Record Requirements When Using Extra-Label Drugs
- Identify the animals, either as individuals or a group.
- Animal species treated.
- Numbers of animals treated.
- Condition being treated.
- The established name of the drug and active ingredient.
- Dosage prescribed or used.
- Duration of treatment.
- Specified withdrawal, withholding, or discard time(s), if applicable, for meat, milk, eggs, or animal-derived food.
- Keep records for two years.
- FDA may have access to these records to estimate risk to public health.

Label Requirements When Using Extra-Label Drugs
- Name and address of the prescribing veterinarian.
- Established name of the drug.
- Any specified directions for use including the class/species or identification of the animal or herd, flock, pen, lot, or other group; the dosage frequency and route of administration; and the duration of therapy.
- Any cautionary statements.
- The specified withdrawal, withholding, or discard time for meat, milk, eggs, or any other food.

Some drugs are classified as **controlled substances**, or drugs considered dangerous because of their potential for human abuse or misuse. The FDA studies the controlled substance and determines its abuse potential, and the **Drug Enforcement Administration (DEA)** enforces the control of these drugs through the Comprehensive Drug Abuse Prevention and Control Act of 1970 (commonly known as the Controlled Substances Act). The original law regulating drugs defined *drug abuse* as the illicit use of an illegal drug or the improper use of a legal prescription drug. However, in 1970, Congress decided that the law should reflect a much more detailed understanding of psychoactive (affecting the mind or behavior) controlled substances. The Controlled Substances Act classifies a drug into one of five schedules based upon the drug's potential for harm and abuse relative to its medical benefit. The higher the drug's schedule, the lower the risk of abuse potential. For example, schedule V drugs (C-V) have less abuse potential than schedule II (C-II) drugs. The Act applies to veterinarians largely because they and their employees have access to controlled substances. Table 1-3 describes the schedules, lists their definitions, and gives examples of drugs in each category.

Table 1-3   Controlled Substances Categories and Examples

DRUG SCHEDULE	DEFINITION OF SCHEDULE	EXAMPLES
Schedule I (C-I)	Substance has high potential for abuse and has no currently accepted medical use; there is a lack of accepted safety for use (considered most dangerous, with virtually no medical benefit)	heroin, LSD, marijuana
Schedule II (C-II)	Substance has high potential for abuse but has currently accepted medical use, with severe restrictions	cocaine, morphine, amphetamines, pentobarbital, etorphine, fentanyl, codeine
Schedule III (C-III)	Substance has potential for abuse less than schedule I and II drugs and has accepted medical uses	acetaminophen/codeine combinations, ketamine, thiamylal, thiopental, hydrocodone
Schedule IV (C-IV)	Substance has low potential for abuse relative to drugs in schedule III and has accepted medical uses	diazepam, phenobarbital, butorphanol
Schedule V (C-V)	Substance has low potential for abuse relative to drugs in schedule IV and has accepted medical uses	buprenophrine, diphenoxylate, codeine cough syrups

Clinical Que

Some controlled substances may be found at multiple schedule levels. For example, codeine is labeled as C-II, C-III, and C-V. This multiple schedule labeling is due to the level of controlled substance found in each product. Codeine-only products are C-II controlled substances; combination products with aspirin or acetaminophen are C-III controlled substances; cough syrups with codeine are either C-V or C-III controlled substances depending on their state. When a controlled substance is combined with another drug it is typically classified as a lower-level controlled substance than when it is the sole active ingredient in a drug.

The prescription, distribution, storage, and use of controlled substances are closely monitored by the DEA in an attempt to decrease substance abuse of prescribed medications. Veterinarians who wish to use or prescribe controlled substances must register triennially (every three years) with the DEA, keep it informed of all address changes, and receive a registration number to be used on all prescriptions and supply order forms. Controlled substances must be stored in a locked cabinet or, preferably, in a safe attached to a concrete floor. The registered veterinarian and other authorized handlers must keep records of orders, receipts, uses, discards, and thefts of controlled substances for two years following each transaction. This record is often in the form of a controlled substance log (Figure 1-3). The veterinarian must keep records of controlled substances on hand and perform an inventory of these substances every two years. Records of controlled substance inventory are kept on file at the clinic so that they are available to the DEA if the facility is audited or has unannounced inspections by DEA agents.

Both the veterinarian and the pharmacist are responsible for the proper prescribing and dispensing of controlled substances. The manufacturer and distributor must identify a controlled substance on the label of the original container by a symbol corresponding to the schedules specified by the DEA (Figure 1-4). Though federal, state, and municipal regulations governing purity, manufacture, sale, and dispensing of drugs may differ, the most stringent regulation

Drug Name: Ketamine Strength: 100 mg/ml Bottle Number: 34

Date	Owner's Name	Animal's Name	Amount Drawn	Amount Used	Amount Discarded	Amount Remaining	Signature
01/10/10	Smith	Gilbert	1.0cc	1.0cc	0	9.0cc	*Janet Romich, DVM*

Figure 1-3 Example of a controlled substance log

takes precedence. The federal regulations of the Controlled Substances Act for recording and storage of controlled substances are often minimum requirements; therefore, state regulations may need to be followed if they are more stringent than the federal guidelines.

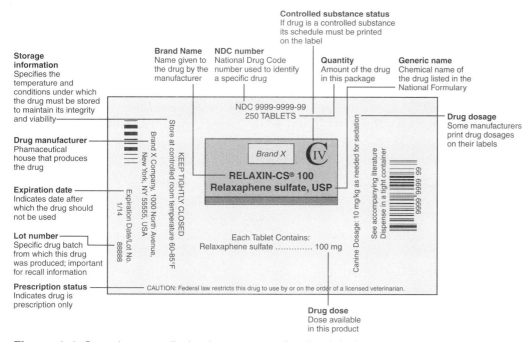

Figure 1-4 Sample controlled substance medication label. Delmar/Cengage Learning

SUMMARY

The history of treating sick animals can be traced to ancient times; however, the use of drug treatment in animals began in the late 1700s in Europe and mid-1800s in the United States. Veterinary pharmacology involves the study of pharmacotherapeutics, pharmacokinetics, and pharmacodynamics. Studying drug therapies (pharmacotherapeutics), drug movement (pharmacokinetics), and drug mechanisms of action (pharmacodynamics) in animals allows veterinarians to use drugs effectively.

Drug regulation, for both people and animals, is accomplished through the FDA. The FDA is a government agency formed to enforce the federal Pure Food and Drugs Act of 1906 and carries out varying levels of drug monitoring. Its oversight responsibilities include prescription drugs, extra-label drugs, over-the-counter drugs, and controlled substances. The FDA's CVM ensures that approved veterinary medicines will not harm animals, or at least that the harm a drug produces will be outweighed by its benefit. The FDA-CVM prohibits the sale and use of a drug that would cause animals to suffer serious health problems. Prescription drugs are limited to use under the supervision of the veterinarian or physician because of their potential dangers. Extra-label drugs are drugs used in a manner not specifically described on the FDA-approved label. OTC drugs are available without a prescription for treating a variety of conditions. Controlled substances are considered dangerous because of their potential for human abuse or misuse and are the most heavily regulated substances. Controlled substances are studied by the FDA and their use is regulated and enforced by the DEA.

It's a Wrap

The answers to the questions in this chapter's Setting the Scene should be understood after reading this chapter.

- How can this owner be sure that the drugs prescribed for his dog are safe and effective?
- How can the veterinary technician convince this owner that this is true?
- Can an explanation of how drugs are manufactured and how their safe use in animals is monitored be given to the owner of the sick dog?
- What agencies regulate drugs?
- Are all drugs regulated in the same way?

The ability to assure clients that the medication they are giving their animals is safe is a fundamental skill needed in the veterinary profession. The CVM at the FDA ensures that approved veterinary drugs will not harm animals, or at least that the harm a drug produces is outweighed by its benefit.

Understanding the FDA approval process will help the veterinary technician assure this client that the medication administered to his dog is safe. Key points in explaining the role of the FDA in the drug approval process will be covered in Chapter 2 and include the following:

- All companies that are attempting to get FDA approval for their drug must submit an Investigational New Animal Drug (INAD) application. Scientific support (experimental data, literature searches, etc.) is presented to support the INAD application.
- Clinical trials are performed on drugs after the FDA approves the INAD application.
- Once clinical data is collected, drug manufacturers can apply for a New Animal Drug Application (NADA) for consideration for marketing approval.
- New drugs are either approved or dismissed.

(continued)

- Monitoring of FDA-approved drugs continues as long as the drug is on the market so that adverse reactions can be reported and documented for further investigation.

Remember, drugs are approved by a different government agency than the one that approves biological agents. It is important to note that the FDA approves pharmaceutical compounds, the U.S. Department of Agriculture (USDA) approves biological agents (such as vaccines) for use, and the Environmental Protection Agency (EPA) approves pesticides.

CHAPTER REVIEW

Matching

Match the term or phrase with its proper definition.

1. _____ drugs that can be purchased without a prescription

2. _____ drugs considered dangerous because of their potential for human abuse or misuse

3. _____ drugs that can be obtained only through a veterinarian or via a prescription

4. _____ drugs used in a manner not specifically described on the FDA-approved label

5. _____ study of a drug's mechanism of action and its biological and physiological effects

6. _____ study of absorption, blood levels, distribution, metabolism, and excretion of drugs

7. _____ the treatment of disease with medicines

8. _____ the study and use of drugs in animal health care

9. _____ the law that allows extra-label use of a drug under certain conditions

10. _____ agency that ensures that approved veterinary medicines are relatively safe for animals

a. pharmacodynamics
b. controlled substances
c. pharmacokinetics
d. over-the-counter drugs
e. pharmacotherapy
f. prescription drugs
g. extra-label drugs
h. veterinary pharmacology
i. FDA-CVM
j. Animal Medicinal Drug Use Clarification Act of 1994

Multiple Choice

Choose the one best answer.

11. The FDA became a government agency after the passage of the
   a. federal Food and Drugs Act of 1906.
   b. Controlled Substances Act of 1970.
   c. Food, Drug, and Cosmetic Act of 1938.
   d. 1972 New Animal Drugs amendment to the Food and Drugs Act.

12. A person studying how the body absorbs, uses, and gets rid of codeine is engaged in the pharmacological specialty called
   a. pharmacotherapeutics.
   b. pharmacodynamics.
   c. pharmacokinetics.
   d. pharmaconeurology.

13. Controlled substances must
   a. be kept in a locked cabinet or safe.
   b. have orders, receipts, uses, and thefts recorded.
   c. be ordered by veterinarians who register triennially with the DEA.
   d. All of the above are true.

14. The higher the schedule number (for example, V versus I) of a controlled-substance drug,
   a. the higher the risk for human abuse potential.
   b. the more questionable its manufacture is.
   c. the lower the risk for human abuse potential.
   d. the less medical value it has.

True/False

Circle a. for true or b. for false.

15. Prescription drugs are limited to use under the supervision of a veterinarian or physician.
   a. true
   b. false

16. The majority of veterinary drugs in use during the early 1900s were found naturally in plants.
   a. true
   b. false

17. The major requirement of the Food, Drug, and Cosmetic Act of 1938 is the requirement of drug safety.
   a. true
   b. false

18. Diazepam is an example of a schedule I drug.
   a. true
   b. false

19. Over-the-counter drugs are approved for human use only by the FDA.
   a. true
   b. false

20. All drugs are utilized on sick animals.
   a. true
   b. false

Case Studies

21. An owner of a 12-year-old male/neutered (M/N) German Shepherd calls the clinic because her dog has been vomiting blood. She says the dog was fine yesterday and has been more active since she began giving aspirin to relieve the pain associated with the dog's arthritis. You explain to the owner that aspirin can cause gastrointestinal upset, and that some signs the animal may show are vomiting and diarrhea. The owner says that it is impossible for the aspirin to be causing the dog to vomit blood, because aspirin can be purchased without a prescription.

    a.   What do you tell this owner?

    b.   What advice can you give this owner?

22. A large animal veterinarian wants to administer flunixin meglumine intramuscularly (IM) to a dairy cow to control her fever. Flunixin meglumine is approved for use intravenously (IV) in beef and dairy cattle to control fever and inflammation. The veterinarian feels that it is easier to give this drug IM versus IV and that the convenience of administration route is a valid reason to use a drug extra-label. Is the veterinarian correct? Why or why not?

Critical Thinking Questions

23. Why would a veterinary technician need or want a clear understanding of the historical development and current practices of drug development and usage?

24. Why are controlled substances an issue in veterinary practice? The controlled substance rating is based on the potential for human abuse, and the veterinary community is not treating humans.

ADDITIONAL REVIEW MATERIAL

Additional review material can be found on the StudyWARE™ CD included with this text.

BIBLIOGRAPHY

American Veterinary Medical Association. *Animal Medicinal Drug Use Clarification Act*, http://www.avma.org, accessed February 2, 2008.

American Veterinary Medical Association. *Minor Use and Minor Species Animal Health Act*, http://www.avma.org, accessed February 28, 2005.

Drug Enforcement Administration. *Drug Scheduling*, http://www.dea.gov, May 30, 2009.

Food and Drug Administration. *Milestones of Drug Regulation in the United States*, http://www.fda.gov, June 18, 2009.

Food and Drug Administration. CVM Update: FDA publishes final rule on extralabel drug use in animals, http://www.fda.gov, May 30, 2006.

Food Animal Residue Avoidance Databank Homepage, www.farad.org, accessed June 3, 2008.

Morens, D. M. (June 2003). Characterizing a "New" Disease: Epizootic and Epidemic Anthrax, 1769–1780, *Am J Public Health*, 93(6): 886–893.

Reiss, B and Evans, M. E. (2007). *Pharmacological Aspects of Nursing Care* (7th ed.). Clifton Park, NY: Delmar Cengage Learning.

Wynn, R. (1999). *Veterinary Pharmacology*. Cambridge, MA: Harcourt Learning Direct.

CHAPTER 2

VETERINARY DRUG DEVELOPMENT AND CONTROL

OBJECTIVES

Upon completion of this chapter, the reader should be able to:

- outline the stages of drug development in the United States.
- differentiate between effective dose and lethal dose.
- define margin of safety, calculate it, and understand how it relates to the appearance of toxicity signs.
- describe how side effects related to reproduction, carcinogenicity, and teratogenicity are monitored.

KEY TERMS

carcinogenicity	preliminary studies
chronic studies	shelf life
clinical trials	short-term tests
effective dose-50 (ED_{50})	short-term toxicity test
lethal dose-50 (LD_{50})	special tests
long-term tests	systems-oriented screen
long-term toxicity test	teratogenicity
margin of safety	therapeutic index
parameter	toxicity evaluation
preclinical studies	

Setting the Scene

A horse is diagnosed with a type of cancer that requires long-term treatment. The owner of this horse knows that drugs used to treat cancer have many side effects and may even be lethal. The owner wants to know how such drugs can be approved and how the drug's side effects are monitored. The owner is also concerned about the cost of this treatment. As the veterinary technician, explain why drugs are expensive and how their side effects are monitored. Also explain the drug approval process to the client in clear and concise terms that she can understand.

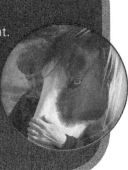

Source: Courtesy of iStockphoto

THE STAGES OF VETERINARY DRUG DEVELOPMENT

Discovering and developing safe and effective medicines is a long, difficult, and expensive process. Developing a new veterinary drug for use in the United States requires evaluation of its effects on animals through a series of tests mandated by the FDA. These mandatory evaluations require the drug company to spend a great deal of resources, both time and money, to prove that the new drug is safe and effective. It can typically take an average of seven years of testing and millions of dollars to bring a new veterinary drug to market. Table 2-1 identifies the regulatory agencies involved in animal health products. Figure 2-1 presents an overview of the animal test phases of drug development.

The four stages in drug development, as illustrated in the figure, are

1. Synthesis/discovery of a new drug compound (pre-FDA phases)

   • Preliminary studies

2. Safety/effectiveness evaluation (preclinical studies and phases I and II)

   • Preclinical studies

   • Clinical trials

3. Submission and review of the New Animal Drug Application (NADA) (phase III)

   • Review by FDA (EPA or USDA)

   • Approval or rejection of the drug based on clinical trial data

4. A postmarketing surveillance stage (phase IV)

   • Product monitoring for safety and effectiveness

   • Listing in *Green Book*, a list of animal drug products published and updated monthly by the Drug Information Laboratory. Information on the *Green Book* can be found at http://www.fda.gov; go to the CVM section and use key search term: green book

Clinical Que

The U.S. system for new drug approvals is perhaps the most rigorous in the world.

Clinical Que

The FDA Center for Veterinary Medicine (CVM) is responsible for ensuring the safety and efficacy of animal drugs and medicated feeds (www.fda.gov/cvm).

Table 2-1	Regulatory Agencies Involved in Animal Health Product Approval	
GOVERNMENT AGENCY	**AREA OF REGULATION**
FDA (Food and Drug Administration)	Development and approval of drugs
EPA (Environmental Protection Agency)	Development and approval of topical pesticides
USDA (United States Department of Agriculture)	Development and approval of biologics such as vaccines and antitoxins

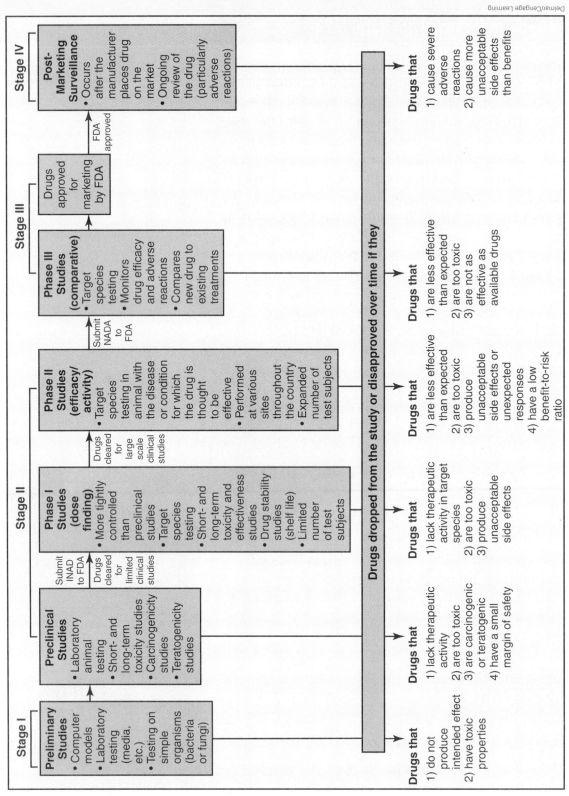

Figure 2-1 Animal test phases of drug development.

Stage One

When scientists synthesize or discover a substance with potential therapeutic value, it must be tested in a series of **preliminary studies**. Preliminary studies are performed to determine if the drug produces the intended effect(s) and whether it has toxic properties. These tests may include simulated testing via computer models, testing in laboratory media, or testing on simple organisms such as bacteria or fungi (Figure 2-2).

Stage Two

If the preliminary studies produce favorable results, **preclinical studies** begin. Preclinical studies are a series of tests performed on laboratory animals to determine safety and effectiveness of the drug. *Target species*, those species in which the drug is intended for use, may also be used in the preclinical studies (Figure 2-3). Safety and effectiveness tests include short-term and long-term toxicity studies and special tests of immediate drug reactions, organ system damage, reproductive effects, **carcinogenicity** (the ability or tendency to produce cancer), and **teratogenicity** (the capacity to cause birth defects).

When sufficient animal data demonstrate the new drug's relative safety and effectiveness, researchers submit an Investigational New Animal Drug (INAD) application to the FDA. If the product is a pesticide, the scientists must file for an Experimental Use Permit (EUP) with the EPA; if the product is a biologic, the scientists must file with the Animal and Plant Health Inspection Services (APHIS) of the USDA. The FDA reviews the INAD application and responds within 30 days.

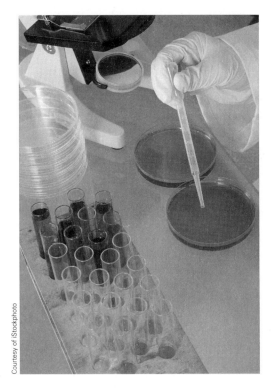

Courtesy of iStockphoto

Figure 2-2 Preliminary studies, such as those using bacteria, are conducted in a laboratory to determine whether or not the drug produces the desired effect.

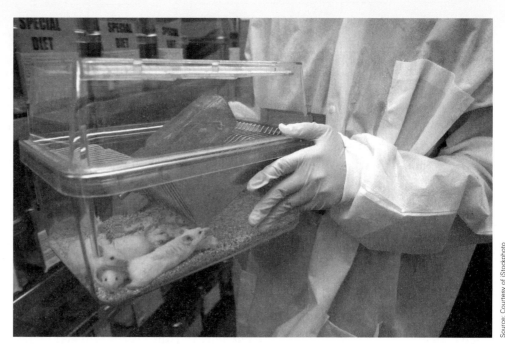

Source: Courtesy of iStockphoto

Figure 2-3 Preclinical studies are performed in laboratory animals to determine the safety and effectiveness of a drug.

Clinical Que

Preclinical studies involve laboratory testing to show the biological activity of the compound against the targeted disease and evaluation of the compound for safety.

Clinical trials are conducted in different phases, each assessing a different aspect of the drug. Phase I examines toxicity effects and may include absorption, distribution, metabolism, and excretion of the drug and its duration of action. Phase I also determines the appropriate dose of the drug. Phase II monitors the drug's effectiveness and helps determine the therapeutic index. Phase III monitors efficacy and adverse reactions and is used to compare the new drug to existing treatments.

If the FDA approves the application, researchers may proceed with **clinical trials** on the target species. Clinical trials are conducted in the target species and are done to evaluate the drug's safety and effectiveness in that species. Any toxic or adverse side effects in the target species are determined in the clinical trials as are tissue residue and withdrawal time information. Studies on **shelf life** (i.e., stability studies), to show how long a drug remains stable and effective for use, are also conducted.

Stage Three

Satisfactory clinical trial results allow the scientists to file a New Animal Drug Application (NADA) with the FDA (or other agencies for pesticides and biologics). All of the research studies are then submitted to the FDA, USDA, or EPA for review. If the results of testing are favorable, an approval and license for manufacture are granted. A new animal drug is deemed unsafe unless there is an approved NADA for that particular drug.

Stage Four

Following approval of the drug, the company and the government monitor the product as long as the drug is manufactured. This monitoring ensures product safety and efficacy. Adverse drug reactions should be reported to drug manufacturers (most have technical service veterinarians who accept reports of drug reactions and discuss problems or concerns of their drugs), the FDA (1-888-FDA-VETS), and the United States Pharmacopeia's Veterinary Practitioner's Reporting Program (1-800-487-7776).

SAFETY AND EFFECTIVENESS EVALUATION

All drugs that are approved by the FDA must be tested to determine if they are effective and whether they cause side effects. If the drug is intended for use in food-producing animals, it must also be tested for safety to human consumers, and the edible animal products must be free of drug residues. The company developing the drug must also devise analytical methods to detect and measure drug residues in edible animal products. Evaluation of a drug's safety and effectiveness begins in the early stages of drug testing and continues past the drug's approval. **Short-term tests** occur in the hours following a test dose, to check the animal for such obvious adverse reactions as seizures, paralysis, depressed breathing, depressed heart rate, and death. **Long-term tests** typically run for 3 to 24 months of repeated dosing, to check the animal's various organ systems for toxicity damage. **Special tests**, such as tests done to determine reproductive effects, carcinogenicity, and teratogenicity, are both short and long term— hence the "special test" designation. Reproductive tests evaluate the drug's effect on conception, fertilization, and pregnancy. Carcinogenicity tests check to see if the drug causes cancerous tumors in such soft tissues as the urinary bladder or brain. Teratogenic tests check for development of fetal defects in pregnant test animals.

Toxicity Evaluation

Drugs must also be tested to determine if they cause toxic side effects. Typically conducted on mice, the **toxicity evaluation** is done to determine the dose at which a drug induces organ or tissue damage that may, at a high-enough dose, result in permanent injury or death. Figure 2-4 shows a typical data sheet researchers would use to collect this information.

The top areas of the sheet are for the compound (drug) number, the animal's number, the animal's weight, the dose of the drug given to the animal, the date of the report, and the animal's vital signs (blood pressure, heart rate, and respiration rate) before receiving the drug. The left-hand column lists the **parameters** (the intensity of effect measured on a subjective scale of either 1 through 3 or 1 through 10) observed and the times of observation. The bottom is reserved for investigator notes and the signatures of the investigator and a witness. The data sheet records information vital to determining everything researchers want to know about a drug, including the key questions: "Does it do anything at all?" and "How much does it take to do it?"

A **short-term toxicity test** that produces adverse reactions may prompt the manufacturer to terminate the drug testing. Further tests sometimes exonerate the drug as the cause of short-term adverse reactions, but usually the study of the drug is terminated because so many other drug candidates await testing. In any case, the next tests do not happen at all unless the bulk of the toxicity tests confirm a degree of safety and effectiveness in the test drug. These tests are too expensive and time-consuming to perform on any drug that does not show significant promise.

```
┌─────────────────────────────────────────────────────────┐
│                      Data sheet                           │
│               Short-term toxicity evaluation              │
│               General profile screen in animal            │
├───────────────────────────────────────────────────────────┤
│                                                           │
│   Compound # _____     Dose _____     Date _____ │
│       Animal # _____                                   │
│       Animal wt. _____         Normal HR _____      │
│                                   Normal BP _____      │
│                                   Normal RR _____      │
│                                                           │
│          Parameter        Time in minutes after drug      │
│                        ┌──┬──┬──┬──┬──┬──┬──┬───┬───┐      │
│                        │1 │5 │10│15│30│45│60│120│180│      │
│      Respiration rate  │  │  │  │  │  │  │  │   │   │      │
│      Heart rate        │  │  │  │  │  │  │  │   │   │      │
│      Blood pressure    │  │  │  │  │  │  │  │   │   │      │
│      Temperature       │  │  │  │  │  │  │  │   │   │      │
│      Sedation/sleep    │  │  │  │  │  │  │  │   │   │      │
│      Seizures          │  │  │  │  │  │  │  │   │   │      │
│                        ├──┼──┼──┼──┼──┼──┼──┼───┼───┤      │
│      Other (list)      │  │  │  │  │  │  │  │   │   │      │
│                        └──┴──┴──┴──┴──┴──┴──┴───┴───┘      │
│   Notes/comments                                          │
│                                                           │
│                              _____             │
│                                Investigator               │
│                              _____             │
│                                  Witness                  │
│                              _____             │
│                                   Date                    │
└───────────────────────────────────────────────────────────┘
```

Source: Delmar/Cengage Learning

Figure 2-4 Data sheet used to collect toxicity evaluation information.

Effective and Lethal Dose Evaluation

Researchers must also determine the amount of a drug, or *dose*, that produces a desired effect. This is called the *effective dose.* A dose can be called effective only if the amount of the test drug causes a defined effect in 50 percent of the animals that receive it. For example, if 50 out of 100 mice given a drug to treat tachycardia (an abnormally rapid heart rate) display a slower heart rate at a certain dose of the drug, that dose is the amount of drug that causes the desired effect. This term is abbreviated ED_{50} for **effective dose-50**.

In addition to finding a drug's effective dose, researchers must also determine the lethal dose. The *lethal dose* is the dose of a test drug that kills 50 percent of the animals that receive it. This figure is important for two reasons: (1) Drugs that kill with a small dose are rarely evaluated further, and (2) every drug marketed must include lethal dose information in the accompanying official drug monograph (a written account of a single drug). The 50 percent figure means, for example, that if 26 mice receive the same dose of a test drug and 13 of them die, then the dose that all 26 received is the test drug's lethal dose. The lethal dose is abbreviated LD_{50} for **lethal dose-50**. Table 2-2 lists some typical LD_{50} values.

Table 2-2   Examples of LD$_{50}$ Values in Some Drugs

DRUG	LD$_{50}$ IN MICE
Valium® (tranquilizer)	720 mg/kg given orally
Cardizem® (blood pressure)	500 mg/kg given orally
Vasotec® (blood pressure)	2000 mg/kg given orally

The Therapeutic Index

Another important part of the short-term toxicity phase is determining the test drug's **margin of safety**, also referred to as the **therapeutic index**. The therapeutic index is the drug dosage or dose that produces the desired effect with minimal or no signs of toxicity. This value is determined by comparing the drug's LD$_{50}$ and its ED$_{50}$. A wide or greater therapeutic index or margin of safety means that the drug can produce its desired effect without approaching toxicity. The wider this therapeutic index or margin of safety is, the better. A narrow or lesser therapeutic index or margin of safety means that the drug levels in plasma need to be monitored more closely because of the small safety range between the effective dose and the lethal dose. The lethal dose is used as the measure of toxicity because death is an easily observed side effect, whereas other toxic effects can be ambiguous.

If the LD$_{50}$ and ED$_{50}$ values are known for a given drug, its therapeutic index is determined by dividing the LD$_{50}$ value by the ED$_{50}$ value. For example, if Drug A has an LD$_{50}$ of 100 milligrams per kilogram and an ED$_{50}$ of 2 milligrams per kilogram, the therapeutic index is

Drug A LD$_{50}$/ED$_{50}$ = 100 mg per kg/2 mg per kg = 50 = TI

If Drug B has an LD$_{50}$ value of 10 milligrams per kilogram and an ED$_{50}$ value of 5 milligrams per kilogram, the therapeutic index is

Drug B LD$_{50}$/ED$_{50}$ = 10 mg per kg/5 mg per kg = 2

Clinical Que

The greater the therapeutic index, the safer the drug. A greater therapeutic index is represented by a larger number.

Drug A's therapeutic index, 50, means that an animal would have to be given 50 times the effective dose to ingest the lethal dose. Drug B's therapeutic index, 2, means that an animal would have to be given only twice the effective dose to ingest the lethal dose. Therefore, Drug A has a substantially larger therapeutic index or margin of safety.

Systems-Oriented Screen

After the toxicity evaluation, researchers conduct a **systems-oriented screen**, a test of the drug's effect on a particular physiological system. This test evaluates effects of interest from the toxicity evaluation as they affect a specific body system. For example, a systems-oriented screen of the cardiovascular system might ask the following questions: How far did blood pressure drop, and with what dose? How much was the heart rate reduced? Did any dose produce no effect on blood pressure or heart rate? Body systems typically tested include the cardiovascular, respiratory, muscular, nervous, and endocrine systems.

Evaluation of Long-Term (Chronic) Effects

As systems-oriented screens begin, researchers may also start tests to find out if the drug has any long-term side effects. Over the years, drugs have found their way to the marketplace before anyone knew of their dangerous long-term effects. Researchers conduct **long-term toxicity tests**, also called **chronic studies**, that run anywhere from three months to two years, to prevent disastrous consequences of a drug therapy. Throughout long-term toxicity studies, test animals are given regular doses of the test drug. At the end of the study, researchers observe the animals for toxic effects that did not appear in the short-term studies. Researchers check each animal's behavior and general appearance, as well as collect blood and urine samples. They then euthanize the animal and perform histological analysis of various tissues (examine tissues microscopically). The liver, heart, and nervous tissue are typically examined for toxic effects. Many of these toxic effects appear only after repeated dosing over a longer period of time.

Evaluating Reproductive Effects, Carcinogenicity, and Teratogenicity

Just as a drug that lowers the heart rate of a mouse may lower a dog's heart rate, a drug that impairs the fertility of a laboratory rat may do the same to other animal species. Because drugs linked to cancer and birth defects have found their way into human and veterinary medicine in the past, researchers and the FDA want to do everything possible to make sure that drugs with the potential to cause serious disease and side effects are not marketed and used on animals in a clinical setting.

Reproductive tests are carried out in test animals to answer questions such as the following: Does the test drug prevent ovulation? Does the test drug prevent fertilization of the egg by sperm? Does the test drug prevent uterine growth of the fertilized egg? Does the drug cause early expulsion of the embryo from the uterus? If the answer to any of these questions is yes, testing on the drug usually ends.

FDA-approved drugs must be free of any cancer-causing potential. Researchers give large daily doses of a test drug to thousands of test animals for six months to see if cancerous tumors appear. After six months, each animal is inspected for tumors. If any tumors or abnormal growths are found, testing on the drug usually ends.

The FDA requires that all drugs be tested to see if they cause fetal defects in animals. Researchers give the test drug to hundreds of pregnant laboratory animals. Researchers then inspect the embryos for such defects as abnormal bone formation, shortened limbs, cleft palate, open skulls, and open spinal columns. Because these defects could be produced in all pregnant animals taking the drug, the FDA will typically not approve it for general use.

SUMMARY

The development of new veterinary drugs in the United States takes time, money, and extensive testing to ensure drug effectiveness and low-level or zero toxicity. Agencies involved in drug testing include the FDA (for drug approval), the EPA (for pesticide approval), and the USDA (for biologics approval). The stages of drug approval include synthesis/discovery of a new drug compound (preliminary studies), safety/effectiveness evaluation (preclinical studies and clinical trials), submission and review of the new animal drug application (FDA [or other agency] review and product approval or rejection), and post-marketing surveillance (product monitoring after the drug is marketed).

Special tests are performed on drugs to determine if there are undesirable properties of the test drug. An effective dose and lethal dose are determined in the process of developing drugs. The ED_{50} is the amount of a drug that produces a desired effect in 50 percent of the population. The LD_{50} is the dose of drug that kills 50 percent of the animals that take it. Both the effective dose and the lethal dose are used to calculate the drug's therapeutic index or margin of safety, which is simply a comparison between the lethal and effective doses. A greater therapeutic index or margin of safety means that the drug can produce the desired effect without approaching toxicity. A narrower therapeutic index means that there is a small safety range between the effective dose and the lethal dose of the drug. Systems-oriented screens are used to test a drug's effect on a particular physiological system. Long-term studies, including examination of reproductive effects, carcinogenicity, and teratogenicity, are also conducted and are important in monitoring the safety of individual drugs.

Clinical Que

Once the FDA approves a drug, the company must continue to submit periodic reports assessing adverse reactions, as well as the appropriate manufacturing quality-control records. Adverse reactions should be reported to the manufacturer, the FDA, and the U.S. Pharmacopeia.

Clinical Que

Drugs intended for administration to animals fall under the same current good manufacturing practices (cGMPs) requirements as do drugs for humans.

It's a Wrap

The answers to the questions in this chapter's Setting the Scene should be understood after reading this chapter. Why are some drugs expensive? Based on the information provided in Chapters 1 and 2, the veterinary technician should be able to explain the lengthy and expensive process of drug approval by the FDA. This process includes preliminary studies, preclinical studies, clinical trials, and postmarketing surveillance. This process is not successful for every drug, and many companies take this into account when determining the price for their medications that receive FDA approval.

How are drug side effects monitored? Many drugs have side effects; the decision to approve or reject a drug based on these side effects depends on the level of adversity that occurs. If the benefit of using that particular drug outweighs the adverse reaction, it may be acceptable. However, certain adverse reactions are so severe that they are contraindicated in certain animals or not approved at all. The classic example is the approval of antineoplastic agents (or anticancer drugs). The adverse effects of vomiting and diarrhea are acceptable in patients if the drug gives them the chance of survival. The same side effects of vomiting and diarrhea may not be acceptable in a drug used to treat gastrointestinal disease.

CHAPTER REVIEW

Matching

Match the term or abbreviation with its proper definition.

1. _____ NADA
2. _____ FDA
3. _____ EPA
4. _____ USDA
5. _____ INAD
6. _____ clinical trials
7. _____ preclinical studies
8. _____ therapeutic index
9. _____ systems-oriented screen
10. _____ effective dose

a. studies conducted in the target species that are done to prove that the drug is safe and effective in that species
b. new animal drug application
c. series of tests performed on laboratory animals to determine safety and effectiveness of the drug
d. government agency that develops and approves drugs
e. government agency that develops and approves biologics such as vaccines and antitoxins
f. government agency that develops and approves topical pesticides
g. test of a drug's effect on a particular physiological system
h. dose that produces desired effect in 50 percent of animals that receive it
i. drug dosage or dose that produces the desired effect with minimal or no signs of toxicity
j. investigational new animal drug

Multiple Choice

Choose the one best answer.

11. Which therapeutic index is the safest of those listed below?
    a. 2
    b. 10
    c. 20
    d. 30

12. The margin of safety is often referred to as
    a. the effective dose.
    b. the lethal dose.
    c. the safety parameter.
    d. the therapeutic index.

13. The LD_{50}/ED_{50} is the mathematical expression of what value?
    a. the lethal dose
    b. the effective dose
    c. the margin of safety
    d. the mortality dose

14. A drug that has a margin of safety of 75 is
    a. safer than a drug whose margin of safety is 5.
    b. less safe than a drug whose margin of safety is 5.
    c. more likely to cause toxic side effects.
    d. not marketable in the United States.

15. How long a drug remains stable and effective for use is known as its
    a. half-life.
    b. shelf life.
    c. effective life.
    d. special test life.

16. The term used to describe the capacity to cause birth defects is
    a. reproductivity.
    b. carcinogenicity.
    c. teratogenicity.
    d. theriogenicity.

True/False

Circle a. for true or b. for false.

17. The FDA is responsible for approval of all chemicals dispensed by veterinarians.
    a. true
    b. false

18. Once the FDA approves a drug, it is no longer monitored for safety and effectiveness because it has already undergone extensive testing prior to approval.
    a. true
    b. false

19. Satisfactory clinical trial results allow scientists to file a NADA with the FDA.
    a. true
    b. false

20. A drug with a narrow margin of safety means less of the drug is needed to produce the lethal dose in comparison to a drug with a wide margin of safety.
    a. true
    b. false

Case Study

21. A client calls your office to ask a question regarding his animal's medication. He is currently giving his dog one antibiotic tablet twice a day to treat a skin infection. The client is planning to go on vacation and is wondering if his dog sitter could give four antibiotic tablets once every two days instead of one antibiotic tablet four times over two days. Because the total dose over the two days would be the same, the client thinks that this would save the dog sitter some time and trouble.

Is it recommended for this dog to receive its entire two-day dose at one time? Why or why not? Relate your answer to testing or test results performed by drug companies to get FDA approval.

Critical Thinking Questions

22. Is there a correlation between onset of drug action and duration of drug action?

23. What is the significance of a drug's therapeutic index to a veterinary technician?

ADDITIONAL REVIEW MATERIAL

Additional review material can be found on the Study StudyWARE™ CD included with this text.

BIBLIOGRAPHY

Drug Approval Process, http://www.fda.gov/animalveterinary/developmentapprovalprocess/newanimaldrugapplications/ucm146121.htm, May 21, 2009.

Beary, J. (Fall 1997). The drug development and approval process. Newsletter of Aplastic Anemia Foundation of America, Inc.

Vail, D. (November 2007). Cancer clinical trials: Development and implementation. In: *Veterinary Clinics of North America Small Animal Medicine*, Volume 37, Issue 6, pp. 1033–1057.

Wynn, R. (1999). *Veterinary Pharmacology*. Cambridge, MA: Harcourt Learning Direct.

Zaret, E. (August/September 2001). Equating Fido's drugs with human quality standards. *Pharmaceutical Formulation & Quality*.

CHAPTER 3

PRINCIPLES AND METHODS OF DRUG ADMINISTRATION

OBJECTIVES

Upon completion of this chapter, the reader should be able to:

- describe the therapeutic range and its role in drug efficacy.
- describe graphically the components of therapeutic range, subtherapeutic range, and toxicity levels.
- describe the three components of the therapeutic range.
- compare and contrast the different routes of administration with regard to rate of onset, formulation, and use.
- differentiate between bolus administration, intermittent therapy, and infusion of fluid IV techniques.
- describe concepts of drug dosing.
- describe dosage and its use in drug dosing.

KEY TERMS

bolus	intravenous (IV)
dosage	loading dose
dosage interval	maintenance dose
dosage regimen	nebulize
dose	nonparenteral
emulsion	oral
epidural	parenteral
immiscible	repository (depot) preparation
inhalation	solution
intra-arterial (IA)	subcutaneous (SC or SQ)
intra-articular	subdural
intracardiac (IC)	suspension
intradermal (ID)	therapeutic range
intramammary	topical
intramedullary	total daily dose
intramuscular (IM)	volatilize
intraosseous (IO)	
intraperitoneal (IP)	
intrathecal	

Setting the Scene

A cat presents to the clinic with extreme dyspnea (difficulty breathing). The cat is breathing with its mouth open, and the veterinary technician notices that its gums and mucous membranes are not as pink as they should be. The veterinarian quickly begins a physical exam to assess the situation. The respiratory sounds are difficult to hear because of fluid in the cat's chest. The veterinarian asks the technician to get a drug that will remove some fluid from the cat's chest to help it breathe easier. The technician rushes over to the drug cabinet and notices that there are different forms of the drug the veterinarian asked for. Which one does the technician reach for, and why? Which route of drug administration would the veterinarian choose, and how should the technician be prepared to assist in the drug administration? In emergency situations, what are the most rapid routes of drug administration? What drug forms are available for these routes of administration?

Courtesy of iStockphoto

SAFE DRUG USE

Drugs are capable of producing a wide variety of effects in animals. All drugs should be considered potential poisons and should be dispensed and given with great care. Only appropriate administration of the appropriate drug at the appropriate dosage determines whether a compound benefits an animal or causes harm. Appropriate administration of a drug includes administration of the appropriate amount of drug (based on dosage) into the animal's body by the appropriate route of administration.

Prior to drug administration it is important to understand that veterinarians and veterinary technicians are accountable for the safe administration of medication. In preparing to administer medication, it is important for the veterinarian or veterinary technician to ensure cleanliness of all materials used. The hands, work surface, and supplies must all be clean. In addition, all supplies needed for administration should be on hand. Medication should be prepared in an area with good lighting and a minimal number of distractions.

Veterinarians and veterinary technicians must know all the components of a drug order and question orders that are not complete, are unclear, or that give a dosage outside the recommended range. The drug order should include the date and time the order was written; drug name; drug dosage; dose, route, frequency, and duration of administration; and signature of prescriber. Sometimes drug orders are given verbally; however, a written order should be documented in the medical record as soon as possible. Verbal orders over the telephone should also be written in the medical record and signed by the prescriber as soon as possible.

Once the order has been examined for completeness, the medication is prepared for administration. To provide safe drug administration, the six rights of proper drug administration should be followed. These six rights include the right drug, the right dose, the right time, the right route, the right patient, and the right documentation (Figure 3-1).

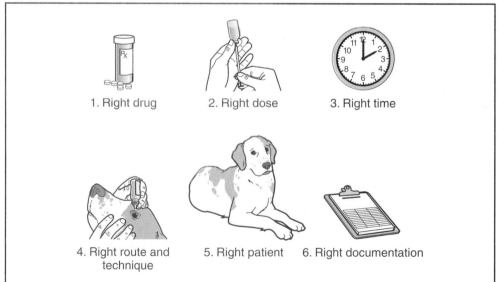

1. Right drug   2. Right dose   3. Right time

4. Right route and   5. Right patient   6. Right documentation
   technique

Delmar/Cengage Learning

FIGURE 3-1 The six rights of medication administration.

The Right Drug

The right drug means that the patient receives the drug that was prescribed. The label on the container should be read three times: when taking the container from its location, when removing the medication from the container, and when returning the container to its storage place. When preparing to administer the medication, it should also be checked three times: when removing it from its location (refrigerator, drawer, or bin), to compare it to the patient's medical record, and before administering it to the patient. Keep in mind that some drug names are similar (e.g., digoxin and digitoxin); therefore, understanding the reason for which the patient is receiving the medication is important. Never give a medication from a container that is unlabeled. Be sure the dose remains packaged until immediately before it is administered.

The Right Dose

The right dose is the dose prescribed for this particular patient. Veterinarians and veterinary technicians must calculate each drug dose accurately. Prior to calculating a drug dose, an estimate of the answer based on knowledge of proportions and ratios should be known. For example, if a 10-pound dog gets 1 tablet of drug A, it would not make sense to give a 50-pound dog 50 tablets of drug A. When in doubt of a drug calculation, have the dose recalculated and checked by another veterinary professional.

The Right Time

The right time is the time at which the prescribed dose should be administered. Daily drug doses are given at specified times during the day (e.g., bid or twice daily, and tid or three times daily) to keep the plasma levels of the drug at the proper level to cause the desired effect without causing signs of toxicity.

The Right Route and Technique

The right route is the proper route of administration and administration in such a way that the patient is able to take the entire dose of the drug at one time. It is important to use aseptic technique when administering drugs (sterile technique is needed for drugs given by injection) and to administer drugs at the appropriate site.

The Right Patient

The right patient means that the veterinarian or veterinary technician takes every opportunity to be certain that the medication is administered to the proper patient. If the patient is in a cage, check the cage card and identification band. If the patient is in the examination room, identify the patient by name so that the owner can verify the name of the animal receiving the medication. When dispensing medication, read the patient's name to the owner to ensure that the proper medication is dispensed to the proper animal.

Clinical Que

According to the Journal of the AVMA, part of owner consent means that client information sheets provided to the veterinarian should be handed out with the drug(s) they pertain to. These client information sheets have information on risks associated with that particular drug.

Clinical Que

The rights of proper drug administration serve as a checklist of activities to be followed by those giving medication. These rights include:

- right drug
- right dose
- right time
- right route and technique
- right patient
- right documentation

The Right Documentation

The right documentation requires that the person administering the drug immediately record the appropriate information about the drug administered (the drug, dosage administered, time and date administered, the route and site if given by injection, the patient's response, and the veterinarian's or veterinary technician's initials or signature). The right documentation is not only a legal obligation, but also a safety obligation. Documentation is the primary method used to communicate drug administration from one professional to another. The basic principle of documentation is "if it is not documented, it was not done." Not documenting drug administration could lead to the animal receiving multiple drug doses, which could result in potentially serious consequences for the animal receiving the drug. It should also be documented if an owner refuses drug treatment for his/her animal, including the reason for this refusal. The client has the right to refuse medical treatment for his/her animal. Without the client's consent, the medication (or treatment) cannot be given. Most refusals by clients are the result of the client's knowledge deficit about what the medication does, past problems with a particular drug, the inability of the drug to provide a cure, or financial limitations. Clients have a right to *owner consent* (formerly known as informed consent), which is based on the client having the knowledge necessary to make a decision. Providing clients with the information needed to make proper decisions for their animals is one way to ensure that optimal medical care is given to patients. Obtaining owner consent typically means the veterinarian explains both the risks and benefits associated with a particular treatment method. Then the client signs a document, which states that the client understands the risks and benefits. By giving owner consent to a procedure or treatment, it is assumed that the client both read and understood all of the terms in the statement. Once owner consent has been given, the animal may be treated according to the conditions listed within the statement.

Clinical Que

If a client questions the appearance, dose, or method of administering the medication, always recheck the order and the medication prior to administration of the dose.

THE SAFE ZONE: THE THERAPEUTIC RANGE

The goal of drug therapy is to deliver the desired concentration of a drug within the target area of the body to ultimately achieve the desired effect. The **therapeutic range** of a drug is the drug concentration in the body that produces the desired effect in the animal with minimal or no signs of toxicity (Figure 3-2 A and B). Onset of action begins when the drug enters the plasma and lasts until it reaches the minimum effective concentration (MEC). Peak action occurs when the drug reaches its highest blood or plasma concentration. Duration of action is the length of time the drug produces the desired effect. Laboratory and clinical evaluation determine the drug dosage used to get and keep the drug level in the therapeutic range. The concept that "more is better" does not hold true in pharmacology: More drug—either an increased amount or increased frequency of dosing—can produce damage to body organs. Similarly, the concept that "less is better" does not hold true in pharmacology: Less drug will not produce toxic side effects, but it will not achieve proper levels in the body to produce the beneficial effects of the drug.

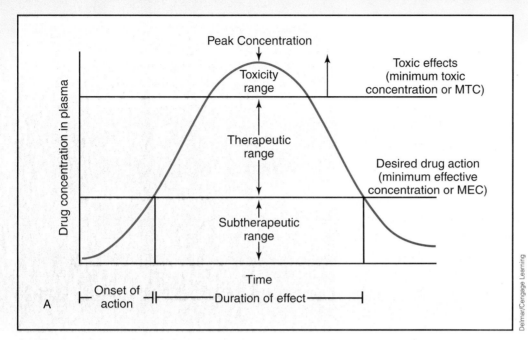

FIGURE 3-2A The therapeutic range of a drug concentration in plasma should be between the lowest drug concentration in plasma for obtaining the desired drug action and the lowest drug concentration in plasma for producing a toxic effect.

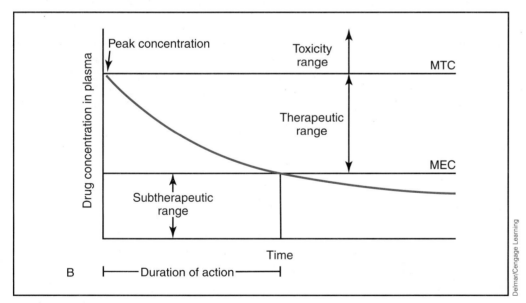

FIGURE 3-2B Therapeutic range for a drug given IV. Drugs administered IV achieve a concentration in the therapeutic range more rapidly than other routes of drug administration.

Many factors play a role in getting and keeping drugs in the therapeutic range. Some factors involve properties of the drugs; other factors involve the health and physiology of the animal. Maintaining drugs in the therapeutic range involves maintaining a balance among the rate of drug entry into the body, absorption of the drug, distribution of the drug, metabolism of the drug, and excretion of the drug (see Chapter 4).

STAYING IN THE SAFE ZONE

Three major drug factors involved in keeping the drug concentration within its therapeutic range include route of administration, drug dose, and dosage interval.

Route of Administration

The route of administration is how the drug is given or the manner in which a drug enters the body. Drugs can enter the body parenterally (Figure 3-3), orally, and locally. **Parenteral** drugs are given by a route other than the gastrointestinal tract, **oral** (also known as **nonparenteral**) drugs are given through the gastrointestinal tract, and local drugs are applied or given directly where the action of the drug is desired.

Many factors are considered before the route of administration for a drug is selected. Some factors are based on the drug itself; other factors are based on the animal being treated. Drug factors that influence the route of administration include the following:

- Some drugs cause one effect when given parenterally and another effect when given orally. For example, magnesium sulfate causes muscle relaxation when given intravenously (IV) and diarrhea when given orally.

- Some drugs are insoluble in water and can be given intramuscularly (IM) but cannot be injected IV.

- Some drugs are destroyed by stomach acid and cannot be given orally.

Animal factors also influence the choice of route of drug administration. Animals that are actively vomiting cannot absorb drugs given orally and must receive drugs through another route of administration. Critically ill patients need to get therapeutic levels of drug in their bodies more rapidly than moderately ill patients do; therefore, critical patients tend to receive IV medication, while moderately ill patients may be treated with oral medications.

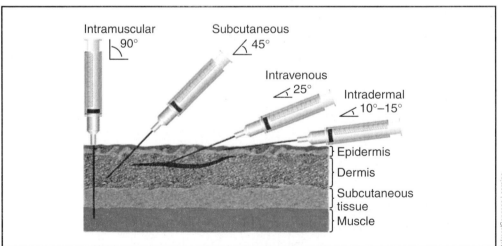

FIGURE 3-3 Some examples of parenteral routes of drug administration.

Injectable Route/Parenteral Drug Forms

Drugs administered by injectable routes are types of parenteral drugs. *Parenteral* literally means "excluding the intestines" (*para-* = apart from, *entero* = small intestine). Injectable drugs are administered by needle and syringe. A syringe consists of a barrel, plunger, flange, and tip (Figure 3-4). The barrel holds the medication and has graduated markings (calibrations) on its surface for measuring medication. The plunger is a movable cylinder that is inserted within the barrel and forms a tight-fitting seal. The plunger is the mechanism by which a medication is drawn into (aspirated) and pushed out of the barrel. The flange is at the end of the barrel where the plunger is inserted. It forms a rim around the end of the barrel against which one places the index and middle fingers when drawing up solution for injection. The tip is the end of the barrel where the needle is attached. Tips may be slip-lok (without grooves in the tip that allow the needle to be held on by friction) or Luer-lok (with grooves in the tip that lock the needle in place). Syringes come in many sizes (1 cc to 60 cc) and are chosen on the basis of the animal, route of administration, and amount of drug to be given (Figure 3-5). Many syringes are now equipped with safety devices to prevent needlestick injuries (Figure 3-6). Needles also come in a variety of sizes and lengths (Figure 3-7a). The parts of a needle are illustrated in Figure 3-7b. The proper use of needles and syringes is described in Appendix A.

Injectable drugs are usually supplied as sterilized solutions, prepackaged syringes with needles for injection, powders that must be reconstituted with sterile solution (called a diluent), or in vials to be drawn up into syringes for injection. Reconstituting a powder medication for administration is illustrated

Clinical Que

To help remember routes of administration, remember key word parts. *Intra-* means within, *inter-* means between, *sub-* means under, *epi-* means above, and *trans-* means across.

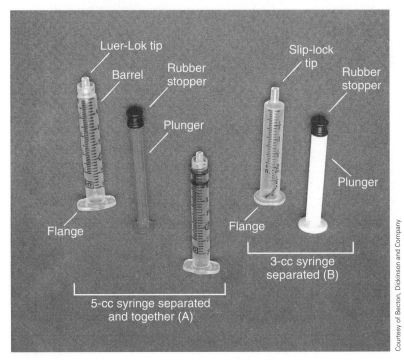

Courtesy of Becton, Dickinson and Company

FIGURE 3-4 Parts of a syringe: (A) A 5-cc syringe separated and unseparated, with Luer-lok tip. (B) A 3-cc syringe separated with slip-lock tip.

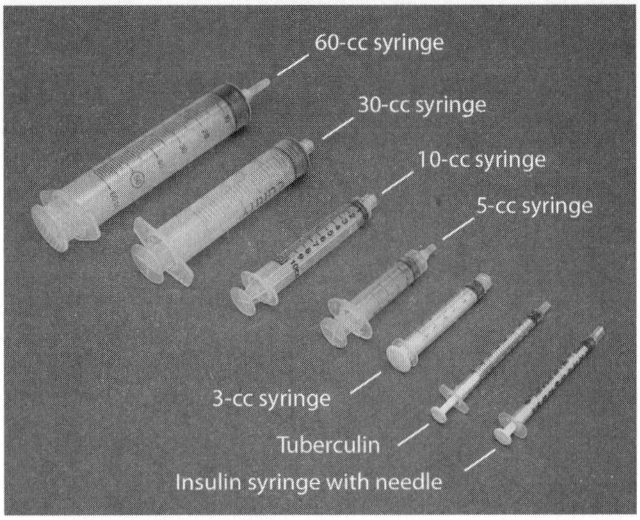

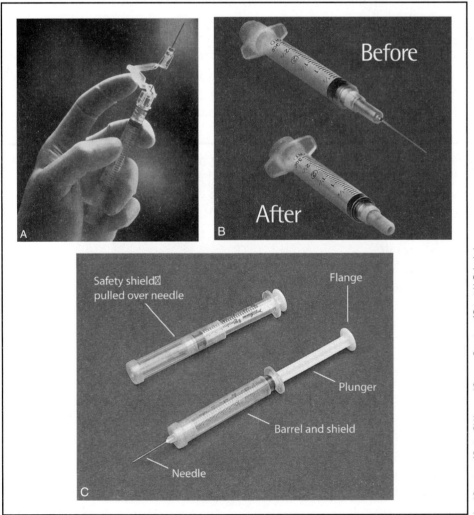

FIGURE 3-5 Various sizes of disposable syringes.

60-cc syringe
30-cc syringe
10-cc syringe
5-cc syringe
3-cc syringe
Tuberculin
Insulin syringe with needle

Courtesy of Becton, Dickinson and Company

Before

After

A

B

Safety shield pulled over needle
Flange
Plunger
Barrel and shield
Needle
C

FIGURE 3-6 Types of safety syringes.

A and C courtesy of Becton, Dickinson and Company; B courtesy of Retractable Technologies

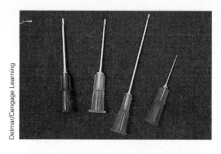

FIGURE 3-7A Various lengths and sizes of needles.

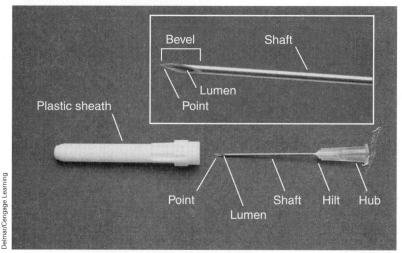

FIGURE 3-7B Parts of a needle and needle sheath. The insert shows point, lumen, bevel, and shaft.

in Appendix B. A vial is a small bottle with a rubber stopper, through which a sterile needle is inserted to withdraw a dose of the medication inside. Vials that store injectable drugs can be multidose, meaning that the vials contain more than one dose, or single dose, meaning that the vials contain only one drug dose. Aspirating medication from a vial is described in Appendix C. Injectable drugs can also be stored in ampules, which are small, sterile, prefilled glass containers containing medication for injection (Figure 3-8). Aspirating medication from an ampule is described in Appendix D.

The three most common injectable administration routes in animals are intravenous (IV), or within a vein; intramuscular (IM), or within the muscle; and subcutaneous (SC or SQ), or under the skin (dermis). Other less common routes of injectable drug administration are intraperitoneal (IP), intradermal (ID), intra-arterial (IA), epidural/subdural/intrathecal, intracardiac (IC), intra-articular, intramammary, and intramedullary/intraosseous (IO).

Intravenous

One of the fastest means of getting drugs into the bloodstream is intravenous injection (the drug is given directly in the vein). Figure 3-9 shows a dog receiving an IV injection. The IV route of administration gives a predictable concentration of drug and usually produces an immediate response. Most IV injections are aqueous solutions (solutions of drugs dissolved in liquid, usually water), but a few are emulsions. An emulsion is a mixture of two immiscible (incapable of

Clinical Que

When giving medication IV, remember that it should be given slowly in most cases. Also, make sure to remove all air bubbles in the substance to be injected before giving an IV injection. Injecting air bubbles can cause them to lodge in areas of the body, forming air emboli (an obstruction of a blood vessel caused by the entrance of air into the bloodstream) and causing tissue damage (and potentially death in small animal species).

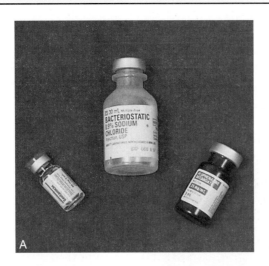

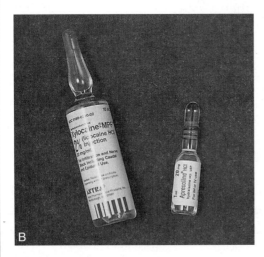

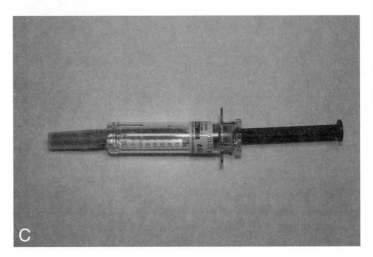

Courtesy of Roche Laboratories, Inc.

FIGURE 3-8 Vials and ampules store injectable drugs: (A) Single-dose and multidose vials, (B) ampules, and (C) prepackaged syringe with needle for injection.

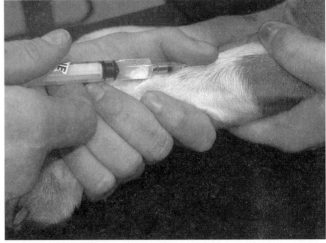

FIGURE 3-9 Dog receiving an IV injection.

Delmar/Cengage Learning

mixing) liquids, one being dispersed throughout the other in small droplets. It is important that no foreign matter or particles be injected along with the medication, because particles can collide with the blood cells and cause blood clots. Filter needles, with filters in the needle hub, may be recommended when emulsions are given. IV drugs that are accidentally injected out of the vein (or other blood vessel) are said to be given *perivascularly*. Some drugs that are inadvertently given perivascularly can cause tissue inflammation and necrosis.

Veterinarians use three intravenous injection techniques: bolus administration, intermittent therapy, and continuous infusion of fluid. Bolus intravenous administration involves injecting a drug in a small amount of fluid, with a syringe and needle only. A **bolus** is a concentrated mass of pharmaceutical preparation. Veterinary professionals use this technique to achieve immediate high concentrations of drugs. Intermittent intravenous therapy involves diluting a drug dose in a small volume of fluid and administering it during a period of 30 to 60 minutes multiple times daily by means of an indwelling catheter (a catheter designed to remain in place). Intermittent intravenous therapy is the best way to maintain blood levels of antibiotics. Continuous infusion of fluid involves the administration of large volumes of fluid continuously over extended periods of time. Veterinary professionals use this method to administer electrolytes and nutritional agents like amino acids.

Intramuscular

Drugs for intramuscular injection come in aqueous (prepared in water) solutions, aqueous suspensions, oily suspensions, and injectable pellets. The body absorbs intramuscular aqueous solution injections rapidly; significant blood levels of the drug may appear within 5 minutes, but typically take about 30 minutes. The body absorbs aqueous suspensions more gradually, so introduction of the drug into the bloodstream is prolonged. Placing an injectable drug in a substance that delays absorption is called a **repository** or **depot preparation**. One means of prolonging drug action and further slowing the body's absorption of an intramuscular injection is to mix the drug with oil, creating an oily suspension. Examples of this are procaine penicillin G and methylprednisolone (Depo-Medrol®). Another way to prolong drug action is to add an ingredient that has limited absorption, which slows absorption of both drugs. An example of this is protamine zinc insulin injection.

When giving medication IM, remember:

1. Aspirate by pulling back on the plunger of the syringe to verify the syringe is not injecting directly into a blood vessel (Figure 3-10). If blood appears in the syringe, pull the needle out of the injection site, pick a new location to inject, and start again.

2. Give the injection within the muscle to make sure the drug will be absorbed properly. If given too shallowly, the drug may not actually have been given IM.

3. Keep in mind that some drugs cause pain when given IM.

Subcutaneous

A third injectable route of administration is subcutaneous (SC, SQ, or subQ). Subcutaneous injections place the drug into the connective tissue underneath the

Clinical Que

A *solution* is a clear liquid preparation that contains one or more solvents and one or more solutes. A *suspension* is a liquid preparation that contains solid drug particles suspended in a suitable medium.

Clinical Que

Repository forms of drugs include sustained-release (SR) products, implants, and spansules.

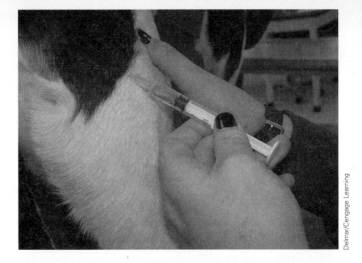

FIGURE 3-10 Aspirating when giving an IM injection ensures that the medication is not being injected directly into a blood vessel.

Delmar/Cengage Learning

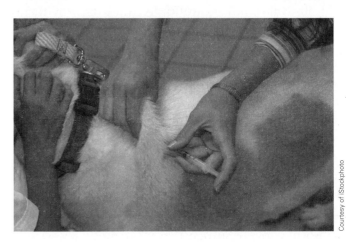

FIGURE 3-11 A dog receiving an SQ injection.

Courtesy of iStockphoto

dermis of the skin (Figure 3-11). When the onset of the drug effect desired is faster than oral administration and slower than intramuscular injection, subcutaneous injection is used. The vascularity of the subcutaneous space is less than that of skeletal muscle; hence, drugs given subcutaneously tend to be absorbed more slowly than IM administrations. Irritating or hyperosmotic solutions should not be given SQ. Larger amounts of solutions can be given SQ; however, the amount given should be based on the animal species involved and should not be so much as to cause tissue death or sloughing of skin. One way to further extend the duration of action of a particular drug given subcutaneously is to implant a pellet into the subcutaneous space where the drug is absorbed slowly and gradually. Hormonal implants are examples of drugs that are administered with injectable pellets.

Intramammary

Drugs given by the intramammary route typically demonstrate fast and even distribution of the drug and a low degree of binding to udder tissue. These properties result in lower concentrations of drug residues in the milk (which are desirable in lactating cows). Particle size is an important factor in intramammary drugs because it affects both the rate of release of the active ingredient and the amount of irritation to the udder tissue. In nonlactating cows treated with intramammary formulations, it is desirable to have prolonged drug release and a high degree of binding to mammary secretions and udder tissues. Drug particle size in nonlactating intramammary formulations is usually smaller than in those for lactating cows, which reduces udder irritation and results in prolonged drug retention in the udder. Thickening agents are added to modify the rate of release of the suspended particles from oil formulations, and antioxidants are commonly incorporated in the formulations to prevent spoilage.

All injectable routes of administration are summarized in Table 3-1.

Table 3-1 Parenteral Routes of Drug Administration	
ROUTE	GENERAL RULES
Intravenous (IV): within the vein	• rapid onset of action • higher initial body levels of drug • shorter duration of activity (need to be redosed more frequently) • larger volumes can be given • irritating drugs or drugs that are painful via other routes can be given IV (e.g., oxytetracycline) • increased risk of adverse effects (if drug given too rapidly, not sterile, or not properly mixed) • drug must be pure, sterile, and free of particles • drug must be water soluble
Intramuscular (IM): within the muscle	• relatively rapid onset of action (generally about 30 minutes) • rate of absorption depends on formulation (oil-based is slow, while water-based is fast) • provides reliable blood levels • longer duration of action than IV (can dose less frequently) • shorter duration of action versus oral (generally) • absorption may be altered by vehicle present in preparation • limited use for giving irritating solutions • convenient route in fractious animals
Subcutaneous (SQ, subQ, or SC): beneath the skin into the subdermis	• slower onset of action than IM • less reliable blood levels (similar to oral) • longer duration of action than IM (can be given less frequently) • absorption may be altered by vehicle in preparation • cannot use irritating solutions • generally used for giving larger volumes of nonirritating, water-soluble solution

(Continued)

ROUTE	GENERAL RULES
Intramammary: within the teat and udder sinuses	• fast and even distribution into mammary tissue • particle size important (in nonlactating cows particle size is small to reduce udder irritation and provide prolonged drug retention) • may contain thickening agents to modify rate of drug release
Intraperitoneal (IP): within the abdominal body cavity	• variable onset of action • variable blood levels • provides large surface area for drug absorption • irritating solutions may cause peritonitis • care must be taken so that the needle does not penetrate any organs; this could lead to peritonitis • first passes via portal system through liver, which could inactivate (or enhance) the drug's action
Epidural/Subdural/ Intrathecal: above the dura mater of the meninges, under the dura mater of the meninges, or into the subarachoid space of the meninges	• rapid onset of action localized to the central nervous system (CNS) • used for diagnostic procedures and for administering some types of anesthetic agents • disadvantages include potential for misperformance resulting in spinal injection or drug moving cranially in the CNS
Intra-arterial (IA): within the artery	• used for treating a specific organ only, because very high drug levels are delivered to a specific site • may be accidental (e.g., given in carotid artery instead of jugular vein)
Intradermal (ID): within the skin	• injection given between dermis and epidermis • very slow absorption • low blood levels obtained • used for local treatments or allergy testing
Intracardiac (IC): within the heart	• rapid drug levels attained because drug passes from heart to systemic circulation • may be used in emergency situations
Intra-articular: within the joint	• injection given in the synovial space of joints • sterile technique is critical • drug can be absorbed systemically
Intramedullary or Intraosseous (IO): within the medullary cavity of bone	• provides rapid blood levels • not commonly used and is painful • route of rapid fluid administration in smaller animals and birds • usually administered in the femur/humerus

Clinical Que

When seeing the term *depo* associated with an injectable drug, think long-acting.

Local Routes

Locally administered drugs are those that are applied or given directly where the action of the drug is desired. Local administration of drugs includes inhalation, topical application (skin, eye, or ear), and other routes (rectal, vaginal, and transdermal patches).

Inhalation

Inhalation administration introduces the drug to the animal by breathing the drug into the respiratory tract. After the animal breathes in the gas, the gas particles enter the alveoli of the lung. Here the particles diffuse (move from an area of high concentration to an area of low concentration) across the alveolar membrane. From the alveolar membrane, the drug molecules enter the blood via capillaries surrounding the alveoli. Once in the capillary, the drug travels through the bloodstream to cause its effect. Drugs inhaled are rapidly absorbed into the bloodstream. Gas anesthesia is one of the most common medications delivered by inhalation.

The inhalation route is also used to treat local respiratory conditions. Inhalation drugs include bronchodilators (drugs that open the airway for better breathing), mucolytic enzymes (drugs designed to liquefy thick mucus for better removal from the lungs), antibiotics (for lung infections), and glucocorticoids (for inflammatory lung conditions).

Drugs that are administered by inhalation must be **volatilized** (turned into gases) in order to be inhaled into the lungs. Anesthetic gases are volatilized from liquids using a gas anesthetic vaporizer. Other drugs, such as bronchodilators, mucolytics, antibiotics, and glucocorticoids, are **nebulized** (turned into a fine spray) and then inhaled into the lungs (Figure 3-12).

Topical

A **topical** medication goes on the surface of the skin or mucous membranes. Topical medications come in ointment, cream, gel, liniment, paste, lotion, powder, and aerosol dose forms (Table 3-2) (Figure 3-13). Topical drugs must first dissolve and then penetrate the skin by diffusion. Topical applications are used for localized treatment of conditions such as localized skin infections (lesions), abrasions, and localized skin allergies. The body as a whole absorbs topical medication more slowly than any other application route; the localized site, however, benefits from a high concentration of the drug. Topical application is good for drugs needed externally that are toxic if injected. The disadvantage of veterinary topical application is that fur and feathers inhibit good skin contact (although clipping or plucking can minimize the problem). Topical treatment also includes eye and ear medications. Eye preparations are available as liquid drops and ophthalmic ointments. Ear preparations are available as drops, ointments, and creams.

Rectal

The rectal route of drug administration may be a practical alternative for delivering such drugs as anticonvulsants, analgesics, and antiemetics in animals due to the danger these animals present to the veterinary staff or due to the inability

Clinical Que

Drugs that constrict blood vessels could delay drug absorption because they affect perfusion to the administration site. Blood vessels that are constricted or narrowed cannot absorb as much drug as rapidly as blood vessels that are dilated or widened. This delay in absorption can prolong drug action. An example of this altered perfusion occurs when injecting lidocaine with epinephrine for local anesthesia. The lidocaine is absorbed more slowly because of the presence of epinephrine, thus lengthening its duration of action.

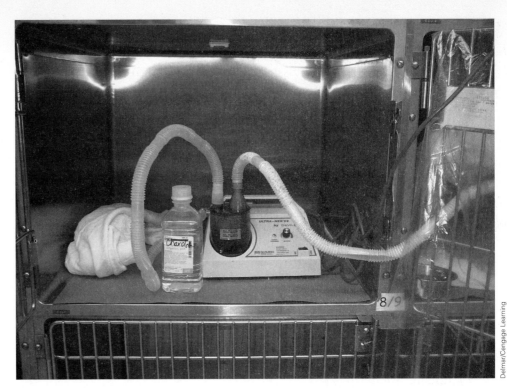

Delmar/Cengage Learning

FIGURE 3-12 Nebulized drugs are delivered as a fine spray that are inhaled into the lungs.

Table 3-2	Topical Drug Forms
TOPICAL DRUG FORM	**DESCRIPTION**
Aerosol	Drug suspended in solvent and packaged under pressure
Cream	Drug suspended in water–oil emulsion
Gel	Drug suspended in semisolid or jelly-like substance
Liniment	Drug suspended in oily, soapy, or alcohol-based substance applied with friction
Lotion	Drug suspended in liquid for dabbing, brushing, or dripping on skin without friction
Ointment	Drug suspended in semisolid, lipid-based preparation that melts at body temperature
Paste	Drug suspended in semisolid preparation that retains its state at body temperature
Powder	Drug suspended in powder for external lubrication or absorption

Figure 3-13 Some examples of topical drug forms. From left to right are a gel, ointment, cream, and paste.

to successfully administer these types of drugs because of the animal's condition. The rate and extent of rectal drug absorption are often lower than with oral drug absorption, possibly due to the relatively small surface area available for drug uptake. In addition, the composition of the rectal formulation (solid versus liquid or nature of the semisolid suppository base) is an important factor in the absorption of rectally administered drugs. Suppositories are used to deliver drugs rectally in animals and are bullet-shaped drug forms intended to be inserted into a body orifice and contain drugs intended for a local effect at the site of insertion (Figure 3-14). Local irritation is a possible side effect of rectal drug therapy.

Vaginal

The vaginal route of drug delivery is a potential administration route for therapeutically important macromolecules. Successful delivery of drugs through the vagina may be challenging due to the poor absorption of drug molecules across the vaginal epithelium. The rate and extent of drug absorption after

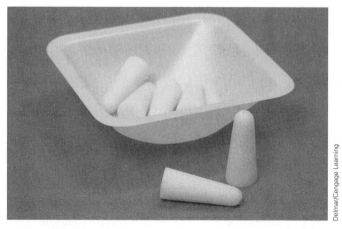

FIGURE 3-14 Suppositories maintain their shape at room temperature but melt or dissolve when inserted into a body orifice.

intravaginal administration may vary depending on drug formulation factors, vaginal physiology, age of the patient, and the phase of the estrous cycle of the patient. Vaginal drug delivery systems used in animals include those used for controlled internal drug release (CIDR) devices, progesterone-releasing intravaginal devices (PRID), and vaginal sponges. These systems are used for estrus synchronization in sheep, goats, and cattle. The active ingredients in these systems are synthetic or natural hormones, which are slowly absorbed through the vaginal wall. Retention of the vaginal drug delivery systems in the vagina may be the limiting factor in animals, and in the systems mentioned, it depends on the entire device (sponges and PRID), or the wings (CIDR device), expanding.

Transdermal

Transdermal drugs are delivered systemically through a patch on the skin. Drugs administered transdermally may be mixed with a chemical (such as alcohol) to enhance skin penetration (Figure 3-15). Through a transdermal patch, the drug passes through the skin to the bloodstream allowing the drug to be delivered slowly and continuously for many hours, days, or even longer. Plasma levels of a drug given transdermally can be relatively constant. Patches are particularly useful for drugs that are quickly eliminated from the body because such drugs, if taken in other forms, would have to be given frequently. Skin irritation is one side effect of this route of drug administration. In addition, patches are limited by how quickly the drug can penetrate the skin, and

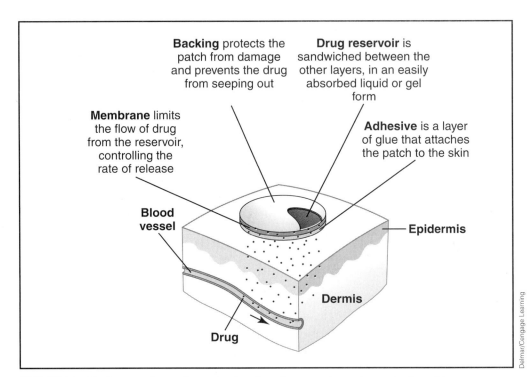

FIGURE 3-15 A transdermal patch slowly releases drug through the skin to the bloodstream. The patch works by diffusion, and the rate of drug release is controlled by an intervening membrane or by suspending the drug in another material called a matrix, which lowers its initial concentration.

only drugs needed in relatively small daily doses can be given through patches. Examples of drugs given transdermally in animals include nitroglycerin and fentanyl.

All forms of local routes of administration are summarized in Table 3-3.

Oral Route

Oral administration delivers the medicine directly to the animal's gastrointestinal (GI) tract (Table 3-4). Oral medications are the most convenient to give and are less likely to cause adverse reactions. They do not have to be sterile because they enter the nonsterile environment of the gastrointestinal tract.

Table 3-3 Local Routes of Drug Administration	
Inhalation: inhaled into the respiratory system	• examples include gas-masking of animals, endotracheal administration of gas anesthesia, and nebulization of drugs • establishes rapid blood levels because the alveoli of the lung provide a large surface area for absorption • may be used for anesthesia, emergency procedures, and treatment of respiratory disease
Topical: applied on top of a surface	• topical routes of administration include skin, conjunctival, and subconjunctival • used mainly in dermatology and ophthalmology • may or may not be absorbed systemically • drug must first dissolve and then penetrate the skin by diffusion • good local effect • may be irritating • easy to administer • animal may chew/lick/rub off
Rectal	• rectal drug absorption may be lower than with oral drugs due to small surface area or composition of the rectal formulation. Local irritation is possible with rectally administered drugs
Vaginal	• vaginal drug delivery varies depending on drug formulation factors, vaginal physiology, age of the patient, and the phase of the estrous cycle of the patient
Transdermal	• transdermal drugs may be mixed with a chemical to enhance skin penetration, which allows the drug to pass through the skin to the bloodstream • transdermal drugs are delivered slowly and continuously for an extended period making plasma levels of a drug relatively constant; skin irritation is possible

Table 3-4   Oral Route of Administration

ROUTE	GENERAL CONCEPTS
Oral (po)	• most convenient route of administration for owner • slower onset of action • longer duration of activity • sometimes erratic and incomplete absorption because drug is affected by gastric fluids (acid) • absorption may be affected by gastrointestinal disease • relatively safe • drug must be able to get through gastrointestinal mucosa • drug need not be sterile • generally causes fewer adverse drug reactions • absorption in ruminants may be questionable with some medications given orally

Before entering the bloodstream, an oral drug must go through a series of events, including release from the dose form (tablet or liquid), transport across the gastrointestinal tract, and passage through the liver. Each of these events can decrease the amount of drug that reaches the bloodstream.

The first event, dissolving the dose form, varies with each type of dose form. Oral dose forms are either solid or liquid. The animal either swallows the intact drug (the most accurate dose method) or eats food laced with the dose (less accurate, as the animal may leave some food uneaten).

Solid oral drug forms include tablets, capsules, boluses, lozenges, and powders (Figure 3-16). Tablets, capsules, and boluses can be given to an animal by hand or pilling/balling gun (Figures 3-17 and 3-18). Tablets are medicines

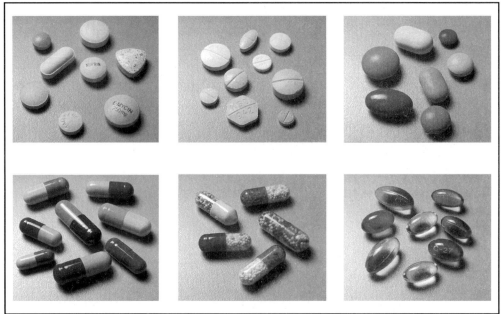

FIGURE 3-16 Examples of solid drug preparations.

FIGURE 3-17 A capsule or tablet is one way to administer oral medication to an animal. This image shows a cat being given a tablet.

FIGURE 3-18 Balling gun used to administer a magnet to a Holstein cow.

Clinical Que

Only scored tablets should be divided, because these are the only ones guaranteed by the manufacturer to distribute equal amounts of drug throughout the tablet (Figure 3-19).

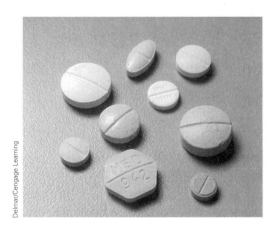

FIGURE 3-19 Scored tablets are the only tablets that should be divided. Unless tablets are scored, the manufacturer cannot guarantee equal distribution of drug in each part.

mixed with an inert binder and molded or compressed into a hard mass. The tablet disintegrates in stomach liquid, releasing the drug for absorption into the bloodstream. Enteric-coated tablets are covered with a special coating that prevents the drug from dissolving in the stomach but permits it to dissolve in the small intestine. These drugs are designed to carry drugs that could irritate the stomach or be chemically destroyed by the acid environment of the stomach. The coating of enteric-coated tablets is designed to dissolve in neutral or alkaline pH. These drugs should not be given with antacids or other alkaline substances. Coated tablets are believed to cause less stomach irritation.

Sustained-release tablets have a coating that allows drugs to be released from tablets in a controlled fashion. For example, some tablets have crystals of potassium chloride embedded in a wax coating that cause small amounts of drug to leak through the channels of wax when the tablet is exposed to stomach acid. This promotes the gradual release of drug over several hours; other tablets may contain microencapsulated drug in which small drug particles are coated with a polymer coating. When the tablet disintegrates, the microencapsulated drug particles are released. This allows for drug release over varying periods.

Molded tablets are soft, chewable tablets that contain drug mixed with lactose, sucrose, dextrose, or a flavoring agent to enhance the taste of the tablet in animals. An example of a molded tablet is heartworm preventative (ivermectin) in a chewable treat.

Capsules are gelatin shells holding a powdered or liquid form of the drug. Most capsules are colored and may bear identifying product markings. The gelatin shell dissolves in the stomach liquids, releasing the drug for absorption into the bloodstream. Boluses are large, compressed, rectangular tablets that are typically used to dose large animals. Lozenges are drugs incorporated into a hard candy tablet that allows slow release of the drug. Lozenges are not utilized in veterinary medicine because of the obvious problems associated with getting animals to suck, rather than chew, the lozenges. Powders are dry, granulated versions of the drug mixed with inert bulking and flavoring agents (such as lactose) to enable dilution. Powders are easily mixed into the animal's food or drinking water.

Liquid oral drug forms include solutions, suspensions, and emulsions (Figure 3-20). **Solutions** are drug preparations dissolved in liquid; they do not settle out if left standing. Solutions often have flavoring added to mask the taste of the drug. Examples of solutions are syrups (drugs dissolved in 85 percent sucrose), elixirs (drugs dissolved in sweetened alcohol), and tinctures (alcohol solutions). An example of a solution is phenobarbital liquid (it is actually an elixir).

A finely divided, undissolved substance dispersed in water is called a **suspension**. Drugs that will not easily dissolve can be dispersed and suspended in liquid so that shaking the container distributes them uniformly. An example of a suspension is pyrantel pamoate oral suspension (a deworming medication).

An emulsion consists of fine droplets of oil in water or water in oil. They separate into layers after standing for long periods of time and must be shaken vigorously before they are given to an animal. An example of an emulsion is castor oil.

Oral liquid drugs may be given by dropper, syringe, and drench (given by forcing the animal to drink) (Figure 3-21 A and B). Solutions, suspensions,

Delmar/Cengage Learning

FIGURE 3-20 Examples of solutions from left to right are a solution (drug preparations dissolved in liquid that do not settle out if left standing), emulsion (consists of fine droplets of oil in water or water in oil that separate into layers after standing for long periods of time), and suspension (finely divided, undissolved substance dispersed in water that will not easily dissolve).

Delmar/Cengage Learning

FIGURE 3-21 (A) A liquid medication is administered orally to this dog. (B) Drenching in a cow.

and emulsions can be mixed with food. Although mixing drugs into food is not as accurate as a direct dose, it offers some advantages if the animal eats all the food. Medications mixed with food are less likely to irritate the stomach and intestinal tract, as capsules and tablets can settle in a small area where any irritants in the drug then concentrate. Mixing drugs in food is also a safe way to medicate an unpredictable or aggressive animal. Getting the animal to eat all the food, however, can be a challenge, especially if the drug smells or tastes bad or the animal's appetite has decreased.

The second event that determines whether an orally administered drug gets into the bloodstream is transport across the gastrointestinal tract. Anatomical differences of the various animal species can affect drug absorption. A ruminant usually takes longer to respond to oral medication than does a monogastric animal. A drug mixed with food can remain in the rumen for up to three days, where it is continually exposed to ruminal secretion. Consequently, some medications that are effective orally in monogastric animals biodegrade and become ineffective in ruminants. A herbivore that grazes, such as a horse, may also have altered transport across the gastrointestinal tract due to the constant presence of food in its intestines. In animals that continuously feed, some drugs may be absorbed more rapidly, while other drugs will be absorbed more slowly. In general, the more complex the digestive tract, the longer it will take to attain therapeutic blood levels of an orally administered drug.

The third event, drug passage through the liver, affects drug concentration in the blood following oral administration, because the liver can alter the drug and render it less (or more) active. Animals with liver disease will need special consideration when receiving oral medications. General guidelines for drug administration are summarized in Table 3-5.

Drug Dose

A **dose** of a drug can also affect the ability to maintain a drug in the therapeutic range. A *dose* is the amount of a drug administered at one time to achieve the desired effect. Examples of doses include milliliters (mL), cubic centimeters (cc), milligrams (mg), grams (g), or tablets (T). When recording a dose in an animal's record, it is more accurate to state the dose in mass units, such as grams or milligrams, because each manufacturer makes tablets in different sizes and liquids in different concentrations from other manufacturers.

Two terms used in reference to dose are **loading dose** and **maintenance dose**. The *loading dose* is the initial dose of drug given to get the drug concentration in the therapeutic range in a very short period of time. This is accomplished by giving a larger amount of drug initially or a normal amount of drug more frequently. A *maintenance dose* is the dose of drug that maintains or keeps the drug in the therapeutic range. The **total daily dose** is the total amount of drug delivered in 24 hours; for example, 30 mg given qid (four times daily) = 120 mg total.

Another term used in medicating animals is **dosage**, the amount of drug per animal species' body weight or measure. Examples of a drug dosage are 5 mg/kg or 25 g/lb.

Table 3-5   General Guidelines for Drug Administration (adapted from *Pharmacological Aspects of Nursing Care* 6th ed.)

1. Enteric-coated tablets should not be administered with antacids, milk, or other alkaline substances, because enteric-coated agents require the acid environment of the stomach to be effective.
2. Enteric-coated tablets should not be crushed before administration; crushing will alter absorption.
3. Suspensions and emulsions must be thoroughly shaken immediately before use, because the separation that occurs after standing for a short period will alter the dose if used in the separated form.
4. Suspensions should never be administered IV.
5. Solutions administered parenterally or in the eye must be sterile to prevent causing infection.
6. Solutions administered IV must be free of particulate matter that could serve as an embolus.
7. Proper storage of solutions is very important to prevent contamination and evaporation.
8. Skin integrity should be assessed for rashes or open areas before applying topical medications, as these conditions will alter absorption time of the medication.
9. Transdermal therapeutic systems or patches allow drugs to pass through intact skin and care should be taken when applying these to animals to prevent self-medication.
10. Proper disposal of transdermal patches is important to prevent their ingestion by animals or improper exposure to people.

Clinical Que

Always use the drug name first when describing the dosage regimen (drug name, dose, route, frequency, and duration).

Dosage Interval

Dosage is involved in the third factor in maintaining a drug in the therapeutic range, the **dosage interval**. The dosage interval is how frequently the dosage is given; for example, sid (*semel in die*) or once daily, bid (*bis in die*) or twice daily, tid (*ter in die*) or three times daily, or qid (*quarter in die*) or four times daily. The dosage interval, dosage, administration route, and duration of treatment together represent the **dosage regimen**. An example of a dosage regimen is 30 mg/kg tid po × 7 days.

OUT OF THE SAFE ZONE

Keeping the level of drug in the therapeutic range involves extensive studies of the drug's efficacy and toxicity signs. Drug toxicities may also occur due to human error and/or accident. Some causes of drug toxicity are summarized in Table 3-6.

Some ways to ensure that drugs stay within the therapeutic range and benefit the animal include the following:

*   Use the proper dosage, frequency, and duration of treatment for each drug prescribed.
*   Avoid combination drug treatment if possible, or allow a margin of error in dosing multiple drugs.

Table 3-6	Causes of Drug Toxicity

CAUSE OF DRUG TOXICITY	EXAMPLE
Outright Overdose	Dosing too frequently, dosing too high, administering too long
Relative Overdose	Recommended dose was too much for this animal, due to individual variation, impaired metabolism or excretion of drug by this individual animal, and improper route of administration
Side Effects	Normal side effects associated with drug may occur at a higher level in this individual
Accidental Exposure	Exposure of animal to drug that is absorbed through skin or inhalation or accidental ingestion
Interaction with Other Drugs	If drug A is highly protein bound and drug B is protein bound as well, the competition for protein binding will affect the amount of free drug (active drug) available to the animal
Incorrect Treatment	Misdiagnosis: treatment causes toxicity levels of drug because the animal does not have the disease; for example, hormone replacement raises hormone levels above normal in healthy animal

- Use less toxic drugs if available (e.g., some antifungals are more toxic than others).

- Be aware of potential hazards and precautions. Sometimes fluid therapy may be given to an animal prior to and during treatment to enhance renal excretion of a drug.

- Use high-quality drugs, check expiration dates, check handling requirements, understand contamination possibilities, and make sure the drug is thoroughly mixed and not precipitating out of solution.

- Follow label directions carefully.

- Know the patient's history and keep in contact with the animal's owner.

If an animal develops drug toxicity, the veterinary staff needs to act quickly to counteract any problems caused by the drug treatment. Some ways to treat drug toxicities include the following:

- removal of the offending drug (e.g., washing off flea products that are causing neurologic signs)

- enhancing drug removal by the animal (e.g., making the animal vomit or administering fluids to enhance drug excretion by the kidney)

- counteracting with an antidote (e.g., giving naloxone for an overdose of morphine)

- providing symptomatic care or nursing care for the animal until the toxicity signs have diminished.

SUMMARY

Appropriate administration of a drug includes administration of the appropriate amount of drug (based on dosage) into the animal's body by the appropriate route of administration. To provide safe drug administration, the six rights of proper drug administration should be followed. These six rights include the right drug, the right dose, the right time, the right route, the right patient, and the right documentation. The dos and don'ts of drug administration are described in Appendix E.

Maintaining drugs within the therapeutic range is important in obtaining optimal drug efficacy. The therapeutic range of a drug is the drug concentration in the body that produces the desired effect in the animal with minimal or no signs of toxicity. Too little drug will keep the drug level in the subtherapeutic range; hence, there is little or no benefit to using the drug. Too much drug will push the drug level into the toxicity range, which will have adverse effects on the animal.

Three key aspects to keeping drug levels within the therapeutic range are route of administration, dose, and dosage interval. Route of drug administration plays a role in speed of drug absorption and utilization of the drug. Routes of drug administration include parenteral, oral, and local means. Various drug forms exist for each type of drug, and the drug form affects its absorption and distribution in the animal's body. Dose is the amount of drug given to the animal and is determined based on the dosage of drug recommended by the manufacturer and the weight of the animal. Dosage is the amount of drug per animal species' body weight or measure. Dosage interval is how frequently we give the dosage. Dosage regimen is the dosage interval, dosage, administration route, and duration of treatment.

Drug toxicities do develop in animals for a variety of reasons. If an animal experiences drug toxicities, the signs of these reactions must be identified quickly and treated with a variety of options, depending on the drug and what toxicities it has caused.

It's a Wrap

The answers to the questions in this chapter's Setting the Scene should be understood after reading this chapter. Which drug form should be used in this case study and why? Which route of drug administration would the veterinarian choose, and how should the technician be prepared to assist in the drug administration? In emergency situations, what are the most rapid routes of drug administration? What drug forms are available for these routes of administration?

(continued)

The route of administration of a drug is important in determining how quickly a response to a drug occurs. In emergency situations, medications and fluids tend to be given IV. Intravenous injections give predictable concentrations of drug and usually give an immediate response because the drug is available to the tissues once it is in the circulation. Other routes of administration such as IM and SQ may be used, but drug absorption will be slower.

An example of a drug used to remove fluid from the chest cavity is furosemide. It is a diuretic (a chemical that increases fluid loss through increased urine production) that helps remove excess fluid from the body. If given IV, the diuretic effect of furosemide occurs within five minutes. If given orally, the diuretic effect of furosemide occurs within one hour. In emergency situations, an hour may be too long to wait for a drug effect to occur.

CHAPTER REVIEW

Matching

Match the definition with its proper term.

1. _____ injectable drug that is in a substance that delays its absorption

2. _____ small, sterile, prefilled glass containers containing medication for injection

3. _____ concentrated mass of drug

4. _____ administration of a large volume of fluid continuously over extended periods of time

5. _____ dilution of a drug dose in a small volume of fluid that is administered over a period of time (typically 30 to 60 minutes) via an indwelling catheter

6. _____ solution types that should not be given subcutaneously

7. _____ term for turned into a gas

8. _____ drug suspended in water–oil emulsion

9. _____ drug suspended in a semisolid preparation that retains its state at body temperature

10. _____ suspended in liquid for dabbing, brushing, or dripping on skin without friction

a. irritating or hyperosmotic solutions
b. ampule
c. continuous infusion of fluids
d. cream
e. intermittent IV therapy
f. lotion
g. volatilized
h. repository or depot preparation
i. paste
j. bolus

Multiple Choice

Choose the one best answer.

11. The dose or dosage of a drug that produces the desired effect in the animal with minimal or no signs of toxicity is the
    a. toxicity subrange
    b. therapeutic range
    c. route of administration
    d. subtherapeutic range

12. An example of a parenteral drug route of administration is
    a. by mouth (po)
    b. rectally
    c. IV
    d. sublingual

13. In general, which route of drug administration has a longer duration of action than IV, yet a shorter duration of action than oral?
    a. rectal
    b. transdermal
    c. inhalation
    d. IM

14. Which route of drug administration do gastric fluids affect?
    a. IM
    b. po
    c. IV
    d. SQ

15. An injectable drug placed into a substance that delays absorption is called a
    a. parenteral drug
    b. repository preparation
    c. suspension
    d. solution

16. Drugs administered via nebulization are
    a. volatilized for inhalation
    b. mucolytic and administered orally
    c. depot prepared and injected
    d. turned into a fine mist for inhalation

True/False

Circle a. for true or b. for false.

17. The loading dose of a drug is the initial dose given to get the drug concentration up to the therapeutic range in a very short period of time.
    a. true
    b. false

18. The dosage of a drug is the total amount of drug delivered in 24 hours and is based on the animal's weight.
    a. true
    b. false

19. 100 mg/kg bid is an example of a dosage regimen.
    a. true
    b. false

20. bid is the Latin abbreviation for twice daily.
    a. true
    b. false

Case Study

21. An elderly client on a limited budget brings her cat in for an examination. The owner tells you that the cat has been vomiting for two days and hiding under the bed. On physical examination, the cat appears lethargic, but its vital signs are normal. Blood is collected from the cat for a chemistry panel and a complete

blood count. Because the blood test results are not available immediately, the owner wants to give the cat something to make it stop vomiting. The owner states that cost is the only concern in medicating this cat.

Cost may be an important factor determining treatment; however, it is not the only concern. What are some other concerns regarding treatment of this cat?

Critical Thinking Questions

22. What is the significance of a drug's therapeutic range to a veterinary technician?

23. What factors must be considered before determining the correct route of administration used to deliver any drug into an animal's body?

ADDITIONAL REVIEW MATERIAL

Additional review material can be found on the StudyWARE™ CD included in this text.

BIBLIOGRAPHY

AVMA (May 15, 2007). AVMA adopts policy on informed consent, http://www.avma.org/onlnews/, accessed February 10, 2009.

McGee, A. R. N. (2006). Overview of Veterinary Client Issues, Animal Legal and Historic Center, http://www.animallaw.info/articles/ovusveterinaryclient.htm, accessed February 10, 2009.

Broyles, B., Reiss, B., and Evans, M. (2007). *Pharamcological Aspects of Nursing Care* (7th ed., pp. 2–77). Clifton Park, NY: Delmar Cengage Learning.

Rice, J. (2006). *Principles of Pharmacology for Medical Assisting* (4th ed.). Clifton Park, NY: Delmar Cengage Learning.

Veterinary Pharmaceuticals and Biologicals (12th ed.). Lenexa, KS: Veterinary Healthcare Communications.

Wynn, R. (1999). *Veterinary Pharmacology.* Cambridge, MA: Harcourt Learning Direct.

Nordenberg, T. Pharmacy compounding: Customizing prescription drugs. *FDA Consumer Magazine,* July–August 2000.

Margolis, M. FDA may not prevent pharmacists from promoting compounded drugs, www.law.uh.edu, accessed August 30, 2002.

CHAPTER 4
PHARMACOKINETICS

OBJECTIVES

Upon completion of this chapter, the reader should be able to:

- outline the four components of pharmacokinetics.
- characterize the four mechanisms that allow drugs to move across cell membranes.
- describe factors that affect drug absorption.
- describe ion trapping.
- characterize factors that affect drug distribution.
- describe the role protein binding plays in drug availability.
- describe factors that affect drug biotransformation.
- list sites of drug biotransformation and excretion.
- describe factors that affect drug excretion.
- describe the role of receptors in pharmacology.
- differentiate between agonist and antagonist.

KEY TERMS

absorption	induce
active transport	ion trapping
affinity	ionized
agonist	lipophilic
antagonist	metabolism
bioavailability	nonionized
biotransformation	passive diffusion
distribution	phagocytosis
drug residue	pharmacokinetics
elimination	pinocytosis
excretion	receptor
facilitated diffusion	steady state
half-life	tolerance
hydrophilic	withdrawal time

Setting the Scene

A horse is experiencing front-limb lameness from an injury that occurred during a race. The owner of this horse calls the veterinarian out to the ranch to have the horse examined. The veterinarian determines that the horse needs an anti-inflammatory drug to decrease some of the front-limb joint swelling noted on the physical exam. The treatment includes an IM injection of an anti-inflammatory drug. The owner wonders why the injection is being given in the horse's muscle when the horse has pain in its joint. How does the medication get from a muscle to another location in the body? How do medications get to where they are needed?

Courtesy of iStockphoto.

59

DRUG MOVEMENT

Pharmacokinetics is the physiological movement of drugs within the body (*pharmaco-* means drug and *kine-* means motion) and includes the mechanisms by which drugs move into, through, and out of the body. There are four steps in pharmacokinetics: absorption, distribution, metabolism or biotransformation, and excretion.

Pharmacokinetics also includes the movement of substances across cell membranes. Four basic mechanisms allow drugs to move across cell membranes: passive diffusion, facilitated diffusion, active transport, and pinocytosis/phagocytosis. These mechanisms are summarized in Table 4-1.

Table 4-1   Mechanisms of Drug Movement

TYPE OF MOVEMENT	ENERGY	CARRIER	NOTES
Passive diffusion: random movement of molecules from an area of high concentration of molecules to an area of low concentration of molecules	– –	– –	• Rapid for nonionic, lipophilic, small molecules • Slow for large, ionic, hydrophilic molecules • Lower temperatures slow the rate of diffusion; higher temperatures speed the rate of diffusion • Thick cell membranes slow the rate of diffusion; thin cell membranes do not slow the rate of diffusion
Facilitated diffusion: passive movement with special molecules within the membrane that carry the molecules through the membrane	– –	++	• Like passive diffusion except uses a carrier molecule • Cannot concentrate molecules on one side or the other • Entry of glucose into the cells occurs by facilitated diffusion with the help of insulin molecules
Active transport: movement of molecules across membranes involving a carrier molecule that pumps the molecule against a concentration gradient	++	++	• Like facilitated diffusion but needs energy (ATP) • Movement of strongly acidic or basic substances into urine usually occurs by active transport • pH gradient in body systems usually occurs by active transport
Pinocytosis/phagocytosis: cell drinking or cell eating in which cellular membrane surrounds the molecule and takes it into the cell	++	– –	• Large drug molecules, such as proteins, are usually involved in these processes

Passive Diffusion

Passive diffusion is the movement of drug molecules (particles) from an area of high concentration of molecules to an area of low concentration of molecules. Drug molecules move from an area where there are many such molecules (high concentration) to an area where there are few such molecules (low concentration) until they start to even out between the two sides (Figure 4-1A). When both sides are nearly equal in concentration, the molecules continue to exchange at an even rate to keep equal numbers of molecules both inside and outside the cell. Passive diffusion does not require energy, nor does it expend energy.

For passive diffusion to occur, the drug must dissolve in the cell membrane (which is made primarily of phospholipids) and must pass readily through the cell membrane. Drugs that move through cell membranes by passive diffusion must be small in size, lipophilic, and nonionic. Keep in mind that pores in the cell membrane are small; therefore, drugs that pass through these pores must also be small. Large drugs cannot pass through small pores and are left outside the cell.

Lipophilic means fat loving (*lipo-* means fat and *-phil* means loving). Lipophilic drugs are chemicals that dissolve in fats or oils. Lipophilic drugs are able to dissolve in the phospholipid or fat-containing cell membrane because like molecules dissolve in like molecules. Recall that oil and water do not mix; therefore, lipophilic and hydrophilic substances do not mix. The intestinal mucosa has lipid-rich cell membranes, allowing lipophilic drugs to be well absorbed from the gut via passive diffusion. **Hydrophilic** means water loving (*hydro-* means water and *-phil* means loving). Hydrophilic drugs are chemicals that dissolve in water. Hydrophilic drugs do not pass through lipid-rich cell membranes as easily as lipophilic drugs. Drugs that are already in fluid (body water) do not need to move through phospholipid membranes. For example, drugs injected intramuscularly are deposited in the fluid that surrounds the cells and must diffuse through fluid to reach the capillaries. Intramuscular drugs should ideally be in a hydrophilic form for rapid absorption. Intramuscular drugs may be in a lipophilic form to slow the rate of absorption.

Drug ionization also affects movement of the drug. The ionization or charge of a drug depends on the pH of the liquid in which it is immersed and the pH of the drug. **Ionized** drugs have either a positive or a negative charge. **Nonionized** drugs have no charge or are considered neutral. Ionization or neutrality also affects the hydrophilic and lipophilic properties of a drug. Nonionized drugs tend to be in lipophilic form; ionized drugs tend to be in hydrophilic form. One goal of passive diffusion is to move molecules through a lipid-containing cell membrane. The molecules that move most effectively through the cell membrane are lipophilic; therefore, nonionized or neutrally charged particles would pass through the cell membrane more easily, as they tend to be lipophilic. When a drug is given, some of the drug molecules go into the hydrophilic form and other drug molecules go into the lipophilic form. The amount of drug that goes into a particular form depends on the chemistry of the drug and the environmental pH in which the drug is found (e.g., oral drugs act in the acid environment of the stomach, whereas intravenous medications are placed in the more neutral environment within the blood vessels).

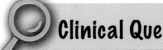

Clinical Que

Lipophilic drugs tend to pass through phospholipid cell membranes readily. Hydrophilic drugs have difficulty passing through phospholipid cell membranes.

Clinical Que

Ionized = charged; Nonionized = uncharged.

Hydrophilic drugs are usually ionized; lipophilic drugs are usually nonionized.

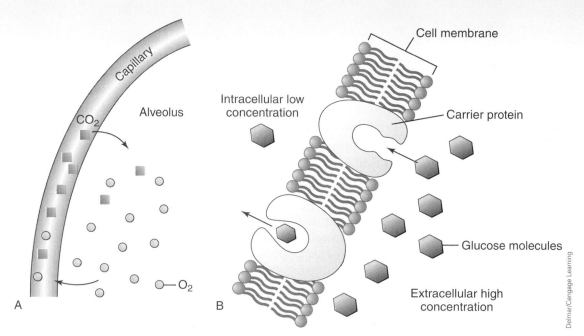

Figure 4-1(A) Diffusion is the movement of atoms, ions, or molecules from an area of high concentration to an area of low concentration. In this example, O_2 and CO_2 diffuse in the lung. **(B)** Facilitated diffusion is a special kind of diffusion that utilizes a carrier molecule. Glucose moving from blood into cells is an example. Glucose molecules are too big and are not lipid soluble enough to diffuse through the cell membrane; therefore, they combine with special carrier proteins so that they can pass through the cell membrane.

Facilitated Diffusion

The second type of cellular movement is **facilitated diffusion,** passive diffusion that utilizes a special carrier molecule (Figure 4-1B). This carrier molecule helps the drug across the cell membrane. When both sides are nearly equal in concentration, the molecules continue to exchange at an even rate to keep equal numbers of molecules both inside and outside the cell. No energy is needed for facilitated diffusion; molecules still move from an area of high concentration to an area of low concentration.

Active Transport

Active transport, the third type of cellular movement, involves both a carrier molecule and energy (Figure 4-1C). Energy is needed in active transport because drug molecules move against the concentration gradient, from areas of low concentration of molecules to areas of high concentration of molecules. Active transport allows drugs to accumulate in high concentration within a cell or body compartment. Sodium, potassium, and some other electrolytes move via active transport and accumulate in higher amounts on one side of a membrane than the other.

Pinocytosis and Phagocytosis

Pinocytosis and **phagocytosis** are mechanisms of molecule movement in which molecules are physically taken in or engulfed by a cell (Figure 4-1D). Pinocytosis (cell drinking) occurs when the cell membrane surrounds and engulfs liquid

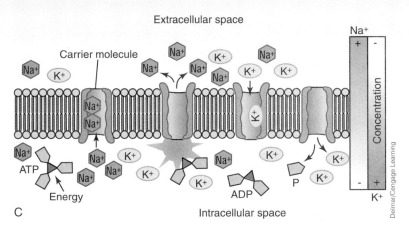

Figure 4-1(C) The active transport process moves particles against the concentration gradient from a region of low concentration to a region of high concentration. Active transport requires energy and a carrier molecule. An example is the Na⁺-K⁺ pump, which keeps sodium levels high outside the cell and potassium levels high inside the cell.

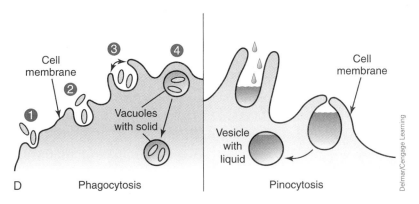

Figure 4-1(D) The processes of phagocytosis (engulfing solids) and pinocytosis (engulfing liquids) move very large particles into the cell by surrounding the particle and forming a vesicle.

particles. Insulin is taken into animal cells by pinocytosis. Phagocytosis (cell eating) occurs when the cell membrane surrounds and engulfs solid particles. Animal cells acquire nutrients by phagocytosis. Engulfing liquid or solid particles into the cell requires energy. Pinocytosis and phagocytosis are important for the movement of large molecules, such as proteins, that cannot enter a cell by passing intact through the cell membrane.

PHARMACOKINETICS PART I: GETTING IN

Drug **absorption** is the movement of drug from the site of administration into the fluids of the body that will carry it to its site(s) of action. Absorption is the first step in the passage of a drug through the body, unless it is introduced directly into the blood via intravenous administration. Most drugs are carried in the systemic blood circulation. Drug factors and patient factors can affect the degree of drug absorption within an animal. Drug factors such as solubility, drug pH, and molecular size can affect drug absorption. Patient factors, such as the animal's age and health status, greatly affect how that individual animal can absorb a drug. Keep in mind that we do not want some drugs to be absorbed, such as topical anesthetics and activated charcoal. Factors affecting absorption include the amount of drug in the body, drug pH and ionization, ion trapping, drug form, and patient factors.

Amount of Drug in the Body

The degree to which a drug is absorbed and reaches the circulation is an important component of drug absorption. This component is termed **bioavailability**. Bioavailability is the percent of drug administered that actually enters the systemic circulation. Intravenous or intra-arterial drugs immediately enter the systemic blood circulation; therefore, drugs given IV or IA are 100 percent bioavailable (they have a bioavailability number of 1). Drugs can be 100 percent bioavailable even if they do not enter the systemic blood circulation immediately. Drugs that are partially absorbed have a bioavailability of less than one. The lower the bioavailability of a drug, the less of it there is in the circulation and subsequently the less of it there is in tissue. Factors that affect bioavailability of a drug include blood supply to the area (muscle has a greater blood supply than the subcutaneous space, so drugs given IM tend to have a higher bioavailability than those given SQ), surface area of absorption (increased surface area means more space for absorption to take place), mechanism of drug absorption (diffusion versus active transport versus pinocytosis/phagocytosis), and dosage form of the drug (IV and IM routes of administration should yield higher bioavailabilities than oral routes of administration).

Some drugs do not go directly into the systemic circulation following absorption but instead pass from the intestinal lumen to the liver via the portal vein. In the liver most of the drug is metabolized to an inactive form of the drug for excretion. This reduces the amount of active drug in the systemic circulation. The process by which the drug passes to the liver first is called the *first-pass effect* or hepatic first pass. Lidocaine and nitroglycerin are drugs that are not given orally because they have extensive first-pass metabolism, so the majority of the drug given orally is destroyed.

pH and Ionization

The pH of the drug and the pH of the environment where the drug is administered both play a role in drug absorption. *pH* is the measurement of the acidity or alkalinity of a substance. pH is based on a scale of 0–14, with the lower numbers indicating acid and the higher numbers indicating alkaline or base. Neutrality on the pH scale is 7.

Drugs exist in both ionized (charged) and nonionized (uncharged) forms. When weakly acidic drugs, such as aspirin, are swallowed, they enter the acidic environment of the stomach. Some of the drug molecules are ionized, and some of the molecules are nonionized in the stomach. The drug particles then progress to the small intestine where the pH is higher (more alkaline). Most of the aspirin molecules are in an ionized form in the small intestine. Recall that hydrophilic drugs are usually ionized (charged) and that lipophilic drugs are usually nonionized (uncharged). The hydrophilic form of aspirin is the charged portion

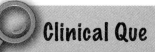

Clinical Que

For a drug to be effective, it must get from where it is administered to the bloodstream. The percent of drug that does this determines the bioavailability of the drug.

in the small intestine, and the lipophilic form of aspirin is the uncharged portion in the stomach. Because the lining of the gastrointestinal tract is composed mainly of phospholipid, the lipophilic or nonionized form is needed for the drug to be readily absorbed through the gastrointestinal mucosa. Weakly acidic drugs are more likely to be absorbed in the stomach, as there is a greater amount of nonionized drug in the stomach than in the small intestine. The more alkaline environment of the small intestine favors the hydrophilic or ionized form of a weakly acidic drug and therefore limits its absorbance in the small intestine.

The acidic or alkaline nature of the drug itself also helps determine whether a drug is predominately in lipophilic or hydrophilic form. Weakly acidic drugs tend to be in hydrophilic form in an alkaline environment, whereas weakly alkaline drugs tend to be in hydrophilic form in an acid environment. Acidic drugs tend to take ionized form in an alkaline environment, whereas alkaline drugs tend to take ionized form in an acidic environment.

The pH at which a drug has a 1:1 ratio of ionized drug to nonionized drug is its pKa. pKa links the drug with the location in which it is best absorbed. By knowing a drug's pH and pKa, the amount of drug in the ionized or nonionized form can be determined. An acid drug with a pKa of 7 will have the same number of molecules in the ionized and nonionized form when it is in an environment with a pH of 7. Acid drugs become more nonionized when placed in an acid environment; therefore, any acid drug placed in any acidic pH environment will produce more drug molecules in the nonionized (lipophilic) form. On the other hand, placing an acid drug in any alkaline pH environment will produce more molecules in the ionized (hydrophilic) form. For an alkaline drug with a pKa of 7, the opposite is true (alkaline drugs placed in acid environments will produce more drug molecules in the ionized [hydrophilic] form and alkaline drugs placed in alkaline environments will produce more drug molecules in the nonionized [lipophilic] form) (Figure 4-2). The ratio of ionized to nonionized molecules changes by a factor of 10 for each pH unit change.

Ion Trapping

Ion trapping occurs when a drug molecule changes from ionized to nonionized form as it moves from one body compartment to another. Drugs can pass from one body compartment to another body compartment with a different pH. When the drug changes compartments, it may change its ionization and become trapped in that new environment. For example, aspirin taken orally enters the acidic environment of the stomach. The majority of the acidic aspirin molecules are in nonionized form in the stomach (remember, some stay in ionized form).

<aside>
Clinical Que

Acid drugs in an alkaline environment tend to be charged. Alkaline drugs in an acid environment tend to be charged. Acid drugs become more lipophilic (and are more readily absorbed across membranes) at a pH more acidic than its pKa. Alkaline drugs become more lipophilic at a pH more alkaline than its pKa.
</aside>

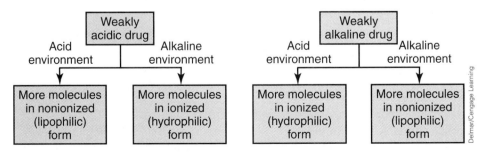

Figure 4-2 The effect of drug pH and environmental pH on the ionization of a drug.

The phospholipid layer of the stomach wall readily absorbs the nonionized form. The aspirin molecules enter the stomach cells and are then in an environment with a pH near neutrality (approximately 7). The pH of the environment that the aspirin is in is becoming more alkaline, and thus the aspirin shifts to ionized form (however, some stay in nonionized form). When the aspirin shifts to ionized form, the molecules become trapped in the stomach cells. The equilibrium between ionized and nonionized molecules within the stomach cells continues to be maintained, and some nonionized molecules pass out of the stomach cell and into the blood. Nonionized molecules are converted to their ionized forms in the blood. This process helps keep the absorbed aspirin molecules in the bloodstream where they can then be distributed to the body cells (Figure 4-3). Ion trapping is especially important in drug excretion, as alterations in the urine pH can allow drugs to be trapped in the urine and excreted.

Oral Versus Parenteral Drug Forms

Absorption is further affected by the choice of oral or parenteral route of drug administration. Solid drugs administered orally must first be broken down and then dissolved in gastric or intestinal fluids before they can be absorbed (Figure 4-4). Disintegration is the breakdown of a solid drug into smaller particles, and dissolution is the dissolving (passing from a solid into solution) of the smaller particles in body fluid for absorption. Administering fluids with solid drugs generally increases the rate at which a drug dissolves and may speed its rate of absorption. Food in the gastrointestinal tract can also interfere with the

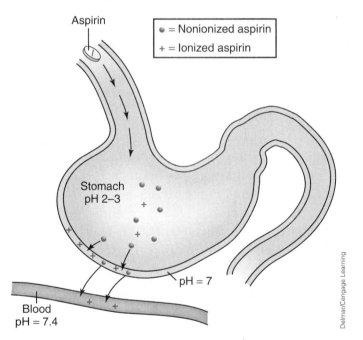

Figure 4-3 Ion trapping occurs when a drug molecule changes from ionized to nonionized form as it moves from one body compartment to another. Aspirin taken orally is nonionized in the stomach and is absorbed. When it enters the stomach cells, the aspirin shifts to its ionized form and cannot pass back into the lumen of the stomach. As a result of equilibrium, some nonionized aspirin molecules pass from the stomach cells into the blood, where they become ionized. These ionized aspirin molecules cannot pass effectively from the blood to the stomach cells.

dissolution and absorption of a drug. The presence of food can raise the levels of stomach acid, which may enhance or hinder drug dissolution. Food in the gastrointestinal tract may increase or decrease transit time, which may alter drug absorption. Food may also inhibit or enhance drug-metabolizing or drug-transportation time, which will affect drug dissolution and absorption. Keep in mind that some drugs are irritating to the gastrointestinal mucosa; so fluid or food is necessary to dilute the drug concentration. Drug size, gastric motility, and the lipophilic nature of the drug also affect orally administered drugs.

Keep in mind that tablets are not 100 percent pure drug, but rather contain filler and inert substances to allow the drug to be a particular size and shape and to enhance dissolution of the drug. Some drug additives, such as the ions potassium and sodium in penicillin potassium and penicillin sodium, increase the absorbability of the drug. Penicillin is poorly absorbed from the gastrointestinal tract because of gastric acid. By adding potassium or sodium to the penicillin, more drug can be absorbed.

Parenterally administered drugs are more affected by tissue blood flow and the hydrophilic nature of the drug than orally administered drugs. For example, drugs given intramuscularly may be absorbed faster when given in muscles that have more blood vessels (thus better tissue blood flow). Drugs given subcutaneously are absorbed more slowly than those given intramuscularly because subcutaneous tissue has fewer blood vessels than muscles. These effects are summarized in Table 4-2.

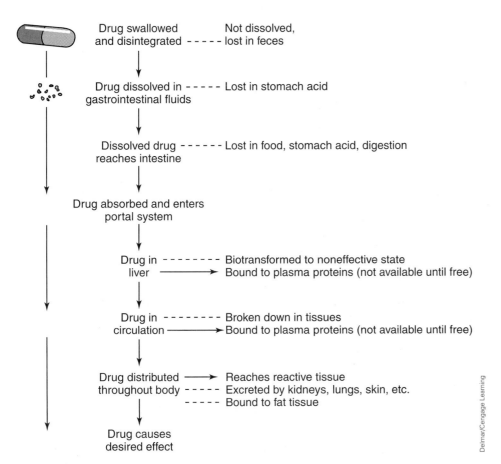

Figure 4-4 Oral drugs need to be disintegrated, dissolved, and absorbed before they can be distributed throughout the body to cause their effect.

Table 4-2 Drug Factors that Affect Drug Absorption

Drug Chemistry	• Lipophilic drugs dissolve in oil-based fluids. Lipophilic drugs are absorbed well across phospholipid-based cell membranes. • Hydrophilic drugs dissolve readily in water (tissue, fluid, and lymph). Tissue fluid is water soluble; therefore, drugs that are hydrophilic dissolve in and diffuse through tissue fluid quite well.
Drug Size	• Molecular size of the drug: Small molecules can pass more readily through cell membranes.
Ionization of the Drug	• Nonionized or neutral drugs are lipophilic and can pass through phospholipid cell membranes. • Ionized or charged drugs are hydrophilic and dissolve in and diffuse through tissue fluid. • Ionization of the drug depends on the drug pH and the environmental pH.
Acid-Base Characteristics	• pH of drug: Drugs may change their form when the pH of the environment changes. Weak acids become more ionized as the pH of the environment increases. Weak bases become more ionized as the pH of the environment decreases.
Ion Trapping	• Drugs can pass from one compartment to another compartment with a different pH. When the drug changes compartments, it may become ionized and become trapped in its new environment
Drug Form	*Oral* drugs must be in lipophilic form to penetrate the GI mucosa. They must be small to dissolve in the membrane. • Tablets must dissolve into smaller particles. Liquid drugs do not have a dissolution step; therefore, oral liquid drugs have a quicker onset of action than pills. • The drug (enteric coating) or special construction of the tablet (sustained-release) may alter dissolution and/or absorption. • Decreased gastric motility lengthens the time it takes for the drug to reach the absorption site. • Increased gastric motility shortens the time the drug remains in the GI tract. This time may not be long enough to allow drug dissolution, and the drug may pass unused in the feces. • The drug must be able to survive first-pass effect or detoxification by the liver. Remember that a drug is absorbed from the intestine and passes through the liver before it enters the systemic circulation. • The presence of food may interfere with the dissolution and absorption of certain drugs. *Parenteral* drugs must be in hydrophilic form. • Anything that interferes with diffusion of the drug from the administration site or alters blood flow to the injection site will delay absorption. • Some drugs are formulated to have a delayed absorption (repository or depot injections). • If there is limited blood flow in the injection site, absorption will be slowed (fat is poorly perfused, muscle is richly perfused). • Temperature may result in vasoconstriction or vasodilation, and affect blood flow to the administration site. • Other drugs may affect blood flow as well (e.g., lidocaine).

Table 4-3 Patient Factors that Affect Drug Absorption

ANIMAL FACTOR	EXAMPLE OF EFFECT
Age	• Young animals may not have well-developed gastrointestinal tracts. • Young animals may have less active enzyme systems.
Health	• Fever may cause molecules to move faster and increase absorption. • Disease signs like diarrhea may speed the drug through the gastrointestinal tract and not allow enough time for proper absorption. • Disease signs like vomiting may affect the time the drug is in the stomach, hindering absorption. • Poor circulation due to a variety of disease conditions hampers drug absorption.
Metabolic rate	• Animals with a high basal metabolic rate may metabolize and/or eliminate drugs more rapidly than those with a normal metabolic rate.
Genetic factors	• Individual variation in response to drugs may occur because of genetic differences between animals. For example, one animal may metabolize a drug more slowly due to a genetically based enzyme deficiency.
Sex	• Male and female animals have different body compositions. The proportion of fat to lean body mass may influence the action and distribution of drugs throughout the animal's body.
Species	• Drugs and food may stay in the rumen for longer periods of time, which may delay drug absorption. • Herbivores that continuously graze may have altered drug absorption due to the presence of food in the gastrointestinal tract.

Patient Factors

Patient factors affecting drug absorption include blood flow, pain, stress, hunger, fasting, food consumption, and pH. Poor tissue perfusion due to altered blood flow or the use of vasoconstricting drugs limits drug absorption. Exercise can affect blood flow by causing more blood to flow to muscles and causing less blood to flow to the gastrointestinal tract. Pain, stress, and the presence of food in the stomach can slow gastric emptying time; so the drug remains in the stomach longer (increasing absorption). Differences in the digestive systems of various species can also affect drug absorption. Patient factors affecting drug absorption are summarized in Table 4-3.

PHARMACOKINETICS PART II: MOVING AROUND

Drug **distribution** is the physiological movement of drugs from the systemic circulation to the tissues. The goal of drug distribution is for the drug to reach the target tissue or intended site of action. Factors that affect drug distribution include membrane permeability, tissue perfusion, protein binding, and volume of distribution.

Membrane Permeability

Membrane permeability has a great effect on drug distribution. Blood capillaries are only one–cell layer thick and have *fenestrations* or small holes between cells to allow drug molecules to move in and out of the capillaries (Figure 4-5A and B). Large molecules usually cannot pass through these fenestrations and are trapped in the bloodstream. One exception to this permeability is the blood-brain barrier. Capillaries in the central nervous system (brain and spinal cord) have no fenestrations. There are also supporting cells (glial cells) in the CNS that surround the capillaries, creating an extra barrier. The lack of fenestrations in the CNS capillaries and the glial cell barriers allow only the most lipophilic drugs to enter an undamaged CNS. Disease manifestations such as fever and inflammation may alter this blood-brain barrier, thus allowing drugs to penetrate the CNS.

Another potential barrier to membrane permeability is the placenta. However, the existence of the placental barrier does not prevent all drugs from passing from mother to fetus. Many drugs can pass through the permeable placental capillaries. Drugs that pass through the placental barrier will affect the fetus.

Tissue Perfusion

Tissue perfusion affects how rapidly drugs will be distributed. Tissue perfusion is the relative amount of blood supply to an area or body system. The level of tissue perfusion varies among body systems and animal species. Distribution will be rapid to well-perfused tissues like the brain, heart, liver, and kidneys and slow to poorly perfused tissues like fat and skin. Some areas

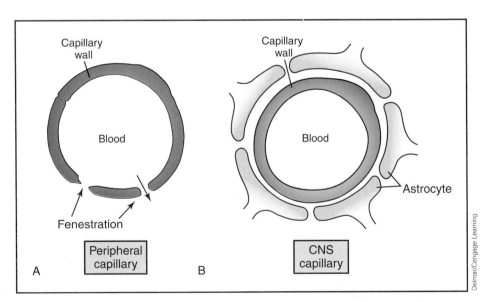

Figure 4-5 (A) Capillaries found in the body have fenestrations (small holes) that allow some molecules to move in and out of the capillaries. (B) Capillaries in the CNS do not have fenestrations; therefore, under normal conditions, only the most lipophilic drugs can enter the CNS.

that are well perfused (such as the brain) may initially achieve high levels of drug, because of the rapid distribution of a particular drug from the bloodstream to the well-perfused area; other poorly perfused areas (such as fat) maintain lower levels of drug. An example of this occurs with the anesthetic agent thiopental. The physical and chemical characteristics of thiopental (and all drugs) usually determine precisely how the drug will be distributed. Thiopental is lipophilic and is stored in fat. It would seem reasonable that overweight patients given thiopental should have high levels of the drug distributed to fat and less of the drug in the brain, resulting in delayed onset of anesthesia. Clinically, however, patients given thiopental IV become anesthetized quickly. Why is this? Remember that tissue perfusion as well as the drug's affinity for a type of tissue plays a role in drug distribution. The well-perfused tissues, like the brain, rapidly receive high concentrations of thiopental when it is given IV. Thiopental in the circulation is simultaneously being distributed to the poorly perfused fat tissue. The concentration of thiopental in the blood decreases as the drug moves from blood to fat. At some point, the concentration of thiopental becomes lower in the blood than its concentration in the brain. Lower drug levels in the bloodstream result in drug being moved to the bloodstream from the brain. The drug in the bloodstream then goes to the fat. Fat becomes a reservoir for thiopental. As thiopental continues to be redistributed from the brain to the bloodstream to the fat, the animal regains consciousness. Typically, more thiopental is given to keep the animal anesthetized. This could eventually lead to high levels of drug in the fat, due to redistribution of thiopental between the brain, bloodstream, and fat. The animal will have too much anesthetic in fat storage, and over time the drug will continue to be redistributed to the brain from the fat stores as the brain levels of thiopental decrease. This phenomenon may keep the animal anesthetized too long. This is especially true of overweight patients given lipophilic drugs, because they have more fat available to store larger amounts of the lipophilic drugs (Figure 4-6 A–D).

Tissue perfusion can also be affected by alterations in blood flow rates, which may occur because of disease conditions or other drug treatments that cause blood vessel constriction or dilation. For instance, decreased blood flow to tissues, as a result of a patient being in heart failure, will decrease the rate and amount of drug delivered to various tissues.

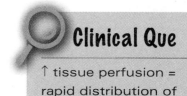

Clinical Que

↑ tissue perfusion = rapid distribution of drug.

Protein Binding

Protein binding also affects drug distribution. Some drugs bind to proteins (particularly albumin) in the blood. Proteins are large and cannot leave the capillaries, so the drug–protein complexes become trapped in the circulation (Figure 4-7). Free or unbound drugs are able to leave the capillaries. In most cases, equilibrium is established between the concentration of bound and unbound drug. This allows bound drug to be released from its binding sites when plasma concentrations of unbound drug diminish. When two protein-bound drugs are given concurrently, they compete for

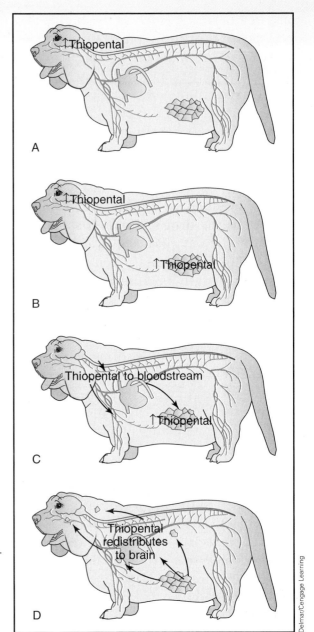

FIGURE 4-6 (A) Thiopental levels are high in the brain following IV administration. (B) Thiopental in circulation is distributed to fat (thus lowering the levels in blood). (C) Thiopental moves from the brain to the blood and then to fat. (D) Over time thiopental will redistribute from fat to the brain.

protein-binding sites. Displacement of one protein-bound drug by another may increase the pharmacological response to the displaced drug since more of it may be circulating in the blood in the free, unbound state. Patients receiving two drugs capable of competing for the same binding sites need to be closely monitored for an enhanced or diminished drug response.

Albumin is the principal protein in systemic circulation and is produced in the liver. An animal that has liver disease or another protein-losing disease will have less protein available for protein binding. Less protein binding of drug will result in more free drug, which means more drug available to the target

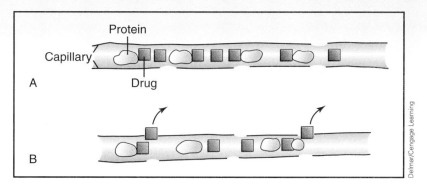

FIGURE 4-7 Protein binding of drugs. Drugs that are bound to protein are unable to leave circulation due to the size of the protein molecule to which they are bound. As free (unbound) drug leaves the capillaries, equilibrium between the concentration of bound and unbound drug is reestablished as bound drug is released from its binding sites.

tissue. Liver disease and protein-losing diseases should alert the veterinary staff to closely monitor drug dosing, because of the decreased potential for drug–protein binding. Less protein binding of drug results in more drug available to the target tissue(s), and potential toxic side effects from high tissue levels of the drug could occur.

Volume of Distribution

Volume of distribution is another factor in drug distribution. *Volume of distribution* is how well a drug is distributed throughout the body based on the concentration of drug in the blood. Volume of distribution assumes that the drug concentration in the blood is equal to the drug concentration dispersed throughout the rest of the body. Drug concentration in blood will be lower if the drug has a large volume to distribute itself through. The larger the volume of distribution, the lower the drug concentration in the blood after distribution.

An example of how volume of distribution affects drug concentration is in a dog with ascites (fluid in its abdomen) (Figure 4-8). A dog with ascites has a greater volume to distribute the drug in (the abdominal fluid), and hence there is a lower concentration of drug in the blood and other tissues. Distributing 10 mg of drug in a small volume (normal dog) will provide a certain concentration of drug in all body compartments. Distributing 10 mg of drug in a large volume (dog with ascites) will provide a decreased concentration of drug in all body compartments. Think of it as 10 mg of drug in a volume of 100 mL (normal dog) versus 10 mg of drug in a volume of 1000 mL (dog with ascites). The larger volume is less concentrated than the smaller volume. A less concentrated situation (dog with ascites) may keep a drug out of the therapeutic range, thus altering the effectiveness of the drug. Drug dosing may have to be altered to achieve drug levels in the therapeutic range in cases with increased volumes of distribution.

Clinical Que

Increased (↑) protein binding = less free drug available to tissue.

Decreased (↓) protein binding = more free drug available to tissue.

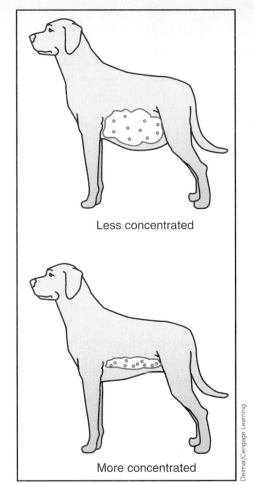

FIGURE 4-8 Volume of distribution is greater in an animal with ascites resulting in a decreased concentration of drug in all its body compartments. Volume of distribution is less in a normal animal than one with ascites resulting in an increased concentration of drug in all its body compartments.

Less concentrated

More concentrated

PHARMACOKINETICS PART III: CHANGING

Biotransformation is also called drug **metabolism**, *drug inactivation*, and/ or *drug detoxification*. Biotransformation is the chemical alteration of drug molecules by the body cells of patients to a metabolite that is in an activated form, an inactivated form, and/or a toxic form. Usually the metabolite is more hydrophilic (therefore more ionized, less likely to be stored in fat, and less likely to pass through membranes) than the parent compound, and is more readily excreted in the urine or bile. The metabolite may also have less affinity for plasma proteins, resulting in wider distribution to tissues or excretion.

There are four main pathways by which drugs undergo biotransformation: oxidation reactions (loss of electrons); reduction reactions (gaining of electrons); hydrolysis (adding of water molecules to a drug, causing it to split); and conjugation (addition of the glucuronic acid molecule, which makes the drug more water soluble).

The primary site of biotransformation is the liver. Other sites of biotransformation include the kidneys, small intestine, brain and neurologic tissue, lungs, and skin. The liver is the primary organ where a microsomal enzyme called *cytochrome P450* is located. Cytochrome P450 is found in the hepatocytes (liver cells) and is actually a family of detoxifying enzymes that alter the structure of drug molecules. Many types of drug interactions are the result of either inhibition or induction of cytochrome P450. Inhibition of cytochrome P450 generally involves competition with another drug for enzyme-binding sites, which can prolong the activity of a particular drug. Cytochrome P450 induction occurs when one drug stimulates the production of more enzyme, thereby enhancing its metabolizing capacity. The anticonvulsant drug phenobarbital is an example of a drug that is capable of inducing or stimulating the release of cytochrome P450 from the liver in a greater quantity than would normally be excreted. This increase in cytochrome P450 causes phenobarbital to be more rapidly biotransformed, thus reducing the patient's response to the drug. This phenomenon explains why the dose of some drugs must be continuously increased to cause the same pharmacological response.

It is important to understand biotransformation because of drug interactions (how one drug's effect on the body affects other drugs). If the administration of two or more drugs produces a pharmacological response that is greater than what would be expected by the individual effects of each drug together, the drugs are said to be acting synergistically. If one drug diminishes the action of another drug, the drugs are said to act antagonistically. Some drug interactions are desirable, and others are undesirable. Ways drugs interact with each other include the following.

- *Altered absorption:* One drug may alter the absorption of other drugs. For example, antacid medication alters the pH of the stomach and may affect how other drugs are absorbed through the gastrointestinal tract. Antacids may also form complexes with other drugs such as tetracycline, thereby rendering it incapable of being absorbed from the gastrointestinal tract.

- *Competition for plasma proteins:* Drug A and Drug B may both bind to plasma proteins. Drug A may have a higher affinity for binding to the plasma protein and so may displace drug B. Drug B may then reach toxic levels because it is free and able to cause its effect(s). Both digitoxin and furosemide are cardiac drugs that are highly protein bound, and if given together, one drug response may be seen at an increased level because it is not able to bind to protein.

- *Altered excretion:* Some drugs may act directly on the kidney and decrease the excretion of other drugs. Diuretics increase the production of urine and may affect drugs excreted via the kidneys.

- *Altered metabolism:* The same enzymes may be needed for biotransformation of two drugs that are prescribed at the same time for an animal. This may cause the enzyme system to become saturated and decrease the rate of metabolism of both drugs. This altered metabolism may cause the parent drug to be broken down more slowly, resulting in lower levels of

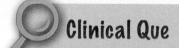

Clinical Que

The goal of biotransformation is to make drugs more water soluble so that they can be more easily excreted from the body.

the active metabolite. The metabolite is produced upon metabolism of the parent drug and is then absorbed to cause the desired effect. If this process is slowed, the desired effect may not be seen. Veterinary staff may then inadvertently increase the animal's drug dose.

Another way some drugs may alter metabolism is by causing the liver enzymes to become more efficient. These drugs induce the enzyme system and hence are known as microsomal enzyme inducers (phenobarbital is one example that was previously described). The enzyme system is referred to as induced if the rate of biotransformation by the enzyme system is increased. If other drugs that are metabolized by this same system are present, their biotransformation will also be increased. An increased rate of drug biotransformation by induction may require the dose to be increased to maintain adequate therapeutic levels.

Under some circumstances, the liver's ability to metabolize drugs may be impaired. Neonates with immature livers may not yet secrete adequate levels of microsomal enzymes. Old animals with liver damage may have diminished production of microsomal enzymes. Drug doses may have to be decreased in animals with altered liver function to avoid toxicity in these patients.

Another way metabolism may be altered is by animals developing tolerance. Tolerance is decreased response to a drug resulting from repeated use. An animal that develops tolerance requires a larger dose of drug to achieve the effect originally obtained by a smaller dose. There are two types of tolerance in animals: metabolic (drug is metabolized more rapidly with chronic use) and cellular ("down regulation" or decreased cellular receptors with repeated use). An example of metabolic tolerance occurs with phenobarbital; as it perfuses the liver, it induces the production of cytochrome P450. This results in faster metabolism of the drug, and more drug must be given to maintain its level in the body. Cellular tolerance occurs when receptors adapt to the continued presence of the drug either by reducing the number of receptors available to the drug or reducing their sensitivity to the drug. This reduction in numbers or sensitivity is called down-regulation, and higher levels of drug are needed to maintain the same pharmacological effect. Prolonged exposure to high concentrations of epinephrine causes a decrease in the number of receptors and in the affinity of the receptors on the target cells for epinephrine.

Drug metabolism can also vary depending on the species of animal. Cats have a slow rate of biotransformation for drugs that depend on glucuronyl transferase for conjugation in the liver. Cats have substantially reduced ability to metabolize drugs such as aspirin that utilize glucuronyl transferase because this enzyme is less effective in cats. As a result, hepatic clearance of aspirin is prolonged in cats (aspirin has a half-life of 37.5 hours in cats in comparison to 8.5 hours in dogs). Aspirin can be used in cats as long as the dosage interval is extended. Another drug, acetaminophen, cannot be given safely in cats. Glucuronyl transferase is needed to metabolize one of the toxic metabolites of acetaminophen. Cats cannot conjugate and clear the toxic metabolites quickly enough with the limited amount of glucuronyl

Table 4-4   Factors Affecting Biotransformation

FACTOR	HOW BIOTRANSFORMATION IS AFFECTED
Plasma protein binding	Less plasma protein binding allows excretion of drug
Storage in tissue and fat depots	Fat and tissue storage decrease the rate of metabolism
Liver disease	Affects cytochrome P450 production
Age of patient	Young animals have decreased metabolic pathways (except horses), a blood-brain barrier that is not yet well established, and higher percent of body water that affects volume of distribution
Nutritional status of patient	Poor nutrition yields inadequate plasma proteins
Species and individual variation	Cats have a reduced ability to biotransform aspirin
Body temperature	Increased body temperature increases rate of drug metabolism
Route of administration	Some drugs given parenterally have an effect, but when given orally have no effect (e.g., apomorphine)

transferase they produce, which results in liver damage and red blood cell breakdown.

Table 4-4 summarizes the primary factors affecting biotransformation.

PHARMACOKINETICS PART IV: GETTING OUT

Drug **elimination** is removal of a drug from the body. Drug elimination is also known as drug **excretion**. Routes of drug elimination include the kidneys, liver, intestine, lungs, milk, saliva, and sweat. The most important routes of drug elimination are the kidneys and liver.

Renal elimination of drugs involves a couple of mechanisms associated with urine production. Glomerular filtration occurs at the level of the nephron and works by pushing water and small molecules through a tuft of capillaries called the *glomerulus*, similar to pushing things through a sieve (Figure 4-9). The rate of glomerular filtration depends on blood pressure. The higher the volume of blood going through the glomerulus, the higher the blood pressure of the capillaries of the glomerulus, and hence the greater the number of particles that could potentially be filtered. Glomerular filtration is

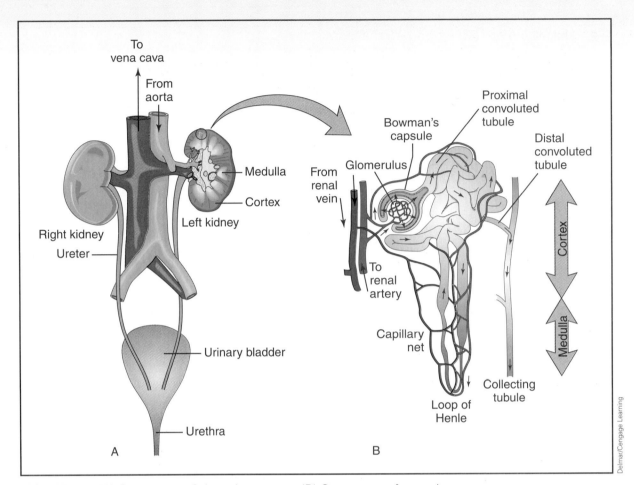

FIGURE 4-9 (A) Structures of the urinary tract. (B) Structures of a nephron.

nonselective. If a particle is small, nonionic, and free (nonprotein bound), it goes through.

The second mechanism by which drugs are eliminated by the kidney is tubular secretion. *Tubular secretion* is active transport across the convoluted tubule membrane that moves certain molecules from blood into the urine filtrate. Drugs such as penicillin are secreted into the tubule, allowing concentrations of these drugs to accumulate in the nephron and urine filtrate. Tubular secretion is generally a more rapid process of elimination than glomerular filtration and results in rapid elimination of drugs. Tubular secretion of drugs requires energy, and any disease of the animal may hinder this function.

Another mechanism that influences how drugs are eliminated by the kidney is tubular reabsorption. Tubular reabsorption of drugs occurs in the loop of Henle of the nephron. Tubular reabsorption depends on lipid solubility and molecule size. A drug that is highly lipid soluble will have increased reabsorption (nonionized particles will have increased reabsorption). Ionization of molecules results in poor reabsorption. In some cases, drugs that enter the tubule via glomerular filtration may be partially reabsorbed through the wall of the tubule back into the bloodstream. This delays the complete elimination of some drugs from the body. Figure 4-10 illustrates the renal elimination of drugs.

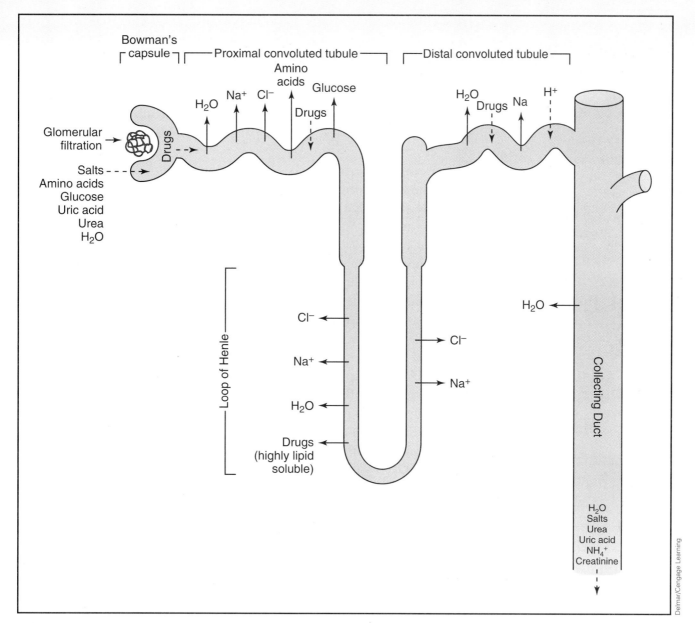

FIGURE 4-10 Renal elimination of drugs involves mechanisms associated with urine production. Glomerular filtration pushes small, nonionized, nonprotein-bound molecules through Bowman's capsule. Tubular reabsorption actively moves molecules from within the tubules into the capillaries surrounding the tubules. Tubular secretion actively moves molecules from capillaries into the tubules.

Urine pH can also affect the rate of drug excretion by changing the chemical form of a drug to one that is more readily excreted or to one that can be reabsorbed into the bloodstream. Drugs that are weak acids, such as barbiturates, penicillins, and other drugs available as sodium or potassium salts, tend to be better excreted in more basic urine, as this increases the proportion of drug in the ionized hydrophilic form. Weak bases, such as atropine, morphine, and other drugs available as sulfate, hydrochloride, or nitrate salts, are better excreted in more acidic urine.

Clinical Que

Drugs available as sodium or potassium salts are typically acidic, while drugs available as sulfate, hydrochloride, or nitrate salts are typically basic.

Clinical Que

A disease condition of the liver and/or kidneys will affect elimination of drug from the body.

Hepatic elimination of drugs is also important. Drugs excreted by the liver usually move by passive diffusion from the blood into the liver cell. Once in the liver cell, they are secreted into the bile. The bile is then secreted into the duodenum (either via the gall bladder or directly to the duodenum if the animal does not have a gall bladder). Lipophilic drugs entering the duodenum will reenter the bloodstream and go back to the liver. Hydrophilic drugs entering the duodenum will most likely become part of the feces and be eliminated by the body.

Intestinal elimination of drugs occurs when drugs are given orally and are not absorbed (allowing them to pass through the feces), when drugs are excreted in bile (allowing them to pass in the feces), or when drugs are actively secreted across mucous membranes into the gut. Pulmonary routes of drug excretion involve the movement of drug molecules out of blood and into the alveoli of the lungs to be eliminated in the expired air. Saliva and sweat also play a minor role in elimination of drugs from the body.

Another route of drug elimination is milk. Drugs or their metabolites, as well as nutrients, may pass directly from the blood to the milk via the mammary glands. Passage of nutrients into milk is important for the nourishment of offspring and human consumers of milk and milk products. Passage of drugs through the milk is an important consideration for veterinarians, as these drugs will affect the suckling offspring and also raise the issue of drug residues for consumers. A **drug residue** is the amount of drug that can be detected in tissues at specific times after administration of the drug ceases. Residues found in food products such as milk, meat, and eggs are important to people who have allergic reactions to the drug, for the development of antibiotic resistance, and for the development of disease secondary to drugs passing through these food sources. Wherever drugs are used to treat sick animals or prevent disease, there is a potential that residues may be incurred. The FDA establishes tolerances for drug residues to insure food safety. The FDA also establishes "withholding periods," which are times after drug treatment when milk and eggs are not to be used for food, and during which animals are not to be slaughtered. This allows time to the animals to eliminate the drug residues. This "withholding period" is called the **withdrawal time** and is the period of time after drug administration during which the animal cannot be sent to market for slaughter and the eggs or milk must be discarded because of the potential for drug residues to persist in these products. Withdrawal times have been established for approved drugs that produce drug residues. The withdrawal time for drugs approved for use in food-producing animals is calculated using the drug's half-life. The **half-life** (abbreviated $T_{1/2}$) is the time required for the amount of drug in the body to be reduced by half of its original level. For example, if the half-life of a drug is four hours, the amount of drug remaining after four hours is 50 percent of the original concentration. With each half-life (four hours in this example), the drug concentration is further reduced by 50 percent of what it was at the beginning of the interval. In this example, after 8 hours, 25 percent of the original concentration would remain; after 12 hours, 12.5 percent would remain; and after 16 hours, 6.25 percent of the original concentration would remain. The half-life of a drug provides an idea of how quickly a drug is eliminated by the body and the drug's steady state. The **steady state** of a drug is the point at which drug accumulation and elimination are balanced. For most drugs it takes about four to five half-lives for the steady state to be reached. This drug level will remain fairly constant as long as the dose of the drug or its frequency of administration is not altered.

The patient's physiology also plays a role in the efficacy of drug excretion. Hydration status, age, and the disease or health status of the animal can all affect drug excretion. Hydration status affects the animal's blood volume and thus affects glomerular filtration of drug molecules excreted through the renal process. Dehydrated animals have less vascular fluid volume and so have decreased pressure in the vessels being filtered by the glomerulus. Age and disease may alter the levels of blood protein in an animal and affect the level of the drug's protein binding. Changes in protein binding can affect elimination rates. Animals with low blood protein levels will have less protein-bound drug (more free drug), which may be filtered through the glomerulus before the drug has had time to cause its effect.

MEASURING DRUG ACTION

Another way to describe drug action is by the use of a graphic depiction of the plasma concentration of the drug versus time (Figure 4-11A). The horizontal line (x-axis) represents time, and the vertical line (y-axis) represents drug concentration in plasma. The zero point on the x-axis represents the time at which the drug is first administered. With an orally administered dose, the drug concentration in the plasma increases from a zero level as the drug is absorbed into the plasma from the gastrointestinal tract. Onset of action begins when the drug enters the plasma. The drug concentration continues to rise until the elimination rate of the drug is equivalent to its rate of absorption. This point is known as the peak plasma level of the drug. The time elapsed from the time of administration to the time that the peak plasma level is reached is known as the time to peak and is important in making clinical judgments about the use of a drug. From the peak plasma level, the concentration declines since the amount of drug being eliminated exceeds the amount being absorbed.

When a drug is administered by rapid intravenous injection, the plasma level versus time plot is different from that observed with oral drug administration since the drug is introduced directly into the bloodstream without requiring the absorption step (Figure 4-11B). With intravenous injection, the drug is introduced directly into the bloodstream without requiring the absorption step; therefore, peak drug level is achieved immediately at the time of administration. Only a decline of plasma concentration of the drug is observed, reflecting the elimination of the drug.

When most drugs are prescribed, an attempt is made to choose a dose and dosage interval, which will permit the plasma level of the drug to remain above the minimal level required to elicit a pharmacological response (the minimal effective concentration or MEC) (see Figure 3-2A and B). The plasma level of the drug must also remain below the plasma level at which toxic effects

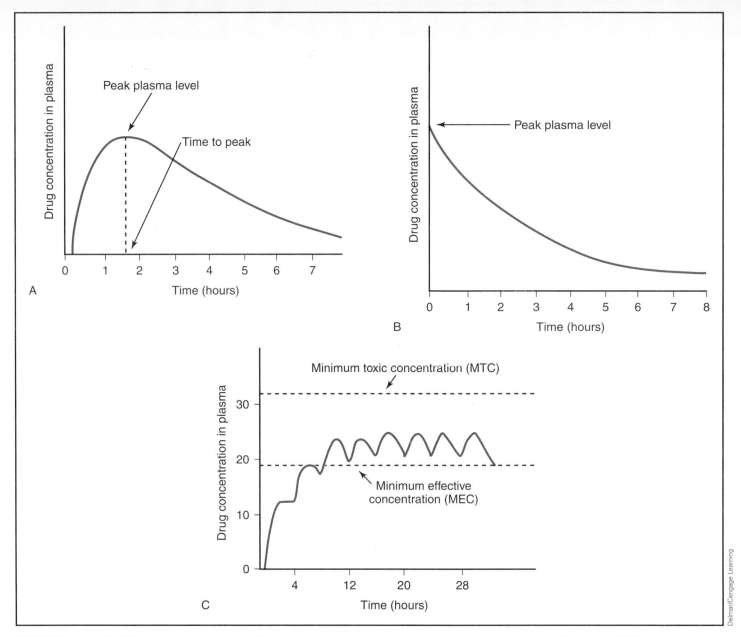

FIGURE 4-11 (A) Plot of drug concentration in plasma versus time after a single oral dose of a drug. (B) Plot of drug concentration in plasma versus time after a single intravenous injection. (C) Plot of drug concentration in plasma versus time after multiple oral administrations.

are observed (the minimum toxic concentration or MTC). Figure 4-11C demonstrates a plot of plasma level versus time of a drug administered orally at four-hour intervals in order to keep the plasma concentration of the drug between the MEC and MTC. With the first oral administration, the MEC was not reached. In situations requiring rapid achievement of therapeutic plasma levels of a drug, a loading dose of a drug may be administered to produce effective plasma levels of the drug quickly. Often multiple drug administrations must be given before a plateau or steady-state concentration of the drug is achieved in plasma. This level will remain fairly constant as long as the dose of the drug and frequency of administration are not changed.

HOW DO DRUGS WORK?

Once a drug is in the animal's body, it is absorbed and distributed by the bloodstream to the tissues. Drugs are capable of causing a variety of effects in animals.

Drugs alter existing cellular functions; they do not create new ones. For example, antibiotics slow the growth (reproduction) of bacteria. Laxatives increase the rate of peristaltic action of the gastrointestinal tract. Drug action is described relative to the physiological state that existed when the drug was administered.

Drugs may interact with the animal's body in several different ways. Some alter the chemical composition of body fluids (antacids alter the acidity of the stomach). Other drugs accumulate in tissues because of their affinity for a tissue component (gas anesthetics have an affinity for the lipid portion of the nerve cell membranes).

Drugs can form a chemical bond with specific cell components on target cells within the animal's body (this is the most common way in which drugs act). These specific cell components are called receptors. **Receptors** are three-dimensional proteins or glycoproteins that are usually located on the cell membrane, but may also be found in the cytoplasm, or within the nucleus of cells (Figure 4-12). Animals have a finite or limited number of receptors. Not all cells have receptors for all substances, but some cells may have multiple receptor sites.

This binding between drug and receptor will occur only if the drug and its receptor have a compatible chemical shape. This interaction between a drug and its receptor is often compared to the relationship between a lock and key. When the drug binds to the receptor, a series of reactions results in some cellular change, which may include the opening or closing of an ion channel; activation of secondary messengers such as cAMP, cGMP, or calcium; inhibition of normal cellular function; and/or activation of normal cellular function.

Affinity is the strength of binding between a drug and its receptor. The better the drug fits at the receptor site, the more biologically active it is. Drugs whose molecules fit precisely into a given receptor can be expected to elicit a good response (e.g., most penicillins); drugs that do not perfectly fit the receptor shape may produce only a weak response or no response at all. The better the fit of the drug with its receptor, the stronger the drug's affinity will be for the receptors and the lower the dose required to produce a pharmacological response. For example, hormone receptors are highly specific. Hormone response may often be elicited by the presence of only minute concentrations of an appropriate hormone, because it has a strong affinity for the receptor. The measure of a drug's affinity for its receptor is known as the dissociation constant K_D.

An **agonist** is a drug that binds to a cell receptor and causes action (usually, more than one molecule of drug must bind to the receptor to cause action). Epinephrine is a drug that stimulates alpha and beta receptors to cause vasoconstriction and an increased heart rate; therefore, it is an agonist (epinephrine is also described as nonselective because it binds to multiple

Clinical Que

Drugs act through receptors, by binding to the receptor to produce (initiate) a response or to block (prevent) a response.

Clinical Que

High-affinity drugs bind more tightly to a receptor than do low-affinity drugs.

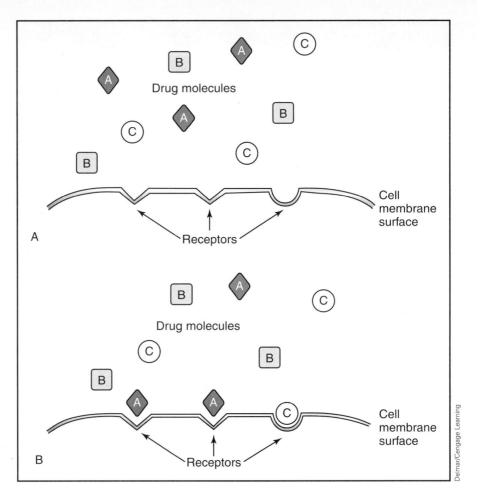

FIGURE 4-12 (A) Receptors are three-dimensional proteins located on the cell membrane surface. Receptors can only bind substances specific for that site. (B) This cell membrane can bind substances A or C.

receptors). A strong agonist is a drug that has an exact fit at the receptor site and causes maximal cellular effect with only a few drug molecules occupying a few receptors. A weak agonist is a drug that does not have an exact fit with the receptor site and has to have many receptors bound with the drug before the effect occurs. A partial agonist is a substance that creates only a partial effect even if the drug occupies all of a cell's receptors.

An **antagonist** is a drug that inhibits or blocks the response of a cell when the drug is bound to the receptors. Cimetidine is an antacid that blocks the H_2 receptor of stomach cells to prevent excessive stomach acid secretion; therefore, it is an antagonist. A competitive antagonist competes with the agonist for the same receptors. Competitive antagonism can be overcome by high doses of agonist (known as *surmountable* or *reversible antagonism*). A noncompetitive antagonist binds to a site different than the agonist's binding site and mechanically changes the agonist's receptor, resulting in a roadblock to the action of the agonist (known as *insurmountable* or *irreversible antagonism*). Figure 4-13 shows the different types of antagonism.

Some drugs work in ways other than binding to a receptor. One example is mannitol, an osmotic diuretic (a drug that causes increased urination).

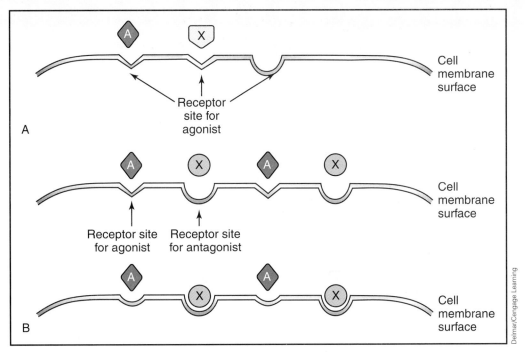

FIGURE 4-13 A = agonist, X = antagonist (A) Competitive antagonism: The receptor is specific for A; however, X can compete for the same receptor because of its shape. (B) Noncompetitive antagonism: When the antagonist binds to its receptor site, it changes the receptor so that the agonist can no longer bind to that site.

Mannitol prevents resorption of water back into the renal tubule by its presence in urine. Because mannitol is a large molecule, it sets up a concentration gradient (one side has more molecules than the other side) in the nephron and water moves into the nephron to try to equalize the concentration of fluid between the nephron and blood. Another example is milk of magnesia, a laxative, that works by raising the concentration of dissolved substances in the gastrointestinal tract, thereby osmotically attracting fluid into the gut in an attempt to dilute the concentration of particles within the gastrointestinal tract.

SUMMARY

Pharmacokinetics describes the physiological movement of drugs within an animal's body. Passive diffusion, facilitated diffusion, active transport, or pinocytosis/phagocytosis are ways drugs move. Physiological drug movement involves four steps: absorption, distribution, biotransformation, and excretion (Figure 4-14). Absorption of drugs is affected by the drug's bioavailability (the percentage of drug administered that actually enters the systemic circulation). Drugs with a bioavailability of 1 are 100 percent bioavailable. Other factors that affect drug absorption include the lipophilic or hydrophilic form of the drug, the degree of ionization, and size of the drug molecule. Acid-base status of the drug, the pH of the surrounding environment, and the drug's pKa all play a role in ionization of the drug. Ion trapping occurs when a drug molecule

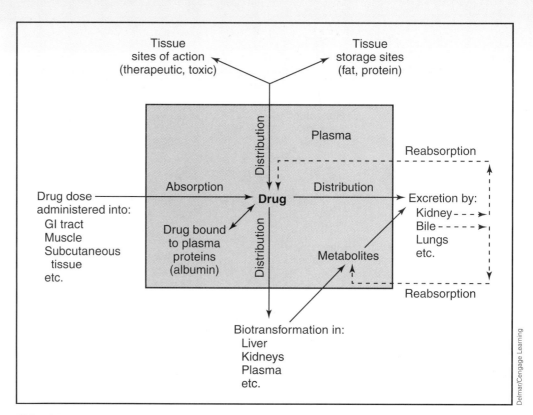

FIGURE 4-14 Summary of pharmacokinetics. The dashed lines indicate that some of the drug and its metabolites may be reabsorbed from the excretory organs.

changes from ionized to nonionized form as it moves from one body compartment to another.

Drug distribution is the physiological movement of a drug to reach the target tissue or intended site of action. Membrane permeability, tissue perfusion, protein binding of the drug, and volume of distribution can affect drug distribution within an animal's body.

Biotransformation, or drug metabolism, occurs mainly in the liver and is affected by a variety of patient and drug parameters. Biotransformation occurs via oxidation reactions, reduction reactions, hydrolysis, and conjugation. Drugs may interact with each other through mechanisms such as altered absorption, competition for plasma proteins, altered excretion, or altered metabolism. Some animals whose metabolism is altered from drug use may develop tolerance.

Drug elimination or excretion is the removal of a drug from the body. Routes of excretion include the kidney, liver, intestine, lungs, milk, saliva, and sweat. The two main routes of drug excretion are through the kidney and the liver.

The majority of drugs work by binding to receptors. Receptors are protein based, may be located in a variety of places on the cell (primarily on the cell membrane), and are three dimensional. Agonists bind to a cell receptor and cause action. Antagonists inhibit or block the response of a cell when bound to its receptors. Antagonists may be competitive or noncompetitive. Some drugs work without the use of receptors.

It's a Wrap

The answers to the questions in this chapter's Setting the Scene case study should be understood after reading this chapter. How does the medication administered in a muscle get from the muscle to another location in the body? How do medications get to where they are needed?

The drug given in the muscle must diffuse from the injection site (the extracellular fluid of the muscle) into the blood. From the blood, it is distributed to the front leg, where it can be effective. Drugs from the blood bind to receptors of certain tissues, thereby causing the desired reactions. Remember, once a drug gets into the blood, it can travel to a variety of areas. When animals are given antibiotics, bacteria may be killed in many areas of the body, not just the area with the bacterial infection. Normal flora bacteria in the gastrointestinal tract are often affected by systemic antibiotics used to treat skin, urinary, or respiratory infections.

CHAPTER REVIEW

Matching

Match the definition with its proper term.

1. _____ type of molecule or drug that easily passes through the cell membrane

2. _____ type of cellular movement in which molecules move against their concentration gradient

3. _____ type of cellular movement that allows intake of large molecules

4. _____ percent of drug administered that enters the systemic circulation

5. _____ term for charged

6. _____ term for water loving

7. _____ physiological movement of drugs from the systemic circulation to the tissues

8. _____ how well a drug is distributed throughout the body based on the concentration of drug in the blood

9. _____ primary site of biotransformation

10. _____ term used to describe an increase in the rate of biotransformation by an enzyme system

a. bioavailability
b. hydrophilic
c. pinocytosis/phagocytosis
d. active transport
e. volume of distribution
f. lipophilic, nonionized, small
g. liver
h. drug distribution
i. induce
j. ionized

Multiple Choice

Choose the one best answer.

11. Which drug has a greater percent that actually enters the systemic circulation?
    a. a drug with a bioavailability of 0.8
    b. a drug with a bioavailability of 0.2

12. The movement of molecules from an area of high concentration to an area of low concentration is known as
    a. passive diffusion
    b. facilitated diffusion
    c. active transport
    d. pinocytosis/phagocytosis

13. What type of drug is well absorbed from the gastrointestinal tract?
    a. hydrophilic
    b. lipophilic
    c. water based
    d. charged

14. In what form do the majority of acidic drugs exist in an acidic environment?
    a. nonionized
    b. ionized

15. Based on blood perfusion, which body compartment will get adequate drug levels more quickly?
    a. fat
    b. subcutaneous tissue
    c. skeletal muscle
    d. smooth muscle

16. Which animal has a greater volume of distribution?
    a. a 7 percent dehydrated cat
    b. a normal cat

17. A dog has decreased renal perfusion. What will this do to the blood levels of a drug excreted through the kidneys?
    a. It will decrease
    b. It will increase
    c. It will remain normal
    d. It will totally stop

18. Drug affinity is the
    a. strength of binding between a drug and its receptor.
    b. the measure of the drug's action.
    c. number of receptors that must be occupied by the drug.
    d. binding of the drug to its receptor.

19. Which food-producing animal drug has the shorter withdrawal time?
    a. drug A with a half-life of 60 minutes
    b. drug B with a half-life of 120 minutes

20. Which of the following is true?
    a. Drugs with short half-lives need not be given more frequently
    b. Noncompetitive antagonists are easily reversed in a clinical setting
    c. Thin animals with low plasma protein levels require more drug than animals of normal weight
    d. Giving fluids to an animal will increase the excretion of drug

Case Study

21. A client brings a 13-year-old M/N Siamese cat into the clinic for an examination. The client states that the cat has been drinking more water and urinating more frequently in the litterbox. Because this is an older cat, the veterinarian collects blood to assess the cat's liver and kidney enzymes. A urine sample is also collected to assess whether the increased drinking and urinating are due to an infection or systemic disease. The blood results indicate that the cat has elevated liver and kidney enzymes. The urinalysis indicates that the cat also has a bacterial urinary tract infection. In addition to medication to treat liver and kidney disease, the client is given antibiotics to treat her cat's urinary tract infection.

The client feels that because this cat has multiple health problems, it would be better to give a larger dose of antibiotics so that the infection can be cleared more quickly. Why is this not a good idea?

Critical Thinking Questions

22. How does the application of cold packs to an area affect the rate of absorption of subcutaneous fluids?

23. In human medicine, some drugs (like nitroglycerin) are given sublingually (under the tongue). Why would these drugs not be given orally?

24. Think of ways that an understanding of pharmacokinetics can be applied to the daily practice of veterinary medicine. Give an example of the application of pharmacokinetics and explain your example.

ADDITIONAL REVIEW MATERIAL

Additional review material can be found on the StudyWARE™ CD included in this text.

BIBLIOGRAPHY

Riviere, J. E., and Papich, M.G. (Eds.). (2009). *Veterinary Pharmacology and Therapeutics* (9th ed.). Hoboken, NJ: John Wiley & Sons.

Johns Cupp, M., and Tracy, T. (January 1, 1998). *Cytochrome P450: New nomenclature and Clinical Implications*, Vol. 57, pp. 107–116. American Academy of Family Physicians.

Julien, Robert. (2001). *A Primer of Drug Action* (pp. 21–25). New York: MacMillan Publishing.

Pineda, M. H. (2003). *McDonald's Veterinary Endocrinology and Reproduction* (p. 192). Iowa: Blackwell Publishing.

Reiss, B., and Evans, M. (2002). *Pharmacological Aspects of Nursing Care* (6th ed. revised by Bonita E. Broyles, pp. 2–31). Clifton Park, NY: Delmar Cengage Learning.

Veterinary Pharmaceuticals and Biologicals (12th ed.). (2001). Lenexa, KS: Veterinary Healthcare Communications.

Wynn, R. (1999). *Veterinary Pharmacology*. Cambridge, MA: Harcourt Learning Direct.

CHAPTER 5

VETERINARY DRUG USE, PRESCRIBING, ACQUISITION, AND PHARMACY MANAGEMENT

OBJECTIVES

Upon completion of this chapter, the reader should be able to:

- compare and contrast generic and brand-name drugs.
- describe the methods of drug marketing in the United States.
- explain compounding and its importance in veterinary medicine.
- explain how veterinary drugs are used and prescribed.
- outline the parts of a drug label and explain the rationale for having this information on the drug label.
- describe the process used in prescribing drugs.
- describe proper drug-dispensing techniques.
- list different drug resources and describe how they are used.
- explain drug package inserts and the information they provide.
- differentiate between prescribing and dispensing drugs.
- outline the essential components of a prescription label.
- define abbreviations used in prescription writing.
- describe the role of the veterinary technician in regard to pharmacy economy.

KEY TERMS

bioequivalency	generic companies
brand name	generic name
compounding	nonproprietary name
direct marketing	package insert
distributors/	prescription
wholesalers	proprietary name
expiration date	trade name

Setting the Scene

A woman calls the veterinary clinic and states that she would like some antibiotics to treat her cat's respiratory infection. She works at a discount store that has a pharmacy, and she would like a prescription called in to that pharmacy for a human-labeled antibiotic that can be used in cats. Essentially, she is asking about the use of human-labeled products for veterinary purposes. Can drugs used for animals be dispensed at human pharmacies? What should the veterinary technician tell her?

Delmar/Cengage Learning

BRAND NAME (®) OR NOT?

Drugs are named in a variety of ways (Table 5-1). During the drug's earliest stages of development, the first name it receives is the chemical name. The chemical name provides scientific and technical information because it is a precise description of the substance in accordance with chemical nomenclature rules established by the International Union of Pure and Applied Chemistry (IUPAC). The chemical name describes the chemical structure of the drug. Because they are long and complex, however, chemical names are rarely used in clinical medicine. An example of a chemical name is N-(2,6-dimethylphenyl)-5,6-dihydro-4H-1,3-thiazin-2-amine, which is more commonly known as xylazine.

Drugs are also named and marketed under two names, a **generic name** and a **brand name**. The generic name (sometimes referred to as the **nonproprietary name**) is written using lowercase letters, and it is the same in all countries. The generic name is the official identifying name of the drug, which is assigned by the U.S. Adopted Names (USAN) Council. The generic name is commonly used to describe the active drug(s) in the product and is easier to pronounce and remember than the chemical name. An example of a generic drug name is the animal sedative that goes by the generic name of xylazine. Another example of a generic drug name is the human-approved painkiller acetaminophen.

The brand name, **trade name**, or **proprietary name** of a drug establishes legal proprietary recognition for the corporation that developed the drug. The brand name of a drug is registered by the U.S. Patent Office, is approved by the U.S. Food and Drug Administration (FDA), and may be used only by the company that has registered the drug. For example, the Bayer Corporation calls its xylazine product by the brand name Rompun® (Figure 5-1). The

Table 5-1   Drug Identification

TERM	DEFINITION	EXAMPLE
Chemical structure	Diagram of the chemical arrangement of the drug	
Chemical name	The name of the drug based on its chemical structure	7-chloro-1,3-dihydro-1-methyl-5-phenyl-2H-1,4-benzodiazepin-2-one
Generic or nonproprietary name	The name assigned by the USAN, but not exclusive to a particular company	diazepam
Trade or proprietary name	The name picked by the company that is registered	Valium® or Valium®

Figure 5-1 Drug label showing generic and trade name of a drug.

Clinical Que

The superscript™ that follows some trade names of drugs is the trademark and indicates that the drug is not yet registered.

Clinical Que

Trademark laws in the United States do not allow generic drugs to look exactly like the brand-name preparation; however, the active ingredients must be the same in both preparations. When switching from brand-name drugs to generic drugs, it is helpful to advise clients that the drug will look different than the brand-name drug.

acetaminophen produced by McNeil Pharmaceutical is commonly known by its brand name Tylenol®.

The brand name of a drug is usually written in capital letters or begins with a capital letter (it is considered a proper noun). It may also have the superscript R or the symbol®, both of which stand for "registered," next to the name. The brand name Rompun, for example, starts with a capital letter. With its superscript, the Rompun brand name looks like this: Rompun^R or Rompun®.

Once a manufacturer's patent for a drug has expired (usually 17 years from its registration date), other companies are free to market the drug under their own trademarked names or under the generic name of the drug. The therapeutic equivalence or **bioequivalency** of these products should be the same regardless of whether they are manufactured by the original patent-holding company or by another company. Bioequivalency is the ability to produce similar blood levels after administration as the original patented drug. In most cases, no significant difference in response is noted when competing products are used. In some instances, however, different products containing identical drugs and drug doses have produced different pharmacological responses, even in the same patient. The vast price differences between some brand-name drugs and other brand-name drugs, and between brand-name drugs and generic drugs, have fueled the debate about whether "identical" drugs have the same therapeutic effectiveness as drugs in a formulation different from the original drug. No generalization can be made regarding the therapeutic effectiveness of competing drug products containing the same dose of a drug. The veterinary staff must carefully assess a patient's response to any drug product change, and the veterinarian should address any variation in therapeutic response.

COMPOUNDING

Drug **compounding** (also called extemporaneous compounding) is the preparation, mixing, assembling, packaging, and/or labeling of a drug based on a prescription drug order from a licensed practitioner for an individual patient (Figure 5-2). Drug compounding also includes the preparation of drugs for use in research or teaching and are not intended for sale or dispensing. Basically, drug compounding occurs when health professionals prepare a specialized drug product to fill an individual patient's needs when an approved drug is not available. Prior to the 1950s most medications for both humans and animals were compounded by pharmacists.

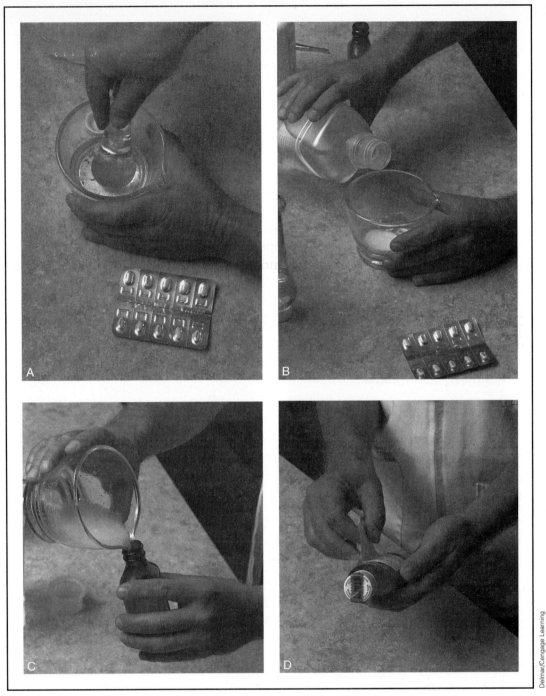

Delmar/Cengage Learning

Figure 5-2 Compounding solutions (liquid dose form in which the active ingredients are dissolved in a liquid) or suspensions (liquid dose form that contains solid drug particles) are common types of drug compounding. (A) The solid form of the drug is ground into a powder; (B) The proper solvent is added to the powder. The proper solvent is chosen based on solubility characteristics of each drug. Flavoring agents may also be added to the compounded drug; (C) The liquid form of the drug is poured into a bottle for dispensing; (D) The bottle is labeled with the patient's information, drug information, and instructions for administration. Suspensions should be dispensed with the "Shake well" label as well.

In human medicine drug compounding has been commonly used to dilute adult formulations into measurable units for pediatric patients, to use in patients allergic to dyes, fillers, preservatives, or flavoring agents in commercially available products, or to make very specific formulations for some specialists. Common examples of compounding in veterinary medicine include the following:

- Creating discontinued drugs that are no longer available commercially yet are still prescribed by veterinarians.

- Tailoring doses and strengths to meet a particular animal's weight and health status.

- Creating alternative dose forms such as liquids, ointments, or chewable tablets to make the medications easier for clients to administer to their animals (Figures 5-3 and 5-4).

- Adding flavoring to unpalatable drugs to make these drugs more appealing to the animals that need to take them.

- Customizing formulations that combine multiple drugs for one-dose administration.

Drug compounding may seem like a way to meet the specific needs of animals, clients, and practitioners; however, it is not always as simple as it appears. Crushing a pill into a powder and mixing it into a liquid is an example of drug compounding. Other types of drug compounding are not so simple and may involve complicated scientific procedures and may need to be performed in a germ-free environment. In these cases, there is no way to ensure that good manufacturing techniques were followed. Some concerns regarding drug compounding include the following:

- Drugs are approved by the FDA based on studies performed on a very specific drug formula and dosage. Small compounding changes may convert an approved drug into an unapproved drug.

- Compounded drugs are made without FDA oversight, thus involving an extra risk for the patients taking them compared with those using FDA-approved drugs.

- Compounded drugs may not be sterile and can cause infection in patients who use them.

- Errors in preparing compounded drugs may result in disease or death in patients who use them.

To ensure the safety of drug compounding, the FDA Modernization Act of 1997 (FDAMA) defines the limits of legitimate compounding in humans. The main provisions of the FDAMA that offer protection against unsafe and ineffective compounded products include that the compounded product must be individually prescribed for an identified patient; the drug's active ingredient must be either an FDA-approved drug, listed in the *United States Pharmacopoeia* (USP), or listed in an FDA rule as acceptable for pharmacy compounding (based on the agency's evaluation of the medical literature);

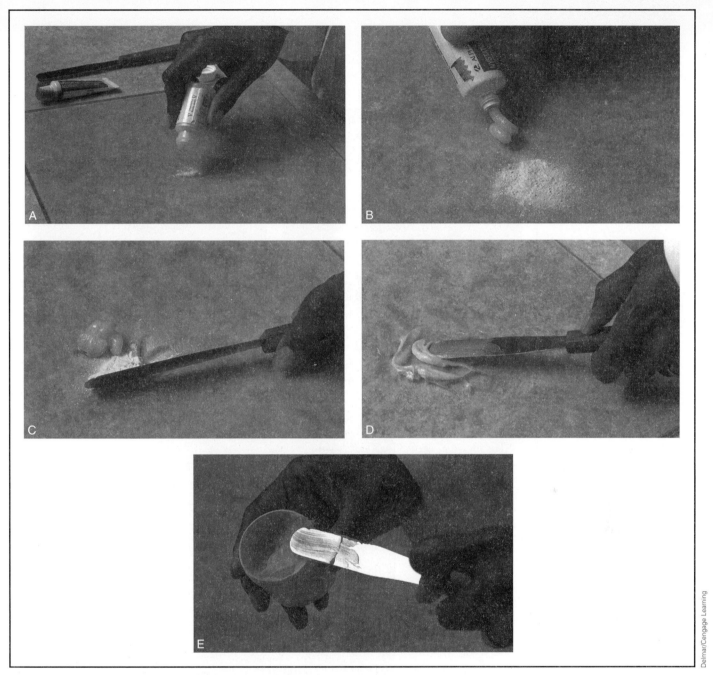

Figure 5-3 Compounding ointments, creams, pastes, and gels are drug forms intended for topical application to skin or mucous membranes. (A) In this example, the proper amount of powder to compound into a cream is placed on a clean surface; (B) A partial amount of cream is applied to the work area; (C) The powder is mixed into the cream; (D) The powder and cream are blended; (E) The blended cream is put into a container for dispensing. The container is labeled with the patient's information, drug information, and instructions for administration.

previously marketed drugs found unsafe or ineffective and removed from the market may not be compounded; and drugs listed by the FDA as difficult to compound drugs may not be compounded. In 2002, the Supreme Court found that section 503A of FDAMA was unconstitutional. Section 503A

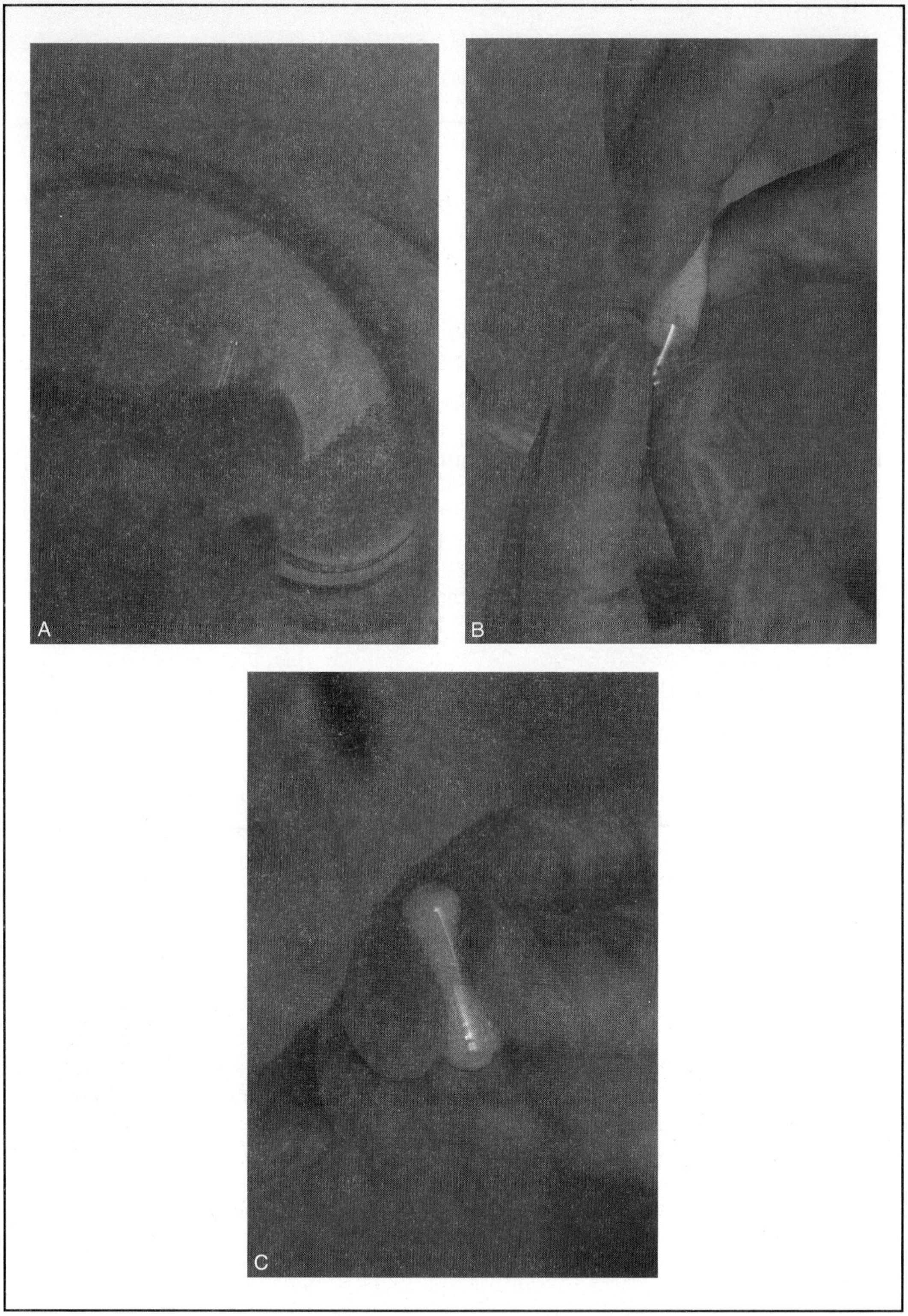

Figure 5-4 Powdered drugs may also be placed in capsules to make administration easier. Powdered drug may need to be weighed with a prescription balance. (A) The powdered drug is punched into the empty capsule shell. Capsule shells come in different sizes and are chosen by trying different sizes; (B) The two capsule halves are put together; (C) The capsule is checked to make sure the powder is properly secured within the capsule.

places restrictions on the advertising of drug-compounding products and services. The FDA believes that compounded drugs for individuals should not be advertised and is also concerned that manufacturing compounded drugs may be seen by some as a way to avoid the FDA testing process, which is expensive and time consuming.

Concern surrounding drug compounding in veterinary medicine revolves around the restrictions for compounded drugs for non-food-producing animals and the use of compounded drugs for food-producing animals versus protecting public health and safety by requiring drugs to be subjected to rigorous FDA testing. In 1996, a task force consisting of veterinarians, pharmacists, and regulators offered comments that the FDA-CVM used in developing their Compliance Policy Guide (CPG). The FDA uses the CPG section 608.400 entitled "Compounding of Drugs for Use in Animals" to provide guidance for veterinarians and pharmacists in their use of compounded drugs. Concerns about the use of compounded drugs in animals include the following:

- When drugs are compounded there is an absence of adequate and controlled safety and effectiveness data as well as lack of oversight for adherence to the guides of good manufacturing practices. This absence of regulation may contribute to the potential of compounded drugs to cause harm to public health and to animals when drugs are compounded, distributed, and used (Figure 5-5).

- The use of compounded drugs in animals may result in adverse reactions, animal deaths, drug residues in food-producing animals, unknown withdrawal times in food-producing animals, and unknown reactions from inactive ingredients used in the compounding process.

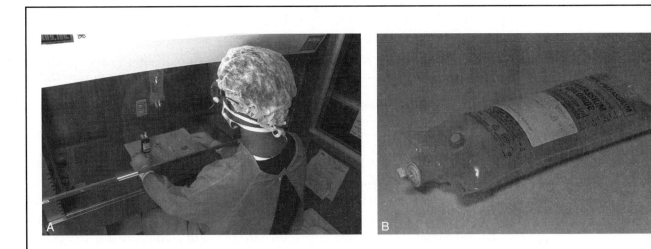

Figure 5-5 Some drugs need to be compounded under sterile conditions. (A) Sterile compounding of IV fluids should be done under a laminar airflow hood; (B) The compounded IV solution should be labeled with the patient's information, drug information, and instructions for administration.

- The FDA is concerned that veterinarians and pharmacies that produce and use compounded drugs may be trying to circumvent the drug approval process and provide the mass marketing of products that have been produced with little or no quality control or manufacturing standards to ensure the purity, potency, and stability of the product.

Since 1996, the CPG has been updated and the currently used version was published in July 2003. The current CPG describes the FDA's present thinking on what types of veterinary compounding might be subject to enforcement action. There is concern and disagreement that the current CPG lacks clarity on the circumstances in which veterinary compounding would be permitted. There are attempts by the American Veterinary Medical Association (AVMA), the International Academy of Compounding Pharmacists (IACP), and the FDA itself to draft a revised CPG. Until that time, the FDA will enforce section 608.400 of the CPG at its discretion and will not take regulatory action when the following are true:

- A legitimate medical need is identified.

- There is a need for an appropriate dose regimen for the particular species, size, age, or medical condition.

- There is no marketed approved animal drug that can treat the condition or there are some extenuating circumstances.

All of the agencies involved (AVMA, IACP, and FDA) agree that there is a need to use compounded drugs in veterinary medicine within a veterinarian/client/patient relationship (VCPR). The goal of coming to an agreement in regard to drug compounding use and regulation will continue to encourage debate among all concerned organizations.

GETTING INFORMATION

Veterinary professionals need to remain aware of the current lawful pharmaceutical use as it is in a constant state of flux and can be the cause of legal action. To remain informed about pharmaceuticals and their lawful use, it is important to know the sources of drug information and understand how these resources are used.

Drug Standards

In the United States, the FDA of the Department of Health and Human Services is responsible for enforcement of the Food, Drug, and Cosmetic Act and its amendments, including the 1972 New Animal Drugs amendment. The FDA oversees adherence to drug standards and regulations established by physicians, pharmacists, dentists, and veterinarians. The standards for drugs are found in the USP. This publication is the legally recognized drug standard of the United States. The USP is revised and published periodically by a committee of the U.S. Pharmacopeial Convention, composed of delegates from all the major American health and medical science organizations. This official compendium describes the source, appearance, properties, standards of purity, and other requirements of the most important pure drugs. A drug may include the designation "USP" on its label if it meets the standards described in the compendium.

The FDA requires that all drugs meet USP standards of purity, quality, and uniformity. The FDA also requires that all drug containers be correctly and completely labeled; each container must identify the manufacturer or distributor, and it must give directions for intended use. All accompanying advertisements and information must be true and correct statements of the indication, toxicity, and general usage of the drug.

Package Inserts

Drugs are packaged or supplied in bottles or vials that are labeled with specific information: the drug names (generic and trade), drug concentration and quantity, name and address of the manufacturer, the manufacturer's control or lot number, the **expiration date** (date the drug is no longer effective) of the drug, withdrawal times (if warranted), and the controlled substance status of the drug (if warranted). The label on the drug bottle or vial usually does not have enough space to list all the information required by FDA regulations. Manufacturers may reference **package inserts** provided with their drugs in order to meet regulatory requirements (Figures 5-6 and 5-7). Much of the information on package inserts is also found in drug references such as the *Physicians' Desk Reference (PDR), Compendium of Veterinary Products (CVP), Veterinary Pharmaceuticals and Biologicals (VPB)*, and other drug reference books. Package inserts provide the following information (some package inserts may vary slightly from the terms listed here):

- Bold-faced letters with a circled R superscript indicate a registered trade name of the product. This is the name, approved by the federal government, which a specific manufacturer uses exclusively for the drug. The generic and/or chemical name of the drug follows the trade name. This is the name, approved by the federal government, which is used to describe the active ingredient of the drug and may be used by anyone. If there is no registered brand name (i.e., the product is a generic drug made by a manufacturer), the generic or chemical name will be given at the top of the insert. An Rx symbol may appear to the right of the brand name to signify that this drug is available through prescription only. The C symbol with an enclosed Roman numeral indicates that the drug is a controlled substance; the Roman numeral indicates one of several varying degrees of addiction potential. The manufacturer's name and trademark may also be on the insert.

- A *description or composition statement* usually follows the product name. This section describes the physical and chemical properties of the active drug. This section may include information about the drug's appearance, solubility, chemical formula and structure, and melting point. Other ingredients added to the product may also be included in this section. It is important to note any additional ingredients because these

NADA 141-108, Approved by FDA

FORT DODGE®

Trade name

EtoGesic®

Generic name

(etodolac)

TABLETS FOR ORAL USE IN DOGS ONLY

Caution: Federal (U.S.A.) law restricts this drug to use by or on the order of a licensed veterinarian.

Description or Composition Statements

DESCRIPTION

Etodolac is a pyranocarboxylic acid, chemically designated as (±) 1,8-diethyl-1,3,4,9-tetrahydropyrano-[3,4-b] indole-1-acetic acid. The structural formula for etodolac is shown:

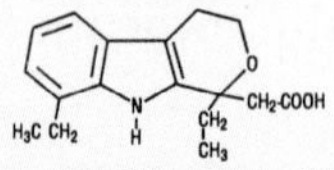

The empirical formula for etodolac is $C_{17}H_{21}NO_3$. The molecular weight of the base is 287.37. It has a pKa of 4.65 and an *n*-octanol:water partition coefficient of 11.4 at pH 7.4. Etodolac is a white crystalline compound, insoluble in water but soluble in alcohols, chloroform, dimethyl sulfoxide, and aqueous polyethylene glycol. Each tablet is biconvex and half-scored and contains either 150 or 300 mg of etodolac.

Clinical Pharmacology

PHARMACOLOGY

Etodolac is a non-narcotic, nonsteroidal anti-inflammatory drug (NSAID) with anti-inflammatory, anti-pyretic, and analgesic activity. The mechanism of action of etodolac, like that of other NSAIDs, is believed to be associated with the inhibition of cyclooxygenase activity. Two unique cyclooxygenases have been described in mammals[1]. The constitutive cyclooxygenase, COX-1, synthesizes prostaglandins necessary for normal gastrointestinal and renal function. The inducible cyclooxygenase, COX-2, generates prostaglandins involved in inflammation. Inhibition of COX-1 is thought to be associated with gastrointestinal and renal toxicity, while inhibition of COX-2 provides anti-inflammatory activity. In *in vitro* experiments, etodolac demonstrated more selective inhibition of COX-2 than COX-1[2]. Etodolac also inhibits macrophage chemotaxis *in vivo* and *in vitro*[3]. Because of the importance of macrophages in the inflammatory response, the anti-inflammatory effect of etodolac could be partially mediated through inhibition of the chemotactic ability of macrophages.

Pharmacokinetics in healthy beagle dogs: Etodolac is rapidly and almost completely absorbed from the gastrointestinal tract following oral administration. The extent of etodolac absorption (AUC) is not affected by the prandial status of the animal. However, it appears that the peak concentration of the drug decreases in the presence of food. As compared to an oral solution, the relative bioavailability of the tablets when given with or without food is essentially 100%. Peak plasma concentrations are usually attained within 2 hours of administration. Though the terminal half-life increases in a nonfasted state, minimal drug accumulation (less than 30%) is expected after repeated dosing (i.e., steady-state). Pharmacokinetic parameters estimated in a crossover study (fed vs. fasted) in eighteen 5-month old beagle dogs are summarized in the following table:

Mean Pharmacokinetic Parameters Estimated in 18 Beagle Dogs After Oral Administration of 150 mg of Etodolac (approximately 12-17 mg/kg)

Pharmacokinetic Parameter	Tablet/ Fasted	Tablet/ Nonfasted
C_{max} (µg/mL)	22.0±6.42	16.9±8.84
T_{max} (hours)	1.69±0.69	1.08±0.46
$AUC_{0-\infty}$ (µg•hours/mL)	64.1±17.9	63.9±28.9
Terminal half-life, $t_{1/2}$ (hrs)	7.66±2.05	11.98±5.52

Pharmacokinetics in dogs with reduced kidney function: In a study involving four beagle dogs with induced acute renal failure, there was no observed change in drug bioavailability after administration of 200 mg single oral etodolac doses. In a study evaluating an additional four beagles, no changes in electrolyte, serum albumin/total protein and creatinine concentrations were observed after single 200 mg doses of etodolac. This was not unexpected since very little etodolac is cleared by the kidneys in normal animals. Most of etodolac and its metabolites are eliminated via the liver and feces. In addition, etodolac is believed to undergo enterohepatic recirculation[4].

EFFICACY

A placebo-controlled, double-blinded study demonstrated the anti-inflammatory and analgesic efficacy of EtoGesic (etodolac) tablets in various breeds of dogs. In this clinical field study, dogs diagnosed with osteoarthritis secondary to hip dysplasia showed objective improvement in mobility as measured by force plate parameters when given EtoGesic tablets at the label dosage for 8 days.

Indications and usage

INDICATIONS

EtoGesic is recommended for the management of pain and inflammation associated with osteoarthritis in dogs.

Dosage and administration

DOSAGE AND ADMINISTRATION

The recommended dose of etodolac in dogs is 10 to 15 mg/kg body weight (4.5 to 6.8 mg/lb) administered once daily. Due to tablet sizes and scoring, dogs weighing less than 5 kg (11 lb) cannot be accurately dosed. The effective dose and duration should be based on clinical judgment of disease condition and patient tolerance of drug treatment. The initial dose level should be adjusted until a satisfactory clinical response is obtained, but should not exceed 15 mg/kg once daily. When a satisfactory clinical response is obtained, the daily dose level should be reduced to the minimum effective dose for longer term administration.

Contraindications

CONTRAINDICATIONS

EtoGesic is contraindicated in animals previously found to be hypersensitive to etodolac.

Precautions

PRECAUTIONS

Treatment with EtoGesic tablets should be terminated if signs such as inappetence, emesis, fecal abnormalities, or anemia are observed. Dogs treated with nonsteroidal anti-inflammatory drugs, including etodolac, should be evaluated periodically to ensure that the drug is still necessary and well tolerated.

EtoGesic, as with other nonsteroidal anti-inflammatory drugs, may exacerbate clinical signs in dogs with pre-existing or occult gastrointestinal, hepatic or cardiovascular abnormalities, blood dyscrasias, or bleeding disorders.

As a class, cyclooxygenase inhibitory NSAIDs may be associated with gastrointestinal and renal toxicity. Sensitivity to drug-associated adverse effects varies with the individual patient. Patients at greatest risk for renal toxicity are those that are dehydrated, on concomitant diuretic therapy, or those with renal, cardiovascular, and/or hepatic dysfunction. Since many NSAIDs possess the potential to induce gastrointestinal ulceration, concomitant use of etodolac with other anti-inflammatory drugs, such as other NSAIDs and corticosteroids, should be avoided or closely monitored.

Studies to determine the activity of EtoGesic tablets when administered concomitantly with other protein-bound drugs have not been conducted in dogs. Drug compatibility should be monitored closely in patients requiring adjunctive therapy.

The safety of EtoGesic has not been investigated in breeding, pregnant or lactating dogs or in dogs under 12 months of age.

INFORMATION FOR DOG OWNERS

EtoGesic, like other drugs of its class, is not free from adverse reactions. Owners should be advised of the potential for adverse reactions and be informed of the clinical signs associated with drug intolerance. Adverse reactions may include decreased appetite, vomiting, diarrhea, dark or tarry stools, increased water consumption, increased urination, pale gums due to anemia, yellowing of gums, skin or white of the eye due to jaundice, lethargy, incoordination, seizure, or behavioral changes. **Serious adverse reactions associated with this drug class can occur without warning and in rare situations result in death (see Adverse Reactions). Owners should be advised to discontinue EtoGesic therapy and contact their veterinarian immediately if signs of intolerance are observed.** The vast majority of patients with drug related adverse reactions have recovered when the signs are recognized, the drug is withdrawn, and veterinary care, if appropriate, is initiated. Owners should be advised of the importance of periodic follow-up for all dogs during administration of any NSAID.

Warnings

WARNINGS

Keep out of reach of children. Not for human use. Consult a physician in cases of accidental ingestion by humans. **For use in dogs only.** Do not use in cats.

All dogs should undergo a thorough history and physical examination before initiation of NSAID therapy. Appropriate laboratory tests to establish hematological and serum biochemical baseline data prior to, and periodically during, administration of any NSAID should be considered. **Owners should be advised to observe for signs of potential drug toxicity (see Information for Dog Owners and Adverse Reactions).**

Adverse reactions

ADVERSE REACTIONS

In a placebo-controlled clinical field trial involving 116 dogs, where treatment was administered for 8 days, the following adverse reactions were noted:

Adverse Reaction	EtoGesic Tablets % of Dogs	Placebo % of Dogs
vomiting	4.3%	1.7%
regurgitation	0.9%	2.6%
lethargy	3.4%	2.6%
diarrhea/loose stool	2.6%	1.7%
hypoproteinemia	2.6%	0
urticaria	0.9%	0
behavioral change, urinating in house	0.9%	0
inappetence	0.9%	1.7%

Following completion of the clinical field trial, 92 dogs continued to receive etodolac. One dog developed diarrhea following 2-1/2 weeks of treatment. Etodolac was discontinued until resolution of clinical signs was observed. When treatment was resumed, the diarrhea returned within 24 hours. One dog experienced vomiting which was attributed to treatment, and etodolac was discontinued. Hypoproteinemia was identified in one dog following 11 months of etodolac therapy. Treatment was discontinued, and serum protein levels subsequently returned to normal.

Post-Approval Experience:

As with other drugs in the NSAID class, adverse responses to EtoGesic tablets may occur. The adverse drug reactions listed below are based on voluntary post-approval reporting. The categories of adverse event reports are listed below in decreasing order of frequency by body system.

Gastrointestinal: Vomiting, diarrhea, inappetence, gastroenteritis, gastrointestinal bleeding, melena, gastrointestinal ulceration, hypoproteinemia, elevated pancreatic enzymes.

Hepatic: Abnormal liver function test(s), elevated hepatic enzymes, icterus, acute hepatitis.

Hematological: Anemia, hemolytic anemia, thrombocytopenia, prolonged bleeding time.

Neurological/Behavioral/Special Senses: Ataxia, paresis, aggression, sedation, hyperactivity, disorientation, hyperesthesia, seizures, vestibular signs, keratoconjunctivitis sicca.

Renal: Polydipsia, polyuria, urinary incontinence, azotemia, acute renal failure, proteinuria, hematuria.

Dermatological/Immunological: Pruritus, dermatitis, edema, alopecia, urticaria.

Cardiovascular/Respiratory: Tachycardia, dyspnea.

In rare situations, death has been reported as an outcome of some of the adverse responses listed above. To report suspected adverse reactions, or to obtain technical assistance, call (800) 477-1365.

SAFETY

In target animal safety studies, etodolac was well tolerated clinically when given at the label dosage for periods as long as one year (see Precautions).

Oral administration of etodolac at a daily dosage of 10 mg/kg (4.5 mg/lb) for twelve months or 15 mg/kg (6.8 mg/lb) for six months, resulted in some dogs showing a mild weight loss, fecal abnormalities (loose, mucoid, mucosanguineous feces or diarrhea), and hypoproteinemia. Erosions in the small intestine were observed in one of the eight dogs receiving 15 mg/kg following six months of daily dosing.

Elevated dose levels of EtoGesic (etodolac), i.e.≥40 mg/kg/day (18 mg/lb/day, 2.7X the maximum daily dose), caused gastrointestinal ulceration, emesis, fecal occult blood, and weight loss. At a dose of ≥80 mg/kg/day (36 mg/lb/day, 5.3X the maximum daily dose), 6 of 8 treated dogs died or became moribund as a result of gastrointestinal ulceration. One dog died within 3 weeks of treatment initiation while the other 5 died after 3-9 months of daily treatment. Deaths were preceded by clinical signs of emesis, fecal abnormalities, decreased food intake, weight loss, and pale mucous membranes. Renal tubular nephrosis was also found in 1 dog treated with 80 mg/kg for 12 months. Other common abnormalities observed at elevated doses included reductions in red blood cell count, hematocrit, hemoglobin, total protein and albumin concentrations; and increases in fibrinogen concentration and reticulocyte, leukocyte, and platelet counts.

In an additional study which evaluated the effects of etodolac administered to 6 dogs at the labeled dose for approximately 9.5 weeks, the incidence of stool abnormalities (diarrhea, loose stools) was unchanged for dogs in the weeks prior to initiation of etodolac treatment, and during the course of etodolac therapy. Five of the dogs receiving etodolac, versus 2 of the placebo-treated dogs, exhibited excessive bleeding during an experimental surgery. No significant evidence of drug-related toxicity was noted on necropsy.

Storage

STORAGE CONDITIONS

Store at controlled room temperature, 15-30°C (59-86°F).

How supplied

HOW SUPPLIED

EtoGesic (etodolac) is available in 150 and 300 mg single-scored tablets and supplied in bottles containing 7, 30 and 90 tablets.

NDC 0856-5520-03 – 150 mg – bottles of 7
NDC 0856-5520-04 – 150 mg – bottles of 30
NDC 0856-5520-05 – 150 mg – bottles of 90
NDC 0856-5530-03 – 300 mg – bottles of 7
NDC 0856-5530-04 – 300 mg – bottles of 30
NDC 0856-5530-05 – 300 mg – bottles of 90

REFERENCES

1. Vane, JR, RM Botting. Overview - mechanisms of action of anti-inflammatory drugs. In Improved Non-steroid Anti-inflammatory Drugs - COX-2 Enzyme Inhibitors. J Vane, J Botting, R Botting (ed.) 1996. Kluwer Academic Publishers. Dordrecht, The Netherlands.
2. Glaser, KB. Cyclooxygenase selectivity and NSAIDs: cyclooxygenase-2 selectivity of etodolac (Lodine®). *Inflammopharmacol* (1995) 3:335-345.
3. Gervais, F, RR Martel, E Skamene. The effect of the non-steroidal anti-inflammatory drug etodolac on macrophage migration *in vitro* and *in vivo*. *J. Immunopharmacol* (1984) 6:205-214.
4. Cayen, MN, M Kraml, ES Ferdinandi, EL Greselin, D Dvornik. The metabolic disposition of etodolac in rats, dogs, and man. *Drug Metab. Revs.* (1981) 12:339-362.

Fort Dodge Animal Health
Fort Dodge, Iowa 50501 USA

5530D Revised March 2001 00781

Figure 5-6 An example of a package insert.

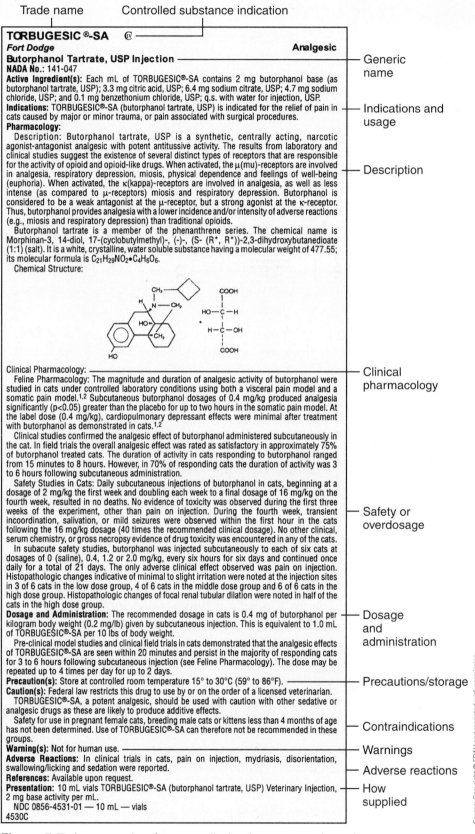

Trade name   Controlled substance indication

TORBUGESIC®-SA Ⓒ
Fort Dodge   **Analgesic**
Butorphanol Tartrate, USP Injection ————————————— Generic name
NADA No.: 141-047
Active Ingredient(s): Each mL of TORBUGESIC®-SA contains 2 mg butorphanol base (as butorphanol tartrate, USP); 3.3 mg citric acid, USP; 6.4 mg sodium citrate, USP; 4.7 mg sodium chloride, USP; and 0.1 mg benzethonium chloride, USP; q.s. with water for injection, USP.
Indications: TORBUGESIC®-SA (butorphanol tartrate, USP) is indicated for the relief of pain in ——— Indications and usage cats caused by major or minor trauma, or pain associated with surgical procedures.
Pharmacology:

 Description: Butorphanol tartrate, USP is a synthetic, centrally acting, narcotic agonist-antagonist analgesic with potent antitussive activity. The results from laboratory and clinical studies suggest the existence of several distinct types of receptors that are responsible for the activity of opioid and opioid-like drugs. When activated, the μ(mu)-receptors are involved ——— Description in analgesia, respiratory depression, miosis, physical dependence and feelings of well-being (euphoria). When activated, the κ(kappa)-receptors are involved in analgesia, as well as less intense (as compared to μ-receptors) miosis and respiratory depression. Butorphanol is considered to be a weak antagonist at the μ-receptor, but a strong agonist at the κ-receptor. Thus, butorphanol provides analgesia with a lower incidence and/or intensity of adverse reactions (e.g., miosis and respiratory depression) than traditional opioids.

 Butorphanol tartrate is a member of the phenanthrene series. The chemical name is Morphinan-3, 14-diol, 17-(cyclobutylmethyl)-, (-)-, (S- (R*, R*))-2,3-dihydroxybutanedioate (1:1) (salt). It is a white, crystalline, water soluble substance having a molecular weight of 477.55; its molecular formula is $C_{21}H_{29}NO_2 \bullet C_4H_6O_6$.
 Chemical Structure:

Clinical Pharmacology: ————————————————————————— Clinical pharmacology
 Feline Pharmacology: The magnitude and duration of analgesic activity of butorphanol were studied in cats under controlled laboratory conditions using both a visceral pain model and a somatic pain model.[1,2] Subcutaneous butorphanol dosages of 0.4 mg/kg produced analgesia significantly ($p<0.05$) greater than the placebo for up to two hours in the somatic pain model. At the label dose (0.4 mg/kg), cardiopulmonary depressant effects were minimal after treatment with butorphanol as demonstrated in cats.[1,2]

 Clinical studies confirmed the analgesic effect of butorphanol administered subcutaneously in the cat. In field trials the overall analgesic effect was rated as satisfactory in approximately 75% of butorphanol treated cats. The duration of activity in cats responding to butorphanol ranged from 15 minutes to 8 hours. However, in 70% of responding cats the duration of activity was 3 to 6 hours following subcutaneous administration.

 Safety Studies in Cats: Daily subcutaneous injections of butorphanol in cats, beginning at a dosage of 2 mg/kg the first week and doubling each week to a final dosage of 16 mg/kg on the fourth week, resulted in no deaths. No evidence of toxicity was observed during the first three weeks of the experiment, other than pain on injection. During the fourth week, transient ——— Safety or overdosage incoordination, salivation, or mild seizures were observed within the first hour in the cats following the 16 mg/kg dosage (40 times the recommended clinical dosage). No other clinical, serum chemistry, or gross necropsy evidence of drug toxicity was encountered in any of the cats.

 In subacute safety studies, butorphanol was injected subcutaneously to each of six cats at dosages of 0 (saline), 0.4, 1.2 or 2.0 mg/kg, every six hours for six days and continued once daily for a total of 21 days. The only adverse clinical effect observed was pain on injection. Histopathologic changes indicative of minimal to slight irritation were noted at the injection sites in 3 of 6 cats in the low dose group, 4 of 6 cats in the middle dose group and 6 of 6 cats in the high dose group. Histopathologic changes of focal renal tubular dilation were noted in half of the cats in the high dose group.
Dosage and Administration: The recommended dosage in cats is 0.4 mg of butorphanol per ——— Dosage and administration kilogram body weight (0.2 mg/lb) given by subcutaneous injection. This is equivalent to 1.0 mL of TORBUGESIC®-SA per 10 lbs of body weight.

 Pre-clinical model studies and clinical field trials in cats demonstrated that the analgesic effects of TORBUGESIC®-SA are seen within 20 minutes and persist in the majority of responding cats for 3 to 6 hours following subcutaneous injection (see Feline Pharmacology). The dose may be repeated up to 4 times per day for up to 2 days.
Precaution(s): Store at controlled room temperature 15° to 30°C (59° to 86°F). ——————— Precautions/storage
Caution(s): Federal law restricts this drug to use by or on the order of a licensed veterinarian.

 TORBUGESIC®-SA, a potent analgesic, should be used with caution with other sedative or analgesic drugs as these are likely to produce additive effects.

 Safety for use in pregnant female cats, breeding male cats or kittens less than 4 months of age ——— Contraindications has not been determined. Use of TORBUGESIC®-SA can therefore not be recommended in these groups.
Warning(s): Not for human use. ————————————————————————— Warnings
Adverse Reactions: In clinical trials in cats, pain on injection, mydriasis, disorientation, ——— Adverse reactions swallowing/licking and sedation were reported.
References: Available upon request.
Presentation: 10 mL vials TORBUGESIC®-SA (butorphanol tartrate, USP) Veterinary Injection, ——— How supplied 2 mg base activity per mL.
 NDC 0856-4531-01 — 10 mL — vials
4530C

Figure 5-7 An example of a controlled substance package insert.

may make the product unsafe for some patients. This section may also be referred to as the *chemistry section.*

- *Clinical pharmacology* or *actions* or *mode of action statements* describe the activities and various properties of the main drug. The toxicology study data on the drug may be represented here. Metabolism and excretion of the drug may also be described in this section.

- The *indications and usage* section of the insert lists the specific uses for which the drug has been approved (indications) and describes how and for how long the drug is generally used (usage). Species-specific information will be listed in this section. Use in species not listed in the indications section and/or uses other than those described in this section are considered extra-label use (refer to Chapter 1 for a description of extra-label drug use).

- The *contraindications* section describes the situations in which the drug should not be used (e.g., hypersensitivity to the drug) and the animals that should not receive the drug.

- The *precautions* section includes reasons to use the drug carefully. This does not imply that the drug use is contraindicated. Precautions usually describe conditions in which the drug is more likely to cause a problem (i.e., use of sulfa antibiotics in dehydrated animals) and suggests steps that should be taken in order for safe drug use.

- The *warnings* section relates situations in which potentially serious problems may occur if the drug is used (e.g., during pregnancy or renal disease). The warnings section may lead into an animal toxicity section if additional studies were completed during development or if market use of the drug has indicated potential toxicity in certain cases.

- *Adverse reactions* or *side effects* are undesirable reactions to drugs that are significant enough to warrant extreme caution in use of the drug. A side effect is any normally occurring effect of the drug other than the intended therapeutic effect. Cautions are sometimes included in this section of the insert as well.

- *Overdosage* information lists the dangers of using excessive quantities of the drug. This section also describes the signs of overdose and how they should be handled.

- *Dosage and administration* information describes the parameters for use of the drug. Dosage refers to the amount of drug (usually per unit of body weight) that will produce the desired effect. Administration refers to the route that can be employed to deliver the drug into the animal's body.

- *Storage* information specifies the temperature and conditions under which the drug must be stored to maintain its integrity and viability.

- *How supplied* information states the dosage forms, strengths, and container size in which the drug is sold.

Additional information found in package inserts may include any or all of the following: how to prepare a suspension if mixing is needed, application methods for topical products, special disposal information, expiration dates on reconstituted powders, residue and withdrawal information, name and address of the manufacturer and distributor, and a *references* section for information on how the drug was tested and dosing information obtained.

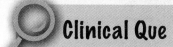

Clinical Que

Drugs packaged or supplied in brown bottles are light sensitive and should be stored out of direct light.

Drug Reference Material

There are a variety of good drug references available for use in veterinary medicine. Formularies are books that contain drug dosages and are available for specific institutions (like veterinary medical colleges or research facilities) or from publishers. Larger references are also available such as *VPB, CVP,* and *Plumb's Veterinary Drug Handbook.* Information on human drugs is not found in veterinary references; therefore, a human drug reference book such as the *PDR* needs to be available in clinical practice for information on human drugs that may be used in animals or accidently ingested by animals. Some of these materials are found in electronic media forms in addition to traditional published format.

What's in the PDR?

- Manufacturer's Index (white pages): Alphabetical listing of pharmaceutical manufacturers included in the *PDR.*

- Brand & Generic Name Index (pink pages): Gives the page number of each product by brand and generic name. The drugs listed in the *PDR* are chosen by the drug manufacturers because they pay a fee to have their drugs in this reference. Therefore, not all drugs manufactured in the United States are included in the *PDR.*

- Product Category Index (blue pages): Lists all fully described products by prescribing category.

- Product Identification Guide (gray pages): Full-color, actual-size photos of tablets, capsules, and other dose forms, arranged alphabetically by manufacturer.

What's in the VPB and CVP?

- Manufacturer/Distributor Index (white pages): Contains alphabetical listing of pharmaceutical manufacturers included in the *VPB* and *CVP.*

- Brand Name and Ingredient Index (green pages): Gives the page number of each product by brand name and ingredients found in these products.

- Therapeutic Index (pink pages): Lists all fully described products by prescribing category.

- Biological Charts (blue pages): List groups of vaccines and bacterins by appropriate species, delineated by their antigens.

- Anthelmintic and Parasiticide Charts (buff pages): List groups of internal parasiticides and endectocides by appropriate species, delineated by the corresponding parasite.
- Withdrawal Time Charts (yellow pages): Alphabetically summarize information on withdrawal times for food products according to the species and route of administration.

DISPENSING VERSUS PRESCRIBING

Veterinary drugs are those approved for use only in animals. Veterinary drugs are not subject to regulatory action by the FDA; however, FDA regulations require that the drug label clearly describe the approved use of the drug in these ways: for use in certain indications, in certain species, by a certain route of administration, at a certain dose, and over a certain length of time. Any use of an approved label drug outside of its "approved use" is subject to regulatory action.

Some drugs, such as antibiotics, are utilized in both human and veterinary medicine. The FDA specifies in its CPG that "food-producing animals" (e.g., cattle, sheep, pigs, and chickens) generally should not receive drugs that are utilized in both human and veterinary practice. Illegal drug residues may be present in milk, meat, and/or eggs from these animals and may present an unknown health risk to the consumer of these products. Most diseases in food-producing animals are treatable with approved veterinary-use-only drugs that can be selected and employed as appropriate by the veterinarian. Any manufacturer, distributor, or pharmacy that suggests the use of human-labeled drugs for animals must comply with FDA regulations on the administration of these drugs.

In the case of non-food-producing animals, the CPG makes an exception. Generally, veterinarians can prescribe human-labeled drugs for any non-food-producing animal without violating any FDA regulations. If such human-labeled drugs harm the animal's health, however, the CPG warns of FDA regulatory action or referral to the state veterinary licensing authority for investigation.

Veterinarians must establish a VCPR prior to prescribing any medication for an animal (Table 5-2). If an animal requires a drug, the veterinarian can either write a prescription to be filled at a human pharmacy, or dispense the drug from the veterinary clinic. Instructions for drug use for the patient are part of the medical record and are often written in abbreviated Latin. Veterinarians abbreviate instructions because Latin terminology is universally used in medical professions and is concise, convenient, and time saving. Table 5-3 provides examples of common Latin abbreviations.

Some veterinarians order drugs by prescription because keeping an extensive pharmaceutical inventory ties up working capital and runs the risk of outdating, spoilage, and obsolescence. A **prescription** is an order to a pharmacist, written by a licensed veterinarian (or physician), to prepare the prescribed

Table 5-2 Guidelines for Veterinary Prescribed Drugs and Veterinarian/Client/Patient Relationship

- Veterinary prescription drugs are labeled for use only by or on the order of a licensed veterinarian. Incidents involving the sale and use of prescription drugs without a prescription should be reported to the proper state authority and the U.S. FDA.

- Veterinary prescription drugs are to be used or prescribed only within the context of a VCPR.

  A VCPR exists when all of the following conditions have been met:

  - The veterinarian has assumed responsibility for making clinical judgments regarding the health of the animal(s) and the need for medical treatment, and the client has agreed to follow the veterinarian's instructions.

  - The veterinarian has sufficient knowledge of the animal(s) to initiate at least a general or preliminary diagnosis of the medical condition of the animal(s). This means that the veterinarian has recently seen and is personally acquainted with the keeping and care of the animal(s) by virtue of an examination of the animal(s) or by medically appropriate and timely visits to the premises where the animal(s) are kept.

  - The veterinarian is readily available for follow-up evaluation, or has arranged for emergency coverage, in the event of adverse reactions or failure of the treatment regimen.

- Veterinary prescription drugs must be properly labeled before being dispensed.

- Appropriate dispensing and treatment records must be maintained.

- Veterinary prescription drugs should be dispensed only in quantities required for the treatment of the animal(s) for which the drugs are dispensed. Avoid unlimited refills of prescriptions or any other activity that might result in misuse of drugs.

- Any drug used in a manner not in accordance with its labeling (extra-label use) should be subjected to the same supervisory precautions that apply to veterinary prescription drugs.

Adapted from the AVMA Guidelines for Veterinary Prescription Drugs on www.avma.org

medicine, to affix the directions, and to sell the preparation to the client. The prescription is a legally recognized document and the writer is held responsible for its accuracy. A pharmacist may dispense human-labeled drugs, and the veterinary staff may dispense veterinary-use-only drugs. Veterinarians cannot legally fill another veterinarian's or physician's prescription because they are not board-certified pharmacists. The dispensing veterinarian must only dispense to those animals with whom he/she has a legal VCPR.

Through a written prescription, the veterinarian gives the pharmacist all the information necessary to dispense safe and effective drugs for the animal. Prescriptions may be written on standard forms, sent electronically, or telephoned in to the pharmacy. Figure 5-8 shows the seven parts of a written prescription.

Clinical Que

Latin abbreviations for pharmacology terms and directions make medical record and prescription writing concise and efficient.

Table 5-3   Pharmacology Abbreviations

ABBREVIATION	MEANING
%	Percent
®	registered trade name when superscript by drug name
ac	before meals (ante cibum)
ad lib	as much as desired (ad libitum)
bid	twice daily (bis in die)
BSA	body surface area
$\bar{c}$	with
cal	calorie
cap	capsule
cc	cubic centimeter (same as mL)
cm	centimeter
conc	concentration
D_5W	5% dextrose in water
dr	dram; equal to 1/8 oz or 4 mL
ED	effective dose
ED_{50}	median effective dose
fl oz	fluid ounce
g	gram
gal	gallon
gr	grain; unit of weight approximately 65 mg
gt	drop (gutta)
gtt	drops (guttae)
hr	hour
IA	intra-arterial
IC	intracardiac
ID	intradermal
IM	intramuscular
IP	intraperitoneal
IT	intratracheal
IU	international units
IV	intravenous
kg	kilogram
km	kilometer

(Continued)

Abbreviation	Meaning
L	liter
lb or # behind a number	pound(s)
LD	lethal dose
LD_{50}	median lethal dose
LRS	lactated Ringer's solution
m	meter
mcg or μg	microgram
mED	minimal effective dose
mEq	milliequivalent
mg	milligram
MIC	minimum inhibitory concentration
MID	minimum infective dose
mL	milliliter (same as cc)
MLD	minimum lethal dose
mm	millimeter (also used to mean muscles)
npo	nothing by mouth (non per os)
NS	normal saline
OTC	over the counter
oz	ounce
p̄	after
pc	after meals (post cibum)
PDR	*Physician's Dest Reference*
pH	hydrogen ion concentration (acidity and alkalinity measurement)
po	orally (per os)
ppm	parts per million
PR	per rectum
prn	as needed (pro re nata)
pt	pint
q	every
q12h	every twelve hours
q4h	every four hours
q6h	every six hours
q8h	every eight hours
qd	every day

(Continued)

ABBREVIATION	MEANING
qh	every hour
qid	four times daily (quarter in die)
qn	every night
qod or eod	every other day
qp	as much as desired
qt	quart
Rx	prescription
s̄	without
sid	once daily (qd or q24h are preferred abbreviations)
sig	let it be written as (used when writing prescription)
sol'n or soln	solution
SQ, SC, subQ, or subc	subcutaneous
T	tablespoon or tablet (or temperature)
tab	tablet
tid	three times daily (ter in die)
tsp	teaspoon
vol	volume
VPB	Veterinary Pharmaceuticals and Biologicals

1. The veterinarian's name, address, and telephone number, all usually preprinted at the top of the form. (If the prescription is for a controlled substance, the DEA number of the prescriber must be given to the pharmacist. DEA numbers may be listed on the preprinted form, but may also be kept confidential and excluded from the preprinted prescription pad.)

2. The client's name and address, as well as the species and name of the patient.

3. The name of the drug, the strength of the drug, and the quantity to be given to the patient.

4. The instructions for giving the drug to the patient; these must include the amount to be given, the route of administration, the frequency of administration, and the duration of administration.

5. The number of refills permitted.

6. The veterinarian's signature.

7. The date on which the prescription was written.

Other things that may be on the written prescription include cautionary statements to be included on the label ("keep out of reach of children", "use gloves when applying", etc.) and withdrawal times.

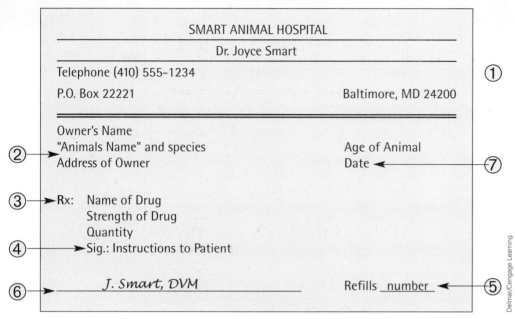

Figure 5-8 The seven parts of a written prescription. Written prescriptions are given to pharmacists so that they can fill medication orders.

Figure 5-9 shows two typical prescriptions with parts 1 through 7 indicated. Prescription A is an order for a cream to be applied to the skin to treat an infection, and prescription B is for insulin injection.

All veterinary prescription drugs should be properly labeled when dispensed. The prescription label should contain the information that appears on the prescription, with the exception of the veterinarian's signature. Remember that veterinary staff members cannot refill or dispense medications without veterinarian approval. All medications should be dispensed in childproof containers; however, it is not currently illegal to dispense veterinary prescriptions in paper envelops or non-childproof containers (in humans, the Poison Prevention Packaging Act of 1970 allows the FDA to require dispensing certain drugs in childproof containers). Labels with cautionary statements may also be used on the prescription bottle to help clients remember key facts about the drugs they are giving their animals. Sample warning labels are illustrated in Figure 5-10.

The label on the prescription bottle dispensed to clients must be complete and contain directions that clients can understand and follow. The complete label must have the following:

A. the name and address of the dispenser (animal hospital and veterinarian's names are usually preprinted on the label)

B. the client's name (address is optional)

C. the animal's name and species

D. the drug name, strength, and quantity to be given

E. the date of the order

F. the directions for use

G. any refill information (may be included if warranted)

Clinical Que

Ways to avoid errors in prescription writing include writing legibly, using a zero in front of any numbers with a decimal point (e.g., 0.02 instead of .02), and specifying both dose and strength of drug required.

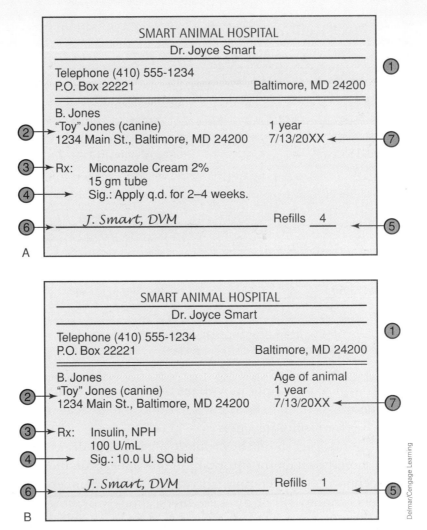

Figure 5-9 Two example prescriptions.

Figure 5-10 Warning labels are placed on prescription medication containers to emphasize key facts about the drug.

Figure 5-11 shows the components of a drug label. Table 5-4 summarizes information required for prescriptions, labels, and medical records.

Medication orders for use in the clinic are recorded in the patient's medical record. The design of the medical record may vary from clinic to clinic. Veterinarians write the medication order in a paper file or type it into an electronic record. These medication orders are then filled, and as the patient is given the medication, it is recorded in the medical record with the date, time, and initials of the person giving the drug (Figure 5-12A). Clients should be given pharmaceutical data sheets when drugs need to be administered to their animals at home (Figure 5-12B).

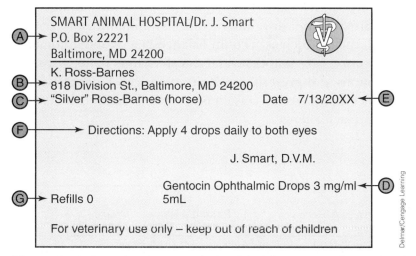

Figure 5-11 Sample of a prescription label for use on dispensing bottle.

Table 5-4   Basic Information for Records, Prescriptions, and Labels

INFORMATION	RECORDS	PRESCRIPTIONS	LABELS
• Name, address, and telephone number of the veterinarian	+	+	+
• Name, address, and telephone number of the client	+	+	Name only
• Identification of the animal(s) treated, species, and numbers of animals treated, when possible	+	+	+
• Date of treatment, prescribing, or dispensing of drug	+	+	+
• Name and quantity of the drug to be prescribed or dispensed	+	+	+
• Dose and duration directions for use	+	+	+
• Number of refills authorized	+	+	May be included

The number of veterinary hospitals that utilize a completely paperless electronic medical record system is increasing. In human medicine, there is a strong push to move into a paperless medical record system because of decreased labor costs, decreased litigation costs, and better physician/client communication. For veterinary facilities, there are similar advantages including improved efficiency (no lost records, immediate access to records from anywhere in the facility, and the ability to access all prescriptions filled or laboratory work done in the past on one screen); space saving (no file cabinets, folders, storage boxes, and less room dedicated to records and record keeping); cost saving (less filing; no time needed to retrieve records; the ability to access records on a computer at a phone station to answer client questions; and automatic download of in-house laboratory results, digital images, and drug records into client records); avoidance of errors (prompts for patients with allergic reactions, information on drug interactions, and identification of clients for special considerations such as account balances, etc.); and automated input (in-house laboratory data automatically transfers into patient record, and departing instructions and prescription instructions can be entered into the computer in advance and linked to certain procedures or billing items that can be personalized for each client).

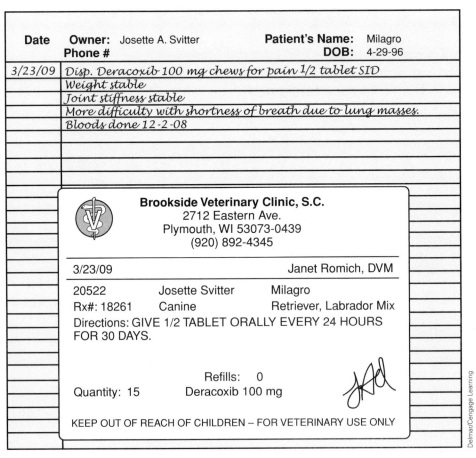

A

Figure 5-12a An example of a paper medical record with the proper documentation for administration of a drug.

Veterinary Pharmaceutical Data Sheet
Brookside Veterinary Clinic, S.C.
2712 Eastern Ave.
Plymouth, WI 53073-0439

Client: Josette Svitter
Patient: Milagro   Canine/Retriever, Labrador Mix/Neutered Male

Your animal has been prescribed deracoxib. Please read the information below to make sure you fully understand the purpose and proper use of deracoxib. Your veterinarian has requested that you give the prescription (deracoxib) to your animal as follows:

GIVE 1/2 TABLET ORALLY EVERY 24 HOURS FOR 30 DAYS.

Pharmaceutical: Deracoxib

–General
INDICATIONS FOR USE: Your veterinarian has prescribed this medication for the relief of pain in dogs.

PRECAUTIONS: As with any medication, there are precautions and side effects associated with their use. This is sometimes complicated by the fact that many of the drugs used in veterinary patients are human drugs that have not been studied for use in animals.
Some of these so called "extra label" drugs have been used for years in veterinary medicine and side effects and precautions listed are those that have been compiled by experience with these drugs.
In cases where veterinary experience is limited, human precautions and side effects have been listed.
In any event, many of the side effects listed have occurred very rarely but have been included for your information and awareness.
The potential benefit of the use of the drug must be weighed against the information known about the risk of the use of the drug.
Your veterinarian has the best interest of your animal in mind and will help you make an informed decision about their treatment.

*It should not be used in patients with known sensitivity to this type of drug–non steroidal anti inflammatory drug. Dogs with bleeding disorders, GI ulcers, liver disease or pre existing renal impairment may be at a greater risk for serious side effects.
Use only when benefit outweighs the risk during pregnancy as safety studies have not been done.
It is recommended to give the lowest effective dose as infrequently as is necessary to control pain.
Baseline bloodwork and urine is recommended to assess the general health of the patient prior to starting this drug.
Periodic bloodwork to monitor this drug's effect on the body is advised if long term therapy is required.

SIDE EFFECTS: GI effects are the most common side effect–nausea, loss of appetite, vomiting, stomach ulceration, soft stool or diarrhea.
Other possible side effects include lethargy, restlessness, aggression, itchiness, hair loss, skin reactions, hives, redness of skin, low albumen, increased water consumption and urination, neurologic signs, seizures and balance disorders.
Liver damage, jaundice and/or kidney disease can occur and dogs with pre existing conditions and geriatric dogs seem to be at greater risk.
Serious blood changes such as immune mediated hemolytic anemia and low platelet count are also possible.
If any reaction becomes severe it may result in the death of the animal.
If your animal experiences any of the above side effects, any unusual bleeding or bruising or any unusual symptoms discontinue use and contact your veterinarian.
Notify your veterinarian if your animal's condition does not improve or worsens despite this treatment.

APPROVED SPECIES: The use of this medication in animals other than dogs is considered extra label as there is no product available for other species.

DRUG INTERACTIONS: If your animal has been prescribed any other medication, including aspirin, steroids or other anti inflammatory medication such as Etogesic®, Rimadyl® or Zubrin please discuss this with your veterinarian as interactions may occur.

OVERDOSE: Keep all drugs out of the reach of children. In case of accidental exposure contact your local health professional immediately.
Keep all drugs out of the reach of pets. Overdose of any medication can cause serious adverse effects.
In case of accidental ingestion or overdose, contact your veterinarian or the National Animal Poison Control Center at 1-800-548-2423 immediately.

B

Figure 5-12b An example of an electronic medical record with the proper documentation for administration of a drug.

PHARMACY ECONOMICS

Maintaining a pharmacy is a business that depends on charging and collecting a fee for services to continue providing medical care. Although veterinary technicians deal primarily with the formation and delivery of medications, they must also pay attention to the fiscal management policies that keep the practice in business.

Often, the responsibility for purchasing and pricing medications is delegated to the veterinary technician. In order to be comfortable in such a position, the veterinary technician must be familiar with pharmacy economics, purchasing terminology, and inventory control.

Inventory Control and Maintenance

Maintaining a veterinary pharmacy requires planning and continuous monitoring. Understanding drug use and patient needs helps prevent drug shortages, makes efficient use of time, decreases costs, and reduces stress. Time invested in maintenance of appropriate stock levels benefits the overall business health of the veterinary practice.

The goal of maintaining drug inventory is to stock quantities of each item as low as possible to reduce overhead and inventory costs yet not so low as to cause drug shortages between ordering periods. The cost of stocking a case of medication is more than simply the price listed on the invoice. Inventory is unproductive until it is used or sold. The longer inventory sits on a shelf, the more it costs the practice in hidden costs such as storage, expiration or spoilage, bookkeeping, breakage, and the appearance that the product is undesirable because it sits on the shelf too long. Too much inventory also ties up money that could be invested and earning interest or used for other purchases that in turn can make money for the practice. Some states also tax inventory, which costs the practice additional money.

Most people err on the side of ordering more drugs than less to make sure that the practice does not have a shortage of drugs and supplies. With the use of computerized inventory software, monitoring daily and annual drug dispensing history can be determined.

One way to assess whether too much money is tied up in inventory is to calculate turnover. The turnover rate is defined as the number of times inventory is depleted ("turned over") and replenished each year. The turnover rate varies greatly with each product, but the average value is important. To calculate the turnover rate for the practice, divide the total inventory expenses for the entire year by the average cost of inventory on hand on any given day. The average cost of inventory on hand is used to average out seasonal variations in some drugs. The average cost of inventory on hand is determined by the year's beginning inventory plus the year's ending inventory divided by two.

$$\text{Turnover rate} = \frac{\text{Yearly inventory expense}}{\text{Average cost of inventory on hand}}$$

An example of calculating the turnover rate is as follows. A veterinary practice spends $100,000 on drugs and supplies during the year. At the beginning of the year, wholesale value of the inventory on hand was $20,000 and $13,000 at the end of the year. The turnover rate would be approximately 6 ($20,000 + $13,000 ÷ 2 = $16,500; $100,000 ÷ $16,500 = 6). This means that the average drug inventory item was replaced 6 times annually or every 2 months. The goal is to turn over inventory often as a high turnover rate means that drugs are being sold (making money for the practice) and not sitting on the shelf (costing the practice money). If the turnover rate is near 12, it would be possible to operate without investing much capital since inventory would be turned over once a month and drug inventories would often be used up prior to the invoice due date. A turnover rate of 12 might be difficult to achieve without being labor intensive; however, the absolute minimum turnover rate should be 4. The turnover rate should be used as a benchmark for monitoring inventory; a rate that works for one practice may not work for another practice.

Some practices, especially those without computerized inventory, do not assess their entire inventory, but rather concentrate on the drugs that are used most often. In this case, one rule to keep in mind is the 80/20 rule, which states that 20% of drugs stocked should account for 80% of annual expenses. This rule implies that most of the practice's drug profits and inventory management center around the most commonly used medications.

Even though it is best to keep inventory at a minimum, sometimes there are advantages for purchasing more of certain products. Special deals offered by vendors may save money; however, it is difficult to accurately predict the use of some products making buying in quantity difficult to justify. When ordering in bulk to take advantage of special deals, make sure that the product will be used within a reasonable period, before the drug expires and that the savings merit the commitment of money. Another advantage for purchasing more of certain products is that processing small orders takes time and the time committed to process a small order is not much different from that of processing a large order. If the items ordered are inexpensive, it may be worthwhile to order more items per order to reduce ordering frequency and minimize handling fees charged by vendors if the minimal required dollar amount per order is not sufficient. Another point to keep in mind when ordering is that some items are hard to obtain or may not be available without back order unless they are ordered early or if ordered in bulk. This is especially true of items such as heartworm preventative or flea and tick preventative that tend to sell more during select times of the year.

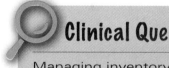

Clinical Que

Managing inventory makes the practice operate more efficiently while reducing expenses.

Inventory Purchasing

Drugs purchased by veterinarians can be marketed in a variety of ways. **Direct marketing** is when a drug is purchased directly from the company that manufactures it. Direct marketing eliminates the "middleman" and added handling fees, but requires a larger commitment of time (purchases need to be made from many vendors if each company only sells its products), larger inventories (purchases are made in larger quantities to get the best deal), and more storage space (to hold the larger quantities ordered). Another way to purchase pharmaceuticals is through distributors. **Distributors** or **wholesalers** are agencies that purchase the drug from the manufacturing company and resell it to veterinarians. Purchasing from distributors means that many items are purchased from one source. Distributors can purchase products from many different companies (they maintain the larger portion of the inventory) and make these products available to veterinarians via sales representatives or telephone/computer order systems. Distributors enable veterinary clinics to reduce inventory costs and the need for a large commitment of personnel to support the purchasing process. The primary disadvantage of distributors is the higher costs of pharmaceuticals. Generic companies sell drugs that are no longer under patent protection. Once a drug patent expires, other drug companies can apply to the FDA to sell a generic equivalent drug, a drug determined to be the therapeutic equal of a drug on which the patent has expired. The application for generic equivalent approval is called an Abbreviated New Drug Application (ANDA). Although no preclinical or clinical data need be included on this application if the generic drug is identical to the original in potency, dose form, and product labeling, the applicant must demonstrate the generic drug's bioequivalency.

Some companies advertise prescription products for sale to nonveterinarians. These companies do require a prescription from a veterinarian before allowing clients to purchase drugs for their animals. When a client prefers a prescription to a dispensed drug, the AVMA recommends that the veterinarian honor the request. While the AVMA recommends that veterinarians honor such client requests for written prescriptions, some state regulations actually require veterinarians to provide a prescription rather than dispense a drug when requested by the client. The prescription must be medically indicated and within a VCPR. In addition, states might have specific requirements or guidelines on the various forms (oral, written, or electronic) in which prescriptions may be offered. Veterinarians should be knowledgeable of state-specific requirements by contacting the board of veterinary medicine in the state(s) they are licensed. Depending on the state, board of pharmacy regulations may also apply to veterinarians as handlers of prescription drugs; so veterinarians should be aware of any pertinent board of pharmacy regulations in their state. State veterinary medical associations monitor state issues that affect veterinary medicine and are a good resource for veterinary professionals.

Other sources of drugs include other veterinary practices (helpful in a crisis), buying groups of several veterinary practices (advantage is decreased drug cost), and pharmacies (good for prescribing drugs not commonly used or human-approved drugs).

Inventory Management

Effectively managing inventory takes time and organization. Pharmaceutical management includes maintaining an adequate stock of all products used, dispensed, and sold; organizing inventory so that items are easy to locate; identifying products that need to be reordered; maintaining accurate purchasing and inventory records; ordering; receiving and inspecting shipments; establishing and updating pricing for all inventory items; rotating stock and monitoring expiration dates; and assessing new or updated products and considering specials offered by suppliers.

Establishing an Inventory System

The first step in establishing an inventory system is record keeping whether the system is manual or computerized. Five types of inventory records should be kept: a reorder log (commonly called a "want list"), purchase order records, individual inventory records, an inventory master list, and vendor files. For inventory records to be properly utilized, invoices and client/patient records must be properly maintained as well. Recording inventory supplies that are used, dispensed, or sold on all invoices and patient records is key to tracking supply levels.

Day-to-day monitoring of drug depletion must be monitored by all staff members. This is accomplished by maintaining a reorder log or "want list" of needed items. The inventory manager should only need to check inventory once weekly if all staff members record when items need to be reordered. By keeping a reorder log (or chalkboard, memoboard, or box that holds the empty drug packages), the minimum quantity of an item that should be kept on hand can be determined and used to predict ordering frequency and sets the reorder point. The reorder point is the level to which an item can be depleted before it is reordered. If orders are placed weekly, a minimum of a three-week supply should be kept in stock (one week for the order period, one week for order processing, and a one-week reserve). A buffer period also allows some protection against deleting the stock during sudden increases in use levels. If the item is ordered monthly, the reorder point should be a six-week supply (one week for the ordering, one week for processing, and one month's reserve). Once the reorder point is determined, the reorder quantity can be determined. The reorder quantity is the amount that should be ordered, which is typically a one-month supply. This amount can be altered as the practice becomes more comfortable with the system. When supplies are stocked, a reorder tag can be placed in front of the first unit of the minimum quantity or secured around the entire drug package.

Computerized inventory records also indicate when the minimum quantity has been reached as the drugs are charged out to the client. In the 1980s, automated dispensing systems were introduced into human health care facilities and are now being seen in some larger veterinary facilities. Automated dispensing systems are drug storage devices or cabinets that electronically dispense medications in a controlled fashion and track medication use (Figure 5-13). Most systems require user identifiers and passwords, and internal electronic devices track which veterinary professionals access the system, track the patients for whom medications are administered, and provide usage data for billing purposes. These automated dispensing systems can be stocked by centralized (pharmacy areas that prepare and distribute medications from a central location) or decentralized (pharmacy areas located near patients and are located throughout a facility) pharmacies. More advanced systems provide additional information support through integration into other external systems, databases, and the Internet. Some models use machine-readable code for medication dispensing and administration. Some examples of automated dispensing systems are the Baxter ATC-212® dispensing system and the Pyxis® Medstation.

Figure 5-13 An example of an automated dispensing unit.

Maintaining records of drugs ordered for a veterinary practice is another part of maintaining an inventory system. Purchase order records may be the actual purchase order form or computerized printouts of inventory control reports. Order records include the date ordered, quantity ordered, size ordered, product name, vendor, product order number (catalog number), unit cost, total cost, date received, quantity received, purchase order number, and discounts if given (Figure 5-14). If phoning in the order, the contact person's name and time and date ordered should be recorded. Computerized practices can also benefit from an order notebook for checking the status of orders.

Individual inventory records are the part of an inventory system that provides information about drug usage, order status, and price changes. An individual card or computer file is made for every supply item. The card or computer file should list product name (trade name, generic name, unit size, color, strength, minimum order quantity such as case or box, expiration date, and serial number), manufacturers and suppliers (names, phone numbers, addresses, and practice account numbers), order information (date and number ordered, date and number received, item cost, reorder point, and amount purchased year-to-date), unit price, location of item, and formula for calculating markup.

An inventory master list is a written tally of all drugs kept in stock (Figure 5-15). The list should contain the most commonly used name, whether that name is generic or trade, cross reference of other names used for the drug, and a notes section for such things as expiration date, location index, or price list.

Vendor files are kept to hold all ordering information pertaining to each supplier. Packing slips, invoices, promotional flyers, price lists, and company correspondence are kept in this file for future reference.

In addition to maintaining an inventory system, it is necessary to maintain drug records and reference materials. The Drug Enforcement Agency (DEA) requires separate inventory and use records to be kept for all controlled

Brookside Veterinary Clinic, S.C.
2712 Eastern Ave.
Plymouth, WI 53073-0439
(920) 892-4345

PURCHASE ORDER

Vendor Name:	Merial Limited	**PO Number:**	363
Contact:	Phone1	**Status:**	Partial
Account No.:	49348	**Date:**	3/3/20XX
Vendor Phone:	(888) 999-9999 Ext.:	**Reference No.:**	

Item ID	Description	Quantity	Unit	Vendor Item ID	Unit Cost	Extended Cost
RABVAC	Rabies Vaccine	100.00	dose	ST D	$1.8900	$189.00
LYMEVAC	Lyme Vaccine	100.00	dose	ST D	$12.9200	$1,292.00
					Total Cost:	$1,481.00

A

2010 Olympic Drive
Athens, GA 30601

Packing List
NOT AN INVOICE

Page 1 of 1

PURCHASE ORDER NUMBER: N070404

SHIP FROM	ROUTING	ORDER NO.	ORDER DATE	SHIP DATE
MEM	036 / UPS Ground - BIO	11200598-2542815	02/24/20XX 12:44	02/24/20XX

SHIP TO: 49348
Brookside Veterinary Center SC
2712 Eastern Ave.
Plymouth, WI 53073-

BILL TO: Brookside Veterinary Center SC
2712 Eastern Ave.
Plymouth, WI 53073-

PRODUCT NO./DESCRIPTION	LOT NO.	QTY ORDERED	QTY SHIPPED	LIST PRICE	EXTENSION	DISCOUNT APPLIED	NET AMOUNT
✓221650 6|10 RECOMBITEK LYME (50X1DS)	42140	2	2	646.00	1292.00		1292.00
✓231150 3|9|10 IMRAB 3 TF (50X1DS)	18090A	2	2	94.50	189.00		189.00

REMARKS		TOTAL VALUE OF SHIPMENT	$1,481.00
		TOTAL PIECES:	04

Please report damages or shipping errors immediately upon receipt at 1-888-991-9237

B

Delmar/Cengage Learning

Figure 5-14 An example of (A) purchase order; (B) packing list; (C) invoice. *(Continued)*

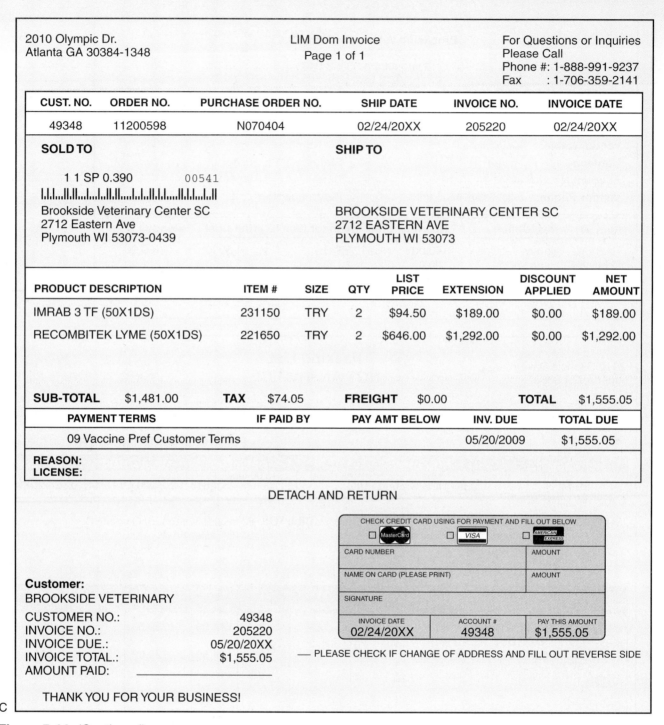

2010 Olympic Dr.
Atlanta GA 30384-1348

LIM Dom Invoice
Page 1 of 1

For Questions or Inquiries
Please Call
Phone #: 1-888-991-9237
Fax   : 1-706-359-2141

CUST. NO.	ORDER NO.	PURCHASE ORDER NO.	SHIP DATE	INVOICE NO.	INVOICE DATE
49348	11200598	N070404	02/24/20XX	205220	02/24/20XX

SOLD TO

1 1 SP 0.390   00541

Brookside Veterinary Center SC
2712 Eastern Ave
Plymouth WI 53073-0439

SHIP TO

BROOKSIDE VETERINARY CENTER SC
2712 EASTERN AVE
PLYMOUTH WI 53073

PRODUCT DESCRIPTION	ITEM #	SIZE	QTY	LIST PRICE	EXTENSION	DISCOUNT APPLIED	NET AMOUNT
IMRAB 3 TF (50X1DS)	231150	TRY	2	$94.50	$189.00	$0.00	$189.00
RECOMBITEK LYME (50X1DS)	221650	TRY	2	$646.00	$1,292.00	$0.00	$1,292.00

SUB-TOTAL $1,481.00   **TAX** $74.05   **FREIGHT** $0.00   **TOTAL** $1,555.05

PAYMENT TERMS	IF PAID BY	PAY AMT BELOW	INV. DUE	TOTAL DUE
09 Vaccine Pref Customer Terms			05/20/2009	$1,555.05

REASON:
LICENSE:

DETACH AND RETURN

CHECK CREDIT CARD USING FOR PAYMENT AND FILL OUT BELOW
☐ MasterCard   ☐ VISA   ☐ AMERICAN EXPRESS

CARD NUMBER   AMOUNT

NAME ON CARD (PLEASE PRINT)   AMOUNT

SIGNATURE

INVOICE DATE	ACCOUNT #	PAY THIS AMOUNT
02/24/20XX	49348	$1,555.05

— PLEASE CHECK IF CHANGE OF ADDRESS AND FILL OUT REVERSE SIDE

Customer:
BROOKSIDE VETERINARY

CUSTOMER NO.:   49348
INVOICE NO.:   205220
INVOICE DUE.:   05/20/20XX
INVOICE TOTAL.: $1,555.05
AMOUNT PAID: _____

THANK YOU FOR YOUR BUSINESS!

C

Figure 5-14 *(Continued)*

substances (see Chapter 1). Computer software is available to monitor controlled substance drug use that interfaces with automated delivery systems such as the Pyxis® Medstation. In addition, the Occupational Safety and Health Administration requires the cataloging of Material Safety Data Sheets (MSDS) for all hazardous substances (Figure 5-16). There should also be a place in the veterinary facility to store pharmacology textbooks, drug and product information, suppliers' catalogs, and state and federal regulatory handbooks.

Inventory – Evaluation Report
Sorted by Item Description

Item:	ANAMAX	Animax Ointment 15 Ml	Class:	I	Dermatology Meds.		
Buy/Sell Ratio:	1 tube to 1.00 tube		Dispensing Fee:	$9.00		Sell UOM:	tube
On Hand Loc/Total:	19.00/19.00 tube		Lead Time:	0		Prices Auto Calculate:	Yes
On Order:	24.00 tube		Reorder Point:	12.00 tube		Minimum Price:	$0.00
On Backorder:	0.00 tube		Reorder Quantity:	24.00 tube		Primary Vendor:	Midwest Veterinary Supply
Buy UOM:	tube		Overstock Point:	37.00 tube		Vendor Item ID:	ST D
Last Purchase Date:	3/4/2009					Location:	PHARM
Base Pricing Info:	$13.88		Markup:	154%			

Item:	ANTISED	Antisedan	Class:	Y	Sm. Ani. Anesthetic Agents		
Buy/Sell Ratio:	1 bottle to 10.00 cc		Dispensing Fee:	$0.00		Sell UOM:	cc
On Hand Loc/Total:	44.00/44.00 cc		Lead Time:	0		Prices Auto Calculate:	Yes
On Order:	0.00 bottle		Reorder Point:	9.00 cc		Minimum Price:	$0.00
On Backorder:	0.00 bottle		Reorder Quantity:	1.00 bottle		Primary Vendor:	Pfizer Animal Health
Buy UOM:	bottle		Overstock Point:	20.00 cc		Vendor Item ID:	ST D
Last Purchase Date:	1/2/2009					Location:	TREATME
Base Pricing Info:	$24.95		Markup:	135%			

Item:	APOMOR	Apomorphine 3 Mg	Class:	Y	Sm. Ani. Anesthetic Agents		
Buy/Sell Ratio:	1 bottle to 15.00		Dispensing Fee:	$0.00		Sell UOM:	capsule
On Hand Loc/Total:	11.00/11.00 capsule		Lead Time:	0		Prices Auto Calculate:	Yes
On Order:	0.00 bottle		Reorder Point:			Minimum Price:	$0.00
On Backorder:	0.00 bottle		Reorder Quantity:			Primary Vendor:	Island Pharmacy
Buy UOM:	bottle		Overstock Point:			Vendor Item ID:	STD
Last Purchase Date:	4/7/2004					Location:	TREATME
Base Pricing Info:	$5.83		Markup:	150%			

Item:	TEARS	Artificial Tears 3.5 Gm	Class:	Q	Ophthalmics		
Buy/Sell Ratio:	1 tube to 1.00 tube		Dispensing Fee:	$7.50		Sell UOM:	tube
On Hand Loc/Total:	58.00/58.00 tube		Lead Time:	0		Prices Auto Calculate:	Yes
On Order:	25.00 tube		Reorder Point:	6.00 tube		Minimum Price:	$0.00
On Backorder:	0.00 tube		Reorder Quantity:	6.00 tube		Primary Vendor:	Midwest Veterinary Supply
Buy UOM:	tube		Overstock Point:	13.00 tube		Vendor Item ID:	
Last Purchase Date:	3/11/2009					Location:	PHARM
Base Pricing Info:	$7.03		Markup:	191%			

Item:	ATROPIN	Atropine Ophthalmic 1%so	Class:	Q	Ophthalmics		
Buy/Sell Ratio:	1 bottle to 1.00 bottle		Dispensing Fee:	$7.50		Sell UOM:	bottle
On Hand Loc/Total:	22.00/22.00 bottle		Lead Time:	0		Prices Auto Calculate:	Yes
On Order:	0.00 bottle		Reorder Point:	3.00 bottle		Minimum Price:	$0.00
On Backorder:	0.00 bottle		Reorder Quantity:	6.00 bottle		Primary Vendor:	Butler Co
Buy UOM:	bottle		Overstock Point:	10.00 bottle		Vendor Item ID:	
Last Purchase Date:	3/24/2008					Location:	PHARM
Base Pricing Info:	$6.58		Markup:	143%			

Date: 3/23/20XX

Figure 5-15 An example of an inventory master list.

Pharmacy Organization

The amount of storage space available in the veterinary facility will determine how drugs and supplies are organized in a practice pharmacy. Most drugs are located in a central pharmacy; however, some bulky items and high-volume purchases need to be placed in other storage areas. Inventory can be arranged alphabetically (by generic or trade name or a combination of both), according to drug classification (antibiotic, antiparasitic, etc.), numeric (each drug is assigned a number), or by dose category (such as oral solids, oral liquids, injectables, ointments, etc.). Most practices use a hybrid form of several inventory

AIR PRODUCTS

Material Safety Data Sheet
Version 1.2
Revision Date 08/12/2003

MSDS Number 300000000099
Print Date 10/01/2003

1. PRODUCT AND COMPANY IDENTIFICATION

Product name	:	Nitrogen
Chemical formula	:	N2
Synonyms	:	Nitrogen, Nitrogen gas, Gaseous Nitrogen, GAN
Product Use Description	:	General Industrial
Company	:	Air Products and Chemicals, Inc 7201 Hamilton Blvd. Allentown, PA 18195-1501
Telephone	:	800-345-3148
Emergency telephone number	:	800-523-9374 USA 01-610-481-7711 International

2. COMPOSITION/INFORMATION ON INGREDIENTS

Components	CAS Number	Concentration (Volume)
Nitrogen	7727-37-9	100 %

Concentration is nominal. For the exact product composition, please refer to Air Products technical specifications.

3. HAZARDS IDENTIFICATION

Emergency Overview

High pressure gas.
Can cause rapid suffocation.
Self contained breathing apparatus (SCBA) may be required.

Potential Health Effects

Inhalation	:	In high concentrations may cause asphyxiation. Asphyxiation may bring about unconsciousness without warning and so rapidly that victim may be unable to protect themselves.
Eye contact	:	No adverse effect.
Skin contact	:	No adverse effect.
Ingestion	:	Ingestion is not considered a potential route of exposure.
Chronic Health Hazard	:	Not applicable.

Exposure Guidelines

Air Products and Chemicals, Inc   1/7   Nitrogen

Material Safety Data Sheet
Version 1.2
Revision Date 08/12/2003

MSDS Number 300000000099
Print Date 10/01/2003

Primary Routes of Entry	:	Inhalation
Target Organs	:	None known.
Symptoms	:	Exposure to oxygen deficient atmosphere may cause the following symptoms: Dizziness. Salivation. Nausea. Vomiting. Loss of mobility/consciousness.

Aggravated Medical Condition
None.

Environmental Effects
Not harmful.

4. FIRST AID MEASURES

General advice	:	Remove victim to uncontaminated area wearing self contained breathing apparatus. Keep victim warm and rested. Call a doctor. Apply artificial respiration if breathing stopped.
Eye contact	:	Not applicable.
Skin contact	:	Not applicable.
Ingestion	:	Ingestion is not considered a potential route of exposure.
Inhalation	:	Remove to fresh air. If breathing is irregular or stopped, administer artificial respiration. In case of shortness of breath, give oxygen.

5. FIRE-FIGHTING MEASURES

Suitable extinguishing media	:	All known extinguishing media can be used.
Specific hazards	:	Upon exposure to intense heat or flame, cylinder will vent rapidly and or rupture violently. Product is nonflammable and does not support combustion. Move away from container and cool with water from a protected position. Keep containers and surroundings cool with water spray.
Special protective equipment for fire-fighters	:	Wear self contained breathing apparatus for fire fighting if necessary.

6. ACCIDENTAL RELEASE MEASURES

Personal precautions	:	Evacuate personnel to safe areas. Wear self-contained breathing apparatus when entering area unless atmosphere is proved to be safe. Monitor oxygen level. Ventilate the area.
Environmental precautions	:	Do not discharge into any place where its accumulation could be dangerous. Prevent further leakage or spillage if safe to do so.

Air Products and Chemicals, Inc   2/7   Nitrogen

Material Safety Data Sheet
Version 1.2
Revision Date 08/12/2003

MSDS Number 300000000099
Print Date 10/01/2003

Methods for cleaning up	:	Ventilate the area.
Additional advice	:	If possible, stop flow of product. Increase ventilation to the release area and monitor oxygen level. If leak is from cylinder or cylinder valve, call the Air Products emergency telephone number. If the leak is in the user's system, close the cylinder valve, safely vent the pressure, and purge with an inert gas before attempting repairs.

7. HANDLING AND STORAGE

Handling

Protect cylinders from physical damage; do not drag, roll, slide or drop. Do not allow storage area temperature to exceed 50°C (122°F). Only experienced and properly instructed persons should handle compressed gases. Before using the product, determine its identity by reading the label. Know and understand the properties and hazards of the product before use. When doubt exists as to the correct handling procedure for a particular gas, contact the supplier. Do not remove or deface labels provided by the supplier for the identification of the cylinder contents. When moving cylinders, even for short distances, use a cart (trolley, hand truck, etc.) designed to transport cylinders. Leave valve protection caps in place until the container has been secured against either a wall or bench or placed in a container stand and is ready for use. Use an adjustable strap wrench to remove over-tight or rusted caps. Before connecting the container, check the complete gas system for suitability, particularly for pressure rating and materials. Before connecting the container for use, ensure that back feed from the system into the container is prevented. Ensure the complete gas system is compatible for pressure rating and materials of construction. Ensure the complete gas system has been checked for leaks before use. Employ suitable pressure regulating devices on all containers when the gas is being emitted to systems with lower pressure rating than that of the container. Never insert an object (e.g. wrench, screwdriver, pry bar, etc.) into valve cap openings. Doing so may damage valve, causing a leak to occur. Open valve slowly. If user experiences any difficulty operating cylinder valve discontinue use and contact supplier. Close container valve after each use and when empty, even if still connected to equipment. Never attempt to repair or modify container valves or safety relief devices. Damaged valves should be reported immediately to the supplier. Close valve after each use and when empty. Replace outlet caps or plugs and container caps as soon as container is disconnected from equipment. Do not subject containers to abnormal mechanical shocks which may cause damage to their valve or safety devices. Never attempt to lift a cylinder by its valve protection cap or guard. Do not use containers as rollers or supports or for any other purpose than to contain the gas as supplied. Never strike an arc on a compressed gas cylinder or make a cylinder a part of an electrical circuit. Do not smoke while handling product or cylinders. Never re-compress a gas or a gas mixture without first consulting the supplier. Never attempt to transfer gases from one cylinder/container to another. Always use backflow protective device in piping. When returning cylinder install valve outlet cap or plug leak tight. Never use direct flame or electrical heating devices to raise the pressure of a container. Containers should not be subjected to temperatures above 50°C (122°F). Prolonged periods of cold temperature below -30°C (-20°F) should be avoided.

Storage

Full containers should be stored so that oldest stock is used first. Containers should be stored in a purpose build compound which should be well ventilated, preferably in the open air. Stored containers should be periodically checked for general condition and leakage. Observe all regulations and local requirements regarding storage of containers. Protect containers stored in the open against rusting and extremes of weather. Containers should not be stored in conditions likely to encourage corrosion. Containers should be stored in the vertical position and properly secured to prevent toppling. The container valves should be tightly closed and where appropriate valve outlets should be capped or plugged. Container valve guards or caps should be in place. Keep containers tightly closed in a cool, well-ventilated place. Store containers in location free from fire risk and away from sources of heat and ignition. Full and empty cylinders should be segregated. Do not allow storage temperature to exceed 50°C (122°F). Return empty containers in a timely manner.

Air Products and Chemicals, Inc   3/7   Nitrogen

Figure 5-16 An example of a MSDS. *(Continued)*

Material Safety Data Sheet
Version 1.2
Revision Date 08/12/2003

MSDS Number 300000000099
Print Date 10/01/2003

Technical measures/Precautions

Containers should be segregated in the storage area according to the various categories (e.g. flammable, toxic, etc.) and in accordance with local regulations. Keep away from combustible material.

8. EXPOSURE CONTROLS / PERSONAL PROTECTION

Personal protective equipment

Respiratory protection	:	Self contained breathing apparatus (SCBA) or positive pressure airline with mask are to be used in oxygen-deficient atmosphere. Air purifying respirators will not provide protection. Users of breathing apparatus must be trained.
Hand protection	:	Sturdy work gloves are recommended for handling cylinders. The breakthrough time of the selected glove(s) must be greater than the intended use period.
Eye protection	:	Safety glasses recommended when handling cylinders.
Skin and body protection	:	Safety shoes are recommended when handling cylinders.
Special instructions for protection and hygiene	:	Ensure adequate ventilation, especially in confined areas.
Remarks	:	Simple asphyxiant.

9. PHYSICAL AND CHEMICAL PROPERTIES

Form	:	Compressed gas.
Color	:	Colorless gas
Odor	:	No odor warning properties.
Molecular Weight	:	28 g/mol
Relative vapor density	:	0.97 (air = 1)
Density at 70 °F (21 °C)	:	0.075 lb/ft3 (0.0012 g/cm3) Note: (as vapor)
Specific Volume at 70 °F (21 °C)	:	13.80 ft3/lb (0.8615 m3/kg)
Boiling point/range	:	-320.8 °F (-196 °C)
Critical temperature	:	-232.6 °F (-147 °C)
Melting point/range	:	-346.0 °F (-210 °C)
Water solubility	:	0.02 g/l

4/7

Air Products and Chemicals, Inc   Nitrogen

Material Safety Data Sheet
Version 1.2
Revision Date 08/12/2003

MSDS Number 300000000099
Print Date 10/01/2003

10. STABILITY AND REACTIVITY

| Stability | : | Stable under normal conditions. |
| Hazardous decomposition products | : | None. |

11. TOXICOLOGICAL INFORMATION

No known toxicological effects from this product.

12. ECOLOGICAL INFORMATION

Ecotoxicity effects

Aquatic toxicity	:	No data available.
Toxicity to other organisms	:	No data available.
Mobility	:	No data available.
Bioaccumulation	:	No data available.

Further information

No ecological damage caused by this product.

13. DISPOSAL CONSIDERATIONS

| Waste from residues / unused products | : | Contact supplier if guidance is required. Return unused product in orginal cylinder to supplier. |
| Contaminated packaging | : | Return cylinder to supplier. |

14. TRANSPORT INFORMATION

CFR
Proper shipping name	:	Nitrogen, compressed
Class	:	2.2
UN/ID No.	:	UN1066

IATA
Proper shipping name	:	Nitrogen, compressed
Class	:	2.2
UN/ID No.	:	UN1066

IMDG
Proper shipping name	:	NITROGEN, COMPRESSED
Class	:	2.2
UN/ID No.	:	UN1066

5/7

Air Products and Chemicals, Inc   Nitrogen

Material Safety Data Sheet
Version 1.2
Revision Date 08/12/2003

MSDS Number 300000000099
Print Date 10/01/2003

CTC
Proper shipping name	:	NITROGEN, COMPRESSED
Class	:	2.2
UN/ID No.	:	UN1066

Further Information

Avoid transport on vehicles where the load space is not separated from the driver's compartment. Ensure vehicle driver is aware of the potential hazards of the load and knows what to do in the event of an accident or an emergency.

15. REGULATORY INFORMATION

OSHA Hazard Communication Standard (29 CFR 1910.1200) Hazard Class(es)
Compressed Gas

Country	Regulatory list	Notification
USA	TSCA	Included on Inventory.
EU	EINECS	Included on Inventory.
Canada	DSL	Included on Inventory.
Australia	AICS	Included on Inventory.
South Korea	ECL	Included on Inventory.
China	SEPA	Included on Inventory.
Philippines	PICCS	Included on Inventory.
Japan	ENCS	Included on Inventory.

EPA SARA Title III Section 312 (40 CFR 370) Hazard Classification:
Sudden Release of Pressure Hazard

US. California Safe Drinking Water & Toxic Enforcement Act (Proposition 65)
This product does not contain any chemicals known to State of California to cause cancer, birth defects or any other harm.

16. OTHER INFORMATION

NFPA Rating

Health	:	0
Fire	:	0
Instability	:	0
Special	:	SA

HMIS Rating

Health	:	0
Flammability	:	0
Physical hazard	:	3

| Prepared by | : | Air Products and Chemicals, Inc. Global EH&S Product Safety Department |

For additional information, please visit our Product Stewardship web site at

http//www.airproducts.com/productstewardship/

6/7

Air Products and Chemicals, Inc   Nitrogen

Figure 5-16 *(Continued)*

styles and also keep some drugs in multiple areas such as examination and surgery rooms, pharmacy, and boarding areas. Dispensing of drugs is easier if tablets and capsules are kept near the counting area, liquids are kept near the sink, and injectables are kept near the needles and syringes. Regardless of how inventory is arranged, special storage conditions such as refrigeration and a securely locked safe or cabinet will be required.

SUMMARY

Drugs are marketed under generic and brand names. The generic name is the official name of the drug and may be used by anyone to identify that drug; the brand name can be used for identification purposes only by the corporation that produces the drug.

Drug compounding is the preparation, mixing, assembling, packaging, and/ or labeling of a drug resulting from a prescription drug order from a practitioner. Drug compounding occurs when health professionals prepare a specialized drug product to fill an individual patient's needs when an approved drug is not available. Common examples of the use of compounding in veterinary medicine include creating discontinued drugs that are no longer available commercially yet are still prescribed by veterinarians; tailoring doses and strengths to meet a particular animal's weight and health status; creating alternative dose forms such as liquids, ointments, or chewable tablets to make the medications easier for clients to administer to their animals; adding flavoring to unpalatable drugs to make these drugs more appealing to the animals that need to take them; and customizing formulations that combine multiple drugs for one-dose administration.

Drug information is available through many resources: the *USP*, drug bottles, package inserts, the *PDR*, *CVP*, *VPB*, and other references.

A veterinarian or physician must order prescription drugs. Prescription drugs may be dispensed by pharmacists (human-labeled drugs) or trained veterinary staff. Specific information is required on prescriptions and medication labels.

Veterinary technicians have many responsibilities in a veterinary practice including pharmacy management. The responsibility for purchasing and pricing medications is typically delegated to the veterinary technician. Maintaining a veterinary pharmacy requires planning and continuous monitoring. The goal of maintaining drug inventory is to stock quantities of each item as low as possible to reduce overhead and inventory costs yet not so low as to cause drug shortages between ordering periods. Drugs may be purchased directly from manufacturers, through veterinary distributors or wholesale suppliers, or through generic companies. Distributors or wholesalers carry a complete line of many products, and dealing with them will reduce the number of small orders required when making purchases. Direct marketed drugs are sold directly to veterinarians rather than working with suppliers. Other sources of drugs include generic companies, other veterinary practices (helpful in a crisis), buying groups of several veterinary practices (advantage is decreased drug cost), and pharmacies (good for prescribing drugs not commonly used or human-approved drugs).

Pharmaceutical management includes maintaining an adequate stock of all products used, dispensed, and sold; organizing inventory so that items are easy to locate; identifying products that need to be reordered; maintaining accurate purchasing and inventory records; ordering; receiving and inspecting shipments; establishing and updating pricing for all inventory items; rotating stock and monitoring expiration dates; and assessing new or updated products and considering specials offered by suppliers. Inventory records provide information about drug usage, order status, and price changes. An individual card or computer file is made for every supply item. An inventory master list is a written tally of all drugs kept in stock. The list should contain the most commonly used name, whether that name is generic or trade, cross reference of other names used for the drug, and a notes section for such things as expiration date, location index, or price list. Vendor files are kept to hold all ordering information pertaining to each supplier. Packing slips, invoices, promotional flyers, price lists, and company correspondence are kept in this file for future reference. Inventory management is summarized in Appendix F.

It is also necessary to maintain drug records and reference materials. The DEA requires separate inventory and use records to be kept for all controlled substances. In addition, the Occupational Safety and Health Administration requires the cataloging of MSDS for all hazardous substances. The amount of storage space available will determine how drugs and supplies are organized in a practice pharmacy. Most drugs are located in a central pharmacy; however, some bulky items and high-volume purchases need to be placed in other storage areas. Regardless of how inventory is arranged, special storage conditions such as refrigeration and a securely locked safe or cabinet will be required.

It's a Wrap

The answers to the questions in this chapter's Setting the Scene case study should be understood after reading this chapter. Can drugs used for animals be dispensed at human pharmacies? What should the veterinary technician tell her?

Veterinarians can call prescriptions into human pharmacies (pharmacists cannot dispense drugs to animals without a veterinarian's orders). Many drugs used in animals are the same as or similar to drugs used in people. Some differences between drugs manufactured for animals and those for people include varying concentrations of drug per unit or drug formulation. The client still needs to bring the cat into the clinic for an examination by the veterinarian to establish a veterinarian/patient/client relationship. This client should be advised to provide follow-up information once the prescription is given in its entirety to her animal.

CHAPTER REVIEW

Matching

Match the English words to the pharmacological
*   abbreviation.*

1. _____ one tablet by mouth twice a day
2. _____ one tablet every 12 hours by mouth
3. _____ one tablet every eight hours by mouth
4. _____ two units subcutaneously after meals
5. _____ one milliliter by mouth every 12 hours
6. _____ one tablet by mouth every day for five days
7. _____ two tablets by mouth once a day
8. _____ one tablet by mouth three times a day
9. _____ apply every day for two to four weeks
10. _____ one tablet by mouth every day

a. Apply qd 2–4 weeks
b. 1 ml q12h po
c. 1 T tid po
d. 1 T q12h po
e. 1 T qd po
f. 1 T qd po X 5d
g. 1 T q8h po
h. 2 T sid po
i. 2U SQ pc
j. 1 T bid po

Multiple Choice

Choose the one best answer.

11. What term describes the date before which a drug meets all specifications and after which the drug can no longer be used?
    a. dispensing date
    b. prescribing date
    c. expiration date
    d. termination date

12. What is the order given or written to a pharmacist by a licensed veterinarian to prepare the medicine, to affix the directions, and to sell the preparation to the client?
    a. dispensatory note
    b. prescription
    c. drug order
    d. package insert

13. On a drug label, which part is usually in capital letters with a superscript R by it?
    a. generic name
    b. drug concentration
    c. manufacturer's name
    d. trade name

14. Direct marketing of veterinary drugs
    a. is done by agencies that purchase the drugs from manufacturers and resell them to veterinarians.
    b. is done by generic companies that sell generic drugs under their own companies's names.
    c. occurs when a drug is purchased by the veterinarian directly from the company that manufactures it.
    d. is not utilized in the veterinary community.

15. What is the preparation, mixing, assembling, packaging, and/or labeling of a drug resulting from a prescription drug order from a practitioner?
    a. compounding
    b. inventory management
    c. prescription
    d. pharmacy management

16. Which of the following is an example of drug compounding?
    a. splitting a pill along its score line
    b. dispensing an unopened bottle of pills to a client
    c. using expired drugs to treat an animal's illness
    d. adding flavoring to formulated drugs

17. What is the number of times inventory is depleted and replenished each year?
    a. turnover rate
    b. inventory management
    c. order record
    d. dispensing history

18. What types of inventory records should be kept?
    a. reorder log and vendor file
    b. purchase order records
    c. individual inventory records and inventory master list
    d. all of the above

19. The Drug Enforcement Agency requires separate inventory and use records to be kept for
    a. all oral medications.
    b. all injectable medications.
    c. schedule I and II controlled substances.
    d. all controlled substances.

20. Organization of pharmaceuticals within a clinic should be
    a. designed so that all drugs are within reach.
    b. arranged alphabetically by drug category.
    c. designed to take into account special refrigeration and security needs.
    d. designed so that liquid medication is near a sink for easier reconstitution.

Fill in the Blank

Write the English meanings for the following prescription abbreviations:

21. bid _____

22. npo _____

23. q4h _____

24. tid _____

25. prn _____

26. po _____

27. gt _____

28. T _____

29. c̄ _____

30. s̄ _____

Case Study

31. A client brings in a puppy for its first examination at your clinic. The routine physical exam is normal; however, the puppy has intestinal parasites. The puppy is prescribed an oral liquid dewormer that is prepared and dispensed in a syringe to the client. The client is advised to bring in a follow-up stool sample two to three weeks after the last dose of dewormer is given to the puppy.

    When the client gets home, he attempts to give the puppy its dewormer. He is having trouble restraining the puppy to insert the syringe rectally. The client believes that the medication has to be given rectally because the worms were present in the puppy's stool. How could this confusion have been avoided?

Critical Thinking Questions

32. Since animals cannot open vials, why are veterinary drugs dispensed in childproof containers?

33. Why is it important to have a veterinary/patient/client relationship before prescribing or dispensing drugs?

ADDITIONAL REVIEW MATERIAL

Additional review material can be found on the StudyWARE™ CD included with this text.

BIBLIOGRAPHY

Margolis, M. (2002). *FDA May Not Prevent Pharmacists from Promoting Compounded Drugs*, www.law.uh.edu.

Nordenberg, T. (2000). *Pharmacy Compounding: Customizing Prescription Drugs*, FDA Consumer Magazine.

Reiss, B., and Evans, M. (2007). *Pharmacological Aspects of Nursing Care* (7th ed. revised by Bonita E. Broyles, pp. 2–31). Clifton Park, NY: Delmar Cengage Learning.

Rice, J. (2006). *Principles of Pharmacology for Medical Assisting* (4th ed.). Clifton Park, NY: Delmar Cengage Learning.

Sanders, M. (2004). *Pharmaceutical Compounding for the Pharmacy Technician*, The University of Mississippi School of Pharmacy, Pharmacy Technician Continuing Education Program (#032-000-02-080-H04).

White-Shim, L. (August 15, 2008). *AVMA Answers: Prescriptions*, JAVMA, Vol. 233, No. 4, p. 539.

Wynn, R. (1999). *Veterinary Pharmacology*. Cambridge, MA: Harcourt Learning Direct.

CHAPTER 6

SYSTEMS OF MEASUREMENT IN VETERINARY PHARMACOLOGY

OBJECTIVES

Upon completion of this chapter, the reader should be able to:

- apply basic math concepts.
- differentiate between the household, apothecary, and metric systems of measurement.
- list the base units in the metric system.
- define the commonly used metric prefixes.
- convert between and within the household, apothecary, and metric systems.
- perform dose calculations.
- perform solution calculations.
- perform reconstitution calculations.

KEY TERMS

apothecary system	meter
centi-	metric system
conversion factor	micro-
dram	milli-
grain	minim
gram	miscible
household system	solute
immiscible	solution
kilo-	solvent
liter	

Setting the Scene

On a bright sunny day, Valerie decides to take a brisk walk with her 55-pound dog Petey. While walking, she notices she is sweating quite a bit and wonders if she and her dog drank enough water today to prevent becoming dehydrated. Valerie knows she should drink about eight (8 oz) glasses of water per day. Valerie recalls that she drank 1 L of water this morning. Did Valerie drink enough water today to replace normal water loss? How can this be mathematically determined? Petey drank one 8-oz bowl of water. If dogs should drink 30 mL of water per pound per day, did he drink enough water today?

Courtesy of iStockPhoto

THE BASICS

Veterinary professionals use mathematical concepts on a daily basis to determine drug doses, fluid rates, and dilution of chemicals. The more comfortable the veterinary technician is with math, the better the technician will be able to perform these calculations.

Basic mathematical concepts include the use of percents, decimals, and ratios. Reducing fractions; interpretation of values; and conversion of fractions, decimals, percents, and ratios are things the veterinary technician should already feel comfortable with. The technician should be able to perform simple algebra to solve for an unknown and use proportions to solve for unknowns. For review of these concepts, refer to Appendix J and the StudyWARE® material provided with this text.

HOW DO WE MEASURE?

Measurement is the use of standard units to determine the weight, length, or volume of substances. Without standard units, pharmacology would be an inaccurate science; everything flows from our ability to standardize weight, length, and volume.

The units of measurement used in pharmacology are the **household system**, the **apothecary system**, and the **metric system**. The household system uses household measures when an approximate dose is acceptable. The household method uses a system of weights and measures based on 1 pound containing 16 fluid ounces. Most household measuring devices lack standardization, but are calibrated in units that most people are familiar with such as teaspoon and cup. The drop, tablespoon, and teaspoon are the only household measures still used in pharmacology; however, ounces, cups, quarts, and gallons may be used when preparing disinfectants and other solutions. The apothecary system uses weights and measures based on 480 grains equal to 1 oz and 12 oz equal to 1 pound (this is different from the household system where 16 fluid ounces equal 1 pound). Today, the grain is the only apothecary weight unit used and the ounce, dram, and minim are the only apothecary volume units used. Pharmacology mainly uses the metric system, and the veterinary technician must become an expert at metric conversion to avoid harmful or fatal mistakes in preparation and dosing of drugs. The metric system is based on a standard to which other factor-of-ten units are compared. The kilogram is a metric weight unit used in veterinary medicine to weigh animals, and milliliter and liter are metric volumes used to administer parenteral drugs and solutions. It is necessary to be able to convert among the three systems, as some drugs and dosages vary in their representation in the veterinary community.

The Household System

Household units are more likely to be used by clients to prepare medications for their animals at home, where veterinary measuring devices are not usually available. Household measures are approximate measurements

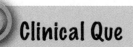

Clinical Que

The metric system is a decimal system, which means it is based on powers of 10.

Table 6-1	Household System Measures	
UNIT	ABBREVIATION	EQUIVALENTS
cup	cup	1 cup = 8 fl oz
drop (s)	gt (gtt)	no standard equivalent
ounce (fluid)	fl oz	2 T = 1 fl oz
ounce (weight)	oz	16 oz = 1 lb
pint	pt	1 pt = 2 cups = 16 fl oz
pound	lb	1 lb = 16 oz
quart	qt	1 qt = 2 pt = 4 cups = 32 fl oz
tablespoon*	T (or tbs)	1 T = 3 t
teaspoon	t (or tsp)	3 t = 1 T

*tablespoon is the larger unit and is expressed with the capital or "large" T, while teaspoon is the smaller unit and is expressed with the lowercase or "small" t.

that lack standardization and are not accurate for measuring medicines. For example, there may be three or four teaspoons per tablespoon depending on the reference source or size of each utensil. Though inaccurate, the household system may be used in some circumstances in which standardization is not crucial. The veterinary technician should be familiar with the household system of measurement in order to properly explain take-home prescriptions to clients at the time of patient discharge or dispensing (Table 6-1). There is no standardized system of notation, but the preferred way to express the quantity is in Arabic numbers and common fractions with the abbreviation following the amount (such as 3 t, ½ oz, or 2 lb).

The inch and foot are common standard units for measuring length in the household system. In the United States, a standard piece of paper is 8½ inches by 11 inches. This figure communicates a length most people understand and can reproduce because of familiarity.

The pound is a common standard unit used to measure weight in the household system. If a box of sugar has a weight of 5 pounds, it is called a 5-pound box of sugar. If a box of table salt contains one of these pounds, it is called a 1-pound box of table salt. Anyone familiar with a pound can purchase sugar or salt and know the amount being purchased because of familiarity.

The fluid ounce is a common standard unit used to measure liquid volume in the household system. If a water glass has a volume of 8 fluid ounces, it is called an 8-oz water glass. Eight ounces is its final measurement. If a bottle of soda holds 12 of these fluid ounces, it is called a 12-oz bottle of soda. Twelve ounces is its final measurement. These figures communicate a volume most people understand and can reproduce because of familiarity.

Clinical Que

Clients need to be aware that when using a liquid medicine at home that is dosed in teaspoons they should be using a "standardized" measuring spoon rather than a flatware spoon.

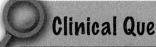

Clinical Que

The ounce in the household system is used for both volume (fluid ounces) and weight (ounce).

The Apothecary System

The apothecary system (also called the *common system*) is derived from the British apothecary system of measures, and is a system of liquid units of measure used chiefly by pharmacists. There is a historic connection between the household system and the apothecary system of measure. The Greeks were the first people to use the apothecary system. *Apothecary* comes from the Greek word for storehouse, which can be loosely translated into "drugstore." The apothecary system made its way from Greece to spread to the rest of Europe via Rome and France. In late-seventeenth-century England, "druggists" agreed that they would stock only drugs and that grocers would sell only food and not drugs. The colonists then brought this system to America. A modified system of measurement for everyday use evolved, which was the household system. Large liquid volumes were based on familiar trading measurements such as pints, quarts, and gallons, which originated as apothecary measurements. Vessels were made by craftspeople to accommodate each measurement and were widely circulated in colonial America.

Units of weight, such as the grain, ounce, and pounds, are also rooted in the apothecary system of measurement. The **grain** (gr) is the basic unit of weight measurement in the apothecary system. The grain originated as the standard weight of a single grain of wheat, which is approximately 60 mg.

A **minim** is the liquid volume of a drop of water from a standard medicine dropper. Sixty minims (or drops) make up a fluid **dram**. Eight fluid dram units, the equivalent of 480 minims (or drops), make up one fluid ounce. Prescription vials to dispense oral medications are available in a variety of dram weight sizes (Figure 6-1). Eight drams are equal to 1 oz. Many medicines are dispensed in quantities of fluid ounces, and liquid medicine bottles are available in standard

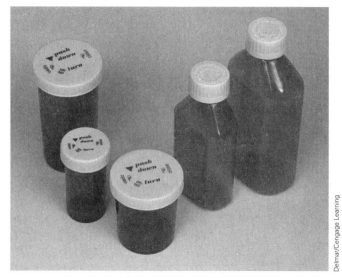

Figure 6-1 Prescription vials are typically available in the apothecary unit of drams, while prescription bottles are typically available in the household unit of fluid ounces. Being familiar with all systems of measure will enable the veterinary technician to convert from one system to another.

sizes ranging from 1-oz bottles to 16-oz bottles. The most frequently dispensed fluid-ounce quantities are 4 oz and 6 oz. Table 6-2 shows the apothecary units of measure and their relationship to other apothecary units.

Traditional apothecary notation uses lowercase Roman numerals to express whole numbers, fractions to designate amounts less than 1, and the unit of measure precedes the amount. For example, gr iii is three grains.

Table 6-2	Apothecary System Weight Measures	
UNIT	ABBREVIATION	EQUIVALENTS
dram (weight)	dr	8 dr = 480 gr = 1 oz
grain	gr	60 gr = 1 dr
ounce (weight)	oz	12 oz = 1 lb
pound	lb	1 lb = 12 oz
minim	ℳ	60 minim = 1 fl dr
dram (volume)	fl dr	8 fl dr = 480 minim = 1 fl oz
ounce (volume)	fl oz	16 fl oz = 1 pt
pint	pt	2 pt = 1 qt
quart	qt	4 qt = 1 gal
Gallon	gal	1 gal = 4 qt = 8 pt

The Metric System

The metric system of measure was formulated by the French government in 1875 as a product of the French Revolution. The term *metric* is borrowed from the Greek term *metron*, which means measure or standard. The metric system is also known as the international system or SI (from the French *Systeme International d'Unites*). The metric system was developed to standardize measures and weights for European countries. It has been used in the United States since 1890 and is the most widely used system of measure. The strength of the metric system is its simplicity, because all units of measure differ from each other in powers of ten. Conversions between units in the system are accomplished by simply moving a decimal point. In the metric system, prefixes are used to denote the size of a metric unit. These metric units are all based on factors of 10, and each prefix in the metric system changes the value of the basic unit by a factor of 10 over the preceding prefix. Table 6-3 lists the fundamental units of the metric system and their prefixes.

The **meter** (m) is the metric standard unit used to measure length. The expression *centimeter* (cm) is an important term in measuring length in the metric system. One meter consists of 100 cm. In other words, it takes one hundred centimeters of length to equal 1 m. The millimeter (mm) is used to measure smaller distances. One meter consists of 1000 mm. In other words, it takes

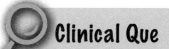

Clinical Que

Some units are used in both the apothecary and household systems (such as ounce, pint, and quart). The conversions for these measures may differ between each system, so make sure to have an understanding of which system is being utilized.

Table 6-3   Common Metric Prefixes and Their Value

PREFIX*	VALUE	DECIMAL REPRESENTATION	FRACTIONAL REPRESENTATION
micro- (mc or μ)	one millionth of a unit	0.000001 base unit	1/1,000,000 base unit
milli- (m)	one thousandth of a unit	0.001 base unit	1/1,000 base unit
centi- (c)	one hundredth of a unit	0.01 base unit	1/100 base unit
kilo- (k)	one thousand units	1000 times base unit	

*Additional metric prefixes can be found in Appendix G

Clinical Que

In the metric system:

- The fundamental unit of length = meter
- The fundamental unit of liquid volume = liter
- The fundamental unit of weight = gram

Clinical Que

1 mL = 1 cc

1000 mm of length to equal 1 m. Keep in mind that 1000 mm = 100 cm = 1 m, 0.001 m = 1 mm, and 0.01 m = 1 cm.

Some drug doses are based on surface area of an animal (m² of body surface area for calculation of chemotherapy drugs for cancer patients). Charts are available for conversion from weight to surface area. This information is found in Appendix H and is covered in Chapter 20 on antineoplastic drugs.

The **liter** (L) is the metric standard unit used to measure liquid volume. The expression *milliliter* (mL) is the other important term in measuring liquid volume in the metric system. One liter consists of 1000 mL. In other words, it takes 1000 ml of liquid to equal 1 L. Keep in mind that 0.001 L = 1 mL.

If water were placed in a cube measuring one centimeter on every side, the amount in the cube would be a cubic centimeter of water. The term *cubic centimeter* is abbreviated as cc. A very important relationship exists between cubic centimeters and milliliters: a cubic centimeter (cc) of liquid is the amount of space occupied by 1 mL volume. It is approximated that 1 cc = 1 mL. In human medicine, the interchangeable use of cc and mL is being discontinued to avoid transcribing errors.

The **gram** (g) is the metric standard unit of weight measurement. The weight of the water in the one cubic centimeter is 1 g. Thus, 1 cc = 1 mL = 1 g, which is read as the volume of one cubic centimeter of water equals one milliliter of water, which in turn weighs one gram.

The denominations most commonly used in measuring weights of medications and powders used in preparing medications are milligrams and grams. One milligram equals one-thousandth of a gram (1 mg = 0.001 g). Keep in mind that 1 g = 1000 mg. Another common denomination is *kilogram*. Animal weights and drug dosages may be represented in kilogram units. One kilogram equals 1000 grams (1 kg = 1000 g). Keep in mind that 1 g = 0.001 kg. Metric weight also includes the microgram, which is one-thousandth of a milligram or one-millionth of a gram (1 mcg = 0.001 mg = 0.000001 g). Keep in mind that 1,000,000 mcg = 1000 mg = 1 g. The relationships of metric units are summarized in Table 6-4.

Table 6-4   Important Metric Relationships

Mass and/or Weight Unit	Equivalents
1 microgram (mcg, μg)	0.000001 g = 0.001 mg = 1 mcg
1 milligram (mg)	0.001 g = 1 mg = 1, 000 mcg
1 gram (g)	1 g = 1,000 mg = 1,000,000 mcg
1 kilogram (kg)	1 kg = 1000 g

Volume Unit	Equivalents
1 milliliter (mL)	0.001 L = 1 mL
1 liter (L)	1 L = 1,000 mL

Length	Equivalents
1 micrometer (mcm, μm)	0.000001 m = 0.01 cm = 0.001 mm = 1 mcm
1 millimeter (mm)	0.001 m = 0.1 cm = 1 mm = 1000 mcm
1 centimeter (cm)	0.01 m = 1 cm = 10 mm = 10,000 mcm
1 meter (m)	1 m = 100 cm = 1000 mm = 1,000,000 mcm
1 kilometer (km)	1 km = 1000 m

A fundamental operation using the metric system is converting to lower or higher denominations. When calculating weights or liquid volumes of medications, it is often necessary to convert between grams, milligrams, micrograms, and kilograms, and between milliliters and liters. Metric system guidelines are summarized in Table 6-5.

Clinical Que

Rounding numbers that will alter drug doses is ultimately up to the discretion of the veterinarian. When performing calculations, it is best not to perform any rounding until the end product is achieved. A decision about rounding can be made at that time.

Table 6-5   Metric System Guidelines

1. Arabic numbers are used to designate whole numbers: 1, 25, 2500, etc.
2. Decimal fractions are used for quantities less than one: 0.1, 0.01, 0.001, etc.
3. To insure accuracy, place a zero before the decimal point: 0.1, 0.01, etc.
4. The Arabic number precedes the metric unit of measurement: 1 g, 30 mL, etc.
5. Abbreviations are generally used.
6. If the term is used, the prefixes are written in lowercase letters: milli, centi, etc.
7. Capitalize the measurement and symbol when it is named after a person: Celsius (C)
8. Periods are not used with most abbreviations or symbols.
9. Abbreviations for units are the same for singular and plural forms. An "s" is not added to indicate the plural form.
10. Separate the amount from the unit so the number and unit of measure do not run together because the unit can be mistaken as zero or zeros, risking a 10-fold to 100-fold overdose.
11. Place commas for amounts at or above 1000. For example, 10,000 mcg, not 10000 mcg.
12. Decimals are used to designate fractional amounts. For example, 1.5 mL not 1½ mL.

CONVERTING WITHIN SYSTEMS

After learning the systems of measurement used in veterinary medicine, it is important to learn how to use them. First, it is necessary to be able to convert or change from one unit to another within the same measurement system. To accomplish this, the veterinary technician needs to be able to recall the equivalents and multiply or divide. Conversion between one unit and another involves the use of a conversion factor. A **conversion factor** is a number used with either multiplication or division to change a measurement from one unit of measurement to its equivalent in another unit of measure. A conversion factor always has a value of 1. To convert from a larger to a smaller unit of measure, multiply by the conversion factor (e.g., if converting 3 L to milliliters, it is necessary to multiply 3 times 1000 because it takes more of the smaller units of measure to represent the same amount that the larger units of measure represented). It takes more parts of a smaller unit to make an equivalent amount of a larger unit. To get more parts, multiply. To convert from a smaller unit of measure, divide by the conversion factor (e.g., if converting 3000 mL to liters, it is necessary to divide 3000 by 1000 because it takes less of the larger units of measure to represent the same amount that the smaller units of measure represented). It takes fewer parts of the larger unit to make an equivalent amount of the smaller unit. To get fewer parts, divide.

Converting Within the Metric System

The most common conversions in dose calculations are within the metric system.

Converting Grams to Milligrams

One method used to convert between metric units is by dimensional analysis (commonly referred to as unit cancellation). Use the example of converting 1.5 g to mg. Solve the equation 1.5 g = _____ mg.

Step 1: Because the unknown factor in the given formula is the number of milligrams contained in 1.5 g, determine the metric equivalents. In this example, the equivalent is 1000 mg = 1 g.

Step 2: Knowing that 1000 mg = 1 g, create a conversion factor. *Conversion factors are used to move between units and always have a value of one.* Because they have a value of one, conversion factors do not change the value of the end product. Write this conversion factor as 1000 mg/1 g.

Step 3: The next step is to determine in which format to write the conversion factor. There are two choices: 1000 mg/1 g or 1 g/1000 mg. The choice is made based on which unit of measure to cancel out and which unit of measure is desired at the end. In this case, cancel out g to end up with mg. Therefore, g is on the bottom, and mg is on the top of the conversion factor.

Step 4: Set up the conversions in an equation. Start with what is known (1.5 g), and use the conversion factor to set up the equation.

$$\left(1.5\,g\right)\left(\frac{1000\,mg}{1\,g}\right) = X$$

Step 5: Now perform the calculation. Remember, values on the bottom of the conversion factors are divided into the numbers on the top. Values on the top of the conversion factors are multiplied by the numbers on the top. In this case, take 1.5 g, multiply it by 1000 mg, and then divide by 1 g.

$$\frac{(1.5\,g \times 1000\,mg)}{1\,g} = 1500\,mg$$

Keep in mind that units should cancel as well to give the desired unit at the end.

Step 6: To make sure the correct answer is determined, prove the work by looking at the units involved. Smaller units (in this case, mg) should have a larger number in front of them versus larger units (in this case, g), which should have a smaller number in front of them. In this case, mg has 1500 and g has 1.5. Always double-check the answer to avoid errors in calculations.

Clinical Que

An oversimplified way to remember how units cancel can be stated as "any unit divided by itself becomes 1 or disappears."

A shortcut method used to convert from grams to milligrams is as follows: When given a value for grams, obtain the equivalent amount of milligrams by moving the decimal point three places to the right of that value (Figure 6-2). Moving the decimal point three places to the right is essentially multiplying the value by 1000, which is the equivalent difference between grams and milligrams. There are 1000 milligrams in each gram of weight. Please note that from now on the abbreviations mg and g will be used for milligram and gram units.

Converting Milligrams to Grams

Sometimes drug preparations will require conversion of milligrams to grams. Using the dimensional analysis method, convert 2500 mg to g. Solve the equation 2500 mg = _____ g.

Step 1: Because the unknown factor in the given formula is the number of grams contained in 2500 mg, determine the metric equivalents. In this example, the metric equivalent is 1000 mg = 1 g.

Step 2: Knowing that 1000 mg = 1 g, create a conversion factor. Write this conversion factor as 1000 mg/1 g.

Figure 6-2 Converting from grams to milligrams.

Step 3: The next step is to determine in which format to write the conversion factor. There are two choices: 1000 mg/1 g or 1 g/1000 mg. In this case, cancel out mg to end up with g. Therefore, mg is on the bottom, and g is on the top of the conversion factor.

Step 4: Set up the conversions in an equation. Start with what is known (2500 mg), and use the conversion factor to set up the equation.

$$\left(2500\ mg\right)\left(\frac{1\ g}{1000\ mg}\right) = X$$

Step 5: Now perform the calculation. Remember, values on the bottom of the conversion factors are divided into the numbers on the top. Values on the top of the conversion factors are multiplied by the numbers on the top. In this case, take 2500 mg, multiply it by 1 g, and then divide by 1000 mg.

$$\frac{(2500\ mg \times 1\ g)}{1000\ mg} = 2.5\ g$$

Keep in mind that units should cancel as well to give the desired unit at the end.

Step 6: To make sure the correct answer is determined, prove the work by looking at the units involved. Smaller units (in this case, mg) should have a larger number in front of them versus larger units (in this case, g) which should have a smaller number in front of them. In this case, mg has 2500 and g has 2.5. Always double-check the answer to avoid errors in calculations.

To use the shortcut method of converting milligrams to grams, apply the following rule: When given the value of the milligram quantity, obtain the equivalent amount of grams by moving the decimal point three places to the left (Figure 6-3). Doing this is essentially dividing the milligram value by 1000. This is because 1000 milligrams is equivalent to 1 gram. The movement of the decimal point to the left is the exact opposite of what was done to convert grams to milligrams.

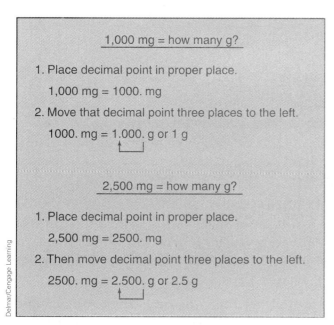

Figure 6-3 Converting from milligrams to grams.

Converting Kilograms to Grams

Some drug dosages for small animals or laboratory species may be expressed in grams. Because some of these animals are weighed on a kilogram scale, it may be necessary to convert between kg and g. Using the dimensional analysis method, convert 45 kg to g. Solve the equation 45 kg = _____ g.

Step 1: Because the unknown factor in the given formula is the number of grams contained in 45 kg, determine the metric equivalents. In this example, the metric equivalent is 1000 g = 1 kg.

Step 2: Knowing that 1000 g = 1 kg, create a conversion factor. Conversion factors are used to move between units and always have a value of one. Because they have a value of one, conversion factors do not change the value of the end product. Write this conversion factor as 1000 g/1 kg.

Step 3: The next step is to determine in which format to write the conversion factor. There are two choices: 1000 g/1 kg or 1 kg/1000 g. The choice is made based on which unit of measure to cancel out and which unit of measure to end up with. In this case, cancel out kg to end up with g. Therefore, kg is on the bottom, and g is on the top of the conversion factor.

Step 4: Set up the conversions in an equation. Start with what is known (45 kg), and use the conversion factor to set up the equation.

$$\left(45\text{ kg}\right)\left(\frac{1000\text{ g}}{1\text{ kg}}\right) = X$$

Step 5: Now perform the calculation. Remember, values on the bottom of the conversion factors are divided into the numbers on the top. Values on the top of the conversion factors are multiplied by the numbers on the top. In this case, take 45 kg, multiply it by 1000 g, and then divide by 1 kg.

$$\frac{(45\text{ kg} \times 1000\text{ g})}{1\text{ kg}} = 45,000\text{ g}$$

Keep in mind that units should cancel as well to give the desired unit at the end.

Step 6: To make sure the correct answer is determined, prove the work by looking at the units involved. Smaller units (in this case, g) should have a larger number in front of them versus larger units (in this case, kg), which should have a smaller number in front of them. In this case, g has 45,000 and kg has 45. Always double-check the answer to avoid errors in calculations.

If using the shortcut method, apply the following rule: When given a value for kilograms, obtain the equivalent amount of grams by moving the decimal point three places to the right of that value (Figure 6-4). Moving the decimal point three places to the right is essentially multiplying the value by 1000, which is the equivalent difference between kilograms and grams. There are 1000 grams in each kilogram of weight.

45 kg = how many g?

1. Place decimal point in proper place.

   45 kg = 45. kg

2. Move that decimal point three places to the right. Remember to add zeros to mark each movement.

   45. kg = 45.000. g or 45,000 g

Figure 6-4 Converting from kilograms to grams.

Converting Grams to Kilograms

Some drug dosages for small animals or laboratory species may be expressed in kilograms. Because some of these animals are weighed on a gram scale, it may be necessary to convert between g and kg. Using the dimensional analysis method, convert 25 g to kg. Solve the equation 25 g = _____ kg.

Step 1: Because the unknown factor in the given formula is the number of kilograms contained in 25 g, determine the metric equivalents. In this example, the metric equivalent is 1000 g = 1 kg.

Step 2: Knowing that 1000 g = 1 kg, create a conversion factor. Write this conversion factor as 1000 g/1 kg.

Step 3: The next step is to determine in which format to write the conversion factor. There are two choices: 1000 g/1 kg or 1 kg/1000 g. The choice is made based on which unit of measure to cancel out and which unit of measure to end up with. In this case, cancel out g to end up with kg. Therefore, g is on the bottom and kg is on the top of the conversion factor.

Step 4: Set up the conversions in an equation. Start with what is known (25 g), and use the conversion factor to set up the equation.

$$\left(25\,g \right)\left(\frac{1\,kg}{1000\,g} \right) = X$$

Step 5: Now perform the calculation. Remember, values on the bottom of the conversion factors are divided into the numbers on the top. Values on the top of the conversion factors are multiplied by the numbers on the top. In this case, take 25 g, multiply it by 1 kg, and then divide by 1000 g.

$$\frac{(25\,g \times 1\,kg)}{1000\,g} = 0.025\,kg$$

Keep in mind that units should cancel as well to give the desired unit at the end.

Step 6: To make sure the correct answer is determined, prove the work by looking at the units involved. Smaller units (in this case, g) should have a larger number in front of them versus larger units (in this case, kg), which should have a smaller number in front of them. In this case, g has 25 and kg has 0.025. Always double-check the answer to avoid errors in calculations.

If using the shortcut method, apply the following rule: When given the value of the gram quantity, obtain the equivalent amount of kilograms by moving the decimal point three places to the left (Figure 6-5). Moving the decimal point three places to the left is essentially dividing the value by 1000, which is the equivalent difference between grams and kilograms. There are 1000 grams, which is equivalent to 1 kilogram. This movement of the decimal point to the left is the exact opposite of what was done to convert kilograms to grams.

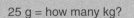

> ## Clinical Que
>
> When converting from larger units to smaller units, the quantity gets larger. When converting from smaller units to larger units, the quantity gets smaller.

25 g = how many kg?

1. Place decimal point in proper place.

    25 g = 25. g

2. Move that decimal point three places to the left. Remember to add zeros to mark each movement.

    25. g = 0.025. kg or 0.025 kg

Figure 6-5 Converting from grams to kilograms.

Delmar/Cengage Learning

Converting Liters to Milliliters

Use of liquids and liquid medications may sometimes require conversion of liters to milliliters. Using the dimensional analysis method, convert 2.525 L to mL. Solve the equation 2.525 L = _____ mL.

Step 1: Because the unknown factor in the given formula is the number of milliliters contained in 2.525 L, determine the metric equivalents. In this example, the metric equivalent is 1000 mL = 1 L.

Step 2: Knowing that 1000 mL = 1 L, create a conversion factor. Write this conversion factor as 1000 mL/1 L.

Step 3: The next step is to determine in which format to write the conversion factor. There are two choices: 1000 mL/1 L or 1 L/1000 mL. The choice is made based on which unit of measure to cancel out and which unit of measure to end up with. In this case, cancel out L to end up with mL. Therefore, L is on the bottom, and mL is on the top of the conversion factor.

Step 4: Set up the conversions in an equation. Start with what is known (2.525 L), and use the conversion factor to set up the equation.

$$\left(2.525\,L \right) \left(\frac{1000\,mL}{1\,L} \right) = X$$

Step 5: Now perform the calculation. Remember, values on the bottom of the conversion factors are divided into the numbers on the top. Values on the top of the conversion factors are multiplied by the numbers on the top. In this case, take 2.525 L, multiply it by 1000 mL, and then divide by 1 L.

$$\frac{(2.525\,L \times 1000\ mL)}{1\,L} = 2525\ mL$$

Keep in mind that units should cancel as well to give the desired unit at the end.

Step 6: To make sure the correct answer is determined, prove the work by looking at the units involved. Smaller units (in this case, mL) should have a larger number in front of them versus larger units (in this case, L), which should have a smaller number in front of them. In this case, mL has 2525 and L has 2.525. Always double-check the answer to avoid errors in calculations.

If using the shortcut method, apply the following rule: When given a value for liters, to obtain the equivalent amount of milliliters, simply move the decimal point three places to the right of that value (Figure 6-6). Moving the decimal point to the right is multiplying the value by 1000, which is equivalent to the difference between liters and milliliters. There are 1000 mL in each liter of a liquid.

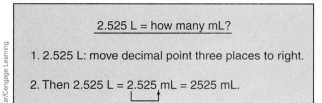

Figure 6-6 Converting from liters to milliliters.

Converting Milliliters to Liters

Sometimes drug preparation will require conversion of milliliters to liters. Using the dimensional analysis method, convert 500 mL to L. Solve the equation 500 mL = _____ L.

Step 1: Because the unknown factor in the given formula is the number of liters contained in 500 mL, determine the metric equivalents. In this example, the metric equivalent is 1000 mL = 1 L.

Step 2: Knowing that 1000 mL = 1 L, create a conversion factor. Write this conversion factor as 1000 mL/1 L.

Step 3: The next step is to determine in which format to write the conversion factor. There are two choices: 1000 mL/1 L or 1 L/1000 mL. The choice is made based on which unit of measure to cancel out and which unit of measure to end up with. In this case, cancel out mL to end up with L. Therefore, mL is on the bottom and L on the top of the conversion factor.

Step 4: Set up the conversions in an equation. Start with what is known (500 mL), and use the conversion factor to set up the equation.

$$\left(500 \text{ mL}\right)\left(\frac{1 \text{ L}}{1000 \text{ mL}}\right) = X$$

Step 5: Now perform the calculation. Remember, values on the bottom of the conversion factors are divided into the numbers on the top. Values on the top of the conversion factors are multiplied by the numbers on the top. In this case, take 500 mL, multiply it by 1 L, and then divide by 1000 mL.

$$\frac{(500 \text{ mL} \times 1 \text{ L})}{1000 \text{ mL}} = 0.500 \text{ L}$$

Keep in mind that units should cancel as well to give the desired unit at the end.

Step 6: To make sure the correct answer is determined, prove the work by looking at the units involved. Smaller units (in this case, mL) should have a larger number in front of them versus larger units (in this case, L), which should have a smaller number in front of them. In this case, mL has 500 and L has 0.500. Always double-check the answer to avoid errors in calculations.

If using the shortcut method, apply the following rule: When given a value for milliliters, to obtain the equivalent amount of liters, simply move the decimal point three places to the left of that value (Figure 6-7). Moving the decimal point to the left, is dividing the value by 1000, which is equivalent to the difference between liters and milliliters. There are 1000 mL in each liter of a liquid.

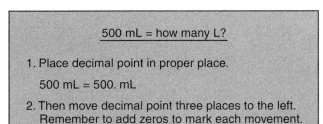

Figure 6-7 Converting from milliliters to liters.

Conversions Within the Apothecary System

The veterinary technician will rarely be required to convert units within the apothecary system; however, an example is the conversion of drams to ounces. Using the dimensional analysis method, convert 64 dr to oz. Solve the equation 64 dr = _____ oz.

Step 1: Because the unknown factor in the given formula is the number of oz contained in 64 dr, determine the apothecary equivalents. In this example, the equivalent is 8 dr = 1 oz.

Step 2: Knowing that 8 dr = 1 oz, create a conversion factor. Write this conversion factor as 8 dr/1 oz.

Step 3: The next step is to determine in which format to write the conversion factor. There are two choices: 8 dr/1 oz or 1 oz/8 dr. The choice is made based on which unit of measure to cancel out and which unit of measure to end up with. In this case, cancel out dr to end up with oz. Therefore, dr is on the bottom, and oz is on the top of the conversion factor.

Step 4: Set up the conversions in an equation. Start with what is known (64 dr), and use the conversion factor to set up the equation.

$$\left(64\ dr\right)\left(\frac{1\ oz}{8\ dr}\right) = X$$

Step 5: Now perform the calculation. Remember, values on the bottom of the conversion factors are divided into the numbers on the top. Values on the top of the conversion factors are multiplied by the numbers on the top. In this case, take 64 dr, multiply it by 1 oz, and then divide by 8 dr.

$$\frac{(64\ dr \times 1\ oz)}{8\ dr} = 8\ oz$$

Keep in mind that units should cancel as well to give the desired unit at the end.

Step 6: To make sure the correct answer is determined, prove the work by looking at the units involved. Smaller units (in this case, dr) should have a larger number in front of them versus larger units (in this case, oz), which should have a smaller number in front of them. In this case, dr has 64 and oz has 8. Always double-check the answer to avoid errors in calculations.

Conversions Within the Household System

The veterinary technician will occasionally be required to convert units within the household system. Using the dimensional analysis method, convert 8 cups to quarts. Solve the equation 8 cups = _____ qt.

Step 1: Because the unknown factor in the given formula is the number of qt contained in 8 cups, determine the household equivalents. In this example, the equivalent is 1 qt = 4 cups.

Step 2: Knowing that 1 qt = 4 cups, create a conversion factor. Write this conversion factor as 1 qt/4 cups.

Step 3: The next step is to determine in which format to write the conversion factor. There are two choices: 1 qt/4 cups or 4 cups/1 qt. The choice is made based on which unit of measure to cancel out and which unit of measure to end up with. In this case, cancel out cups to end up with qt. Therefore, cups are on the bottom, and qt is on the top of the conversion factor.

Step 4: Set up the conversions in an equation. Start with what is known (8 cups), and use the conversion factor to set up the equation.

$$\left(8\ \text{cups}\right)\left(\frac{1\ \text{qt}}{4\ \text{cups}}\right) = X$$

Step 5: Now perform the calculation. Remember, values on the bottom of the conversion factors are divided into the numbers on the top. Values on the top of the conversion factors are multiplied by the numbers on the top. In this case, take 8 cups, multiply it by 1 qt, and then divide by 4 cups.

$$\frac{(8\ \text{cups} \times 1\ \text{qt})}{4\ \text{cups}} = 2\ \text{qt}$$

Keep in mind that units should cancel as well to give the desired unit at the end.

Step 6: To make sure the correct answer is determined, prove the work by looking at the units involved. Smaller units (in this case, cups) should have a larger number in front of them versus larger units (in this case, qt), which should have a smaller number in front of them. In this case, cup has 8 and qt has 2. Always double-check the answer to avoid errors in calculations.

CONVERTING AMONG SYSTEMS

The use of the apothecary and household systems is becoming less frequent; however, the veterinary technician will need to be familiar with conversions among the metric, apothecary, and household systems. Approximate equivalents are used for conversions from one system to another. Table 6-6 contains conversion factors among systems.

Conversions Between Metric and Household Systems of Measure

Sometimes calculating drug doses or preparing solutions calls for conversions between the metric and household systems. Teaspoons may not be added to milliliters, nor grams subtracted from pounds. The differences must be bridged with conversion to a single system. Using the dimensional analysis method, convert 66 lb to kg. Solve the equation 66 lb = _____ kg.

Step 1: Because the unknown factor in the given formula is the number of kg contained in lb, determine the conversion factor between the metric and household systems. The relationship between metric and household weight is 1 kg = 2.2 lb. This bridge converts kilograms to pounds.

Step 2: Knowing that 1 kg = 2.2 lb, create a conversion factor. Write this conversion factor as 1 kg/2.2 lb.

Step 3: The next step is to determine in which format to write the conversion factor. There are two choices: 1 kg/2.2 lb or 2.2 lb/1 kg. The choice is made based on which unit of measure to cancel out and which unit of measure to end up with. In this case, cancel out kg to end up with lb. Therefore, pounds are on the bottom, and kilograms are on the top of the conversion factor.

Step 4: Set up the conversions in an equation. Start with what is known (66 lb), and use the conversion factor to set up the equation.

$$\left(66\ \text{lb}\right)\left(\frac{1\,\text{kg}}{2.2\,\text{lb}}\right) = X$$

Step 5: Now perform the calculation. Remember, values on the bottom of the conversion factors are divided into the numbers on the top. Values on the top of the conversion factors are multiplied by the numbers on the top. In this case, take 66 lb, multiply it by 1 kg, and then divide by 2.2 lb.

$$\frac{(66\ \text{lb} \times 1\ \text{kg})}{2.2\ \text{lb}} = 30\ \text{kg}$$

Keep in mind that units should cancel as well to give the desired unit at the end.

Step 6: To make sure the correct answer is determined, prove the work by looking at the units involved. Smaller units (in this case, pounds) should have a larger number in front of them versus larger units (in this case, kilograms), which should have a smaller number in front of them. In this case, lb has 66 and kg has 30. Always double-check the answer to avoid errors in calculations.

Table 6-6	Equivalents for the Metric, Household, and Apothecary Systems of Measure

Length

1 meter = 1.0936 yards

1 centimeter = 0.39370 inch

1 inch = 2.54 centimeters

1 kilometer = 0.62137 mile

1 mile = 5280 feet or 1.6093 kilometers

1 foot = 0.3048 meter

Weight

1 kilogram = 2.2 pounds

1 pound = 453.59 grams

1 pound = 16 ounces

1 grain = 65 milligrams or 60 milligrams

1 dram = 3.888 grams

1 ounce = 28.35 grams

1 ton = 2000 pounds

1 gram = 0.035274 ounces

1 gram = 15.4 grains

Volume

1 liter = 1.0567 quarts

1 gallon = 4 quarts

1 gallon = 8 pints

1 pint = 2 cups = 16 fluid ounces

Volume

1 cup = 8 fluid ounces

1 gallon = 3.7854 liters

1 quart = 32 fluid ounces

(Continued)

Table 6-6   *(Continued)*

1 quart = 0.94633 liter

1 minim = 0.06 milliliter

1 fluid dram = 3.7 milliliter

1 ounce = approximately 30 milliliters

1 milliliter = 1 cubic centimeter

1 teaspoon = 5 milliliters

Conversions Between Metric and Apothecary Systems of Measure

Sometimes calculating drug doses or preparing solutions calls for conversions between the metric and apothecary systems. Grains may not be added to grams, nor milliliters subtracted from fluid drams. The differences must be bridged with conversion to a single system. Using the dimensional analysis method, convert 240 gr to g. Solve the equation 240 gr = _____ g.

Step 1: Because the unknown factor in the given formula is the number of g contained in gr, determine the conversion factor between the metric and apothecary systems. The relationship between metric and apothecary weight is 1 g = 15.4 gr. This bridge converts grams to grains.

Step 2: Knowing that 1 g = 15.4 gr, create a conversion factor. Write this conversion factor as 1 g/15.4 gr.

Step 3: The next step is to determine in which format to write the conversion factor. There are two choices: 1 g/15.4 gr or 15.4 gr/1 g. The choice is made based on which unit of measure to cancel out and which unit of measure to end up with. In this case, cancel out grains to end up with grams. Therefore, grains are on the bottom and grams are on the top of the conversion factor.

Step 4: Set up the conversions in an equation. Start with what is known (240 gr), and use the conversion factor to set up the equation.

$$\left(240\ gr\right)\left(\frac{1\ g}{15.4\ gr}\right) = X$$

Step 5: Now perform the calculation. Remember, values on the bottom of the conversion factors are divided into the numbers on the top. Values on the top of the conversion factors are multiplied by the numbers on the top. In this case, take 240 gr, multiply it by 1 g, and then divide by 15.4 gr.

$$\frac{(240\ gr \times 1\ g)}{15.4\ gr} = 31.2\ g$$

Clinical Que

Technically, 1 gr = 64.9 mg (rounded to 65 mg); however, both 1 gr = 65 mg and 1 gr = 60 mg conversion factors are still used.

Keep in mind that units should cancel as well to give the desired unit at the end.

Step 6: To make sure the correct answer is determined, prove the work by looking at the units involved. Smaller units (in this case, grains) should have a larger number in front of them versus larger units (in this case, grams), which should have a smaller number in front of them. In this case, gr has 240 and g has 31.2. Always double-check the answer to avoid errors in calculations.

Table 6-6 lists the most useful bridges among the metric, household, and apothecary (common) systems of measure. Table 6-7 shows metric-to-metric conversions. If extra practice is needed with systems of measurement and unit conversions, refer to Appendix J and the StudyWARE® material provided with this text.

Table 6-7	Review of Metric-to-Metric Conversions

Linear Measure: base unit is meters (m)

1 m = 100 centimeters (cm)

1 m = 1000 millimeters (mm)

1 m = 1,000,000 micrometers or microns (mcm, μm or μ)

1000 m = 1 kilometer (km) or 0.001 m = 1 km

Volume Measure: base unit is liter (L)

1 L = 100 centiliters (cL)

1 L = 1000 milliliters (mL)

1000 L = 1 kiloliter (kL)

Weight Measure: base unit is gram (g)

1 g = 100 centigrams (cg)

1 g = 1000 milligrams (mg)

1 g = 1,000,000 micrograms or 0.000001 g = 1 microgram (mcg or μg)

0.001 mg = 1 microgram (mcg or μg)

1 mg = 1000 micrograms (mcg or μg)

TEMPERATURE CONVERSIONS

In the veterinary field, the technician will need to take and record an animal's body temperature accurately. The type of thermometer used will determine whether the reading will be in the Fahrenheit or Celsius scale. The veterinary technician needs to understand both these scales.

Gabriel Fahrenheit developed the mercury thermometer. He used a mixture of salt and ice to experiment with temperature. The coldest mixture he could make he called "zero." He noted that water froze at 32° and boiled at 212° on this scale. The Fahrenheit system is used in the United States and uses the references of water freezing at 32°F and boiling at 212°F.

Anders Celsius suggested a temperature scale based on the freezing and boiling points of water. He felt that the point at which water freezes should be 0 and the point at which water boils should be 100. The Celsius temperature scale, which is used in the metric system, states that water freezes at 0°C and boils at 100°C.

Conversion between the Fahrenheit and Celsius scales is based on the relationship between scales as to the freezing and boiling points of water. In the Celsius scale, there is a difference of 100 between the freezing and boiling points of water (100°C − 0°C = 100°C). In the Fahrenheit scale, there is a difference of 180 between the freezing and boiling points of water (212°F − 32°F = 180°F). The ratio between these two differences is 1.8 (180 ÷ 100 = 1.8) (Figure 6-8). Using this ratio and the variation in freezing point, conversions between Celsius and Fahrenheit can be calculated.

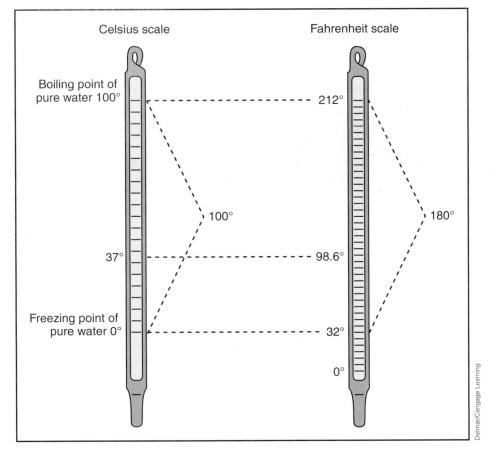

Figure 6-8 Comparison of Celsius and Fahrenheit temperature scales. Note that there is 180° difference between the boiling and freezing points on the Fahrenheit thermometer and 100° difference between the boiling and freezing points on the Celsius thermometer. The ratio of the difference between the Fahrenheit and Celsius scales can be expressed as 180:100 or 180/100. When reduced, this ratio is equivalent to 1.8.

To convert a temperature reading from Fahrenheit to Celsius, subtract 32 from the Fahrenheit reading and divide by 1.8.

$$°C = \frac{(°F - 32)}{1.8}$$

For example, convert 98.6°F to Celsius.

Step 1: Take the Fahrenheit reading (98.6) and subtract 32. This gives 66.6.

Step 2: Divide the result (66.6) by 1.8. This gives 37.

Therefore, 98.6°F converts to 37°C.

To convert a temperature reading from Celsius to Fahrenheit, multiply the Celsius reading by 1.8 and add 32.

$$°F = 1.8°C + 32$$

For example, convert 100°C to Fahrenheit.

Step 1: Multiply the Celsius reading (100) by 1.8. This gives 180.

Step 2: Take the result (180) and add 32. This gives 212.

Therefore, 100°C converts to 212°F.

Clinical Que

For converting temperatures, use the following equations

$$°F = 1.8°C + 32$$

$$°C = \frac{(°F - 32)}{1.8}$$

DOSE CALCULATIONS

Dose calculations are performed daily in veterinary practice. The following are examples of how dose calculations are performed in a veterinary setting.

Determining the Dose and the Amount of Drug Dispensed

Calculating the total dose of a drug is a common calculation performed in veterinary medicine. Once the animal's weight and the amount of a drug needed per unit of the animal's weight (the drug dosage) are known, the dose of drug can be determined. The dose is the amount of drug given to the animal in one administration.

Dose in mg

An example of calculating a dose in mg is as follows: The dosage of a drug is 2 mg per kilogram body weight of the animal. How many mg should be given to an animal weighing 22 lb?

First convert lb to kg. From Table 6-6, choose the appropriate conversion factor of 1 kg = 2.2 lb. The animal weighs 22 lb. Therefore, use the conversion factor 2.2 lb = 1 kg to get the weight into kg units. Use the original weight, and set up a unit conversion to switch from one system of measure to the other.

$$\left(22\text{ lb}\right)\left(\frac{1\text{ kg}}{2.2\text{ lb}}\right) = 10\text{ kg}$$

If the animal needs 2 mg of drug for every 1 kg of body weight, and the animal weighs 10 kg, use the weight in kilograms and set up a cancellation method to incorporate the dosage.

$$\left(10\text{ kg}\right)\left(\frac{2\text{ mg}}{1\text{ kg}}\right) = 20\text{ mg dose}$$

The formula used is animal weight (in lb) divided by 2.2 (to convert to kg), then multiplied by amount of drug (in mg) per kilogram.

$$\frac{\text{Wt (lb)}}{2.2} \times \text{dose (units/kg)} = \text{dose to animal}$$

Remember, the amount of medicine and the strength of dose must be measured in the same units.

This calculation has determined the strength of each dose to be given to the animal and the total amount of medicine to dispense/administer. To measure the dose, the dose needs to be in a measurable unit. For example, tablets can be dispensed as a unit, and milliliters can be dispensed as a unit.

Dose in Tablets

Using the example of the 22-lb animal needing 20 mg of drug, how many tablets will the owner give per dose? To find this, it is necessary to know the strength of tablets available. If the clinic inventory contains 40 mg, 80 mg, and 100 mg tablets of this drug, how many tablets per dose would this owner give the animal?

It was already determined that the animal needs 20 mg per dose, so the technician would divide the 20 mg dose by 40 mg (the tablet strength closest to the dose).

$$\left(20\text{ mg}\right)\left(\frac{1\text{ tablet}}{40\text{ mg}}\right) = \frac{1}{2}\text{ tablet per dose (assuming that the tablets are scored)}$$

Clinical Que

Sometimes the same dose of a drug may be cheaper when a scored tablet of a higher dose is divided in half rather than using a full tablet of the lower dose. For example, splitting a 40 mg tablet in half to get a 20 mg dose may be cheaper than using one 20 mg tablet.

Dose in mL

Using the example of the 22-lb animal needing 20 mg of drug again, how many mL of drug will the owner give? If the drug vial states the concentration of drug is 10 mg/mL, how many mL would the owner give?

It was determined that the animal needs 20 mg per dose, so the technician would divide 20 mg by the concentration of 10 mg/mL.

$$\left(20\ mg\right)\left(\frac{1\ mL}{10\ mg}\right) = 2\ mL$$

A 3-mL syringe is used to measure and administer 2 mL of drug (Figure 6-9).

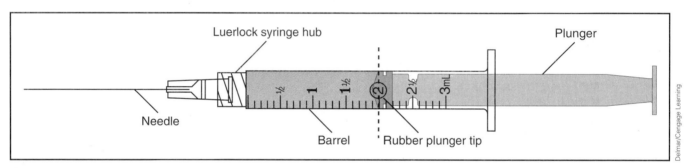

Figure 6-9 3-mL syringe with needle unit measuring 2 mL. The calibrations are read from the top of the plunger ring, not the raised middle section and not the bottom of the plunger ring.

Dose in U

Some liquid medications, including insulin, heparin, and penicillin, are measured in units (U) or international units (IU) (Figure 6-10). The international unit or unit is a standardized amount of drug needed to produce the desired effect rather than based on the drug's weight. These medications are standardized in units based on their strengths. The strength varies from one medicine to another, depending on their source, their condition, and the method by which they are obtained. It is not necessary to learn conversion for the international unit or unit because medications prescribed in these measurements are also prepared and administered in the same system.

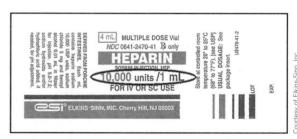

Figure 6-10 Some drugs, such as the anticoagulant heparin, are measured in units.

An example of calculating a dose in units is as follows: It is necessary to give a 1000-lb cow 50,000 U/kg of penicillin G IM, how many mL would be given based on a concentration of 300,000 U/mL?

First, convert 1000 lb to kg, using the conversion factor 2.2 lb = 1 kg

$$\left(1000 \text{ lb}\right)\left(\frac{1 \text{ kg}}{2.2 \text{ lb}}\right) = 454.54 \text{ kg}$$

Next, take the weight in kg, and multiply it by the dosage

$$\left(454.54 \text{ kg}\right)\left(\frac{50,000 \text{ U}}{\text{kg}}\right) = 22,727,272.73 \text{ U}$$

Then take the units and divide by the concentration to determine a dose in mL

$$\left(22,727,272.73\right)\left(\frac{1 \text{ mL}}{300,000 \text{ U}}\right) = 75.8 \text{ mL}$$

Clinical Que

Veterinary professionals should always question dose calculations that seem unreasonably large or small for the size animal for which the drug is prescribed.

Calculating Total Dose

Using the example of the animal that needs 20 mg per dose, how much drug would need to be dispensed if this animal needed the drug bid for seven days? How many tablets would be dispensed? It was previously determined that each dose in tablet form was ½ tablet. Because the animal needs the drug bid (twice daily), the animal will get 1 tablet per day (½ tablet × 2 doses per day = 1 tablet per day). The animal needs to take the medication for one week. Multiply 1 tablet per day by 7 days: 1 tablet × 7 days = 7 tablets total to be dispensed.

Calculating Number of Doses

To find the number of doses that can be given from a total dose, use this formula

Number of doses = total amount of medicine divided by strength of each dose

Examples of calculating number of doses are as follows:

Example 1: The veterinarian prescribes 200 mg of medicine to an ill animal. Knowing the strength of each dose (20 mg), the number of doses that can be given from the total dose can be calculated.

$$\frac{200 \text{ mg}}{20 \text{ mg}} = 10$$

Therefore, the animal would need to be given 10 doses.

Example 2: A single dose of the drug to be given is 1 g. How many doses are contained in 10 g?

The total amount of medicine is 10 g. The strength of each dose is 1 g. The units are already in the same unit (grams). Calculate the problem as follows:

$$\text{Number of doses} = \frac{10 \text{ g}}{1 \text{ g}} = 10 \text{ doses}$$

Sometimes it is necessary to convert between units of measure within the same dose calculation.

Example 3: The dose of a drug given to an ill sow is 200 mg. How many doses are there in 10 g?

The total amount of medicine is 10 g. The strength of each dose is 200 mg. The units need to be converted to similar units; therefore, convert 10 g to mg so that the common unit is mg. From Table 6-2, the conversion factor 1 g = 1000 mg is found. Using this conversion factor, mg can be determined by taking (10 g) (1000 mg/1 g) = 10,000 mg.

Now that the total amount of medicine and each dose of medicine are in the same units, the number of doses can be calculated.

$$\text{Number of doses} = \frac{10,000 \text{ mg}}{200 \text{ mg}} = 50 \text{ doses}$$

Therefore, there are 50 doses (20 mg each) in 10 g.

Determining the Amount in Each Dose

Now that the total amount of medicine prescribed and the number of doses in the total amount is known, the quantity of each individual dose can be determined using the following formula:

$$\text{Quantity in each dose} = \frac{\text{quantity in total amount}}{\text{the number of doses}}$$

The following example demonstrates a determination of the amount in each dose. If the total amount of drug is 100 mg in a solution of 100 mL, which represents 20 doses, how much drug is in one dose?

$$\text{quantity in each dose} = \frac{100\ \text{mg}}{20\ \text{doses}} = 5\ \text{mg in each dose}$$

For additional problems on calculating doses, refer to Appendices G, H, I, and J, and the StudyWARE® material provided with this text.

SOLUTIONS

Solutions are mixtures of substances not chemically combined with each other. The dissolving substance of a solution is referred to as the **solvent** and is usually a liquid. The dissolved substance of a solution is referred to as the **solute** and is usually the solid or particulate part of the mixture. Substances that form solutions are called **miscible** and those that do not are called **immiscible**.

When working with solutions, the amount of solute (particles) in the solvent is important to know. The amount of solute dissolved in solvent is known as the *concentration*. Concentrations may be expressed as parts (per some amount), weight per volume, volume per volume, and weight per weight. They are usually reported as percents or percent solution. Remember that a percent is the parts per the total times 100. Here are some rules of thumb for working with concentrations.

Clinical Que

Ratio strengths (parts per some amount) are used primarily in solutions. They represent parts of drug per parts of solution.

- *Parts:* Parts per million means 1 mg of solute in a kg (or L) of solvent. Ratios or fractions must be translated into percents of solution to perform many of the necessary calculations. For example, to determine the percent concentration of 1:1000 epinephrine (Figure 6-11), divide the 1 by 1000 and multiply by 100.

$$\left(\frac{1}{1000}\right)\left(100\right) = 0.1\%$$

- *Liquid in liquid:* The percent concentration is the volume per 100 volumes of the total mixture. The two volumes may be expressed in any unit as long as they are the same unit within the percent. For example, 1 mL/100 mL, 5 oz/100 oz, and 15 L/100 L.

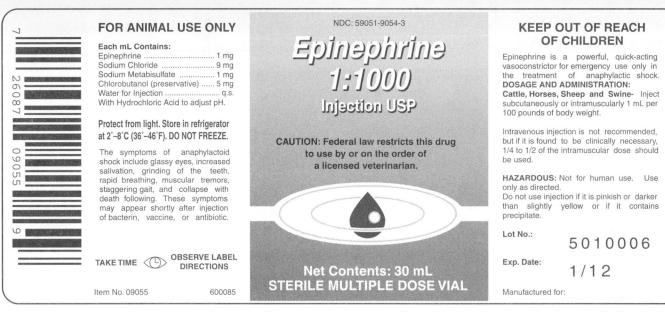

Figure 6-11 Epinephrine label.

- *Solids in solids:* The percent concentration is the weight per 100 weights of total mixture. The two weights may be expressed in any units as long as they are the same unit within the percent. For example, 60 mg/100 mg, 55 oz/100 oz, and 4.5 g/100 g.

- *Solids in liquid:* The percent concentration is the weight in grams per 100 volume parts in milliliters. The weight must be in grams and the volume must be in milliliters. For example, dextrose 5% = 5 g/100 mL. Dextrose 5% can also be expressed as 5000 mg/100 mL or 50 mg/mL, as long as the number is based originally on grams.

Percent Concentration Calculations

Occasionally the veterinary technician may have to prepare a drug solution from a pure drug or stock solution. *Pure drugs* are substances, in solid or liquid form, that are 100 percent pure. A *stock solution* is a relatively concentrated solution from which more dilute (weaker) solutions are made.

One method of determining the amount of pure drug needed to make a solution is the ratio-proportion method. The formula used to determine the amount of pure drug needed to make the solution is

$$\frac{\text{Amount of drug}}{\text{Amount of finished solution}} = \frac{\text{percentage of finished solution}}{100\% \text{ (based on a pure drug)}}$$

For example, to determine how much sodium chloride is needed to make 500 mL of a 0.9% solution (0.9% NaCl = 0.9 g NaCl per 100 mL solution), the following calculation would be performed:

> ## Clinical Que
>
> When using ratios, it is important to understand what the notation represents. A 1:2 ratio means of the total amount (2 parts), 1 part is the medication. A 1:1000 ratio means of the total amount (1000 parts), 1 part is medication. Therefore, a 1:1 ratio means of the total amount (1 part), the entire 1 part is medication. In other words, a 1:1 ratio contains all medication.

$$\frac{X\text{ g}}{500\text{ mL}} = \frac{0.9\text{ g}}{100\text{ mL}}$$

To solve for X, cross-multiply to get

$$100(\text{mL})\ X\ (\text{g}) = 450(\text{mL})(\text{g})$$

Then divide each side by 100 percent to isolate X.

X = 4.5 g of sodium chloride
(The answer technically is 4.5 mL; however, 4.5 mL of sodium chloride can be multiplied by the density of solution which is 1 g/mL. This gives a final answer of 4.5 g)

The amount of drug used to prepare a solution adds to the total volume of the solvent. The amount of volume contributed by a dry drug varies by its individual structure and is usually not accounted for. However, it is fairly easy to determine the volume a liquid drug will add to the solvent volume. To determine the amount of solvent needed to make the finished solution, subtract the volume of liquid drug from the amount of finished solution.

For example, to prepare a liter of 4 percent formaldehyde fixative solution from a 37 percent stock solution, use the ratio-proportion method as follows:

$$\frac{X\text{ mL}}{1000\text{ mL}} = \frac{4\ \%}{37\ \%}$$

To solve for X, cross-multiply to get

$$37(\%)\ X(\text{mL}) = 4000(\%)(\text{mL})$$

Then divide each side by 37 percent to isolate X.

X = 108 mL of stock solution

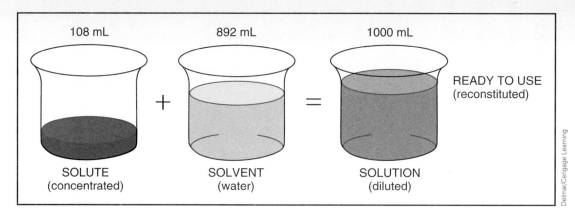

Figure 6-12 Solvent is added to concentrated liquid solute to make diluted solution.

To determine the amount of solvent to add to the stock solution, take the total desired amount (in this case, 1 L), convert 1 L to 1000 mL (since the volume of stock solution is in mL, this volume also needs to be in mL), and subtract the amount of stock solution in mL

1000 mL – 108 mL = 892 mL of solvent should be added (Figure 6-12)

Another way to determine volume for a desired final volume is via the volume concentration method. The equation for this is as follows:

$$V_s \times C_s = V_d \times C_d$$

V_s = volume of the beginning or stock solution
C_s = concentration of the beginning or stock solution
V_d = volume of the final solution
C_d = concentration of the final solution

An example using this equation is as follows: How much water must be added to a liter of 90 percent alcohol to change it to a 40 percent solution?

To solve this, consider what is known. There is 1 L of 90 percent alcohol. Now consider what is wanted. A 40 percent solution is wanted, and whatever amount of water needed to make this happen is used.

$$1000 \text{ mL} \times 90\% = V_d \times 40\%$$

$$90,000 \text{ mL } (\%) = V_d \, 40\%$$

$$2250 \text{ mL} = V_d$$

It is also possible to convert the percents to decimals before solving the preceding equation.

$$1000 \text{ mL} \times 0.90 = V_d \times 0.40$$

$$900 \text{ mL} = 0.40 \, V_d$$

$$2250 \text{ mL} = V_d$$

Remember that 2250 mL is the final volume. Therefore, it is necessary to subtract the original amount of 90 percent alcohol.

$$2250 \text{ mL} - 1000 \text{ mL} = 1250 \text{ mL}$$

Another example is as follows: How much of a 1:25 solution of NaCl is needed to make 3 L of 1:50 solution?

To solve this, convert 1:25 and 1:50 to percent

$$\left(\frac{1}{25}\right)\left(100\right) = 4\%$$

$$\left(\frac{1}{50}\right)\left(100\right) = 2\%$$

Next, use the volume concentration equation to complete the problem.

$$V_s \times 4\% = 3 \text{ L} \times 2\%$$

$$V_s \, 4(\%) = 6(\text{L})(\%)$$

$$V_s = 1.5 \text{ L}$$

Therefore, 1.5 L of 4 percent NaCl and 1.5 L of solvent are needed to make a 2 percent NaCl solution.

A final example is as follows: How much water is needed to add to a 1 percent solution to make a 10 percent solution? The answer is that it cannot be done. It is not possible to make a more dilute solution more concentrated simply by adding water. The veterinary technician should always check to make sure the answer makes sense before mixing or dispensing a product.

Sometimes drug concentrations are listed in percents. The percent concentrations are usually found on the front of the drug vial or bottle. Occasionally, these containers also have the concentration listed in mg/mL.

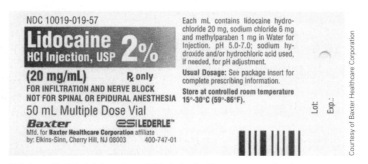

Figure 6-13 Lidocaine 2% label. Drugs supplied as percent solutions may also have the mg/mL concentration on the label.

An example is as follows: Lidocaine is a drug used as a topical anesthetic and as an antiarrhythmic drug. The dosage for a dog is 3 mg/kg. Calculate the dose of lidocaine in mL for a 15-lb dog.

First, change lb to kg by using the 2.2 lb = 1 kg conversion factor.

$$\left(15 \text{ lb}\right)\left(\frac{1 \text{ kg}}{2.2 \text{ lb}}\right) = 6.81 \text{ kg}$$

Next calculate the dose needed in mg.

$$(6.81 \text{ kg}) \, (3 \text{ mg/kg}) = 20.45 \text{ mg}$$

The front of the vial lists the concentration of lidocaine as 2 percent (Figure 6-13). Because percents are parts per the total, 2 percent is 2 parts per 100. In solutions, percent represents the number of grams of drug per 100 mL of solution. Remembering that percents are g/100 mL, lidocaine 2 percent equals 2 g/100 mL. In this example, the units of weight (mg and g in this case) need to be the same to perform the calculation.

Change 20.45 mg to g

$$\left(20.45 \text{ mg}\right)\left(\frac{1 \text{ g}}{1000 \text{ mg}}\right) = 0.02045 \text{ g}$$

Then calculate the dose based on 2% = 2 g/100 mL

$$\left(0.02045 \text{ g}\right)\left(\frac{100 \text{ mL}}{2 \text{ g}}\right) = 1.02 \text{ mL}$$

This dog would get 1.02 mL per dose (which would be rounded to 1.0 mL).

Additional problems concerning concentration calculations can be found in Appendix J and the StudyWARE® material provided with this text.

RECONSTITUTION PROBLEMS

Some parenteral medications are not stable when suspended in solution. Such drugs are usually stored in a powder form because in time (hours, days, or weeks) the drug begins to deteriorate in solution. For this powdered drug to be administered parenterally, it must first be dissolved or reconstituted in sterile water, saline, or dextrose solution. Usually a vial label or package insert indicates the amount of solution necessary to dissolve the powder in the vial, as well as the solvent that should be used when dissolving the powder (Figure 6-14).

In multiple-dose vials, the powder to be dissolved often adds to the total final volume of the liquid being reconstituted. The label in these cases indicates the amount of solvent to use to reconstitute the powder and the total volume this reconstitution will produce. This total volume will include the volume added by both the powder and the liquid. Consider the following example.

Cefazolin sodium comes in a 500 mg vial strength. The label instructions state that when 2.0 mL of sterile saline is added to the vial, the total final volume will be 2.2 mL (Figure 6-15). The label also states that the approximate average concentration when cefazolin sodium is reconstituted as directed will yield a concentration of 225 mg/mL. The dose calculation is based on this concentration.

Some labels or package inserts allow a choice of dilution amounts, which will provide different concentrations of the drug based on which dilution is chosen. The amount of diluent (liquid) to use is based on the strength most appropriate for the dose ordered (to yield the smallest possible volume of drug to be given) and personal preference. Consider the following example.

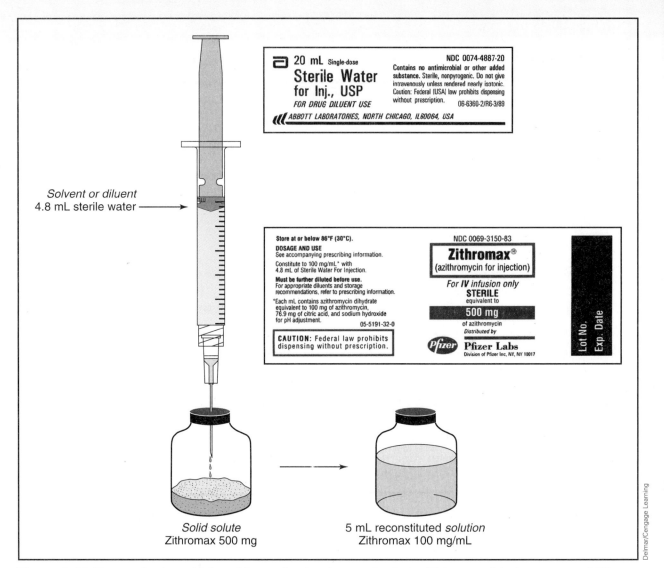

Figure 6-14 Reconstitution of injectable medication in powder form (Zithromax label courtesy of Pfizer Labs, Sterile Water label courtesy of Abbott Laboratories).

A brand name of ampicillin comes in a vial containing 25 g of ampicillin. It can be diluted with sterile water for injection in volumes of 104.5 mL (final concentration 200 mg/mL), 79.0 mL (final concentration 250 mg/mL), or 41.0 mL (final concentration 400 mg/mL). If the 25 g ampicillin vial is diluted with 41.0 mL of sterile water, how many mL would be given to an animal needing an 800 mg dose of ampicillin?

The concentration of ampillicin in this case is 400 mg/mL. If this animal needs 800 mg per dose, the following calculation is made

$$\left(800 \text{ mg}\right)\left(\frac{1 \text{ mL}}{400 \text{ mg}}\right) = 2 \text{ mL}$$

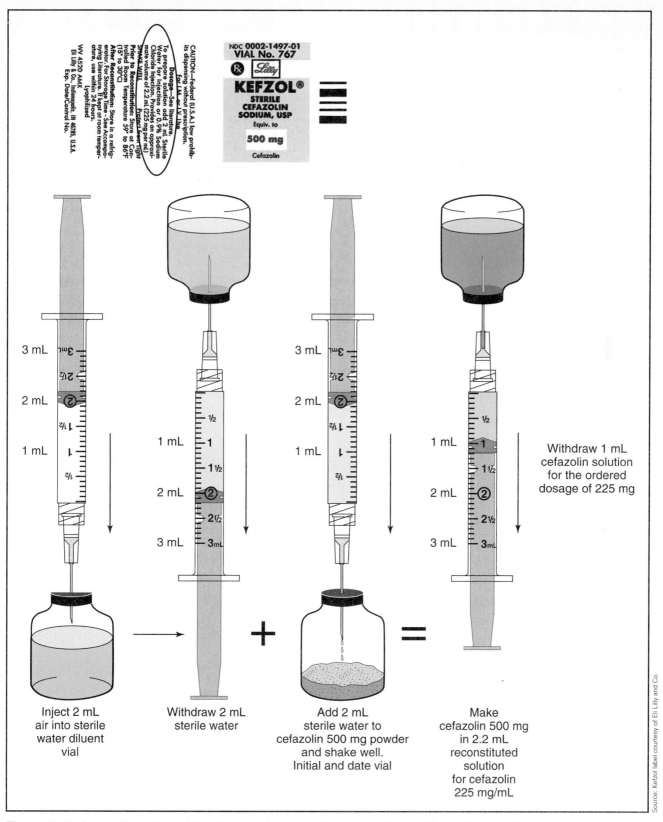

Figure 6-15 In multiple-dose vials, the powder to be dissolved often adds to the total final volume of the liquid being reconstituted. This cefazolin label indicates the amount of solvent to use to reconstitute the powder and the total volume this reconstitution will produce.

Inject 2 mL air into sterile water diluent vial

Withdraw 2 mL sterile water

Add 2 mL sterile water to cefazolin 500 mg powder and shake well. Initial and date vial

Make cefazolin 500 mg in 2.2 mL reconstituted solution for cefazolin 225 mg/mL

Withdraw 1 mL cefazolin solution for the ordered dosage of 225 mg

Clinical Que

When there is a
choice as to the
amount of diluent to
use for reconstitution
of a drug
(Figure 6-16), the
volume used and the
final concentration
should be circled or
written on the vial
to avoid any dose
calculation errors.
The date and time of
the reconstitution or
expiration date, as
well as the initials
of the person who
reconstituted the
drug, should also be
written on the vial.

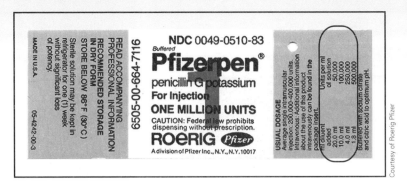

Figure 6-16 Multiple-strength solution label.

Some drugs that require reconstitution can be prepared for various parenteral routes. The dilutions for these routes of administration may vary. Use caution to ensure that the proper reconstitution for the desired route of administration is performed!

SUMMARY

The ability to perform mathematical calculations in a veterinary setting is critical for patient care. Three systems of measurement are used in determining drug doses and solution concentrations: household, apothecary, and metric. The metric system is the most commonly used. The base units of the metric system are meter (for length), liter (for volume), and gram (for weight). Prefixes applied to the base units denote the size of the metric unit. Common prefixes include micro- (0.000001), milli- (0.001), centi- (0.01), and kilo- (1000). Conversions can be done within a system or between systems using conversion factors. Appendix G contains conversion tables for reference when doing mathematical conversions.

Veterinary technicians need to be able to properly convert or change from one unit to another within the same measurement system or between measurement systems to perform dose calculations. Converting between one unit and another involves the use of a conversion factor. A conversion factor is a number used with either multiplication or division to change a measurement from one unit of measurement to its equivalent in another unit of measure.

Dose calculations are performed daily in veterinary practice. Veterinary technicians need to be proficient at calculating doses in milligrams, tablets, milliliters, and units. Veterinary technicians need to be proficient at calculating the total dose and the number of doses needed for a patient, and the amount in each dose for their patients.

Solutions are mixtures of substances not chemically combined with each other. The dissolving substance of a solution is referred to as the solvent and is usually a liquid. The dissolved substance of a solution is referred to as the solute and is usually the solid or particulate part of the mixture. Substances that form solutions are called miscible and those that do not are called immiscible.

When working with solutions, the amount of solute (particles) in the solvent is important in determining a solution's concentration. The amount of solute dissolved in solvent is known as the concentration. Concentrations may be expressed as parts (per some amount), weight per volume, volume per volume, and weight per weight. Solution calculations are used to determine amounts of solute or solvent to add to dilute or concentrate a solution.

Some parenteral medications are not stable when suspended in solution and are stored in a powder form because in time, the drug begins to deteriorate in solution. For a powdered drug to be administered parenterally, it must first be dissolved or reconstituted in sterile water, saline, or dextrose solution. The vial label or package insert indicates the amount of solution necessary to dissolve the powder in the vial, as well as the solvent that should be used when dissolving the powder. In multiple-dose vials, the powder to be dissolved often adds to the total final volume of the liquid being reconstituted.

It's a Wrap

The answers to the questions in this chapter's Setting the Scene case study should be understood after reading this chapter. Did Valerie drink enough water today to replace normal water loss? No, she did not. How can this be mathematically determined? Valerie needed to drink 8 glasses of water that contain 8 oz of water; therefore, she needed to drink 64 oz of water. When converting 64 oz to liters, it is determined that Valerie should have consumed 1.92 L of water (using dimensional analysis [64 oz] [30 mL/oz] [L/1000 mL] = 1.92 L). Since she drank only 1 L of water, she needs to drink almost another liter of water to meet her needs. Petey drank one 8 oz bowl of water. If dogs should drink 30 mL of water per pound per day, did he drink enough water today? Since Petey is 55 pounds, he needs to drink 1650 mL of water per day (using dimensional analysis [55#][30 mL/#] = 1650 mL). Drinking one 8-oz bowl of water provides only 240 mL of water (using dimensional analysis [8 oz] [30 mL/oz] = 240 mL). Petey needs to drink 1410 additional mL of water to meet his daily needs.

CHAPTER REVIEW

Matching

Match the abbreviation or term to its definition.

1. _____ centi-

2. _____ milli-

3. _____ kilo-

4. _____ micro-

5. _____ pt

6. _____ qt

7. _____ cc

8. _____ g

9. _____ L

10. _____ m

a. liter
b. gram
c. mL
d. 2 pints
e. 1000 times the base unit
f. meter
g. 1/100th of the base unit
h. 1/1000th of the base unit
i. 1/1,000,000th of the base unit
j. 16 fl. oz or 16 fluid ounces

Multiple Choice

Choose the one best correct answer. You may refer to the unit conversion chart in Appendix G.

11. In the metric system, the fundamental unit of liquid volume is the
    a.   meter
    b.   gram
    c.   liter
    d.   inch

12. In the metric system, the fundamental unit of weight is the
    a.   meter
    b.   gram
    c.   liter
    d.   inch

13. One gram is equivalent to how many milligrams?
    a.   1 mg
    b.   10 mg
    c.   100 mg
    d.   1000 mg

14. 38 fluid ounces are equal to how many pints?
    a.   0.421 pint
    b.   2.375 pints
    c.   38 pints
    d.   608 pints

15. 7.6 quarts are equal to how many pints?
    a.   14 pints
    b.   15.2 pints
    c.   28.8 pints
    d.   30.4 pints

16. How many quarts are equal to 17 gallons of liquid?
    a.   4.25 quarts
    b.   34 quarts
    c.   8.5 quarts
    d.   68 quarts

17. 3.5 fluid ounces equal how many milliliters?
    a.   7 mL
    b.   105 mL
    c.   56 mL
    d.   1655.5 mL

18. If the dose of a drug is 5 mg, how many doses are contained in 10 mg?
    a.   1
    b.   10
    c.   2
    d.   20

19. If the dose of a drug is 16 g, how many doses (in g) are contained in 50,000 mg?
   a. 0.0032
   b. 31.25
   c. 3.125
   d. 312.5

20. If each dose of medicine is 400 mg, how many doses are in 86 g of medicine?
   a. 0.215
   b. 34.4
   c. 4.6
   d. 215

21. The dose of a drug is 1 mg per kilogram of body weight. How many mg should be given to a rabbit weighing 10 pounds?
   a. 4.5 mg
   b. 10 mg
   c. 5.5 mg
   d. 45 mg

22. The dose of a drug is 0.05 mg per kg of body weight. How many mcg should be given to a cat weighing 17.6 kg?
   a. 0.008 mcg
   b. 880 mcg
   c. 0.88 mcg
   d. 8880 mcg

23. If the total amount of 18 L of medicine represents 240 doses, how many mL are in each dose?
   a. 0.75 mL
   b. 75 mL
   c. 1.33 mL
   d. 432 mL

24. If one dose is 0.005 g, how many doses (in mg) are in 100 mg of medicine?
   a. 0.05
   b. 20
   c. 20,000
   d. 200

25. Translate 1:70 into a percent.
   a. 10%
   b. 1.4%
   c. 1%
   d. 2.5%

26. How much water needs to be added to a 5 g vial of drug powder to make a 2.5% solution?
   a. 200 mL
   b. 50 mL
   c. 100 mL
   d. 150 mL

27. How much of a 1:20 solution of NaCl is needed to make 1 L of 1:50 solution of NaCl?
   a. 2000 mL
   b. 1000 mL
   c. 500 mL
   d. 400 mL

Fast Conversions

Convert the following.

28. 1.550 g = _____ mg

29. 674 mg = _____ g

30. 2.55 L = _____ mL

31. 600 cc= _____ mL = _____ L

32. 15 mg = _____ g

33. 25 mcg = _____ mg

34. 400 cc = _____ L

35. 0.003 g = _____ mg

36. 0.015 mg = _____ mcg

37. 176 lb = _____ kg

38. 1000 g = _____ kg

39. ¼ L = _____ mL

40. 51 cm = _____ inches

41. Which is larger, 0.125 mg or 0.25 mg?

42. 10 lb = _____ kg

43. 13 mm = _____ m

44. 125 mL + 3 L = _____ mL

45. How many 250 mg tablets are needed to equal 1 g? _____

46. How would 5 mg of a drug be dispensed that comes in 20 mg tablets (assuming the tablet is properly scored)? _____

47. 32 oz = _____ mL

48. 2.5 cm = _____ inches

49. 32 kg = _____ lb

50. 100 mm = _____ cm

Temperature Conversions

51. 98°F = _____ °C

52. 102°F = _____ °C

53. 212°F = _____ °C

54. 25°C = _____ °F

55. −40°C = _____ °F

Parenteral Doses of Drugs

Calculate the amount needed for each dose. The labels provided represent the drugs available. Draw an arrow to the syringe calibration that corresponds to the amount of drug needed for administration.

56. Bicillin C-R 900/300 1,200,000 units IM

Give _____ mL

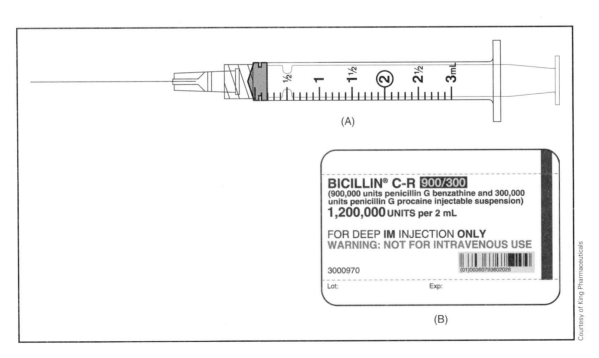

(A)

BICILLIN® C-R 900/300
(900,000 units penicillin G benzathine and 300,000 units penicillin G procaine injectable suspension)
1,200,000 UNITS per 2 mL

FOR DEEP **IM INJECTION ONLY**
WARNING: NOT FOR INTRAVENOUS USE

3000970

Lot: Exp:

Courtesy of King Pharmaceuticals

(B)

57. Dexamethasone sodium phosphate 1.5 mg IV q12 h

Give _____ mL

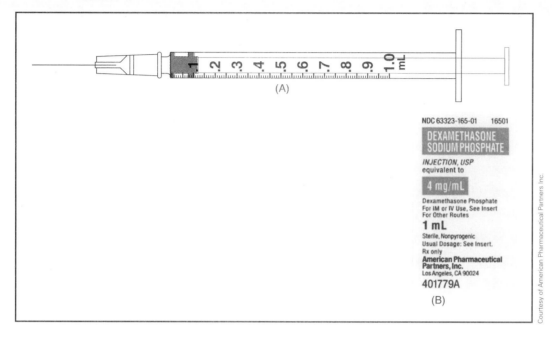

58. Heparin 4000 units SQ daily

Give _____ mL

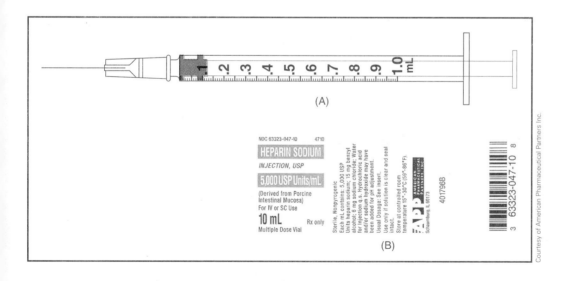

59. Furosemide 15 mg IV daily

Give _____ mL

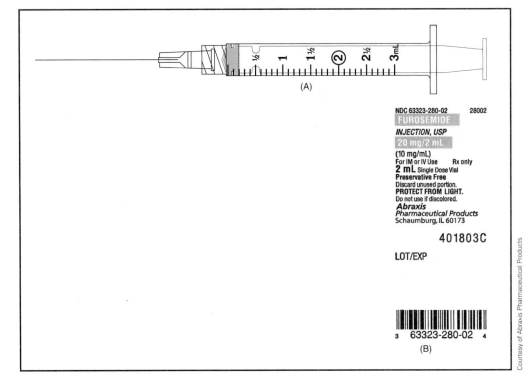

(A)

NDC 63323-280-02   28002

FUROSEMIDE

INJECTION, USP

20 mg/2 mL

(10 mg/mL)
For IM or IV Use   Rx only
2 mL Single Dose Vial
Preservative Free
Discard unused portion.
PROTECT FROM LIGHT.
Do not use if discolored.
Abraxis
Pharmaceutical Products
Schaumburg, IL 60173

401803C

LOT/EXP

3 63323-280-02 4

(B)

Courtesy of Abraxis Pharmaceutical Products

60. Morphine sulfate gr ¼ IV for pain

Give _____ mL

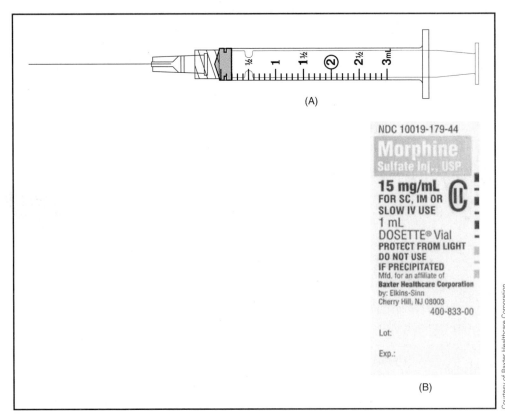

(A)

NDC 10019-179-44

Morphine
Sulfate Inj., USP

15 mg/mL
FOR SC, IM OR
SLOW IV USE
1 mL
DOSETTE® Vial
PROTECT FROM LIGHT
DO NOT USE
IF PRECIPITATED
Mfd. for an affiliate of
Baxter Healthcare Corporation
by: Elkins-Sinn
Cherry Hill, NJ 08003
400-833-00

Lot:

Exp.:

(B)

Courtesy of Baxter Healthcare Corporation

Reconstitution of Solutions

Calculate the amount of diluent needed to prepare each dose. The labels provided are the drugs available. Prepare a reconstitution label if needed.

61. Penicillin G 150,000 units IM q12h

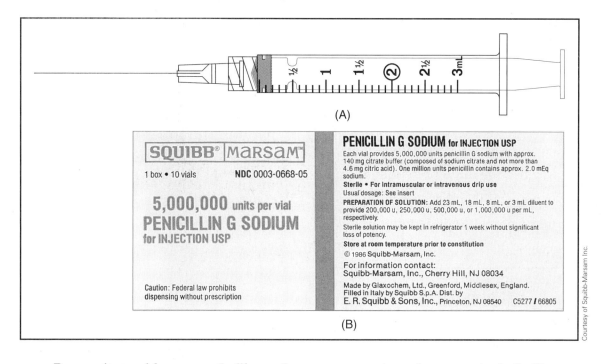

(A)

SQUIBB® MARSAM™

1 box • 10 vials   NDC 0003-0668-05

5,000,000 units per vial
PENICILLIN G SODIUM
for INJECTION USP

Caution: Federal law prohibits
dispensing without prescription

PENICILLIN G SODIUM for INJECTION USP

Each vial provides 5,000,000 units penicillin G sodium with approx.
140 mg citrate buffer (composed of sodium citrate and not more than
4.6 mg citric acid). One million units penicillin contains approx. 2.0 mEq
sodium.

Sterile • For intramuscular or intravenous drip use
Usual dosage: See insert

PREPARATION OF SOLUTION: Add 23 mL, 18 mL, 8 mL, or 3 mL diluent to
provide 200,000 u, 250,000 u, 500,000 u, or 1,000,000 u per mL,
respectively.

Sterile solution may be kept in refrigerator 1 week without significant
loss of potency.

Store at room temperature prior to constitution

© 1986 Squibb-Marsam, Inc.

For information contact:
Squibb-Marsam, Inc., Cherry Hill, NJ 08034

Made by Glaxochem, Ltd., Greenford, Middlesex, England.
Filled in Italy by Squibb S.p.A. Dist. by
E. R. Squibb & Sons, Inc., Princeton, NJ 08540   C5277 / 66805

(B)

Courtesy of Squibb-Marsam Inc.

Reconstitute with _____ mL diluent for a concentration of _____ units/mL. Give _____ mL.

Reconstitute with _____ mL diluent for a concentration of _____ units/mL. Give _____ mL.

Reconstitute with _____ mL diluent for a concentration of _____ units/mL. Give _____ mL.

Reconstitute with _____ mL diluent for a concentration of _____ units/mL. Give _____ mL.

Indicate the concentration you would choose, and explain the rationale for your selection. Select _____ units/mL and give _____ mL. Rationale: _____

How many full doses are available in this vial? _____ dose (s)

Prepare a reconstitution label for the remaining solution

62. Ceftriaxone sodium 750 mg IM daily

Reconstitute with _____ mL diluent for an initial concentration of _____ mg/mL. Give _____ mL.

How many full doses are available in this vial? _____ doses

Roche Laboratories Inc.
Nutley, New Jersey 07110

1 gram   Single-Use Vial
ROCEPHIN®
(ceftriaxone sodium)
For I.M. or I.V. Use
equivalent to 1 gram ceftriaxone

℞ only

For I.M. Administration: Reconstitute
with 2.1 mL 1% Lidocaine Hydrochloride
Injection (USP) or Sterile Water for
Injection (USP). Each 1 mL of solution
contains approximately 350 mg
equivalent of ceftriaxone.
For I.V. Administration: Reconstitute
with 9.6 mL of an I.V. diluent specified
in the accompanying package insert.
Each 1 mL of solution contains
approximately 100 mg equivalent
of ceftriaxone. **Withdraw entire** contents
and dilute to the desired concentration
with the appropriate I.V. diluent.
USUAL DOSAGE: See package insert.
Storage Prior to Reconstitution:
Store powder at room temperature
77°F (25°C) or below.
Protect From Light.
Storage After Reconstitution:
See package insert.
        LOT   EXP.

27888829

(01) 103 0004 1964 04 8

Courtesy of Roche Laboratories Inc.

Understanding Drug Labels

63. Examine this lidocaine drug label, and answer the following questions.

What percent of lidocaine is present in this vial? _____ %

How much pure drug in g per 100 mL is present in this vial? _____ g/100 mL

How much pure drug in mg per 100 mL is present in this vial? _____ mg/100 mL

How much pure drug in mg per mL is present in this vial? _____ mg/mL

64. Examine the potassium chloride label, and answer the following questions.

Is this a single-dose or multi-dose vial? _____

What is the name of the drug manufacturer? _____

What is the lot number of this drug? _____

What is the total volume of medication in this vial? _____

65. Examine the propranolol label, and answer the following question.

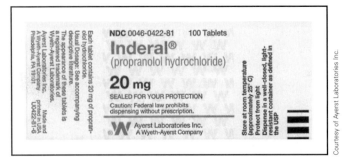

What is the dose strength of the tablets in this vial? _____

How should this drug be stored? _____

Is this a prescription or over-the-counter drug? _____

How many tablets are in this vial? _____

Case Studies

66. A 13-year-old, M/N domestic short hair (DSH) named Buttons has a bite wound in the right side of the mandible. The owner does not know when the cat got the bite wound because Buttons is an outside cat. He has not been eating or drinking the past few days, is lethargic, and on physical examination (PE) has temperature (T) = 103.5°F, heart rate (HR) = 180 bpm, and respiration rate (RR) = 45 breaths/min. Other than the mandibular wound, he is healthy. He is currently on vaccinations. The veterinarian decides to sedate Buttons so that the wound can be clipped, cleaned, and debrided. You must calculate the dose of injectable anesthetic for Buttons, who weighs 12.5 lb. The dosage of ketamine anesthetic is 22 mg/kg IM. The concentration listed on the vial of ketamine is 100 mg/mL. What volume of ketamine should be given to this cat?
    a.   What is his weight in kg?
    b.   What dose of ketamine should he get in mg?
    c.   What dose of ketamine should he get in mL?

67. Eliza, a three-year-old, female spayed (F/S) German Shepherd, was presented for signs of weakness and joint pain of three months' duration. On PE, vital signs were within normal limits (WNL). The hips were palpated because of this breed's predilection for developing hip dysplasia. Pelvic radiographs showed no signs of hip dysplasia. Because Eliza is a young, active dog that spends a lot of time outside, a Lyme titer was drawn. Her blood test returned positive, and she was started on doxycycline antibiotic. The dosage is 5 mg/kg po qd for 30 days. Eliza weighs 100 lb. Doxycycline comes in 50 mg and 100 mg tablets.
    a.   What is Eliza's weight in kg?
    b.   What dose of doxycycline should she get in mg?
    c.   How many 50 mg tablets would she get per dose?
    d.   How many 100 mg tablets would she get per dose?

68. A 13-year-old, M/N Golden Retriever named Chaska (45 kg) has not been able to bear weight on his rear limbs for the past 2½ days. On PE, moderate pain was elicited on palpation of the pelvic joints. Temperature, pulse, and respiration (TPR) were WNL; the rest of the PE was normal. Pelvic radiographs revealed a shallow acetabulum on both right (R) and left (L) sides and narrowing of the joint space. Chaska was diagnosed (dx'd) as having hip dysplasia with degenerative joint disease. The owner was given both surgical and medical options, but because of the dog's age, the owner opted for medical treatment. This dog was put on carprofen 1 mg/lb bid. Before treatment was started, a blood sample was collected to assess liver enzymes. Carprofen comes in 25 mg, 75 mg, and 100 mg tablets. Calculate the amount of medication needed for 14 days of treatment.
    a.   How many tablets do you dispense to this owner?

69. Kyra, a seven-year-old, F/S mixed-breed dog (107 lb), is presented to the clinic with signs of increased vocalization and urine leaking. She does not appear to have PU/PD (increased urination/increased drinking) or having accidents due to lack of training. She usually leaks urine after she has lain down (her bed is damp). A urinalysis (UA) is WNL, as is the PE. Scout radiographs of the urinary bladder are unremarkable. Based on her history and pattern of urine leaking, the veterinarian determines that she might have estrogen-responsive incontinence (she was spayed at an early age). She is prescribed DES (diethylstilbestrol) at a dose of 1 mg po sid for three days, followed by maintenance therapy of 1 mg po per week. DES comes in 1 mg and 5 mg tablets.
    a.   How many mg does this dog receive per dose?
    b.   How many mg does this dog receive for the first three days?
    c.   How many tablets does this dog receive for the first three days?
    d.   How many mg does this dog receive for three weeks of treatment?
    e.   How many tablets does this dog receive for three weeks of treatment?

70. Fenbendazole is an antiparasitic drug used in the treatment of roundworms, hookworms, whipworms, and some species of tapeworms. For dogs the dosage is 50 mg/kg po for three days.
   a.   How many mg would you give a 50 lb dog?
   b.   Fenbendazole comes in a 10% suspension. How many mg/mL is that?
   c.   How many mL would you give this dog per dose?
   d.   How many mL would you give this dog per treatment regimen?

Critical Thinking Questions

71. While performing a dose calculation, a value of 6.0 cc of drug A is calculated to be given to an animal. The calculation is double-checked by a coworker, who finds that the value is actually 0.6 cc of drug A. The statement is made that the original calculation was close and was only off by a decimal point. Is this statement a true reflection of the calculation error made?

72. When someone describes a 1 to 10 dilution, what exactly does this mean?

ADDITIONAL REVIEW MATERIAL

Additional review material can be found on the StudyWARE™ CD included in this text.

BIBLIOGRAPHY

Curran, A. M. (2006). *Dimensional Analysis for Meds* (3rd ed.). Clifton Park, NY: Delmar Cengage Learning.

Pickar, G. D. (2008). *Dosage Calculations* (8th ed.). Clifton Park, NY: Delmar Cengage Learning.

Rice, J. (2002). *Medications and Mathematics for the Nurse* (9th ed.). Clifton Park, NY: Delmar Cengage Learning.

Rice, J. (2006). *Principles of Pharmacology for Medical Assisting* (4th ed.). Clifton Park, NY: Delmar Cengage Learning.

Stumpt, E., Fritz, F., and Bradford, W. (2006). *Mathematics for Veterinary Medical Technicians* (2nd ed.). Durham, NC: Carolina Academic Press.

CHAPTER 7

DRUGS AFFECTING THE NERVOUS SYSTEM

OBJECTIVES

Upon completion of this chapter, the reader should be able to:

- explain the basic anatomy and physiology of the nervous system.
- explain the role of neurotransmitters at a synapse.
- describe the function of the autonomic nervous system (ANS).
- describe the branches of the ANS and how they affect body function.
- differentiate between the types of sympathetic receptors.
- differentiate between the types of parasympathetic receptors.
- describe the types of CNS drugs, including anticonvulsants, tranquilizers/sedatives, analgesics, opioid antagonists, neuroleptanalgesics, anesthetics, CNS stimulants, and euthanasia solutions.
- describe the types of ANS drugs, including cholinergic drugs, anticholinergic drugs, adrenergic drugs, and adrenergic blocking agents.

KEY TERMS

adrenergic blocking agent
adrenergic drug
analgesic
anesthetic
antianxiety drug
anticholinergic drug
anticonvulsant
cholinergic drug
CNS stimulant

euthanasia solution
narcotic antagonist
neuroleptanalgesic
opioid antagonist
parasympatholytic
parasympathomimetic
sedative
sympatholytic
sympathomimetic
tranquilizer

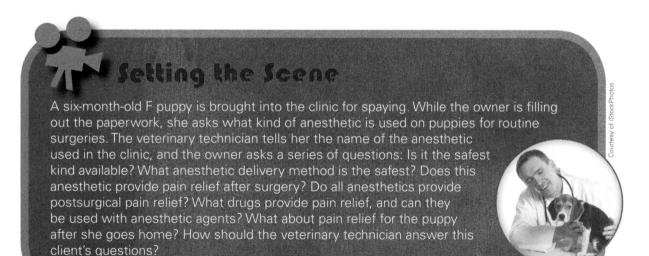

Courtesy of iStockPhotos

Setting the Scene

A six-month-old F puppy is brought into the clinic for spaying. While the owner is filling out the paperwork, she asks what kind of anesthetic is used on puppies for routine surgeries. The veterinary technician tells her the name of the anesthetic used in the clinic, and the owner asks a series of questions: Is it the safest kind available? What anesthetic delivery method is the safest? Does this anesthetic provide pain relief after surgery? Do all anesthetics provide postsurgical pain relief? What drugs provide pain relief, and can they be used with anesthetic agents? What about pain relief for the puppy after she goes home? How should the veterinary technician answer this client's questions?

BASIC NERVOUS SYSTEM ANATOMY AND PHYSIOLOGY

The nervous system is the main regulatory and communication system in the animal's body. The function of the nervous system is to receive stimuli and transmit information to nerve centers to initiate an appropriate response.

The basic unit of the nervous system is the *neuron* (Figure 7-1). There are three types of neurons: sensory (carry impulses toward the central nervous system [CNS]), associative (carry impulses from one neuron to another), and motor (carry impulses away from the CNS). The parts of the neuron include the cell body, one or more dendrites, one axon, and terminal end fibers. The cell body has a nucleus and is responsible for maintaining the life of the neuron. The *dendrites* are rootlike structures that receive impulses and conduct them toward the cell body. The *axon* is a single process that extends away from the cell body and conducts impulses away from the cell body toward muscle cells, glands, or other nerves. The *terminal end fibers* are the branching fibers that lead the impulse away from the axon and toward the synapse.

The space between two neurons or between a neuron and its receptor is the *synapse* (Figure 7-2). A chemical substance called a *neurotransmitter* allows the impulse to move across the synapse from one neuron to another. There are different neurotransmitters for different functions.

Figure 7-3 shows the two divisions of the nervous system: the central and the peripheral. The brain and spinal cord constitute the portion of the nervous

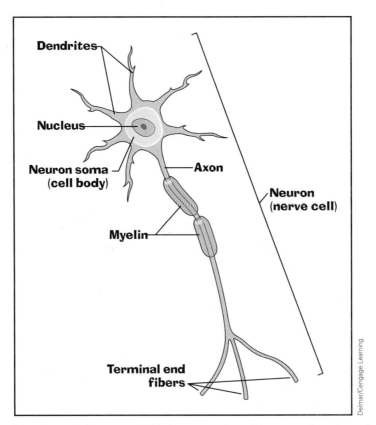

Figure 7-1 Structures of a neuron. Dendrites conduct impulses toward the neuron cell body; axons conduct impulses away from the cell body.

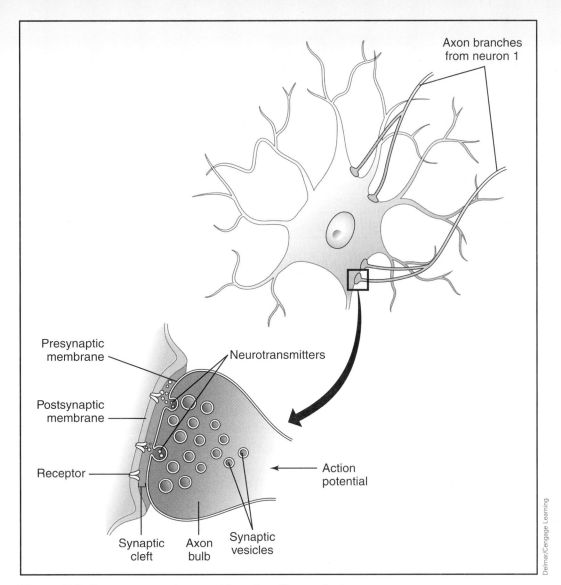

Axon branches from neuron 1

Presynaptic membrane

Neurotransmitters

Postsynaptic membrane

Receptor

Action potential

Synaptic cleft

Axon bulb

Synaptic vesicles

Delmar/Cengage Learning

Figure 7-2 Synapse structure and function. Transmission across a synapse occurs when a neurotransmitter is released by the presynaptic neuron, diffuses across the synaptic cleft, and binds to a receptor in the postsynaptic membrane.

system known as the CNS. The function of the CNS is to interpret information sent by impulses from the peripheral nervous system (PNS) and return instructions through the PNS for appropriate cellular actions. CNS stimulation may increase nerve cell activity or may block nerve cell activity. The CNS is encased in a multilayered connective tissue membrane called the *meninges*. The three layers of the meninges from superficial to deep are the dura mater, the arachnoid membrane, and the pia mater. The CNS is cushioned and nourished by the cerebrospinal fluid (CSF). The CSF is clear, colorless fluid produced by special capillaries within the ventricles of the brain.

The PNS consists of cranial nerves, spinal nerves, and the autonomic nervous system (ANS). Twelve *cranial nerves* originate from the undersurface of the brain. They provide a variety of functions depending on the different areas they serve. The *spinal nerves* arise from the spinal cord and are paired. The dorsal roots of spinal nerves carry sensory impulses from the periphery to the

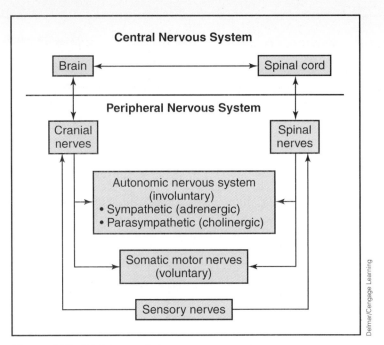

Figure 7-3 Divisions of the nervous system.

spinal cord. The ventral root carries motor impulses from the spinal cord to muscle fibers or glands. The ANS is an involuntary response system that innervates smooth muscle, cardiac muscle, and glands. The two divisions of the ANS are the sympathetic nervous system and the parasympathetic nervous system.

Sympathetic Nervous System or Adrenergic System

The sympathetic nervous system is known as the *fight-or-flight system*. It is responsible for increasing heart rate, respiratory rate, and blood flow to muscles. It also decreases gastrointestinal function and causes pupillary dilation. The sympathetic nervous system is found in the thoracic and lumbar regions between T1 and L3. It has short preganglionic fibers and long postganglionic fibers. *Acetylcholine* is the neurotransmitter released at the preganglionic synapse, and *epinephrine* or *norepinephrine* is the neurotransmitter released at the postganglionic synapse (Figure 7-4).

Sympathetic receptors include the following:

- alpha 1, found in the smooth muscles of blood vessels. Stimulation of alpha-1 receptors causes constriction of the arterioles (except in the gastrointestinal tract), increasing the blood pressure.

- alpha 2, found in the postganglionic sympathetic nerve endings. Stimulation of alpha-2 receptors causes inhibition of norepinephrine release in the brain, resulting in sedation and analgesia. Side effects of alpha-2 receptor stimulation are initial hypertension (due to vasoconstriction) followed by a reflex bradycardia that causes the central alpha-2 action of decreased blood pressure and cardiac output.

- beta 1, located in the heart. These cause increased heart rate, conduction, and contractility.

- beta 2, found mainly in smooth muscles of the lung. Stimulation of these receptors causes bronchodilation and dilation of skeletal blood vessels.

- dopaminergic, located in the renal, mesenteric, and cerebral arteries. Stimulation of dopaminergic receptors causes dilation of the coronary vessels, dilation of the blood vessels of the kidney, and dilation of mesenteric blood vessels.

Parasympathetic Nervous System or Cholinergic System

This branch of the ANS is known as the *homeostatic system*. Parasympathetic nervous system effects are generally opposite to sympathetic nervous system

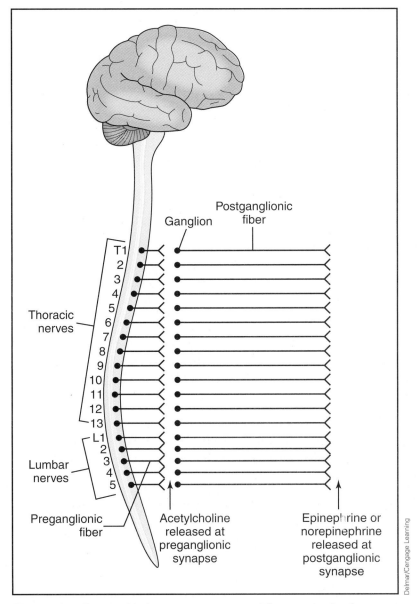

Figure 7-4 Sympathetic nervous system. The sympathetic nervous system, also known as the fight-or-flight system, has acetylcholine released at the preganglionic synapse and epinephrine or norepinephrine released at the postganglionic synapse.

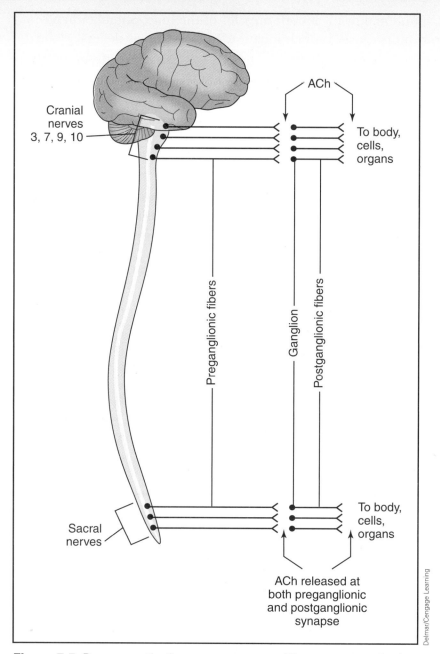

Figure 7-5 Parasympathetic nervous system. The parasympathetic nervous system, also known as the homeostatic system, has acetylcholine released at both the preganglionic and postganglionic synapse.

effects. It is responsible for returning heart rate, respiratory rate, and blood flow to normal levels. It also returns gastrointestinal function to normal and causes the pupils to constrict to normal size. The parasympathetic nervous system is found in the brain stem region and sacral segments. It has long preganglionic fibers and short postganglionic fibers. Acetylcholine (ACh) is released at both the pre- and postganglionic synapses (Figure 7-5).

Parasympathetic receptors include muscarinic receptors, which stimulate smooth muscles and slow the heart rate; and nicotinic receptors, which affect skeletal muscles.

Clinical Que

Because acetylcholine is the only neurotransmitter present in the parasympathetic nervous system, it makes sense that this system is referred to as the cholinergic system.

CENTRAL NERVOUS SYSTEM DRUGS

Central nervous system drugs include anticonvulsants, tranquilizers/sedatives, barbiturates, dissociatives, opioids, opioid antagonists, neuroleptanalgesics, stimulants, and analgesics.

Anticonvulsants

Seizures are periods of altered brain function due to recurrent abnormal electrical impulses; they are characterized by loss of consciousness, increased muscle tone and movement, and altered sensations (Figure 7-6). Seizures occur in animals for a variety of reasons, such as traumatic, idiopathic (unknown), infectious, toxic, and metabolic factors. Seizures result from abnormal electric discharges by the cerebral neurons.

Anticonvulsants are drugs that help prevent seizures. Ongoing seizures, known as *status epilepticus*, are treated as an emergency situation because they can cause elevated body temperatures, hypoxia, or acidosis. Periodic, recurring seizures are treated with long-term preventative therapy using oral drugs. Many types of anticonvulsants work by suppressing the abnormal electric impulses from the seizure focus to other areas of the cerebral cortex, thus preventing the seizure but not eliminating the cause of the seizure. The goal of anticonvulsant therapy is to obtain the greatest degree of control over the seizures without causing severe side effects. All anticonvulsants are classified as CNS depressants and may cause ataxia, drowsiness, and hepatotoxicity (especially phenobarbital and primidone). Anticonvulsants include the following:

Barbiturates

Barbiturates are drugs that are chemical derivatives of barbituric acid and act as CNS depressants. They are classified into groups based on their duration of action: ultra-short-acting, short-acting, and long-acting (Table 7-1).

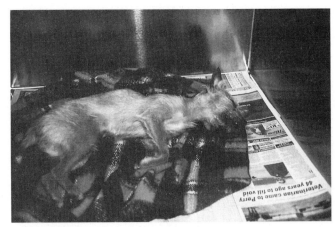

Figure 7-6 Dog suffering from a seizure.

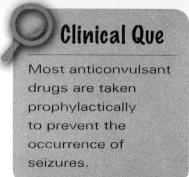

Table 7-1   Barbiturate Classifications (by duration of action)

CLASSIFICATION	DURATION OF ACTION	EXAMPLES
Long-acting	6–8 hours	phenobarbital
Short-acting	1–2 hours	pentobarbital
Ultrashort-acting	10–15 minutes	thiopental methohexital

Phenobarbital

Phenobarbital is a long-acting barbiturate that depresses the motor centers of the cerebral cortex and is the anticonvulsant of choice in dogs and cats. Barbiturates work by impairing chemical transmission of impulses across synapses in the brain stem. Phenobarbital is a C-IV controlled substance that is used orally (tablet and elixir) and parenterally (IV) to control seizures. Examples include Solfoton®, Luminal®, Barbita®, and a variety of generic forms. Induction of liver enzymes (the animal develops tolerance) may increase the rate of phenobarbital and other drug metabolism. An increase in liver enzymes, especially alkaline phosphatase (alk phos) and alanine aminotransferase (ALT), is commonly seen in patients on phenobarbital. Side effects include ataxia (which is usually temporary), drowsiness, liver damage, respiratory depression, PU/PD, and polyphagia. Blood samples should be taken to determine phenobarbital levels and doses should be altered accordingly.

Pentobarbital

Pentobarbital is a short-acting barbiturate that is administered IV to control seizures. Pentobarbital is especially useful when controlling seizures caused by toxins (such as strychnine). It is a C-II controlled substance that lasts about one to two hours. It can also be given orally, but this administration route is less common. Examples include Nembutal®, Somnotol®, and generic products. Side effects include respiratory depression and hypothermia. It is very irritating when given SQ or perivascularly, so these administration routes are avoided. Pentobarbital is also used as an euthanasia solution (covered later in this chapter).

Primidone

Primidone is structurally similar to phenobarbital and is broken down to phenobarbital and phenylethylmalondiamide. It is given orally; examples include Mylepsin®, Mysoline®, Neurosyn®, and generic brands. It can induce liver enzymes that increase its own and other drug metabolism. Side effects include ataxia, PU/PD, and polyphagia.

Benzodiazepines

Benzodiazepines are drugs that potentiate the effects of gamma-aminobutyric acid (GABA), an inhibitory neurotransmitter that stabilizes nerve cell membranes. These drugs also cause muscle relaxation and relieve anxiety.

Diazepam

Diazepam (Valium®) is a benzodiazepine anticonvulsant that is a C-IV controlled substance used intravenously for the treatment of status epilepticus (rapid succession of epileptic seizures). Once the seizures are brought under control with IV diazepam, oral anticonvulsant therapy is initiated. Diazepam can also be given orally as an adjunct to seizure treatment. It has a shorter duration of action (three to four hours) than phenobarbital. Diazepam works by increasing GABA, a substance that inhibits impulse transmission in nerve cells (Figure 7-7). Side effects include CNS excitement (a paradoxical or contradictory response) and weakness.

Clinical Que

When diazepam is given IV, it is given slowly to avoid vascular and cardiac problems.

Clinical Que

Benzodiazepines can often be recognized by the -epam suffix such as in the drugs diazepam and lorazepam.

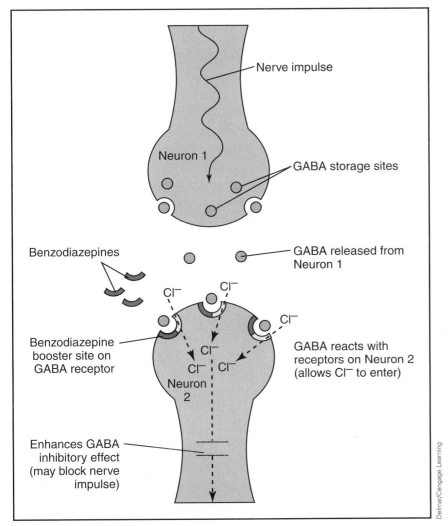

Figure 7-7 GABA is an inhibitory neurotransmitter on nerve cells and GABA receptors are found throughout the CNS. When a nerve impulse causes the release of GABA from storage sites on neuron 1, GABA is released into the synaptic cleft. GABA then reacts with receptors on neuron 2. This reaction allows chlorine to enter the neuron which inhibits further progress of the nerve impulse. Benzodiazepines react with GABA receptors to enhance the inhibitory effects of GABA.

Lorazepam

Lorazepam (Ativan®) is another benzodiazepine anticonvulsant that is a C-IV controlled substance with a longer duration of action than diazepam and can be easier to administer (IN, IM, and buccal) to patients with status epilepticus. Side effects of lorazepam include increased appetite, ataxia, and vocalization.

Clorazepate

Clorazepate is a C-IV controlled substance in the same drug classification as diazepam and lorazepam. It is used orally as an adjunct anticonvulsant (usually used with phenobarbital) and for behavioral phobias. Trade names include Cloraze®Caps, Tranxene®-SD, and GenENE®. Side effects include sedation and ataxia.

Potassium Bromide

Potassium Bromide (KBr) is used as an adjunct to anticonvulsant therapy when seizures cannot be controlled by phenobarbital or primidone alone. Potassium bromide's antiseizure activity is due to its depressant effects on neuron excitability because bromide ions compete with chloride ion transport across cell membranes resulting in membrane hyperpolarization. Membrane hyperpolarization raises the seizure threshold and limits the spread of seizures. It is either sprinkled on the food or squirted in liquid form into the mouth. KBr has a long half-life (in dogs, its half-life is up to 24 days) and a narrow therapeutic range. Because of KBr's long half-life, it takes several days to months to reach a steady-state concentration (when drug concentration reaches a plateau where the peak concentrations and trough (lowest) concentrations are the same after repeated doses of the drug) of KBr in blood. To avoid the time needed to achieve the maximum anticonvulsant effects of this drug, many dosing regimens include an initial oral bolus loading dose to reduce the amount of time needed to attain therapeutic concentrations. The loading dose may be divided over a number of days at which time a maintenance dose is administered daily. Blood serum levels of KBr should be monitored to assure that the concentration is within the therapeutic range. If KBr is used in conjunction with phenobarbital, the phenobarbital dose should be reduced as the KBr concentrations rise and begin to show clinical response to avoid sedation or ataxia in the animal. This chemical must be compounded and is ordered USP grade from chemical supply houses and specialty pharmacies. The powder form is usually mixed with corn syrup (Karo® syrup) to make it more palatable to animals. People are advised to wear gloves while handling KBr because it is toxic to people. A side effect of KBr use is electrolyte imbalance and vomiting.

Phenytoin

The anticonvulsant *phenytoin* (Dilantin®) is used in human medicine, but has undesirable side effects (including rapid blood pressure decreases) that have severely limited its use in animals. Anticonvulsant activity occurs only after accumulation of several doses of phenytoin, and this presumed lack of efficacy has also limited its use.

Clinical Que

Anticonvulsants help prevent seizures but do not eliminate the cause of the seizures.

Add-ons

Treatment of seizures is not always successful for a variety of reasons including age at the time of first seizure (later onset typically has a better response), concentration of GABA in cerebrospinal fluid (low GABA levels in CSF do not respond well to treatment), and resistance to medication. A patient is labeled refractory or is considered to have refractory epilepsy when a single anticonvulsant medication has been used at the high end of the therapeutic blood level and the patient continues to have the same or an increased number of seizures, the patient has developed side effects from the anticonvulsant medication that now limits its use, or the patient has been well controlled for a period of time and recently has developed a significant increase in seizure frequency. A category of drugs called "add-on" drugs is used for animals with refractory seizures. These drugs include levetiracetam, zonisamide, gabapentin, and felbamate. In general, these drugs have fewer side effects than the standard anticonvulsants and are being considered as first-line anticonvulsant treatments.

Levetiracetam

Levetiracetam (Keppra®) is an oral add-on anticonvulsant drug in dogs and cats with refractory epilepsy. Levetiracetam binds to a synaptic vesicle protein (SV2A) that affects neurotransmission, preventing propagations of seizure activity. Levetiracetam is 100 percent bioavailable in the oral form (tablet and solution) and has limited side effects, which include sedation, ataxia, and anorexia. It is not metabolized by the liver making its use valuable in animals with phenobarbital hepatic toxicity. Levetiracetam is also available in an injectable form that can be given as an IV bolus for dogs with status epilepticus.

Zonisamide

Zonisamide (Zonegran®) is an oral add-on anticonvulsant that works by blocking calcium and sodium channels in the brain and facilitating dopamine and serotonin neurotransmission. Zonisamide has a half-life of 15 hours allowing for bid dosing. Intrarectal dosing of zonisamide is used for treating status epilepticus in dogs. Side effects include ataxia and sedation.

Gabapentin

Gabapentin (Neurontin®) is an oral add-on anticonvulsant that works by inhibiting calcium channels resulting in decreased excitatory neurotransmission. Side effects include sedation, ataxia, and the potential for hepatotoxicity. Gabapentin is primarily excreted by the kidneys and should be used with caution in animals with renal insufficiency. It is available in tablet and oral solution form; however, the oral solution contains xylitol and should not be used in dogs due to toxicity seen with ingestion of xylitol in this species.

Felbamate

Felbamate (Felbatol®) is an oral add-on anticonvulsant used in dogs that controls seizures by potentiating GABA-mediated neuronal inhibition and inhibiting neuronal calcium channels. Side effects of felbamate include hepatotoxicity,

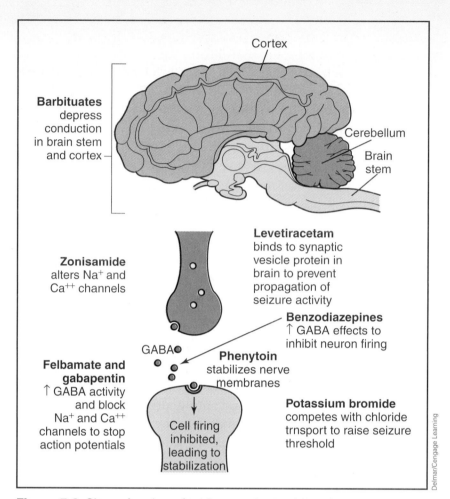

Figure 7-8 Sites of action of various anticonvulsant drugs.

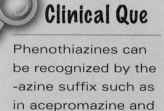

reversible blood dyscrasias (thrombocytopenia, lymphopenia, and leucopenia), and keratoconjunctivitis sicca.

The various types of anticonvulsants and their sites of action are summarized in Figure 7-8.

Calming (Tranquilizers, Sedatives, and Antianxiety Agents)

Tranquilizers are drugs that calm animals and are used to reduce anxiety and aggression in animals. **Sedatives** are drugs that decrease irritability and excitement in animals and are used to quiet excited animals. **Antianxiety drugs** lessen anxiousness but do not make the animal drowsy. Drugs that fall into one or more of these categories are phenothiazine derivatives, butyrophenones, benzodiazepines, and alpha-2 agonists (Table 7-2).

Phenothiazine Derivatives

Phenothiazine derivatives work via an unknown mechanism, but are believed to block dopamine and the alpha-1 receptors found in the smooth muscle cells of peripheral blood vessels. This group of drugs causes sedation and relieves fear and anxiety, but does not produce analgesia. The primary uses of phenothiazines include calming animals for physical examination

Table 7-2   Properties of Sedatives, Tranquilizers, and Antianxiety Drugs

	PHENOTHIAZINES	BUTYROPHENONES	BENZODIAZEPINES	ALPHA-2 AGONISTS
Classification:				
Sedation	X	X		X
Antianxiety	X		X	
Antiemetic	X	X		
Analgesic				X (some of short duration)
Antiarrhythmic	X			
Antihistimine effect	X			
Peripheral vasodilation	X			
Seizure threshold reducer	X			
Muscle relaxant			X	X
Emetic				X

(restraint) or transport and sedation prior to minor procedures or as preanesthetic agents. Phenothiazines depress the chemoreceptor trigger zone in the brain and therefore are also used as antiemetics, usually to prevent motion sickness in animals. Clinically, giving more phenothiazine drug to an animal does not necessarily mean greater effectiveness. The main side effects of this drug group are the development of hypotension, the lowering of the seizure threshold in animals with a seizure history, protrusion of the nictitating membrane or third eyelid (Figure 7-9), and the potential development of paraphimosis (retraction of the prepuce, causing swelling of the penis that prevents it from being retracted) in horses. Phenothiazines can also cause inappropriate aggressive behaviors in animals. This drug group does not cause respiratory depression or pronounced adverse effects on the heart. Examples in this group include *acepromazine* (Promace® and generic), *chlorpromazine* (Thorazine®), *prochlorperazine/isopropamide* (Compazine®), and *promazine* (Sparine®).

Butyrophenones

Butyrophenones are tranquilizers and sedatives that have similar effects as phenothiazines. A drug in this category is *azaperone* (Stresnil® is available in the United Kingdom for use in swine) and is labeled for control of aggressiveness in pigs. Azaperone causes less sedation than the phenothiazine drugs. Side effects include salivation, panting, and shivering.

Delmar/Cengage Learning

Figure 7-9 Protrusion of the third eyelid in a cat treated with acepromazine.

Benzodiazepines

Benzodiazepines are antianxiety drugs; this group includes *diazepam* (Valium®), *zolazepam* (found in Telazol®), *midazolam* (Versed®), and *clonazepam* (Klonopin®). These drugs have anticonvulsant activity, produce muscle relaxation, and reduce anxiousness in the animal while it remains alert. They work by increasing GABA, an inhibitory neurotransmitter in the brain. They are sometimes used as appetite stimulants in cats and in combination with ketamine or tiletamine for short-term anesthesia. These drugs have the benefit of causing minimal cardiovascular and respiratory depression. These drugs are C-IV controlled substances and do not provide any analgesic effects. Side effects include CNS excitement and weakness. The reversal agent for benzodiazepines is flumazenil (Romazicon®).

Alpha-2 Agonists

Alpha-2 agonists are drugs that bind to alpha-2 receptors on neurons that normally release the neurotransmitter norepinephrine. When molecules of this group of drug bind to the alpha-2 receptor, norepinephrine production is decreased. Because norepinephrine maintains alertness, its absence produces sedation. Alpha-2 agonists produce a calming effect, some analgesia and muscle relaxation, and decrease the animal's ability to respond to stimuli. Ruminants are very sensitive to this group of drugs and need to be dosed with caution. Side effects include bradycardia and heart block; so premedication with anticholinergics like atropine is recommended. The following are examples of alpha-2 agonists.

Xylazine

Xylazine (Rompun®, AnaSed®, and Gemini®) may be combined with ketamine for short-term procedures such as castration in horses and cats and surgical wound repair. It also causes vomiting in cats and some dogs; so it is used as an emetic especially in cats. It produces analgesia and is used in horses for

Clinical Que

Alpha-2 agonists produce cardiovascular side effects.

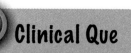

Clinical Que

Because xylazine causes vomiting, it is contraindicated in animals with bloat, gastric torsion, or other conditions in which vomiting is dangerous.

Each mL contains / chaque mL contient:
Xylazine: 20 mg
(Present as xylazine hydrochloride /
présent comme chlorhydrate de xylazine)
Veterinary use only /
usage vétérinaire seulement

Bayer, Bayer Cross, and Rompun are
registered trademarks of Bayer AG, used
under licence.
Bayer, la Croix Bayer et Rompun sont des
marques déposées de Bayer AG utilisées
sous licence.

08937020, R.0   08937551   12784

Each mL contains / chaque mL contient:
Xylazine: 100 mg
(Present as xylazine hydrochloride /
présent comme chlorhydrate de xylazine)
Veterinary Use Only /
usage vétérinaire seulement

Bayer, Bayer Cross, and Rompun are
registered trademarks of Bayer AG, used
under licence.
Bayer, la Croix Bayer et Rompun sont des
marques déposées de Bayer AG utilisées
sous licence.

08939929, R.0   08939953   12794

Figure 7-10 Xylazine (Rompun®) is available in both small and large animal concentrations. Technicians need to take care when calculating the dose of xylazine and when choosing the bottle of xylazine to use when preparing the animal's dose (Courtesy of Bayer Corporation).

pain associated with colic. (Note that horses may appear sedate, but can still respond to stimuli by kicking!) It is used extra-label in cattle for surgical procedures such as cesarean (C-) sections. Small-animal (20 mg/mL) and large-animal (100 mg/mL) concentrations are available; so care must be taken when determining drug doses so that the proper concentration is used (Figure 7-10). A main side effect is profound cardiovascular effects, especially when given IV or without an accompanying anticholinergic drug. Respiratory side effects include hypoventilation and cyanosis. Xylazine can also slow insulin secretion by the pancreas, resulting in transient hyperglycemia. This should be kept in mind when interpreting blood sample results of animals given xylazine.

Reversal agents for xylazine include *yohimbine* (Yobine®, Antagonil®) and *tolazoline* (Tolazine®), which are alpha-adrenergic blocking agents. Yohimbine is used to reverse xylazine administration or overdose. Yohimbine also reverses the toxic effects of the antiparasitic drug amitraz, an alpha-2 agonist, and may be used prophylactically before amitraz dips. It is available in an injectable form for IV and IM use. Side effects include transient CNS stimulation, muscle tremors, and salivation. Tolazoline is approved for reversal of effects associated with xylazine in horses; it is not approved for use in food-producing animals. Side effects of tolazoline include transient tachycardia, peripheral vasodilatation (seen as sweating and injected mucous membranes of the gingival), and piloerection.

Detomidine

Detomidine (Dormosedan®) produces better analgesia than xylazine. It is labeled for use in horses for sedation, but the sedated animal can still respond to stimuli. Detomidine is also used for treating horses for pain associated with colic and when trimming foot abscesses in horses. It also causes severe cardiovascular and respiratory side effects.

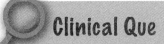

Clinical Que

When using xylazine, it is important to know the concentration per milliliter (100 mg/mL versus 20 mg/mL), to prevent overdosing or underdosing a patient: 10 percent solution is 100 mg/mL, and 2 percent solution is 20 mg/mL.

Medetomidine

Medetomidine (Domitor®) is labeled for use in dogs older than 12 weeks as a sedative and analgesic. It is used for minor surgical procedures, and as a restraint for diagnostic and dental procedures. Muscle twitching can be seen in dogs sedated with medetomidine. Blood pressure is initially increased, followed by decreased heart rate due to a vagal response in the animal; therefore, anticholinergic use is recommended with this drug. Side effects include bradycardia and decreased respiration rates. Medetomidine has a reversal agent called atipamezole (Antisedan®). *Atipamezole* is an alpha-2 antagonist. Dosing charts for both atipamezole and medetomidine are provided in the package insert. The volumes given per animal weight are the same for both drugs. Care must be taken to accurately read the IM versus IV doses in these charts.

Dexmedetomidine

Dexmedetomidine (Dexdomitor®) is labeled for use in dogs older than six months of age and cats older than five months of age as a preanesthetic agent, for sedation, and as an analgesic. It is given IM and provides 40 minutes of sedation in dogs and 60 minutes of sedation in cats. It is contraindicated in animals with cardiac, liver, or kidney disease, seizures, or very young or very old animals. Dexmedetomidine is dosed on body surface area and not weight. Dexmedetomidine can be reversed with atipamezole (Antisedan®).

Pain Relieving (Analgesics)

Pain is an unpleasant sensory event of the peripheral and central nervous systems and an emotional and cognitive experience associated with actual or potential tissue damage. Pain involves a complex series of physical and chemical responses whose terminology and basic neurophysiology need to be understood so that it can be identified, treated, and prevented in animals.

Pain is classified as physiologic (the body's protective mechanism to avoid tissue injury) or pathologic (that which arises from tissue injury and inflammation or from damage to a portion of the nervous system). Pathologic pain can be further divided into categories such as nociceptive (peripheral tissue injury) or neuropathic (damage to the peripheral nerves or CNS) (Table 7-3 and Figure 7-11). Pain can also be defined as acute (arising from a sudden stimulus such as surgery) or chronic (persisting beyond the time normally associated with tissue injury).

Nociceptive pain is transient pain in response to a noxious stimulus. *Nociception* is the processing of a noxious stimulus, such as mechanical (like cuts and bruises), chemical (like acid), and thermal changes (like heat), resulting in the perception of pain by the brain. The nociceptive system originates in peripheral tissues, spans the spinal cord, traverses the brain stem and thalamus, and terminates in the cerebral cortex where pain sensation is perceived. Peripheral tissues are innervated by nociceptors, which are highly specialized primary sensory neurons, which contain specific receptors or ion channels at their peripheral terminals. Activation of these receptors or ion channels by noxious stimuli generates an action potential that is then relayed to the brain

Table 7-3   Nociceptive Pain vs. Neuropathic Pain

NOCICEPTIVE PAIN		NEUROPATHIC PAIN	
Normal processing of stimuli that damages normal tissues or has the potential to do so		Abnormal processing of sensory input by the peripheral or central nervous system	
SOMATIC PAIN	VISCERAL PAIN	CENTRALLY GENERATED PAIN	PERIPHERALLY GENERATED PAIN
• Aching or throbbing in quality • Well localized • Arises from bone, joint, muscle, skin, or connective tissue	• Arises from visceral organs • May be aching in quality and fairly well localized (tumor involvement) • May cause intermittent cramping and poorly localized (obstruction to hollow organs)	• May be due to either peripheral or central nervous system injury • May be associated with dysregulation of the autonomic nervous system	• Pain is felt along the distribution of peripheral nerves (polyneuropathies such as diabetic neuropathy) • Pain is associated with a known peripheral nerve injury (mononeuropathies such as nerve root compression)

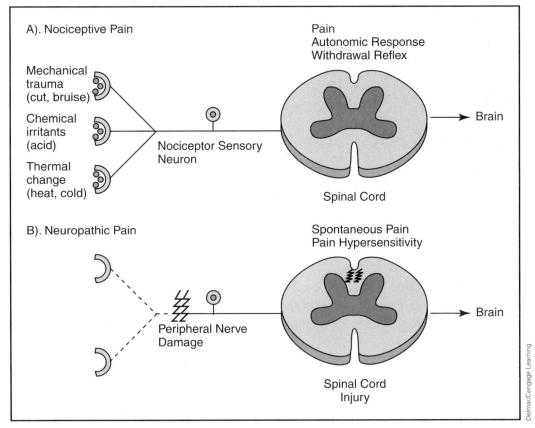

Figure 7-11 Nociceptive and neuropathic pain pathways.

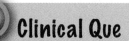

Clinical Que

Nociceptors are pain receptors from somatic sources (found primarily in the skin and joints) and visceral sources (walls of organs).

Clinical Que

Some signs of pain in animals include changes in personality or attitude; abnormal vocalization; licking, biting, scratching, or shaking of a painful area; changes in the appearance of the haircoat; changes in posture or ambulation; changes in activity level; changes in appetite; changes in facial expression; excessive sweating or salivation; oculonasal discharge; teeth grinding; and changes in bowel movements or urination.

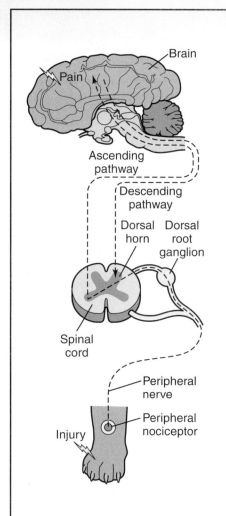

Perception: Brain response to nociceptive signals. Inhibited by general anesthesia, opiods, and alpha-2 agonists.

Modulation: Secondary neurons recieve excitatory and inhibitory inputs from neurotransmitters released by peripheral neurons and neuropathways. Action potential is transmitted further or inhibited by local anesthetics, alpha-2 agonists, opiods, NSAIDs, tricyclic antidepressants, serotonin selective reuptake inhibitors and dissociative.

Transmission: Action potential is transmitted along the length of the neuron to cells in the spinal cord and brain. Inhibited by local anesthetics and alpha-2 agonists.

Transduction: Chemical, mechanical, or thermal inputs act on nociceptors and are translated into electrical activity at sensory nerve endings. Inhibited by local anesthetics, opioids, and NSAIDs.

Delmar/Cengage Learning

Figure 7-12 Nociceptive pain pathway.

for pain perception. The nociception pain pathway consists of the four processes of transduction, transmission, modulation, and perception, which are described as follows and in Figure 7-12:

- Transduction is the conversion of a noxious stimulus (mechanical, chemical, or thermal) into electrical energy by an afferent nerve ending (afferent nerves carry information toward the CNS). Basically, transduction refers to the conversion of a stimulus into an action potential.

- Transmission is the propagation through the PNS to the spinal cord and then to the brain. Basically, the action potential produced during transduction is transmitted along the length of the neurons to cells in the spinal cord and brain. Transmission occurs at three junctions: first, between nociceptor and dorsal horn of the spinal cord; second, between the spinal cord and the thalamus and brain stem; and third, from the thalamus into the cerebral cortex. Nerve fibers involved in transmission include A-delta (fast) fibers responsible for the initial sharp pain, C (slow) fibers that cause the secondary dull, throbbing pain, and A-beta (tactile) fibers, which have a lower threshold of stimulation.

- Modulation is the alteration of nociceptive information by endogenous mechanisms. The modulation of pain involves changing transmission of pain impulses in the spinal cord. After transmission, various neurotransmitters are released from the afferent fiber into the synapse, which facilitates and/or inhibits the further transmission of nociceptive impulses. Some neurotransmitters are excitatory, such as glutamate and substance P, while others are inhibitory, such as opioid and serotonin. Basically modulation is what the CNS does with the information (pain signals).

- Perception is the brain response to nociceptive signals that are projected from neurons to the brain. Once the impulse reaches the brain stem, thalamus, and cortex, the animal becomes aware of pain. Perception is the emotional and physical experience of pain. Basically perception is what the brain does with the information.

Neuropathic pain arises from an abnormality in processing sensations by the CNS or PNS. Examples of neuropathic pain include phantom pain following limb amputation, pain from cancer chemotherapy, and pain associated with diabetes mellitus. Neuropathic pain typically produces shooting pain, burning pain, numbness, and tingling. Peripheral nerve damage may lead to abnormal nerve regeneration leading to the formation of a neuroma. Neuromas can generate spontaneous discharge, and these abnormal charges are thought to induce pain.

Analgesics are prescribed for pain relief and are administered in a variety of ways. Preemptive analgesia is the administration of analgesics prior to a painful stimulus to decrease the anticipated poststimulus pain. Administering analgesics prior to inducing a painful stimulus is more effective than giving the same drug after induction of the stimulus. Once pain is established, it becomes more difficult to control. Multimodal analgesia is the use of analgesic drugs from multiple classes. By using drugs from more than one analgesic class, the chance of successfully preventing and treating pain is increased. Multimodal analgesia use also blocks the painful impulse from a variety of points in the pathway to achieve a synergistic effect and potentially lowering doses of individual drugs and the risks of their potential side effects. Analgesics are also categorized as either nonnarcotic or narcotic. The choice of analgesic prescribed depends on the severity of pain. Mild to moderate pain of skeletal muscles and joints is frequently relieved with the use of nonnarcotic analgesics. Nonnarcotic analgesics are not addictive and are less potent than narcotic analgesics. Nonnarcotic analgesics act on PNS receptor sites, whereas narcotic analgesics act mostly on the CNS. Nonnarcotic analgesics are covered in Chapter 16 on anti-inflammatory and pain-reducing drugs.

Narcotic analgesics are usually used for moderate to severe pain in smooth muscles, organs, and bones. Narcosis is a state of disorientation, and narcotics are named after the reversible state of drug-induced CNS depression they produce. *Narcotic* refers to opiate (natural from opium poppy seeds) or opioidlike (synthetic) products that were required to become prescription drugs in 1914 (under the Harrison Narcotic Act of 1914).

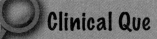

Clinical Que

Pain from a stimulus originates via two mechanisms: (1) Pain signals, which are transduced in peripheral pain receptors, and (2) inflammation, modulated by leukotrienes, nerve growth factor, bradykinin, and other cytokines, as well as prostaglandins, histamine, and hydrogen ions and neurosignaling chemicals such as norepinehrine and neuropeptides.

Clinical Que

It can be difficult to differentiate pain from anxiety as physical signs such as excessive panting, vocalization, and increased heart rate and respiratory rate can be seen with both conditions.

Opioids

Opioids produce analgesia and sedation and relieve anxiety. They do not produce anesthesia; animals still respond to sound and sensation when taking opioids. Side effects of opioids may include respiratory depression and excitement if given rapidly. They affect the regulatory centers in the brain for body temperature control and may cause panting in animals. They cross the placenta very slowly and are used in C-sections. They have central parasympathetic effects, producing salivation, defecation, and vomiting associated with the GI tract and slowed heart rate and hypotension associated with the cardiovascular system. Opioids are used as preanesthetics and postanesthetics for analgesia and sedation, in combination with other drugs for surgical procedures, for restraint, and as antitussives and antidiarrheals. Cats and horses are extremely sensitive to opioids.

The mechanism of action of opioids involves complex interactions of different types of receptors. Opioids produce their effect by the action of opioid receptors, which are located in high concentrations in the nervous tissue but are also found in the gastrointestinal tract, urinary tract, and smooth muscle. Opioids mimic the action of endogenous opioids, which are peptides produced in the nervous and endocrine systems that act on different opioid receptors. Three opioid receptors have been identified in animals: mu (μ), kappa (κ), and delta (δ). The mu, kappa, and delta receptors may be distributed to different parts of the nervous system in animals with this distribution varying among species. Mu receptors are found in the pain areas of the brain and spinal cord. They cause strong analgesia, minimal to mild sedation, euphoria, respiratory depression, bradycardia, and physical dependence. If a drug affects the mu receptor, it is classified as a controlled substance. Stimulation of the mu receptors produces sedation associated with narcosis. Mu receptors are also found in the respiratory center of the brain, and their stimulation can suppress coughing and produce respiratory depression. Mu-receptor stimulation results in the most significant analgesic effect of any opioid receptor.

Kappa receptors are found in the cerebral cortex and spinal cord. They produce spinal analgesia, minimal sedation, and miosis with limited respiratory depression. Delta receptors are found in the spinal cord and cortex of the brain and may provide some spinal cord analgesia, but are believed to have a minimal effect in animals.

Opioids are described by their action at the various opioid receptors. *Affinity* is the drug's ability to bind with a receptor, and a drug with a high affinity for a receptor will bind readily and strongly to that receptor. Potency of an opioid drug is related to its affinity. An opioid may be very potent (have strong affinity) yet have no effect when bound to that receptor. *Activity* of a drug is the ability of that drug to cause action in or on a cell in which the receptor is found. The opioids are categorized based on their affinity and activity into full (or pure) agonists, partial agonists, full antagonists, and partial antagonists (Figure 7-13). Agonists are drugs that bind to a receptor and produce an effect. Full (or pure) agonists have a strong affinity for the receptor and bind tightly with the receptor to produce a potent effect. Morphine

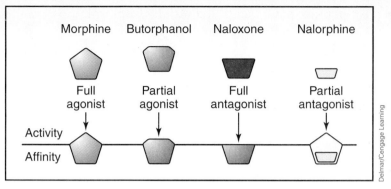

Figure 7-13 Opioids are categorized as full agonists, partial agonists, full antagonists, and partial antagonists.

is a strong agonist for mu receptors. Partial agonists have an affinity for some receptors (but not others) and have potent activity for those they bind to. Partial agonists produce some effect when they bind to a receptor, thus producing a milder effect. Butorphanol is a partial mu agonist. Antagonists do not produce an effect when bound to a receptor. Full antagonists have a strong affinity for but no activity at receptors and are used as reversal agents. Naloxone is an example of a mu antagonist. Partial antagonists have some affinity for receptors and are used as reversal agents. Nalorphine is an example of a partial mu antagonist. Agonist–antagonist drugs have some agonist activity at one type of opioid receptor and some antagonistic activity at another opioid receptor. Buprenorphine is a mu-receptor agonist and kappa-receptor antagonist.

Morphine is the opioid to which all others in this category are compared, with regard to activity. The following are examples of opioids.

Opium

Opium is a naturally occurring opioid that is known as paregoric when it is the camphorated tincture of opium. It is a C-III controlled substance and is used as an antidiarrheal in calves and foals.

Morphine Sulfate

Morphine sulfate (Duramorph® and Astramorph® PF are injectables; Roxanol®, MS Contin®, and generic tablets are oral forms) is a naturally occurring opium derivative that affects mu and kappa receptors and is a C-II controlled substance. It is used to treat severe pain, as a preanesthetic, and as an anesthetic. The dosage used for dogs can cause mania in cats, so a lower dosage rate is used in cats. Side effects include gastrointestinal stimulation (vomiting) and severe respiratory depression.

Hydromorphone

Hydromorphone (Dilaudid®) is a semisynthetic opioid that is five to seven times more potent than morphine. It is a mu agonist and a C-II controlled substance. It is used IV, IM, or SQ preoperatively because it produces more sedation and causes less vomiting in animals than morphine. Hydromorphone is also used to offset moderate to severe postoperative pain. Pain relief with hydromorphone typically lasts for four hours. Side effects include respiratory depression and bradycardia.

Oxymorphone

Oxymorphone (Numorphan®) is a semisynthetic opioid that is ten times more potent than morphine. It is a mu agonist and a C-II controlled substance. It is used IM, IV, and SQ for sedation, as a restraining agent for diagnostic procedures, analgesic, and preanesthetic agent mainly in dogs and occasionally cats. It is also used in horses as an analgesic but may cause CNS excitement in them. Oxymorphone may be used alone or in combination with neuroleptic agents or barbiturates. Oxymorphone should not be mixed in the same syringe with barbiturates as this causes precipitates to form.

Butorphanol

Butorphanol (Torbugesic®, Torbutrol®) is a synthetic opioid with kappa- and mu-receptor activity. It is a C-IV controlled substance that provides two to five times more analgesia than morphine. It is a potent antitussive and is labeled as an antitussive agent in dogs. It is also used as an analgesic and preanesthetic in dogs, cats, and horses. Side effects include sedation and anorexia; however, anxiety and excitation have been noted in some animals. It produces less respiratory depression than other opioids. Butorphanol is available in two concentrations: 0.2 mg/mL (Torbutrol®, Torbugesic®-SA), which is approved for use as an antitussive and analgesic in dogs, and 10 mg/mL (Torbugesic®), which is approved as an analgesic for horses.

Hydrocodone

Hydrocodone (Hycodan®, Tussigon®) is a synthetic opioid and C-III controlled substance. It is used primarily as an antitussive in dogs and it is a more effective antitussive than codeine. Hydrocodone is also used for treating some behavior problems in dogs and cats such as compulsive licking. Side effects include sedation and vomiting.

Fentanyl

Fentanyl (transdermal fentanyl patches are marketed under the name Duragesic®) is a synthetic opioid and C-II controlled substance. It is about 200 times more potent than morphine and had been used IV, IM, and SQ as an analgesic/tranquilizer for minor surgical and dental procedures, and for chemical restraint in dogs, as a combination fentanyl/droperidol product under the trade name Innovar-Vet®. This product is no longer available. Fentanyl is now available in a transdermal patch delivery system, as a tablet, or as an injectable drug. Transdermal fentanyl patches must be applied with gloves. The fur is clipped, and the skin is cleaned and dried prior to application of the patch over the dorsal neck area. The animal should not be allowed to lick or to eat the patch. Patches should not be cut or torn as this may allow fentanyl to pass into the skin too quickly. Exposure of the patch to heat (such as a heating blanket) will enhance its absorption. Side effects include defecation, respiratory depression, and pain after injection. Most side effects are associated with higher doses. It is not approved for use in food-producing animals.

Alfentanil

Alfentanil (Alfenta®) is a synthetic opioid similar to fentanyl, but is less potent and has a shorter half-life. Its short duration of action makes it ideal for infusion that may reduce the use of other anesthetic concentrations (especially in cats) or as an induction agent. It is a full mu agonist and C-II controlled substance. It is metabolized in the liver, and its dosage may need to be adjusted in animals with liver disease.

Etorphine

Etorphine (M-99®) is a synthetic opioid and C-II controlled substance. It has analgesic effects 1000 times more potent than morphine and is used in zoo and exotic animals for immobilization. Diprenorphine (Diprenorfin®) is its antagonist if people accidentally inject themselves with this potent—and potentially fatal—drug.

Buprenorphine

Buprenorphine (Buprenex®, Subutex®) is an opioid agent that provides long-term analgesia (8 to 10 hours) and is used postsurgically in veterinary medicine. It is considered 30 times more potent than morphine and has a high affinity for mu receptors. Side effects are rare but include respiratory depression. In the United States, it is available only in low-concentration doses and injectable form. In October 2002, the DEA rescheduled buprenorphine from a C-V to a C-III narcotic under the Controlled Substances Act.

Pentazocine

Pentazocine (Talwin®, Talwin®-V) is a synthetic opioid and C-IV controlled substance. It acts as an antagonist at the mu receptor. It has a short duration of action that has limited its use in animals. It is used for analgesia in horses and dogs.

Methadone

Methadone (Dolophine®) is a synthetic opioid and C-II controlled substance. It is used for treatment of colic in horses. Side effects include respiratory depression and sedation.

Codeine

Codeine (generic) is a synthetic opioid and C-III controlled substance. Its analgesic effects are less than morphine. It is used as an antitussive in dogs. It may be combined with acetaminophen for pain relief (especially in humans). Side effects are rare at low to moderate dosages; sedation is seen with higher dosages.

Tramadol

Tramadol (Ultram®) is a centrally acting opioid agonist that has primarily mu-receptor activity and also inhibits the reuptake of serotonin and norepinephrine. It is not a controlled substance (it demonstrates little respiratory depression or abuse potential) and is used as an oral analgesic and antitussive in dogs and cats. Tramadol is also used to control chronic cancer pain, and its side effects include sedation and gastrointestinal signs such as anorexia and vomiting.

Diphenoxylate, Loperamide, and Apomorphine

Diphenoxylate (Lomotil®) is a synthetic opioid and C-V controlled substance. It is combined with atropine and used as an antidiarrheal. *Loperamide* (Immodium®) is a synthetic opioid that is sold over the counter as an antidiarrheal for dogs weighing more than 10 kg. *Apomorphine* (generic) is a synthetic opioid that stimulates the chemoreceptor trigger zone in the brain stem to induce vomiting. These drugs are covered in Chapter 11 on GI drugs.

Opioid Blocking (Opioid Antagonist)

Opioid antagonists, also known as **narcotic antagonists**, block the binding of opioids to their receptors. The opioid antagonist has a higher affinity for the opioid receptor site than the narcotic does. These drugs both displace bound molecules and prevent binding of new molecules. Opioid antagonists are used to treat the respiratory and CNS depression of opioid use. These drugs are given IV, and effects are rapid. Examples include the following:

- *naloxone* (Narcan®, Naloxone® Injection), an opioid antagonist given IV, IM, or SQ that has a high affinity for mu receptors. It does not reverse the analgesic effects of the original drug. Naloxone is used to reverse respiratory depression following narcotic overdose with oxymorphone and morphine. It is relatively free of side effects. The duration of action of naloxone may be shorter than that of the narcotic being reversed; so the patient should be closely monitored, in case additional doses and/or ventilatory support are needed.

- *naltrexone* (Trexan®), an opioid antagonist given SQ or po. It is more often used for behavior disorders and excessive licking of pruritic dermatitis than it is as a reversal agent. It has rare side effects.

Pain Relieving and Anxiety Calming (Neuroleptanalgesics)

Neuroleptanalgesics are a combination of an opioid and a tranquilizer or sedative. These drugs cause CNS depression and analgesia and may or may not produce unconsciousness. The opioid antagonists can reverse the opioid portion of a neuroleptanalgesic. Examples include fentanyl and droperidol (Innovar-Vet®), formerly the only commercially available neuroleptanalgesic. It is no longer available as a veterinary or human (Innovar®) product. Combinations prepared by veterinarians include *xylazine and butorphanol, acepromazine and morphine*, and *acepromazine and oxymorphone*. Side effects include panting, flatulence, bradycardia, and increased sensitivity to sound.

No Pain (Anesthetics)

Anesthesia means without sensation. **Anesthetics** are drugs that interfere with the conduction of nerve impulses and are used to produce loss of sensation, muscle relaxation, and/or loss of consciousness. General anesthetics affect the CNS and produce loss of sensation with partial or complete loss of consciousness; local anesthetics block nerve transmission in the area of application, causing loss of sensation without loss of consciousness (Figure 7-14).

Clinical Que

Anesthesia means lacking or without sensation; it does not mean unconsciousness.

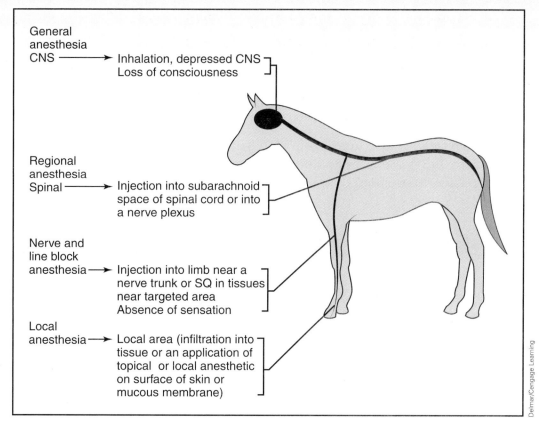

General
anesthesia
CNS ——→ Inhalation, depressed CNS
Loss of consciousness

Regional
anesthesia
Spinal ——→ Injection into subarachnoid
space of spinal cord or into
a nerve plexus

Nerve and
line block
anesthesia ——→ Injection into limb near a
nerve trunk or SQ in tissues
near targeted area
Absence of sensation

Local
anesthesia ——→ Local area (infiltration into
tissue or an application of
topical or local anesthetic
on surface of skin or
mucous membrane)

Delmar/Cengage Learning

Figure 7-14 Sites and mechanisms of action of drugs used for anesthesia.

Local Anesthetics

Local anesthetics block pain at the site of administration or application in the PNS and spinal cord (part of the CNS). They can be used as nerve blocks in equine lameness exams to localize lesions, as nerve blocks in cattle to allow surgical and medical procedures, as an aid to help endotracheal tube placement in cats by preventing laryngeal spasm, and to ease skin irritation. Local anesthetics work by preventing the conduction of nerve impulses in the peripheral nerves. They may be applied topically to mucous membranes and the cornea, by infiltration of a wound or joint, by IV, and around nervous tissue. Local anesthesia usually lasts from 5 to 30 minutes. Their duration of action can be lengthened by use of epinephrine, which causes vasoconstriction and thus prolongs absorption time. Side effects include restlessness and hypotension from the added epinephrine. Local anesthetics can usually be recognized by their *-caine* ending. Types of local anesthetics are listed in Table 7-4, and their use is illustrated in Figure 7-15.

Clinical Que

There are no safe anesthetics, just safe anesthetists.

- *Lidocaine* (Xylocaine®) is available in 0.5 to 2 percent solution for injection and 2 to 4 percent for topical use. It provides immediate onset and if mixed with epinephrine has a two-hour duration.

- *Proparacaine* (Ophthaine®, Ophthetic®) is a rapid-acting topical anesthetic used for ophthalmic procedures. It comes in 0.5 percent solution that is applied in drop doses every 5 to 10 minutes for a maximum of 7 doses. It provides 5 to 10 minutes of anesthesia to the cornea with limited drug penetration to the conjunctiva.

- *Tetracaine* (LiquiChlor®, Neo-Predef® with tetracaine, Pontocaine®) is used topically on the skin and as an ophthalmic or otic solution.

- *Mepivacaine* (Carbocaine®) is available in 1 to 2 percent injectable form. It has an immediate onset of action and lasts 90 to 180 minutes.

- *Bupivacaine* (Marcaine®) is available in 0.25 to 0.5 percent for injection and 0.75 percent for epidural use. It provides immediate onset and lasts for four to six hours.

Table 7-4   Types of Local Anesthesia

TYPE	DESCRIPTION	BENEFIT	USE	DRUG EXAMPLE
Infiltration anesthesia	Small amounts of anesthetic solution are injected into the tissue surrounding the site to be worked on (surgical site, wound repair site, etc.)	Because small amounts of anesthetic are used, there is reduced danger of systemic side effects	• Wound suturing • Wound debriding • Skin biopsies	• lidocaine • mepivacaine • bupivacaine
Topical anesthesia	Anesthetic agent is applied directly onto the surface of the skin or eye; also used to aid in diagnostic procedures and intubation in cats	Systemic absorption is limited from these sites	• Eye examinations • Minor skin irritation • Catheter passing (gel is applied to the catheter tip) • Larynx is sprayed to liquid applied to prevent spasming during intubation in cats	• tetracaine • proparacaine
Nerve block anesthesia	Anesthetic solution is injected along the course of a nerve so that the area it innervates is desensitized	Localization of pain relief Ability to determine the source of pain	• Helps locate areas of injury • Provides local desensitization	• lidocaine • mepivacaine • bupivacaine
Line block anesthesia	A continuous line of local anesthetic is given SQ in the tissues proximal to the targeted area	Allows for larger area of desensitization, but limits systemic effects	• Surgical procedures	• lidocaine • mepivacaine • bupivacaine
Regional (epidural)	Anesthetic agent is injected into a nerve plexus or area of the spinal cord (subarachnoid space)	Provides adequate restraint and may prevent movement (can affect respiratory muscles if given in the cranial parts of the spinal cord)	• Surgeries like C-sections, tail amputations, anal sac removal, and surgery of the rear limb	• lidocaine • mepivacaine • bupivacaine

General Anesthetics

General anesthetics covered in the following sections are either used strictly as anesthetics or as induction agents to produce general anesthesia. Induction agents are used to provide enough sedation for animals to allow inhalant anesthetics to be given. Induction agents are given by injection and tend to have a rapid onset of action. There are two major categories of general anesthetics: injectable and inhalant. Figure 7-16 summarizes agents used in general anesthesia.

Injectable Anesthetics

Injectable anesthetics include the barbiturates, dissociatives, and miscellaneous.

Barbiturates

Barbiturates are CNS depressants that are derived from barbituric acid. Barbiturates are used mainly as anticonvulsants, anesthetics, and euthanasia solutions. They are easy and inexpensive to administer; however, they can cause potent cardiovascular and respiratory depression. They are highly protein-bound drugs, and plasma proteins can serve as reservoirs for these drugs. Acidotic animals (e.g., animals in shock, with diabetes mellitus, and having other systemic diseases) show less binding of barbiturates to plasma proteins. In animals with hypoproteinemia and acidosis, barbiturate doses must be decreased to avoid side effects associated with overdosing.

Barbiturates are classified according to their duration of action: long-acting, short-acting, or ultrashort-acting (refer to Table 7-1). They may also be classified according to the side chain on the barbituric acid: either *oxybarbiturate* (a side chain connected by oxygen) or *thiobarbiturate* (a side chain connected by sulfur). Oxybarbiturates are usually long-acting or short-acting, whereas thiobarbiturates are usually ultrashort-acting. Thiobarbiturates have an ultrashort duration of activity because they are very fat soluble and move out of the CNS rapidly to fat stores within the body. The following are examples of barbiturates.

Phenobarbital (generic) is a long-acting oxybarbiturate that lasts 8 to 12 hours. It is a C-IV controlled substance and is used as an anticonvulsant (discussed in previous sections).

Pentobarbital (generic, Nembutal®) is a short-acting oxybarbiturate that lasts one to two hours. It is a C-II controlled substance that used to be employed as a general anesthetic, but is now utilized mainly as an anticonvulsant and euthanasia solution.

Thiopental (Pentothal®) is an ultrashort-acting thiobarbiturate that lasts 5 to 30 minutes. It is available in a vial as a sterile powder that must be reconstituted with sterile water for use. Thiopental is a C-III controlled substance that is used as an anesthetic induction agent and can only be given IV because it is very alkaline. If thiopental accidentally is injected perivascularly, severe inflammation, tissue swelling, and tissue necrosis may result. Care must be taken when administering thiobarbiturates to thin animals (like sighthounds), because they lack fat stores. Following IV injection,

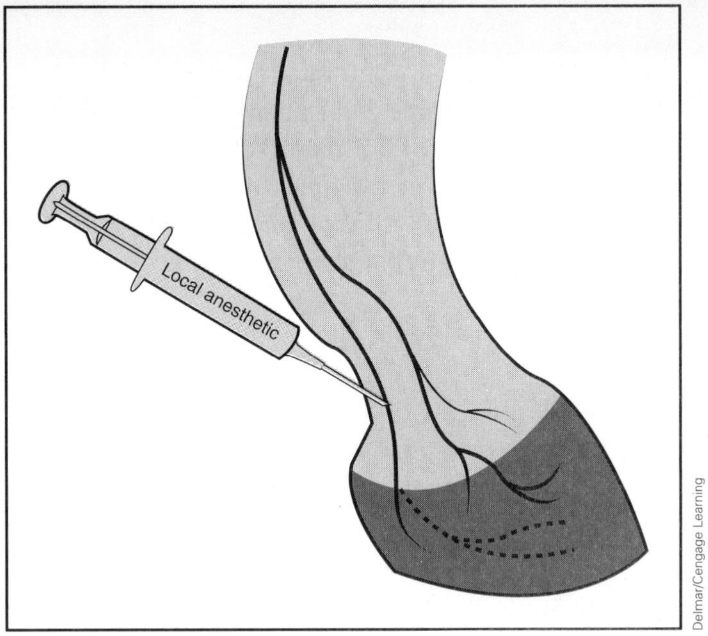

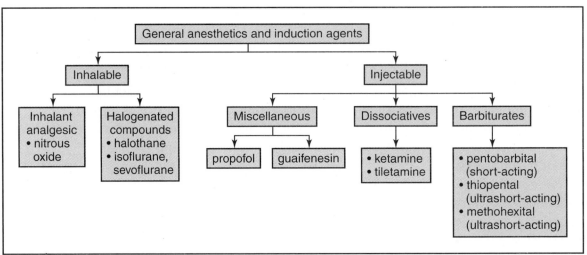

Delmar/Cengage Learning

Figure 7-15 Local anesthesia (peripheral nerve block) is used to provide anesthesia to a surgery site and aid in precise diagnosis, ideal therapy, and accurate prognosis of equine lameness. This figure illustrate a palmar (or plantar) digital nerve block in a horse.

Figure 7-16 General anesthetics and induction agents.

thiobarbiturates rapidly enter the CNS (highly perfused) and then redistribute to fat (poorly perfused). The duration of action of thiobarbiturate is short because of this rapid redistribution of drug out of the CNS. However, the thiobarbiturate does not leave the body: It just leaves the CNS and goes to fat. The animal may not appear anesthetized; so another dose may be given. This second dose may not be able to be redistributed to the fat (as the fat already has drug in it) and thus stays in the CNS. Because the thiobarbiturate has nowhere to go, the amount available in the CNS increases, and brain concentrations of drug may increase to extremely high levels. Likewise, obese animals may accumulate excessive stores of thiobarbiturates, which leave poorly perfused tissue like fat very slowly and result in prolonged anesthesia (see Figure 4-6 A–D). Thiobarbiturates used in obese animals must be dosed based on lean body weight. Thiobarbiturates can also cause apnea if given too fast (resulting in the need for artificial respirations to be given) and CNS excitement if given too slowly. Typically animals are given one-third to one-half of the calculated dose rapidly, and then the rest is given to effect. Side effects include cardiac arrhythmias and transient apnea (absence of breathing).

Methohexital (Brevane®) is a methylated oxybarbiturate similar in structure to the ultrashort-acting thiobarbiturates. It is used in sighthounds because of its rapid redistribution and metabolism by the liver. Its duration of action is short (5 to 10 minutes). It is a C-III controlled substance. Methohexital can cause profound respiratory depression.

Dissociatives

Dissociatives belong to the cyclohexamine family (which includes the street drug PCP or angel dust) and cause muscle rigidity (catalepsy), amnesia, and mild analgesia by altering neurotransmitter activity. Dissociatives are used only for restraint, diagnostic procedures, and minor surgical procedures because dissociatives do not relieve deep pain. They are usually used in combination with other agents for surgical procedures. They cause minor cardiac stimulation, respiratory depression, and exaggerated reflexes. During induction and recovery, tremors, spasticity, and convulsions may occur. Examples of dissociatives include the following.

Ketamine (Ketaset®, Ketalar®, Vetalar®) is a C-III controlled substance usually used in combination with acepromazine, xylazine, and/or diazepam to provide muscle relaxation and deepen anesthesia. It is given IM or IV. It is approved for cats and primates and is used extra-label in other species. Animals keep their eyes open when given ketamine; therefore, ocular lubricants are applied to the eye when using this drug. Spastic muscle jerking and increased salivation may also be seen with this drug. Pain at the injection site is frequently noted with this drug, due to its low pH. Apneustic breathing is seen with ketamine use. Anesthetic depth in animals given ketamine may be difficult to assess because palpebral reflexes are altered.

Tiletamine (found in combination with zolazepam in the product Telazol®) is a C-III controlled substance manufactured as an injectable anesthetic approved for dogs and cats. Tiletamine provides better analgesia than ketamine. In cats tiletamine is metabolized first, whereas in dogs zolazepam is metabolized first. Therefore, in dogs a premedication or postsurgical sedative is usually recommended. Ocular lubricants are also needed for animals given this drug. Pain at the injection site is frequently noted with this drug, due to its low pH.

Ketamine-diazepam mixtures are prepared in clinics and used for IV induction. This mixture is made by combining equal volumes of diazepam (5 mg/mL) and ketamine (100 mg/mL). It may begin to precipitate if stored for

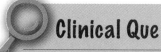

Clinical Que

Unlike other general anesthetics, dissociatives cause CNS stimulation.

Clinical Que

Apneustic breathing is an inspiration followed by a long pause and short expiration. It seems like the animal is holding its breath.

more than one week. It may be mixed in a syringe as needed to prevent problems with precipitation.

Miscellaneous

Miscellaneous injectable anesthetics include the following:

Guaifenesin (Guailaxin®, Gecolate®) is a skeletal muscle relaxant used in combination with an anesthetic drug to induce general anesthesia in the horse. It comes in a 5 percent or 10 percent solution. Generally, large volumes are given to horses to induce general anesthesia, with the addition of small increments to maintain or extend anesthesia.

Propofol (Rapinovet®, PropoFlo®) is a short-acting injectable hypnotic anesthetic agent that produces rapid and smooth induction when given slowly IV. Propofol is used as an induction agent (especially before endotracheal intubation or an inhalant anesthetic) and as an anesthetic for minor procedures such as laceration repair, minor biopsies, and endoscopy. Propofol can also be used to treat status epilepticus because it tends to cause less cardiovascular depression and produce smoother recoveries than pentobarbital. A single bolus of propofol lasts two to five minutes. Propofol is typically administered as 25 percent of the calculated dose every 30 seconds until the desired effect is achieved. If given too rapidly, propofol can produce apnea. Recovery time after a single bolus injection is approximately 20 minutes in dogs and 30 minutes in cats. Propofol provides sedation and minimal analgesic activity; so other agents may be given to the animal as well. It is also used as a treatment for refractory epilepsy as it causes less cardiovascular depression and smoother recoveries than pentobarbital. It crosses the placenta and may be excreted in maternal milk, making its use in pregnant and nursing animals less desirable. Repeated use of propofol in cats may produce Heinz body anemia because cats cannot metabolize it well due to their low concentrations of the enzyme glucoronyl transferase. Drugs containing -OH, -COOH, -NH_2, -HN, and -SH (such as propofol and other phenolic compounds, acetaminophen, morphine, chloramphenicol, and salicylic acid) require conjugation prior to renal excretion. The lower concentrations of the glucoronyl transferase in cats mean prolonged exposure of their red blood cells to the drug and/or its metabolites. Propofol is a white emulsion made of egg lecithin, soybean oil, and glycerol without preservatives; this allows bacteria to grow if the emulsion is contaminated. Unused portions of propofol should be discarded because of the potential for contamination. It should not be used if there is evidence of separation of the phases of the emulsion. Side effects include cardiac arrhythmias and apnea, but these are less severe than with other agents.

Inhalant Agents

Inhalant anesthetics are brought into the body via the lungs and are distributed by the blood into different tissues. The main target of inhalation anesthetics, also called *volatile anesthetics*, is the brain. Inhalant anesthetics are volatile liquids that are purchased in a liquid form (they are liquid at room temperature) and are administered by inhalation in combination with air or oxygen. Advantages of inhalant anesthetics include the ability to alter the depth of anesthesia, because

Table 7-5   Methods of Administering Anesthetics by Inhalation

METHOD	DESCRIPTION	DESCRIPTION/EXAMPLE
Open-drop	Liquid anesthetic is dropped onto a cloth and extended over the animal's nose and mouth	• This method is not used anymore because there is lack of control over the amount of anesthetic delivered, and there is no respiratory assistance. It is also not currently used because these agents (such as ether) are flammable.
Semi-closed	Anesthetic is provided through a mask connected to a reservoir (usually a gas anesthetic machine)	• Need to use a rapid-acting anesthetic like halothane or isoflurane and related anesthetics • Provides greater control of anesthesia delivered to the patient as compared to the open-drop method • Exhaled gases still leak into the environment
Closed	Anesthetic is delivered by anesthesia machine after the liquid has vaporized to the inhalant (gas) form	• Animal is intubated with an endotracheal tube and inhalant anesthetic is delivered directly to the respiratory system • Examples include isoflurance and related anesthetics; halothane

anesthetic continually enters and leaves the body; the constant delivery of oxygen; and the emergency access for ventilation in intubated animals. Methods of administering anesthetics by inhalation are summarized in Table 7-5.

How do inhalant anesthetics get into the blood? The liquid form of the drug is vaporized (converted to gas phase) through a vaporizer as oxygen passes over it in the anesthetic machine. The gas travels in the respiratory system until it gets to the alveoli of the lung, where it diffuses across the alveolar membrane (gas particles move from an area of high concentration in the alveoli to an area of low concentration in the blood capillary). The concentration of gas is high in the alveoli and low in the blood when anesthesia is induced; therefore, there is rapid diffusion of anesthetic into the blood. The anesthetic goes from the blood to the areas of the body that are well perfused (like the brain). Diffusion occurs in the opposite direction when the vaporizer is turned off after a procedure. The concentration of gas in the alveoli goes down when the vaporizer is turned off, although the concentration of anesthetic is high in the blood; therefore, there is diffusion of particles from the blood into the alveoli. The gas is expired from the alveoli. If the animal is provided with pure oxygen when turning off the vaporizer, this diffusion will happen more rapidly because of the greater difference in concentration.

Inhalant anesthetics are compared on a value called *minimum alveolar concentration* (MAC). MAC is the lowest concentration of an anesthetic that produces no response to painful stimuli in 50 percent of patients. It is often referred to as the "strength of anesthetic." Inhalant anesthetics with a lower MAC are more potent, and those with a higher MAC are less potent. Halothane has a lower

 **Clinical Que**

Volatile means that the substance evaporates quickly. Anesthetic agents that have a high vapor pressure will evaporate easily.

Clinical Que

A single vaporizer should not be used for multiple drugs, even if those drugs have similar vapor pressures. Vaporizers should be temperature-, flow-, and back-pressure-compensated.

Clinical Que

↓MAC = more potent anesthetic = less anesthetic needed to produce the effect.

↑MAC = less potent anesthetic = more anesthetic needed to produce the effect.

Clinical Que

The higher the blood-to-gas solubility of an inhalant anesthetic, the longer the induction and recovery times.

MAC than isoflurane; therefore, it is necessary to provide the patient a higher concentration or percent of isoflurane to get the animal under anesthesia.

Another factor used in comparing inhalant anesthetics is *blood-to-gas solubility*, the measure of the inhalation anesthetic to distribute between the blood and gas phases in the body. It is a measure of the tendency of an inhalation anesthetic to exist as a gas or to dissolve in blood. Anesthetic gases that are highly soluble must saturate the blood before molecules entering the blood are "available" to distribute to the tissues. Because it takes time to saturate the blood, it takes longer for highly soluble anesthetics to "fill" the blood so that "leftover" molecules can then distribute to the tissue. Less soluble anesthetics need not saturate the blood, and therefore pass into the blood from the alveolus (absorption of drug) and then more readily move to the tissues (distribution of drug).

An analogy of sugar and tea can be used to represent anesthetic gas and its absorption and distribution in blood. Hot tea represents a solution that is highly soluble; cold tea represents a solution with low solubility. If the same volume of sugar (anesthetic gas) is dissolved in both hot and cold tea, the sugar will remain in solution in the hot tea and will precipitate out of solution in the cold tea. The precipitated (leftover) sugar at the bottom of the cup of cold tea can be used for other purposes. Anesthetic gas that is poorly soluble quickly saturates the blood so that additional gas is "leftover" for distribution to tissues.

An inhalant anesthetic with high blood-to-gas solubility will have a longer induction time and longer recovery time than an inhalant anesthetic with low blood-to-gas solubility. Inhalant anesthetics with low blood-to-gas solubility stay in the alveoli (gas phase), and a large concentration difference between the agent in blood and in alveoli builds. This large concentration difference between the agent in blood and in alveoli allows rapid entry of the inhalant anesthetic into the circulation and then distribution into the tissues (like the brain). In contrast, an inhalant anesthetic with high blood-to-gas solubility is rapidly absorbed into the blood and tissues, and thus no large concentration difference between the agent in blood and in alveoli occurs. This small concentration difference between the agent in blood and in alveoli allows wider and more even distribution of the agent throughout the body. This results in slower induction and recovery rates. Examples of inhalant agents include the following.

Inhalant Analgesics

Nitrous oxide, or laughing gas, is an inhalant analgesic that is stored in blue cylinders in the United States. It diffuses rapidly throughout the body and can enter gas-filled body compartments (like the stomach and bowels), thus increasing the pressure in these compartments. Therefore, nitrous oxide is contraindicated in cases of gastric dilatation, pneumothorax, and twisted intestines. At the end of surgery in which nitrous oxide had been used, it is recommended to leave the animal on 100 percent oxygen for about 10 minutes and keep the endotracheal tube in place. This is to prevent diffusion hypoxia, which occurs when nitrous oxide rapidly diffuses out of the tissues back to the blood and then back to the alveoli. The alveoli are flooded with nitrous oxide, which dilutes the oxygen in the lung and causes hypoxia (low levels of oxygen).

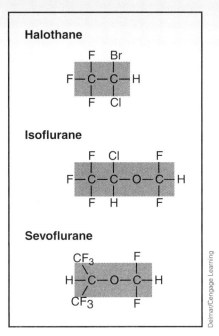

Figure 7-17 Chemical structures of volatile anesthetic agents.

Inhalant Anesthetics

The volatile anesthetic agents are halogenated hydrocarbons (carbon- and hydrogen-based molecules that have fluorine, chlorine, bromine, and/or iodine attached; Figure 7-17). In general, increasing the halogenation of the molecule increases its potency and reduces its flammability. The addition of fluoride atoms increases the agent's stability.

Halothane (Fluothane®) is a nonflammable inhalant anesthetic administered via a precision vaporizer, because of its high vapor pressure (precision vaporizers limit the concentration of vaporization). Halothane is susceptible to decomposition; therefore, it is stored in amber-colored bottles, and thymol is added as a preservative. Hepatic problems often occur with halothane use because about 25 percent of the halothane that is delivered to the patient is metabolized by the liver (the metabolites are hepatotoxic). Other problems that may occur with halothane are malignant hyperthermia, cardiac arrhythmias, bradycardia, and tachypnea (that will eventually increase the amount of anesthetic that the animal receives). Animals on halothane need close monitoring because they can change planes of anesthesia quickly.

Isoflurane (Isoflo®, Isosol®, Forane®) is a nonflammable inhalant anesthetic that causes rapid induction of anesthesia and short recoveries following anesthetic procedures. These drugs must be administered via a precision vaporizer (Figure 7-18). They do not cause the cardiac arrhythmia problems that halothane does. However, vigilant monitoring is needed because of the animal's ability to change planes of anesthesia quickly and the very short recovery period. Using a tranquilizer or sedative prior to induction can smooth recovery. Masking an animal with isoflurane and related anesthetics may be difficult due to the ability of these drugs to irritate the respiratory system.

Clinical Que

Advantages of inhalant anesthesia include easily and rapidly controlled anesthetic depth, increased inspired oxygen to the patient with the delivery of inhalation agents, provision of respiratory support via endotracheal tube placement, and minimal metabolism requirements for recovery. Disadvantages of inhalant anesthesia include waste gas exposure and dose-dependent cardiopulmonary depression.

Delmar/Cengage Learning

Figure 7-18 An isoflurane precision vaporizer.

A side effect of isoflurane is respiratory depression (greater than that of halothane) and, like halothane, it can trigger malignant hyperthermia.

Sevoflurane (Ultane®, SevoFlo®) is another nonflammable inhalant anesthetic and is the main isomer of isoflurane used in veterinary medicine (isomers vary in the amount and type of chemicals that are attached to the base molecule). Like isoflurane, it produces rapid induction and rapid recoveries. In general, sevoflurane produces fewer cardiovascular side effects than the other inhalant anesthetics. It is a profound respiratory depressant, and close monitoring of animals receiving sevoflurane is needed. Patients should be continuously monitored, and facilities for maintenance of patient airway, artificial ventilation, and oxygen supplementation must be immediately available. Sevoflurane quickly enters the bloodstream and escapes to the brain, making it good for mask inductions. It is also quickly eliminated making it good for C-sections because any sevoflurane absorbed by the fetus is quickly eliminated. Sevoflurane undergoes temperature-dependent degradation by the soda lime and barium lime crystals used in carbon dioxide absorber canisters; therefore, sevoflurane cannot be used in low-flow or closed-system anesthesia. Sevoflurane has low tissue solubility, resulting in rapid elimination of the drug by the body and rapid awakening making the judicious use of tranquilizers important. Sevoflurane provides limited analgesia; therefore, pre- and postoperative analgesics must be administered to alleviate pain in surgical patients.

CNS Stimulants (↑ CNS)

CNS stimulants are used to reverse CNS depression caused by CNS depressants. *Doxapram* (Dopram-V®, Respiram®) stimulates the brain stem to increase respiration in animals with apnea or bradypnea. It is commonly used when animals have C-sections. It is given sublingually or via the umbilical cord to the neonates

---

who received CNS depressants in the form of anesthetics from the dam through the placenta. It should be used with caution in animals with a seizure history.

Methylxanthines, a group of drugs that inhibits an enzyme that normally breaks down cyclic adenosine monophosphate (cAMP), include substances such as caffeine, *theophylline*, and *aminophylline*. Methylxanthines are typically used as bronchodilators; however, one of their adverse effects is CNS stimulation. *Theophylline* (generic, Theo-dur®) and *aminophylline* (generic) are very similar compounds that are available in oral and injectable forms. *Aminophylline* contains about 80 percent *theophylline* and is better tolerated by the gastrointestinal tract; however, *theophylline* comes in sustained-release forms that can be given less frequently. Methylxanthines cause CNS stimulation, gastrointestinal irritation, and bronchodilation.

Euthanasia Solutions

Euthanasia solutions are used to humanely end an animal's life. They usually contain pentobarbital. When pentobarbital is the only narcotic agent present, it is a C-II controlled substance (Sleep Away®); it is a C-III controlled substance when in combination with other agents (Beuthanasia®-D Special, which also has phenytoin in it). T-61 is a nonnarcotic, nonbarbiturate, noncontrolled-substance, general anesthetic euthanasia solution that causes muscle paralysis. Appendix K reviews euthanasia procedures and provides useful resources for the reader to assess regarding euthanasia.

AUTONOMIC NERVOUS SYSTEM DRUGS

Autonomic nervous system drugs work either by acting like neurotransmitters or by interfering with neurotransmitter release. They may affect either the parasympathetic or the sympathetic nervous systems (Figure 7-19).

Parasympathetic Nervous System Drugs

The parasympathetic nervous system is the homeostatic system that releases ACh at both the pre- and postganglionic fibers. Two groups of drugs affect the parasympathetic nervous system: the cholinergics and the anticholinergics.

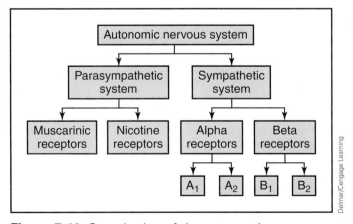

Figure 7-19 Organization of the autonomic nervous system.

Cholinergic Drugs/Parasympathomimetics

Cholinergic drugs, also known as **parasympathomimetics**, mimic the action of the parasympathetic nervous system. They work either by mimicking the action of acetylcholine (direct acting) or by inhibiting acetylcholine breakdown (indirect acting) (Figure 7-20). Direct-acting cholinergics are selective for muscarinic receptors and affect the smooth muscles of the urinary and gastrointestinal tracts (Figure 7-21). Indirect-acting cholinergics work by decreasing the inactivation of ACh in the synapse by AChE. Table 7-6 summarizes the effects of cholinergic drugs. Side effects of cholinergic drugs include bradycardia, diarrhea and vomiting, and increased secretions (intestinal, bronchial, and ocular). Examples include the following:

- *bethanechol* (Urecholine®), a direct-acting cholinergic used to treat gastrointestinal and urinary atony.

- *metoclopramide* (Reglan®), a direct-acting cholinergic used to control vomiting and aid in gastric emptying.

- *pilocarpine* (Akarpine®, Pilocar®, IsoptoCarpine®), a direct-acting cholinergic used as an ophthalmic solution to decrease the intraocular pressure seen in glaucoma. Local irritation is an additional side effect.

- *edrophonium* (Tensilon®), an indirect-acting cholinergic used to diagnose myasthenia gravis.

- *neostigmine* (Prostigmine®, Stiglyn®) and *physostigmine* (Antilirium®, Eserine®), indirect-acting cholinergics used to treat rumen atony, intestinal atony, and urine retention.

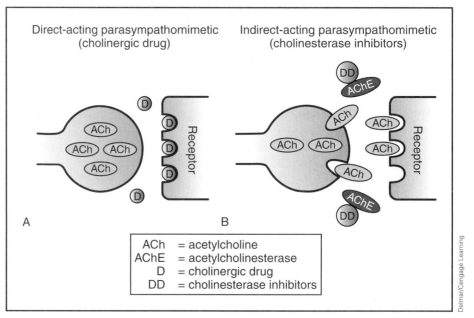

Figure 7-20 Direct-acting versus indirect-acting parasympathomimetic drugs. (A) Direct-acting cholinergic drugs resemble acetylcholine and act directly on the receptor. (B) Indirect-acting cholinergic drugs are known as cholinesterase inhibitors because they inactivate the enzyme acetylcholinesterase thus permitting acetylcholine to react with the receptor.

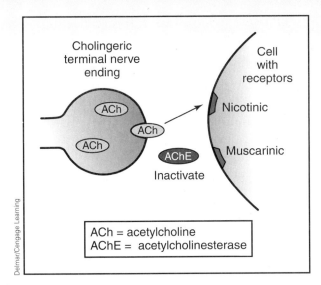

Delmar/Cengage Learning

Figure 7-21 Cholinergic receptors at organ cells are either muscarinic or nicotinic, meaning that they are stimulated by the substances muscarine and nicotine respectively. Nicotinic receptors are located in motor nerves and skeletal muscles. Stimulation of nicotinic receptors results in muscle contraction. Muscarinic receptors are located in most internal organs. Stimulation of muscarinic receptors may result in excitation or inhibition depending on the organ involved.

Table 7-6   Effects of Cholinergic Drugs

BODY TISSUE		EFFECT OF CHOLINERGIC DRUGS
Cardiac		Decreases heart rate and slows conduction of the AV node
Vascular		Causes vasodilation (lowers blood pressure)
Lung (bronchi)		Stimulates bronchial smooth muscle contraction and increases bronchial secretions
Gastrointestinal		Increases motility of the smooth muscles of the stomach, increases peristalsis, and relaxes sphincter muscles
Urinary		Contracts urinary bladder muscles, relaxes sphincter muscles of the urinary bladder, and stimulates urination
Ocular		Causes miosis (pupillary constriction)
Skeletal muscle		Maintains muscle strength and tone
Glandular		Increases salivation, perspiration, and tear production

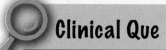

Clinical Que

Cholinergic receptors are found mainly in the heart, blood vessels, gastrointestinal tract, bronchi, and eye.

- *demecarium* (Humorsol®) and *isoflurophate* (Floropryl®), indirect-acting cholinergics used to manage glaucoma.

- *organophosphates,* indirect-acting cholinergics used in antiparasitic products. If used improperly or in debilitated animals, toxicity may be seen that can be reversed with 2-PAM (Protopam®). 2-PAM reactivates cholinesterase.

Anticholinergic Drugs/Parasympatholytics

Anticholinergic drugs inhibit the actions of acetylcholine by occupying the acetylcholine receptors (Figure 7-22). These drugs are also referred to as **parasympatholytics**, antimuscarinic agents, or antispasmodics. The major body tissues affected by the anticholinergic drugs are the heart, respiratory tract, gastrointestinal tract, urinary bladder, eye, and exocrine glands. By blocking the parasympathetic nerves, the sympathetic or adrenergic nervous system dominates. Anticholinergic and cholinergic drugs have opposite effects. Table 7-7 summarizes the effects of anticholinergic drugs. Side effects of anticholinergics may include tachycardia, constipation, dry mouth, dry eye, and drowsiness. Examples of anticholinergics include the following:

- *atropine* (generic), used as a preanesthetic agent to prevent bradycardia and to decrease salivation, to dilate pupils for ophthalmic examination, to control ciliary spasm of the eye, to decrease gastrointestinal motility, to treat bradycardia, and as an antidote for organophosphate poisoning.

- *glycopyrrolate* (Robinul-V®), which is similar to atropine but with a longer duration of action and is mainly used as a preanesthetic agent.

- *aminopentamide* (Centrine®), used to control diarrhea and vomiting.

- *propantheline* (Pro-Banthine®), used to decrease gastric secretions and gastrointestinal spasms, to treat urinary incontinence, and to decrease colonic peristalsis in horses.

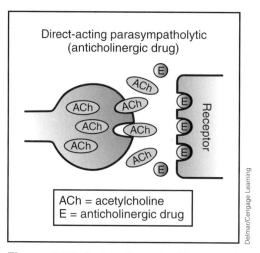

Figure 7-22 Anticholinergic drugs work by occupying the receptor sites, thus blocking acetylcholine.

Table 7-7   Effects of Anticholinergic Drugs

BODY TISSUE		EFFECT OF ANTICHOLINERGIC DRUGS
Cardiac		Increases heart rate
Vascular		Causes vasoconstriction to digestive organs and skin and vasodilation to skeletal muscles
Bronchi		Dilates bronchi and decreases bronchial secretions
Gastrointestinal		Relaxes smooth muscle tone of the gastrointestinal tract, decreases gastrointestinal motility and peristalsis, and decreases gastric and intestinal secretions
Urinary		Relaxes urinary bladder muscles, increases constriction of the internal sphincter muscle of the urinary bladder, and causes urine retention
Ocular		Causes mydriasis (pupillary dilation) and paralyzes the ciliary muscle
CNS/Muscular system		Decreases muscle rigidity and can cause drowsiness and disorientation
Glandular		Decreases salivation, perspiration, and tear production

Sympathetic Nervous System Drugs

The sympathetic nervous system is the fight-or-flight system that releases ACh at the preganglionic fibers and NE or epinephrine at the postganglionic fibers. Two groups of drugs affect the sympathetic nervous system: the adrenergics and the adrenergic blocking agents.

Adrenergic Drugs/Sympathomimetics

Adrenergic nerves have either alpha (α) or beta (β) receptors (Figure 7-23). Drugs that simulate the action of the sympathetic nervous system are called **adrenergic drugs** or **sympathomimetics**. They act on one type (either alpha or beta receptors only; these are called selective drugs) or both types (both alpha and beta receptors; these are called nonselective drugs) of adrenergic receptors located on the cells of smooth muscles (Figure 7-24). Catecholamines are chemicals that can cause a sympathomimetic response. Examples of naturally occurring catecholamines are epinephrine, norepinephrine, and dopamine. Table 7-8 summarizes the effects of adrenergic (and adrenergic blocking)

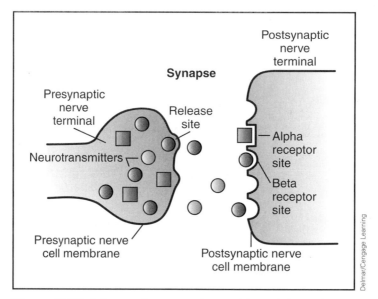

Figure 7-23 Adrenergic nerves have either alpha or beta receptors. Norepinephrine released by the presynaptic nerve crosses the synapse and binds with alpha and beta receptors in the cell membrane of the postsynaptic nerve, resulting in the transmission of the nerve impulse.

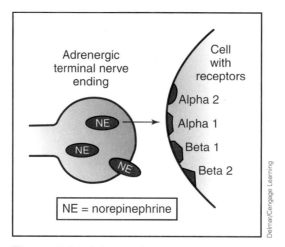

Figure 7-24 Adrenergic receptors are either alpha (1 or 2) or beta (1 or 2). Whether the neurotransmitter acts on an alpha or beta receptor or both accounts for the variation of responses seen with adrenergic stimulation.

Table 7-8	**Effects of Adrenergics and Adrenergic Blocking Agents**	

RECEPTOR	ADRENERGIC DRUG EFFECT	ADRENERGIC BLOCKING DRUG EFFECT
alpha 1	Increases force of heart contraction, increases blood pressure, and causes mydriasis	Vasodilation and miosis
alpha 2	Inhibits release of norepinephrine and dilates blood vessels, producing hypotension	None
beta 1	Increases heart rate and force of heart contraction	Decreases heart rate
beta 2	Dilates bronchioles and relaxes gastrointestinal tract	Constricts bronchioles

agents. Side effects include tachycardia, hypertension, and cardiac arrhythmias. Synthetic catecholamines include the following:

- *epinephrine* (generic, Adrenalin®), a nonselective injectable adrenergic drug that affects alpha-1, beta-1, and beta-2 receptors. It increases heart rate, cardiac output, constriction of blood vessels to the skin, dilation of bronchioles, and dilation of blood vessels to muscles. It is used in emergency situations for cardiac resuscitation and treatment of anaphylaxis.

- *norepinephrine* (Noradrenalin®, Levophed®, Levarterenol®), a nonselective injectable adrenergic drug that affects alpha-1 and beta-1 receptors and is used mainly to increase blood pressure.

- *isoproterenol* (Isuprel®), a selective adrenergic drug that affects beta-1 and beta-2 receptors and is used mainly to cause bronchodilation.

- *dopamine* (Intropin®), a selective adrenergic drug that affects beta-1 receptors and is used to treat shock and congestive heart failure.

- *dobutamine* (Dobutrex®), a selective adrenergic drug that affects beta-1 receptors and is used to treat heart failure.

- *phenylpropanolamine* (Prion®, Propalin®), a nonselective adrenergic drug that affects alpha-1 and beta-1 receptors and is used to treat urinary incontinence in dogs.

- *isoetharine* (Bronkosol®), *albuterol* (Proventil®), and *terbutaline* (Brethine®), selective adrenergic drugs that affect beta-2 receptors and are used to treat bronchospasm by producing bronchodilation.

- *ephedrine* (generic), a selective adrenergic drug that affects alpha-1, beta-1, and beta-2 receptors and is used to produce bronchodilation. In some states, ephedrine may be a controlled substance or have restrictions on its sale because it is used to make methamphetamine. Be cautious of people wanting to purchase ephedrine for their animals.

- *xylazine* (Rompun®, AnaSed®), a selective adrenergic drug that affects alpha-2 receptors and was covered previously in this chapter under tranquilizers/sedatives.

Adrenergic Blocking Agents/Sympatholytics

Drugs that block the effects of the adrenergic neurotransmitters are called **adrenergic blocking agents** or **sympatholytics**. They act as antagonists to the adrenergic agonists by blocking the alpha- and beta-receptor sites (Figure 7-25). They can block the receptor site either by occupying the receptor or by inhibiting the release of the neurotransmitter.

Alpha blockers usually promote vasodilation and a decrease in blood pressure. Side effects of alpha blockers include tachycardia and hypotension. Examples of drugs used in veterinary practice include the following:

- *phenoxybenzamine* (Dibenzyline®), used in small animals to decrease urethral sphincter tone and in horses to prevent or treat laminitis. It should not be used in horses with colic.

- *prazosin* (Minipress®), used to treat animals with heart failure and hypertension.

- *yohimbine* (Yobine®), *atipamezole* (Antisedan®), and *tolazoline* (Tolazine®), reversal agents already discussed in this chapter.

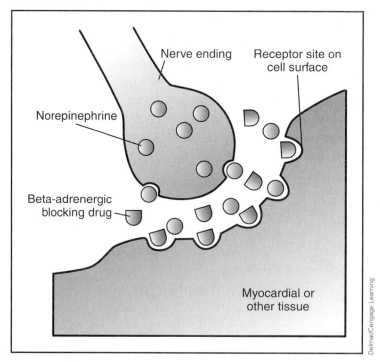

Figure 7-25 Adrenergic blocking agents work by blocking the receptor site either by occupying the receptor (as shown in this image) or by inhibiting the release of the neurotransmitter.

Beta blockers decrease heart rate and blood pressure. Side effects of beta blockers include bradycardia and hypotension. Examples of beta blockers used in veterinary practice include the following:

- *propranolol* (Inderal®), which blocks beta-1 and beta-2 receptors and is used to treat cardiac arrhythmias and cardiac disease in animals.

- *metoprolol* (Lopressor®) and *atenolol* (Tenormin®), which block beta-2 receptors and are used to treat hypertension.

- *timolol* (Timoptic®), which blocks beta-1 and beta-2 receptors and is used as an ophthalmic preparation to treat glaucoma.

SUMMARY

The nervous system has a basic unit called a neuron, which consists of a cell body, dendrites, and an axon. The space between two neurons or between a neuron and its receptor is the synapse. Neurotransmitters move across the synapse to cause action.

There are two branches of the nervous system: the CNS and the PNS. The PNS consists of cranial nerves, spinal nerves, and the ANS. The ANS is divided into two branches: the sympathetic (fight-or-flight pathway) and the parasympathetic (the homeostatic pathway).

CNS drugs include anticonvulsants that control seizures, tranquilizers/ sedatives/antianxiety drugs that reduce anxiety and excitement in animals, analgesics that minimize pain, opioid antagonists that treat respiratory and CNS depression, neuroleptanalgesics that cause CNS depression and analgesia, anesthetics that remove pain sensation, CNS stimulants that counteract CNS depression, and euthanasia solutions that humanely end an animal's life.

Drugs that work on the ANS include cholinergics, anticholinergics, adrenergics, and adrenergic blocking agents. ANS drugs affect the cardiovascular system, bronchi, gastrointestinal tract, urinary tract, CNS/muscular system, glands, and ocular system. Table 7-9 summarizes the types of drugs covered in this chapter.

Clinical Que

Alpha blockers mainly are used to treat hypertension and heart failure (also laminitis and to decrease urethral tone). Beta blockers are mainly used to treat hypertension and cardiac arrhythmias (also glaucoma).

Table 7-9	Types of Drugs Covered in this Chapter
DRUG CATEGORY	EXAMPLES
Anticonvulsants	• phenobarbital • pentobarbital • primidone • diazepam • lorazepam • clorazepate • potassium bromide • phenytoin • levetiracetam • zonisamide • gabapentin • felbamate

(Continued)

Table 7-9   *(Continued)*

DRUG CATEGORY	EXAMPLES
Tranquilizers/sedatives/ antianxiety	• phenothiazine derivatives: acepromazine, chlorpromazine, prochlorperazine/isopropamide, promazine • butyrophenones: azaperone • benzodiazepines: diazepam, zolazepam, midazolam, clonazepam • alpha-2 agonists: xylazine, detomidine, Medetomidine, dexmedetomidine
Narcotic analgesics	• opium • morphine • hydromorphone • oxymorphone • butorphanol • hydrocodone • fentanyl • alfentanil • etorphine • buprenorphine • pentazocine • methadone • codeine • tramadol • diphenoxylate • apomorphine • loperamide
Opioid antagonists	• naloxone • naltrexone
Neuroleptanalgesics	• xylazine and butorphanol • acepromazine and morphine • acepromazine and oxymorphone
Local anesthetics	• lidocaine • proparacaine • mepivacaine • tetracaine • bupivacaine
General anesthetics (injectable)	• barbiturates: phenobarbital, pentobarbital, thiopental, methohexital • dissociatives: ketamine, tiletamine • miscellaneous: guaifenesin, propofol
Analgesic (inhalant)	• nitrous oxide
General anesthetic (inhalant)	• halothane • isoflurane, sevoflurane
CNS stimulants	• doxapram • methylxanthines: aminophylline, theophylline
Euthanasia solutions	• pentobarbital combinations • T-61

DRUG CATEGORY	EXAMPLES
Cholinergics	• bethanechol • metoclopramide • pilocarpine • edrophonium • neostigmine • demecarium • organophosphates
Anticholinergics	• atropine • glycopyrrolate • aminopentamide • propantheline
Adrenergics	• epinephrine • norepinephrine • isoproterenol • dopamine • dobutamine • phenylpropanolamine • isoetharine • albuterol • terbutaline • ephedrine • xylazine
Adrenergic blocking agents	• alpha blockers: phenoxybenzamine, prazosin, yohimbine, atipamezole, tolazoline • beta blockers: propranolol, metoprolol, timolol

It's a Wrap

The answers to the questions in this chapter's Setting the Scene case study should be understood after reading this chapter. Is it the safest kind available? What anesthetic delivery method is the safest? The safest kind of anesthetic tends to be that available in inhalant form, since the drug can be taken away quickly by disconnecting the anesthesia tubing. However, some inhalant anesthetics are safer than others (halothane causes more cardiovascular problems than isoflurane). Injectable anesthetics that have reversal agents may be considered safer than those that do not have reversal agents. Examples of injectable anesthetics with reversal agents include xylazine, whose reversal agents include yohimbine and tolazoline, and detomidine, whose reversal agent is atipamezole.

Does this anesthetic provide pain relief after surgery? Do all anesthetics provide postsurgical pain relief? What drugs provide pain

(continued)

relief, and can they be used with anesthetic agents? Some anesthetics provide some degree of postsurgical analgesia; however, analgesics are administered to extend pain relief after surgical procedures. The choice of analgesic prescribed depends on the severity of pain. Nonnarcotic analgesics are used for mild to moderate pain (covered in Chapter 16), while narcotic analgesics are used for moderate to severe pain. Some analgesics can cause respiratory depression and when given in addition to the anesthetic, they may cause problems in diseased or older animals. The type of analgesic used depends on the procedure performed: Oxymorphone, an opioid, is a more potent analgesic than xylazine for visceral pain, whereas a nonsteroidal anti-inflammatory drug like phenylbutazone or carprofen may be sufficient for musculoskeletal pain and inflammation in minor musculoskeletal surgical procedures. Butorphanol, another opioid, causes less respiratory depression than other opioids and may be used for geriatric patients.

What about pain relief for the puppy after she goes home? Analgesics may be dispensed for the client to give her puppy for pain relief after she goes home. Clients are usually dispensed oral analgesics to give their animals for postsurgical pain relief.

CHAPTER REVIEW

Matching

Match the drug name with its action.

1. _____ phenobarbital
2. _____ xylazine
3. _____ acepromazine
4. _____ phenylpropanolamine
5. _____ diazepam
6. _____ morphine
7. _____ ketamine
8. _____ glycopyrrolate
9. _____ bethanecol
10. _____ methohexital

a. analgesic drug
b. anticonvulsant drug
c. phenothiazine derivative tranquilizer
d. adrenergic drug
e. cholinergic drug
f. anticholinergic drug
g. thiobarbiturate
h. dissociative anesthetic
i. alpha-2 agonist
j. anticonvulsant and tranquilizer that causes muscle relaxation

Multiple Choice

Choose the one best answer.

11. Which neurotransmitter is released at the postganglionic synapse of parasympathetic nerves?
   a. acetylcholine
   b. epinephrine
   c. dopamine
   d. mu

12. Which neurotransmitter is released at the postganglionic synapse of sympathetic nerves?
   a. acetylcholine
   b. epinephrine
   c. dopamine
   d. mu

13. Why do some animals become "resistant" to their dose of phenobarbital?
   a. They excrete this drug through the kidneys.
   b. They cannot biotransform this drug.
   c. Liver enzymes are induced causing the animal to develop tolerance.
   d. The half-life of the drug is shortened due to the high level of protein binding.

14. Which of the following sedatives produces analgesia in the horse, but still allows the horse to respond to stimuli by kicking?
   a. xylazine
   b. ketamine
   c. diazepam
   d. yohimbine

15. Obese animals that are anesthetized with thiobarbiturate would
   a. only be anesthetized for a short time.
   b. be anesthetized for a long time.
   c. not become anesthetized because of their fat stores.
   d. have severe perivascular inflammation.

16. Which group of anesthetics causes muscle rigidity and mild analgesia?
   a. barbiturates
   b. dissociatives
   c. inhalants
   d. propofol

17. For which inhalant analgesic is it recommended to leave the animal on 100 percent oxygen for about 10 minutes following the procedure to prevent diffusion hypoxia?
   a. nitrous oxide
   b. halothane
   c. isoflurane
   d. sevoflurane

18. Which anesthetic is a white emulsion, the effects of which last about two to five minutes, when given slow IV?
   a. barbiturate
   b. dissociative
   c. propofol
   d. guaifenesin

True/False

Circle a. for true or b. for false.

19. All euthanasia solutions are C-II controlled substances.
   a. true
   b. false

20. Cholinergic drugs are mainly used to improve the cardiovascular system.
   a. true
   b. false

Case Studies

21. A 13-year-old F/S Poodle with a 5-year history of epilepsy has been treated with phenobarbital. Her liver enzymes are checked every six months, and this time her liver enzymes are increasing.

a.   Is an increase in liver enzymes normally seen with phenobarbital treatment?

b.   This poodle's brother is also being treated with phenobarbital for seizures. He has been on ¼ grain of phenobarbital bid but is now having seizures again. What phenomenon is occurring?

c.   These dogs are going on vacation, and the owner would like to receive an antianxiety drug for each of them. What antianxiety drug would *not* be used?

d.   These dogs need refills on their phenobarbital. Before dispensing the medication that has been approved by the veterinarian, what needs to be done?

22. A four-year-old, 85#, M/N Brittany Spaniel is presented to the clinic for a laceration he received on his morning walk. His PE is normal except for his obesity and the laceration that must be sutured. The veterinarian intends to induce this dog with thiobarbiturate and needs his dose calculated.

a.   What needs to be considered in determining a dose for this dog?

b.   Why would this dog have an anesthetic overdose if the dose were calculated based on his 85-lb weight?

c.   When giving thiobarbiturate IV, how is overdosing the patient avoided?

Critical Thinking Questions

23. Why would animals prescribed phenobarbital for seizure control need to have blood collected to determine phenobarbital blood levels about three to six weeks after starting treatment?

24. What are some instances in which use of certain inhalant anesthetics would be contraindicated?

ADDITIONAL REVIEW MATERIAL

Additional review material can be found on the StudyWARE™ CD included in this text.

BIBLIOGRAPHY

Allen, D. G. (2005). *Handbook of Veterinary Drugs* (3rd ed.). Baltimore, MD: Lippincott Williams & Wilkins.

Amundson Romich, J. (2008). *An Illustrated Guide to Veterinary Medical Terminology* (3rd ed.). Clifton Park, NY: Delmar Cengage Learning.

Beckman, B. (August 2007). Pain management for oral surgery in dogs and cats. *DVM magazine.*

Broyles, B., Reiss, B., and Evans, M. (2007). *Pharmacological Aspects of Nursing Care* (7th ed., pp. 358–400). Clifton Park, NY: Delmar Cengage Learning.

Hendrix, P. (May 2001). Perioperative pain management in cats and dogs. *Practice Builder, DVM: The Newsletter of Veterinary Medicine,* http://www.veterinarynews.dvm360.com

Hoskins, J. D. (March 1, 2008). New anticonvulsant drugs show promise in dogs, cats. *DVM magazine.*

Plumb, D. C. (2008). *Plumb's Veterinary Drug Handbook* (6th ed.). Ames, IA: Blackwell Publishing.

Smith Bailey, K., Dewey, C., et al. (March 15, 2008). Levetiracetam as an adjunct to Phenobarbital treatment in cats with suspected idiopathic epilepsy. *Journal of the American Veterinary Medical Association,* 232(6): 867–872.

Spratto, G. R. and Woods, A. L. (2008). *PDR Nurse's Drug Handbook.* Clifton Park, NY: Delmar Cengage Learning.

Thompson, D. (2005). Alpha-2 agonists. Veterinary Anesthesia and Analgesia Support Group. http://www.vasg.org

CHAPTER 8
CARDIOVASCULAR DRUGS

OBJECTIVES

Upon completion of this chapter, the reader should be able to:

- describe the anatomy and physiology of the cardiovascular system, including blood flow through the heart and the electrical conduction system.
- describe the components of the ECG.
- differentiate between preload and afterload.
- describe the basic pathophysiology of the various types of heart disease.
- outline compensatory mechanisms by which the heart can improve its workload in heart disease.
- describe the clinical usages of the following drugs in the treatment of heart disease: positive inotropes, antiarrhythmics, vasodilators, diuretics, anticoagulants, hemostatic drugs, and blood-enhancing drugs.

KEY TERMS

afterload	catecholamine
angiotensin I	diuretic
angiotensin II	erythropoietin
antiarrhythmic drug	hemostatic drug
anticoagulant	inotropy
arrhythmia	preload
benzimidazole-pyridazinone	renin
blood-enhancing drug	vasodilator
cardiac glycoside	

Setting the Scene

A 14-year-old F/S Cocker Spaniel is brought into the clinic because her owner feels that the dog is not as energetic as usual. The Spaniel is up to date on her vaccinations and has been heartworm-tested annually. On PE, the veterinarian notes that her mucous membrane color is not as pink as it has been in the past, hears a murmur upon auscultation of her heart, and finds that some fluid appears to be present in the abdominal cavity. These signs indicate possible heart disease, so the appropriate diagnostic tests are performed on the dog. Chest radiographs show an enlarged heart. The dog's ECG is normal, and her blood values are normal. The dog is diagnosed with congestive heart failure, and the veterinarian prescribes a diuretic and a mixed vasodilator. The veterinarian also prescribes a sodium-restricted diet without any treats. The owner asks how all these drugs are going to help her dog. Why can't she keep her dog on the same dog food that she has been feeding for many years? What should the veterinary technician tell her? What is the rationale behind low-sodium diets for animals with congestive heart failure? How is the success or failure of cardiac treatment assessed? Is there any way to determine if the drugs and diet restriction are working? When should the owner bring the dog in again for reexamination?

Clinical Que

The cardiovascular system is a closed system that relies on pressure differences to ensure delivery of blood to the tissues and the return of that blood to the heart.

BASIC CARDIAC ANATOMY AND PHYSIOLOGY

The functions of the cardiovascular system include delivery of oxygen, nutrients, and hormones to the various body tissues and the transport of waste products to the appropriate waste removal system. There are three major parts of the cardiovascular system: the heart, the blood vessels, and the blood.

The Heart

The heart is a hollow, muscular organ that provides the power to move blood through the body. The heart consists of a thick muscular wall called the *myocardium*; an inner membrane that lines the chambers and valves called the *endocardium*; and the outer membrane of the heart called the *epicardium*. A double-walled membrane called the *pericardium* surrounds the heart. The heart consists of chambers: the *atria* are the cranial (upper) chambers and the *ventricles* are the caudal (lower) chambers. In animals with four-chambered hearts (mammals and birds), the right atrium receives deoxygenated blood from the venous circulation. Blood is pumped to the right ventricle, which in turn pumps blood to the lungs to receive oxygen via diffusion. The left atrium receives oxygenated blood and pumps it to the left ventricle, which in turn pumps it to the aorta. Separating these chambers are four valves that help control the direction of blood flow in the heart. There are two atrioventricular (AV) valves (tricuspid on the right and mitral on the left) and two semilunar valves (pulmonic on right and aortic on the left). Figures 8-1 and 8-2 illustrate the structures of the heart. Table 8-1 describes blood flow through the heart.

Function of the Heart

The primary function of the heart is twofold: to pump fresh, oxygenated blood to the tissues of the body and to take waste products such as carbon dioxide away from the tissues. Blood carries oxygen to tissues via arteries and arterioles and takes wastes away from tissues via venules and veins. Diffusion of oxygen and carbon dioxide occurs in the single-cell-layer-thick capillaries.

The Conduction System

The conduction of electrical impulses of the heart originates in the sinoatrial (SA) node. The SA node is located in the wall of the right atrium and regulates heart rate. When the cells in the SA node depolarize, a wavelike sequence of depolarization passes through the right and then the left atrium. *Depolarization* occurs when sodium channels open, allowing sodium ions into the cell, which then allows the cells of the myocardium to contract. Depolarization continues cell to cell very quickly, so that it seems as though both the right atrium and the left atrium are contracting at the same time. As the atria contract, pressure in the atria increases and blood is pushed through the AV valves into the ventricles. The wave of depolarization spreads from the atria, but is prevented from entering the ventricles by a cellular barrier. The depolarization wave can enter the ventricles only through the AV node. The AV node is located

Clinical Que

Cardiac rhythms initiated by the SA node are sinus rhythms; those initiated by the AV node are nodal rhythms.

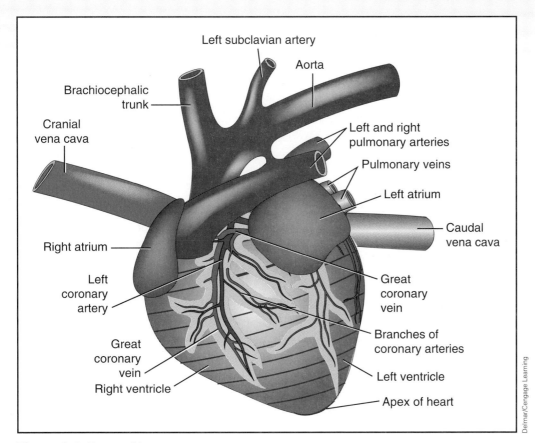

Figure 8-1 External heart structures.

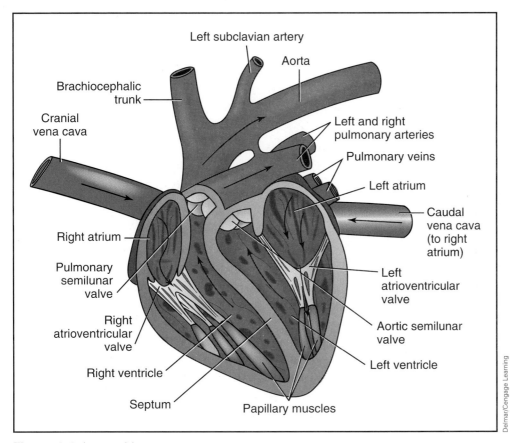

Figure 8-2 Internal heart structures.

Table 8-1   Blood Flow Through the Heart

- The right atrium receives blood from all tissues, except the lungs, through the cranial and caudal venae cavae. Blood flows from here through the tricuspid valve into the right ventricle. (This is systemic circulation.)

- The right ventricle pumps the blood through the pulmonary semilunar valve and into the pulmonary artery, which carries it to the lungs. (This is pulmonary circulation.)

- The left atrium receives oxygenated blood from the lungs through the four pulmonary veins. The blood flows through the mitral valve into the left ventricle. (This is pulmonary circulation.)

- The left ventricle receives blood from the left atrium. From the left ventricle, blood goes out through the aortic semilunar valve and into the aorta and is pumped to all parts of the body, except the lungs. (This is systemic circulation.)

- Blood is returned by the venae cavae to the right atrium, and the cycle continues.

in the interatrial septum. As the depolarization wave enters the AV node, it is delayed for a fraction of a second; then it enters the ventricles. The delay allows the atria to complete contraction before the ventricles contract. After the depolarization wave passes through the AV node, it goes through the interventricular septum, along a conduction system known as the *bundle branches*. Bundle branches conduct electrical impulses to the Purkinje fibers at the apex of the heart. From the Purkinje fibers, an impulse travels rapidly through the ventricular muscle cells, causing contraction from the apex toward the heart valves. At this point, blood is pushed from the ventricles through the pulmonic and aortic semilunar valves to the lung and body. Figure 8-3 summarizes the conduction system of the heart.

Conduction of electrical impulses can be seen on an electrocardiogram (ECG). The P-wave of the ECG corresponds to the depolarization of the atria (and the contraction of the atrial cells). A short delay occurs between the time when the depolarization wave enters the AV node and the time when the ventricles contract. This delay is represented on the ECG by the flat line after the P-wave known as the P-R interval. Depolarization of the ventricles (and the contraction of the ventricular cells) and repolarization of the atria are represented as the large QRS complex. After the ventricles contract, the ventricular muscle cells relax and repolarize (sodium channels close, potassium channels open, and potassium moves into the cell). This repolarization of the ventricles is represented by the T-wave. Figure 8-4 shows the appearance of the ECG.

Heart Rate

Heart rate is controlled primarily by the autonomic nervous system. Parasympathetic (cholinergic) nerve endings (vagal fibers) are located close to the SA node. When parasympathetic nerves are stimulated, the neurotransmitter

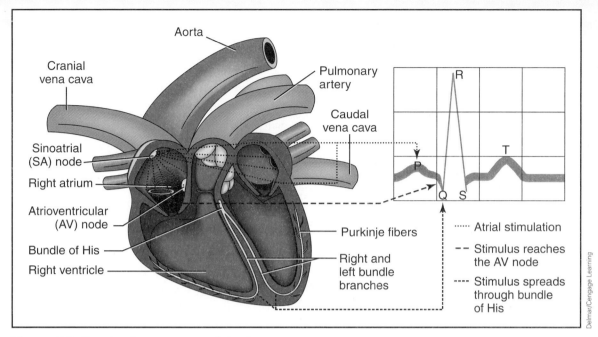

Figure 8-3 Conduction systems of the heart.

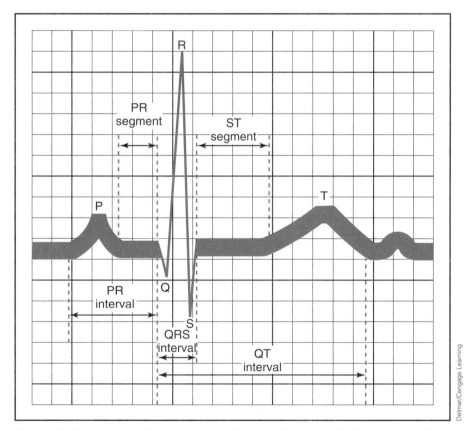

Figure 8-4 Anatomy of an electrocardiogram. The first deflection, the P wave, represents excitation (depolarization) of the atria. The PR interval represents conduction through the atrioventricular valve. The QRS complex results from excitation of the ventricles. The QT interval represents ventricular depolarization and repolarization. The ST segment represents the end of ventricular depolarization to the onset of ventricular repolarization. The T wave results from recovery (repolarization) of the ventricles.

acetylcholine is released at the junction of the nerve and the cardiac muscle. This acts to slow the heart rate by inhibiting impulse formation and electrical conduction in the heart.

Sympathetic (adrenergic) nerve fibers also innervate part of the heart. When sympathetic nerves are stimulated, the neurotransmitter norepinephrine is released. This acts to increase heart rate by promoting impulse formation and electrical conduction in the heart. Sympathetic stimulation also decreases the time between impulses, thus reducing the duration of the refractory period (time between consecutive muscle contractions).

Heart Rhythm

A *rhythm* is a recurrence of an audible action at regular intervals. The heart's contractions are supposed to be rhythmic. Cardiac contractions are divided into the states of systole and diastole. *Systole* refers to contraction of the heart chambers; diastole refers to relaxation of the chambers, during which time the chambers are filling with blood. The force of contraction is known as **inotropy**.

The rate and regularity of the heart rhythm, or *heartbeat,* is innate to the myocardial cells as directed by the SA node. Normal heart rhythm is called the *sinus rhythm* because it starts in the SA node. If the SA node does not function properly and is unable to send the impulse to the rest of the heart, other areas of the conduction system can take over and initiate a heartbeat. The resulting abnormal rhythm is called an **arrhythmia**. Depending on the type of arrhythmia present, drugs may be used to control heartbeat irregularities.

Auscultation (listening) of the heart allows for determination of heart rate and rhythm.

Cardiac Output

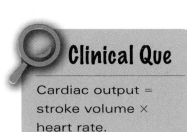

Clinical Que

Cardiac output = stroke volume × heart rate.

The workload of the heart is divided into preload and afterload. **Preload** is the volume of blood entering the right side of the heart, or the ventricular end-diastolic volume. **Afterload** is the force needed to push blood out of the ventricles, or the impedance to ventricular emptying presented by aortic pressure. Preload problems are usually associated with right-sided heart disease, whereas afterload is associated with left-sided heart disease. Preload and afterload, along with contractility (the force of ventricular contraction), make up *stroke volume,* the amount of blood ejected from the left ventricle with each heartbeat. Stroke volume multiplied by heart rate is the cardiac output. *Cardiac output* is the volume of blood expelled from the heart in one minute.

If the heart is not working properly, compensatory mechanisms to improve the workload of the heart may take over. Some examples of compensatory mechanisms of the heart include the following:

- *increasing heart rate* to increase cardiac output, so long as the heart rate is not so fast as to limit the chambers' filling with blood.

- *increasing stroke volume,* because increased force of contraction should result in improved ventricular emptying during systole, thus allowing more blood to be pumped from the heart.

- *maximizing efficiency* of the heart muscle by reducing vascular tone to decrease the amount of work done by the heart thus improving cardiac output.

- *enlarging the heart* in an attempt to pump out more blood. This can be achieved either by dilating the chambers (allowing more blood to collect in the ventricle, thus increasing stroke volume) or by thickening the myocardium (allowing more forceful contractions to move blood more efficiently out of the ventricle).

Blood Vessels

There are three major types of blood vessels: arteries, veins, and capillaries. All blood vessels have a lumen, which is the opening in a vessel through which fluid flows. The diameter of the lumen is affected by constriction (narrowing of the vessel diameter) and dilation (widening of the vessel diameter). The pumping action of the heart drives blood into the arteries.

Arteries

An artery is a blood vessel that carries blood away from the heart. Blood in the arteries is usually oxygenated (the main exception is the pulmonary artery, which carries deoxygenated blood) and is bright red. The aorta is the main trunk of the arterial system and begins from the left ventricle of the heart. After leaving the left ventricle, the aorta arches dorsally and then progresses caudally. The aorta branches into other arteries that supply many muscles and organs of the body.

The arterioles are smaller branches of arteries. Arterioles are smaller blood vessels that carry blood away from the heart. Arterioles are smaller and thinner than arteries and carry blood to the capillaries.

Veins

Veins form a low-pressure collecting system that returns blood to the heart. Veins have thinner walls and are less elastic than arteries, which have muscular walls to allow contraction and expansion to move blood throughout the body (Figure 8-5). Because the veins do not have muscular walls, contractions of the skeletal muscles cause the blood to flow through the veins toward the heart. Veins also have valves that permit blood flow toward the heart and prevent blood from flowing away from the heart (Figure 8-6).

Capillaries

The capillaries are single-cell-thick vessels that connect the arterial and venous systems. Blood flows rapidly through arteries and veins; however, blood flow is slower through the capillaries due to their smaller diameter. This slower flow allows time for the diffusion of oxygen, nutrients, and waste products. Blood in the alveolar capillaries picks up oxygen and gives off carbon dioxide. In the rest of the body, oxygen diffuses (passes through) from the capillaries into tissue, and carbon dioxide diffuses from tissue into the capillaries. Capillaries connect with venules, which are tiny blood vessels that carry blood to the veins.

Figure 8-5 Cross section of a large vein (left) and artery (right). Compare the differences between the thin-walled vena cava on the left and the thick-walled aorta on the right.

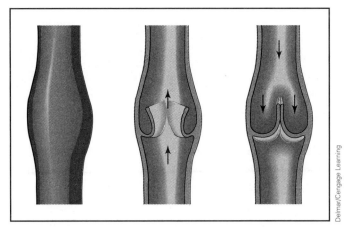

Figure 8-6 Veins contain valves to prevent the backward flow of blood. (A) External view of the vein shows wider area of the valve. (B) Internal view with the valve open as blood flows through. (C) Internal view with the valve closed.

Blood Pressure

The blood in the cardiovascular system flows from areas of higher pressure to areas of lower pressure. The area of highest pressure in this system is the left ventricle during systole. The pressure in the left ventricle propels blood out of the aorta and into the system. The lowest pressure in this system is the right atrium, which collects deoxygenated blood from the body. The maintenance of this pressure system is controlled by areas of the brain and various hormones.

Blood pressure in the cardiovascular system is determined by three things: heart rate, stroke volume (amount of blood pumped out of the ventricle with each heart beat), and peripheral resistance (resistance of the muscular arteries to the blood being pumped through). Systems that influence blood pressure include the following:

- sympathetic nervous system: Arterioles are the most important factor in determining peripheral resistance because they have the smallest diameter and are able to decrease blood flow to the capillaries when they constrict. The arterioles are very responsive to the sympathetic nervous system (alpha-1 receptors), and they constrict when the sympathetic system is stimulated, which increases peripheral resistance and blood pressure.

- baroreceptors. As blood leaves the left ventricle, it enters the aorta and influences specialized pressure receptors in the aortic arch called baroreceptors (similar cells are located in the carotid arteries, which deliver blood to the brain). If the blood pressure is high, sensory input is sent to the medulla (vasomotor center) of the brain, which stimulates vasodilation and a decrease in heart rate and output, causing the blood pressure to decrease. If the blood pressure is low, the medulla stimulates an increase in heart rate and output and vasoconstriction. The medulla mediates these effects through the autonomic nervous system.

- renin-angiotensin system. The kidneys require a constant perfusion of blood to function properly; therefore, they have a compensatory mechanism to ensure that blood flow is maintained. This mechanism is the renin-angiotensin system (sometimes called the renin-angiotensin-aldosterone system) (Figure 8-7A). Poor oxygenation or low blood pressure of a nephron is detected by juxtaglomerular cells (groups of specialized cells located in the afferent arteriole delivering blood to the glomerulus that monitor blood pressure and flow into the glomerulus), which signal **renin** to be released into the blood. Renin is delivered to the liver and converts angiotensinogen (produced in the liver) to **angiotensin I**. Angiotensin I travels in the bloodstream to the lungs, where alveolar cells convert it, using angiotensin-converting enzyme (ACE), to **angiotensin II**. Angiotensin II binds to receptor sites on blood vessels to cause intense vasoconstriction. This raises blood pressure and peripheral resistance, which restore blood flow to the kidneys and decrease the release of renin. Angiotensin II also stimulates the adrenal cortex to release aldosterone, which acts on the nephrons to cause sodium and water retention. This retention of sodium and water increases blood volume, which should also increase blood pressure. The sodium-rich blood also signals the hypothalamus to release antidiuretic hormone (ADH), which causes additional water retention, further increasing blood pressure. As blood pressure increases and blood flow to the kidneys increase, release of renin should decrease causing the compensatory mechanism to stop (Figure 8-7B).

Blood

Blood supplies body tissues with oxygen, nutrients, and various chemicals. Blood transports waste products to various organs for removal from the body. Blood cells also play important roles in the immune and endocrine systems. Blood is formed in the bone marrow and is composed of 55 percent liquid plasma (which contains clotting proteins such as fibrinogen and prothrombin) and 45 percent formed elements (cells such as red blood cells, white blood cells, and clotting cells). An erythrocyte is a mature red blood cell (oxygen-carrying cell). Erythrocytes have a biconcave disc shape and contain hemoglobin, a blood protein that transports oxygen. Heme is the nonprotein, iron-containing portion of hemoglobin. A leukocyte is a white blood cell that functions primarily in fighting disease in the body.

Clotting cells called thrombocytes or platelets are also produced in the bone marrow and play a part in the clotting of blood.

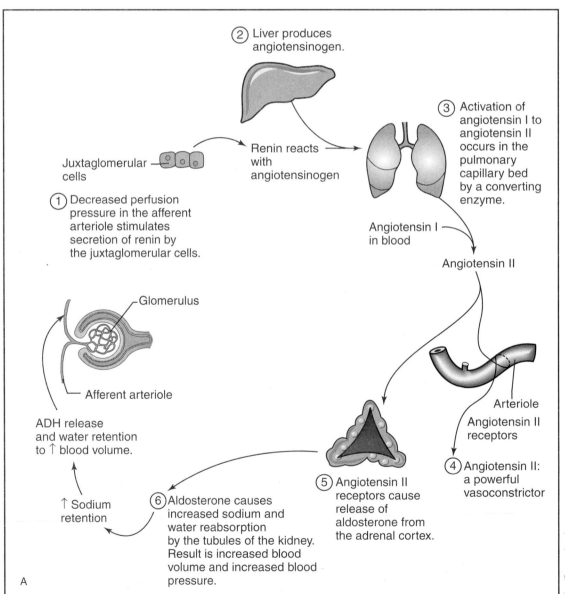

Figure 8-7 (A) The renin-angiotensin system.

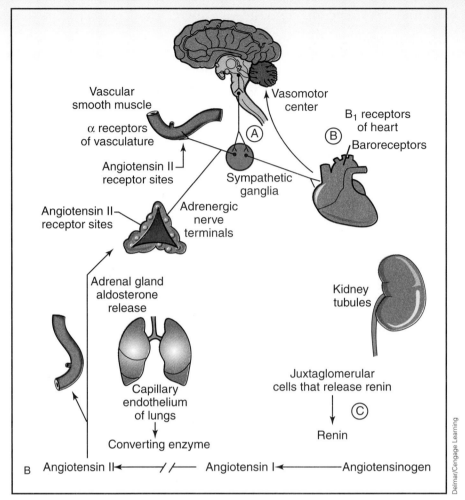

Figure 8-7 *(Continued)* (B) Control of blood pressure. (A) The sympathetic nervous system controls blood pressure through α_1 receptors. When α_1 receptors present on arterioles are activated, they constrict, which increases peripheral resistance and blood pressure. (B) The vasomotor center in the brain responds to stimuli from aortic baroreceptors to cause vasodilation if blood pressure is high or vasoconstriction if blood pressure is low. (C) The kidneys release renin to activate the renin-angiotensin system, causing vasoconstriction and increased blood volume.

CARDIOVASCULAR CONDITIONS

Cardiovascular conditions are those that affect the heart, blood vessels, and blood. Understanding conditions that affect the heart and circulatory system will aid in understanding how cardiovascular drugs work.

Congestive Heart Failure

Congestive heart failure (CHF) is a syndrome that can occur with any disorder that damages or overworks the heart muscle and results in the heart failing to effectively pump blood around the body. Because there is a balance between the pumping of the right and left sides of the heart, any failure of the heart muscle to pump blood out of either side of the heart may result in a backup of blood. Backup of blood may cause the blood vessels to become congested; in time the cells of the body will become deprived of oxygen and nutrients and exposed to excessive amounts of waste products.

The basic unit of cardiac muscle is the sarcomere, which contains two contractile proteins, actin and myosin, which are highly reactive with each other but at rest are kept apart by troponin. When a cardiac muscle cell is stimulated, calcium enters the cell and inactivates the troponin, allowing actin and myosin to form cross-bridges (Figure 8-8). The formation of these cross-bridges allows muscle fibers to slide together or contract. The contraction process requires energy, oxygen, and calcium to allow the formation of the cross-bridges. The degree of shortening (strength of contraction) is determined by the amount of calcium present (more calcium means more cross-bridges will be formed) and by the stretch of the sarcomere before contraction begins (the farther apart the actin and myosin are before cell stimulation, the more cross-bridges will be formed and the stronger the contraction). Therefore, the more blood in the heart, the greater the contraction will be to empty the heart (up to a point). If the cross-bridges are stretched too far apart, they will not be able to reach each other, and no contraction will occur.

Conditions that can lead to CHF include cardiomyopathy (caused by infections, genetic disorders, or degeneration), hypertension, or valvular disease. Cardiomyopathy causes muscle alterations and ineffective contraction and

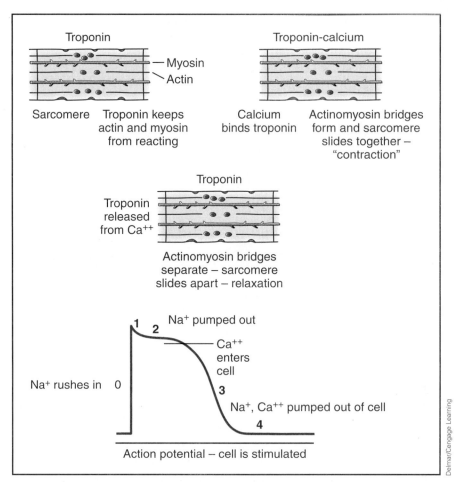

Figure 8-8 The sliding filament theory of heart muscle. Calcium entering the cell deactivates troponin and allows actin and myosin to reaction (causing contraction). Calcium pumped out of the cell frees troponin to separate actin and myosin; the sarcomere filament slides apart and the cell relaxes.

pumping. Hypertension leads to enlarged cardiac muscle because the heart has to work harder than normal to pump against the high pressure in the arteries. Hypertension puts constant, increased demands for oxygen on the system because the heart is continually pumping so hard. Valvular disease leads to an overload of the ventricles because the valves do not close tightly, which allows blood to leak backward into the ventricles. This overloading causes muscle stretching and increased demand for oxygen and energy as the heart muscle has to constantly contract harder. The end result of all of these conditions is that the heart muscle cannot pump blood effectively throughout the vascular system. When the left ventricle pumps inefficiently, blood backs up into the lungs, causing pulmonary vessel congestion and fluid leakage into the alveoli and lung tissue. In severe cases, pulmonary edema can occur (Figure 8-9A). When the right side of the heart is the primary problem, blood backs into the venous system leading to the right side of the heart. Liver congestion and ascites (fluid in the abdomen) are seen with right-sided heart failure (Figure 8-9B). Because the cardiovascular system works as a closed system, one-sided heart failure, if left untreated, eventually leads to failure of both sides.

In CHF, the heart is unable to provide blood with adequate levels of oxygen to the body. Without adequate levels of oxygen, the body's cells trigger a series of compensatory responses (Figure 8-10). Decreased cardiac output stimulates baroreceptors in the aortic arch and carotid arteries causing sympathetic

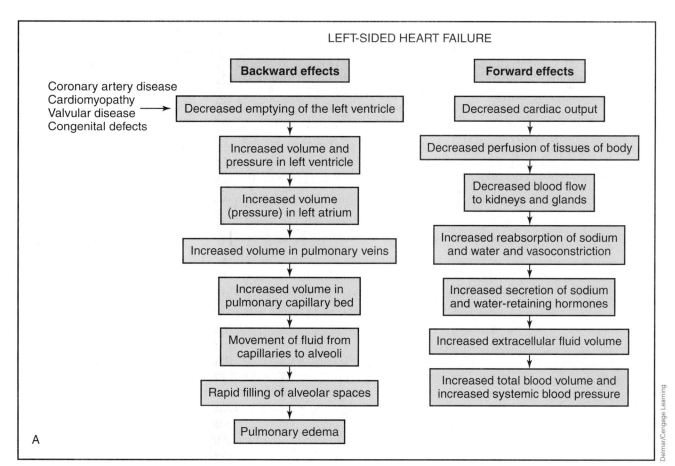

Figure 8-9 (A) Pathophysiology of left-sided heart failure; (B) pathophysiology of right-sided heart failure.

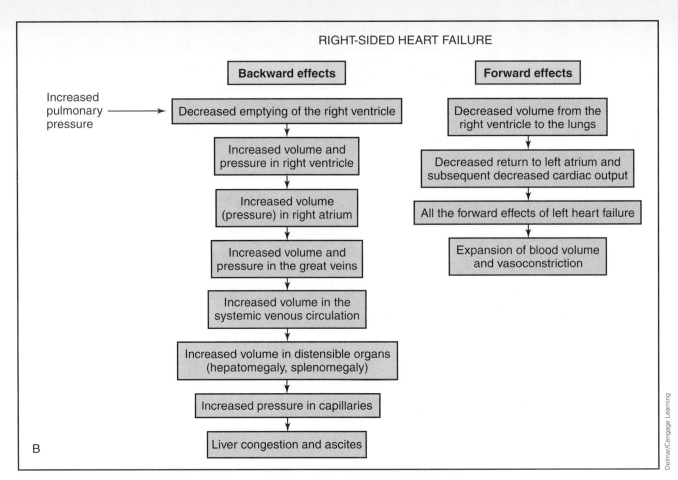

Figure 8-9 *(Continued)*

stimulation. Sympathetic stimulation causes an increase in heart rate, blood pressure, and force of contraction. Various hormones such as aldosterone are released in an attempt to correct the problem by conserving and retaining fluids in an effort to increase blood volume and the output of blood. Decreased cardiac output also stimulates the release of renin from the kidneys and activates the renin-angiotensin system, which further increases blood pressure and blood volume. These compensatory responses help the situation temporarily; however, increased fluid retention eventually becomes harmful, causing fluid to leak out of the capillaries, in turn causing coughing and reduced stamina. Eventually, the heart muscle stretches from overwork, and the chambers of the heart dilate secondary to the increased blood volume that they have to handle. This hypertrophy (enlargement) leads to inefficient pumping.

Cardiac Arrhythmias

All cardiac cells have some degree of automaticity, and these cells undergo a spontaneous depolarization during rest because they decrease the flow of potassium ions out of the cells and probably leak sodium into the cell, causing an action potential. There are five phases of the action potential of cardiac cells:

- Phase 0 occurs when the cell reaches a point of stimulation, sodium gates open along the cell membrane, and sodium rushes into the cell, resulting in an electrical potential. This is called depolarization.

Clinical Que

The underlying problem with CHF usually involves muscle function.

- Phase 1 is a very short period during which sodium ion concentration equalize inside and outside the cell.

- Phase 2 occurs as the cell membrane becomes less permeable to sodium, calcium slowly enters the cell, and potassium begins to leave the cell. This is repolarization.

- Phase 3 is a time of rapid repolarization as sodium gates are closed and potassium flows out of the cell.

- Phase 4 occurs when the cell comes to a rest; the sodium-potassium pump returns the membrane to its resting membrane potential, and spontaneous depolarization begins again.

A disruption in the cardiac rate or rhythm is an arrhythmia. Arrhythmias interfere with the work of the heart and can disrupt cardiac output. Arrhythmias can be caused by changes in the rate, stimulation from an ectopic focus, or by alterations in conduction through the muscle. The underlying causes of

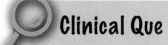

Clinical Que

Fluid in the lungs is called pulmonary edema, fluid below the skin is called peripheral or limb edema, and fluid in the abdomen is called ascites.

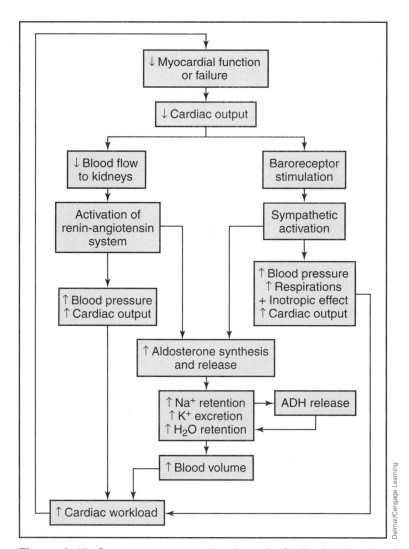

Figure 8-10 Compensatory mechanisms in CHF which lead to increased cardiac workload and further CHF.

arrhythmias can arise from changes to the automaticity or conductivity of the heart cells. These changes can be caused by electrolyte disturbances, decreases in oxygen delivered to cells, structural damage that changes the conduction pathway, accumulation of waste products, and acidosis. In some cases, arrhythmias may be the result of drugs that alter the action potential or cardiac conduction.

Types of arrhythmias include the following:

- *Sinus arrhythmias* occur when the autonomic nervous system changes the rate of firing in order to meet the animal's oxygen demand. A faster-than-normal heart rate with a normal-appearing ECG is called sinus tachycardia (Figure 8-11). Sinus bradycardia is a slower-than-normal heart rate with a normal-appearing ECG.

- *Supraventricular arrhythmias* originate above the ventricles but not in the SA node. These arrhythmias have an abnormally shaped P-wave, because the site of origin is not the sinus node. The QRS complex is normally shaped because the ventricles are still conducting impulses normally. Types of supraventricular arrhythmias include premature atrial contractions (an ectopic focus in the atria generating an impulse out of the normal rhythm), paroxysmal atrial tachycardias (sporadically occurring runs of rapid heart rate originating in the atria), atrial flutter (a single ectopic focus generating a regular, fast atrial rate), and atrial fibrillation (many ectopic foci firing in an uncoordinated manner throughout the atria).

- *Atrioventricular block* is a slowing or lack of conduction at the AV node due to structural damage, hypoxia, or injury to the heart muscle. There is first-, second-, and third-degree heart block that vary in their appearance on ECG.

- *Ventricular arrhythmias* originate below the AV node from ectopic foci that do not use the normal conduction pathways. The QRS complexes are wide and prolonged, and the T-waves are inverted. An example of a ventricular arrhythmia is premature ventricular complexes or PVCs; PVCs can arise from a single ectopic focus or many ectopic foci.

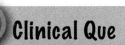

Clinical Que

Supraventricular arrhythmias originate above the ventricles (e.g., the SA node or AV node). Ventricular arrhythmias originate in the ventricles (e.g., premature ventricular complexes or PVCs).

Alterations of Blood Pressure

Alterations of blood pressure may result in hypertension (increased blood pressure) or hypotension (decreased blood pressure). Hypertension results in prolonged force being put on the vessels of the vascular system, which eventually could lead to left ventricle thickening because the muscle must constantly work harder to pump blood out at a greater force. Damage to the arteries and arterioles could lead to loss of kidney function if affecting the vessels to the glomeruli. Hypotension results in the tissues of the body not receiving sufficient oxygenated blood to function, which may allow waste products to accumulate and cells to die from lack of oxygen.

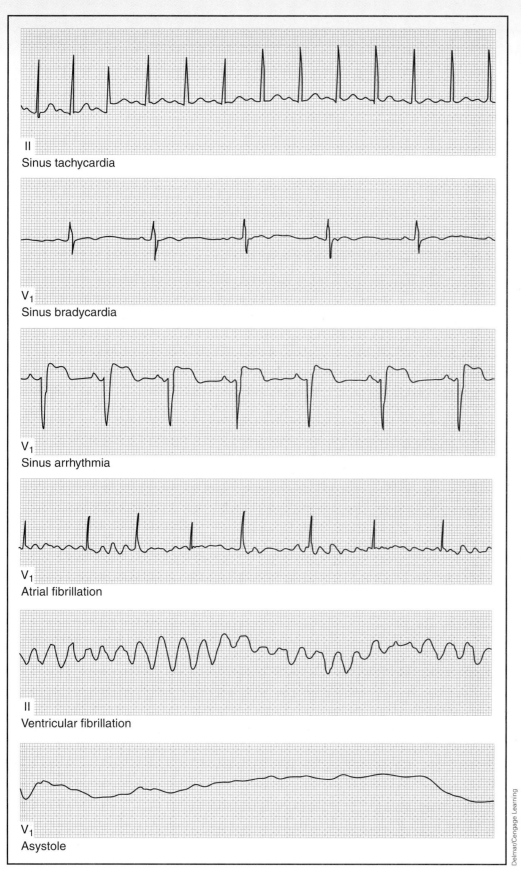

II
Sinus tachycardia

V$_1$
Sinus bradycardia

V$_1$
Sinus arrhythmia

V$_1$
Atrial fibrillation

II
Ventricular fibrillation

V$_1$
Asystole

Delmar/Cengage Learning

Figure 8-11 ECG tracings of common arrhythmias.

CARDIOVASCULAR DRUGS

Cardiovascular drugs can influence aspects of cardiac performance in the following ways:

- Increasing or decreasing the force of myocardial contraction. Drugs that increase the force of contraction are called *positive inotropic drugs*. Drugs that reduce the force of contraction are called *negative inotropic drugs*.

- Increasing or decreasing heart rate by altering the rate of impulse formation at the SA node. Drugs that increase heart rate are called *positive chronotropic drugs*. Drugs that decrease heart rate are called *negative chronotropic drugs*.

- Increasing or decreasing the conduction of electrical impulses through the myocardium. Drugs that increase the rate of electrical conduction are called *positive dromotropic drugs*. Drugs that decrease the rate of electrical conduction are called *negative dromotropic drugs*.

Increasing Force

Positive inotropic drugs aim to improve the strength of myocardial contraction. Although these drugs may increase the heart's demand for oxygen, they are commonly used to increase the strength of contractions by failing heart muscle. Examples of positive inotropic drugs are cardiac glycosides, catecholamines, and benzimidazole-pyridazinones.

Cardiac Glycosides

Cardiac glycosides, or digitalis drugs, are derived from natural sources (the foxglove plant) and have been in use for hundreds of years to treat heart problems in humans. Digitalis drugs increase the strength of cardiac contractions, decrease heart rate, have an antiarrhythmic effect, and decrease signs of dyspnea. Digitalis increases the strength of contraction by inhibiting the sodium-potassium pump, which increases intracellular calcium concentrations. This increase in intracellular calcium concentration causes the myocardial fibers to contract more efficiently. Cardiac glycosides are used in animals to treat such heart problems as CHF, atrial fibrillation, and supraventricular tachycardia. Side effects include anorexia, vomiting, diarrhea, and cardiac arrhythmias. Blood levels must be monitored to prevent toxicity. Digitalis compounds have interactions with many other drugs, so multiple drug dosing must be monitored very closely. Examples of digitalis drugs include *digoxin* (Lanoxin®, Lanoxicaps®), which comes in elixir, injectable, tablet, and capsule formulations, and *digitoxin* (Crystodigin®), which comes in tablet form. Digoxin has a shorter duration of action than digitoxin, and thus it is less likely to result in cumulative toxic effects. Because of toxicity issues associated with cardiac glycoside use, blood must be routinely collected to determine digoxin levels.

Catecholamines

Catecholamines are chemicals that cause a sympathomimetic response and have the following influences: They increase the force and rate of myocardial contraction,

Clinical Que

All cardiac glycosides have a low therapeutic index; that is, the therapeutic dose is very close to the toxic dose. Careful patient monitoring for signs of toxicity (vomiting, lethargy, and arrhythmias) is a very important part of the animal care protocol for patients prescribed any cardiac glycoside.

which increases cardiac output; they constrict peripheral blood vessels, which increases blood pressure; and they increase blood glucose levels. Receptors in the sympathetic nervous system are referred to as *adrenergic receptors* and are classified as alpha-1 (α_1), alpha-2 (β_2), beta-1 (β_1), and beta-2 (β_2). Alpha-1 receptors are found primarily in smooth muscle tissue of peripheral blood vessels and in the sphincters of the gastrointestinal system and urinary tract. When α_1 receptors are stimulated, contraction occurs in the smooth muscles associated with them. Alpha-2 receptors seem to function as controllers of neurotransmitters in the synaptic space. Stimulation of β_2 receptors results in reduction of neurotransmitter release. Beta-1 receptors are located primarily in heart muscle and fatty tissue. Stimulation of β_1 receptors in the heart increases heart rate and causes more forceful heart contractions. Stimulation of β_1 receptors in fatty tissue promotes the breakdown of stored fat to fatty acids, which can be used by the body as energy sources. Beta-2 receptors are located primarily in bronchial smooth muscle and in the walls of blood vessels of skeletal muscle, the brain, and the heart. Stimulation of β_2 receptors in bronchial smooth muscle produces muscle relaxation, thereby causing bronchodilation. Table 8-2 summarizes the effects of stimulation of catecholamine receptors. The following are the primary catecholamines (sympathomimetic drugs).

Epinephrine

Epinephrine has both alpha and beta activity. It causes smooth muscle relaxation in the bronchioles (bronchodilation), raises blood glucose levels, and increases heart rate and contractility. It is also used as a cardiac stimulant following cardiac arrest. It may be given intracardiac, intratracheal, intramuscular, or subcutaneous. It is manufactured in two concentrations (1:1000 solution and 1:10,000 solution), so care should be taken when ordering, performing dose calculations, or measuring drug doses to make sure that the proper concentration is chosen. The 1:10,000 solution is the preferred concentration for use in veterinary medicine. Side effects include the possible development of arrhythmias and hypertension.

Dopamine

Dopamine (Inotropin®, Dopastat®, Dopamine® HCl, and Dopamine® HCl in 5 percent dextrose) is a precursor to norepinephrine and works on both α_1 and β_1 receptors. Dopamine causes increased heart contractility, increased heart

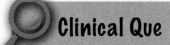

Clinical Que

Digoxin and digitoxin are positive inotropes and negative chronotropes. They cause the heart to pump harder and the heart rate to slow.

Clinical Que

Sympathetic response through β_1 receptors produces positive inotropic and chronotropic effects on the heart. Parasympathetic response through cholinergic receptors produces negative inotropic and chronotropic effects on the heart.

Table 8-2	Catecholamine Receptors	
RECEPTOR	EFFECT WHEN STIMULATED	EXAMPLE OF CLINICAL USE AFFECTING RECEPTOR
α_1	Constriction of peripheral blood vessels	Constricts blood vessels within the area and prevents rapid diffusion of drug away from the injection site (e.g., epinephrine can be injected with a local anesthetic).
α_2		Affects CNS by causing drowsiness and sedation.
β_1	Increases heart rate and force of heart contractions; increases blood glucose	Treats animals with decreased cardiac function.
β_2	Causes bronchodilation	Treats diseases where bronchodilation is needed.

rate, and increased blood pressure. It is used in treating acute heart failure, severe shock, and oliguric (scanty amount of urine) renal failure. Side effects include tachycardia, dyspnea, and vomiting.

Dobutamine

Dobutamine (Dobutrex®) is a direct β_1 agonist, with slight β_2 and alpha activity. It produces increased cardiac output without the dilation of blood vessels seen with dopamine. It is given as a constant-rate IV infusion.

Isoproterenol

Isoproterenol (Isuprel®) has beta activity and is used in the treatment of cardiac arrhythmias and bronchial constriction. Side effects include tachycardia, weakness, and tremors; therefore, it is no longer frequently used in veterinary practice.

Benzimidazole-pyridazinones

Benzimidazole-pyridazinones are inodilators because they increase force of contraction (positive inotrope) and cause widening of the blood vessels (vasodilation). The drug in this category used in veterinary medicine is *pimobendan* (Vetmedin®), which is approved for use in dogs to treat signs of heart failure due to atrioventricular valvular insufficiency or dilated cardiomyopathy. Pimobendan increases ventricular contractility and reduces preload and afterload (it also usually decreases heart rate). A benefit of pimobendan treatment is that it does not increase oxygen consumption in patient with heart disease. It is available as a chewable tablet. Side effects include anorexia, lethargy, diarrhea, and dyspnea.

Fixing the Rhythm

Antiarrhythmic drugs are divided into categories according to the types of arrhythmia they treat and the actions they have on the cells of the heart. Antiarrhythmic drugs can have the following influences on the heart: decreasing automaticity, altering the rate of electrical impulse conduction, or altering the refractory period of the heart muscle between consecutive contractions. Table 8-3 summarizes types of arrhythmias, Table 8-4 summarizes antiarrhythmic drugs, and Figure 8-12 summarizes action potentials.

Table 8-3 Types of Arrhythmias

ARRHYTHMIA	DESCRIPTION
Atrial flutter	Rapid contraction of the atria at a rate too rapid for the ventricles to pump efficiently.
Atrial fibrillation	Irregular and rapid atrial contraction resulting in a quivering of the atria and inefficient and irregular ventricular contraction.
Premature ventricular contractions	Beats originating from the ventricles instead of the SA node, causing the ventricles to contract before the atria, and resulting in a decrease in the amount of blood pumped to the body.
Ventricular tachycardia	Rapid heartbeat originating in the ventricles.
Ventricular fibrillation	Rapid, disorganized contractions of the ventricles resulting in the inability of the heart to pump blood to the body.

Table 8-4   Antiarrhythmic Drugs

CLASS OF ANTIARRHYTHMIC	DRUG NAME	ACTION
Class IA: Local Anesthetics This group works as local anesthetics to the nerves and myocardial membrane (prolonging the action potential).	A. *quinidine* (Cin-Quin®, Cardioquin®, Quinidex®) B. *procainamide* (Pronestyl®, Procamide® SR, Procan® SR)	A. Suppresses myocardial excitability and increases conduction times. Used orally to treat atrial and ventricular arrhythmias. Side effects include vomiting, diarrhea, and weakness. B. Suppresses myocardial excitability and increases conduction times. Used orally to treat premature ventricular contractions (PVC), ventricular tachycardia, and some atrial tachycardias. Side effects are vomiting, diarrhea, and weakness.
Class IB: Membrane Stabilization This group works by blocking the influx of sodium into the cell, thus stabilizing the myocardium and preventing depolarization (shortens the duration of the action potential).	A. *lidocaine* (Xylocaine®, Lidocaine® for Injection 1% and 2%) B. *tocainide* (Tonocard®) C. *mexiletine* (Mexitil®)	A. Depresses myocardial excitability. Used IV to control PVCs and treat ventricular tachycardia. Side effects are rare. B. Action similar to lidocaine, but is only given orally. Used orally to treat ventricular arrhythmias (ventricular tachycardia and PVCs). Side effects include ataxia, vomiting, and hypotension. C. Action is similar to lidocaine, but is used orally. Used to treat ventricular arrhythmias (ventricular tachycardia and PVCs). Side effects include vomiting and unsteadiness.
Class II: Beta-Adrenergic Blockers This group works by blocking beta-adrenergic receptors or by preventing release of norepinephrine from the adrenergic neuron (depresses the depolarization phase).	*propranolol* (Inderal®, Intensol®)	Reduces automaticity by blocking β_1 and β_2 receptors, thus reducing oxygen demand by the myocardium. Used to treat hypertrophic cardiomyopathy and some ventricular arrhythmias. Side effects include bradycardia, lethargy, and depression.
Class III: Potassium Channel Blockers This group works by lengthening the time between action potentials, which decreases the sinus rate (prolongs repolarization).	A. *bretylium* (Bretylol®) B. *amiodarone* (Cordarone®) C. *sotalol* (Betapace®)	A., B., C. Increases the duration of the action potential. Both are used for emergency treatments of ventricular tachycardia and fibrillation. These drugs are used in animals that are resistant to other drugs. Side effects include hypotension and vomiting.
Class IV: Calcium Channel Blockers This group works by blocking channels that allow calcium to enter the myocardial cell. The entry of calcium into the cell facilitates muscle contractility (depresses depolarization and lengthens repolarization).	A. *verapamil* (Isoptin®, Calan®) B. *nifedipine* (Procardia®, Adalat®) C. *diltiazem* (Cardizem®)	A. Blocks calcium passage, relaxing vascular smooth muscle to lower blood pressure and inhibit the cardiac conduction system. Given orally and IV to treat supraventricular tachycardia and atrial flutter. B., C. Blocks calcium passage, dilating coronary and peripheral blood vessels, which results in the reduction of cardiac workload. Both are used orally to treat atrial fibrillation and supraventricular tachycardia. Side effects of calcium channel blockers include hypotension and edema.

Polarization

When the nerve cell is polarized, positive ions (+) are on the outside of the cell membrane and the negative ions (–) are on the inside of the cell membrane.

Depolarization

In response to a stimulus, the positive ions move from the outside to the inside of the cell membrane, while the negative ions move to the outside.

Repolarization

After the stimulus has passed along the nerve fiber, the ions move back to their original place until another stimulus occurs.

Figure 8-12 Polarization, depolarization, and repolarization.

The following points highlight special concerns about antiarrhythmic drugs.

- Digoxin levels may increase in animals taking quinidine; the dosage of digoxin may have to be lowered.

- Reactions to procainamide are likely in animals that are sensitive to procaine and other "-caine" local anesthetics.

- Be certain not to use the lidocaine product with epinephrine when giving lidocaine IV. Lidocaine is not effective when given orally because of its high first-pass effect (metabolism of orally administered drugs by hepatic enzymes after absorption from the gastrointestinal tract resulting in significant reduction of the amount of unmetabolized drug that reaches the systemic circulation). Cats are sensitive to lidocaine, so dosing must be carefully monitored.

- Because propranolol blocks β_1 and β_2 activity, it affects the heart (lowering heart rate and blood pressure) and bronchioles (causing bronchoconstriction).

Correcting Constriction

As heart failure occurs, the animal's body tries to compensate for the loss of cardiac function. The first reaction to a failing heart is to increase heart rate. Next, blood vessels are constricted because of a nervous system reaction and renin production by the kidney. Renin converts angiotensinogen to angiotensin I, and angiotensin I is converted by ACE to angiotensin II, a potent vasoconstrictor. Angiotensin II stimulates aldosterone secretion by the adrenal cortex. Aldosterone influences the tubules of the nephron to reabsorb sodium ions and thus retain water. The retained water is physiologically utilized to expand the circulating blood volume, which improves tissue perfusion.

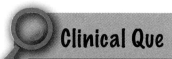

Clinical Que

Peripheral resistance to blood flow is primarily controlled by constriction or relaxation of the arterioles. Constricted arterioles raise blood pressure; dilated arterioles lower blood pressure.

Vasodilators are drugs used to dilate arteries and/or veins, which alleviates vessel constriction and improves cardiac output. The actions of vasodilators often greatly improve the condition of an animal with CHF. Types of vasodilators include the following.

Angiotensin-Converting Enzyme Inhibitors

ACE inhibitors are combined vasodilators (both venous and arterial) used in the treatment of heart failure and hypertension. They prevent the conversion of angiotensin I to angiotensin II (Figure 8-13). Examples include *enalapril* (Enacard®, Vasotec®), *lisinopril* (Zestril®), *benazepril* (Lotensin®), *captopril* (Capoten®), and *ramipril* (Altace®, Vasotop®). Side effects include hypotension and gastrointestinal problems.

Arteriole Dilators

Hydralazine (Apresoline®) is an arteriole dilator used to reduce the afterload associated with CHF. Side effects include hypotension and gastrointestinal signs.

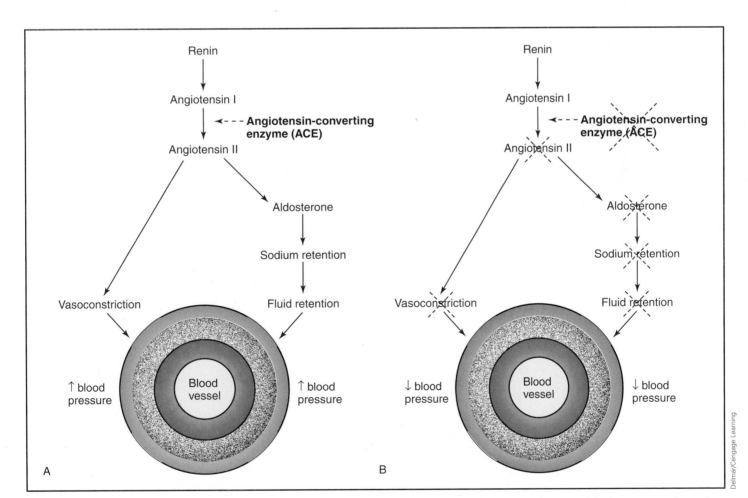

Figure 8-13 ACE inhibitors act as antagonists of the renin-angiotensin system by interfering with the conversion of angiotensin I to angiotensin II. This ultimately results in vasodilation.

Venodilators

Nitroglycerin ointment (Nitro-Bid®, Nitrol®) is a venodilator used to improve cardiac output and reduce pulmonary edema. It is applied as a dose in inches to the skin (wear gloves when applying nitroglycerin). It is also available in transdermal system form (patch). Care must be taken so that the animal does not lick or chew the patch if this method of delivery is used. Side effects include rashes and irritation at the application site.

Combined Vasodilators

Prazosin (Minipress®) is a combined vasodilator (both venous and arterial). It is used to treat CHF, dilated cardiomyopathy, hypertension, and pulmonary hypertension. Side effects include hypotension and gastrointestinal problems. *Nitroprusside* (Nitropress®) is another combined vasodilator that is used to treat CHF secondary to mitral valve regurgitation and hypertension. Side effects include hypotension, retching, and restlessness.

Calcium Channel Blockers

Amlodipine (Norvasc®), *verapamil* (Isoptin®, Calan®), *nifedipine* (Procardia®, Adalat®), and *diltiazem* (Cardizem®) are calcium channel blockers that are used to treat CHF and hypertension. Calcium channel blockers inhibit the movement of calcium through channels across the myocardial cell membranes and vascular smooth muscle. Cardiac and vascular smooth muscle depends on the movement of calcium ions into the muscle cells through specific ion channels.

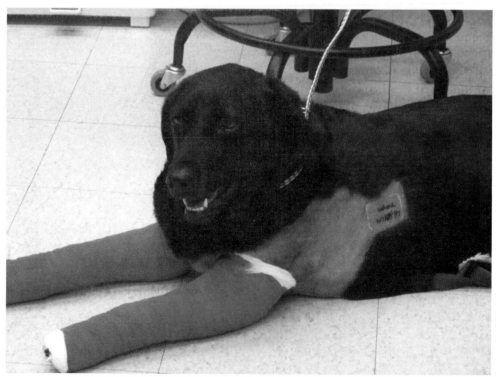

Figure 8-14 Dog with a transdermal patch.

Courtesy of Teri Raffel, CVT

When this movement is inhibited, the coronary and peripheral arteries dilate, which in turn decreases the force of cardiac contraction. Some calcium channel blockers are also used as antiarrhythmic drugs. Side effects of calcium channel blockers include hypotension and anorexia.

Losing Fluid

Diuretics are drugs that increase the volume of urine excreted by the kidneys and thus promote the release of water from the tissues. This process, called *diuresis*, lowers the fluid volume in tissues (extracellular fluid volume). Diuretics are used in the treatment of hypertension because they promote fluid loss and thus reduce blood pressure. These drugs are covered in Chapter 12, which is on the urinary system. Table 8-5 summarizes diuretic drugs.

DRUGS AFFECTING BLOOD

Drugs affecting blood are categorized by their effects and goals for use: to stop clots from forming, to stop bleeding, or to enhance blood and its products.

Clot Stopping

Anticoagulants are used to inhibit clot formation by inactivating one or more clotting factors (Figure 8-15). They are used clinically to inhibit clotting in catheters, to prevent blood samples from clotting, to preserve blood transfusions, and to treat emboli that may occur. Examples of anticoagulants include the following.

Heparin

Heparin prevents the conversion of prothrombin to thrombin (Figure 8-16). Mainly used in the process of testing blood and in transfusion, heparin is used to treat heart disorders such as thromboembolism, where a clot, or thrombus, is constricting an artery of the heart. Heparin does not break down clots, but prevents them from getting bigger. It is also used to treat disseminated intravascular coagulation (DIC) and laminitis in horses. Side effects include bleeding and thrombocytopenia (decrease in blood clotting cells). Heparin overdose is treated with protamine sulfate. The strongly basic protamine combines with the strongly acidic heparin to form a stable complex with no anticoagulant activity.

Clopidogrel Bisulfate

Clopidogrel bisulfate (Plavix®) is an oral platelet aggregation inhibitor that may prevent thrombi formation in cats and may improve circulation in cats following an embolic event. Clopidogrel bisulfate prevents platelet aggregation by a different mechanism (inhibits adenosine diphosphate binding to platelets) than aspirin (see the following section). Side effects are gastrointestinal in nature and include vomiting and anorexia. Giving the drug with food may alleviate these side effects.

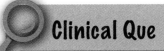

Clinical Que

Anticoagulants do not alter the size of an existing thrombus or clot.

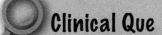

Clinical Que

Heparin is not effective when administered orally because it is rapidly inactivated by HCl in the stomach.

Table 8-5   Diuretic Drugs

Diuretic Category	Action	Examples	Side Effects
Thiazide diuretics	Inhibit sodium and chloride reabsorption from the distal convoluted tubule of the nephron	• *hydrochlorothiazide* (HydroDIURIL®) • *chlorothiazide* (Diurel®) • *hydroflumethiazide* (Saluron®) • *bendroflumethiazide* (Corzide®)	Hypokalemia (low blood potassium levels)
Loop diuretics	Inhibit reabsorption of sodium and chloride in the loop of Henle, reducing the ability of the kidneys to concentrate urine; more potent than thiazides in promoting sodium and water excretion	• *furosemide* (Lasix®, Disal®, Diuride®) • *ethacrynic acid* (Edecrin®)	Hypokalemia
Potassium-sparing diuretics	Work in a variety of ways 1. Inhibit aldosterone (mineralocorticoid that normally increases sodium retention and thus water in the kidney) 2. Block sodium reabsorption in the distal convoluted tubule 3. Inhibit sodium reabsorption in the distal convoluted tubule	1. *spironolactone* (Aldactone®) 2. *trimterene* (Dyazide®) 3. *amiloride* (Midamor®)	Hyperkalemia
Osmotic diuretics	Large molecules that can be filtered by the glomerulus but have limited capability of being reabsorbed into the blood. High concentration of osmotic agent is left in the kidney tubule, which carries large amounts of fluid with it	• *mannitol* (Osmitrol®) • *glycerin* (Osmoglyn®)	Vomiting, electrolyte imbalance
Carbonic anhydrase inhibitors	Block the action of carbonic anhydrase, an enzyme that normally is involved in the reabsorption of sodium, potassium, bicarbonate, and water	• *acetazolamide* (Diamox®) • *dichlorphenamide* (Daranide®)	Gastrointestinal disturbances, CNS signs (depressed CNS activity)

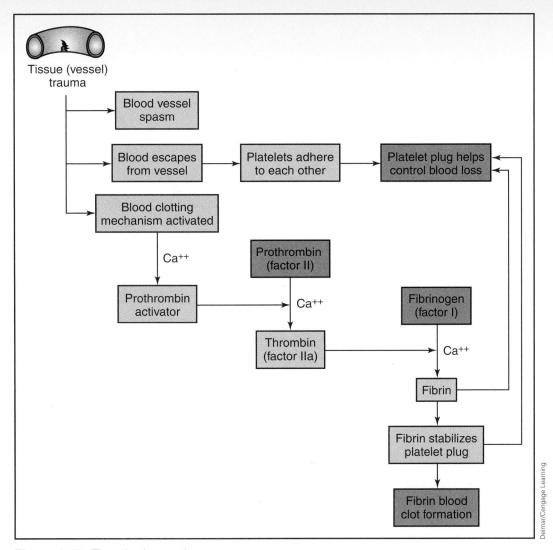

Figure 8-15 The clotting pathway.

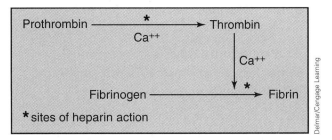

Figure 8-16 Heparin exerts its anticoagulant activity by interfering with the conversion of prothrombin to thrombin.

Ethylenediamine Tetraacetic Acid (EDTA)

EDTA (Meta-Dose®) chelates calcium, which prevents clots from forming. It is used in lavender-topped blood tubes to prevent clotting of blood samples and as an injectable chelating agent in the treatment of lead poisoning. Side effects include vomiting, diarrhea, and renal toxicity.

Coumarin Derivatives

This type of anticoagulant binds vitamin K_1, thereby inhibiting the formation of prothrombin. *Coumarin derivatives* are used for the long-term treatment of thromboembolisms and include the trade-named drugs Coumadin®, Dicumarol®, and Sofarin®. Overdose of coumarin can be treated with vitamin K_1. Side effects include bleeding and weakness.

Aspirin

Aspirin has antiplatelet activity because it inhibits the stickiness of platelets through the inhibition of thromboxane. It is used to prevent thromboembolism associated with heartworm disease in small animals and cardiomyopathies in cats. Side effects include bleeding, gastrointestinal bleeding, and vomiting. Caution should be exercised when using aspirin in cats (see Chapter 16).

Blood Transfusion Anticoagulants

Acid citrate dextrose (ACD®) and citrate phosphate dextrose adenine (CPDA-1®) are chemicals found in preservative bags used to collect blood for preservation. Both work by chelating calcium.

Bleeding Control

Hemostatic drugs help promote the clotting of blood. This category of drugs can be divided into parenteral and topical.

Parenteral hemostatic drugs include the following:

- *vitamin K_1* or *phytonadione*, a synthetic version of vitamin K_1. Vitamin K is involved in the clotting cascade; if it is not available, formation of some clotting factors is hindered. It is used in veterinary medicine for the treatment of rodenticide poisoning and for bleeding disorders. Brand-name formulations include AquaMEPHYTON®, Mephyton®, and Konakion®. Side effects include anaphylactic reactions and bleeding from the injection site.

- *protamine sulfate*, a basic protein used to treat heparin overdose and bracken fern poisoning in cattle. It is given slowly IV and is marketed generically. Side effects include hypotension and bradycardia.

Topical hemostatic drugs are used to control capillary bleeding. They include the following:

- *silver nitrate sticks*, which has astringent (an agent that constricts tissue) properties to stop capillary bleeding. Staining of skin and clothing is seen with silver nitrate use.

- *hemostat powder* (ferrous sulfate), to control capillary bleeding from superficial cuts and wounds and after dehorning of cattle, sheep, or goats. Irritation, redness, and swelling may be seen with this product, and if seen, use of this product should be discontinued.

- *gelfoam gelatin sponge*, which provides hemostasis following surgical procedures. This sponge, available in a variety of sizes and shapes, is applied to the bleeding tissue to quickly stop the flow of blood. The gelatin material is gradually absorbed by the body without inducing excessive scar formation.

- *thrombogen topical thrombin solution*, which acts directly to promote the conversion of fibrinogen to fibrin. Topical thrombin is derived from animal sources and is available as a sterile powder, which is usually reconstituted with sterile distilled water or isotonic saline before use. The speed with which the drug promotes clotting of blood is proportional to its concentration.

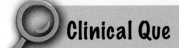

Clinical Que

The synthesis of new clotting factors after giving vitamin K_1 takes 6 to 12 hours; therefore, emergency needs for clotting factors must be met by administering blood products.

Blood-Enhancing Drugs

Blood-enhancing drugs include drugs that affect red blood cells. Erythrocytes or red blood cells (RBCs) are made in the bone marrow in response to a substance produced in the kidneys called **erythropoietin**. Red blood cells carry oxygen to tissues. Hemoglobin is the portion of the RBC that transports oxygen. Hemoglobin is made up of a heme protein and an iron-containing compound. Adequate amounts of iron and other vitamins and minerals are essential to the formation of RBCs. Drugs that affect the production or quality of RBCs include the following.

Iron

Iron compounds include iron dextran, ferrous sulfate, ferric hydroxide, and others. Iron compounds are used clinically in the treatment of baby pig anemia and as a nutritional supplement. Iron comes in both oral and injectable forms; examples of brand-name forms are Iron Dextran Complex Injection (Armedexan®, Ferrextran®, Pigdex 100®, and Imposil®), Iron-Gard 100®, and Purina Oral Pigemia®. Side effects include discoloration at the injection site and muscle weakness.

Erythropoietin

Erythropoietin is a protein made by the kidneys that stimulates the differentiation of bone marrow stem cells to form RBCs. It is used to treat anemia in animals with chronic renal failure. It is marketed as a human recombinant genetically engineered hemopoietin (epoetin alpha) under the trade names Epogen® and Procrit®. While human recombinant erythropoietin works in dogs and cats, it is not the same protein, and antibodies eventually develop in response to its exposure. Because of this, allergic reactions are sometimes seen with erythropoietin products. At the present time, the amino acid sequence of the canine and feline versions of erythropoietin are known but commercial products are not available. Erythropoietin must be refrigerated.

Cyanocobalamine

Cyanocobalamine is vitamin B_{12} and is used to treat vitamin B_{12} deficiencies. Cyanocobalmine is a cobalt-containing, water-soluble vitamin that serves as an important cofactor for many enzymatic reactions in mammals that are required for normal cell growth and erythropoiesis. Brand names include Cyanoject® and Rubesol® and are available in injectable and oral forms; however, oral forms are not appropriate for small animal medicine due to limited gastrointestinal absorption.

Folic Acid

Folic acid is a B vitamin needed for normal erythropoiesis, and deficiency of folic acid may be seen in dogs, cats, and horses due to small intestinal disease. Folate® and Folacin® are brands of folic acid. There are few side effects from folic acid administration with either the oral or injectable form.

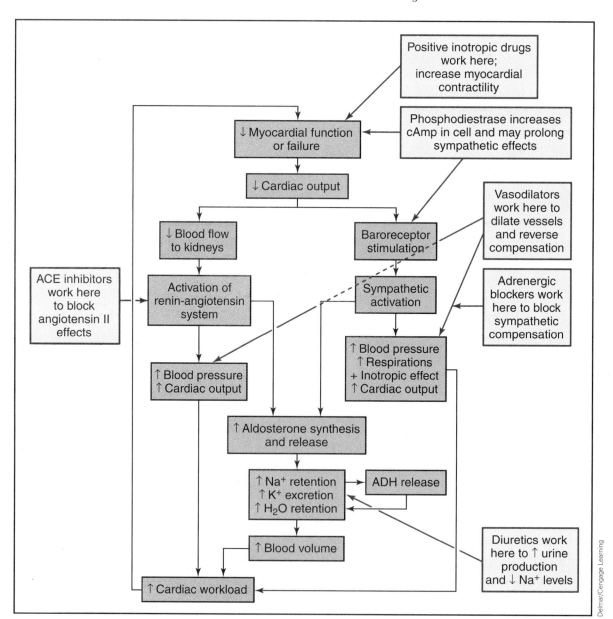

Figure 8-17 Summary of select drugs used to treat cardiovascular disease.

SUMMARY

The functions of the cardiovascular system include the delivery of oxygen, nutrients, and hormones to body tissues and the transportation of wastes to waste removal systems of the body. Abnormalities that can affect the heart include decreased myocardial contraction, changes in heart rate, and changes in heart rhythm.

Positive inotropes are drugs that improve the strength of myocardial contraction. Examples of positive inotropes are cardiac glycosides (digoxin and digitoxin), catecholamines (such as epinephrine, dopamine, dobutamine, and isoproterenol), and benzimidazole-pyridazinones (pimobendan). Antiarrhythmics are drugs that correct abnormal heart rhythms and include local anesthetics, membrane stabilizers, beta-adrenergic blockers, potassium channel blockers, and calcium channel blockers. Vasodilators counteract the vasoconstriction associated with heart disease and may dilate veins only, arterioles only, or both veins and arterioles (called combined or mixed vasodilators). Examples of vasodilators are ACE inhibitors, arteriole dilators, venodilators, combined vasodilators, and calcium channel blockers. Diuretics decrease the fluid volume that a diseased heart is required to pump, thereby decreasing the stress on the heart. Figures 8-17 and 8-18 summarize the sites where some of these drugs work.

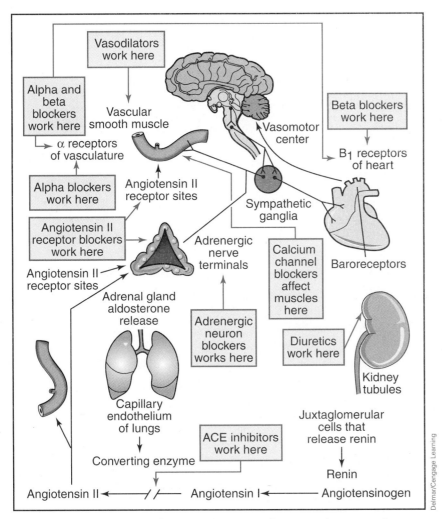

Figure 8-18 Summary of select drugs used to treat hypertension.

Anticoagulants inhibit clot formation; they include heparin, clopidogrel bisulfate, EDTA, coumarin derivatives, aspirin, and chemicals used in blood preservation for transfusion. Hemostatic drugs are substances that help promote clotting of blood. This category of drugs includes vitamin K_1 and protamine sulfate (parenteral) and silver nitrate sticks, hemostat powder (ferrous sulfate), gelfoam gelatin sponge, and thrombogen topical thrombin solution (topical). Blood-enhancing drugs affect red blood cells. Drugs that affect the production or quality of RBCs include iron, erythropoietin, cyanocobalmin (vitamin B12), and folic acid.

It's a Wrap

The answers to the questions in this chapter's Setting the Scene case study should be understood after reading this chapter. When an owner asks how drugs are going to help her pet, it is important for the veterinary technician to be able to answer these questions in a manner that is clear to the client. In this case, the dog is prescribed diuretics and vasodilators. Diuretics decrease preload (volume of blood transported by the heart) and vasodilators decrease afterload (resistance to blood flow). If the heart has less fluid to move, it does not have to work so hard (theory behind diuretic use). If the heart does not have so much resistance to overcome to allow movement of blood, it does not have to work so hard (theory behind vasodilator use). Why can't she keep her dog on the same dog food that she has been feeding for many years? What is the rationale behind low-sodium diets for animals with CHF? The sodium-restricted diet also decreases preload. Remember that where sodium goes, water follows. By decreasing sodium in the diet, the volume of "water" in the blood decreases, reducing the volume of blood that the heart has to move (similar to the theory behind diuretic use). How is the success or failure of cardiac treatment assessed? Is there any way to determine if the drugs and diet restriction are working? Two ways to determine if the drugs and diet restrictions are working are by clinical assessment (decreased fluid in the abdomen, decreased coughing, increased energy levels, etc.) and by diagnostic tests (chest radiographs, blood tests, ECG, etc.). It is wise to recheck animals that are diagnosed with cardiovascular disease. When should the owner bring the dog into the clinic again for reexamination? If there is no improvement with therapy, the dog should be rechecked right away; if there is some improvement, the time before recheck examination depends on the animal's status and veterinarian recommendation. In nonemergency situations, this may be one week from the original diagnosis and treatment.

CHAPTER REVIEW

Matching

Match the drug name with its action.

1. _____ digoxin

2. _____ propranolol

3. _____ prazosin

4. _____ coumarin

5. _____ epinephrine

6. _____ protamine sulfate

7. _____ hydralazine

8. _____ silver nitrate

9. _____ heparin

10. _____ lidocaine

a. cardiac glycoside that increases the strength of contractions and decreases heart rate

b. catecholamine that increases force of myocardial contraction

c. membrane-stabilizing antiarrhythmic used to control PVCs and ventricular tachycardia

d. beta-adrenergic blocker used to treat hypertrophic cardiomyopathy

e. arteriole dilator

f. combined vasodilator

g. anticoagulant that prevents the conversion of prothrombin to thrombin

h. hemostatic drug used to treat heparin overdose

i. topical hemostatic drug

j. anticoagulant that binds vitamin K_1

Multiple Choice

Choose the one best answer.

11. Coumarin toxicity, found when animals ingest rat poisoning, is treated with
   a. heparin.
   b. vitamin K_1.
   c. EDTA.
   d. silver nitrate sticks.

12. Erythropoietin is used to
   a. increase WBC count in infectious disease.
   b. increase urine loss to decrease fluid retention.
   c. stop bleeding.
   d. increase RBC production.

13. Which group of drugs prevents the conversion of angiotensin I to angiotensin II?
   a. arteriole dilators
   b. venodilators
   c. ACE inhibitors
   d. combined vasodilators

14. Which antiarrhythmic drug is used to treat ventricular fibrillation?
   a. propranolol
   b. bretylium
   c. verapamil
   d. lidocaine

15. Catecholamines function as what autonomic nervous system response?
   a. sympathetic
   b. parasympathetic
   c. cholinergic
   d. sympathetic-blocking

16. What drug is manufactured in both 1:1000 and 1:10,000 concentrations?
   a. epinephrine
   b. isoproterenol
   c. dopamine
   d. dobutamine

17. Which cardiovascular drug should be monitored for toxicity levels through blood testing?
    a. digoxin
    b. epinephrine
    c. dopamine
    d. lidocaine

18. The use of diuretics in treating heart disease is believed to
    a. increase preload.
    b. decrease preload.
    c. increase afterload.
    d. decrease afterload.

19. The use of vasodilators in treating heart disease is believed to
    a. increase preload.
    b. decrease preload.
    c. increase afterload.
    d. decrease afterload.

20. Which of the following are false regarding the ECG?
    a. The P-wave represents atrial depolarization.
    b. The QRS complex represents ventricular depolarization.
    c. The T-wave represents ventricular repolarization.
    d. None of the above are false.

Case Studies

21. A 12-year-old M/N Boston Terrier (14#) comes into the clinic for exercise intolerance and coughing at night. On PE, his TPR are normal; however, he has developed ascites and is very lethargic. The veterinarian requests chest X-rays and an ECG. The tests reveal that the dog is in heart failure.
    a. What medication group could be given to this dog to reduce preload (think of reducing fluid volume that the heart must pump around)?
    b. What medication group could be given to this dog to reduce afterload (think of reducing the tension against which the heart has to pump)?

22. A seven-year-old F Poodle (10#) presents to the clinic with weakness. On PE, the veterinarian notes an arrhythmia, so she hooks the dog up to an ECG. The ECG reveals that the dog has a ventricular arrhythmia.
    a. Considering that this is an emergency, what drug could be given IV?
    b. What drug could this dog be sent home on once she is stabilized?

Critical Thinking Questions

23. What clinical signs of possible cardiovascular disease can be observed by the client without medical training?

24. Think about the implications of giving an animal the diagnosis of cardiac disease. What major changes in care and lifestyle will follow for the animal and the owner?

ADDITIONAL REVIEW MATERIAL

Additional review material can be found on the StudyWARE™ CD included with this text.

BIBLIOGRAPHY

Allen, D. G. (2005). *Handbook of Veterinary Drugs* (3rd ed.). Baltimore, MD: Lippincott Williams & Wilkins.

Amundson Romich, J. (2008). *An Illustrated Guide to Veterinary Medical vTerminology* (3rd ed.). Clifton Park, NY: Delmar Cengage Learning.

Broyles, B., Reiss, B., and Evans, M. (2007). *Pharmacological Aspects of Nursing Care* (7th ed.). Clifton Park, NY: Delmar Cengage Learning.

Plumb, D. C. (2008). *Plumb's Veterinary Drug Handbook* (6th ed.). Ames, IA: Blackwell Publishing.

Spratto, G. R. and Woods, A. L. (2008). *PDR Nurse's Drug Handbook.* Clifton Park, NY: Delmar Cengage Learning.

Thomason, J. D., Fallaw, T. K., and Calvert, C. (November 2007). *Pimobendan Treatment in Dogs with Congestive Heart Failure.* Veterinary Medicine.

Yin, S. (February 2008). *Cardiology Update: Canine Heart Failure.* Veterinary Forum.

CHAPTER 9
RESPIRATORY SYSTEM DRUGS

OBJECTIVES

Upon completion of this chapter, the reader should be able to:

- describe the basic anatomy and physiology of the respiratory tract.
- differentiate between ventilation and respiration.
- describe the basic pathophysiology of various types of respiratory diseases.
- describe the clinical usages of the following drugs in the treatment of respiratory disease: expectorants, mucolytic drugs, antitussives, decongestants, bronchodilators, antihistamines, respiratory stimulants.

KEY TERMS

antihistamine

antitussive

beta-2-adrenergic agonist

bronchodilator

cholinergic blocking agents

decongestant

expectorant

mucolytic

respiratory stimulant

Setting the Scene

A four-year-old male (M) Beagle is brought into the clinic with a two-day history of a harsh, honking cough. When questioned, the owner says that the dog was boarded the previous week while the family was on vacation. A PE shows that the dog's TPR are normal, and he does not appear dehydrated. While in the clinic, he coughs quite a bit, and the veterinarian determines that this dog probably has kennel cough (a disease that can be caused by the bacterium *Bordetella bronchiseptica* and several viruses, including canine parainfluenza virus). Because this disease is usually self-limiting, no treatment is initiated. The owner, however, tells the veterinary technician that the dog is keeping the family awake all night with the incessant coughing. He wants to know if there is anything he can give the dog to control its cough. Is this a wise thing to do? Should all coughs be suppressed? The desire to suppress or not suppress a cough is based on the type of cough it is. How can the difference be explained to the owner of this beagle? What is the rationale behind the treatment chosen for this dog? What drugs are available for this dog, and what are their potential side effects?

BASIC RESPIRATORY ANATOMY AND PHYSIOLOGY

The *respiratory system* is the body system that brings oxygen from the air into the body for delivery via the blood to the cells. *Respiration* is the exchange of gases (oxygen and carbon dioxide) between the atmosphere and the cells of the body. *Ventilation* is the term used to describe the bringing in of fresh air.

The respiratory system is divided into two parts: the upper respiratory tract, consisting of the nostrils, nose, nasal cavities, pharynx, and larynx, and the lower respiratory tract, consisting of the trachea, bronchi, bronchioles, and alveoli (Figures 9-1 and 9-2). Several important anatomical features play a role in respiratory therapies.

- The upper respiratory tract is lined with epithelial cells that contain cilia (microscopic hairs) and goblet cells. Goblet cells secrete mucus, which can trap foreign particles that are then moved out of the respiratory tract via the movement of cilia (Figure 9-3).

- The bronchioles are lined with smooth muscle that can be dilated with sympathetic nervous stimulation and constricted with parasympathetic nervous stimulation.

- The alveoli (small, terminal, saclike structures where oxygen and carbon dioxide diffuse) produce a lipoprotein called *surfactant*. Surfactant keeps the alveoli open by reducing surface tension of these small sacs. Absence of surfactant leads to alveolar collapse.

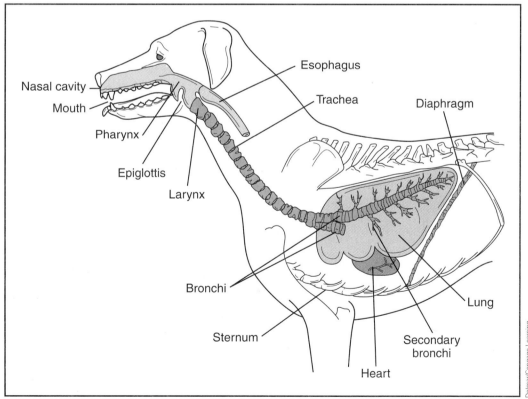

Figure 9-1 Structures of the respiratory system.

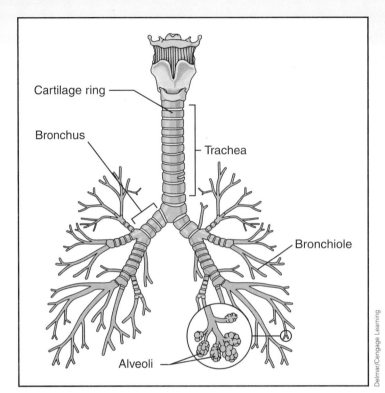

Figure 9-2 Lower respiratory tract structures.

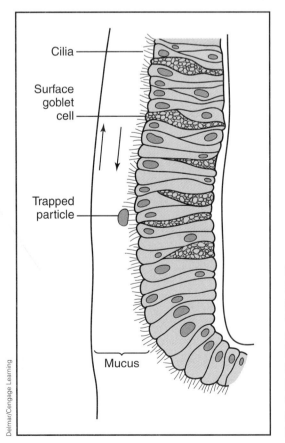

Figure 9-3 The mucociliary escalator. Respiratory epithelium contains goblet cells that secrete mucus, which traps dust, microorganisms, and other foreign substances. Respiratory epithelial cells also contain cilia (microscopic, hairlike projections) which are constantly moving and directing the mucus and any trapped substances toward the throat.

- Irritation to the respiratory tract is sensed by cough receptors in the respiratory tract that send a signal to the cough center of the brain stem. The brain stem in turn sends impulses to the respiratory muscles to produce a forceful expiration or cough. Nonproductive or dry coughs are not accompanied by the expulsion of fluid or material from the respiratory tract. Productive coughs help expel mucus and foreign material from the respiratory tract. It is usually recommended that only nonproductive coughs be suppressed, because productive coughs help clear the respiratory tract.

RESPIRATORY CONDITIONS

Disorders of the respiratory system are typically categorized into upper respiratory and lower respiratory conditions. Upper respiratory conditions in animals are mainly caused by infectious agents that produce clinical signs such as congestion and coughs. Lower respiratory tract conditions in animals include asthma (chronic inflammatory disease of the airways that causes bronchospasms and bronchoconstriction), bronchitis (inflammation and possible infection of the bronchi), pneumonia (inflammation of the lungs caused by microorganism invasion of the tissue or by aspiration of foreign substances into the lower respiratory tract), and chronic obstructive pulmonary disease or COPD (a slowly progressive disease of the airways characterized by a gradual loss of lung function). COPD in horses is commonly called heaves.

RESPIRATORY DRUGS

Drugs that affect the respiratory system work to keep the airways open and gases moving efficiently. Respiratory drugs used to treat upper respiratory conditions include expectorants, mucolytics, antitussives, and decongestants. Respiratory drugs used to treat lower respiratory conditions include bronchodilators and antihistamines. Respiratory stimulants are used to enhance ventilation in animals. The drugs in these categories affect the respiratory tract in different ways.

Cough Causing (Expectorants)

Expectorants increase the flow of respiratory secretions to allow material to be coughed up from the lungs. Expectorants do this by increasing the fluidity of mucus, which is more effectively coughed up than thicker mucus. Expectorants are believed to work by acting on the goblet cells or by reducing the stickiness of the mucus. Increased secretory activity also helps keep dry, irritated tissue moist, thus protecting it from further trauma. *Glyceryl guaiacolate (guaifenesin)* is a secretory expectorant used orally in humans to increase secretions from the airway and clear it of fluid. Guaifenesin is also used in horses as a muscle relaxant, as part of a general anesthesia protocol. Trade names of guaifenesin products include Guailazin® and Gecolate®. Side effects include a mild decrease in blood pressure and increase in heart rate.

Mucus Breaking (Mucolytics)

Mucolytic drugs decrease the viscosity or thickness of respiratory secretions. *Acetylcysteine* (Mucomyst®, Mucosol®) is a mucolytic expectorant used to break up thick mucoid secretions in the airway to promote better respiration. Acetylcysteine is administered orally, intravenously, or as an aerosol through nebulization (Figure 9-4). Acetylcysteine is also used to treat acetaminophen toxicity in cats; it helps metabolize the acetaminophen and decreases its toxicity in the liver.

Cough Controlling (Antitussives)

Antitussives suppress coughing. Antitussives can be centrally acting (working on the cough centers of the brain stem) or locally acting (soothing irritation to the mucosal lining of the respiratory tract that initiates coughing). Locally acting antitussives are usually syrups or lozenges and are not used commonly in veterinary practice.

Butorphanol (Torbugesic®, Torbutrol®) is a centrally acting antitussive that is available by injection (with differing concentrations per mL) and oral tablet form. It is now classified as a C-IV controlled substance; therefore, careful documentation of its use is required. Butorphanol is also used as a preanesthetic and analgesic, because its main side effects are sedation and ataxia. It also causes respiratory depression.

Hydrocodone (Hycodan®, Tussigon®) is a C-III controlled substance that is a centrally acting narcotic. It comes in tablet and syrup forms, and is used primarily for harsh, nonproductive coughs. Side effects include sedation and slowing of gastrointestinal motility that may result in constipation.

Codeine is a C-V, C-III or C-II centrally acting antitussive that may or may not be mixed with aspirin or acetaminophen (C-III) or guaifenesin (C-V or C-III, depending on the concentration of guaifenesin). Codeine-only products are C-II. Codeine is available under many generic and trade names in injectable,

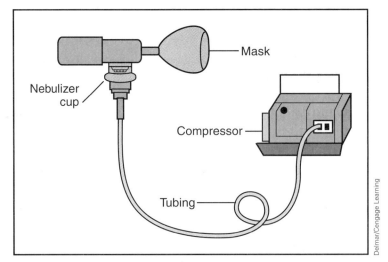

Figure 9-4 Nebulizers are devices that transform solutions or suspensions of medications into aerosols that are optimal for deposition in the lower airway. Liquid medicine is placed in the nebulizer cup.

syrup, and tablet forms. Its side effects also include sedation and decreased gastrointestinal motility.

Dextromethorphan is a nonnarcotic, centrally acting antitussive that is chemically similar to codeine. Dextromethorphan is not addictive like codeine and is sold over the counter. It is available in syrups, but owners should be advised that products containing other "cold products" in the syrup might be harmful to animals. Dextromethorphan is generally not very effective in nonhuman species.

Trimeprazine (a centrally acting antitussive) and prednisolone (a glucocorticoid) are found in the combination drug Temaril-P®. This drug has antitussive and antipruritic effects and is available in spansules and tablets. Side effects include sedation, depression, and hypotension.

Congestion Reducing (Decongestants)

Decongestants decrease the congestion of nasal passages by reducing swelling. Decongestants may be given by spray or orally as a liquid or tablet. These products have limited use in veterinary practice, but have been used to help treat feline upper respiratory tract disease. Decongestants include *phenylephrine* (Neo-Synephrine®) and *pseudoephedrine* (Sudafed®). Phenylephrine has cardiostimulatory properties and should not be used in animals with hypertension or tachycardia (increased heart rate).

Bronchi Widening (Bronchodilators)

Bronchodilators widen the lumen of the bronchioles and counteract bronchoconstriction. Bronchoconstriction occurs when acetylcholine is released by the parasympathetic nerves, causing increased respiratory secretions; histamine is released from mast cells through allergic or inflammatory reactions, causing it to bind to receptors on bronchiole smooth muscle; or beta-2-adrenergic receptors are blocked as a result of other drug treatments. **Cholinergic blocking agents**, more commonly known as *anticholinergics*, produce bronchodilation by counteracting the first mechanism: They bind to acetylcholine receptors and prevent bronchoconstriction (Figure 9-5). Cholinergic blockers include *aminopentamide* (Centrine®), *atropine* (Atropine SA®, Atropine LA®), and *glycopyrrolate* (Robinul-V®). Side effects include dry mouth, dry eyes, and tachycardia. Atropine has additional side effects, including vomiting, constipation, and urinary retention.

Another group of bronchodilators works by stimulating beta receptors and hence are referred to as **beta-2-adrenergic agonists**. Beta-2 receptors are involved in bronchodilation and stabilization of mast cells (reducing histamine release). Bronchodilators in this category are *epinephrine* (which is usually reserved for life-threatening situations and has beta-1 and alpha-1 activity); *isoproteronol* (which also causes beta-1 stimulation and thus affects the heart); and *terbutaline, albuterol,* and *clenbuterol* (which have primarily beta-2 activity and thus little effect on the heart). These drugs go by the following generic and trade names: *epinephrine* (Adrenalin Chloride® and generically labeled forms), *isoproteronol* (Isuprel®), *terbutaline* (Brethine®), *albuterol* (Proventil®, Ventolin®), and *clenbuterol* (Ventipulmin® syrup). Side effects include tachycardia (increased

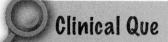
Clinical Que

Spansules are medicine-containing capsules that are coated to slow the dissolution rate, so that the medicine is delivered more slowly.

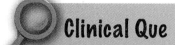
Clinical Que

Bronchodilators ↓ airway resistance and ↑ airflow.

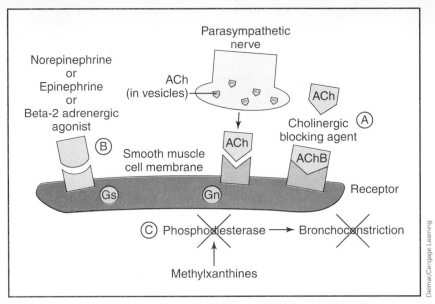

Figure 9-5 Mechanism of action of bronchodilators. (A) Cholinergic blocking agents (AChB) bind to acetylcholine receptors and prevent bronchoconstriction. (B) Beta-2-adrenergic agonists bind to beta receptors and cause bronchodilation. (C) Methylxanthines inhibit an enzyme (phosphodiesterase) in smooth muscle cells that normally promotes bronchoconstriction, but when it is inhibited, bronchodilation occurs.

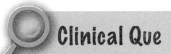

Clinical Que

Some theophylline products are sustained release. The reported release rates for humans may or may not correlate to actual release rates in veterinary patients.

heart rate), and CNS excitement and weakness. Clenbuterol is banned for use in food-producing animals.

A last group of bronchodilators are the methylxanthines. Methylxanthines work by inhibiting an enzyme in smooth muscle cells. The enzyme normally promotes bronchoconstriction, but when it is inhibited, bronchodilation occurs. *Aminophylline* (Aminophyllin®), *theophylline* (Slo-BID Gyrocaps®, Theo-dur®), caffeine, and theobromine (found in chocolate) are examples of methylxanthines. Side effects include CNS stimulation and gastrointestinal irritation.

Histamine Blocking (Antihistamines)

Antihistamines block the effects of histamine, a chemical released from mast cells that combines with H_1 receptors on bronchiole smooth muscle to cause bronchoconstriction (histamine also works on other body systems such as the heart (stimulates), stomach (stimulates gastric secretions), arterioles (causes dilation), and the inflammatory response) (Figure 9-6). Antihistamines are usually used in prevention of respiratory problems such as heaves (in horses) and feline asthma. Generic names for antihistamines usually end with *-amine*. Examples in this category include *diphenhydramine* (Benadryl®) and *chlorpheniramine* (ChloroTrimeton®). Side effects include CNS depression and anticholinergic effects (dry mouth and urinary retention).

Respiratory Stimulants

Respiratory stimulants are drugs that stimulate the animal to increase its respirations. *Doxapram* is a central nervous stimulant that is usually used in neonates to stimulate respiration after Cesarean section or dystocia. Doxapram

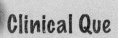

Clinical Que

Some over-the-counter antihistamines may contain other active ingredients. Take care to counsel clients to choose OTC products that contain only the desired antihistamine compound.

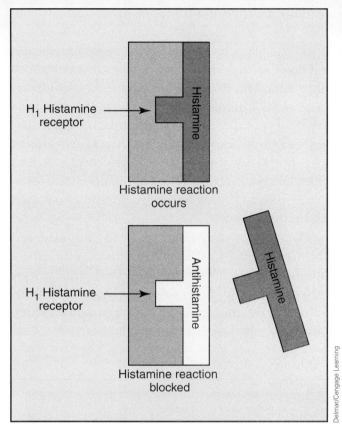

Figure 9-6 Mechanism of antihistamine action. Antihistamines block the effects of histamine, a chemical released from mast cells that combines with H₁ receptors on bronchiole smooth muscle to cause bronchoconstriction. By blocking histamine vasoconstriction does not occur.

is also used to restore reflexes after anesthesia. It is available for injection as Dopram-V®. Side effects include hypertension, arrhythmias, and seizures.

Asthma Drugs

Feline asthma is a chronic, noninfectious inflammatory disease of the lower airways characterized by intermittent, reversible airway obstruction and heightened airway sensitivity. Feline asthma is a condition seen in some cats when inhaled allergens cause sudden contraction of the smooth muscles around airways. Common allergens that can trigger feline asthma include grass and tree pollens, cigarette or fireplace smoke, various sprays (hair sprays, deodorants, flea sprays, and deodorizers), and dust from cat litter. The most common clinical signs in cats with asthma are wheezing and a dry, hacking cough that may be confused with gagging. In mildly affected cats, coughing and wheezing may occur only occasionally; however, severely affected cats have daily coughing and wheezing and increased airway constriction, leading to open-mouth breathing and panting that can be life threatening.

In humans, dogs, and horses, histamine is the key mast cell mediator leading to bronchial smooth muscle contraction, but in feline asthma, seratonin plays the more important role. Both seratonin and histamine cause bronchoconstriction and increased vascular permeability allowing the influx of inflammatory cells. Traditionally, feline asthma has been managed with oral glucocorticoids to decrease underlying inflammation and bronchodilators to reverse smooth muscle contraction. The use of metered dose inhalers (MDIs) for delivery of medications directly into the airways is another option for delivery of both glucocorticoids and bronchodilators that are available as MDI preparations (Figure 9-7). Aerosolized therapy produces increased drug concentration at the site of action and decreased systemic absorption, leading to fewer side effects. Additionally, the faster onset of action facilitated by the aerosolized delivery of bronchodilators provides owners a home treatment for acute asthmatic attacks, thus reducing emergency trips to the veterinarian and reducing patient stress during transport. Specific drugs used in the treatment of asthma include the following:

Glucocorticoids are anti-inflammatory drugs that may be administered orally or by inhalation for the treatment of asthma. Traditional oral glucocorticoids used in tvhe treatment of feline asthma include oral *prednisone* or *prednisolone* for moderately affected cats and IV *dexamethasone* for severely affected cats. Glucocorticoids decrease the formation of cytokines, which decreases production of inflammatory prostaglandins and leukotrienes. Cats

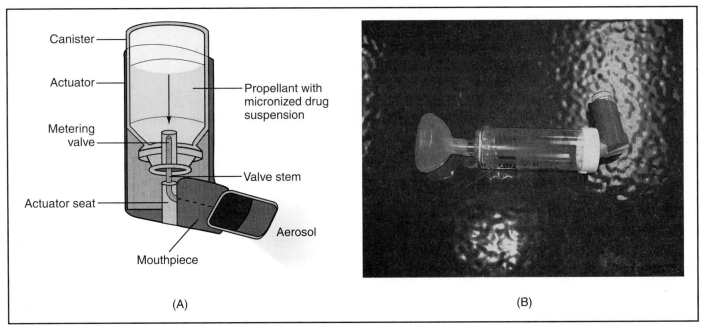

(A) (B)

Figure 9-7 (A) The MDI includes a pressurized canister, a metering chamber, and an actuator. The canister contains a liquid propellant that serves as a delivery and dispersal medium for the dissolved or suspended medication. Depressing the canister releases a measured amount of the aerosolized combination of propellant and medication. As the patient inhales, the propellant readily evaporates, leaving the particles of medication to be deposited in the bronchial tree. (B) Adapting MDIs for cats requires the use of spacers and facemasks. A spacer is an extension add-on device that permits the aerosol from the MDI to expand and slow down, turning it into a very fine mist instead of a high-pressure actuation spray. The fine drug particles are carried deep into the lung where they are most efficacious, instead of hitting the tongue or the back of the throat the way a blast from an MDI sometimes does. Spacers also eliminate the need to coordinate inhalation with actuation of the device. The MDI fits into one end of the spacer and the facemask on the other. A one-way valve at the facemask connection ensures that medication only leaves the spacer during inhalation. (B. Courtesy of Kimberly Kruse Sprecher, CVT).

on chronic high doses of oral glucocorticoids may develop polyuria, polydipsia, polyphagia, cystitis, pancreatitis, and insulin resistance. Inhaled glucocorticoids are used to decrease swollen and narrowed airways that are caused by inflammation. Inhaled glucocorticoids increase air flow and facilitate respiration by decreasing airway inflammation. *Fluticasone* (Flovent®) has potent anti-inflammatory activity and is available as an aerosol and powder for inhalation. Typically, cats with asthma are initially treated with oral prednisone until a beneficial response is achieved (usually a few weeks), and then inhaled glucocorticoids are started while the animal is being weaned off oral prednisone. Cats are gradually weaned off oral glucocorticoids to avoid acute adrenal insufficiency (see Chapter 10). Side effects of fluticasone include upper respiratory infections and irritation to the upper respiratory tract. Side effects of inhaled glucocorticoids are significantly fewer than with oral or injectable ones.

Bronchodilators may be used in addition to glucocorticoids in the treatment of feline asthma for control of clinical signs. Selective beta-2-adrenergic agonists such as *terbutaline* (Brethine®) and *albuterol* (Proventil®, Ventolin®) cause bronchial smooth muscle relaxation and reduced airway resistance. Side effects include tachycardia and CNS excitement. Tolerance may develop with chronic therapy with beta-2-adrenergic agonists. Short-acting beta-2-agonists such as terbutaline and albuterol can produce a rapid onset of action when administered by inhalation. *Ipratropium* (Atrovent®), an inhalational anticholinergic agent, may be useful if clinical signs are not controlled adequately with beta-2-adrenergic agonists alone. Side effects include tracheal or bronchial irritation.

Methylxanthines, such as *theophylline* (Theo-Dur®, Slo-BID®), are another group of bronchodilators used in the treatment of feline asthma and have been previously discussed. Selective beta-2-adrenergic agonists are generally preferred over methylxanthines in the treatment of feline asthma because methylxanthines have a narrow therapeutic index.

Other oral treatments for feline asthma include cyproheptadine and T-cell modulators. They are not first-line drugs, and are prescribed as adjunct therapies for patients whose signs cannot be controlled with more traditional medications. *Cyproheptadine* (Periactin®) is more commonly used as an appetite stimulant in cats and is covered in Chapter 11. It is an antihistamine and antiseratonin medication, but is used in feline asthma for its antiseratonin effects. It blocks seratonin receptors in airway smooth muscle cells, preventing bronchoconstriction secondary to mast cell degranulation.

Cyclosporin (Sandimmune®) is a fungal derived protein that modulates T lymphocyte activity, and may be useful in the late phase of the allergic response in feline asthma (it is covered in Chapter 20). The oral absorption of cyclosporine is unpredictable in cats, and blood levels should be monitored during therapy. Side effects include nephrotoxicity and vomiting.

COPD Drugs

Equine COPD or heaves is a chronic, noninfectious respiratory disease in which inflammation in the small airways of the lung leads to impaired ventilation due to enlargement and destruction of airspaces secondary to chronic generalized bronchiolitis (similar to asthma). COPD is seen in horses in cold climates, where horses are kept in barns for prolonged periods of time and where hay is often moldy. It is believed that COPD is caused by an allergic reaction to these molds when horses inhale them, causing an inflammation in the terminal bronchioles. Other allergens such as pollens, chemicals, and microorganisms may also trigger the condition. Signs of COPD include coughing, tachypnea, labored breathing, and yellow nasal discharge. Horses with severe disease appear listless and dyspnic, and develop a muscular "heave line" along the barrel from taking a double exhalation.

COPD is treated by improving management practices (keeping horses outside as much as possible, reducing dust, storing hay in a dry place away from horses, and improving ventilation where horses are stabled) and the use of some medications. Glucocorticoids such as oral *prednisone* or injectable *dexamethasone* help decrease smooth muscle contraction, suppress inflammation, and reduce mucus production. Inhaled glucocorticoids are used to decrease swollen and narrowed airways that are caused by inflammation. Inhaled glucocorticoids provide a high dose of medication within the airways with minimal systemic side effects but a special mask is necessary for administration.

Bronchodilators such as the beta-2-adrenergic agonists *clenbuterol* (Ventipulmin®) and *albuterol* (Alupent®, Proventil®) relax the smooth muscle of the airways. They may be safely combined with anti-inflammatory drugs to treat severely affected horses. Bronchodilators may be given orally, by injection, or by inhalation.

Other Drugs Used in the Respiratory System

Other drugs used in the respiratory system include glucocorticoids (Chapter 16), antimicrobials (Chapter 14), and diuretics (Chapter 12). These drugs are addressed under different chapters, and information regarding these drug categories is found in those chapters. The respiratory drugs covered in this chapter are summarized in Table 9-1.

Table 9-1	Respiratory Drugs Covered in This Chapter	
DRUG CATEGORY	ACTION	EXAMPLES
Expectorants	Increase the flow of respiratory secretions to allow the coughing up of material from lungs.	• guaifenesin (Guailaxin®, Gecolate®, Robitussin Cough Syrup®, and Triaminic®)
Mucolytics	Decrease viscosity or thickness of respiratory secretions.	• acetylcysteine (Mucomyst®)
Antitussives	Work centrally by affecting cough centers of brain stem. Work locally by soothing irritation to mucous lining of respiratory tract. Limited use in veterinary practice.	• butorphanol (Torbugesic®, Torbutrol®) • hydrocodone (Hycodan®) • codeine (generic) • dextromethorphan (Dimetapp®) • trimeprazine and prednisolone (Temaril-P®)

(Continued)

Drug Category	Action	Examples
Decongestants	Decrease the congestion of nasal passages by reducing swelling (limited veterinary use).	• phenylephrine (Neo-Synephrine®) • pseudoephedrine (Sudafed®)
Bronchodilators	Cholinergic blockers (work by counteracting the action of acetylcholine by binding to acetylcholine receptors), beta-2-adrenergic agonists (stimulate beta-2 receptors that cause bronchodilation), or methylxanthines (inhibit an enzyme in smooth muscle cells that normally causes bronchoconstriction)	Cholinergic blockers: • aminopentamide (Centrine®) • atropine (Atropine SA®, Atropine LA®) • glycopyrrolate (Robinul-v®) Beta-2-adrenergic agonists: • epinephrine (Adrenalin Chloride®) • isoproteronol (Ventoline®) • terbutaline (Brethine®) • albuterol (Alupent®, Proventil®) • clenbuterol (Ventipulmin®) Methylxanthines: • aminophylline (Aminophyllin®) • theophylline (Theo-Dur®)
Antihistamines	Block the effects of histamine on bronchiole smooth muscle causing bronchioconstriction	• diphenhydramine (Benadryl®) • chlorpheniramine (ChloroTrimeton®)
Respiratory Stimulants	Act centrally to stimulate CNS	• doxapram (Dopram-V®)
Asthma Drugs	Glucocorticoids reduce inflammation	• prednisone (Deltasone®) • prednisolone (Meticorten®) • fluticasone (Flovent®)
	Bronchodilators (previously covered in this table)	• see above
	Antiseratonin properties prevent bronchoconstriction due to mast cell degranulation	• cyproheptadine (Periactin®)
	T lymphocyte modulator that controls allergic response	• cyclosporin (Sandimmune®)
COPD Drugs	Glucocorticoids (previously covered in this table)	• see above • dexamethasone (Azium® Dexasone®)
	Bronchodilators (previously covered in this table)	• see above

SUMMARY

Respiratory drugs include expectorants, mucolytics, antitussives, decongestants, bronchodilators, antihistamines, respiratory stimulants, asthma drugs, and COPD drugs. Expectorants increase the flow of respiratory secretions to allow material to be coughed up from the lungs. Guaifenesin is an expectorant. Mucolytic drugs decrease the viscosity or thickness of respiratory secretions. Acetylcysteine is a mucolytic drug that is also used for acetaminophen toxicity in cats. Antitussives suppress coughing. Antitussives can be centrally acting (working on the cough centers of the brain stem) or locally acting (soothing irritation to the mucosal lining of the respiratory tract that initiates coughing). Examples of centrally acting antitussives are butorphanol, hydrocodone, codeine, trimeprazine, and dextromethorphan. Locally acting antitussives are rarely used in veterinary medicine. Decongestants decrease the congestion of nasal passages by reducing swelling; they have limited use in veterinary practice. Bronchodilators expand the bronchioles and counteract bronchoconstriction. Bronchodilators are cholinergic blockers, beta-2-adrenergic agonists, or methylxanthines. Cholinergic blockers include atropine, aminopentamide, and glycopyrrolate; beta-2-adrenergic agonists include epinephrine, isoproterenol, terbutaline, albuterol, and clenbuterol; and methylxanthines include aminophylline and theophylline. Antihistamines block the effects of histamine, a chemical released from mast cells that combines with H-1 receptors on bronchiole smooth muscle to cause bronchoconstriction. The generic names of antihistamines usually end with -*amine* and examples include diphenhydramine and chlorpheniramine. A respiratory stimulant is doxapram, which is used mainly in neonates to stimulate respiration after C-section or dystocia.

Drugs used to treat asthma and COPD include glucocorticoids and bronchodilators. Additional oral treatments for feline asthma include cyproheptadine and T-cell modulators.

It's a Wrap

The answers to the questions in this chapter's Setting the Scene case study should be understood after reading this chapter. The client wants to know if there is anything he can give his dog to control its cough. In this particular case, is this a wise thing to do? Should all coughs be suppressed? How can the difference be explained to the owner of this beagle? Coughs are categorized as productive and nonproductive. Productive coughs produce fluid that can be expelled through coughing. Nonproductive coughs are harsh, do not produce fluid to be expelled through coughing, and tend to irritate the respiratory tract. Typically, nonproductive coughs are suppressed with antitussives to limit irritation and further injury to the respiratory tract. What is the rationale behind the treatment chosen for this dog? Antitussives are recommended for severe, continuous coughs that are nonproductive (usually associated with chronic bronchitis). What drugs are available for this dog, and what are their potential side effects? Available antitussives include butorphanol, hydrocodone, codeine, dextromethorphan, and trimeprazine. Side effects of most antitussives include respiratory depression. In this case, antitussive could be prescribed for the patient if the veterinarian feels that the cough is nonproductive.

CHAPTER REVIEW

Matching

Match the drug name with its action.

1. _____ guaifenesin

2. _____ theophylline

3. _____ terbutaline

4. _____ acetylcysteine

5. _____ doxapram

6. _____ butorphanol

7. _____ phenylephrine

8. _____ atropine

9. _____ dextromethorphan

10. _____ diphenhydramine

a. respiratory stimulant
b. methylxanthine
c. beta-2-adrenergic agonist
d. expectorant
e. mucolytic
f. centrally acting antitussive
g. decongestant
h. anticholinergic (cholinergic blocker)
i. nonnarcotic centrally acting antitussive
j. antihistamine

Multiple Choice

Choose the one best answer.

11. What type of respiratory drug inhibits cough production?
   a. antihistamine
   b. antitussive
   c. decongestant
   d. expectorant

12. What drug is a mucolytic and also is used to treat acetaminophen toxicity?
   a. guaifenesin
   b. hydrocodone
   c. theophylline
   d. acetylcysteine

13. Which of the following are controlled substances?
   a. theophylline
   b. aminophylline
   c. butorphanol
   d. guaifenesin

14. Which respiratory drug can be used to prevent respiratory problems?
   a. expectorant
   b. mucolytic
   c. antihistamine
   d. antitussive

15. Which of the following is an antihistamine?
   a. diphenhydramine
   b. hydrocodone
   c. Temaril-P®
   d. butorphanol

16. Which group of bronchodilators works by inhibiting an enzyme in smooth muscle cells that normally causes vasoconstriction?
   a. anticholinergics
   b. beta-2-adrenergic agonists
   c. methylxanthines
   d. all of the above work in a similar fashion

17. Which of the following drugs are used to stimulate respiration in neonates after a cesarean section?
   a. naloxone
   b. doxapram
   c. yohimbine
   d. theophylline

18. Which respiratory drug category decreases the viscosity of respiratory secretions?
   a. expectorant
   b. mucolytic
   c. antitussive
   d. decongestant

19. The cough center is located in what part of the brain?
   a. cerebrum
   b. cerebellum
   c. brain stem
   d. meninges

20. Which antitussive combination drug also has a corticosteroid in it?
   a. acetaminophen with codeine
   b. hydrocodone
   c. Temaril-P®
   d. butorphanol

Case Study

21. A two-year-old F/S Chihuahua (10 lb) presents to the clinic with a dry, harsh cough. On PE, T = 104°F, HR = 120 bpm, and RR = 24 breaths/min. Upon auscultation of the lung fields, harsh referred sounds are heard. The trachea is palpated to determine if the dog has a collapsing trachea, but she does not. The owner states that this dog was boarded while the owner was on vacation. The veterinarian suspects that this dog has kennel cough because she was not vaccinated before she was kenneled.
   a. What category of drug might the veterinarian prescribe for this dog?
   b. The veterinarian knows that most cases of kennel cough resolve in three to seven days. She decides to prescribe butorphanol for this dog, as well as antibiotics to treat a secondary bacterial infection she believes the dog has. What type of antitussive is butorphanol?
   c. Using your answer to question b, what side effect of butorphanol do you think this owner should be warned about?
   d. Is butorphanol a controlled substance?

Critical Thinking Questions

22. If an animal with a nonproductive cough is given fluids because it is dehydrated, what might happen to its cough?

23. Do animals with bronchitis or pneumonia have more severe coughs?

24. Why would epinephrine not be the first-choice bronchodilator for animals?

ADDITIONAL REVIEW MATERIAL

Additional review material can be found on the StudyWARE™ CD included with this text.

BIBLIOGRAPHY

Allen, D. G. (2005). *Handbook of Veterinary Drugs* (3rd ed.). Baltimore, MD: Lippincott Williams & Wilkins.

Amundson Romich, J. (2008). *An Illustrated Guide to Veterinary Medical Terminology* (3rd ed.). Clifton Park, NY: Delmar Cengage Learning.

Broyles, B., Reiss, B., and Evans, M. (2007). *Pharmacological Aspects of Nursing Care* (7th ed.). Clifton Park, NY: Delmar Cengage Learning.

Plumb, D. C. (2008). *Plumb's Veterinary Drug Handbook* (6th ed.). Ames, IA: Blackwell Publishing.

Spratto, G. R. and Woods, A. L. (2008). *PDR Nurse's Drug Handbook*. Clifton Park, NY: Delmar Cengage Learning.

CHAPTER 10

HORMONAL AND REPRODUCTIVE DRUGS

OBJECTIVES

Upon completion of this chapter, the reader should be able to:

- describe the basic anatomy and physiology of the endocrine system.
- explain how endocrine function is regulated.
- describe the functions of pituitary gland hormones and how they influence other glands and organs.
- outline the usual disease mechanisms for endocrine disease.
- explain the pharmaceutical management of endocrine disease including regulation of pituitary gland hormones, blood glucose, metabolic rate, adrenal cortex hormones, and reproductive hormones.
- differentiate the role of insulin and glucagon in regulating blood glucose levels.
- describe the role of the thyroid gland in regulating metabolism.
- list the stages of estrous and describe the function of the hormones in each stage.
- describe the functions of FSH, LH, GnRH, estrogen, progesterone, and prostaglandins in the estrous cycle.
- differentiate between the follicular and luteal phases of the estrous cycle.
- describe the role of testosterone as a growth promotant.
- outline the role of endocrine drugs in regulating the reproductive status of animals, including estrous cycle, abortion, and pregnancy maintenance.
- describe uses of reproductive drugs other than influencing reproductive status.

KEY TERMS

adrenocortical insufficiency
anabolic steroids
androgen
anestrus
diabetes insipidus
diabetes mellitus
diestrus
estrogen
estrus
follicular phase
glucagon
glucocorticoid
gonadotropin
growth promotants

hormone
hyperadrenocorticism
hyperthyroidism
hypothyroidism
insulin
luteal phase
metestrus
mineralocorticoid
proestrus
progesterone
prostaglandin
testosterone
thyroxine
tri-iodothyronine

Setting the Scene

A five-year-old M/N Dachshund comes into the clinic for an examination. He has been lethargic lately and seems to be gaining weight. Upon weighing him, the veterinary technician discovers that he has not gained weight since his last appointment, but that his abdomen hangs lower than it used to. He is still a bit overweight, but has been for many years. The owner says he has been drinking a lot and urinating a lot. On PE, the veterinary technician finds that TPR are normal, but note areas of alopecia (abnormal hair loss). The veterinarian orders some blood tests to determine the cause of this dog's clinical condition. What tests are appropriate to run on this dog's blood? What are some examples of endocrine diseases that this dog might have? Are the treatments for these endocrine diseases safe? What information should be given to the owner? A veterinary technician should be able to adequately explain endocrine diseases to clients and how treatment of endocrine disease can help their pet.

BASIC ENDOCRINE SYSTEM ANATOMY AND PHYSIOLOGY

The endocrine system is composed of ductless glands that secrete chemical messengers called *hormones* into the bloodstream. Hormones enter the bloodstream and are carried throughout the body to affect a variety of tissues and organs. Tissues and organs that the hormones act upon are called *target organs*. The glands of the endocrine system (Figure 10-1) include the following:

- one pituitary gland (with two lobes)
- one thyroid gland (right and left lobes fused ventrally)
- four parathyroid glands (in most species)
- two adrenal glands
- one pancreas
- one thymus
- one pineal gland
- two gonads (ovaries in females, testes in males)

REGULATION OF THE ENDOCRINE SYSTEM

The endocrine system is controlled by a feedback mechanism that includes the hypothalamus, pituitary gland, and the other endocrine glands (Figure 10-2). The pituitary gland is located at the base of the brain below the hypothalamus. The hypothalamus secretes releasing and inhibiting factors that affect the release of substances from the pituitary gland. When signaled by releasing factors from the hypothalamus, the pituitary gland secretes hormones that control other endocrine glands.

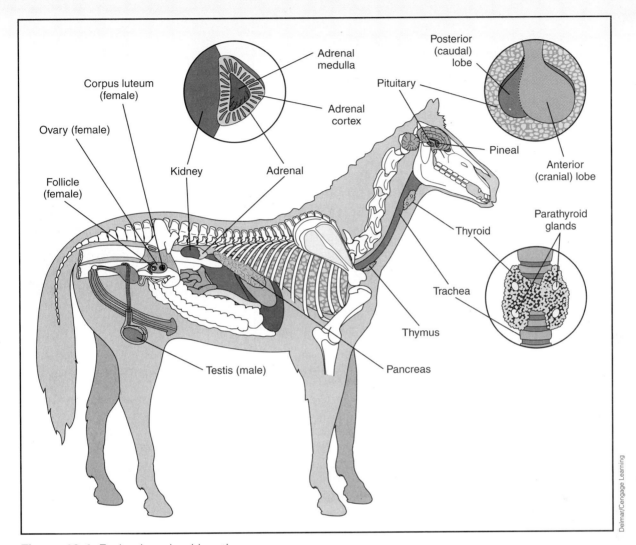

Figure 10-1 Endocrine gland locations.

Two types of feedback loops are involved in regulating the endocrine system. Feedback loops are described as either negative or positive. An example of the negative feedback loop in action is the release of glucocorticoid from the cortex of the adrenal gland. If there are low blood levels of glucocorticoid, a signal is sent to the hypothalamus. The hypothalamus then secretes a releasing factor that signals the anterior pituitary gland to secrete adrenocorticotropic hormone (ACTH). ACTH is delivered via the bloodstream to the adrenal gland and causes the adrenal cortex to produce more glucocorticoid. If too much glucocorticoid is produced, a signal is sent to the hypothalamus to secrete inhibiting factor to the anterior pituitary gland. The anterior pituitary gland will then decrease its production and release of ACTH, thus lowering the amount of glucocorticoid ultimately produced by the adrenal cortex.

Positive feedback loops occur less frequently in the body. An example is the release of oxytocin from the posterior pituitary gland during parturition. The presence of the fetus in the pelvic canal causes oxytocin to be released from the posterior pituitary. Oxytocin signals the uterus to contract. The uterine contraction signals the hypothalamus to produce more releasing factor. The releasing factor signals the posterior pituitary gland to secrete more stored

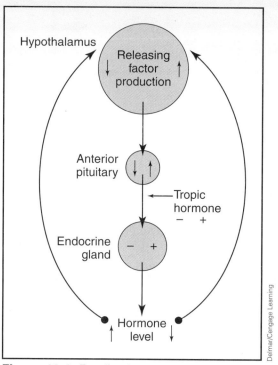

Figure 10-2 Feedback control mechanism. Positive and negative feedback mechanisms control the levels of a particular hormone in the blood by secreting releasing factors or inhibiting factors that affect hormone release. This is an example of a negative feedback loop.

oxytocin (oxytocin is actually produced in the hypothalamus and stored in the posterior pituitary). Continued uterine contractions signal more releasing factor to be released from the hypothalamus, resulting in the release of more oxytocin from the posterior pituitary in an effort to complete labor.

HORMONAL DRUGS

A **hormone** is a chemical substance produced by cells in one part of the body and transported to another part of the body where it influences and regulates cellular activity and organ function. Veterinarians use hormones and hormone-like preparations for two purposes: to replace hormones diminished by disease, or to block excess hormones created by disease.

Master Gland Control

The pituitary gland, also known as the master gland, has two divisions: the anterior (cranial) and the posterior (caudal) (Figure 10-3). Anterior pituitary hormones regulate an animal's growth and the proper functioning of its thyroid, gonads (testes and ovaries), and various endocrine glands. The posterior pituitary secretes antidiuretic hormone (ADH, which is also known as vasopressin) and oxytocin (hormone that induces smooth muscle contraction especially in the uterus and udder) that have been produced by the hypothalamus. Table 10-1 summarizes the pituitary gland hormones and their functions.

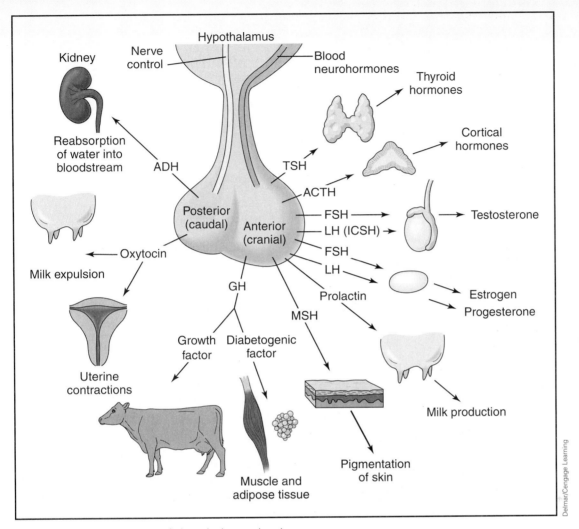

Figure 10-3 Secretions of the pituitary gland.

Anterior pituitary hormones are used in veterinary practice for the following:

- *Thyroid stimulating hormone* (TSH) is used in the diagnosis of primary hypothyroidism.

- *Adrenocorticotropic hormone* (ACTH) is used to stimulate the adrenal cortex to secrete glucocorticoids in the diagnosis of adrenal cortex disease.

- *Follicle stimulating hormone* (FSH) and *luteinizing hormone* (LH) (and its signal from the hypothalamus, GnRH) are discussed in the reproductive section.

- *Growth hormone* (GH), also known as somatotropin, is used to increase growth rate and feed efficiency in livestock, and is believed to increase milk production in dairy cows. In healthy animals, it is released throughout the animal's life to promote tissue building by increasing protein synthesis. The FDA approved a genetically produced *bovine somatotropin* (BST) for use in 1993; the trade name of BST is Posilac®. There is sharp debate about its use because of potential milk residues and potential overproduction of milk that would result in low milk prices for farmers.

Table 10-1   Pituitary Gland Hormones and Functions

PORTION OF PITUITARY GLAND	HORMONE PRODUCED	HORMONE ACTION
Anterior Pituitary	thyroid stimulating hormone (TSH)	Stimulates growth and secretions of the thyroid gland
	adrenocorticotropic hormone (ACTH)	Stimulates the growth and secretions of the adrenal cortex (glucocorticoids and mineralocorticoids)
	follicle stimulating hormone (FSH); FSH is a type of gonadotropic hormone	Stimulates secretion of estrogen and growth of eggs in the ovaries (female) and the production of sperm in the testes (male)
	luteinizing hormone (LH); LH is a type of gonadotropic hormone	Stimulates ovulation and aids in maintenance of pregnancy (females)
	interstitial cell-stimulating hormone (ICSH); ICSH is now considered to be LH and is a type of gonadotropic hormone	Stimulates testosterone secretion (males)
	prolactin (also known as lactogenic hormone or luteotropin)	Stimulates milk secretion and influences maternal behavior
	growth hormone (GH); also known as somatotropin	Regulates body growth
	melanocyte stimulating hormone (MSH)	Causes skin pigmentation
Posterior Pituitary	antidiuretic hormone (ADH); also known as vasopressin	Maintains water balance in the body by augmenting water reabsorption in the kidneys
	oxytocin	Stimulates uterine contractions during parturition and milk letdown from the mammary ducts

The posterior pituitary secretes antidiuretic hormone (ADH) and oxytocic (labor-producing) hormone. ADH is available commercially and has been used in dogs and cats to treat diabetes insipidus. **Diabetes insipidus** is a disease characterized by the inability to concentrate urine, caused by the posterior pituitary's failure to release sufficient ADH or the inability of the kidneys to respond to ADH stimulation. This results in increased thirst and urination (PD/PU). *Vasopressin* (Pitressin®) injection has been used to diagnose and treat diabetes insipidus. Vasopressin is available from both natural and synthetic sources. Side effects include local irritation at the injection site and abdominal pain.

Desmopressin (Stimate®, DDAVP®) is a product used in the conjunctival sac or IV to treat diabetes insipidus. Desmopressin is structurally similar to vasopression but has more antidiuretic activity and does not affect urinary sodium or potassium secretion. Side effects include eye irritation and hypersensitivity reactions.

Oxytocin is available as a synthetic commercial injection whose primary effects are stimulation of the smooth muscle of the uterus and mammary gland. Veterinarians use it to stimulate or enhance uterine contracts at parturition in animals at term. This is especially helpful in the queen, bitch, and sow, where the possibility of manual manipulation is limited. Oxytocin also helps expel the placenta and uterine debris after Cesarian section. Oxytocin can also induce milk letdown by stimulating the smooth muscle that surrounds the milk-secreting cells of the mammary gland. This muscle contraction forces milk into the sinuses, making the milk readily available to the suckling young. Oxytocin is also used as adjunct therapy in animals with an open pyometra, so that the increase in uterine contractions can help expel pus from the uterus. If used in animals with a closed pyometra, the increase in uterine contractions and the inability to dispel pus through the undilated cervix may cause uterine tearing. Oxytocin is marketed generically and does not have a milk or meat withdrawal time.

Blood Glucose Regulation

To remain healthy, animals must maintain blood glucose levels within a narrow, minimally fluctuating range. The two hormones that maintain this range are insulin and glucagon. **Insulin**, which is formed in the pancreas, responds primarily to a rise in blood glucose; it promotes the uptake and utilization of glucose for energy in body cells, and the storage of glucose in the liver as glycogen (the chief source of carbohydrate storage in animals). Insulin's main function is to decrease blood glucose concentrations by distributing glucose to tissue. **Glucagon**, in contrast, increases blood glucose levels by promoting the breakdown of liver glycogen into glucose, which exits the liver and enters the bloodstream.

Diabetes mellitus is a complex disease of carbohydrate, fat, and protein metabolism caused by lack of insulin or inefficient use of insulin in animals. Diabetic animals have elevated blood glucose levels (hyperglycemia), glucose in the urine (glucosuria), frequent thirst and urination (PU/PD), and alterations in fat metabolism that can lead to toxic effects and diabetic coma. Diabetes mellitus results in cell starvation in the presence of hyperglycemia. Cell starvation leads to the animal developing polyphagia with concurrent weight loss. Hyperglycemia leads to polyuria and compensatory polydipsia due to glucose spilling into the urine and osmotically drawing water with it. Diabetes mellitus, though recorded in all species, shows up most frequently (about 1 in 1000) in dogs, especially obese female dogs that are middle aged or older. The exact cause of diabetes mellitus in domestic animals is not known. However, dogs with the condition often have a chronic pancreatic disorder such as chronic inflammation affecting the islets of Langerhans, the pancreatic cells that secrete insulin. Other suspected causes of diabetes mellitus include pancreatic atrophy and immune-mediated destruction of the islets of Langerhans cells.

Veterinarians treating diabetic dogs or cats consider both dietary and medical management. Because a diabetic animal has difficulty using carbohydrates, its diet must be low in carbohydrates and high in protein. The diabetic diet should

Clinical Que

Oxytocin increases both the frequency and force of contractions of uterine smooth muscle.

Clinical Que

Persistent fasting hyperglycemia is the hallmark of diabetes mellitus. In dogs a fasting serum glucose concentration >200 mg/dl indicates diabetes mellitus. Cats may become transiently hyperglycemic when stressed, so a single elevated serum glucose concentration is not enough to diagnose diabetes mellitus. Urinary glucose levels along with serum glucose concentration can be used to confirm or negate a diagnosis of diabetes mellitus. Serum fructosamine concentration (which provides information about serum glucose levels for the previous three weeks) may be needed to make the diagnosis of diabetes mellitus.

also be high in soluble fiber, which slows digestion and reduces postprandial (after eating) hyperglycemia. Feeding habits of animals are also important as dogs tend to eat defined meals, whereas cats are more likely to graze throughout the day. Along with a managed diet, the animal usually receives insulin injections; trial doses establish the final doses that restore the animal's former physical condition and activity. Insulin is not given orally (it is a protein that would be denatured by stomach acid if given orally), so owners need to be shown how to give insulin injections. Owners should also be counseled about signs of insulin overdose, which include lethargy, weakness, ataxia, and potentially seizures. Treatment of insulin overdose includes advising the owner to offer food immediately and to bring the animal into the clinic for supportive veterinary care.

Insulin preparations differ with respect to degree of purity, source (beef, pork, or recombinant human origin), and onset and duration of action. Insulin purity is based on the concentration of proinsulin contamination in the insulin product. *Proinsulin* is a long chain of amino acids that are used by the beta cells of the pancreas to produce insulin. The greater the concentration of proinsulin, the greater the likelihood of local or systemic allergic reactions. Purified insulin products generally do not contain more than 10 parts per million (ppm) of proinsulin contamination.

Insulin has traditionally been extracted from bovine or porcine pancreas or a combination of both. The source of insulin is printed on every label (Figure 10-4). Porcine (pork) insulin is similar in structure to naturally occurring canine and human insulin, whereas bovine (beef) insulin is similar to naturally occurring feline insulin. The more similar the commercial insulin is to the host insulin, the less likely it is that the recipient animal will have problematic immunological reactions. Recombinant human and synthetically processed insulin are now available, but have not been as successful in treatment of veterinary cases of diabetes mellitus as it is in treating human cases. Manufacturers have discontinued bovine (beef) insulin production for humans due to the possible risk of bovine spongiform encephalopathy (mad cow disease) transmission to humans, which has decreased the options veterinarians have for choosing a type of insulin product

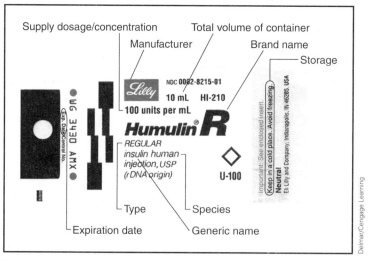

Figure 10-4 Insulin Label (Courtesy of Ely Lilly and Company).

to use. Purified pork insulin products manufactured for human use have also decreased, and most pork insulin products available are veterinary products.

The onset and duration of insulin activity in the body may be controlled by modification of regular insulin. Regular insulin has the most rapid onset and the shortest duration of action. By precipitating insulin with zinc, various modified insulins can be made. Another way of modifying insulin to achieve a longer onset and duration of action is to precipitate insulin with zinc and a large protein, protamine, or by the production of an insulin analog (glargine). These modifications result in NPH (isophane insulin suspension), PZI (protamine zinc), and insulin glargine products (Table 10-2).

The insulin categories, based on onset and duration of action, are seen in Figure 10-5 and are described as follows:

- Short-acting insulin (*regular crystalline insulin*, which is the only insulin given IV, and it can also be given IM or SQ). Short-acting insulin is used initially to treat diabetic ketoacidosis and control blood glucose until the animal is stabilized. Short-acting insulin is not used in animals for long-term glucose management and is not an appropriate choice for at-home therapy. Short-acting insulin has an immediate onset of action when given IV with maximum effects occurring at 0.5 to 2 hours. Its duration of action varies depending on the administration route. Examples include Regular Iletin® I, Humulin® R, Novolin® R, Regular Insulin, and Humalog®.

- Intermediate-acting insulin (*Neutral protamine hagedorn [NPH]* and *lente*, which are given SQ). Intermediate-acting insulin is used for control of blood glucose in uncomplicated diabetes mellitus in cats and dogs. This category of insulin has a longer duration of action than the short-acting insulins and a more rapid onset of action than the long-acting insulins. Its duration of action is between 6 and 24 hours in dogs and 4 and 12 hours in cats. Vetsulin® is an FDA-approved porcine lente insulin for use in dogs and cats that is available in a U-40 concentration. It produces two peaks of activity with one occurring shortly after administration and the second one several hours later. Most dogs require twice daily dosing with lente insulin products such as Vetsulin® (Caninsulin® outside the United States). Examples of intermediate-acting human insulin products include NPH Iletin® I, Lente Iletin® II, Humulin® N, and Novolin® N, which are less expensive than the veterinary products and may be an economically better option for large dogs.

- Long-acting insulin (*protamine zinc insulin* and *ultralente*, which are given SQ). Long-acting insulin is poorly absorbed from tissue; therefore, it maintains a long-lasting blood level. The duration of action for long-acting insulin is 6 to 28 hours in dogs and 6 to 24 hours in cats. PZI is the insulin of choice for cats and is given once or twice daily. PZI was taken off the human market and was manufactured on a much smaller scale under the Medically Necessary Veterinary Products Policy for use in cats until the manufacturing of a 90 percent bovine and 10 percent porcine veterinary product was available (PZI Vet®). Due to limited

Clinical Que

Clinically important terms related to insulin include onset (when insulin first begins to act in the body), peak (when the insulin is exerting maximum action), and duration (length of time the insulin remains in effect).

supplies of bovine pancreas as the source of insulin, new production of PZI Vet® may cease and may not be available. Veterinarians may need to start using ultralente insulin products in cats or switch diabetic cats to another form of insulin such as lente (Vetsulin®). Examples of long-acting insulin include Novolin ge Ultralente®, PZI Vet®, and PZI®.

- Ultralong-acting insulin (glargine, which is given SQ). Glargine is a recombinant human type of insulin that is available in a U-100 concentration. Insulin glargine can take a few days to achieve its maximal effect and overdose can result in up to 72 hours of hypoglycemia; therefore, blood glucose levels may not change significantly for the first three days of initial therapy, and dose increases are not recommended for the first week of therapy. It is appropriate for use in cats because the prolonged duration matches with the tendency of cats to eat small meals frequently. In cats, insulin glargine is typically given twice daily. This insulin cannot be diluted because its effect is pH dependent. An example of insulin glargine is Lantus®.

Table 10-2 Types of Insulin and Their Properties

INSULIN	TYPE	ROUTE OF ADMINISTRATION	ONSET OF ACTION	DURATION OF ACTION
Regular insulin (injection)	Recombinant human	• IV • IM • SQ	• Immediate • 10–30 minutes • 10–30 minutes	• 1–4 hours • 3–8 hours • 4–10 hours
NPH (isophane insulin suspension)	Recombinant human	• SQ	• 20–120 minutes	• 6–18 hours (dog) • 4–12 hours (cat)
Lente (porcine insulin zinc suspension)	Recombinant human Purified pork	• SQ	• 30–120 minutes	• 14–24 hours
PZI (protamine zinc suspension)	Beef-pork	• SQ	• 60–240 minutes	• 6–28 hours (dog) • 6–24 hours (cat)
Insulin glargine (insulin analog)	Recombinant human	• SQ	• 60–120 minutes	• 18–24 hours

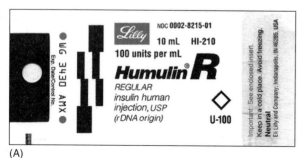

(A)

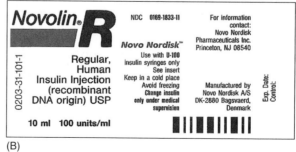

(B)

Figure 10-5 Labels for selected types of insulin grouped by onset and duration of action.

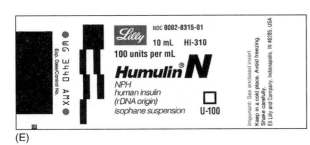

(C)

Lilly
NDC 0002-7510-01
10 mL VL-7510
100 units per mL
Humalog®
insulin lispro injection
(rDNA origin)
Neutral
U-100

CAUTION—Federal (USA) law prohibits dispensing without a prescription.
For parenteral use.
See accompanying literature for dosage.
Eli Lilly and Company, Indianapolis, IN 46285, USA

WG 3550 AMX
Exp. Date/Control No.

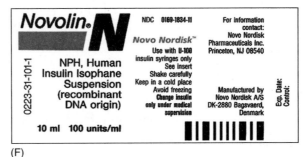

(D)

Lilly
NDC 0002-8415-01
10 mL HI-410
100 units per mL
Humulin® L
LENTE®
human insulin
(rDNA origin)
zinc suspension
U-100

Important: See enclosed circular.
Keep in a cold place. Avoid freezing.
To mix, roll or carefully shake the insulin bottle several times.
If pregnant or nursing, see carton.
Eli Lilly and Co. Indianapolis, IN 46285, USA

WG 2560 AMX
Exp. Date/Control No.

(E)

Lilly
NDC 0002-8315-01
10 mL HI-310
100 units per mL
Humulin® N
NPH
human insulin
(rDNA origin)
isophane suspension
U-100

Important: See enclosed insert.
Keep in a cold place. Avoid freezing.
Shake carefully.
Eli Lilly and Company, Indianapolis, IN 46285, USA

WG 3440 AMX
Exp. Date/Control No.

(F)

Novolin® N
NDC 0169-1834-11
Novo Nordisk™
NPH, Human
Insulin Isophane
Suspension
(recombinant
DNA origin)
10 ml 100 units/ml

Use with U-100
insulin syringes only
See insert
Shake carefully
Keep in a cold place
Avoid freezing
Change insulin
only under medical
supervision

For information
contact:
Novo Nordisk
Pharmaceuticals Inc.
Princeton, NJ 08540

Manufactured by
Novo Nordisk A/S
DK-2880 Bagsvaerd,
Denmark

Exp. Date;
Control:

0223-31-101-1

(G)

Lot: A140A01
Exp: 04.2010

10 mL

See package Insert for directions for use.

Indication: VETSULIN® (porcine insulin zinc suspension) is indicated for the reduction of hyperglycemia and hyperglycemia-associated clinical signs in dogs and cats with diabetes mellitus.

Storage: Store in an upright position under refrigeration at 2° to 8° C (36° to 46° F) Do not freeze. Protect from light.

vetsulin®
(porcine insulin zinc suspension)

40 units per mL (U-40)

CAUTION: Federal law restricts this drug to use by or on the order of a licensed veterinarian.

Distributed by: **INTERVET INC.,**
Millsboro, DE 19966
Made in Germany
www.vetsulin.com

intervet

NADA No. 141-236, Approved by FDA
049544LPU2400

(H)

Lilly
NDC 0002-8615-01
10 mL HI-610
100 units per mL
Humulin® U
ULTRALENTE®
human insulin
(rDNA origin)
extended zinc suspension
U-100

Important: See enclosed circular.
Keep in a cold place. Avoid freezing.
To mix, roll or carefully shake the insulin bottle several times.
If pregnant or nursing, see carton.
Eli Lilly and Co. Indianapolis, IN 46285, USA

WG 2570 AMX
Exp. Date/Control No.

A, C, D, E, H Courtesy of Eli Lilly and Company; B and F Courtesy of Novo Nordisk; J Courtesy of Aventis; and G Courtesy of [JLR33]

(I)

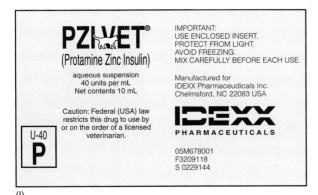

PZI VET®
(Protamine Zinc Insulin)

aqueous suspension
40 units per mL
Net contents 10 mL

Caution: Federal (USA) law restricts this drug to use by or on the order of a licensed veterinarian.

U-40
P

IMPORTANT:
USE ENCLOSED INSERT.
PROTECT FROM LIGHT.
AVOID FREEZING.
MIX CAREFULLY BEFORE EACH USE.

Manufactured for
IDEXX Pharmaceudicals Inc.
Chelmsford, NC 22083 USA

IDEXX
PHARMACEUTICALS

05M678001
F3209118
S 0229144

(J)

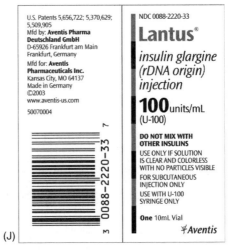

U.S. Patents 5,656,722; 5,370,629; 5,509,905
Mfd by: **Aventis Pharma Deutschland GmbH**
D-65926 Frankfurt am Main Frankfurt, Germany
Mfd for: **Aventis Pharmaceuticals Inc.**
Kansas City, MO 64137
Made in Germany
©2003
www.aventis-us.com
50070004

NDC 0088-2220-33

Lantus®
insulin glargine
(rDNA origin)
injection

100 units/mL
(U-100)

DO NOT MIX WITH OTHER INSULINS

USE ONLY IF SOLUTION IS CLEAR AND COLORLESS WITH NO PARTICLES VISIBLE

FOR SUBCUTANEOUS INJECTION ONLY

USE WITH U-100 SYRINGE ONLY

One 10mL Vial

Aventis

3 0088-2220-33 7

Figure 10-5 *(continued)*

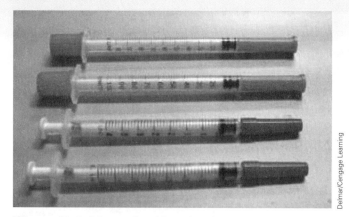

Figure 10-6 Measuring U-40 insulin to the 1 unit mark in a U-40 syringe will contain 1 unit of insulin. Measuring U-100 insulin to the 1 mark in a U-100 syringe will contain 1 unit of insulin. Note that U-40 insulin syringes have red caps and U-100 insulin syringes have orange caps.

Insulin concentration is expressed in units of insulin per milliliter. Insulin is available in 40 units/mL (U-40), 100 units/mL (U-100), and 500 units/mL (U-500). U-40 insulin is used when administering small amounts of insulin to cats and small dogs. U-40 insulin is given with U-40 syringes (Figure 10-6). U-100 insulin is given with U-100 syringes. U-40 insulin syringes are calibrated in 1 unit increments with every 5 units labeled up to 40 units, while U-100 insulin syringes are calibrated on one side in even-numbered, 2 unit increments with every 10 units labeled (10, 20, 30, etc.) and on the reverse side in odd-numbered, 2 unit increments with every 10 units labeled (5, 15, 25, etc.) (Figure 10-7). Drawing up small volumes on a U-40 syringe is easier and more accurate than drawing up small volumes on a U-100 syringe.

Although insulin products are generally stable at room temperature, they are usually stored in the refrigerator and not frozen. Exposure for even short periods to freezing or high temperatures can permanently degrade insulin products. If the product becomes discolored or a precipitate forms, it should be discarded. Insulin should be gently rolled between the palms to mix the product before it is withdrawn from the vial.

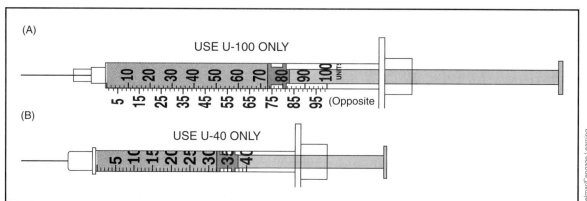

Figure 10-7 (A) Standard U-100 insulin syringe measuring 73 units; (B) standard U-40 insulin syringe measuring 32 units.

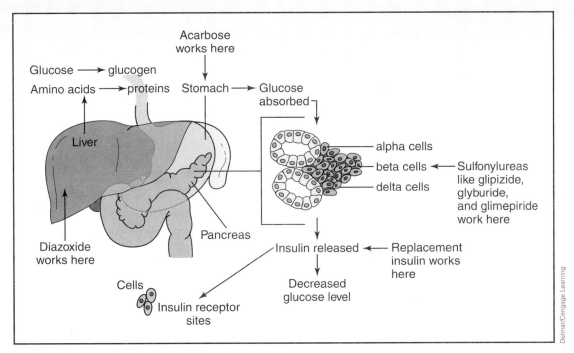

FIGURE 10-8 Sites of action of drugs used to treat blood sugar abnormalities.

Delmar/Cengage Learning

Oral hypoglycemic agents have been used with some success in animals. Oral hypoglycemic agents work by stimulating pancreatic beta cells to secrete insulin (Figure 10-8); therefore, some pancreatic function is required for these drugs to act. The oral hypoglycemics used in veterinary practice are in the chemical class known as the sulfonylureas. The representative drug in this class is *glipizide* (Glucotrol®). Glipizide, a second-generation sulfonylurea, is more potent than older agents and must be dosed less frequently. Because most dogs with diabetes mellitus tend to have the insulin-dependent form of the disease, glipizide has not been effective. Glipizide has been shown to be effective in treating about 25 percent of diabetic cats. Side effects of glipizide include gastrointestinal disturbances, hypoglycemia, and liver toxicity. Other drugs in this group are *glyburide* (DiaBeta®, Micronase®) and *glimepiride* (Amaryl®) that may be used if glipizide is unavailable or if once daily dosing is needed. Side effects are the same as with glipizide.

Another oral hypoglycemic agent is *acarbose* (Predcose®), which reduces the rate and amount of glucose absorbed from the gastrointestinal tract after a meal. Acarbose is given with food and is used to cause mild reduction in blood glucose concentrations in dogs and cats with non–insulin-dependent diabetes mellitus or as an adjunct treatment of insulin-dependent diabetes mellitus. Side effects include diarrhea and flatulence.

In a healthy animal, glucagon is used to regulate blood glucose levels when they become low. Low blood sugar or hypoglycemia is associated with insulinomas (tumors of the pancreas) that cause increased secretion of insulin. *Diazoxide* is a drug that is labeled for oral use in treating hypoglycemia secondary to hyperinsulin secretion by directly inhibiting pancreatic insulin secretion. Clients giving diazoxide to their pets should be instructed to monitor for

 Clinical Que

Insulin syringes are designed for use with a specific strength of insulin, with needles color coded according to strength. U-40 syringes have red tops, while U-100 syringes have orange tops. U-40 syringes have 40 units per ml, and U-100 syringes have 100 units per ml.

signs of hyperglycemia, hypoglycemia, or gastrointestinal problems. It can also cause sodium and water retention and should be used cautiously in animals with heart or kidney disease. Trade names of diazoxide include Proglycem® suspension and Eudemine® tablets.

Other drugs used to increase blood glucose levels include glucocorticoids, epinephrine, and progesterone. Glucose and dextrose can be administered to animals that are hypoglycemic. These drug categories have been covered under other sections.

Regulation of Metabolic Rate

The thyroid gland is an organ located within the neck near the larynx. Hormones produced by the thyroid are necessary for an animal's normal growth and reproduction. *Thyroid hormone* is a collective term for two active hormones found in the thyroid gland: **thyroxine** (T_4) and **tri-iodothyronine** (T_3). The synthesis of these hormones takes place in a series of chemical steps. Iodides consumed in food and water are absorbed through the gastrointestinal tract and enter the blood stream. When blood passes through the thyroid gland, iodide is trapped and converted to iodine. Iodine combines with tyrosine (an amino acid) to form iodotyrosine. Iodotyrosine molecules combine to form T_4 and T_3, which are stored in the thyroid gland until they are released. When T_4 is released into the bloodstream, some of it is converted to T_3. Figure 10-9 summarizes the steps of T_4 and T_3 synthesis.

Thyroid hormones act as catalysts in the body and affect many metabolic, growth, reproductive, and immune functions. They help regulate lipid and

> **Clinical Que**
>
> The "T" designation in T_4 (thyroxine) and T_3 (tri-iodothyronine) refers to the iodine atoms in each hormone: Tyroxine has four, tri-iodothyronine has three.

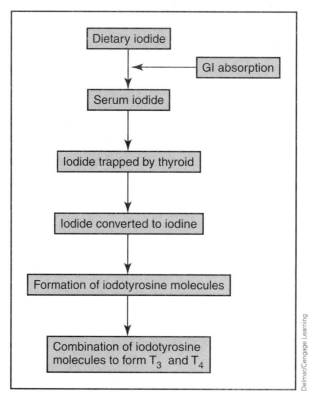

Figure 10-9 Summary of the biosynthesis of T_4 and T_3.

carbohydrate metabolism. Thyroid hormones also affect heat production in the body.

Hypothyroidism, a disease characterized by a deficiency of thyroid hormone, may be caused by many things. Abnormalities of the thyroid gland can result from disorders of iodide trapping, conversion of iodide to iodine, and/or the release of thyroid hormone from its storage sites. Another cause of hypothyroidism is a disorder of the anterior pituitary gland in which inadequate amounts of TSH are released. Whatever the cause, the signs of hypothyroidism reflect thyroxine's necessity to virtually all the cells and organ systems of the animal. These signs include decreased coat luster or hair loss; weight gain without any increase in appetite; listlessness; intolerance to cold; reproductive failure; and skin that is more susceptible to mites, bacteria, and scales (Figure 10-10).

Hypothyroidism is a serious condition in dogs and is diagnosed in a variety of ways. One way to diagnose hypothyroidism is to measure serum total T_4 and T_3. However, hypothyroid dogs may test in the normal range when only serum levels of thyroid hormone are measured. Another diagnostic test is the thyroid stimulation test, in which thyroid hormones are measured before and after the administration of TSH. Blood is drawn prior to the test, and then TSH is given IV. Another blood sample is collected four to six hours after the TSH injection. If the level of thyroid hormone is low following TSH injection, then the dog is hypothyroid. *Thyrotropin* is a purified form of TSH that is collected from bovine anterior pituitary glands and used to make commercial TSH. Trade names for TSH are Dermathycin® and Thyrotropar®. They are available in powder forms that must be reconstituted prior to use. Reconstituted thyrotropin can be stored for up to three weeks before losing its effectiveness.

Another test used to detect hypothyroidism is the thyrotropin-releasing hormone (TRH) response test. The TRH response test is performed by obtaining blood samples before and six hours after IV administration of *TRH* (Relefact®). An increase in serum T_4 of 1.5 times baseline is normal; levels below that are considered abnormal.

The use of thyroid hormone to treat hypothyroidism is called thyroid replacement therapy. The goal in treating hypothyroid animals is to achieve an euthyroid state (normal thyroid state) by supplying the animal with appropriate

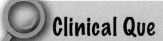

Clinical Que

What is a stimulation test? Stimulation tests evaluate an endocrine gland's responsiveness to exogenous stimulating factors. Stimulation tests measure plasma levels of hormone after administration of a substance that should increase production of the hormone. If this increase does not occur, the endocrine gland is not functioning as it should.

Courtesy of Kimberly Kruse Sprecher, CVT

Figure 10-10 Dog with hypothyroidism.

concentrations of thyroid hormone. Drugs used in thyroid replacement therapy include the following:

- *levothyroxine sodium* (T_4): Levothyroxine is the synthetic isomer of T_4. It is the drug of choice for treating hypothyroidism in all animals. Its advantages are that it is chemically pure, is inexpensive, and has a long half-life, which helps in lowering dosing frequency. Side effects are rare. Examples are Soloxine®, Thyro-Form®, Thyro-Tab®, and Synthroid®.

- *liothyronine sodium* (T_3): Liothyronine is a synthetic form of T_3 that is chemically pure but has a shorter half-life than levothyroxine. It is used in animals that do not respond well to treatment with levothyroxine. Side effects are rare. Examples are Cytobin® and Cytomel®.

Hyperthyroidism is excessive functional activity of the thyroid gland. Excessive secretion of thyroid hormone results in increased levels of metabolism, excessive heat production, and increased nervous system activity. Seen mainly in cats, hyperthyroidism is characterized by increased thirst, weight loss despite increased appetite, increased stool production, restlessness, and tachycardia Treatment includes destruction or removal of the dysfunctional thyroid gland or the administration of antithyroid drugs.

The thyroid can be surgically excised to remove the source of excess thyroid hormone. Prior to surgery, animals are brought to an euthyroid state with drugs that suppress thyroid function.

The thyroid gland can be destroyed by radioactive isotopes of iodine (sodium iodine I-131). Domestic animals require iodine for normal thyroid function, though no one knows the minimum amount of iodine that various species require. It is known that iodine (in the form of iodide salts), dosed substantially above dietary requirements, inhibits thyroid hyperfunction. Veterinarians use radioactive iodine, I-131, to evaluate thyroid gland function and to destroy hyperactive thyroid tissue. Radiation is emitted by the trapped isotope, which destroys thyroid cells without excessively damaging the surrounding tissue. Because I-131 has a half-life of only eight days, more than 99 percent of the radiant energy will be gone in about 56 days.

Antithyroid drugs work by blocking excess hormone production in the thyroid. *Methimazole* (Felimazole®, Tapazole®) is the drug of choice for treating hyperthyroidism. It interferes with the incorporation of iodine in the molecules of T_3 and T_4. Thyroid hormone existing in the blood is not altered. Felimazole® is available as 2.5 mg and 5 mg coated tablets that may be given once or twice daily. Side effects include vomiting, anorexia, and lethargy. *Carbimazole* (Carbazole®) is the antithyroid drug used in Europe. *Propranolol* (Inderal®), a beta-adrenergic blocking agent, has been used to suppress the tachycardia associated with hyperthyroidism. Propranolol is covered in Chapter 8.

Regulation of the Adrenal Cortex

The adrenal glands are paired glands located near the cranial portion of the kidney. Each gland has an outer region (adrenal cortex) and an inner region (adrenal medulla). The hypothalamus regulates both regions of the adrenal

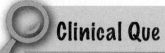

Clinical Que

An animal's response to stress is multidimensional. The sympathetic nervous system releases norepinephrine and/ or epinephrine and works on target organs like the heart (to increase the heart rate). The adrenal medulla responds to sympathetic nervous stimulation by releasing the hormones epinephrine and norepinephrine.

gland. The medulla receives direct nervous stimulation that originates in the hypothalamus, then travels to the brain stem, the spinal cord, and the sympathetic nerves. The adrenal medulla releases epinephrine and norepinephrine when stimulated by the sympathetic nervous system. The hypothalamus regulates the adrenal cortex by secreting releasing hormones for ACTH, which stimulates the anterior pituitary gland to secrete ACTH. ACTH stimulates the adrenal cortex.

The hormones produced by the adrenal cortex are produced in different regions, but all are steroids. **Mineralocorticoids** regulate blood volume and electrolyte concentration in the blood. The principal mineralocorticoid is aldosterone, which acts to conserve sodium ions and water in the body. Aldosterone is secreted in direct response to sodium and potassium ions.

Glucocorticoids, also secreted by the adrenal cortex, help regulate nutrient levels in the blood. They do this by increasing cellular utilization of energy sources (proteins and fats) and conserving glucose, causing blood glucose levels to increase. Glucocorticoids also stimulate liver cells to produce glucose from amino acids and fat. ACTH from the anterior pituitary gland controls glucocorticoid secretion. The main glucocorticoid is cortisol, which increases blood glucose levels.

Diseases of the adrenal gland include **adrenocortical insufficiency** (too little hormone produced) and **hyperadrenocorticism** (too much hormone produced). Adrenocortical insufficiency, also known as Addison's disease, is a progressive condition associated with adrenal atrophy, usually caused by immune-mediated inflammation. Adrenal atrophy results in deficient production of glucocorticoids and mineralocorticoids. Signs of this disease include lethargy, weakness, anorexia, vomiting, diarrhea, and PU/PD. Adrenocortical insufficiency is diagnosed by the ACTH stimulation test. Exogenous ACTH is available as an animal extract ACTH® gel or *corticotropin* (Acthar®) or as synthetic ACTH (*cosyntropin* [Cortrosyn® and Synacthen®]). A blood sample is obtained for determination of a resting plasma cortisol level. ACTH® gel is then given parenterally (cannot be given orally because it is inactivated by gastrointestinal enzymes), and another blood sample is collected one to two hours later. Low levels of cortisol indicate adrenocortical insufficiency. Adrenocortical insufficiency can also be diagnosed by determining ACTH blood levels, which are elevated in primary hypoadrenocorticism.

Treatment of adrenocortical insufficiency includes use of *desoxycorticosterone* (commonly referred to as DOCP) (Percorten-V®), which is a long-acting mineralocorticoid, in conjunction with prednisone or prednisolone, both of which are glucocorticoids. DOCP is thought to act by controlling the rate of protein synthesis. DOCP is given every 25 days and has relatively few side effects. Those side effects, which are related to increased renal absorption of sodium, may include edema and electrolyte imbalance. DOCP should be used with caution in animals with cardiovascular and renal disease. Glucocorticoids are covered in Chapter 16 on anti-inflammatory and pain-reducing drugs.

Another drug used to treat adrenocortical insufficiency is *fludrocortisone acetate* (Florinef®). Fludrocortisone is a potent corticosteroid with both glucocorticoid and mineralocorticoid activity. It is given once or twice daily, but it has

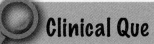

Clinical Que

All hormones produced from the adrenal cortex are steroids. Steroids have a parent molecule that consists of three six-carbon ring structures and one five-carbon ring structure.

Clinical Que

Glucocorticoids are contraindicated in horses with laminitis as they can worsen the condition.

not been as effective as DOCP in the treatment of adrenocortical insufficiency. Side effects include edema and electrolyte imbalance.

Hyperadrenocorticism, also known as Cushing's disease, may be caused by prolonged administration of adrenocortical hormones (iatrogenic), by adrenocortical tumors, or by pituitary disorders. Animals with hyperadreno-corticism have signs of PU/PD, hair loss, and a pendulous abdomen due to abnormal nutrient metabolism (Figure 10-11A and B). Hyperadrenocorticism is diagnosed by a low-dose dexamethasone suppression test and/or ACTH

Courtesy of Mark Jackson DVM, PhD, University of Glasgow

Courtesy of Lynn Paron

FIGURE 10-11 (A) Dog with hyperadrenocorticism. Hyperadrenocorticism (Cushing's disease) results in poluria, polydipsia, and redistribution of body fat; (B) Hyperadrenocorticism in horses presents as a failure to shed out the coat in spring making the coat look long, thick, and matted; excessive sweating; and loss of muscle mass developing a dipped back and pendulous abdomen.

stimulation test. The low-dose dexamethasone suppression test has histori-cally been used as the initial diagnostic test for confirming hyperadrenocorti-cism, but the ACTH stimulation test can provide information on the pituitary gland prior to initiation of treatment. The low-dose dexamethasone test mea-sures plasma cortisol concentrations before and four and eight hours after IV administration of dexamethasone. If plasma cortisol levels are high eight hours after dexamethasone administration, hyperadrenocorticism is present. The ACTH stimulation test measures serum cortisol concentrations before and following the administration of ACTH. This test can differentiate between iat-rogenic (caused by treatment) and naturally occurring hyperadrenocorticism. ACTH gel is commercially made from porcine pituitary glands and stimulates the production and release of glucocorticoids. The availability of ACTH gel is limited in the United States. Corticotropin (Acthar®) is commercially made from animal pituitary glands, while synthetic ACTH (cosyntropin [Cortrosyn® and Synacthen®]) is manufactured in a laboratory.

Treatment of hyperadrenocorticism due to adrenocortical tumors is aimed at destroying part of the adrenal cortex using *mitotane* (Lysodren®), also known as *o,p'*-DDD in veterinary medicine. Side effects of mitotane use include neu-rologic signs, lethargy, vomiting, and diarrhea. Adverse effects are common if there is a very rapid decrease in plasma cortisol levels. To prevent gastrointes-tinal side effects, mitotane should be given with food.

Other treatment options for hyperadrenocorticism include *ketoconazole* (Nizoral®), an antifungal drug that also blocks the enzymes needed to pro-duce steroid compounds; *selegiline* (Anipryl®), a monoamine oxidase inhibi-tor; and *trilostane* (Vetoryl®), a 3-beta hydroxysteroid dehydrogenase inhibitor. Ketoconazole's low toxicity and minimal effect on mineralocorticoid produc-tion make it a viable option in treatment of this disease. Selegiline, also known as L-deprenyl, is used for treating pituitary-dependent hyperadrenocorticism because it increases synthesis and release of dopamine (a monoamine neu-rotransmitter). Dopamine release signals the pituitary gland to shut down, which in turn decreases cortisol production by the adrenal gland. Hypothalamic dopamine deficiency plays a role in the pathology of pituitary-dependent hyperadrenocorticism in dogs. Response to selegiline treatment varies; it may take one to two months after initiation of treatment for signs of improvement to appear. Side effects of selegiline include vomiting, diarrhea, and restless-ness. Trilostane is used to treat pituitary-dependent or adrenal-dependent hyperadrenocorticism in dogs, pituitary-dependent hyperadrenocorticism in cats, and equine hyperadrenocorticism. Trilostane is a competitive inhibitor of 3-beta hydroxysteroid dehydrogenase and works at the level of the adrenal gland to block the production of cortisol, aldosterone, and androgens. This inhibition is reversible and dose dependent. It is given once or twice daily with food. Trilostane should be used with caution in animals with renal or hepatic disease. Side effects include lethargy, mild electrolyte abnormalities, and anorexia. Vomiting and diarrhea may also be seen.

Treatments for equine hyperadrenocorticism include *cyproheptadine* (Periactin®), a serotonin blocker, and *pergolide mesylate* (Permax®), a dopa-mine agonist that can also worsen laminitis (a common sequela of equine

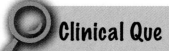

Clinical Que

Iatrogenic Addison's disease occurs when animals are supplemented with glucocorticoids and not tapered off the dose gradually. Sudden withdrawal of exogenous glucocorticoid does not give the adrenal cortex-hypothalamus-anterior pituitary feedback loop time to adjust to the changes in glucocorticoid level. Iatrogenic Addison's disease is more likely to occur when supplementation with high doses of glucocorticoids is done for long periods of time.

hyperadrenocorticism). Cyproheptadine is an antihistamine that blocks sero-tonin production in the hypothalamus and stimulates dopamine release. Side effects include anorexia and lethargy. Pergolide mesylate works by binding with drug receptors in the brain that control the production of dopamine in the hypothalamus (thus it mimics the action of dopamine). Since dopamine concentrations in horses with hyperadrenocorticism are lower than in healthy horses, supplementation of a dopamine agonist results in normal dopamine levels. Pergolide mesylate has been linked to heart valve damage in humans and was withdrawn from the market in 2007, but may still be available from compounding pharmacies. It appears to be well tolerated in horses.

The last set of hormones secreted by the adrenal cortex is the androgens (male sex hormones) and estrogens (female sex hormones). Because their effect is usually masked by the hormones of the testes and ovaries, abnormalities of androgens are usually not seen with adrenal cortex disease.

BASIC REPRODUCTIVE SYSTEM ANATOMY AND PHYSIOLOGY

The reproductive system is responsible for the process of producing offspring. The reproductive organs, whether male or female, are called the *genitals*. The structures that produce sex cells are called *gonads* and consist of the ovaries in females and testes in males. The male reproductive system consists of the testes (located in the scrotum), epididymis, ductus deferens, accessory sex glands, urethra, and penis. Sperm are produced in the seminiferous tubules of the testes. The Leydig's cells in the testes produce testosterone. Figure 10-12 shows the structures of the male reproductive system.

The female reproductive system consists of the ovaries, uterine tubes, uterus, cervix, vagina, and vulva (Figure 10-13). Ova are produced in the graafian follicle of the ovary. Estrogen and progesterone are produced in the ovary (progesterone is also produced in other places in pregnant animals).

Clinical Que

Because mitotane destroys some of the adrenal cortex, clients who give mitotane to their animals need to know the signs of adrenocortical insufficiency: lethargy, weakness, PU/PD, and gastrointestinal problems. Due to the extreme toxicity associated with mitotane, clients should wear gloves when administrating this drug and wash their hands thoroughly after handling this drug.

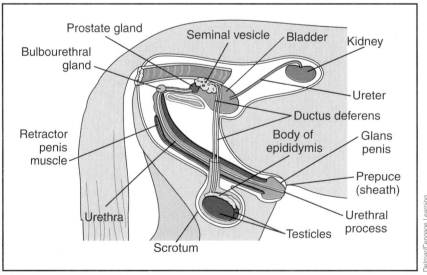

Figure 10-12 Reproduction tract of a stallion.

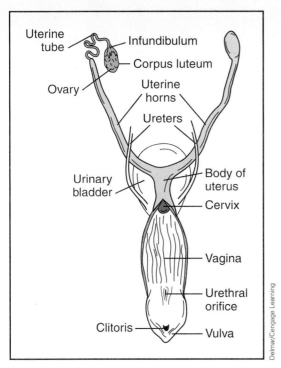

Figure 10-13 Reproduction tract of a bitch.

The estrous cycle in animals consists of

- **proestrus**: The period of the cycle before sexual receptivity. It involves the secretion of FSH by the anterior pituitary gland, which causes the follicles to develop in the ovary. FSH stimulates ovarian release of estrogen, which helps prepare the reproductive tract for pregnancy.

- **estrus**: The period of the cycle in which the female is receptive to the male. During estrus, FSH levels decrease and LH levels increase, causing the graafian follicle to rupture and release its egg (ovulation).

- **metestrus**: The period of the cycle after sexual receptivity. The corpus luteum (CL) forms in the ovary and produces progesterone if the animal is pregnant. In cattle, there is metestrus bleeding.

- **diestrus**: The period of the cycle that is a short phase of inactivity in polyestrous animals. If the animal is pregnant, the CL is fully functional and produces high levels of progesterone. If the animal is not pregnant, the CL decreases in size and become a corpus albicans.

- **anestrus**: The period of the cycle when the animal is sexually quiet; this is a long phase in seasonally polyestrous and monestrous animals.

Another way of categorizing the estrous cycle is by the follicular and luteal phases. The **follicular phase** is the stage of the estrous cycle in which the graafian follicle is present. Estrogen is the predominant hormone during the follicular phase. The **luteal phase** is the stage of the estrous cycle in which the CL is present. Progesterone is the predominant hormone during the luteal phase.

Control of the reproductive system, like the endocrine system, occurs via the hypothalamus and pituitary glands. The hypothalamus makes gonadotropin-releasing hormone (gonadorelin or GnRH) in response to various stimuli such as daylight length and feedback mechanisms. The hormones regulating reproduction are briefly described here and in Figure 10-14.

- GnRH causes the release of FSH and LH from the anterior pituitary gland. GnRH release is controlled by levels of FSH and LH via a negative feedback loop.

- FSH causes the growth and maturation of ovarian follicles. Follicles produce estrogen. Increased estrogen levels signal the hypothalamus to produce less GnRH.

- LH causes ovulation of mature follicles and the formation of the CL. The CL produces progesterone. Increased progesterone levels signal the hypothalamus to produce less GnRH.

- ACTH is released in pregnant animals as parturition approaches. ACTH causes cortisol to be produced by the adrenal cortex. Increased cortisol levels increase production of estrogen and prostaglandin by the uterus.

- Prostaglandin breaks down the CL at the end of pregnancy and at the end of diestrus in nonpregnant animals.

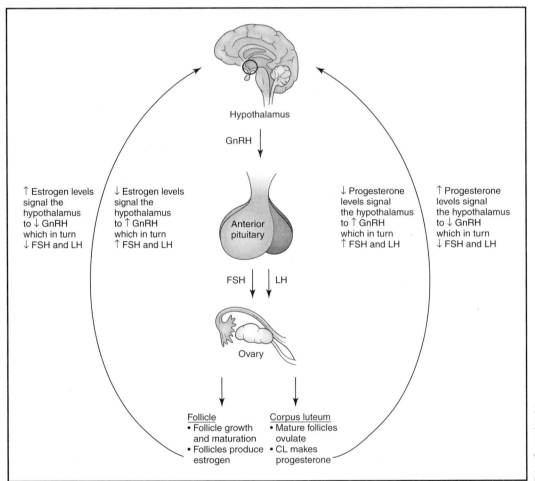

Figure 10-14 Control of the reproductive system through release of GnRH, LH, and FSH.

DRUGS AFFECTING REPRODUCTION

Drugs affecting the reproductive system are used in veterinary medicine to treat infertility; improve gonad production; aide in the treatment of pyometra; cause abortion; and detect, block, synchronize, or delay estrus. Because the reproductive system uses a cyclical balance of hormone to maintain homeostasis, any change in one part of the cycle has a wide variety of effects on the entire animal.

Male Hormone–like Drugs

Androgens are male sex hormones produced in the testes, ovaries, and adrenal cortex. **Testosterone**, the primary male sex hormone, is primarily synthesized in the interstitial cells (Leydig's cells) of the testes. Testosterone production is initiated and controlled by FSH and interstitial cell stimulating hormone. Testosterone has both androgenic (promoting male characteristics) and anabolic (tissue-building) effects. In addition to maturation of the male sex organs and development of secondary male sex characteristics, testosterone helps spermatozoa develop and also helps develop and maintain accessory sex organs (like the prostate gland).

Veterinarians use testosterone to treat conditions such as infertility and hypogonadism (the decreased function and retarded growth and sexual development of the gonads). Testosterone is also used to produce estrus detectors or "teaser animals" in cattle and for testosterone-responsive urinary incontinence in dogs. Testosterone propionate USP in oil can be given subcutaneously, intramuscularly, as subcutaneous implants, or as an oral tablet. Testosterone products are C-III controlled substances. Side effects of testosterone include the development of perianal tumors, prostatic disorders, and behavior changes. Examples of testosterone products include the following:

- *testosterone cypionate* in oil (generic and Depo-Testosterone®).
- *testosterone enanthate* in oil (generic).
- *testosterone propionate* in oil (generic).
- *methyltestosterone* (Android®, Methitest®), an androgen with anabolic effects that may be used to suppress estrus, treat estrogen-dependent mammary tumors, pseudopregnancy, or some hormonal-dependent alopecias.
- *danazol* (Danocrine®), a weak, synthetically produced androgen used primarily for immune-mediated hemolytic anemia in animals.

Mibolerone is an androgenic compound that works by blocking the release of LH from the anterior pituitary gland by negative feedback. When LH release is blocked, the follicle does not fully develop, resulting in failure of ovulation and CL development. Mibolerone is used to prevent estrus in adult female dogs not intended for breeding and for treatment of false pregnancies. An example is Cheque® drops, which should not be used in cats because of its low margin of safety. Side effects include reproductive organ problems (such as vulvovaginitis) and behavior changes. Implants of mibolerone are discussed under growth promotants.

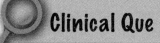

Clinical Que

Testosterone products that are derived naturally from animal testes are not effective when administered orally, because the liver rapidly inactivates them. Synthetic forms of testosterone are effective when administered orally, subcutaneously, or intramuscularly.

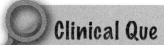

Clinical Que

Teaser animals are animals that are used to detect females in heat, but are altered so that they can not impregnate the females.

Finasteride (Proscar®, Propecia®) is an inhibitor of an enzyme (5-alpha-reductase) that converts testosterone to dihydrotestosterone (DHT) in the prostate, liver, and skin. DHT is responsible for development of the prostate. Finasteride is used orally in dogs to treat benign prostatic hyperplasia and in ferrets as an adjunctive treatment of adrenal disease. Side effects include vomiting and anorexia.

A nonsurgical neutering drug, *zinc gluconate neutralized by arginine* (Neutersol®), is a chemical sterilant approved for use in 3- to 10-month-old male dogs. This drug was approved in 2003 (and discontinued in 2005) and was used as an intratesticular injection administered once in each testicle. This drug is expected to be reintroduced onto the United States market in 2009. It should not be used in cryptorchid animals (those with undescended testicles), animals with testicular disease, or animals with scrotal irritation. Most testicles atrophy to some degree following intratesticular injection of this product; however, variability in size between the left and right testicles may be noticed. Testosterone production is not completely eliminated, so diseases that occur as a result of testosterone production (such as prostatic disease and testicular tumors) may not be prevented. Clients should be advised that this product does not kill sperm present at the time of injection; therefore, treated dogs should be kept away from females in heat for at least 60 days. Side effects include vocalization upon injection, scrotal pain, vomiting, and anorexia.

Female Hormone–like Drugs

The ovaries, testicles, adrenal cortex, and placenta produce estrogens. The ovarian follicle (the ovum and its encasing cells) and placenta in the female are the main sources of **estrogen**, a hormone that promotes female sex characteristics and stimulates and maintains the reproductive tract and accessory reproductive organs, including duct growth of the mammary glands. Estrogen stimulates female reproductive function, and is necessary for the uterus to contract and respond to oxytocin. Estrogen also enhances such female secondary sex characteristics as mammary development, plumage, and beak color. Veterinarians use synthetic estrogens in lieu of the more expensive natural estrogens. Intramuscular injections of synthetic estrogen prevent implantation of fertilized ova in dogs; however, due to serious side effects (such as pyometra and aplastic anemia), the use of estrogen for mismating was not recommended, and it was removed from the market. Estrogens can be helpful in correcting urinary incontinence, vaginitis, and dermatitis in dogs that have had their ovaries removed. Estrogen use has not been shown to be an efficacious therapy in dairy cows. In horses, estrogen is used to induce estrus following withdrawal of progesterone in the nonbreeding season. Side effects of estrogens include bone marrow suppression, endometrial hyperplasia, follicular cyst formation, and pyometra. Examples of estrogen include estradiol cypionate (ECP® injectable and Depo-Estradiol Cypionate® have been removed from the market; implants for cattle are available and covered under the hormonal implant section), and diethylstilbestrol (DES tablets and specialized preparations due to the unavailability of DES). Implants are discussed under growth promotants.

The "Gest"s (Progesterone Drugs)

Progesterone is a female sex hormone produced and secreted after ovulation by the CL, a group of ovarian cells. Progesterone decreases uterine activity when a female is in estrus (heat) or pregnant; progesterone deficiency may cause embryonic death in some animals. Veterinarians use progesterone and progestins (a group of compounds similar in effect to progesterone) with some success in cows (for reducing embryonic death), mares, and other domestic animals, but recommend it selectively. Progestins also block estrus in the bitch. When estrus is blocked, the ovary does not develop follicles and ovulation ceases. Progestins also help cattle breeders by synchronizing the breeding and birth cycles so that newborn care is better programmed to use capital facilities efficiently. Progestins help breeders of other species by delaying estrus (heat) during racing seasons, travel, and livestock shows. Progestins can also be used to treat behavior disorders and some forms of dermatitis. Examples include the following:

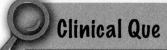

Clinical Que

Progesterone, when administered in conjunction with a luteolytic agent (such as prostaglandin or estradiol), can provide an effective means of resetting the cow's physiological clock.

- *me**gest**rol acetate* (Ovaban®, Megace®). These oral tablets are used in small animals to postpone estrus, alleviate false pregnancies, treat behavior problems (especially in cats), and treat skin problems. Side effects include hyperglycemia (usually a transient diabetes mellitus in cats) and adrenal cortex suppression.

- *medroxypro**gest**erone acetate* (Depo-Provera®, Provera®, Cycrin®). These oral and injectable drugs are used in small animals similarly to megestrol acetate. Side effects include pyometra, mammary changes, and behavior changes.

- *altreno**gest*** (Regu-Mate®). This oral and injectable preparation is used in horses to suppress estrus in mares, for estrus synchronization, and in low levels to maintain pregnancy in mares. When the drug is stopped, GnRH release is stimulated, and the mare will cycle again. Side effects are minimal and include electrolyte imbalance and changes in liver enzyme values. Altrenogest can be absorbed by the skin and should not be handled by pregnant women.

Clinical Que

Progestins or progesterone products can be recognized by the gest in their generic names.

- *pro**ges**terone* (Eazi-Breed Cidr® cattle insert). Another type of estrus synchronization tool is the application of an intravaginal insert containing progesterone. This intravaginal insert has progesterone in elastic rubber molded over a nylon spine that is inserted with a special applicator into the vagina. The insert is placed in the cranial portion of the vagina for seven days and then removed. When the progesterone insert is removed, plasma progesterone rapidly decreases, triggering estrus to occur within three days. One day before the insert is removed, an injection of dinoprost (Lutalyse®) is given to assure estrus synchronization.

- *melen**gest**rol acetate* or MGA (Steakmaker® and MGA 200®). This is a synthetic form of progesterone used to suppress estrus in feedlot heifers intended for breeding. MGA is a premix that is fed for 10 days to feedlot heifers; when it is removed from the feed, the heifers come into heat. MGA increases rate of weight gain, improves feed efficiency, and suppresses estrus.

Clinical Que

When inserting implants, it is important to treat the area with an external antiparasitic spray or repellent to prevent fly infestation.

The "Prost"s (Prostaglandin Drugs)

Many groups of prostaglandins occur naturally in the body (groups include A, B, C, D, E, and F) and are made by almost every cell in the body. In the reproductive system, prostaglandins are made in the uterus. Prostaglandin $F_{2\alpha}$ affects the reproductive system by causing lysis of the CL, which results in lowered progesterone levels in plasma and initiation of a new estrous cycle. Prostaglandin $F_{2\alpha}$ also causes contraction of uterine muscle, facilitating either expulsion of pus (in pyometra) or the mummified fetus (in fetal death), or an abortion. In cattle, prostaglandin $F_{2\alpha}$ is used in estrus synchronization (if given to cattle in diestrus, it will lyse any CL, and all of the cattle will cycle at the same time, in about two to four days), treatment of silent heat, and treatment of pyometra. In small animals, it is used to treat pyometra, cause abortion, and induce parturition. In mares, it is used for estrus synchronization. The CL of mares is resistant to prostaglandins for five days after ovulation; therefore, two injections (13 days apart) are usually given to synchronize estrus. Pregnant women should not handle this medication, as it is absorbed through the skin and can cause uterine contractions. Side effects include bronchoconstriction and elevated blood pressure in animals and humans (through skin absorption). Examples include the following:

- *dino**prost*** tromethamine (Lutalyse®).
- *flu**prost**enol* (Equimate®) (labeled for mares only).
- *clo**prost**enol* sodium (Estrumate®) (labeled for cows only).

Gonad Stimulators (Gonadotropins)

Gonadotropins are hormones that stimulate the gonads. Gonadotropin drugs cause release of LH and FSH or simulate their activity. The three substances that play a role in this category are LH, FSH, and GnRH (Figure 10-15). LH activity is simulated with the use of human chorionic gonadotropin (hCG), FSH activity is simulated by pregnant mare serum gonadotropin (PMSG) or by FSH processed from the pituitary gland, and GnRH is synthetically prepared.

Gonadorelin or GnRH is produced in healthy animals by the hypothalamus. GnRH causes release of FSH and LH by the anterior pituitary gland. GnRH is used IM or IV to treat follicular cysts in cattle, for estrus synchronization in cattle, and to induce estrus in small animals. Side effects are rare. Examples of GnRH are Cystorelin® and Factrel®.

Pregnant mare serum gonadotropin (PMSG) and FSH obtained from the pituitary gland (FSH-P) act like naturally occurring FSH. PMSG is a FSH-like hormone that comes from a pregnant mare's endometrium. Its function in the pregnant mare is to cause follicular development and formation of multiple corpus lutea as a progesterone source during pregnancy. Inactive ovaries that are dormant for various reasons can be effectively stimulated into activity by PMSG. A single subcutaneous injection produces estrus and ovulation within two to five days in cows and within five to eight days in horses. It has been used

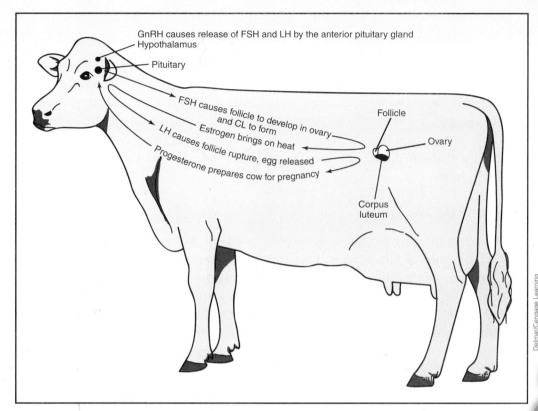

Labels in figure:
GnRH causes release of FSH and LH by the anterior pituitary gland
Hypothalamus
Pituitary
FSH causes follicle to develop in ovary and CL to form
Estrogen brings on heat
LH causes follicle rupture, egg released
Progesterone prepares cow for pregnancy
Follicle
Ovary
Corpus luteum
Delmar/Cengage Learning

FIGURE 10-15 Gonadotropins and their role in the estrous cycle.

as a follicle stimulant in horses, cattle, sheep, swine, dogs, and cats. FSH-P causes growth and maturation of the ovarian follicle.

FSH is also used to promote superovulation (ova release from multiple follicles) in cattle. Superovulation can be achieved by giving 8 to 10 injections of FSH SQ or IM at half-day intervals. Prostaglandin $F_{2\alpha}$ is then given 48 to 72 hours after initiation of treatment with the fifth, sixth, or seventh FSH injection to cause the CLs to lyse at the same time thus inducing ovulation.

Human chorionic gonadotropin (hCG) is a compound that has gonadotropic activity similar to LH. The source of hCG is the human placenta. Human chorionic gonadotropin appears in the urine a few weeks after conception, reaches a peak 50 days into pregnancy, then decreases. The rabbit pregnancy test (in which a pregnant woman's urine causes ovarian changes in the rabbit) employs hCG. Veterinarians use hCG obtained from the urine of pregnant women to treat cystic ovaries in nymphomanic cattle, detect cryptorchidism in dogs, get infertile bitches to cycle, and make breeding mares ovulate. Examples are Follutein®, Chorulon®, A.P.L.®, and LyphoMed®. Veterinarians use IV or IM injections of hCG, because hCG is destroyed in the GI tract after oral administration. Because hCG is a protein, the main side effect is anaphylactic reaction. If an anaphylactic reaction is observed, antihistamines are given.

Clinical Que

Pheromones are substances secreted to the outside of an animal. They are perceived through smell by other members of the same species. Pheromones cause specific behavior, usually sexual or territorial in nature. Pheromones produced synthetically can prevent some of these behaviors, such as urine marking. An example of a commercially produced pheromone product is Feliway®, an analogue of feline facial pheromones. It is labeled to prevent urine marking by cats.

PROMOTING GROWTH

Improved feed conversion efficiency (rate of conversion from food to tissue) promotes growth. A number of chemical compounds, called anabolic agents or **growth promotants**, can improve growth in animals. Antibiotics, typically considered agents to treat bacterial infections, can improve the efficiency of a healthy animal's gastrointestinal tract by changing its microbial population. Antibiotics used in this fashion are classified as growth promotants (Table 10-3).

Hormonal Implants

Growth-promoting implants have been used since the 1940s to increase feed efficiency and weight gain in beef cattle. *Implants* are small pellets or devices placed under the skin at the back of the ear (to avoid the possibility of hormone residue in food products, implants inserted in the ear are removed prior to slaughter). Figure 10-16 shows ear implants. Each pellet contains a growth-promoting hormone that is slowly released into the blood and is subsequently carried to tissues. When growth-promoting implants are placed in the ear, there is a rapid release of hormone from the implant. The level of growth-promoting hormone will begin to fall a few days after implantation, but it remains above the animal's normal level for a varying length of time, depending on the type of implant and the chemical in the implant. Growth promotants must be officially approved and are subject to local regulation. Many implants contain small amounts of antibiotic for local protection.

Diethylstilbestrol (DES) was one of the first growth-promoting compounds; however, it was taken off the market many years ago because of its potential danger for human consumers (increased chance of some forms of cancer and infertility). Veterinarians currently use implants of natural hormones (such as testosterone, progesterone, and estradiol) and synthetic hormones to provide a slow, sustained release of the growth promotant. Examples of growth promotants include the following.

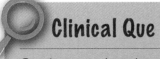

Clinical Que

Cattle must be given adequate nutrition in order for implants to positively affect feed efficiency and weight gain.

Table 10-3   Some Antimicrobial Growth Promotants Used in Livestock

COMPOUND	EFFECTS
bacitracin	Promotes growth in poultry
flavomycin	Increases feed efficiency, promotes growth in poultry and cattle
virginiamycin	Promotes growth in poultry
avoparcin	Increases feed efficiency; promotes growth in poultry, pigs, and cattle
lasalocid	Increases feed efficiency, promotes growth in cattle
monensin	Increases feed efficiency, promotes growth in cattle

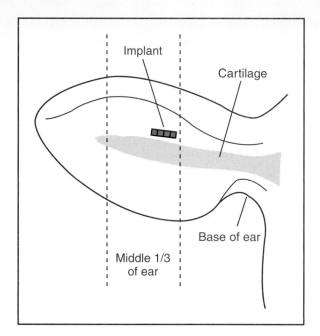

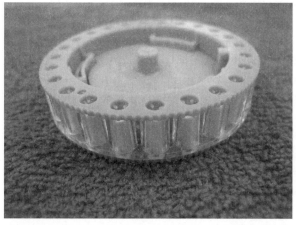

Figure 10-16 (A) Example of ear implant placement; (B) An example of implants.

Estradiol is a potent anabolic agent in ruminants and is administered either as a compressed-tablet implant or a silastic rubber implant. Estradiol given to steers increases growth rate by 10 to 20 percent, lean meat content by 1 to 3 percent, and feed efficiency by 5 to 8 percent. Estradiol is not an effective anabolic agent in pigs. Examples of estradiol products are Compudose® and Encore®.

Testosterone is used in combination with estradiol as a component in compressed-tablet implants to slow the release rate of estradiol. Testosterone is not used on its own as an anabolic agent in farm animals. Examples include Synovex H® (heifers), Synovex S® (steers), and Implus-H®.

Progesterone, like testosterone, is used in combination with estradiol to slow the release rate from implanted compressed tablets. There is no evidence to suggest that progesterone itself is anabolic in livestock. Examples include Synovex C® (calves more than 45 days old), Implus-S®, Component®, and CALF-oid Implant®.

Synthetics are synthetic hormones that generally have more potency and fewer adverse effects than naturally occurring hormones. The synthetic hormones used in veterinary practice include *trenbolone acetate* (TBA), *melengestrol acetate* (MGA), and *zeranol.*

Veterinarians use TBA to promote growth in feedlot heifers and (to a lesser extent) sheep. It works like testosterone, but with greater activity. Unlike natural testosterone, TBA works by itself and with estradiol as a pellet-type implant in heifers and cull cows. It is not intended for use in breeding or dairy animals. Examples of TBA are Finaplix-H®, Finaplix-S®, and Revalor-S®.

Veterinarians use MGA, a synthetic form of progesterone, as a feed supplement to improve rate of weight gain and feed utilization. Examples of MGA are Steakmaker® and MGA 200® (these products were discussed previously with methods to suppress estrus in feedlot heifers). Heifers are fed this supplement daily to improve weight gain and feed efficiency (estrus is also suppressed). This supplement is not effective in steers.

Zeranol is an analog of a naturally occurring plant estrogen and is used as a subcutaneous ear implant to enhance weight gain and feed efficiency in cattle and sheep. Examples of zeranol are Ralgro® beef cattle implant and Ralgro® feedlot lamb implant.

Side effects of growth promotants that are synthetic hormones include mounting behavior, rectal prolapse, ventral edema, and udder development. Most growth promotants should not be given to dairy cattle or breeding animals.

Tissue Building

Anabolic steroids are tissue-building substances. Anabolic steroids cause a positive nitrogen balance and reverse tissue breakdown. These products are labeled for use in dogs, cats, and horses for anorexia, weight loss, debilitation, and to promote red blood cell formation. Anabolic steroids are C-III controlled substances. They should not be used in pregnant animals (they are teratogens) or animals intended for food. Testosterone is a natural anabolic steroid. Other examples include *boldenone undecylenate* (Equipoise®), which comes in injectable form, and *nandrolone decanoate* (Deca-Durabolin®), which comes in injectable form.

SUMMARY

The endocrine system consists of ductless glands that secrete hormones into the bloodstream. Disease of the endocrine system is usually a result of over- or underproduction of these hormones. Drugs used in the management of the endocrine system include substances to regulate pituitary hormones, blood glucose, metabolism via the thyroid gland, and the reproductive system.

The anterior pituitary gland produces and secretes chemicals that affect other endocrine glands and these substances tend to be used in the diagnosis of

Clinical Que

All removed implants or inserts should be placed in a sealed container until they can be properly disposed of in accordance with local, state, and federal regulations.

endocrine diseases. The posterior pituitary gland secretes ADH, which affects urine production; and oxytocin, which affects uterine contractions. Diabetes insipidus is a disease characterized by the inability to concentrate urine due to insufficient ADH release. Drugs used to treat diabetes insipidus include vasopressin (also used to diagnose diabetes insipidus) and desmopressin. Synthetic oxytocin is used to cause uterine contraction and induce labor in animals at term.

Insulin and glucagon regulate blood glucose levels. Insulin is used to treat the hyperglycemia seen in patients with diabetes mellitus. Insulin is categorized as short-, intermediate-, long-, or ultralong-acting. Insulin concentration is expressed in units of insulin per milliliter. Insulin should be refrigerated and not frozen when stored. It is important never to shake insulin, as this results in degradation of the insulin molecule. Oral hypoglycemic agents such as glipizide, glyburide, glimepiride, and acarbose are also used to treat diabetes mellitus.

Hypoglycemia, secondary to hyperinsulin secretion, is treated with oral hyperglycemic agents such as diazoxide. Blood glucose levels may also be increased with the use of glucocorticoids, epinephrine, progesterone, glucose, and dextrose.

Hypothyroidism is a disease in which animals do not produce enough thyroid hormone. Hypothyroidism is diagnosed by measuring total T_4 and T_3 levels, the thyroid stimulation test using thyrotropin (TSH), or the TRH response test. The drug of choice for treating hypothyroidism is levothyroxin, a synthetic form of thyroxine. Liothyronine is used in animals that do not respond to treatment with levothyroxin. Hyperthyroidism is a disease in which animals produce too much thyroid hormone. Hyperthyroidism is treated by surgically removing the dysfunctional thyroid gland, by destroying the dysfunctional thyroid gland with radioactive iodine, or by blocking hormone function with methimazole.

Corticosteroids are produced by the adrenal cortex. Adrenocortical insufficiency results in deficient production of corticosteroids (glucocorticoids and mineralocorticoids) and is diagnosed via the ACTH stimulation test using corticotropin or cosyntropin. Treatment of adrenocortical insufficiency includes use of desoxycorticosterone (DOCP), a long-acting mineralocorticoid; and prednisone, a glucocorticoid. Fludrocortisone acetate has both glucocorticoid and mineralocorticoid activity and has been used to treat adrenocortical insufficiency. Hyperadrenocorticism, a disease that results in overproduction/oversupplementation of glucocorticoids, is treated with mitotane. Mitotane destroys part of the adrenal cortex. Other treatment options for hyperadrenocorticism include ketoconazole, selegiline, trilostane, cyproheptadine, and pergolide mesylate.

Drugs affecting reproduction in animals include testosterone, estrogen, progesterone, prostaglandin, and gonadotropins. Veterinarians use testosterone to treat infertility, hypogonadism, and urinary incontinence, and to produce teaser animals in cattle. Other androgens used in veterinary medicine include danazol (used to treat immune-mediated hemolytic anemia), mibolerone (used to prevent estrus in adult female dogs), and finasteride (used to treat benign prostatic hyperplasia in dogs). Estrogen is used in animals to prevent conception and treat urinary incontinence in dogs and to induce estrus in mares. Progesterone and progestins prevent estrus in dogs, synchronize breeding in cattle, and treat behavior problems and dermatitis in small animals. Progesterones and progestins can be recognized by "gest" in their generic names. Prostaglandins, mainly prostaglandin $F_{2\alpha}$, are used for estrus synchronization in cattle and horses, and for treating pyometra, causing abortion, and inducing parturition in small animals. Prostaglandins can be recognized by "prost" in their generic names. Gonadotropins cause release of LH and FSH or simulate their activity. Gonadotropins include GnRH (used to treat follicular cysts in cattle and induce estrus in small animals), PMSG and FSH-P (used to induce estrus or as a follicle stimulant), and hCG (used to treat cystic ovaries in cattle, to detect cryptorchidism and induce cycling in dogs, and to cause ovulation in mares).

Growth promotants, used to improve growth in animals, include antimicrobials, estradiol, testosterone, progesterone, synthetic hormones, and anabolic steroids. Side effects include mounting behavior, rectal prolapse, ventral edema, and udder development.

Table 10-4 summarizes the drugs covered in this chapter.

Table 10-4 Drugs Covered in This Chapter

ENDOCRINE SYSTEM CONDITION OR ACTION	DRUG(S) USED TO TREAT CONDITION
Diabetes insipidus (insufficient ADH)	• vasopressin • desmopressin
Dystocia (induction of uterine contraction)	• oxytocin
Blood glucose regulation	• insulin • glipizide • glyburide • glimepiride • acarbose • diazoxide
Hypothyroid	• levothyroxine • liothryonine
Hyperthyroid	• methimazole • radioactive iodine I-131
Adrenocortical insufficiency	• desoxycorticosterone (DOCP) • fludrocortisone • prednisolone

(continued)

ENDOCRINE SYSTEM CONDITION OR ACTION	DRUG(S) USED TO TREAT CONDITION
Hyperadrenocorticism	• mitotane • ketoconazole • selegiline • trilostane • cyproheptadine • pergolide mesylate
Testosterone or testosterone-like products	• testosterone cypionate • testosterone enanthate • testosterone propionate • methyltestosterone • danazol • mibolerone • finasteride
Estrogen or estrogen-like products	• estradiol cypionate • diethylstilbesterol
Progesterone or progesterone-like	• megestrol acetate • medroxyprogesterone acetate • altrenogest • progesterone • melengestrol
Prostaglandins	• dinoprost • fluprostenol • cloprostenol
Gonadotropins	• gonadorelin • PMSG • FSH-P • hCG
Growth promotants (antimicrobial)	• bacitracin • flavomycin • virginiamycin • avoparcin • lasalocid • monensin
Growth promotants (hormonal)	• estradiol • testosterone • progesterone • trenbolone acetate • melengestrol acetate • zeranol
Anabolic steroids	• boldenone undecylenate • nandrolone decanoate

It's a Wrap

The answers to the questions in this chapter's Setting the Scene case study should be understood after reading this chapter. In this case study, the veterinarian orders some blood tests to determine the cause of this dog's clinical condition. What tests are appropriate to run on this dog's blood? Recommended diagnostic tests for animals with PU/PD include a complete blood count (CBC), serum biochemistry profile (assesses liver, kidney, and glucose status), and urinalysis. Endocrine tests include a thyroid test, which should be included in the serum biochemistry profile in cats, and suppression tests if hyperadrenocorticism is suspected. What are some examples of endocrine diseases that this dog might have? Some examples of endocrine diseases this dog may have include hyperadrenocorticism, hypothyroidism, and diabetes mellitus. Are the treatments for these endocrine diseases safe? The safety of treating endocrine diseases depends on the disease and the treatment selected. Side effects of insulin used for treating diabetes mellitus include hypoglycemia. Hypoglycemia can result if too much insulin is given, and that can cause seizures, coma, and death. Side effects of treating hyperthyroidism with methimazole include vomiting, diarrhea, and lethargy. Side effects of mitotane for treatment of hyperadrenocorticism include neurologic signs, lethargy, vomiting, and diarrhea. The levels of side effects are related to the severity of the disease and how long the disease has been present in the animal. What information should be given to the owner? Owners should be advised that endocrine diseases may take a long time to regulate and that diagnostic testing and physical examinations should be performed until the condition is regulated. Additional testing to monitor the disease is also warranted to keep the disease and treatment in check. The ability to adequately explain endocrine diseases to clients is an important part in achieving owner compliance with treatment and patient monitoring.

CHAPTER REVIEW

Matching

Match the drug name with its action.

1. _____ levothyroxine
2. _____ vasopressin
3. _____ gonadorelin
4. _____ insulin
5. _____ oxytocin
6. _____ TBA (trenbolone acetate)
7. _____ estradiol
8. _____ methimazole
9. _____ megestrol
10. _____ dinoprost

a. a synthetic hormone used to induce estrus in horses during the nonbreeding season
b. a synthetic hormone used to promote growth in feedlot heifers
c. drug of choice for treating hypothyroidism
d. chemical that has antidiuretic effect
e. GnRH used to treat follicular cysts in cattle
f. drug of choice for treating hyperthyroidism
g. drug used to treat diabetes mellitus in dogs
h. injectable drug used to stimulate uterine contractions
i. progestin used to treat behavior problems in cats
j. progesterone used to synchronize estrus in cattle

Multiple Choice

Choose the one best answer.

11. Which group of growth promotants causes tissue building and promotes red blood cell formation?
    a. estradiol
    b. progesterones
    c. anabolic steroids
    d. antimicrobials

12. Which estrus synchronization drug, if given to a herd of cattle, will lyse any CLs and cause the cattle to cycle all at the same time?
    a. prostaglandins
    b. GnRH
    c. estrogens
    d. testosterone

13. Adrenal cortex dysfunction, including adrenocortical insufficiency and hyperadrenocorticism, can be diagnosed by
    a. supplementing with desoxycorticosterone.
    b. supplementing with o,p'-DDD.
    c. clinical signs.
    d. measuring cortisol levels before and after administration of ACTH.

14. Decreased coat luster, hair loss, decreased appetite, listlessness, intolerance to cold, and reproductive failure are all signs of
    a. diabetes mellitus.
    b. hyperthyroidism.
    c. Addison's disease.
    d. hypothyroidism.

15. Insulin concentration is measured in
    a. units/mm.
    b. units/mL.
    c. cm/lb.
    d. g/mL.

16. Which insulin form is used initially to treat diabetic ketoacidosis and to stabilize glucose levels in newly diagnosed diabetic animals?
    a. short-acting
    b. intermediate-acting
    c. long-acting
    d. ultralong-acting

17. The link between the hypothalamus, pituitary gland, and some endocrine glands—in which low levels of hormone signal the hypothalamus to secrete more releasing factor, which in turn signals the pituitary gland to secrete stimulating hormones, resulting in increased hormone levels—is called a

    a.   feedback stimulation mechanism.
    b.   negative feedback loop.
    c.   positive feedback loop.
    d.   connected feedback loop.

True/False

Circle a. for true or b. for false.

18. Oxytocin is beneficial in treating both open and closed pyometra.
    a.   true
    b.   false

19. Bovine somatotropin (BST) is believed to cause increased milk production in dairy cattle.
    a.   true
    b.   false

20. TSH, ACTH, LH, and FSH are hormones controlled by the posterior pituitary gland.
    a.   true
    b.   false

Fill in the Blanks

Read the following labels. Identify the insulin brand name and its action time (short-acting, intermediate-acting, long-acting, or ultralong-acting).

21. Insulin brand name _____

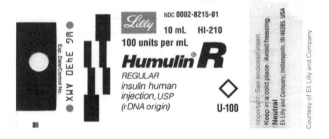

    Insulin type _____

    Action time _____

22. Insulin brand name _____

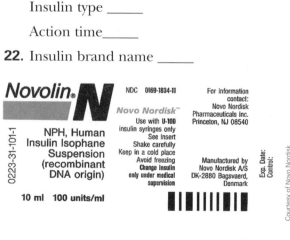

    Insulin type _____

    Action time _____

23. Insulin brand name _____

PZI·VET®
(Protamine Zinc Insulin)

aqueous suspension
40 units per mL
Net contents 10 mL

Caution: Federal (USA) law restricts this drug to use by or on the order of a licensed veterinarian.

U-40
P

IMPORTANT:
USE ENCLOSED INSERT.
PROTECT FROM LIGHT.
AVOID FREEZING.
MIX CAREFULLY BEFORE EACH USE.

Manufactured for
IDEXX Pharmaceuticals Inc.
Chelmsford, NC 22083 USA

IDEXX
PHARMACEUTICALS

05M678001
F3209118
S 0229144

Courtesy of Eli Lilly and Company

    Insulin type _____

    Action time _____

24. Insulin brand name _____

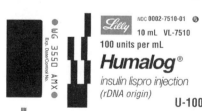

    Insulin type _____

    Action time _____

25. Insulin brand name _____

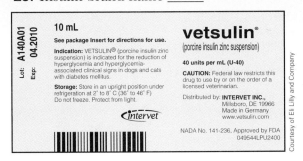

Insulin type _____

Action time_____

26. Insulin brand name _____

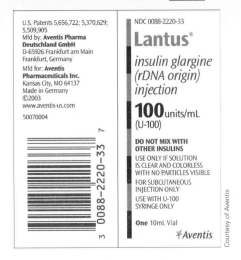

Insulin type _____

Action time_____

Insulin Syringe Reading

Identify which type of insulin syringe is being used and the dose indicated by the colored area of the syringe.

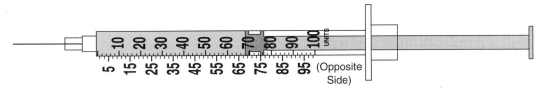

27. _____ type of insulin syringe

_____ units

28. _____ type of insulin syringe

_____ units

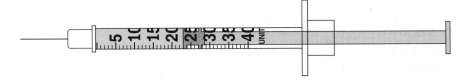

29. _____ type of insulin syringe _____

_____ units

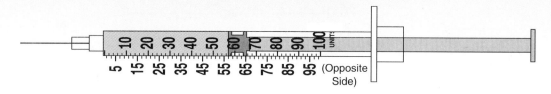

30. _____ type of insulin syringe _____

_____ units

Case Studies

31. A rancher would like his replacement heifers to calve three weeks before his mature cows. He would like to artificially inseminate the heifers with proven "calving ease bulls." Assume that all heifers are cycling females.

    a. What treatment options does this rancher have?

    b. The rancher decides to use 25 mg dinoprost IM. This drug comes in 5 mg/mL concentration. How much does each heifer get?

    c. The rancher can give the calculated dose of dinoprost IM and breed on standing heat. When would he see the heifers in heat?

32. A dairy farmer is having trouble getting cows bred. He reads about a newer protocol for ovulation synchronization called Ovsynch. It uses GnRH (Cystorelin®) and dinoprost (Lutalyse®). Here is the protocol:

- **Day 1: Give 2 cc Cystorelin® (50 mcg/mL GnRH) IM.**

- **Day 8: Give 5 cc Lutalyse® (5 mg/mL dinoprost) IM at 8 A.M.**

- **Day 10: Give 2 cc Cystorelin® IM at 5 P.M.**

- **Day 11: Breed all animals by artificial insemination at 8 A.M.**

    a. On day 1 and day 10, what does the Cystorelin® do?

    b. On day 1 and day 10, how many total mcg does the rancher give each heifer?

    c. On day 8, what does the Lutalyse® do?

    d. On day 8, how many total mg does the rancher give each heifer?

    e. What should you tell this rancher (and remember yourself) about handling dinoprost?

33. A Jersey cow, 120 days in milk, is examined during a normal herd health visit at the farm for being anestrus. PE: The cow is in good body flesh. Rectal palpation reveals a 40-mm soft circular structure on the left ovary. There are no other structures on either ovary. The uterus is normal size for the number of days postpartum. This cow is diagnosed with a cystic follicle (follicular cyst).

    a. What drug do you think the veterinarian will prescribe for this cow?

    b. This cow needs 100 mcg Cystorelin® IM. It comes in 50 mcg/mL concentration. How much should this cow receive?

34. A three-year-old M/N Golden Retriever weighing 90# presents to the clinic with hair loss and weight gain. He has been less active than normal. PE reveals normal TPR, alopecia over the shoulders and hindquarters, and listlessness. Blood tests showed that the dog was hypothyroid.

    a. Explain to the owner what hypothyroidism is.

    b. How do we treat hypothyroidism? What form of hormone is this?

    c. The veterinarian prescribes Soloxine® at a dosage of 0.02 mg/kg daily. How much should this dog get?

    d. Soloxine® comes in 0.1 mg, 0.2 mg, 0.3 mg, 0.4 mg, 0.5 mg, 0.6 mg, 0.7 mg, and 0.8 mg pills. How many pills should this dog get per dose?

Critical Thinking Questions

35. Consider the use of certain reproductive hormones in the practice of animal husbandry. Why would a breeder of herd animals such as dairy cattle be interested in all their breeding females experiencing "heat" at or near the same time (also known as "synchronization")?

36. Why would a breeder of pack animals such as dogs be interested in discouraging synchronized heat in the breeding females?

37. What are "mismating" shots?

ADDITIONAL REVIEW MATERIAL

Additional review material can be found on the StudyWARE™ CD included with this text.

BIBLIOGRAPHY

Allen, D. G. (2005). *Handbook of Veterinary Drugs* (3rd ed.). Baltimore, MD: Lippincott Williams & Wilkins.

Amundson Romich, J. (2008). *An Illustrated Guide to Veterinary Medical Terminology* (3rd ed.). Clifton Park, NY: Delmar Cengage Learning.

Broyles, B., Reiss, B., and Evans, M. (2007). *Pharmacological Aspects of Nursing Care* (7th ed.). Clifton Park, NY: Delmar Cengage Learning.

Cook, A. K. (September 2007). *The Latest Management Recommendations for Cats and Dogs with Nonketotic Diabetes Mellitus.* Veterinary Medicine.

Curran, A. M. (2006). *Dimensional Analysis for Meds* (3rd ed.). Clifton Park, NY: Delmar Cengage Learning.

Feline Diabetes (2006). Cornell Feline Health Center, http://www.vet.cornell.edu/

Pickar, G. D. (2008). *Dosage Calculations* (8th ed.). Clifton Park, NY: Delmar Cengage Learning.

Plumb, D. C. (2008). *Plumb's Veterinary Drug Handbook* (6th ed.). Ames, IA: Blackwell Publishing.

Spratto, G. R. and Woods, A. L. (2008). *PDR Nurse's Drug Handbook.* Clifton Park, NY: Cengage Learning.

CHAPTER 11
GASTROINTESTINAL DRUGS

OBJECTIVES

Upon completion of this chapter, the reader should be able to:

- describe the basic anatomy and physiology of the gastrointestinal tract.
- differentiate between peristalsis and segmentation.
- list the various signs of gastrointestinal disease.
- describe the basic pathophysiology of the various types of gastrointestinal disease.
- describe the use of the following drugs in veterinary medicine: antisialogues, antidiarrheals, laxatives, antiemetics, systemic and nonsystemic antacids, emetics, antiulcer compounds, antifoaming agents, prokinetic agents, and digestive enzymes.
- outline the use of dental prophylaxis and treatment aids.

KEY TERMS

adsorbents	cholinergic drugs
antacids	dentifrices
anticholinergic drugs	disclosing solutions
antidiarrheals	emetics
antiemetics	laxatives
antifoaming agents	probiotics
antisialogues	prokinetic agents
antiulcer drugs	protectants
cathartics	sympathetic drugs

Setting the Scene

A teenager calls the clinic and says that she has been babysitting a child and his dog. The dog got into the cupboard and spilled cleaning supplies on the floor. When the babysitter found the dog, he was lapping up some of the chemicals from the floor. She wants to know what she can give the dog to make him vomit up the chemicals. What should the veterinary technician tell her? Is vomiting warranted in this case? What information is needed before making that decision? What home options does the babysitter have, or is it better to bring the dog into the clinic?

BASIC DIGESTIVE SYSTEM ANATOMY AND PHYSIOLOGY

Gastrointestinal tract, alimentary system, GI tract, and *digestive tract* are all terms that basically describe a long, muscular tube that begins at the mouth and ends at the anus. The anatomic structures therein include the oral cavity, pharynx, esophagus, stomach, small intestine, and large intestine. These structures vary from monogastric animals with simple stomachs to ruminant animals with multichambered forestomachs (Figure 11-1). In either type of animal, the digestive tract plays a role of bringing life-sustaining elements into the body, and

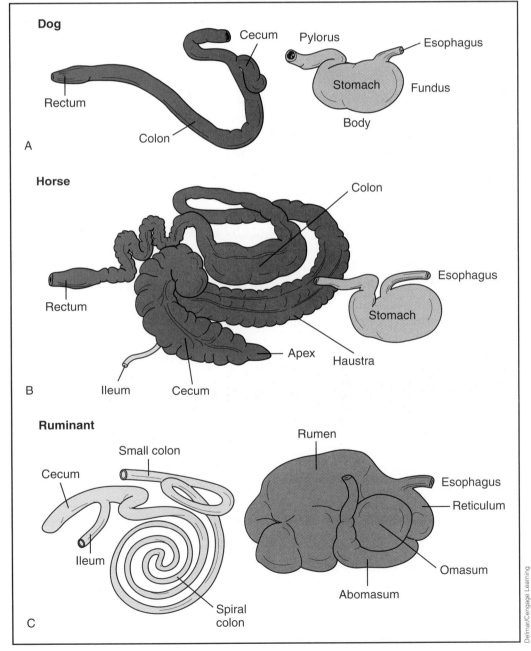

Figure 11-1 Gastrointestinal tracts (small intestinal segments omitted for clarity): (A) Dog (B) Horse (C) Ruminant.

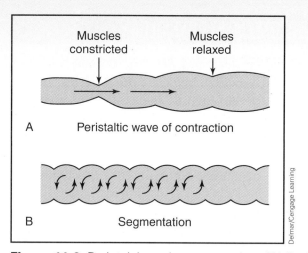

Figure 11-2 Peristalsis and segmentation: (A) Peristaltic contraction moves food through the digestive tract. (B) Segmentation helps break down and mix food through cement-mixer-type action.

taking waste products out of it. Proper functioning of this system depends on an unobstructed and regulated flow of these elements. In addition to the structures of this tube, there are accessory organs (e.g., liver, gall bladder, salivary glands, and pancreas) that aid the digestive tract through the production or secretion of enzymes that help in the digestion of food products.

Flow of material along the gastrointestinal tract occurs by *peristalsis*, a wave-like contraction of longitudinal and circular muscle fibers that moves food through the gut. Another process occurring in the gastrointestinal tract is *segmentation*: periodic, repeating intestinal constrictions that cause churning of the GI contents (Figure 11-2). Regulation of these actions is controlled by many mechanisms. One control mechanism of the GI tract is the autonomic nervous system (ANS), which consists of the sympathetic branch (fight-or-flight response) and the parasympathetic branch (homeostatic response). Parasympathetic stimulation increases intestinal motility and GI secretions and relaxes sphincters. **Cholinergic drugs** simulate these actions. **Anticholinergic drugs** inhibit these actions. Sympathetic stimulation decreases intestinal motility, decreases intestinal secretions, and inhibits the action of sphincters. **Sympathetic drugs** simulate these actions. Stretch receptors in the GI tract also increase peristalsis.

Chemical secretions also control the GI tract. Hormones released from the intestinal cells help control actions such as gall bladder emptying. Chemical substances such as histamine can attach to the H_2 receptors of the gastric cells, causing increased HCl production.

GASTROINTESTINAL DISORDERS

Gastrointestinal disorders are among the most common complaints seen in veterinary medicine. The underlying causes of these disorders range from infectious sources, dietary excess, adverse drug effects, and a variety of systemic diseases. Infections and disease can alter the functioning of the GI tract. Bacterial infections resulting in endotoxin release from gram-negative bacterial cell walls can increase intestinal blood vessel permeability, resulting in increased fluid loss. Disease conditions such as liver disease can affect the production of bile and ultimately fat digestion. Gastrointestinal disorders affect the gastrointestinal tract either directly or indirectly resulting in clinical signs such as diarrhea, constipation, vomiting, bloat, ulcer development, or pain. Some of the clinical signs that an animal develops are protective mechanisms; however, these clinical signs may need to be treated to prevent other complications such as dehydration or electrolyte alterations.

GASTROINTESTINAL DRUGS

Gastrointestinal drugs help maintain the unobstructed and regulated flow of food into the body and waste products out of the body. Although many drugs can affect gastrointestinal tract function, this chapter covers only drugs that directly affect the gastrointestinal tract.

Veterinary gastrointestinal drugs can be administered for a variety of reasons. Some gastrointestinal drugs encourage peristalsis, suppress it, or reduce its undesirable by-products. Other GI drugs decrease the flow of saliva, control vomiting and diarrhea, loosen stool, cause vomiting, protect the GI tract, decrease acid production, or reestablish GI normal flora. Gastrointestinal drugs do these things in a variety of ways, resulting in a variety of outcomes.

Saliva Stopping

Antisialogues are drugs that decrease salivary flow. Veterinarians administer antisialogues intravenously, intramuscularly, or subcutaneously to limit this excess saliva production, which often occurs secondary to anesthetic drug use. *Glycopyrrolate* (Robinul®) and *atropine* (Atropine Injectable-SA®, Atropine Injectable-LA®) are antisialogues via their anticholinergic mechanism of action. Anticholinergic drugs block the effects of acetylcholine at parasympathetic nerve endings. The effect of anticholinergics is to reduce gastrointestinal motility and secretions (including saliva). Antisialogues, although used to decrease salivation, can also affect peristalsis. In addition to decreasing saliva production, anticholinergic drugs are used in the gastrointestinal tract to treat vomiting, diarrhea, and excess gastric secretion by blocking parasympathetic nerve impulses (see section on antidiarrheals and antiemetics). Side effects of anticholinergics include dry mouth, constipation, CNS stimulation, tachycardia, and pupillary dilation.

Diarrhea Stopping

Diarrhea is abnormal frequency and liquidity of fecal material due to failure of the intestinal tract to adequately absorb fluids from the intestinal contents. Diarrhea is not a disease but rather a sign of underlying disease. There are many causes of diarrhea, including parasitic disease, bacterial infection, viral infection, dietary indiscretion, or a systemic nongastrointestinal disease. Diarrhea is a concern because it may cause excessive fluid loss that results in dehydration and electrolyte imbalance. Diarrhea also decreases the uptake of nutrients from the intestine if the contents are moving too quickly through it. The longer diarrhea is allowed to continue, the more likely it is that the animal will experience

Clinical Que

Atropine comes in two concentrations: Atropine Injectable-SA (0.5 mg/mL) and Atropine Injectable-LA (2 mg/mL). Care should be taken when using atropine to ensure that the proper concentration is chosen.

Clinical Que

Glycopyrrolate does not appreciably cross into the CNS or placenta, making it more desirable for use in pregnant animals than atropine, which does cross into the CNS and placenta.

side effects. Side effects of diarrhea include electrolyte imbalances, muscle weakness, acid–base disturbances, and anorexia.

Antidiarrheals are drugs that decrease peristalsis of the gastrointestinal tract, thereby allowing fluid absorption from the intestinal contents. This helps to reverse diarrhea by decreasing the liquidity of stool. Antidiarrheals include anticholinergics, protectants, adsorbents, and narcotic analgesics (Figure 11-3). Probiotics and metronidazole are also used to supplement the treatments of diarrhea.

Anticholinergics

Anticholinergics are used to treat tenesmus (straining to defecate) associated with colitis and vomiting related to colonic irritation. This group of drugs blocks acetylcholine release from the parasympathetic nerve endings, thus decreasing gastrointestinal motility and secretions. It should be kept in mind that in some cases decreased gastrointestinal motility is already associated with

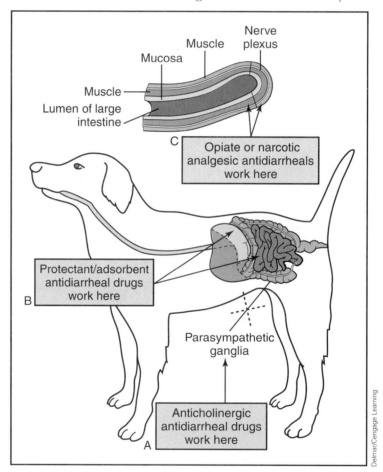

Figure 11-3 Mechanisms of action of antidiarrheal drugs. (a) Anticholinergic antidiarrheal drugs block acetylcholine release from the parasympathetic nerve endings to decrease gastrointestinal motility and secretions. (B) Protectants/adsorbents work either by coating inflamed intestinal mucosa with a protective layer (protectants), or by binding bacteria and/or digestive enzymes and/or toxins to protect intestinal mucosa from their damaging effects (adsorbents bind substances). (C) Opiates or narcotic analgesics control diarrhea by decreasing both intestinal secretions and the flow of feces, and increasing segmental contractions, resulting in increased intestinal absorption.

diarrhea; therefore, these drugs should be used with caution. Side effects of anticholinergics include dry mucous membranes, urine retention, tachycardia, and constipation. Drugs in this category include the following:

- *atropine* (generic, Atropine Injectable-SA®, Atropine Injectable-LA®), which is available as an injection to decrease gastrointestinal motility in many animal species (atropine also has a wide variety of other actions that are covered in the appropriate chapters).

- *aminopentamide* (Centrine®), which is available in tablet and injection form, for treatment of acute abdominal spasm in dogs and cats.

- *propantheline* (Pro-Banthine®), which is available as tablets, is used to treat gastrointestinal spasms and hypersecretions associated with colitis and irritable bowel syndrome in dogs and cats and to reduce rectal contractions in horses.

- *methscopolamine* (Biosol-M®), which is available as a liquid, is a product that has antispasmotic properties and an antibiotic (neomycin) that is used to treat bacterial diarrhea in dogs and cats.

- *N-butylscopolammonium bromide* (Buscopan®), which is available in injection form, is an anticholinergic used as a single IV dose in horses with spasmodic colic, flatulent colic, or simple impactions.

Protectants/Adsorbents

This category of antidiarrheal drugs works either by coating inflamed intestinal mucosa with a protective layer (**protectants**), or by binding bacteria and/or digestive enzymes and/or toxins to protect intestinal mucosa from their damaging effects (**adsorbents** bind substances). Side effects from these drugs are uncommon except for the possible development of constipation. Drugs in these categories include the following:

- *bismuth subsalicylate* (Corrective Mixture®, Pepto-Bismol®). The bismuth portion of this drug coats the intestinal mucosa and has antiendotoxic and weak antibacterial effects. The subsalicylate portion has anti-inflammatory effects because it reduces the production of prostaglandins. Subsalicylate is an aspirin-like product; therefore, this drug should not be used in cats. Corrective Mixture with Paregoric® has an opium tincture in it and is a C-V controlled substance. These products can blacken stool and can also cause opacities on radiographs.

- *kaolin/pectin* (Kao-Forte®, Kaopectolin®, K-P-Sol®). This combination drug has both adsorbent and protective qualities. Bacteria and toxins are adsorbed in the gut and the coating action protects inflamed intestinal mucosa.

- *activated charcoal* (Liqui-Char®, Superchar®, Actidose-Aqua®). Activated charcoal is a fine, black, tasteless powder used to adsorb many chemicals and drugs in the upper gastrointestinal tract. It is used primarily to treat ingestions of certain toxins and is typically administered via stomach tube, dosing gun, or premeasured syringe (Figure 11-4).

Clinical Que

An *adsorbent* is a substance that binds other material to its surface. A gastrointestinal adsorbent adsorbs gases, toxins, bacteria, drugs, and digestive enzymes in the stomach and intestines. Adsorbents lack specificity and may prevent absorption of other drugs the animal was given orally.

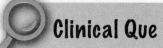

Figure 11-4 Activated charcoal is used primarily to treat ingestions of certain toxins and is typically administered via stomach tube, dosing gun, or premeasured syringe.

Opiate or Narcotic Analgesics

Opiates or narcotic analgesics control diarrhea by decreasing both intestinal secretions and the flow of feces, and increasing segmental contractions, thereby resulting in increased intestinal absorption. Side effects of these drugs include CNS depression (excitement in horses and cats), ileus, urinary retention, bloat, and constipation with prolonged use. Examples include the following:

- *diphenoxylate* (Lomotil®, Lonox®, Diphenatol®), a C-V controlled substance with atropine added. Diphenoxylate is not an opium derivative, but is structurally similar to meperidine.

- *loperamide* (Imodium®, Imodium A-D®). This drug causes less CNS depression than other drugs in this category and can be purchased over the counter. Like diphenoxylate, it is structurally similar to meperidine.

- *paregoric*. This drug is a C-III controlled substance and may be combined with kaolin/pectin. Paregoric is camphorated tincture of opium.

Probiotics

Another approach in treating diarrhea is the use of **probiotics** to seed the gastrointestinal tract with beneficial bacteria such as *Lactobacillus* spp., *Enterococcus faecium*, and *Bifidobacterium* spp. This approach is based on the theory that some forms of diarrhea (like that secondary to antibiotic use) are caused by disruption of the normal bacterial flora of the gastrointestinal tract. The mechanisms of action of probiotics may involve competing with pathogenic bacteria for colonizing sites, production of antimicrobial factors, alteration of the microenvironment, reduction of local inflammation, and alteration of immune responses. Plain yogurt with active cultures is often used to try to repopulate the gastrointestinal tract with beneficial bacteria. Trade names of probiotic products used in veterinary medicine include Fastrack® gel, FortiFlora®, Proviable™-DC, Proviable™-KP (with kaolin and pectin), Prostora™ Max, Probiocin® oral gel for pets, and Probiocin® oral gel for ruminants. These products may have to be refrigerated to maintain the viability of the bacterial culture.

Anaerobic Antibiotic

Metronidazole (Flagyl®) is an antibiotic that is effective against anaerobic bacteria. It is sometimes used as an antidiarrheal on the theory that disruption of the normal-flora environment may increase the number of anaerobic bacteria present in an animal with diarrhea. Because metronidazole is effective against anaerobes, it may help return the animal's stool to its normal consistency.

Additional information on metronidazole and its spectrum of activity appears in Chapter 14.

Stool Loosening

Constipation is a condition in which passage of feces is slowed or nonexistent. A **laxative** is a medicine that loosens the bowel contents and encourages evacuation of stool (Figure 11-5). Veterinarians use laxatives to help animals evacuate stool without excessive straining, to treat chronic constipation from nondietary causes and movable intestinal blockages (such as hair balls), and to evacuate the GI tract before surgery, radiography, or proctoscopy. **Cathartics** are harsher laxatives that result in a soft to watery stool and abdominal cramping.

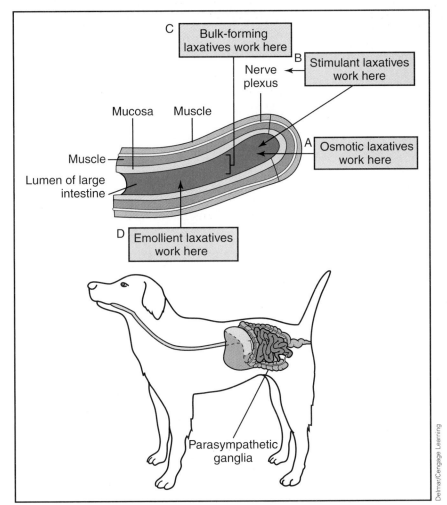

Figure 11-5 Mechanisms of action of laxatives. (A) Osmotic laxatives work by producing a hyperosmolar environment in the intestinal lumen that pulls water into the colon and increase water content in the feces, increasing bulk and stimulating peristalsis. (B) Stimulant laxatives increase peristalsis by chemically irritating sensory nerve endings in the intestinal mucosa. (C) Bulk-forming laxatives are compounds that absorb water into the intestine, increase fecal bulk, and stimulate peristalsis, resulting in large, soft stool production. (D) Emollients reduce stool surface tension and reduce water absorption through the colon, facilitate passage of fecal material, increase water retention in stool, and permit easier penetration and mixing of fats and fluid with the fecal mass, resulting in a softer stool that is more easily passed.

A *purgative* is a harsh cathartic, causing watery stool and abdominal cramping. Categories of laxatives include osmotic, stimulant, bulk forming, and emollients (stool softeners).

Osmotic

Osmotic (or hyperosmolar) laxatives include salts or saline products, lactulose, and glycerin. Hyperosmolar salts pull water into the colon and increase water content in the feces, increasing bulk and stimulating peristalsis. The saline products are composed of sodium or magnesium, and small amounts may be systemically absorbed, causing electrolyte imbalances. Use of saline products should be limited in animals with heart failure and renal dysfunction; some (such as Fleet Enema®) are not recommended for cats because they can cause severe electrolyte imbalance. Prolonged use of osmotic laxatives can also cause dehydration. Examples of osmotics include the following:

- *lactulose* (Cephulac®), also used in liver disease because it eliminates ammonia
- *sodium phosphate with sodium biphosphate* (Fleet Enema®, Gent-L-Tip Enema®)
- *magnesium sulfate* (Epsom Salts)
- *magnesium hydroxide* (Milk of Magnesia®, Carmilax-Powder®, Magnalax®, Poly Ox II Bolus®)
- *polyethylene glycol-electrolyte solution* (OCL® Solutions, CoLyte®, GoLYTELY®)

Stimulants

Stimulant (irritant or contact) laxatives increase peristalsis by chemically irritating sensory nerve endings in the intestinal mucosa. Their site of action may vary from affecting only the large or small intestine to affecting the entire gastrointestinal tract. Although most stimulant laxatives come from natural sources, two agents (bisacodyl and phenolphthalein) are synthetic. Many stimulant laxatives are absorbed systemically and can cause a variety of side effects. Examples in this group include the following:

- *bisacodyl* (Dulcolax®), a cathartic that comes in enteric-coated and suppository forms. Side effects include cramping and diarrhea. Tablets should not be crushed or chewed, as this will result in intense abdominal cramping. Milk or antacids should not be given within one hour of administration of this drug, as they may dissolve the enteric coating.
- *castor oil*, which has an active ingredient (ricinoleic acid) that is liberated in the small intestine. Ricinoleic acid inhibits water and electrolyte absorption, leading to fluid accumulation in the gastrointestinal tract and increased peristalsis. Side effects include diarrhea, abdominal pain, and electrolyte imbalance.

Clinical Que

Enemas are liquids administered directly into the colon. They can elicit a laxative response as well as cleansing the bowel prior to a surgical or diagnostic procedure.

Clinical Que

Mechanoreceptors in the digestive tract detect tension in response to distention and contraction in the gastrointestinal tract. Chemoreceptors in the digestive tract detect chemicals in the intestinal lumen.

Bulk Forming

The bulk-forming laxatives are natural fibrous substances or semisynthetic compounds that absorb water into the intestine, increase fecal bulk, and stimulate peristalsis, resulting in large, soft stool production. Unlike stimulant laxatives, bulk-forming laxatives tend to produce normally formed stools. Most of these products contain psyllium preparations or plantago seed (the seed coating absorbs water and swells). Most bulk-forming laxatives have minimal effects on nutrient absorption and are not systemically absorbed. Examples include the following:

- *psyllium hydrophilic mucilloid* (Metamucil®, Equine Psyllium®, Perdiem®)
- *polycarbophil* (FiberCon®, Fiberall®)
- *bran*

Emollients

Emollients are stool softeners (which reduce stool surface tension and reduce water absorption through the colon), lubricants (which facilitate passage of fecal material, increasing water retention in stool), and fecal wetting agents (detergent-like drugs that permit easier penetration and mixing of fats and fluid with the fecal mass, resulting in a softer stool that is more easily passed). Emollients are not absorbed systemically, and thus have fewer side effects, as well as decreasing straining during defecation. Side effects are rare but include some abdominal cramping and diarrhea. Examples include the following:

- *docusate sodium* (Colace®). This is also known as docusate sodium succinate (DSS).
- *docusate calcium* (Surfak®)
- *docusate potassium* (Dialose®)
- *petroleum products* (solid or liquid). Mineral oil is liquid petroleum, commonly used in horses as a laxative, but also used to decrease absorption of fat-soluble toxins. White petroleum is solid petroleum, commonly used to prevent and treat hair balls in cats (CatLax®, Laxatone®).

Vomit Stopping

Vomiting or emesis is the expulsion of stomach (and sometimes duodenal) contents through the mouth. Vomiting is caused by a variety of things, including viral and bacterial infection, dietary indiscretion, food intolerance, surgery, pain, and other drugs. The act of vomiting is controlled by the vomiting center in the medulla of the brain stem. Acetylcholine is the neurotransmitter for the vomiting center. The vomiting center gets input from many pathways that activate it. These inputs include equilibrium changes in the inner ear, responses of the higher centers of the cerebral cortex due to pain or fear, intracranial pressure changes, vagus nerve stimulation in the gastrointestinal tract, and activity in the chemoreceptor trigger zone (CRTZ). Figure 11-6 shows the areas that affect the vomiting center.

Clinical Que

Animals taking bulk-forming laxatives should have free access to water. Water consumption is necessary to assure that the bulk-forming process goes smoothly.

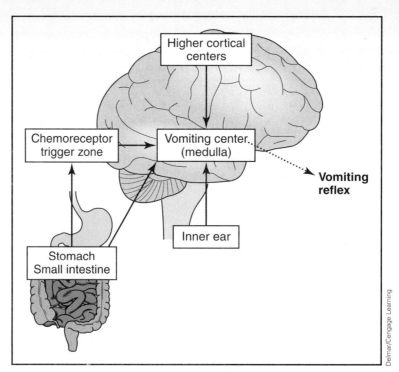

Figure 11-6 The vomiting center and factors that control vomiting.

Vomiting represents a coordinated effort of the gastrointestinal, musculoskeletal, and nervous systems to expel food, fluid, or debris from the gastrointestinal tract. It is initiated centrally by direct stimulation of the vomiting center in the medulla of the brain stem or indirectly via the CRTZ or peripherally by abdominal afferent nerves. Stimulation of receptors in the semicircular canals of the vestibular system in the inner ear, inflammation within the CNS, and increases in intracranial pressure can also cause vomiting. In the CNS, the CRTZ (which lies near the medulla) senses chemical changes. The CRTZ is outside the blood-brain barrier and therefore responds to stimuli (toxins, drugs, etc.) from either the cerebrospinal fluid (CSF) or the blood. This detection of chemical changes stimulates activation of the vomiting center. Dopamine is the neurotransmitter for the CRTZ, and stimulation of the CRTZ results in dopamine release and stimulation of the vomiting center. Serotonin is another mediator of vomiting in the CRTZ, while histamine acts as a secondary neurotransmitter in the CRTZ.

In addition to the CRTZ, the vomiting center is another area of the brain that controls vomiting. Some sensory impulses, such as odor and taste, are transmitted directly to the vomiting center from higher cerebral cortical centers. The main neurotransmitter of the higher centers of the brain is acetylcholine. Impulses from the inner ear related to equilibrium are also sent directly to the vomiting center.

Peripheral receptors involved in the vomiting reflex are found in the abdominal viscera (mainly in the duodenum). Peripheral irritation to the gastrointestinal tract by infectious agents, foreign material, chemicals, and over distention sends impulses to the vomiting center via the vagus nerve. Irritation or inflammation of the liver, pancreas, kidneys, spleen, genitourinary tract, or peritoneum can also send signals to the vomiting center. This mechanism may be protective by inhibiting absorption of toxic materials. Acetylcholine release from cholinergic nerves due to abdominal afferent neural stimulation is the primary neurotransmitter acting on the emetic center. The neurotransmitters involved with vomiting are illustrated in Figure 11-7A.

The concerns with vomiting are dehydration, electrolyte loss (the animal's electrolyte balance will be shifted due to loss of sodium, potassium, and chloride ions), and acid-base changes. Ideally, the cause of any vomiting will be identified and treated; however, **antiemetics** (drugs that control vomiting) may help alleviate discomfort and help control electrolyte balance. Most antiemetics are given parenterally, as the patient may vomit the medication before it can be absorbed through the gastrointestinal tract. Types of antiemetics include the following:

Phenothiazine Derivatives

This group of drugs works by inhibiting dopamine in the CRTZ, thus decreasing the stimulation to vomit. Side effects include hypotension and sedation. Phenothiazines lower the seizure threshold and therefore should not be used in epileptic animals. Hydration is important when using phenothiazine derivatives because they cause vasodilation and hypotension. Phenothiazine derivatives should not be used to treat vomiting animals with abnormal gastrointestinal motility because it may promote ileus and worsen vomiting. Examples in this group include the following:

- *acepromazine* (PromAce®)
- *chlorpromazine* (Thorazine®, Laragactil®, Thor-Prom®)
- *prochlorperzine* (Compazine®; Darbazine® has isopropamide added to the prochlorperzine)
- *perphenazine* (Trilafon®)

Antihistamines

Antihistamines are used to control vomiting in small animals when the vomiting is due to motion sickness, vaccine reactions, or inner ear problems. They work by blocking input from the vestibular system to the CRTZ. Antihistamines can cause sedation. Examples include the following:

- *trimethobenzamide* (Tigan®)
- *dimenhydrinate* (Dramamine®)
- *diphenhydramine* (Benadryl®)
- *meclizine* (Antivert®)

Anticholinergics

Anticholinergics block acetylcholine peripherally, which decreases intestinal motility and secretions. These drugs also decrease gastric emptying, which may increase the tendency to vomit. Side effects include dry mouth, constipation,

Clinical Que

Vomiting can be caused by primary GI disease, kidney or liver failure, electrolyte abnormalities (due to diseases such as hypoadrenocorticism), pancreatitis, or CNS disorders.

Clinical Que

Keep in mind that not all animals vomit; horses, rabbits, and rats do not. In ruminants, abomasal content can go into the forestomach, but ejection of content from the mouth does not occur.

Clinical Que

Many times vomiting can be controlled by withholding food and water (NPO) and correcting dehydration.

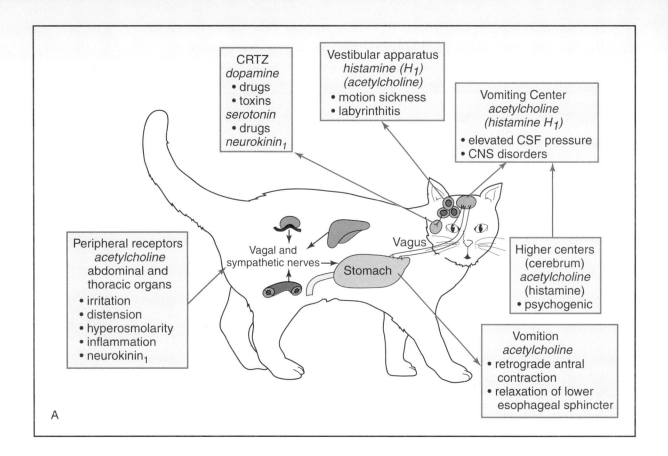

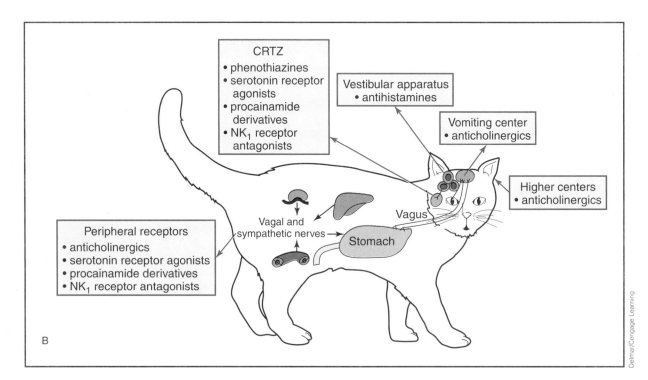

Figure 11-7 (A) Sites and neurotransmitters that mediate vomiting. (B) Sites of action of antiemetic drugs.

Delmar/Cengage Learning

urinary retention, and tachycardia. These drugs are contraindicated in animals with glaucoma or pyloric obstruction. Examples include the following:

- *aminopentamide* (Centrine®) (may also block CRTZ)
- *atropine* (may also block CRTZ)
- *propantheline* (Pro-Banthine®)

Procainamide Derivatives

This group of antiemetics works both centrally, by blocking the CRTZ (as a dopamine antagonist), and peripherally, by speeding gastric emptying, strengthening cardiac sphincter tone (decreasing gastroesophageal reflux), and increasing the force of gastric contractions. These antiemetics are not recommended in animals with GI obstructions, GI perforation, or hemorrhage, because of the peripheral effects (stimulation of gastric motility, which is harmful in these cases). However, this effect also makes them acceptable for the treatment of gastric motility disorders. One example in this group is *metoclopramide* (Reglan®) used widely in human medicine to control vomiting associated with cancer chemotherapy treatment. It stimulates motility of the upper gastrointestinal tract without stimulating the production of gastric, biliary, or pancreatic secretions. It can be given orally or parenterally (SQ, IM, or IV). When given SQ or IM, metoclopramide is rapidly absorbed and removed; therefore, it doesn't have a prolonged effect. It is most effective when given continually by infusion pump. It is contraindicated in animals with a gastrointestinal obstruction or seizure disorder.

Serotonin Receptor Antagonists

Serotonin receptor antagonists work selectively on 5-HT$_3$ receptors, which are located peripherally (on nerve terminals of the vagus nerve) and centrally (in the CRTZ). It is believed that some chemicals (especially chemotherapeutic agents and certain anesthetic agents such as propofol) cause vomiting because they increase serotonin release from cells in the small intestine. Blocking the release of serotonin from these sites controls vomiting. These drugs are expensive, can cause gastrointestinal upset, and are reserved for vomiting associated with specific chemicals or prior to chemotherapy. Drugs in this category include *ondansetron* (Zofran®), *dolasetron* (Anzemet®), and *granisetron* (Kytril®).

Neurokinin (NK$_1$) Receptor Antagonists

Neurokinin (NK$_1$) receptor antagonists work on NK$_1$ receptors, which are located in the vomiting center of the brain. These drugs work by inhibiting substance P, the key neurotransmitter involved in vomiting. An example of a veterinary-approved neurokinin (NK$_1$) receptor antagonist is *maropitant citrate* (Cerenia®), which suppresses both peripherally and centrally mediated vomiting in dogs and is used extralabelly in cats. Maropitant citrate is used to prevent acute vomiting and motion sickness and is administered orally or SQ. Side effects include pain at the injection site, pretravel vomiting and hypersalivation (when given at higher doses for motion sickness), and diarrhea.

Clinical Que

Emetics should not be used in unconscious, seizuring, or compromised animals.

Clinical Que

Ipecac syrup should not be given with activated charcoal, as the charcoal may absorb the emetic portion of the ipecac. However, if vomiting does not occur, it is recommended to give activated charcoal to absorb ipecac and the ingested toxin.

Vomit Producing

Emetics are drugs that induce vomiting. Emetics are used in the treatment of poisonings and drug overdoses. Keep in mind that vomiting should not be induced if caustic substances have been ingested, such as ammonia, lye, bleach, or other damaging products. Always check with poison control before inducing vomiting. Emetics may be centrally acting (working on the CRTZ) or peripherally acting (working on receptors locally).

Centrally acting emetics include *apomorphine* (Apokyn®), which stimulates dopamine receptors in the CRTZ and thus induces vomiting. It is a morphine-derived emetic that is given SQ, IM, or topically in the conjunctival sac. Vomiting occurs rapidly, usually within 5 to 10 minutes when given peripherally or 10 to 20 minutes when given subconjunctively. It is the emetic of choice for dogs and is available generically and through compounding pharmacies. Side effects include protracted vomiting, CNS depression, and restlessness. *Xylazine* (Rompun®, Gemini®, AnaSed®) induces vomiting in cats as a side effect of its use as a sedative. The mechanism of emetic action is not fully understood for xylazine. It is considered the emetic of choice for cats. Xylazine does not induce vomiting in horses, cattle, sheep, or goats, due to their inability to vomit. Side effects include bradycardia and decreased respiratory rate.

The best-known peripherally acting emetic is *ipecac syrup*, which is made from roots and rhizomes of plants. It is an over-the-counter medication that is given orally. Two alkaloids, emetine and cephaeline, cause irritation to the gastric mucosa and centrally stimulate the CRTZ. Contents of both the stomach and small intestine are evacuated within 10 to 30 minutes. It is usually used in dogs and cats. Ipecac can cause cardiovascular problems with higher doses.

Some home remedy emetics that have been used to induce vomiting include hydrogen peroxide, salt and water, mustard and water, and salt followed by food. The results of these home remedies vary from case to case, and they are considered less reliable methods of inducing emesis.

Activated charcoal is given in poisoning cases when emesis is contraindicated. Activated charcoal absorbs many chemicals and drugs in the upper gastrointestinal tract, reducing their absorption. It comes in liquid form or powder form for reconstitution with water. Side effects include constipation or diarrhea and blackening of the feces. Trade names of activated charcoal products include SuperChar® Vet Powder, SuperChar® Vet Liquid, and Toxiban® Granules. (Toxiban® products have activated charcoal and kaolin in them.)

Ulcer Stopping

Ulcers are erosions of mucosa and are named according to the site of involvement: for example, gastric ulcer, duodenal ulcer, and esophageal ulcer. Ulcers form for many reasons, including metabolic disease, drug therapy, and stress. Most cases are due to increased release of hydrochloric acid from the parietal cells of the stomach. Histamine, gastrin, and acetylcholine influence the parietal cells. A thick layer of mucus separates and protects the mucosal lining from gastric secretions. Two sphincter muscles, the cardiac (located at the upper portion of the stomach) and the pyloric (located at the lower portion of the stomach), act as barriers to prevent reflux of acid into the esophagus and the duodenum. Signs of ulcers may include anorexia, melena, abdominal pain, and hematemesis.

Antiulcer drugs help prevent the formation of ulcers. Figure 11-8 summarizes antiulcer drug actions. Categories of antiulcer drugs include antacids, histamine-2 antagonists, mucosal protective drugs, prostaglandin analogs, and proton pump inhibitors.

Antacids

Antacids are substances that promote ulcer healing by neutralizing HCl and reducing pepsin activity. They do not coat the ulcer. Antacids can interact with other drugs by adsorption or binding of the other drugs (decreasing oral absorption of bound drug), by increasing stomach pH (causing a decrease in absorption of certain drugs), and by increasing urinary pH (inhibiting elimination of drugs that are weak bases). There are two types of antacids: systemic (those absorbed into the blood) and nonsystemic (those that remain primarily in the GI tract).

Systemic antacids are readily soluble in stomach fluid and, once dissolved, are readily absorbed. Most systemic antacids have a rapid onset and short

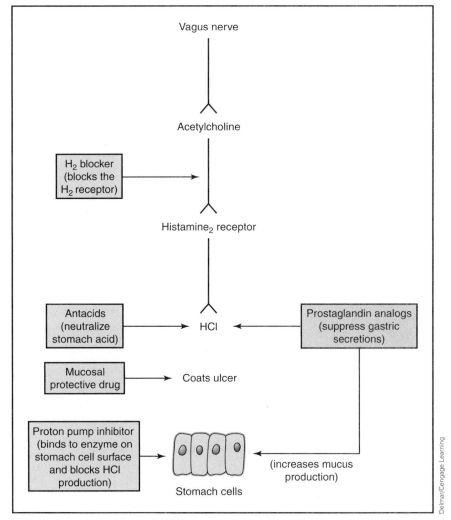

Figure 11-8 Mechanism of action of antiulcer drugs.

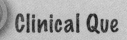

duration of action. Examples of systemic antacids are sodium bicarbonate and calcium carbonate. Sodium bicarbonate can cause sodium excess and water retention and is not frequently used as an antacid in veterinary medicine. Calcium carbonate is more effective than sodium bicarbonate in neutralizing acid, but can result in excessive calcium levels, and is not frequently used as an antacid in veterinary medicine.

Nonsystemic antacids remain primarily in the GI tract. They are composed of alkaline salts of aluminum (aluminum hydroxide) and magnesium (magnesium hydroxide). Magnesium hydroxide has greater neutralizing power than aluminum hydroxide. Magnesium may cause diarrhea, and aluminum may cause constipation. A combination of these two compounds provides acid neutralization with minimal GI side effects. Sometimes simethicone, an antigas agent, is added to these antacids.

Antacid use in animals has decreased due to difficulty of administration and the introduction of histamine-2 blockers (see next section). Examples include the following:

- *magnesium hydroxide* (Magnalax®, Rulax II®, Milk of Magnesia®). In ruminants, magnesium hydroxide is used to increase rumen pH and as a laxative to treat rumen acidosis (grain overload). Magnesium-containing antacids are contraindicated in animals with kidney disease because they may develop electrolyte imbalances

- *aluminum/magnesium hydroxide* (Maalox®, Mylanta®). In dogs and foals, aluminum/magnesium hydroxide may be used as an adjunctive treatment for gastric ulcers.

- *aluminum hydroxide* (Amphojel®). Aluminum hydroxide has also been used to lower high serum phosphate levels because it binds and thereby depletes phosphorus.

Histamine-2 (H$_2$) Receptor Antagonists

Histamine-2 (H$_2$) receptor antagonists prevent acid reflux by competitively blocking the H$_2$ receptors of the parietal cells in the stomach, thus reducing gastric acid secretion. Side effects are rare, but may include diarrhea and inhibition of liver enzymes that affects metabolism of other drugs and waste products. Examples from this group include the following:

- *cimetidine* (Tagamet®), the first H$_2$ blocker developed. It comes in tablet, oral solution, and injectable forms.

- *ranitidine* (Zantac®), which is more potent and has a longer duration of action than cimetidine. It is also available in tablet, oral solution, and injectable forms.

- *famotidine* (Pepcid®), which is more potent and has fewer side effects than ranitidine, but is less bioavailable. It is available in coated tablets, oral powder, and injectable forms.

- *nizatidine* (Axid®), which is an H$_2$-receptor antagonist and also has prokinetic action (see the following section on prokinetic agents) via acetylcholinesterase inhibition (which increases acetylcholine at muscarinic receptors). Its primary use is as a prokinetic agent because it stimulates gastrointestinal motility. Side effects are rare and include anemia and pruritus. Nizatidine is administered orally via tablet, capsule, or solution.

Mucosal Protective Drugs

Mucosal protective drugs, also known as pepsin inhibitors, are typified by the drug *sucralfate* (Carafate®). Sucralfate is a chemical derivative of sucrose that is nonabsorbable and combines with protein to form an adherent substance that covers the ulcer and protects it from stomach acid and pepsin. Sucralfate comes in 1-g tablets, and its only side effect is constipation. Because sucralfate binds to ulcers in an acid environment, it should not be given at the same times as H$_2$ receptor antagonists.

Clinical Que

Sucralfate does little to neutralize stomach acid.

Prostaglandin Analogs

Prostaglandin analogs appear to suppress gastric secretions and increase mucus production in the GI tract. An example of a prostaglandin analog is *misoprostol* (Cytotec®), an oral tablet that is usually given to animals taking non-steroidal anti-inflammatory drugs (NSAIDs). Pregnant women should be cautioned about handling this medication, because of its prostaglandin effects. Side effects include GI signs such as diarrhea, vomiting, and abdominal pain.

Proton Pump Inhibitors

Proton pump inhibitors are drugs that bind irreversibly to the H$^+$-K$^+$-ATPase enzyme on the surface of parietal cells of the stomach. This inhibits hydrogen ion transport into the stomach so that the cell cannot secrete HCl. When this enzyme is blocked, acid production is decreased, and this allows the stomach and esophagus to heal. Drugs in this group include the following:

- *omeprazole* (Prilosec®, Gastrogard®). In horses and foals more than four weeks of age, omeprazole is used to heal gastric ulcers and to prevent recurrence. Gastric ulcers in horses form secondary to feeding problems (too little hay intake), training (intense exercise increases gastric acid production), and changes with growth (a developing stomach may be injured by acid and enzymes). Gastric ulcers in horses interfere with performance.

- *lansoprazole* (Prevacid®), used to treat gastroesophageal reflux disease and to help heal gastric and duodenal ulcers. It blocks the last step of gastric acid production and is used extra-label in animals. Lansoprazole is less likely to react with other drugs than omeprazole.

Foam Stopping

Antifoaming agents are drugs that reduce or prevent the formation of foam. As a gastrointestinal drug, antifoaming agents are used in ruminants whose rumens are subject to acute frothy bloat (Figure 11-9). *Frothy bloat* (also known as pasture bloat or legume bloat) is a condition in which the rumen is distended with

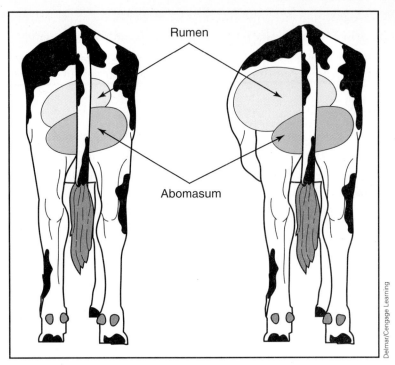

Figure 11-9 In ruminants, bloat is accumulation of gas in the rumen (most common), abomasum, or cecum. Bloat can be prevented with proper management techniques. Some forms of bloat are treated with antifoaming agents.

a gas that mixes with fluid to form a froth; the froth can asphyxiate the animal by blocking its ability to eructate (belch). Antifoaming agents make this foam less stable, breaking it up to promote gas release through belching. *Poloxalene* and *polymerized methyl silicone* accomplish this when administered as solutions by stomach tube directly into the forestomach. Trade names of antifoaming agents include Therabloat®, Bloat Guard®, Bloat-Pac®, and Bloat Treatment®.

Motility Enhancing

Prokinetic agents increase the motility of parts of the GI tract to enhance movement of material through it. Parasympathomimetic agents, dopaminergic antogonists, and serotonergic agents may act as prokinetics.

Parasympathomimetic Agents

Parasympathomimetic agents include acetylcholinesterase inhibitors and cholinergics. The best example of an acetylcholinesterase inhibitor is *neostigmine* (Prostigmin®). It works by competing with acetylcholine for acetylcholinesterase, resulting in increased intestinal tone and salivation. It is used in ruminants for treatment of rumen atony. It is also used for diagnosis of myasthenia gravis in dogs. Side effects are cholinergic in nature (vomiting, diarrhea, and increased salivation).

Cholinergics mimic the parasympathetic nervous system because they make a precursor to acetylcholine, thus assuring adequate production of acetylcholine. An example of a drug in this category is *dexpanthenol* (d-panthenol injectable, Ilopan®). Dexpanthenol is used to treat intestinal distention or atony, paralytic ileus, and colic. It is also used after surgery to help animals increase gastric motility.

Dopaminergic Antagonists

Dopaminergic antagonists stimulate gastroesophageal sphincter, stomach, and intestinal motility by sensitizing tissues to the action of the neurotransmitter acetylcholine. Side effects are behavioral in nature. Examples of dopaminergic antagonists include *metoclopramide* (Reglan®) and *domperidone* (Motilium®).

Serotonergic Agents

Serotonergic agents stimulate motility of the gastroesophageal sphincter, stomach, small intestine, and colon. An example is *cisapride* (Propulsid®), used for the treatment of constipation, gastroesophageal reflux, and ileus. Side effects may include diarrhea, megacolon, and abdominal pain.

Clinical Que

Because metoclopramide interferes with acetylcholine action, anticholinergic drugs completely inhibit the action of metoclopramide.

Enzyme Supplementing

Pancreatic exocrine insufficiency (PEI) is a disease in which the pancreas does not produce digestive enzymes. These enzymes can be supplemented in the diet through the use of pancrealipase. *Pancrealipase* (Viokase®-V Powder, Pancrezyme®) contains primarily lipase (enzyme that digests fats), but also has amylase (enzyme that digests starch) and protease (enzyme that digests proteins) to help with digestion of fats, starch, and protein. Side effects include gastrointestinal problems such as diarrhea and abdominal pain. Care should be taken when handling this drug, as it can be irritating to the skin on contact and to nasal passages upon inhalation.

Appetite Altering Drugs

Appetite disorders are common in animals whether it is obesity as a result of overeating or anorexia secondary to clinical disease. Drugs used to alter the appetite of animals include appetite-stimulating and appetite-controlling drugs.

Appetite Stimulating Drugs

For anorexic animals that do not respond to coaxing with small amounts of palatable foods, drug therapy may be used to stimulate appetite to prevent initiating more invasive procedures, such as nasogastric or gastrostomy tube feeding, or total parenteral nutrition. Some drugs such as B vitamins have side effects of increasing an animal's appetite and will not be covered in this section. Vitamin supplements are listed in Appendix L. Types of appetite stimulants include the following:

- serotonin antagonist antihistamines. Serotonin antagonist antihistamines are drugs that compete with histamine for sites on H$_1$ receptors on effector cells (not by blocking histamine release, but antagonizing its effects). Serotonin antagonist antihistamines promote appetite by inhibition at the serotoninergic receptors, which normally control satiety (fullness). By inhibiting these sites, appetite is stimulated. An example in this group is *cyproheptadine* (Periactin®), which is used orally in cats (and dogs) as an appetite stimulant. Cyproheptadine also has antiserotonin activity that may have limited effects in treating equine Cushing's disease and serotonin syndrome in small animals. Side effects include sedation and dry mouth.

- benzodiazepines. The benzodiazepines are effective appetite stimulants in cats (but not dogs) by induction of aminobutyric acid (GABA) and by central inhibition of the satiety center in the hypothalamus. *Diazepam* (Valium®) is the classic example of a benzodiazepine used to stimulate appetite and can be administered IV, IM, or orally. Cats that respond begin eating within a few seconds of IV administration; therefore, palatable food should be available when giving this drug. *Oxazepam* (Serax®) is a metabolite of diazepam that can be given orally. Diazepam is the more effective appetite stimulant but also has a greater sedative effect than oxazepam. Side effects of both diazepam and oxazepam are sedation and ataxia. Some dogs may exhibit a contradictory CNS excitement with diazepam administration, and some cats may develop liver failure after receiving oral diazepam.

- tetracyclic antidepressants. Tetracyclic antidepressants such as *mirtazapine* (Remeron®) stimulate appetite by antagonizing alpha-2 receptors, which normally inhibit norepinephrine release. By blocking these receptors, there is an increase in norepinephrine, which acts on other receptors to stimulate appetite. Mirtazapine also inhibits serotonin receptors (which gives it antiemetic effects) and histamine receptors (which provides sedative effects). Mirtazapine is used extralabelly in dogs and cats as an appetite stimulant and/or antiemetic (it is only FDA approved for humans). Side effects include sedation, vocalization, hypotension, and tachycardia.

- glucocorticoids. Glucocorticoids increase gluconeogenesis and antagonize insulin resulting in hyperglycemia. Glucocorticoids such as *prednisone* (generic brands) stimulate steroid-induced euphoria, which in turn stimulates appetite. Extended use of glucocorticoids has catabolic effects because as skeletal muscle and collagen proteins are broken down, they provide precursors for the gluconeogenesis process. Side effects of glucocorticoids include polydipsia, polyuria, dull haircoat, weight gain, and behavioral changes. A zero tolerance of residues in milk for glucocorticoids has been established in dairy cattle.

- anabolic steroids. The anabolic steroids are synthetic derivatives of testosterone that have enhanced anabolic effects with reduced androgenic effects. Anabolic steroids antagonize the catabolic effect of glucocorticoids and the negative nitrogen balance associated with conditions such as surgery, illness, trauma, and aging. Improved nitrogen balance depends on adequate intake of protein and calories and on treatment of the underlying disease. Anabolic steroids stimulate hematopoiesis, appetite, and weight gain. *Boldenone undecylenate* (Equipoise®) is an anabolic steroid approved for use in horses (and is used extralabelly as an appetite stimulant in cats). Side effects of anabolic steroid therapy include hepatotoxicity, masculinization, and early closure of the growth plate in young animals. Anabolic steroids are contraindicated in animals with congestive heart failure because of increased sodium and water retention and in stallions (has detrimental effects on testis size and sperm production) or pregnant mares (may cause fetal masculinization). Because of the potential for abuse by people, anabolic steroids are schedule III controlled substances.

- Progestins. *Megestrol acetate* (Ovaban®, Megace®) is a synthetic progestin that has significant antiestrogen and glucocorticoid activity, which results in adrenal suppression. It is used to stimulate appetite and promote weight gain in anorectic cats and dogs. Megestrol acetate is contraindicated in pregnant animals and in animals with uterine disease, diabetes mellitus, or mammary neoplasia. Side effects include behavior change, endometritis, and mammary enlargement. In cats, megestrol acetate can induce adrenocortical suppression, adrenal atrophy, and diabetes mellitus, which may or may not be reversible.

Appetite Controlling

Appetite control in animals is necessary when animals have too much body fat that may put them at risk for developing other health issues. Data suggest that approximately 5 percent of dogs in the United States are obese, and another 20 to 30 percent are overweight. *Dirlotapide* (Slentrol®) is a drug approved by the FDA for management of obesity in dogs. Dirlotapide reduces appetite and fat absorption, which produces weight loss. Dirlotapide is a selective microsomal triglyceride transfer protein inhibitor, which blocks the assembly and release of lipoproteins into the bloodstream. The mechanism for producing weight loss is not completely understood, but seems to result from reduced fat absorption and a satiety (fullness) signal from lipid-filled cells lining the intestine. Dirlotapide is administered to the dog as an initial dose for the first 14 days; then the dog's progress is assessed at monthly intervals and the dose adjusted based on the amount of the dog's weight loss. After the dog has achieved the goal weight, the drug is continued for a three-month period, during which time an optimal level of food intake and physical activity needed to maintain the dog's weight is determined. Side effects associated with dirlotapide include vomiting, diarrhea, lethargy, and anorexia.

Dental Prophylaxis and Treatment

Advances in veterinary dental care have greatly increased both attention to dental care and dental product use (Figure 11-10). The types of products available for dental care include enzymes, antiseptics, and fluoride products. Human dental products should not be used on animals, as human products may have higher concentrations of certain chemicals that can lead to problems in animals. Some veterinary dental products and their uses are listed in Table 11-1.

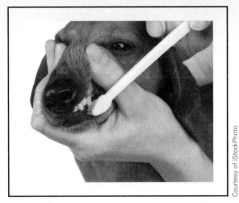

Courtesy of iStockPhoto

Figure 11-10 Brushing an animal's teeth with veterinary products is one way to decrease dental disease in animals.

Table 11-1	Veterinary Dental Products and Their Uses	
PRODUCT TYPE	TRADE NAME	CONSIDERATIONS
Cleansing	• C.E.T.® Dentifrice • Oxydent® • Nolvadent® Oral Cleansing Solution • Oral Dent® • C.E.T.® Oral Hygiene Spray • C.E.T.® Chews • C.E.T.® Toothbrush • CHX® Oral Cleansing Solution • MaxiGuard® Oral Cleansing Gel • Hill's t/d Diet® • C.E.T.® Prophypaste (polishing paste) • C.E.T.® Veggiedent™ Chews • Purina Dental Health Diets® • Del Monte Tartar Check® Dog Biscuit • Friskies Cheweez® Beefhide Treats for Dogs	• These products are used to clean oral surfaces with or without antiseptic components. • These products may help with plaque removal and breath freshening. • Most of these products need not be rinsed. • These products contain either enzymes such as lactoperoxidase or antiseptics such as chlorhexidine. • **Dentifrices** are substances that clean the teeth.

(continued)

Product Type	Trade Name	Considerations
	• Greenies® Edible Dog Treats	
	• Iams Chunk Dental Defense Diet for Dogs®	
	• Science Diet Oral Care Diet for Dogs®	
Fluoride containing	• C.E.T.® Oral Hygiene Spray with Fluoride	• Fluoride-containing products designed for animals have lower concentrations of fluoride than human products.
	• FluroFom®	• Fluoride desensitizes exposed dentin, strengthens tooth enamel, and stimulates remineralization of the enamel.
Disclosing	• First Sight® Disclosing Swab	• Single color systems may have dyes (such as fluorescein or D & C red #28) to stain plaque
	• 2-Tone® Disclosing Solution	• 2-color systems stain new plaque red and old plaque blue.
	• Plak-Check ® Disclosing Solution	• **Disclosing solutions** are used as client education or after dental prophylaxis to check dental technique.
	• Reveal ® Disclosing Solution	
Vaccine	• Porphyromonas denticanis-gulae-salivosa Bacterin ®	• Inactivated *P. denticanis*, *P. gulae*, and *P. salivosa* bacterin
		• Approved for the vaccination of healthy dogs to aid in preventing periodontitis (demonstrated by a reduction in bone changes [osteolysis/osteosclerosis]).

(continued)

Table 11-1	(Continued)	
PRODUCT TYPE	TRADE NAME	CONSIDERATIONS
		• Healthy dogs should receive 2 doses administered 3 weeks apart.
		• Duration of immunity for this product has not been determined; 6 and 12 month intervals are being evaluated.

SUMMARY

The gastrointestinal tract is a long, muscular tube that plays a central role in bringing life-sustaining nutrients into the body and taking waste products out of it. Gastrointestinal drugs play a role in maintaining the unobstructed and regulated flow of elements in and out of the body.

Gastrointestinal drugs include antisialogues (drugs that decrease salivary flow), antidiarrheals (drugs that decrease peristalsis to reverse diarrhea), laxatives (drugs that loosen stool), antiemetics (drugs that control vomiting), emetics (drugs that cause vomiting), antiulcer drugs (drugs that heal or prevent ulcers), antifoaming agents (drugs used to prevent frothy bloat in ruminants), prokinetic agents (drugs that increase motility of parts of the GI tract), digestive enzyme supplements (drugs used to replace digestive enzymes), appetite altering drugs (drugs that either stimulate or depress an animal's appetite), and dental products (products to clean teeth and prevent dental disease). Table 11-2 summarizes the drugs covered in this chapter.

Table 11-2	Drugs Covered in This Chapter
CATEGORY	EXAMPLES
Antisialogues	• Anticholinergics: glycopyrrolate and atropine
Antidiarrheals	• Anticholinergics: atropine, aminopentamide, isopropamide, propantheline, methscopolamine, and N-butylscopolammonium bromide
	• Protectants/adsorbents: bismuth subsalicylate, kaolin/pectin, and activated charcoal
	• Opiate or narcotic agents: diphenoxylate, loperamide, and paregoric
	• Probiotics: beneficial bacteria-containing products
	• Anaerobic antibiotic: metronidazole

(continued)

CATEGORY	EXAMPLES
Laxatives	• Osmotics: lactulose, sodium phosphate with sodium biphosphate, magnesium sulfate, magnesium hydroxide, and polyethylene glycol-electrolyte solution • Stimulant: bisacodyl and castor oil • Bulk forming: psyllium hydrophilic mucilloid, polycarbophil, and bran • Emollients: docusate sodium, docusate calcium, docusate potassium, and petroleum products
Antiemetics	• Phenothiazine derivatives: acepromazine, chlorpromazine, prochlorperazine, and perphenazine • Antihistamines: trimethobenzamide, dimenhydrinate, diphenhydramine, and meclizine • Anticholinergics: aminopentamide, atropine, and propantheline • Procainamide derivatives: metoclopramide • Serotonin receptor antagonists: ondansetron and dolasetron • Neurokinin receptor antagonist: maropitant citrate
Emetics	• Centrally acting: apomorphine and xylazine • Peripherally acting: ipecac syrup and various home remedies
Antiulcer drugs	• Systemic antacids: sodium bicarbonate and calcium carbonate • Nonsystemic antacids: magnesium hydroxide, aluminum/magnesium hydroxide, and aluminum hydroxide • H_2-receptor antagonists: cimetidine, ranitidine, famotidine, and nizatidine • Mucosal protective drugs: sucralfate • Prostaglandin analogs: misoprostol • Proton pump inhibitors: omeprazole and lansoprazole
Antifoaming agents	• Defoaming agents: poloxalene and polymerized methyl silicone
Prokinetic agents	• Parasympathomimetic agents: neostigmine and dexpanthenol • Dopaminergic antagonists: metoclopramide and domperidone • Serotonergic agents: cisapride
Digestive enzyme supplements	• Digestive enzyme replacement: pancrealipase

(continued)

Table 11-2   *(Continued)*

CATEGORY	EXAMPLES
Appetite stimulants	• Serotonin antagonist antihistamine: cyproheptadine • Benzodiazepines: diazepam and oxazepam • Tetracyclic antidepressant: mirtazapine • Glucocorticoids: prednisone • Anabolic steroid: boldenone undecylenate • Progestin: megestrol acetate
Appetite depressants	• Microsomal triglyceride transfer protein inhibitor: dirlotapide (weight management)

It's a Wrap

The answers to the questions in this chapter's Setting the Scene case study should be understood after reading this chapter. Veterinary technicians need to understand what to do when presented with an animal that has ingested a toxic substance. Should the babysitter in this case try to make the dog vomit? What should the veterinary technician tell her? It is important to find out what chemicals the animal ingested to determine whether or not to induce vomiting in this animal. The veterinary technician should ask the client to read the ingredients on the container to determine what chemicals are in the cleaning solution. Is vomiting warranted in this case? What information is needed before making that decision? Vomiting is not recommended for ingestion of caustic substances such as ammonia, lye (drain cleaner), and bleach. If it is unknown whether or not to induce vomiting, contact the Animal Society to Prevent Cruelty to Animals (ASPCA) national animal poison control center to determine the best treatment for animal poisonings (there is a charge for each case). What home options does the babysitter have, or is it better to bring the dog into the clinic? If vomiting is indicated, some home remedies include ipecac syrup (most parents and babysitters are familiar with this product), hydrogen peroxide, salt and water, mustard and water, and salt followed by food. Vomiting is not immediate and can continue for an extended period of time. Ideally, this dog should be brought into the clinic once a home remedy antiemetic is given so that supportive care may be established if needed. The veterinary technician should also recommend bringing towels in the car to collect any vomit produced during the ride to the clinic. If the owner lives close enough to the clinic, it is best to bring the animal into the clinic to receive antiemetics administered by veterinary professionals. Appendix M describes the management of animal toxicities.

CHAPTER REVIEW

Matching

Match the drug name with its action.

1. _____ ipecac syrup
2. _____ apomorphine
3. _____ aminopentamide
4. _____ kaolin/pectin
5. _____ bran
6. _____ poloxalene
7. _____ sodium bicarbonate
8. _____ glycopyrrolate
9. _____ pancrealipase powder
10. _____ docusate calcium

a. centrally acting emetic
b. drug used to reduce salivation
c. digestive enzyme supplement
d. antifoaming agent used to treat frothy bloat in cattle
e. systemic antacid
f. peripherally acting emetic
g. anticholinergic antiemetic
h. bulk-forming laxative
i. emollient laxative
j. protectant/adsorbent antidiarrheal

Multiple Choice

Choose the one best answer

11. What part of the brain controls vomiting?
    a. cerebrum
    b. cerebellum
    c. brain stem
    d. ventricles

12. What category of drug simulates the parasympathetic nervous system or homeostatic function?
    a. dopaminergic
    b. adrenergic
    c. cholinergic
    d. sympathetic

13. Which group of antidiarrheals controls diarrhea by blocking acetylcholine release from parasympathetic nerve endings?
    a. cholinergic
    b. anticholinergic
    c. sympathetic
    d. sympathetic blocking agents

14. What type of laxative loosens stool by pulling water into the colon and increasing water in the feces?
    a. osmotics
    b. contact
    c. bulk forming
    d. emollients

15. Which of the following control vomiting by inhibiting dopamine in the CRTZ?
    a. acepromazine
    b. trimethobenzamide
    c. aminopentamide
    d. metoclopramide

16. Which emetic can be given in the subconjunctival sac?
    a. xylazine
    b. ipecac
    c. apomorphine
    d. hydrogen peroxide

17. Which group of antiulcer medications block receptors on parietal cells in the stomach, thereby reducing gastric acid secretion?
    a. mucosal protectives
    b. H_2 receptor antagonists
    c. systemic antacids
    d. nonsystemic antacids

18. What drug is used for rumen atony?
    a. poloxalene
    b. neostigmine
    c. cisapride
    d. metoclopramide

True/False

Circle a. for true or b. for false.

19. Vomiting should be induced whenever an animal ingests a poison.
    a. true
    b. false

20. Fluoride-containing products designed for animals should have a higher concentration of fluoride than human products.
    a. true
    b. false

Case Studies

21. A two-year-old M Labrador retriever is brought to the clinic after ingesting drain cleaner. The veterinarian wants to sedate the dog to perform a PE.
    a. Xylazine is typically used at your clinic to sedate animals. Is this a good choice in this case? Why or why not?
    b. To prevent this animal from vomiting, what are good choices for sedation?
    c. What drug can be given to this animal via a stomach tube to absorb toxins if this is indicated by poison control?

22. A six-month-old Saddlebred colt has had diarrhea for the past few days. The owner wants to stop this horse's diarrhea.
    a. What antidiarrheal is commonly used in large animals?
    b. What test(s) should be performed on this horse with diarrhea?

Critical Thinking Questions

23. What are the most important questions to ask the owner/client of an animal with vomiting and diarrhea?

24. Some animals may be on long-term glucocorticoids for treatment of certain disease conditions. What are the gastrointestinal implications for a patient on long-term glucocorticoid use? What kinds of preventative measures can be taken to prevent gastrointestinal complications?

ADDITIONAL REVIEW MATERIAL

Additional review material can be found on the StudyWARE™ CD included with this text.

BIBLIOGRAPHY

Allen, D. G. (2005). *Handbook of Veterinary Drugs* (3rd ed.). Baltimore, MD: Lippincott Williams & Wilkins.

Broyles, B., Reiss, B., and Evans, M. (2007). *Pharmacological Aspects of Nursing Care* (7th ed.). Clifton Park, NY: Delmar Cengage Learning.

Plumb, D. C. (2008). *Plumb's Veterinary Drug Handbook* (6th ed.). Ames, IA: Blackwell Publishing.

Spratto, G. R., and Woods, A. L. (2008). *PDR Nurse's Drug Handbook.* Clifton Park, NY: Delmar Cengage Learning. www.thomsonhc.com/pdrel/librarian; PDR Electronic Library, Thomson Health Communications.

CHAPTER 12

URINARY SYSTEM DRUGS

OBJECTIVES

Upon completion of this chapter, the reader should be able to:

- describe the basic anatomy and physiology of the urinary system.
- outline the biological process of urine formation.
- explain how diuretics affect urine production.
- describe the modes of action of various diuretics.
- describe the ways various hypertensive drugs affect urine production.
- describe the use of the following drugs employed in veterinary medicine: cholinergics, anticholinergics, adrenergics, adrenergic antagonists, hormones, skeletal muscle relaxants, urinary acidifiers and alkalinizers, and xanthine oxidase inhibitors.

KEY TERMS

adrenergic antagonists
anticholinergics
antihypertensives
cholinergic agonists

diuresis
diuretics
urinary incontinence
uroliths

Setting the Scene

A client calls the clinic and says that her cat is spending a lot of time in his litterbox. When he is in the litterbox, he squats and occasionally starts meowing loudly. The client says this has been going on since last night. What questions should the veterinary technician ask this client? What concerns are there regarding this cat? Does this cat need immediate veterinary care, or can this problem wait until the cat's next scheduled PE?

Delmar/Cengage Learning

BASIC URINARY SYSTEM ANATOMY AND PHYSIOLOGY

The structures of the urinary system include paired kidneys, paired ureters, a single urinary bladder, and a single urethra. Within each kidney are millions of individual structures that do the actual work of the kidneys. These structures, referred to as the functional units of the kidney, are called *nephrons* (Figure 12-1). The nephron consists of the glomerulus, Bowman's capsule, the proximal convoluted tubule, the loop of Henle, a distal convoluted tubule, and a collecting duct. Urine is formed in the nephron via the processes of glomerular filtration, tubular reabsorption, and tubular secretion.

A glomerulus is a cluster of capillaries that is responsible for glomerular filtration. Basically, the glomerulus filters substances from blood into the glomerular filtrate (water and dissolved substances). Glomerular filtration is affected by blood pressure (and hence blood volume and flow rate), plasma osmotic pressure (related to particles in the blood), and capsule pressure (which can alter resistance to blood flow).

The tubules are responsible for reabsorption or secretion of substances. The substances needed by the body are reabsorbed from the filtrate (fluid within the tubules) through the tubular cell wall, and then they reenter the plasma. These substances include water, glucose, amino acids, and ions (sodium [Na^+], potassium [K^+], calcium [Ca^{+2}], chloride [Cl^-], and bicarbonate [HCO_3^-]). If these substances are in excess or are not useful, they remain in the filtrate to be excreted in the urine. Tubular secretion is the process in which substances

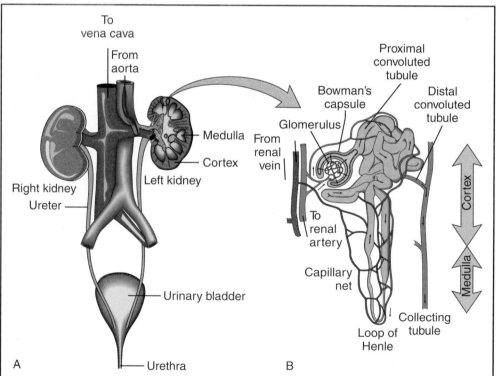

Figure 12-1 (A) Structures of the urinary tract (B) Structures of a nephron.

are carried into the tubular lumen. These substances are secreted by active transport and include K^+, H^+, NH_3^+, creatinine, and drugs. Tubular secretion is helpful in maintaining blood pH and in excreting drugs. Collecting ducts collect urine from the nephron and carry it to the renal pelvis. Figure 4-10 shows elimination of substances from the kidney.

Because the nephron plays an important role in drug excretion, animals with kidney problems will experience diminished function in excretion of drugs. Drug doses for these animals may have to be modified to meet their specific needs.

The main responsibilities of the urinary system are to remove from the body waste products produced during metabolism and to maintain homeostasis, a steady state in the internal environment of the body. The urinary system removes waste products from the body by the constant process of blood filtration. The major waste product of protein metabolism is urea, which is filtered by the kidney and used in some diagnostic tests to determine the health status of the kidney. Uremia is a condition in which we see waste products in the blood (that should normally be excreted in the urine). Uremia may be seen with many kinds of kidney or renal disease.

URINARY SYSTEM DISORDERS

Disorders of the urinary system include urinary tract infections, inflammation and irritation of the urinary tract causing smooth muscle spasms, renal failure, urinary incontinence, and uroliths (urinary stones). Bacterial infections of the urinary system (particularly the urinary bladder) may lead to inflammation, pollakiuria (increased frequency of urination), and dysuria (painful urination), which may result in inappropriate urination for household pets. Crystals in the urine can cause urinary tract inflammation and irritation and may lead to the development of uroliths. Renal failure may lead to an increase of nitrogenous waste products and electrolytes in the blood that may cause severe illness in an animal. Urinary incontinence may be due to a variety of issues (including trauma and systemic disease) and may lead to skin lesions, infection, and the inability of the owner to care for the animal. Urinary system disorders affect the urinary tract resulting in clinical signs such as inappropriate urination, inability to urinate, frequent urination, increased urination, or pain. Some of the clinical signs that an animal develops relating to the urinary system may be due to treatment of other diseases such as heart disease and hypertension. Clinical signs that present with urinary system disease may need to be treated to prevent other complications such as dehydration, electrolyte alterations, or toxic accumulations of nitrogenous waste products in blood.

URINARY SYSTEM DRUGS

Many different types of drugs are used in the management of renal disease and urinary system disorders. Some urinary drugs directly influence urine production and electrolyte balance. Others maintain blood pressure and reduce urinary system disease.

Urine Producing

Diuretics increase the volume of urine excreted by the kidneys and thus promote the release of water from the tissues. This process, called **diuresis**, lowers the fluid volume in tissues. Many disease conditions make fluid reduction desirable. The two main purposes of diuretic use are to decrease edema and to lower blood pressure. Edema (extracellular fluid accumulation) occurs with conditions such as congestive heart failure and chronic liver and kidney disease. Diuretics effectively reduce the edema associated with these conditions, as well as edemas of nonspecific nature, pulmonary edema, pulmonary congestion, and any pathological accumulation of noninflammatory liquid. Diuretics are summarized in Figure 12-2 and Table 8-4.

The old adage used in the discussion of urine formation is "Where sodium goes, water will follow." The kidneys secrete and reabsorb sodium and chloride ions as they make urine. Diuretics block the reabsorption of these ions, so the sodium has nowhere to go but out of the kidneys and into the urinary bladder. Water then follows the sodium out of the kidneys and into the urinary bladder, and diuresis occurs. Though diuretics come from different chemical families, most work by affecting the reabsorption of sodium and chloride. The classes of diuretics include the thiazides, the loop diuretics, potassium-sparing diuretics, carbonic anhydrase inhibitors, and osmotics.

Thiazides

Thiazides are diuretics that act directly on the distal convoluted tubules to block sodium reabsorption and promote chloride ion excretion. Oral administration

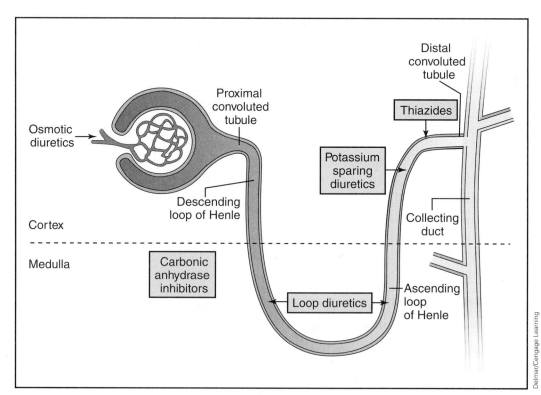

Figure 12-2 Sites of action of diuretics in the nephron.

of thiazides produces diuresis in all animal species, with few documented side effects. This effect remains even with prolonged use, but long-term thiazide use does cause excessive potassium secretion, leading to hypokalemia (potassium deficiency) and cardiac dysfunction. To prevent hypokalemia, veterinarians suggest that potassium-rich diets or potassium supplements accompany thiazide diuretics. Thiazides most often manage edema associated with congestive heart failure. *Hydrochlorothiazide* (HydroDIURIL®), the standard thiazide drug, is given intravenously, intramuscularly, and orally to small animals, horses, and cattle. Other veterinary thiazides include *chlorothiazide* (Diuril®), *hydroflumethiazide* (Saluron®), and *bendroflumethiazide* (Naturetin®).

Loop Diuretics

Loop diuretics get their name from their point of action. The loop of Henle, a U-shaped renal tubule, is the sodium-reabsorbing site that lends its name to this type of diuretic. Loop diuretics influence the reabsorption action at the loop of Henle. *Furosemide* (Lasix®, Disal®, Diuride®) and *ethacrynic acid* (Edecrin®), two loop diuretics, are potent and effective drugs that block absorption of the following ions: chloride, potassium, calcium, hydrogen, magnesium, and bicarbonate. The result of blocking reabsorption of all of these electrolytes is tremendous diuresis. Dogs and cats given oral loop diuretics respond within 30 minutes. Parenteral administration produces diuresis immediately on a fully functional kidney. In small animal practice, furosemide is used to treat congestive heart failure, pulmonary edema, and hypertension. In large animal practice, furosemide is used in dairy cattle with udder edema and in racing horses to control respiratory hemorrhaging, which appears as a nosebleed. The main side effects of loop diuretics are electrolyte imbalances, especially hypokalemia, which can lead to cardiac arrhythmias.

Potassium-Sparing Diuretics

Potassium-sparing diuretics are weaker diuretics than thiazides or loop diuretics. Drugs in this group are used as mild diuretics or in combination with other drugs. Potassium-sparing diuretics act on the distal convoluted tubules to promote sodium and water excretion and potassium retention. These drugs interfere with the sodium-potassium pump that is controlled by aldosterone (a mineralocorticoid produced by the adrenal cortex that affects sodium and potassium levels). Potassium is reabsorbed, and sodium is excreted. Drugs in this category include *spironolactone* (Aldactone®), *triamterene* (Dyazide®), and *amiloride* (Midamor®). The main side effect of these drugs is hyperkalemia.

Carbonic Anhydrase Inhibitors

Carbonic anhydrase inhibitors, such as *acetazolamide* (Diamox®) and *dichlorphenamide* (Daranide®), block the action of the enzyme carbonic anhydrase. This enzyme is used by the body to maintain acid-base balance (mainly between hydrogen and bicarbonate ions). Inhibition of this enzyme causes increased sodium, potassium, and bicarbonate excretion. The main side effect with prolonged use of carbonic anhydrase inhibitors is metabolic acidosis. This group of drugs is mainly used to decrease intraocular pressure with open-angle glaucoma; they are covered in Chapter 18 on ophthalmic and otic drugs.

Osmotic Diuretics

Osmotic diuretics increase the osmolality (concentration) of the filtrate in the renal tubules. This results in excretion of sodium, chloride, potassium, and water. This group of drugs is used to prevent kidney failure, to decrease intracranial pressure, and to decrease intraocular pressure (i.e., glaucoma). *Mannitol* (Osmitrol® and generic) and *glycerin* (Osmoglyn®) are examples of osmotic diuretics. Side effects are uncommon with osmotic diuretics, but can include fluid and electrolyte imbalance and vomiting.

Blood Pressure Lowering

Drugs used to decrease hypertension, called **antihypertensives**, work in a variety of ways. Hypertension is a condition in which an animal's arterial blood pressure is higher than normal. The primary factor in hypertension is increased resistance to blood flow, resulting from the narrowing of peripheral blood vessels. If left untreated, animals with elevated blood pressure are at risk for developing cardiac and renal dysfunction. Some drugs that affect blood pressure include the following.

Diuretics

Diuretics have an antihypertensive effect by promoting sodium and water loss, which causes a decrease in fluid volume and blood pressure. These drugs were covered earlier in this chapter and in Chapter 8.

Angiotensin-Converting Enzyme Inhibitors (ACE Inhibitors)

The kidneys regulate blood pressure via the renin-angiotensin system. Renin, an enzyme released by the kidneys, stimulates the conversion of angiotensin I to angiotensin II (a potent vasoconstrictor). Angiotensin II causes the release of aldosterone (a mineralocorticoid from the adrenal cortex that promotes retention of sodium and water). Retention of sodium and water increases fluid volume, which elevates blood pressure. ACE inhibitors block the conversion of angiotensin I to angiotensin II, which decreases aldosterone secretion (Figure 12-3). Clinically, ACE inhibitors are used to treat hypertension. Examples in this category are *enalapril* (Enacard®, Vasitec®), *captopril* (Capoten®), *lisinopril* (Zestril®), and *benazepril* (Lotensin®).

Calcium Channel Blockers

Calcium channel blockers block the influx of calcium ions into the myocardial cells, resulting in an inhibition of cardiac and vascular smooth muscle contractility. This decreased resistance to blood flow reduces blood pressure, thus affecting glomerular filtration. Side effects include hypotension and edema. Examples include *diltiazem* (Cardizem®), *verapamil* (Isoptin®), and *nifedipine* (Procardia®).

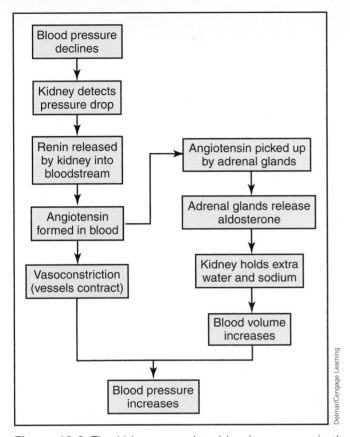

Delmar/Cengage Learning

Figure 12-3 The kidneys regulate blood pressure via the renin-angiotensin system. The renin-angiotensin system also influences aldosterone release from the adrenal cortex.

Direct-Acting Arteriole Vasodilators

Direct-acting arteriole vasodilators relax smooth muscles of the blood vessels, mainly arteries, causing vasodilation. The main side effect of this drug group is edema due to sodium and water retention. Examples include *hydralazine* (Apresoline®) and *minoxidil* (Loniten®).

Beta-Adrenergic Antagonists

Beta-adrenergic antagonists, also known as beta blockers, can affect the heart and bronchi. Beta-1 blockers work on the heart, and beta-2 blockers work on the bronchial receptors. Nonselective beta blockers will inhibit the activity of beta-1 and beta-2 receptors, resulting in bradycardia and bronchoconstriction. Side effects include decreased blood pressure, decreased cardiac output, and bronchospasm. An example of a nonselective beta blocker is *propranolol* (Inderal®).

Alpha-Adrenergic Antagonists

Alpha-adrenergic antagonists block the alpha-1 adrenergic receptors, resulting in vasodilation. This group is covered later in the chapter in Table 12-1.

Table 12-1   Urinary Bladder Drugs and their Mechanisms of Action

CATEGORY OF URINARY BLADDER DRUG	MECHANISM OF ACTION	EXAMPLES
Cholinergic agonists or parasympathomimetic agents	Clinically used to promote voiding of urine from the urinary bladder	bethanechol (Urecholine®, Duvoid®, Urabeth®)
Anticholinergics or parasympatholytic drugs	Used to treat urinary incontinence by promoting urine retention in the urinary bladder	propantheline (Pro-Banthine®), dicyclomine (Bentyl®), butyl hyoscine (Buscpan®)
Adrenergic agonists	Increase the tone of the urethral sphincters via both alpha and beta receptors	phenylpropanolamine (Proin®, Propagest®), ephedrine (available as a generic), pseudoephrine (Equiphed®, Sudafed®)
Adrenergic antagonists 1. Alpha-adrenergic antagonists 2. Beta-adrenergic antagonists	1. Clinically used to decrease the tone of internal urethral sphincters (needed in the treatment of decreased urinary tone due to overextension of the urinary bladder). 2. Clinically used for hypertension (see under blood-pressure lowering drugs in this chapter).	1. phenoxybenzamine (Dibenzyline®), prazosin (Minipress®), nicergoline (Sermion®) 2. propranolol (Inderol®)
Estrogen	Sex hormone; helps maintain urethral muscle tone in some animals	diethylstilbestrol (DES)
Testosterone	Sex hormone; helps maintain urethral muscle tone in some animals	testosterone cypionate (Depo-Testosterone®), testosterone propionate (available as a generic)
Alpha- and beta-adrenergic agonists	Stimulation of these receptors increases urethral tone	phenylpropanolamine (Proin®, Propagest®), ephedrine (available in generic form), pseudoephedrine (Equiphed®, Sudafed®)
Skeletal muscle relaxants	Clinically used to treat animals that have urge incontinence or urethral obstructions due to increased external urethral sphincter tone; may also be used after unobstructing male cats to limit spasticity of the external urethral sphincter	dantrolene (Dantrium®), aminopropazine (Jenotone®), diazepam (Valium®), baclofen (Lioresal®)
Urinary acidifiers	Clinically used to produce acid urine, which dissolves and helps prevent formation of struvite uroliths; use has decreased with increasing use of urinary acidifying diets	methionine (Methio-tabs®, Methigel®), ammonium chloride (Uroeze®)
Urinary alkalinizers	Clinically used in treatment of calcium oxalate, cystine, and ammonium urate uroliths	potassium citrate (Urocit-K®), sodium bicarbonate (generic)
Xanthine oxidase inhibitors	Decreases the production of uric acid, which helps decrease the formation of ammonium urate uroliths	allopurinol (Zyloprim®, Lopurin®)

Urolith Treatment

Uroliths (also known as *urinary calculi*) are abnormal mineral masses in the urinary system. Uroliths are composed of a large amount of crystalline material (organic and inorganic crystalloids) and a small amount of organic matrix (typically mucoid material). The different types of uroliths include struvite, also known as triple phosphate crystals because they contain the three minerals magnesium, ammonium phosphate, and hexahydrate; calcium oxalate; calcium phosphate; urate, also known as ammonium biurate; cystine; and mixed. In some animal species, urolith formation may result from bacterial infections of the urinary tract. The development of urolith formation is not fully understood, but dietary factors are known to be important in some cases. Uroliths in the urinary bladder cause hematuria (blood in the urine) and dysuria. Uroliths that lodge in the urethra may cause obstruction, which is a major concern in male cats and male sheep because of the narrow diameter of the urethra.

Diagnosis of uroliths is done via clinical signs (hematuria, dysuria, and possible urethral obstruction when the animal cannot urinate), urinalysis (Figures 12-4 and 12-5), and radiography and/or ultrasound. Treatment includes antibiotic therapy (if warranted), medical dissolution of the uroliths, and possibly surgical removal of the uroliths.

Medical dissolution of uroliths depends on the type of urolith found and the area of the urinary tract in which it occurs (Figure 12-6). Uroliths found in the urinary bladder can be successfully treated with medical dissolution because the urolith is bathed in urine; changing the composition of the urine can allow the urolith to dissolve. This is not true of kidney uroliths. The pH of the urine in which the urolith is found also plays a role in the type of urolith formed. Animals with uroliths that form in alkaline (basic) urine (struvite uroliths) are fed a urine-acidifying diet that dissolves uroliths (Hill's Prescription Diet s/d®), or are prescribed urinary acidifiers such as *methionine* (Methio-tabs®, Methigel®) and *ammonium chloride* (Uroeze®). Once the uroliths are dissolved

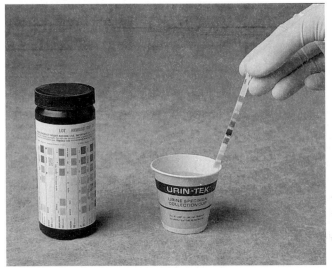

Figure 12-4 In the urinalysis, the chemical properties of urine, such as pH, glucose, ketones, and bilirubin, are tested with a dipstick.

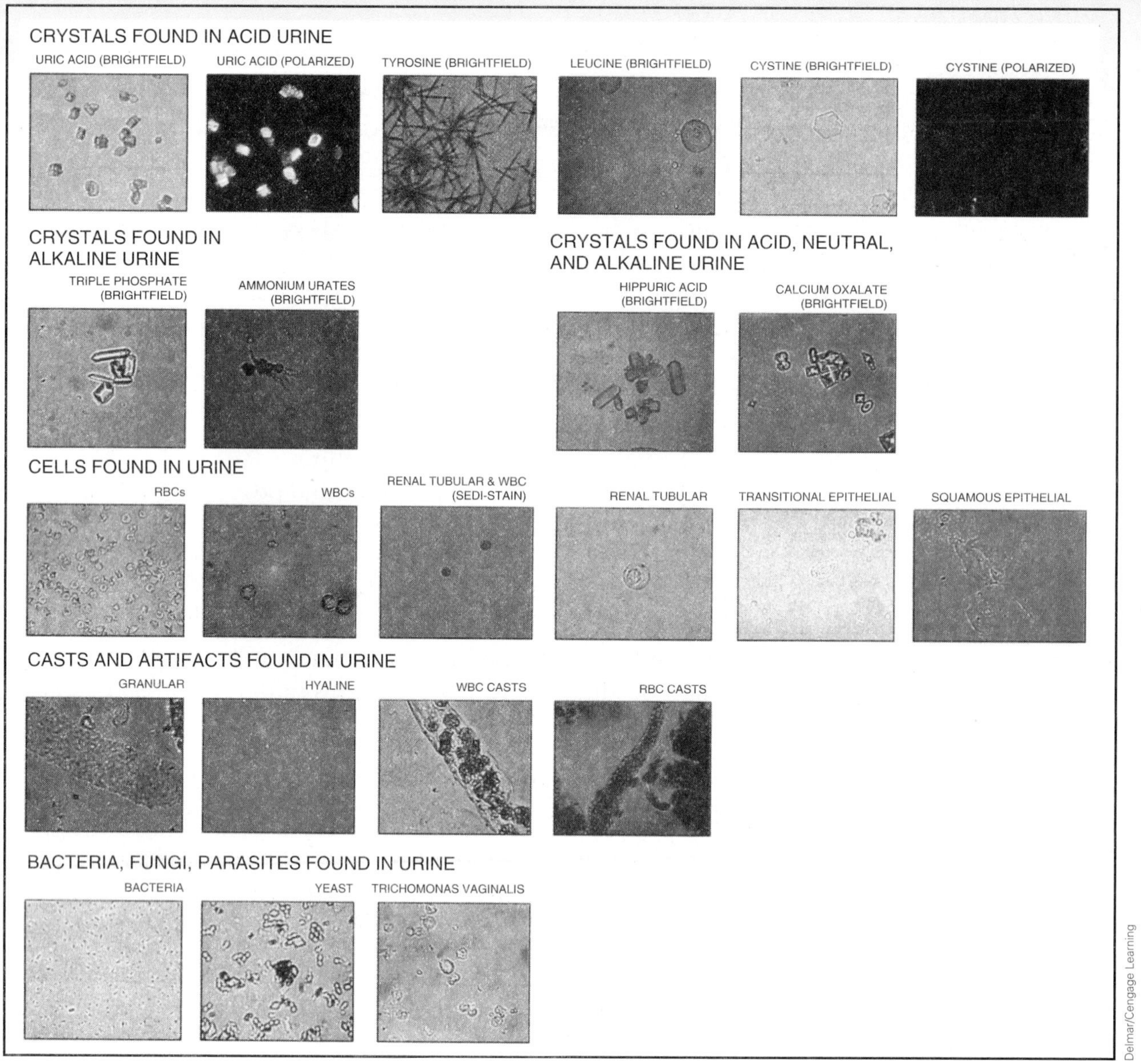

Figure 12-5 Crystals, cells, and casts found in urine.

or removed, animals are then maintained on diets that produce acid urine (Hill's Prescription Diet c/d®, Iams Low pH/s™). Animals with uroliths that form in acid urine (calcium oxalate, cystine, and ammonium urate uroliths) are fed urine-alkalinizing diets (Hills Prescription diet k/d®, Iams Moderate pH/O™ diet, or Purina's NF-Formula® diet), or are prescribed urinary alkalinizers such as *potassium citrate* (Urocit-K®) and *sodium bicarbonate* (generic). Animals that have renal disease should not be fed acidifying diets or prescribed urinary-acidifying drugs. Drugs used in the treatment of uroliths are summarized in Table 12-1.

Some uroliths, such as ammonium urate uroliths, also indicate the need for a low-protein, low-purine, low-oxalate diet to prevent recurrence (Hill's Prescription diet u/d®, Purina's NF Kidney Function® diet). Animals with ammonium urate uroliths are also often prescribed xanthine oxidase inhibitors.

Figure 12-6 Uroliths.

This group of drugs decreases the production of uric acid. Decreasing uric acid production helps prevent the formation of ammonium urate uroliths. *Allopurinol* (Zyloprim®, Lopurin®) is an example of a xanthine oxidase inhibitor. Side effects of allopurinol use are rare.

Urinary Incontinence

Urinary incontinence is the loss of voluntary control of micturition (a two-stage process involving the passive storage of urine and the active voiding of urine). Urinary incontinence can be divided into two main categories: (1) disorders due to neurologic disorders and (2) nonneurologic disorders. The body of the urinary bladder is composed of smooth muscle, the neck of the urinary bladder (also known as the *internal sphincter*) is composed of smooth muscle, and the external urethral sphincter is composed of skeletal muscle (Figure 12-7). Nervous control to the urinary bladder and urethra consists of both autonomic and somatic nervous system components. Autonomic nervous system control is both parasympathetic (the pelvic nerve supplies the smooth muscle of the urinary bladder, called the *detrusor muscle*) and sympathetic (the hypogastric nerve supplies beta-adrenergic fibers to the detrusor muscle and alpha-adrenergic fibers to the smooth muscle of the internal sphincter and part of the urinary bladder). Somatic innervation (pudendal nerve) stimulates the skeletal muscle of the external urethral sphincter. The coordination of these nervous system components results in voluntary micturition.

Neurologic problems may result in urinary incontinence (dribbling) or urine retention (inability to void urine that is in the urinary bladder). Neurologic causes of urinary problems can result from trauma to the spinal cord, tumors affecting the nervous system tracts, or degeneration of the nervous system tracts. Nonneurologic causes of urinary incontinence include hormone-responsive incontinence, stress incontinence, urge incontinence, ectopic ureters (abnormal entry of the ureters into the distal urethra instead of the urinary bladder), or urinary bladder overdistention. The cause of urinary

Clinical Que

Animals that are being fed urinary-acidifying diets should not also be given urinary-acidifying drugs. This combination can cause acid-base disturbances in the animal.

Clinical Que

Detrusor muscle contraction and external urethral sphincter relaxation are both necessary for micturition to occur.

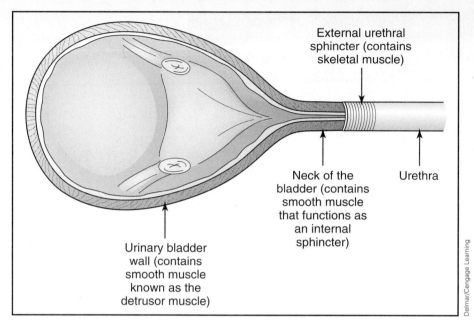

Figure 12-7 Urine pathway: Urine travels from the renal pelvis of the kidney through the ureters via peristalsis and is collected in the urinary bladder. During micturition, urine is directed through the neck of the bladder (known as the internal sphincter) and through the external urethral sphincter toward the urethra. Urine does not escape from the urinary bladder while the bladder is filling because the external urethral sphincter is contracted and urine cannot pass into the urethra. When urine is expelled from the urinary bladder, the external urethral sphincter muscle relaxes, and the urinary bladder muscles contract.

incontinence determines the treatment used in each case. Drugs used to treat urinary incontinence are summarized in Table 12-1.

Neurologically Caused Incontinence

Drugs used to treat animals with neurologic causes of urinary incontinence or urine retention include the following:

Cholinergic agonists (parasympathomimetic agents) are used to treat animals with "spinal cord bladders"—that is, damage to the nerves that control relaxation of the urinary bladder outflow sphincters. This nerve damage results in the retention of urine. Cholinergic agonists promote voiding of urine from the urinary bladder. These drugs simulate the action of acetylcholine by direct stimulation of cholinergic receptors. The cholinergic agonist binds to the receptors on smooth muscles, allowing sodium and calcium to enter the cells. This influx of sodium and calcium in turn allows muscle contraction. Tone of the detrusor muscle of the urinary bladder is increased, which may increase detrusor muscle contractions. An example of a cholinergic agonist is *bethanechol* (Urecholine®, Duvoid®, Urabeth®). Side effects include GI signs such as vomiting and diarrhea.

Anticholinergics (parasympatholytic drugs) are used to treat urinary incontinence by promoting urine retention in the urinary bladder. These drugs work by blocking the binding of acetylcholine to its receptor sites and thereby causing muscle relaxation. Examples of anticholinergics that promote urine retention are *propantheline* (Pro-Banthine®), *dicyclomine* (Bentyl®), and *butyl hyoscine*

(Buscpan®). Side effects of propantheline include decreased gastric motility (may affect absorption of other drugs) and other gastrointestinal problems.

Adrenergic antagonists are divided into alpha and beta categories. Beta-adrenergic antagonists, used in the treatment of hypertension, were covered in Chapter 8. Alpha-adrenergic antagonists are used to decrease the tone of internal urethral sphincters; they are useful in the treatment of decreased urinary tone due to overdistention of the urinary bladder. Alpha-adrenergic antagonists work by blocking circulating epinephrine or norepinephrine from binding to their receptors. These drugs are also used to decrease blood pressure. Examples of alpha-adrenergic antagonists include *phenoxybenzamine* (Dibenzyline®), *prazosin* (Minipress®), and *nicergoline* (Sermion®). The main side effect of these drugs is weakness due to decreased blood pressure.

Nonneurologically Caused Incontinence

Drugs used to treat nonneurologic causes of urinary incontinence include the following.

Estrogen (*diethylstilbestrol* or *DES*) is used to treat hormone-responsive incontinence. Hormone-responsive incontinence is seen primarily in spayed female dogs that are more than eight years old. It is believed that sex hormones contribute to the maintenance of urethral muscle tone; therefore, lack of sex hormones decreases urethral muscle tone and allows urine to dribble from the urethra. This urine dribbling usually occurs when the animal is relaxed or asleep. Side effects of estrogen include bone marrow suppression, endometrial hyperplasia, and pyometra.

Testosterone (*testosterone cypionate, testosterone propionate*) is used to treat hormone-responsive incontinence in castrated male dogs. Males suffer hormone-responsive incontinence less frequently than spayed female dogs. Side effects of testosterone include the development of prostatic disorders and behavior changes (including aggression).

Alpha- and beta-adrenergic agonists increase the tone of the urethral sphincters and include the following:

- *phenylpropanolamine* (Proin®, Propagest®), an oral medication used to treat stress incontinence. It may be used before resorting to hormones like estrogen and testosterone in the treatment of hormone-responsive incontinence. It is an alpha- and beta-adrenergic agonist that increases urethral tone. Side effects of phenylpropanolamine include anorexia, restlessness, and hypertension. In 2000, the FDA began taking steps to remove phenylpropanolamine (PPA) from all drug products, based on evidence that PPA increases the risk of hemorrhagic stroke in humans. This action has limited the availability of PPA in some areas.

- *ephedrine* (generic), an alpha and beta agonist that increases urethral tone. It is used to treat stress incontinence and is available in tablet or injectable form. Side effects are the same as for phenylpropanolamine.

- *pseudoephedrine* (Equiphed®, Sudafed®), an alpha- and beta-adrenergic agonist used to increase urethral sphincter tone and produce closure of the urinary bladder neck. It is used to treat urinary incontinence and comes in tablet and liquid forms. Side effects include restlessness, irritability, hypertension, and anorexia.

Skeletal muscle relaxants, such as *dantrolene* (Dantrium®), *aminopropazine* (Jenotone®), *diazepam* (Valium®), and *baclofen* (Lioresal®) are sometimes used to treat animals that have urge incontinence or urethral obstructions due to increased external urethral sphincter tone. These drugs may also be used after unobstructing male cats to limit spasticity of the external urethral sphincter. Side effects of skeletal muscle relaxants include mild tranquilization and sedation. Dantrolene has the additional side effect of hepatotoxicity.

Miscellaneous Drugs

Other groups of drugs, such as antibiotics, are used in management of urinary disease and infection. These groups of drugs are covered in Chapter 14.

SUMMARY

Urinary system disease and dysfunction can be caused by a variety of factors. Drugs used in the management of urinary system disease include diuretics, urinary bladder drugs, and antihypertensive drugs. Diuretics increase the volume of urine excreted by the kidneys and thus promote the release of water from the tissues. The classes of diuretics include thiazides, loop diuretics, potassium-sparing diuretics, carbonic anhydrase inhibitors, and osmotic diuretics. Antihypertensives decrease blood pressure; this class includes diuretics, ACE inhibitors, calcium channel blockers, direct-acting vasodilators, and beta- and alpha-adrenergic antagonists.

Uroliths or urinary calculi are abnormal mineral masses in the urinary system. Treatment of uroliths includes antibiotic therapy (if warranted), medical dissolution of the uroliths, and possibly surgical removal of the uroliths. Medical dissolution of struvite uroliths in alkaline (basic) urine involves feeding the animal a urine-acidifying diet that dissolves uroliths or using prescribed urinary acidifiers like methionine and ammonium chloride. Animals with uroliths that form in acid urine (calcium oxalate, cystine, and ammonium urate uroliths) are fed urine-alkalinizing diets or are prescribed urinary alkalinizers such as potassium citrate or sodium bicarbonate. A low-protein, low-purine, low-oxalate diet helps to prevent urate uroliths from recurring. Animals with urate uroliths are also prescribed xanthine oxidase inhibitors, which decrease the production of uric acid and thus help decrease the formation of ammonium urate uroliths. Allopurinol is an example of a xanthine oxidase inhibitor.

Urinary incontinence is the loss of voluntary control of micturition (a two-stage process involving the passive storage of urine and the active voiding of urine). Urinary incontinence may be caused by neurologic or nonneurologic disorders. Neurologic causes of urinary incontinence include trauma to the spinal cord, tumors affecting the nervous system tracts, or degeneration of the nervous system tracts. Types of nonneurologic urinary incontinence include hormone-responsive incontinence, stress incontinence, urge incontinence, ectopic ureters, and urinary bladder overdistention. The cause of urinary incontinence determines the treatment used in each case.

Drugs used to treat animals with neurologic causes of urinary incontinence include cholinergic agonists (parasympathomimetic agents) that promote voiding of urine from the urinary bladder; anticholinergics (parasympatholytic drugs) that promote urine retention in the urinary bladder; and alpha-adrenergic antagonists, used to decrease the tone of urethral sphincters and balance decreased urinary bladder tone due to overextension.

Drugs used to treat nonneurologic causes of urinary incontinence include estrogen and testosterone, which are used to treat hormone-responsive incontinence; alpha- and beta-adrenergic agonists, such as phenylpropanolamine, ephedrine, and pseudoephedrine, which increase urethral tone; and skeletal muscle relaxants, such as dantrolene, aminopropazine, diazepam, and baclofen, which are sometimes used to treat animals that have urge incontinence or urethral obstructions due to increased external urethral sphincter tone.

It's a Wrap

The answers to the questions in this chapter's Setting the Scene case study should be understood after reading this chapter. What questions should the veterinary technician ask this client about the male cat that is straining to urinate? The veterinary technician needs to determine whether or not this cat can urinate. Other information needed from this client are whether or not the animal is eating or drinking, is it lethargic, and when was the last time it urinated. What concerns are there regarding this cat? Since this cat is male, the risk of urinary obstruction is higher than in female cats. Male cats are especially prone to urethral obstruction due to the narrow diameter of their urethra. The veterinary technician needs to determine how long this has been going on, and the owner must bring the cat into the clinic. Does this cat need immediate veterinary care, or can this problem wait until the cat's next scheduled PE? Urinary obstruction is an emergency situation since the inability to urinate allows toxins to accumulate in the animal, which can lead to electrolyte imbalance and death. If this animal has a urinary tract infection rather than urinary obstruction, it is still best to treat promptly to avoid any long-term problems.

CHAPTER REVIEW

Matching

Match the drug name with its action.

1. _____ furosemide
2. _____ mannitol
3. _____ propantheline
4. _____ methionine
5. _____ potassium citrate
6. _____ diltiazem
7. _____ enalapril
8. _____ propranolol
9. _____ spironolactone
10. _____ acetazolamide

a. potassium-sparing diuretic
b. urinary acidifier
c. calcium channel blocker
d. carbonic anhydrase inhibitor
e. urinary alkalinizer
f. beta blocker
g. loop diuretic
h. ACE inhibitor
i. osmotic diuretic
j. anticholinergic that promotes urine retention

Multiple Choice

Choose the one best answer.

11. Which group of urinary drugs is also used for treatment of glaucoma?
    a. loop diuretics
    b. ACE inhibitors
    c. carbonic anhydrase inhibitors
    d. cholinergic agonists

12. What part of the nephron is responsible for filtration?
    a. glomerulus
    b. proximal convoluted tubule
    c. loop of Henle
    d. distal convoluted tubule

13. ACE inhibitors work by
    a. blocking electrolyte channels.
    b. causing vasoconstriction.
    c. pulling fluid into the renal tubules.
    d. blocking the conversion of angiotensin I to angiotensin II.

14. What type of drug promotes water release from tissues?
    a. diuretic
    b. urinary acidifier
    c. urinary alkalinizer
    d. cholinergic

15. What type of urine helps prevent formation of and encourages dissolution of struvite crystals?
    a. acidic
    b. basic

16. What type of diuretic acts directly on the renal tubules to block sodium reabsorption and promote chloride ion excretion?
    a. thiazide
    b. loop
    c. potassium sparing
    d. osmotic

17. Which category of urinary drug promotes voiding of urine from the urinary bladder?
a. cholinergics
b. anticholinergics
c. sympathomimetic
d. sympathetic blocking agents

18. Which category of urinary drug decreases internal urethral sphincter tone and is used to treat animals with an overdistended urinary bladder?
a. cholinergics
b. anticholinergics
c. alpha-adrenergic agonists
d. alpha-adrenergic antagonists

19. Which group of drugs is used in animals with ammonium urate uroliths?
a. xanthine oxidase inhibitors
b. loop diuretics
c. osmotic diuretics
d. urinary acidifiers

20. Glomerular filtration is affected by
a. blood pressure.
b. plasma osmotic pressure.
c. capsule pressure.
d. all of the above.

Case Studies

21. A three-year-old M Dalmatian presents to the clinic with what the owner describes as "urinating all the time, but only producing a small amount of urine." The PE reveals T = 102°F and HR = 84 bpm, and the dog is panting. The rest of the PE was unremarkable. The veterinarian requests a clean-catch urine sample on this dog. The urinalysis reveals a urinary tract infection (UTI) and the presence of ammonium urate crystals. The veterinarian prescribes a broad-spectrum antibiotic for this dog to treat the UTI and a low-protein diet to restrict purines.
a. What other drug should be dispensed to the owner of this dog to help decrease the formation of ammonium urate uroliths?
b. This dog would also benefit from a urinary alkalizer. What is an example of a urinary alkalizer?

22. A 15-year-old M/N Llasa Apso (25#) was put on a loop diuretic for his congestive heart failure.
a. What is an example of a loop diuretic?
b. Where do loop diuretics work?
c. What is the main side effect of loop diuretics?
d. Diuresis is rapid and tremendous with loop diuretics. What would you want to warn this owner about?

Critical Thinking Questions

23. Urinalysis is one of the basic fundamental diagnostic tests used in the practice of medicine. Discuss the information that urinalysis provides about the health of an animal.

24. Why do loop diuretics cause electrolyte imbalances?

ADDITIONAL REVIEW MATERIAL

Additional review material can be found on the StudyWARE™ CD included with this text.

BIBLIOGRAPHY

Allen, D. G. (2005). *Handbook of Veterinary Drugs* (3rd ed.). Baltimore, MD: Lippincott Williams & Wilkins.

Amundson Romich, J. (2008). *An Illustrated Guide to Veterinary Medical Terminology* (3rd ed.). Clifton Park, NY: Delmar Cengage Learning.

Broyles, B., Reiss, B., and Evans, M. (2007). *Pharmacological Aspects of Nursing Care* (7th ed.). Clifton Park, NY: Delmar Cengage Learning.

Plumb, D. C. (2008). *Plumb's Veterinary Drug Handbook* (6th ed.). Ames, IA: Blackwell Publishing.

Spratto, G. R., and Woods, A. L. (2008). *PDR Nurse's Drug Handbook.* Clifton Park, NY: Cengage Learning.

Iams Web site www.iams.com

Hill's Pet Nutrition Web site www.hillspet.com

CHAPTER 13

DRUGS AFFECTING MUSCLE FUNCTION

OBJECTIVES

Upon completion of this chapter, the reader should be able to:

- describe the basic anatomy and physiology of muscles
- explain the process of skeletal muscle activation.
- describe the role of the neuromuscular junction.
- describe the use of the following drugs in veterinary medicine: anti-inflammatory drugs (in regard to treating muscle disease), neuromuscular blockers, skeletal muscle spasmolytics, and anabolic steroids.

KEY TERMS

acetylcholine

acetylcholinesterase

anabolic steroid

anti-inflammatory drugs

competitive nondepolarizers

inflammation

neuromuscular blockers

neuromuscular junction

spasmolytics

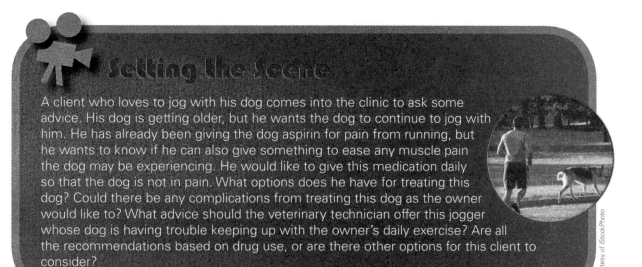

Setting the Scene

A client who loves to jog with his dog comes into the clinic to ask some advice. His dog is getting older, but he wants the dog to continue to jog with him. He has already been giving the dog aspirin for pain from running, but he wants to know if he can also give something to ease any muscle pain the dog may be experiencing. He would like to give this medication daily so that the dog is not in pain. What options does he have for treating this dog? Could there be any complications from treating this dog as the owner would like to? What advice should the veterinary technician offer this jogger whose dog is having trouble keeping up with the owner's daily exercise? Are all the recommendations based on drug use, or are there other options for this client to consider?

Courtesy of iStockPhoto

BASIC MUSCLE ANATOMY AND PHYSIOLOGY

Muscles are tissues that contract to produce movement. Muscles are made up of long, slender cells called *muscle fibers*. Muscle fibers are encased in a fibrous sheath. Muscle cells are categorized into three types, based on their appearance and function: skeletal, smooth, and cardiac. The focus here is on skeletal muscles. Skeletal muscles are those muscles that are attached to the skeleton and are activated by voluntary control.

A motor nerve that originates in the spinal cord and terminates in fibers connected to muscle cells activates skeletal muscle. The point at which a motor nerve fiber connects to muscle cells is known as a **neuromuscular junction** (Figure 13-1). When an electrical impulse of sufficient strength travels from the spinal cord to the neuromuscular junction, it causes release of the cholinergic neurotransmitter acetylcholine. **Acetylcholine** moves across the neuromuscular junction and binds to specialized receptor sites on the muscle opposite the nerve ending. As the electrical charge travels from the receptor sites across the length of the muscle fiber, depolarization of the muscle occurs, calcium is released, and the muscle contracts. The acetylcholine that causes this action is inactivated by **acetylcholinesterase**, an enzyme that breaks down acetylcholine and readies the muscle fibers for the next nerve impulse.

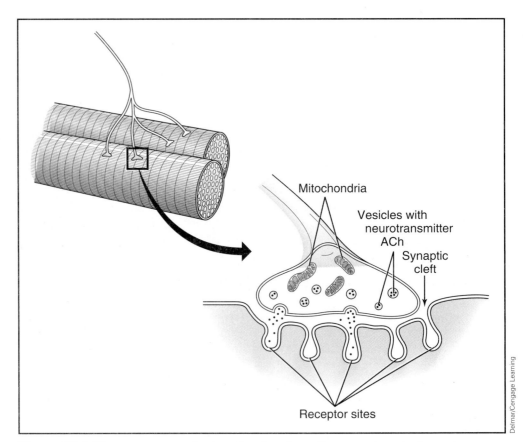

Mitochondria

Vesicles with neurotransmitter ACh

Synaptic cleft

Receptor sites

Delmar/Cengage Learning

Figure 13-1 The neuromuscular junction is where the axon terminal meets a muscle fiber. The axon terminal does not touch the sarcolemma of the muscle, but rather fits into a depression in the cell membrane.

MUSCLE DISORDERS

Muscle disorders in veterinary medicine are usually associated with traumatic injuries, neuromuscular disease such as myasthenia gravis, or chronic debilitating disorders such as those associated with spinal cord disease. Traumatic injury and chronic debilitating disorders may result in muscle spasms and pain. Myasthenia gravis is a disease caused by lack of acetylcholine reaching the cholinergic receptors and is characterized by weakness and fatigue of skeletal muscles.

Clinical signs such as weakness, muscle atrophy, and pain may need to be treated to prevent other complications such as pressure sores, skin infection, and further debilitation so that the animal may lead a normal life. Muscles may also need to be paralyzed so that some types of surgery can be performed.

DRUGS AFFECTING MUSCLE FUNCTION

Drugs that affect skeletal muscle include anti-inflammatories (drugs that counteract inflammation), neuromuscular blockers (drugs that produce paralysis), skeletal muscle spasmolytics (drugs that reduce muscle spasms), and anabolic steroids (whose tissue-building effects can reverse muscular atrophy or wasting).

Inflammation Reducing

Inflammation is a normal response to injury, infection, or irritation of living tissue. Redness, pain, swelling, heat, and decreased range of motion are all signs of inflammation. **Anti-inflammatory drugs** are used to relieve these signs. Anti-inflammatory drugs include steroidal and nonsteroidal varieties. Steroidal anti-inflammatory drugs (glucocorticoids) are chemically related to the naturally occurring hormone cortisone, which is secreted by the adrenal cortex. Glucocorticoids have potent anti-inflammatory effects and cause a variety of metabolic effects, including modification of the body's immune system. Nonsteroidal anti-inflammatory drugs (NSAIDs) are synthetic products that are unrelated to the substances produced by the body. These agents are widely used in the treatment of inflammation and pain reduction. The anti-inflammatory mechanism of action of NSAIDs is the result of inhibition of prostaglandin synthesis. These drugs are covered in more detail as to their mechanisms of action and side effects in Chapter 16.

Muscle Paralyzing

Veterinarians need to relax the muscles of animals that are undergoing surgery and to prevent or treat muscle spasms. To do this, they may use **neuromuscular blockers**. These agents, administered intravenously, paralyze muscles by disrupting the transmission of nerve impulses from motor nerves to skeletal muscle fibers.

All but one of the commonly used neuromuscular blockers are **competitive nondepolarizers**, which are neuromuscular blocking agents that compete with acetylcholine for the same receptor sites, thus inhibiting the effects of

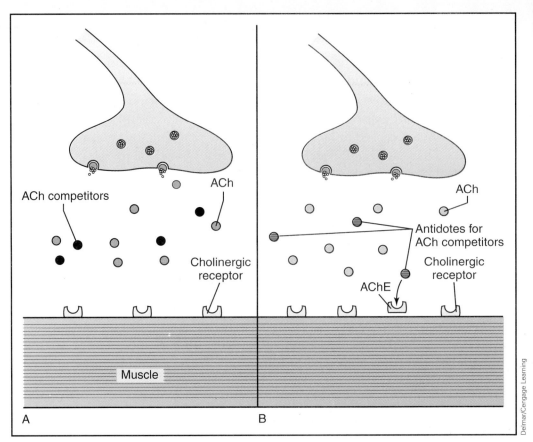

Figure 13-2 Neuromuscular blockers and their antidotes. (A) Competitive nondepolarizers compete with acetylcholine (ACh) for the same receptor sites thus inhibiting the effects of ACh. (B) Antidotes for competitive nondepolarizers compete with ACh for acetylcholinesterase (AChE), which allows ACh to accumulate in the neuromuscular junction and cause its effect.

acetylcholine (Figure 13-2). These drugs are also called *curarizing agents* because they resemble curare alkaloids, toxic botanical extracts originally used as arrow poisons in South America. Examples of competitive nondepolarizers are *pancuronium* (Pavulon®) and *atracurium* (Tracrium®). Pancuronium is used as an adjunct to general anesthesia to produce muscle relaxation and to facilitate endotracheal intubation. When given IV, it causes muscular relaxation in 2 to 3 minutes, and the effect lasts for about 45 minutes. Side effects include increased heart rate and blood pressure. Atracurium is used for the same purposes as pancuronium. It is given IV, and takes 2 to 3 minutes to cause muscle relaxation that lasts about 25 minutes. It causes minimal cardiovascular side effects and can be used in animals with kidney disease.

The competitive nondepolarizers require antidotes once their paralytic effect is no longer needed. Veterinarians administer anticholinesterase drugs such as *neostigmine* (Prostigmin®, Stiglyn®), *pyridostigmine* (Mestinon®), and *edrophonium* (Enlon®, Tensilon®, Reversol®), which allow more acetylcholine into the muscle site and thus reverse the paralyzing effect of competitive nondepolarizers. Neostigmine competes with acetylcholine for acetylcholinesterase, allowing acetylcholine to accumulate in the neuromuscular junction and prolong the cholinergic effect. Neostigmine is also used to treat rumen atony (see Chapter 11) and as a diagnostic agent and treatment for myasthenia gravis. Side effects include vomiting, diarrhea, and excess salivation. Pyridostigmine also competes with acetylcholine for attachment to acetylcholinesterase. This competition allows acetylcholine to accumulate and bind to the cholinergic receptors available. It is used to treat myasthenia gravis since this disease is believed to be caused by the production of autoantibodies to acetylcholine receptors (acquired form) or lack of acetylcholine receptors (congenital form). Pyridostigmine has a longer duration of action than neostigmine, and its side effects include vomiting and diarrhea. Edrophonium is a very short-acting agent that attaches to acetylcholinesterase, thereby hindering its breakdown of acetylcholine. Side effects related to acetylcholine buildup include miosis (pupillary constriction), bronchoconstriction, and excessive salivation. Edrophonium is used for the diagnosis of myasthenia gravis.

Other examples of competitive nondepolarizing neuromuscular blocking agents include *gallamine* (Flaxedil®), *vecuronium* (Norcuron®), and *metocurine* (Metubine®). Because all these agents are cholinergic in nature, atropine (an anticholinergic) is sometimes given with these drugs to decrease the profound cholinergic effects.

One muscle paralyzer that is not a competitive nondepolarizer is *succinylcholine* (Anectine®, Quelicin®, Sucostrin®). It is known as a depolarizing drug or noncompetitive neuromuscular blocker. Rather than competing with acetylcholine, succinylcholine works independently of the transmitter. Depolarizing drugs mimic the action of acetylcholine in muscle fibers, and because they are not destroyed by acetylcholinesterase, their action is prolonged. Succinylcholine binds to the cholinergic receptors to produce depolarizations that continue as long as sufficient amounts of succinylcholine remain. This noncompetitive depolarizer does not need an antidote; it wears off by itself in a short period of time. It is used to reduce the muscle contraction associated with toxicities or pharmacologically induced convulsions. Side effects include excessive salivation, cardiovascular effects, and rash.

Spasm Stopping

Spasmolytics are drugs used to treat acute episodes of muscle spasticity associated with neurological and musculoskeletal disorders, such as malignant hyperthermia, equine postanesthetic myositis, and traumatic injury that results in muscle spasms. Spasmolytics "break down" the muscle spasticity. They work by unknown mechanisms; some are associated with calcium release from the sarcoplasmic reticulum or are responsible for depressing nerve impulse transmission. They include *methocarbamol* (Robaxin-V®), *guaifenesin* (Guailaxin®, Gecolate®), *diazepam* (Valium®), and *dantrolene* (Dantrium®).

Methocarbamol (Robaxin-V®) is used as an adjunctive therapy for inflammatory and traumatic conditions of skeletal muscle (especially intervertebral disc disease and traumatic injuries). It is also used to reduce muscle spasms. It is a potent, centrally acting muscle relaxant with selective action on the CNS that helps reduce muscle spasms but does not decrease muscle tone.

Clinical Que

Neuromuscular blockers do not provide analgesia or sedation. Animals receiving neuromuscular blocking agents must be monitored closely, because they may develop respiratory paralysis and/or cardiac collapse.

Clinical Que

Atracurium and pancuronium should be refrigerated. Do not store these drugs in plastic containers/syringes because the plastic may absorb them.

Clinical Que

Animals on
methocarbamol
may develop darker
urine—clients should
be advised that this
is not a concern.

Methocarbamol is available in injectable and oral forms. Side effects are rare, but include sedation, lethargy, and weakness.

Guaifenesin (Guailaxin®, Gecolate®) is a muscle relaxant that relaxes both laryngeal and pharyngeal muscles, thereby easing intubation. It is used as an adjunct to anesthesia in large and small animals. A benefit of guaifenesin is that it does not affect the diaphragm and thus does not affect respiratory function. Guaifenesin is also covered in Chapter 7.

Diazepam (Valium®) is a centrally acting muscle relaxant that decreases the turnover of acetylcholine in the brain. It is a C-IV controlled substance used for muscle relaxation and anxiety control in animals. It is also covered in Chapter 7.

Dantrolene (Dantrium®) is a peripherally acting muscle relaxant that inhibits calcium release from the muscle, making it less responsive to nerve impulses. Dantrolene is used to prevent and treat malignant hyperthermia in various species, to treat urethral obstruction due to increased external urethral sphincter tone in dogs and cats (see Chapter 12), and to treat postanesthetic myositis in horses. The most significant side effect of dantrolene is liver toxicity, followed by sedation and vomiting.

Tissue Building

Anabolic steroids are steroids with a tissue-building effect—simply put, these drugs increase muscle mass. Veterinarians use them to promote growth, counteract postsurgical debility, and treat diseases such as muscular atrophy and orthopedic conditions (Figure 13-3). These drugs include *nandrolone decanoate* (Deca-Durabolin®) and *boldenone undecylenate* (Equipose®). Nandrolone decanoate is an injectable anabolic steroid used to stimulate erythropoiesis (red blood cell formation) in animals. It is also used as an appetite stimulant. It is a C-III controlled substance and should not be used in animals that are pregnant or have kidney disease.

Figure 13-3 Muscle atrophy may be treated with anabolic steroids.

Boldenone undecylenate is a long-acting injectable anabolic steroid used to treat debilitated horses. Boldenone increases appetite and improves musculature and haircoat appearance. It is a C-III controlled substance. Most horses respond with one to two treatments. This drug should not be used in animals that are pregnant or intended for use as food.

SUMMARY

Skeletal muscles are those muscles that are attached to the skeleton and are activated by voluntary control. Motor nerves that affect the neuromuscular junction activate skeletal muscle. Acetylcholine moves across the neuromuscular junction, binds to muscle receptors, and causes muscular contraction for movement. Acetylcholinesterase stops this action. Drugs used to treat muscle disorders include anti-inflammatories, neuromuscular blockers, skeletal muscle spasmolytics, and anabolic steroids. Anti-inflammatory drugs reduce inflammation. Neuromuscular blockers disrupt transmission of nerve impulses from motor nerves to skeletal muscle fibers. Skeletal muscle spasmolytics work via unknown mechanisms to decrease muscle spasticity. Anabolic steroids build up muscle tissue.

Table 13-1 summarizes the drugs covered in this chapter.

Clinical Que

To obtain optimal results when giving anabolic steroids, adequate and well-balanced dietary intake is essential. Caution is required, however, because anabolic agents can cause electrolyte imbalance, liver toxicity, behavioral changes, and reproductive abnormalities.

Table 13-1 Drugs Covered in This Chapter

DRUG CATEGORY	EXAMPLES
Neuromuscular blockers	pancuronium (Pavulon®)
	atracurium (Tracrium®)
	gallamine (Flaxedil®)
	vecuronium (Norcuron®)
	metocurine (Metubine®)
	succinylcholine (Anectine®, Quelicin®, Sucostrin®)
Antidotes (anticholinesterases)	neostigmine (Prostigmin®, Stiglyn®)
	pyridostigmine (Mestinon®)
	edrophonium (Enlon®, Tensilon®, Reversol®)
Spasmolytics	methocarbamol (Robaxin-V®)
	guaifenesin (Guailaxin®, Gecolate®)
	diazepam (Valium®)
	dantrolene (Dantrium®)
Anabolic steroids	nandrolone decanoate (Deca-Durabolin®)
	boldenone undecylenate (Equipose®)

It's a Wrap

The answers to the questions in this chapter's Setting the Scene case study should be understood after reading this chapter. What options does this owner have for treating an older dog that is experiencing muscle pain? There are a variety of treatment options here. Chapter 16 presents the anti-inflammatory options (nonsteroidal anti-inflammatories, such as carprofen and buffered aspirin; other drugs, such as flunixin meglumine that may not be indicated for long-term use; and glucocorticoid anti-inflammatories). Sometimes animals with musculoskeletal injuries or degeneration are prescribed muscle relaxants, such as methocarbamol. Muscle relaxants are thought to work by decreasing muscle rigidity without affecting normal muscle tone. Conditions that can benefit from muscle relaxant use include intervertebral disk disease, muscle strain, and myositis. Side effects from methocarbamol use are rare but may include muscle weakness, ataxia, and vomiting. Could there be any complications from treating this dog as the owner would like to? One risk in medicating this dog is that the dog may overexert itself and cause more muscle damage or injure another part of his body secondary to the drug side effects of ataxia and weakness. What advice should the veterinary technician offer this jogger whose dog is having trouble keeping up with the owner's daily exercise? Are all the recommendations based on drug use, or are there other options for this client to consider? Weight loss is recommended for dogs that are overweight. The use of padded bedding may benefit dogs with musculoskeletal pain. Some nonmedication options for this owner include physical therapy, chiropractic medicine, and neutraceuticals, such as polysulfated glycosaminoglycans. Polysulfated glycosaminoglycans are a mixture of glycosaminoglycans derived from bovine cartilage that are believed to work by increasing synovial fluid production from the synovial membrane and possibly reducing cartilage damage. Increased amounts of synovial fluid can help lubricate joints and ease joint movement. These options are discussed in Chapter 16.

CHAPTER REVIEW

Matching

Match the drug name with its action. Answers will be used more than once.

1. _____ pancuronium

2. _____ nandrolone

3. _____ guaifenesin

4. _____ dantrolene

5. _____ atracurium

6. _____ edrophonium

7. _____ succinylcholine

8. _____ boldenone undecylenate

9. _____ methocarbamol

10. _____ neostigmine

a. skeletal muscle spasmolytic

b. neuromuscular blocking agent

c. neuromuscular blocking agent antidote

d. anabolic steroid

Multiple Choice

Choose the one best answer.

11. What type of drug overrides the stimulative effect of acetylcholine?
    a. anti-inflammatory
    b. competitive nondepolarizer
    c. spasmolytic
    d. anabolic steroid

12. How is acetylcholine normally deactivated at the neuromuscular junction?
    a. It is competitively overridden.
    b. It is inactivated by acetylcholinesterase.
    c. It is degraded by the warmer temperature at the neuromuscular junction.
    d. It has a half-life of two to four minutes.

13. Anabolic means
    a. breakdown.
    b. building.
    c. homeostasis.
    d. deactivation.

14. What mineral is released from the sarcoplasmic reticulum of muscles?
    a. potassium
    b. sodium
    c. chloride
    d. calcium

15. What type of neuromuscular drug would be used to counteract postsurgical debility?
    a. anti-inflammatory
    b. neuromuscular blocker
    c. spasmolytic
    d. anabolic steroid

Case Study

16. A-10-year-old F/S Dachshund (18#) presents to the clinic with reluctance to walk and lethargy. The owner states that she thought the dog had hurt herself, because she no longer wants to jump on the bed or walk up stairs. Physical examination reveals T = 103°F, RR = panting, HR = 120 bpm. The dog has pain on palpation of its back and proprioceptive deficits of both hind limbs (seen with spinal cord disease). The veterinarian orders an X-ray of the back, which shows the dog has intervertebral disc disease. The veterinarian recommends surgery; however, the owner is reluctant to have surgery done on her dog. She opts for medical management, which includes the use of glucocorticoids to reduce inflammation of the spinal cord, and cage rest.
    a. What muscular drug may also benefit this dog?
    b. The veterinarian prescribes methocarbamol at a dosage of 15 mg/kg po tid. How much would this dog get per dose?
    c. Methocarbamol comes in 500 mg tablets that are scored into quarters. How many would this dog get per dose?
    d. The veterinarian wants this dog on methocarbamol for seven days. How many pills do you dispense?

Critical Thinking Questions

17. The disease known as milk fever can strike lactating female animals under certain conditions. What are these conditions? How does milk fever affect muscle function?

18. Botulism and tetanus toxins both affect the neuromuscular junctions. Perform an Internet search to determine how each of these toxins affects the neuromuscular junctions.

ADDITIONAL REVIEW MATERIAL

Additional review material can be found on the StudyWARE™ CD included with this text.

BIBLIOGRAPHY

Allen, D. G. (2005). *Handbook of Veterinary Drugs* (3rd ed.). Baltimore, MD: Lippincott Williams & Wilkins.

Amundson Romich, J. (2008). *An Illustrated Guide to Veterinary Medical Terminology* (3rd ed.). Clifton Park, NY: Delmar Cengage Learning.

Broyles, B., Reiss, B., and Evans, M. (2007). *Pharmacological Aspects of Nursing Care* (7th ed.). Clifton Park, NY: Delmar Cengage Learning.

Plumb, D. C. (2008). *Plumb's Veterinary Drug Handbook* (6th ed.). Ames, IA: Blackwell Publishing.

Spratto, G. R., and Woods, A. L. (2008). *PDR Nurse's Drug Handbook*. Clifton Park, NY: Delmar Cengage Learning.

Veterinary Pharmaceuticals and Biologicals (12th ed.). (2001). Lenexa, KS: Veterinary Healthcare Communications.

CHAPTER 14

ANTIMICROBIALS

OBJECTIVES

Upon completion of this chapter, the reader should be able to:

- describe the role of antimicrobials in treating animal infections.
- explain the function of antibiotics.
- differentiate between narrow- and broad-spectrum antibiotics.
- differentiate between bactericidal and bacteriostatic antibiotics.
- describe the use of MIC.
- describe various mechanisms by which antibiotics work.
- list key uses, side effects, and differences among the different categories of antibiotics.
- list key uses, side effects, and differences among the different categories of antifungals.
- describe the uses, side effects, and limitations of antiviral drugs.
- differentiate between disinfectants and antiseptics.
- list key uses and differences among the different categories of disinfectants and antiseptics.

KEY TERMS

antibiotics	germicide
antibiotic residues	intermediate
antibiotic resistance	minimum inhibitory
antifungals	concentration (MIC)
antimicrobials	narrow-spectrum
antiseptic	antibiotics
antiviral	potentiated
bactericidal	resistant
bacteriostatic	sanitizing
beta-lactamase	sensitive
broad-spectrum	spectrum of action
antibiotics	sporicidal
disinfectant	tuberculocide
fungicidal	virucidal

Setting the Scene

A cat owner comes into the clinic requesting a refill of an antibiotic because her cat is sneezing again. The client explains that when the cat starts sneezing, she gives him antibiotics. The owner says she only needs about 14 pills, because she only has to give her cat two to three days worth of pills before he stops sneezing. Fourteen pills will give her enough antibiotic to treat the cat for a few "bouts" of sneezing. She would also like the "strongest" antibiotic available. How can the veterinary technician help this client understand that her rationale of antibiotic use is flawed? What is the best way to explain to this client how antibiotics work, what side effects can be expected, and why "stronger" is not a good term to use in describing antibiotics? What are some consequences of this kind of antibiotic use?

Courtesy of iStockPhoto

PATHOGENIC MICROORGANISMS AND ANIMAL DISEASE

Pathogenic microorganisms may cause a wide variety of infections and produce illness in different organs or body systems. The invasion of pathogenic microorganisms that cause infection may be classified as either local or systemic. A localized infection may involve the skin or internal organ and may progress to a systemic infection. A systemic infection affects the whole animal and is typically a more serious infection than localized infections.

Pathogenic microorganisms enter the animal through a variety of ways including the integumentary system (a break in the skin or mucous membrane), respiratory system (by inhaling contaminated droplets), gastrointestinal tract (through ingestion of contaminated food and water), and genitourinary system (through contaminated vaginal secretions, semen, or urine). Preventing a microorganism from entering a host (animal that can be affected by an agent) and keeping an animal's environment properly sanitized are important factors of controlling infection in animals.

ANTIMICROBIAL TERMINOLOGY

Antimicrobials are drugs that counteract infection. An antimicrobial is a chemical substance that has the capacity, in diluted solutions, to kill (biocidal activity) or inhibit the growth (biostatic activity) of microorganisms. Biocidal agents attack something that the microorganisms have that the patient does not have and thereby kill the microorganisms. Biostatic agents attack something that both the microorganism and the patient have, but that the microorganism needs more of. These agents only injure the microorganisms, but rely on the patient's immune system to kill the bacteria. Depending on the drug dosage and serum level of the drug, certain drugs can be both biocidal and biostatic.

Antimicrobials can be further classified as antibiotics, antifungals, antiviral drugs, and antiparasitic drugs. There are many antimicrobials available to treat infections. This chapter describes some of the important agents used widely by veterinarians to treat bacterial, fungal, and viral infections. Chapter 15 covers antiparasitic drugs.

ANTIMICROBIALS FOR BACTERIA

Antibiotics are chemicals that work only on bacteria. Bacteria are prokaryotic cells and have fewer organelles and are smaller than eukaryotic cells (Figure 14-1). They are described by their **spectrum of action**, which refers to the range of bacteria on which the agent is effective. Most bacteria are classified by Gram stain, a staining and decolorizing procedure that divides them into gram-positive bacteria and gram-negative bacteria. Gram-positive bacterial strains resist decolorization by the Gram stain process and have cell walls less complex in chemical composition than gram-negative bacteria, which are

Clinical Que

There are many chemicals (such as cyanide) that are lethal to bacteria, but they cannot be used to treat infections because they are lethal to the host as well. The goal is to find substances that attack a structure or metabolic pathway found in bacteria but not in the host.

Clinical Que

Antimicrobial drugs may be natural compounds (such as *Penicillium* mold that makes the antibiotic penicillin); semisynthetic products (such as rifampin), which have been chemically modified to improve efficacy or reduce side effects; or synthetic products (such as the sulfa antibiotics), which are completely processed.

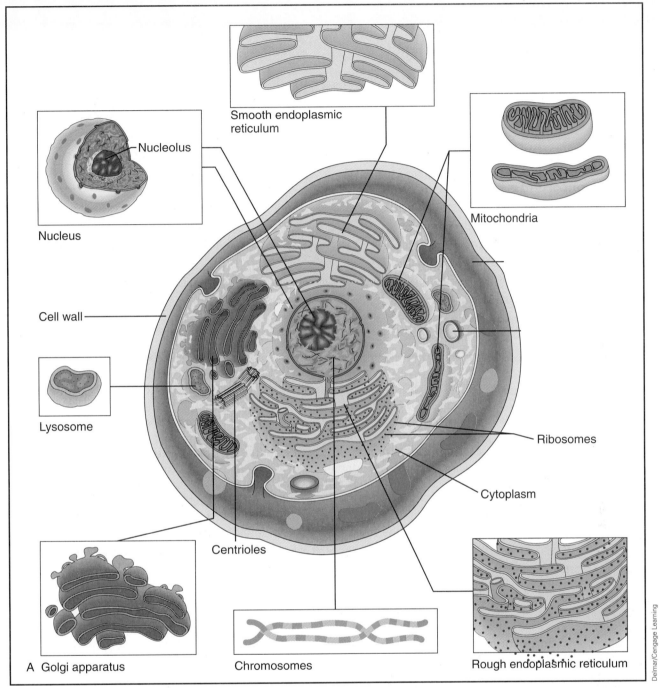

Smooth endoplasmic reticulum

Nucleolus

Nucleus

Mitochondria

Cell wall

Lysosome

Ribosomes

Cytoplasm

Centrioles

A Golgi apparatus

Chromosomes

Rough endoplasmic reticulum

Delmar/Cengage Learning

Figure 14-1 Parts of a eukaryotic versus prokarotic cell. (A) Animal cells are one type of eukaryotic cell. Parts of the eukaryotic cell include the cell or plasma membrane (serves as the cell's boundary and is semipermeable to allow some things in and some things out of the cell), the nucleus (controls cellular activity and contains genetic material of the cell), nucleolus (produces RNA that forms ribosomes), cytoplasm (semifluid medium containing organelles), mitochondria (energy producers of the cell), Golgi apparatus (chemical processor of the cell), endoplasmic reticulum (collection of folded membranes) that may contain ribosomes (known as rough endoplasmic reticulum that is the synthesizer of protein) or may be void of ribosomes (known as smooth endoplasmic reticulum that is the synthesizer of lipids, and some carbohydrates), ribosomes (site of protein synthesis), vacuoles (small, membrane-bound organelles containing water, food, or metabolic waste), and lysosomes (digestive system of the cell).

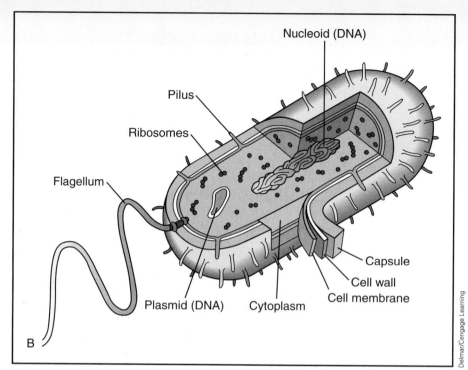

Figure 14-1 (B) Bacteria are prokaryotic cells. Parts of the prokaryotic cell include the cell wall (outermost structure that maintains the overall cell shape), the cell or plasma membrane (serves as the cell's boundary and is semipermeable to allow some things in and some things out of the cell), the nucleoid (region where DNA is generally found though not surrounded by a membrane), cytoplasm (semifluid medium containing organelles), and ribosomes (site of protein synthesis that are smaller than those found in eukaryotic cells). Some bacteria have capsules (protective layer outermost to the cell wall that serves as a barrier against phagocytosis), pili (hollow, hairlike protein extensions that allows the bacterium to attach to other cells or communicate genetic information between cells), plasmids (DNA located outside of the chromosome that can replicate independently and can transfer this information to other cells through pili), and a flagellum (a long appendage that allows mobility). A few bacteria have the ability to produce endospores (resistant structures that allow select bacteria to become dormant during environmental adversity and germinate once environmental conditions improve).

decolorized by the Gram stain process (Figure 14-2). Some of the gram-positive infectious bacteria include *Staphylococcus* sp. and *Streptococcus* sp.; some of the gram-negative types include *Salmonella* sp. and *Proteus* sp. Drugs that act specifically on the gram-positive family or specifically on the gram-negative family of bacteria are referred to as **narrow-spectrum antibiotics**. Those that act on both gram-positive and gram-negative bacteria are called **broad-spectrum antibiotics**.

Antibiotics can also be classified as bactericidal or bacteriostatic. **Bactericidal** means that an agent is capable of killing bacteria, whereas **bacteriostatic** means that an agent is capable of inhibiting the growth or replication of bacteria. Some bactericidal drugs kill bacteria by damaging key bacterial structures during development; therefore, they are effective against actively dividing bacteria. Other bactericidal drugs disrupt cell membranes or protein synthesis of bacteria and cause bacterial death in existing and multiplying bacteria. Bacteriostatic drugs work by preventing the division of bacteria.

Clinical Que

Narrow-spectrum antibiotics act on *either* gram-positive or gram-negative bacteria. Broad-spectrum antibiotics act on *both* gram-positive and gram-negative bacteria (this does not mean that they work on all gram-positive and all gram-negative bacteria).

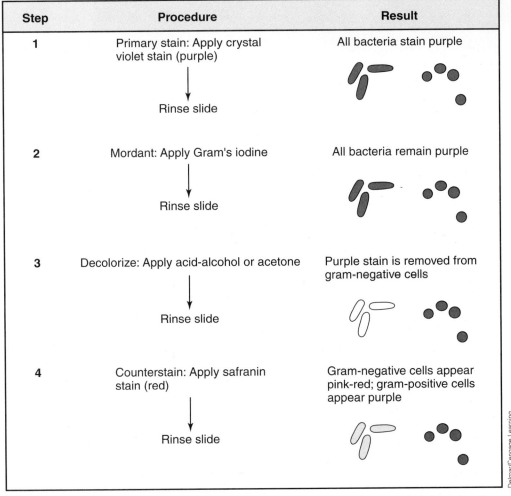

Step	Procedure	Result
1	Primary stain: Apply crystal violet stain (purple) ↓ Rinse slide	All bacteria stain purple
2	Mordant: Apply Gram's iodine ↓ Rinse slide	All bacteria remain purple
3	Decolorize: Apply acid-alcohol or acetone ↓ Rinse slide	Purple stain is removed from gram-negative cells
4	Counterstain: Apply safranin stain (red) ↓ Rinse slide	Gram-negative cells appear pink-red; gram-positive cells appear purple

Delmar/Cengage Learning

Figure 14-2 Gram stain procedure: 1. Smear is covered with crystal violet; the dye is removed. 2. Smear is covered with iodine; the iodine is removed. 3. Smear is washed with a decolorizer; the decolorizer is washed off. 4. Safranin is applied on the smear; the stain is removed.

The goal of antibiotic treatment is to render the bacteria helpless (either by killing them or inhibiting their replication) and not to hurt the animal being treated. This is accomplished by making sure that infecting bacteria are susceptible to the antibiotic, that the antibiotic reaches the infection site, and that the animal can tolerate the antibiotic. One method to determine if a particular antibiotic is effective against a particular bacterium is the *agar diffusion test* (also known as Kirby-Bauer *antibiotic sensitivity testing*). The agar diffusion test uses a variety of antibiotic-impregnated disks that are placed onto agar plates containing the bacterium being tested. The test uses standard media, standard bacterial counts, and impregnated disks with known amounts of antibiotic. After incubation at the proper temperature for the proper time, zones of inhibition or clear zones where the bacterium has failed to grow are measured (Figure 14-3). These measurements are in millimeters and are compared with a standardized chart to determine R (R = **resistant**, meaning the antibiotic does not work against these bacteria), I (I = **intermediate**, meaning

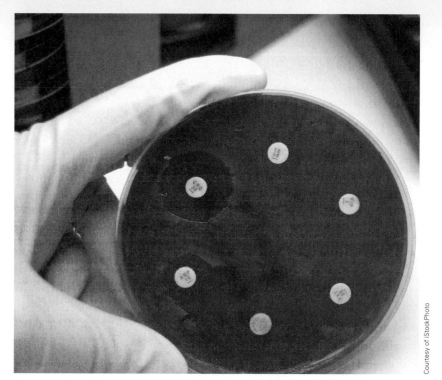

Figure 14-3 Zones of inhibition on an agar diffusion sensitivity test plate

the antibiotic may work against these bacteria), or S (S = **sensitive**, meaning the antibiotic does work against these bacteria). Because the antibiotics diffuse out into the agar at a rate unique to each antibiotic, each antibiotic has its own zone-of-inhibition reference values to determine resistance and sensitivity.

Another method to determine bacterial susceptibility is the *broth dilution method*. In this test, a standard dilution of bacteria is placed into a series of tubes or wells containing different concentrations of a particular antibiotic. The lowest concentration of that particular antibiotic that visually inhibits the growth of bacteria is the **minimum inhibitory concentration** or **MIC** (Figure 14-4). An antibiotic's MIC varies with the bacterial species.

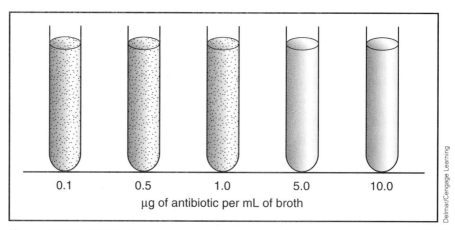

Figure 14-4 MIC is the lowest level of antibiotic that will at least inhibit the growth of bacteria. In this example, the MIC is 5 µg of antibiotic/mL.

In addition to making sure that bacteria are sensitive to the antibiotic being used and using the proper dose of antibiotic, the veterinarian must make sure that the antibiotic gets to the infection site in high-enough concentrations. An antibiotic that is effective against *E. coli* is of no help to an animal with an *E. coli* urinary tract infection if the antibiotic does not reach adequate levels in the urinary tract.

How Antibiotics Work

Antibiotics can work by a variety of mechanisms (Table 14-1 and Figure 14-5). Some of these mechanisms include the following:

- inhibition of cell wall synthesis. Bacterial cells have cell walls, while animal cells do not; therefore, this is an effective way to destroy bacteria without interfering with the host. An intact bacterial cell wall keeps the bacterium from filling with water and bursting. Disruption of cell wall synthesis can occur only when bacteria are growing and dividing. Already developed bacteria would not be affected by this mechanism.

- damage to the cell membrane. Damaging the bacterial cell membrane alters the membrane permeability of the bacterium. This allows substances to enter or leave the cell when they are not supposed to, resulting in cell death.

- inhibition of protein synthesis. Protein synthesis in bacteria is necessary to maintain the cell and allow for cell division. Protein synthesis occurs at the ribosome, and bacteria make protein by sending an RNA copy of the DNA for a specific protein to these ribosomes. Transfer RNA takes different amino acids to the ribosome where they are sequenced. If the amino acids are not linked properly, normal protein production in the bacterium is disrupted.

- interference with metabolism. Some antibiotics block the action of enzymes or the use of essential nutrients by bacteria, leading to an inability to divide and eventually to cell death.

- impairment of nucleic acid production. Some antibiotics interfere with the production of nucleic acids, so that the bacterium cannot divide or function properly.

Additional Considerations in Antibiotic Use

The use of antibiotics has also brought two concerns to the veterinary community: **antibiotic resistance** and **antibiotic residues**. When bacteria are sensitive to an antibiotic, the bacteria are inhibited or destroyed. If bacteria are resistant to an antibiotic, the bacteria survive and continue to multiply, despite administration of that antibiotic. In this instance, antibiotic use promotes the development of antibiotic-resistant bacteria. *Antibiotic resistance* occurs when bacteria change in some way that reduces or eliminates the effectiveness of the agent used to treat or prevent the infection. Bacteria can become resistant to a particular antibiotic when the drug is not used properly (e.g., when

Table 14-1 Mechanisms of Action of Antibiotic Drugs

ACTION	EFFECT	DRUG EXAMPLES
Inhibition of cell wall synthesis	Bactericidal effect by inhibition of an enzyme in the synthesis of the cell wall	penicillins, cephalosporins, bacitracin, vancomycin, carbapenems, monobactams, aminocoumarins
Alteration in cell membrane permeability	Bacteriostatic or bactericidal effect; as the membrane permeability is increased, the loss of cellular substances causes cell lysis	polymyxin B
Inhibition of protein interference	Bacteriostatic or bactericidal effects due to interference with bacterial protein synthesis, but not animal cell protein synthesis or inhibition of the steps of protein synthesis	aminoglycosides, tetracyclines, chloramphenicol, florfenicol, macrolides, lincomycins, aminocoumarins, diterpines
Interference with metabolism	Bacteriostatic effects due to the deprivation of essential material for bacterial metabolism	sulfonamides
Nucleic acid impairment	Bactericidal effects due to inhibition of nucleic acid enzymes	quinolones, fluoroquinolones

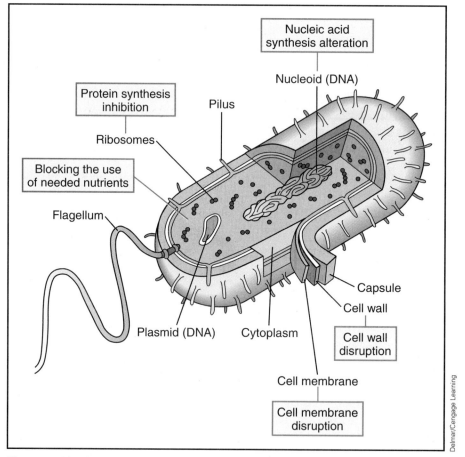

Figure 14-5 Antibiotics can affect bacteria by disrupting their cell wall, damaging their cell membrane, inhibiting protein synthesis, blocking the use of essential nutrients, or interfering with nucleic acid synthesis.

antibiotics are used in an animal with a viral infection) or when the drug is not administered for the proper length of time or at the proper dosage.

One way bacteria can become resistant is through mutation. Even a single random gene mutation can have a large impact on the bacterium's pathogenicity, and since most bacteria replicate very quickly, they can evolve rapidly. Therefore, a mutation that helps a bacterium survive in the presence of an antibiotic will quickly become predominant throughout the bacterial population. For example, some bacteria that were once sensitive to penicillin now produce an enzyme called beta-lactamase (sometimes called penicillinase), which inactivates penicillin before it can affect bacteria.

Another way that bacteria become resistant is by acquiring genes that code for resistance. Such genes can be transferred from members of the same bacterial species or from unrelated bacteria through plasmid transfer (transfer of DNA material from one resistant bacterium to a nonresistant bacterium). Again, the antibiotic-resistant bacterium survives in the presence of the antibiotic and quickly becomes predominant throughout the bacterial population.

Antibiotic resistance forces veterinarians to increase the dose and/or the duration of a course of a particular antibiotic or to abandon use of a particular antibiotic altogether. Cross-resistance can also occur between antibiotics that have similar actions, for example, between penicillins and cephalosporins. Clinics also serve as a fertile environment for antibiotic-resistant bacteria. Close contact among sick patients can lead to the survival of antibiotic-resistant bacteria; therefore, close contact among hospitalized patients should be minimized. Proper handwashing techniques, including interlacing fingers and vigorous rubbing, will also decrease the indirect transfer of potentially resistant microorganisms from patient to patient. To determine if an antibiotic has an effect on a specific bacterium, it is best to perform a culture and sensitivity test (see the preceding description of antibiotic sensitivity testing). Other ways to help prevent the development of antibiotic-resistant bacteria are to avoid administering antibiotics for viral infections, to avoid saving antibiotics used to treat one type of infection in an animal and using it again on another type of infection, to administer antibiotics at the prescribed dose for the prescribed time, and to avoid using one animal's prescription for another animal.

A *residue* is the presence of a chemical or its metabolites in animal tissues or food products. Antibiotic residues in food-producing animals are of great concern because even low levels of these residues can cause allergic reactions in people or can produce resistant bacteria that can be transferred to the people who consume these products. Cooking or pasteurization does not usually degrade antibiotic residues. The FDA approves all drugs marketed for use in animals in the United States and establishes tolerances for drug residues to insure food safety. The FDA also establishes "withdrawal times" or "withholding periods," which are times after drug treatment when milk and eggs are not to be used for food, and during which time animals are not to be slaughtered for their meat. Withdrawal times allow sufficient time for the animals to eliminate the drug residues. Withdrawal times for antibiotics are aimed at

Clinical Que

Natural or intrinsic resistance occurs when an antimicrobial acts on specific enzyme systems or biological processes that are not used by a particular microorganism; hence, that microorganism is resistant to that agent. Acquired resistance occurs when microorganisms that were once sensitive to the effects of a particular drug are no longer altered by this drug. Microorganisms can acquire resistance through mutation or plasmid transfer of resistance coding.

eliminating antibiotic residues in food-producing animals and vary for each drug. Current label information including withdrawal times of all drugs approved for use in food-producing animals in the United States is maintained by the Food Animal Residue Avoidance Databank (FARAD) (see Chapter 4). It also maintains a list of drugs prohibited for use in food-producing animals. The FARAD is administered through the Department of Agriculture, and information about FARAD can be obtained on their Web site (www.farad.org) or by phone (1-888-873-2723).

CLASSES OF ANTIBIOTICS

Antibiotics are often categorized by their mode of action such as cell wall agents, cell membrane agents, protein synthesis agents, antimetabolites, nucleic acid agents, and others.

Cell Wall Agents

Many bacteria have cell walls, while animal cells do not. Finding drugs that are active against these structures is thus advantageous, because they will attack the bacterium and not the host cell. Examples of antibiotics that affect the cell wall are penicillins, cephalosporins, bacitracin, vancomycin, carbapenems, and monobactams.

Penicillins

Penicillin was discovered by Scottish scientist Sir Alexander Fleming in 1928 and began to be commercially available in 1941. Today the penicillins consist of a group of natural and semisynthetic agents and come in a variety of forms that may be active against gram-positive bacteria only or gram-positive and gram-negative cocci (spheres) and bacilli (rods). Penicillin may be either the natural product of mold (such as *Penicillium notatum*) or a semisynthetic derivative of the *Penicillium* molds. Penicillins can be identified by the *-cillin* suffix in their generic names.

The penicillins have a beta-lactam structure that interferes with bacterial cell wall synthesis among newly formed bacterial cells. Because the newly formed bacterial cells are unable to develop rigid cell walls when affected by penicillin, the rapidly developing cells die as a result of an increase of fluid in the cells (bactericidal action).

There are four groups of penicillins: natural penicillins, broad-spectrum penicillins, beta-lactamase-resistant penicillins, and potentiated penicillins.

Natural Penicillins

Penicillin G is a natural penicillin and the penicillin most commonly used in veterinary practice. It is a narrow-spectrum antibiotic used against *Staphylococcus* sp., *Streptococcus* sp., and some gram-positive bacilli. Penicillins are distributed extensively throughout the body to most body fluids and bone except the brain and cerebrospinal fluid unless inflamed. Penicillins are excreted unchanged in the urine, making them good choices for urinary tract infections for bacteria sensitive to penicillin. Penicillin G is inactivated by stomach acid and thus is only used parenterally. Penicillin G is available as a sodium or potassium salt, both of which are rapidly absorbed after IM injection and provide peak levels usually within 20 minutes following injection. The sodium or potassium salt of penicillin G should be the only form administered IV. To produce prolonged blood levels of the antibiotic, veterinarians often inject procaine penicillin G or benzathine penicillin G, two penicillin suspensions. Procaine and benzathine extend the duration of penicillin activity by slowing its absorption from IM sites. Procaine penicillin G usually has a 24-hour duration; benzathine penicillin G usually has a 5-day duration. A combination product, penicillin G benzathine with penicillin G procaine, is also available, which has an extended slaughter withdrawal time in cattle (30 days). Side effects of penicillins include anorexia, vomiting and diarrhea when given orally, and hypersensitivity reactions (especially when given parenterally).

Penicillin V is another natural and narrow-spectrum penicillin that comes in oral tablet and powder forms. Penicillin V is usually commercially available as a potassium salt. Penicillin V potassium is better absorbed from the gastrointestinal tract and is relatively stable in stomach acid; therefore, it is the preferred penicillin for oral administration.

Broad-Spectrum Penicillins

Some other members of the penicillin family are broader-spectrum (have an extended spectrum of activity) than the natural penicillins and have activity against some gram-negative bacteria. These broader-spectrum penicillins are semisynthetic and are the result of chemical treatment of a biological precursor to penicillin allowing them to be more slowly excreted by the kidneys. Broad-spectrum penicillins include drugs such as *amoxicillin, ampicillin, carbenicillin, ticarcillin,* and *piperacillin*. All the penicillins kill susceptible bacteria by interfering with cell wall synthesis.

Beta-Lactamase-Resistant Penicillins

One concern that arises with the use of penicillin and its derivatives is the ability of bacteria to produce beta-lactamase. Beta-lactamase, or penicillinase, is an enzyme that destroys the beta-lactam structure found in penicillin (Figure 14-6). If the beta-lactam ring is destroyed, the penicillin is useless.

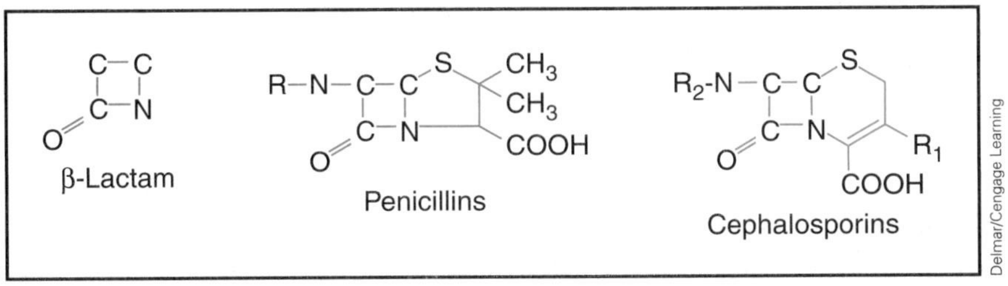

Clinical Que

If refrigerated, reconstituted penicillin sodium and penicillin potassium products typically last 14 days. Any drug left after 14 days should be discarded. Other penicillin products may have a shorter shelf life depending upon their chemical properties and storage conditions; always follow the manufacturer's guidelines.

Figure 14-6 Beta-lactam ring and beta-lactam ring antibiotics

Some forms of penicillin are more resistant to beta-lactamase action than other penicillins. These penicillins are referred to as beta-lactamase-resistant penicillins; examples are *oxacillin*, *dicloxacillin*, and *cloxacillin*. Most penicillins in this group are narrow spectrum.

Potentiated Penicillins

A **potentiated** drug is chemically combined with another drug to enhance the effects of both. The susceptibility of amoxicillin to destruction by beta-lactamase led to the development of a potentiated drug containing amoxicillin and clavulanic acid. Clavulanic acid is a beta-lactam inhibitor that has a beta-lactam ring and competitively binds to beta-lactamase. This binding protects the beta-lactam ring of amoxicillin against beta-lactamase destruction. *Amoxicillin/ clavulanic acid* combinations are Clavamox® for animals and Augmentin® for humans. Other beta-lactam inhibitors include *sulbactam* and *tazobactam*.

Cephalosporins

Cephalosporins are semisynthetic, broad-spectrum antibiotics first introduced in the 1960s and are produced from chemical manipulation of cephalosporin C (a fungus derivative). Cephalosporins are structurally and pharmacologically related to the penicillins. Like the penicillins, they have a beta-lactam ring and also interfere with cell wall synthesis killing susceptible bacteria quickly. Cephalosporins are typically used in veterinary medicine to treat respiratory, skeletal, genitourinary, skin, and soft tissue infections caused by susceptible bacteria. The cephalosporins are also used throughout the perioperative period (preoperative, intraoperative, and postoperative periods) in patients having surgery on a contaminated or potentially contaminated area. Side effects include vomiting and diarrhea following oral administration (which may be avoided when given with food) and phlebitis and myositis after IV or IM injection, respectively. All cephalosporins are potentially nephrotoxic and need to be used with caution if administering concurrently with other potentially nephrotoxic drugs.

Cephalosporins are classified into four generations: first, second, third, and fourth. First-generation cephalosporins tend to have the greatest bacteriostatic or bactericidal activity against gram-positive and some gram-negative bacteria. Generally, they can be inactivated by the beta-lactamase produced by some bacteria. Second-generation cephalosporins have a broader spectrum of activity against gram-negative bacteria, and slightly less activity against gram-positive bacteria, than first-generation cephalosporins. First-and second-generation cephalosporins do not readily cross the blood-brain barrier. Third-generation cephalosporins have an even broader spectrum of activity against gram-negative organisms and are resistant to the beta-lactamase produced by some bacteria. Fourth-generation cephalosporins have an expanded spectrum activity similar to that of the third-generation cephalosporins; however, fourth-generation cephalosporins are active against *Pseudomonas aeruginosa* and certain Enterobacteriaceae (gram-negative rods of the intestinal tract) that are resistant to third-generation cephalosporins. Fourth-generation cephalosporins may be more active against some gram-positive bacteria than some third-generation cephalosporins. Cephalosporins can be recognized by

Clinical Que

Oral penicillin (except amoxicillin) should be given on an empty stomach, as food affects its bioavailability.

Clinical Que

Penicillin use can cause fatal diarrhea in guinea pigs, rabbits, and hamsters because these species have normal flora that consists largely of gram-positive bacteria. Penicillin kills the normal flora of these species, allowing harmful bacteria to populate the gastrointestinal tract.

Clinical Que

Remember that antibiotics given to food-producing animals have withdrawal times that vary with the species, the antibiotic, and the food use (milk or meat). Check the package insert for the most current recommendations.

the "ceph" spelling in older first-generation cephalosporins and the "cef" in newer first-generation and later-generation cephalosporins.

Veterinarians use cephalosporins when first-line antibiotics, such as penicillins, are not effective. Animals allergic to penicillins will probably also be allergic to cephalosporins. Some cephalosporin drugs are longer-acting than others such as *cefpodoxime* (which may be dosed once daily) and *cefovecin* (which may provide up to 14 days of treatment).

Bacitracin

Bacitracin disrupts the bacterial cell wall like penicillin and cephalosporin, but not via the same mechanism. Bacitracin is a polypeptide antibiotic. Polypeptide antibiotics are chemically composed of long chains of amino acids called *polypeptides.* Bacitracin does not have a beta-lactam ring; therefore, it is not susceptible to destruction by beta-lactamase.

Bacitracin works primarily against gram-positive bacteria. Bacitracin is toxic to the kidneys and is most popular as a topical medication for skin, mucous membranes, and ocular surfaces. It is also used as a feed additive to control intestinal pathogens, because it is poorly absorbed from the gastrointestinal tract.

Vancomycin

Vancomycin is a glycopeptide bactericidal antibiotic that is effective against many gram-positive bacteria. It is used primarily for treatment of infections that are resistant to the toxic antibiotics such as the penicillins or cephalosporins. It is especially useful in the treatment of drug-resistant *Staphylococcus aureus* and *Clostridium difficile* in humans. It is rarely used in nonhumans to avoid promoting resistance development. Side effects of vancomycin include ototoxicity, nephrotoxicity, and pain on IV injection.

Carbapenems

The carbapenems are beta-lactam antibiotics that inhibit synthesis of the bacterial cell wall and are therefore bactericidal. Examples include *imipenem-cilastatin* and *meropenem.* Imipenem is administered with cilastatin to decrease renal tubular metabolism. Cilastatin does not affect the antibacterial activity. Imipenem-cilastatin has the widest antibacterial spectrum of any beta-lactam antibiotic whose spectrum includes gram-positive, gram-negative, and anaerobic bacteria. This combination drug is also used to treat bacteria resistant to cephalosporins, penicillins, and aminoglycosides. Meropenem has antibacterial activity equal to, or greater than, imipenem. Its advantage over imipenem is that it is more soluble and can be administered in less fluid volume more rapidly. Meropenem can be administered subcutaneously with almost complete absorption. Disadvantages of carbapenems include inconvenient administration (injectable forms only) and high cost. Side effects of the carbapenems include gastrointestinal upset, pain on injection, hypotension, and induction of seizures.

Monobactams

The monobactam antibiotics have a beta-lactam structure that inhibits bacterial cell wall synthesis and are bactericidal. This group of antibiotics is used

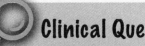

Clinical Que

As the generation of cephalosporin increases, the spectrum of activity and potential side effects also increase.

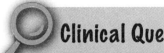

Clinical Que

Cephalosporins can be identified by the ceph- or cef-prefix in their generic and/or brand names.

to treat gram-negative bacteria (it is not effective for gram-positive or anaerobic bacteria), has good penetration into most tissues, and has low toxicity. An example of a monobactam antibiotic is *aztreonam*, which is reserved for treating serious infections that are caused by bacteria not sensitive to aminoglycosides or fluoroquinolones or their use is contraindicated. Aztreonam is available for IM or IV use. Due to the limited use of monobactams, there is not a complete profile of their side effects; however, gastrointestinal upset, pain and/or swelling following IM injection, and phlebitis after IV injection have been documented in people.

Cell Membrane Agents

Attacking a bacterial cell membrane (also known as the plasma membrane) is a more challenging problem than attacking a cell wall, because both bacterial and animal cells have cell membranes. This group of agents can thus cause more toxicity problems in animals. This group of antibiotics includes the *polymyxins*, which act on the phospholipid bilayer of the cell membrane to alter permeability. Polymyxin is also classified as a polypeptide antibiotic, like bacitracin.

Polymyxin B is effective against gram-negative bacteria only. Polymyxin B is not absorbed when taken orally or applied topically. It is most often applied topically as an ointment or a wet dressing for local treatment. Polymyxin B is often combined with neomycin and bacitracin to create a wide-spectrum topical medication.

Protein Synthesis Agents

Within a bacterial cell, information must be transmitted from DNA to the operational parts of the cell. This process starts with the transfer of information from DNA to messenger RNA. This process is followed by translation, which stimulates the formation of 30S and 50S ribosomal units (these ribosomal units are different in animal cells). Agents that interfere with this process, especially the formation of the 30S and 50S ribosomal units, can affect the production of protein in bacterial cells.

Aminoglycosides

The aminoglycosides include any of a group of bactericidal antibiotics derived from various species of *Streptomyces* microbes. Aminoglycosides are specialized antibiotics that work by inhibiting protein synthesis, thereby promoting death of affected bacteria. Veterinarians use them to treat infections caused by gram-negative bacteria. Because they stay within the GI tract when administered orally, veterinarians inject aminoglycosides to treat systemic infections. Aminoglycosides, identified by the *-micin* or *-mycin* endings in their generic names, include *gentamicin, amikacin, dihydrostreptomycin, kanamycin, streptomycin, tobramycin,* and *neomycin.* Aminoglycosides derived from *Streptomyces* bacteria are named with the *-mycin* suffix and those derived from the *Micromonospora* genus are named with the *-micin* suffix (although some drugs such as vancomycin and erythromycin have the *-mycin* suffix and are not aminoglycosides). Veterinarians generally inject these antibiotics for short-term treatment of gram-negative bacterial septicemias (bacteria and their toxins released into the blood) and various skin, soft-tissue, bone, joint, respiratory, and postoperative infections. Topical preparations containing aminoglycosides are also available to treat ocular and otic infections. Aminoglycosides can be extremely nephrotoxic; so monitoring of kidney function via urine for the appearance of casts and serum creatinine levels is recommended. Ototoxicity is another side effect of aminoglycosides. Veterinarians avoid injecting the most toxic aminoglycoside, neomycin, reserving it for topical treatment of localized skin infections caused by gram-negative bacteria. Aminoglycosides are not approved for use in food-producing animals.

Tetracyclines

Discovered in 1947 in bacterial soil samples, tetracyclines are broad-spectrum, bacteriostatic antibiotics effective against many gram-positive and gram-negative bacteria (including *Mycoplasma* and *Chlamydia*) and rickettsial agents (including *Mycoplasma haemofelis [formerly known as Haemobartonella]* and *Ehrlichia*). Veterinarians also use tetracyclines to treat Lyme disease (caused by the spirochete bacterium *Borrelia burgdorferi*), leptospirosis (caused by the spirochete *Leptospira* sp.), and toxoplasmosis in cats (caused by the protozoan *Toxoplasma gondii*). Tetracyclines work by inhibiting protein synthesis and are available in oral and parenteral forms. Tetracycline drugs can be classified as short-acting, water soluble (*tetracycline, oxytetracycline, chlortetracycline*), intermediate-acting (*demeclocycline*), and long-acting, lipid soluble (*minocycline, doxycycline*).

Tetracyclines can be identified by the *-cycline* ending in their generic names. Tetracyclines can bind to calcium and can be deposited in growing bones (slowing bone development) and teeth (causing yellow discoloration of the teeth). Therefore, tetracycline should not be given to growing or pregnant animals or with dairy products. Tetracyclines can also bind to other minerals, such as magnesium, iron, and copper; therefore, they should not be given with iron supplements, antacids, or products containing kaolin, pectin, or bismuth. Side effects of tetracyclines include nausea, vomiting, diarrhea, and renal damage when given orally, and anaphylaxis, hypotension, and shock when given parenterally.

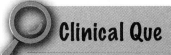

Clinical Que

Tetracycline should not be given to pregnant or growing animals. Tetracycline injectable should be used with caution to prevent cardiac problems related to its calcium-binding properties, which may affect muscle contraction. Injectable tetracycline products are painful when given IM.

Chloramphenicol

Chloramphenicol is a bacteriostatic antibiotic that was originally derived from the bacterium *Streptomyces venezuelae* and introduced commercially in 1949. It is a broad-spectrum antibiotic that works on a wide variety of microorganisms by inhibiting bacterial protein synthesis. Chloramphenicol penetrates tissues and fluids well, including the eye and CNS. This drug is not considered first-line in developed countries because of its toxicity. A *first-line drug* is the veterinarian's first drug of choice. Drugs not considered first-line are those that have toxic side effects or are considered for use only when the first-line choice cannot be used or is not effective. These drugs are called *second-line drugs*.

Veterinarians suspect that chloramphenicol suppresses blood cell formation by depressing bone marrow function. Because of its potential for bone marrow suppression, chloramphenicol is banned from use in food-producing animals. When using or dispensing chloramphenicol, the veterinary staff and owners should wear gloves for protection and avoid inhalation of chloramphenicol powder. Side effects of chloramphenicol include bone marrow suppression with long-term use, anorexia, vomiting, and diarrhea.

Florfenicol

Florfenicol (Nuflor®) is a synthetic, broad-spectrum antibiotic that works by inhibiting bacterial protein synthesis. Florfenicol is available as an injectable solution and is approved for treatment of bovine respiratory disease. Intramuscular injections may result in local tissue reaction and may result in loss of edible tissue at slaughter; therefore, it is recommended to give the injection near the neck area. Withdrawal times vary with the route of administration with IM withdrawal times less than SQ withdrawal times. Other side effects include inappetence, decreased water consumption, and diarrhea. Florfenicol is not approved for use in breeding-age cattle.

Macrolides

Macrolides are antibiotics with large molecular structure and many-membered rings. Macrolides work by inhibiting protein synthesis and can cause stomach upset in animals. One example of a macrolide is *erythromycin*, a broad-spectrum antibiotic produced by a strain of the bacterium *Streptomyces erythreus*. Erythromycin has primary activity against gram-positive bacteria and may be bactericidal or bacteriostatic depending on the concentration of drug given. Erythromycin is used to treat penicillin-resistant infections or in animals that have allergic reactions to penicillin. Erythromycin comes in oral and ointment forms. Side effects of erythromycin include vomiting, abdominal pain, anorexia, and diarrhea.

Tylosin is another macrolide, structurally similar to erythromycin. It is used mainly in livestock for a variety of diseases including those caused by gram-positive and gram-negative bacteria, spirochetes, chlamydiae, and mycoplasma bacteria. It comes in oral and injectable forms. Tylosin may also be found in feed or water additives (these products usually have other ingredients in the formulations as well). It is not recommended for use in horses, as reports of fatal diarrheas have been documented. If an animal has an adverse reaction to tylosin, it will also react to the other macrolide antibiotics.

Tilmicosin, another macrolide, is used to treat bovine and ovine respiratory disease especially those caused by *Mannheimia haemolytica* (formerly known as *Pasteurella haemolytica*). Tilmicosin is only approved for use in cattle and sheep. It is administered subcutaneously and at lower volumes than other antibiotics, so as to cause less muscle damage in meat-producing animals. An advantage to tilmicosin use is that it is administered in a single injection, which remains in the body for approximately three days. Single-injection therapy means less stress on the animal and less labor for the owner. Side effects include swelling at the injection site and tachycardia (increased heart rate). Tilmicosin is fatal if administered IV to cattle and sheep; injectable use of tilmicosin in horses, goats, swine, and primates has been fatal. Extreme care must be used when administering tilmicosin because injection of this antibiotic into people may cause lethal cardiovascular toxicity.

Lincosamides

Lincomycin, and its derivatives *clindamycin* and *pirlimycin*, are lincosamide antibiotics produced from a different strain of *Streptomyces* than the macrolides or aminoglycosides. They work primarily by inhibiting protein synthesis and work against gram-positive aerobic bacteria. Veterinarians typically use erythromycin instead of these antibiotics because they may cause serious gastrointestinal problems. Lincomycin is effective in the treatment of upper respiratory infections, skin infections, nephritis, abscesses, and metritis caused by susceptible microorganisms. Clindamycin has good efficacy against many pathogenic anaerobes and has a greater spectrum of activity than lincomycin. Additional side effects of clindamycin include local pain following injection and the need for cautious use in animals with renal disease.

Aminocoumarins

Novobiocin is an aminocoumarin antibiotic produced from *Streptomyces niveus* that works by a variety of mechanisms including inhibiting protein and nucleic acid synthesis and interferes with bacterial cell wall synthesis. Novobiocin is effective against some gram-positive cocci and is used in dry dairy cattle as a mastitis tube and in dogs as a combination oral product containing tetracycline and prednisolone. Side effects include fever, gastrointestinal disturbances, rashes, and blood abnormalities.

Diterpines

Tiamulin is a diterpine antibiotic that is used in swine to treat pneumonia and as a feed additive to enhance weight gain. It is mainly effective against gram-positive bacteria with the exception of some gram-negative bacteria such as *Hemophilus* sp., *E. coli*, and *Klebsiella* sp. Tiamulin is available as a feed additive, oral powder, and injection. Side effects are unlikely but include causing redness of the skin.

Antimetabolites

Another approach to eradicating bacteria is to deprive them of essential material needed for metabolism. These agents are referred to as antimetabolites.

Sulfonamides

The sulfonamides are a group of drugs that inhibit the synthesis of folic acid, an action that hinders the growth of a wide variety of bacteria. They were the first antibacterial drugs to be used clinically in large patient populations, about seven years before the advent of penicillin. Because they are synthesized in the chemical laboratory, sulfonamides are not considered true antibiotics, but rather synthetic antimicrobials. Some sulfonamides may be designed to stay in the gastrointestinal tract (enteric forms) and are used to treat coccidia protozoal infections. Other sulfonamides are in a form that is absorbed by the gastrointestinal tract and can penetrate tissues including cerebrospinal fluid (systemic forms). Sulfonamides are ineffective in the presence of pus and necrotic tissue. Sulfonamides have a fairly broad spectrum of activity and can treat coccidia and *Toxoplasma* organisms. Sulfonamides can produce such side effects as crystalluria (formation of tiny, sharp-edged crystals in the tubules of the nephron, which can damage the kidneys), keratoconjunctivitis sicca (dry eye), skin rashes, and blood imbalances; so veterinarians need to monitor these animals for possible side effects.

Sulfonamides may be combined with *trimethoprim* and *ormetoprim* to increase their antibacterial effects (these are referred to as *potentiated sulfonamides*). Trimethoprim and ormetoprim inhibit a different step of folic acid synthesis from the sulfonamides. Sulfonamides are bacteriostatic; trimethoprim and ormetoprim are bactericidal; however, if they penetrate tissue sites together, they can have a bactericidal effect.

Nucleic Acid Agents

Nucleic acids of bacteria and animal cells are similar; however, some of the enzymes associated with nucleic acids differ between bacterial and animal cells. This makes interference with the bacterial nucleic acids a viable option for treating bacterial infections. A group of drugs that affects bacterial nucleic acids are the quinolones.

Quinolones are synthetic antimicrobials that work by inhibiting DNA function in the bacterium; they do not harm mammalian DNA because of enzyme differences between mammals and bacteria. Quinolones are bactericidal and are effective against both gram-positive and gram-negative bacteria. *Nalidixic acid*, the first member of this family, has been used to treat human urinary tract infections for many years, but is not used in veterinary medicine to any significant extent. *Flumequine* has been used in many countries to control intestinal infections in livestock. The quinolones are indicated in animals for the treatment of local and systemic infections, particularly against deep-seated infections and intracellular pathogens.

Newer generations of quinolones, referred to as the fluoroquinolones, include *enrofloxacin* (Baytril®), *ciprofloxacin* (Cipro®), *orbifloxacin* (Orbax®), *difloxacin* (Dicural®), *marbofloxacin* (Zeniquin®), and *sarafloxacin* (SaraFlox®). Fluoroquinolones have fluorine bound to the quinolone base structure, which increases the drug's potency and spectrum of activity as well as improves its absorption. The fluoroquinolones are effective against gram-negative and gram-positive bacteria and can be recognized by the *-floxacin* ending in their generic names. Generally, fluoroquinolones are very safe drugs, but they can cause bubble-like cartilage lesions in growing dogs, and crystalluria has been seen with ciprofloxacin use in humans. Fluoroquinolones given at high doses can cause quinolone-induced blindness in cats. These drugs are potentially teratogenic and should not be given to pregnant animals. Indiscriminate use of flurouquinolones, for infections that could be treated with other antibiotics, may result in bacterial resistance. These antibiotics tend to be reserved until other antibiotics have been tried, so as to maintain the effectiveness of fluoroquinolones on superinfections. Food, mineral supplements, and antacids impair the absorption of fluoroquiniolones; therefore, they should not be given after eating or when administering these drugs. The FDA prohibits the extra-label use of fluoroquinolones in food-producing animals, and labels should be consulted prior to administration in these animals.

Miscellaneous Antibiotics

Certain antimicrobials do not fit the preceding categories, but do treat specific veterinary diseases effectively.

Nitrofurans

Nitrofurans are bacteriostatic antibiotics that affect bacterial enzymes. These agents have a broad antimicrobial spectrum, but are less potent than traditional antibiotics. Nitrofurans include *furazolidone, nitrofurazone,* and *nitrofurantoin.* Nitrofurantoin is rapidly eliminated from the body, and the animal may not maintain therapeutic levels for long. Nitrofurantoin is used to treat urinary tract infections because it is filtered unchanged through the kidney, canine infectious tracheobronchitis (kennel cough), and is applied topically to treat wound infections. Furazolidone and nitrofurazone are found in topical products for wounds and ocular infections. Nitrofuran use in food-producing animals is prohibited because there is evidence that this drug may induce carcinogenic residues in animal tissues. Topical application of nitrofuran is extra-label use in food-producing animals and is prohibited by the FDA. This order is based on evidence that extra-label use of topical nitrofuran drugs in food-producing animals may result in the presence of residues that are carcinogenic and have not been shown to be safe. A carbon-14 (C-14) radio-label residue depletion study conducted by the FDA showed that detectable levels of nitrofuran derivatives are present in edible tissues (milk, meat, kidney, and liver) of cattle treated by the ocular route. Side effects of nitrofurans include gastrointestinal and liver disturbances.

Nitroimidazoles

The nitroimidazole group of drugs has both antibacterial and antiprotozoal activity. *Metronidazole* is the representative drug of this group. Although there is no approved veterinary form of metronidazole (Flagyl®), it has been used both orally and intravenously for many years to treat *Giardia* and *Trichomonas* (parasitic protozoans), amoebiasis (infection with amoebae), and anaerobic bacteria. Metronidazole is considered by some the drug of choice for canine diarrhea. Metronidazole is believed to work by disrupting DNA and nucleic acid synthesis, and is considered bactericidal. It is recommended that use of this drug be avoided in pregnant animals. It can also cause anorexia, vomiting, diarrhea, and neurologic signs.

Rifamycin Antibiotic

Rifampin is a rifamycin antibiotic that disrupts RNA synthesis. It is a broad-spectrum antibiotic and is used primarily with erythromycin for the treatment of *Rhodococcus equi* (formerly known as *Corynebacterium equi*) infections in foals. It is bactericidal or bacteriostatic depending on the dose. Owners should be warned that rifampin causes a reddish color to urine, tears, sweat, and saliva.

Table 14-2 summarizes the various groups of antibiotics and gives examples of their trade and generic names. Figure 14-7 summarizes the site of action of various groups of antibiotics.

Table 14-2 Classes of Antibiotics and Their Effectiveness

CLASS OF ANTIBIOTIC	ACTION OF ANTIBIOTIC	CONSIDERATIONS	EXAMPLES
penicillins	• Inhibit cell wall synthesis • Bactericidal • Mainly work on gram+ bacteria; some gram– with amoxicillin, ampicillin, ticarcillin, and carbenicillin • Carbenicillin, ticarcillin, and piperacillin are effective against *Pseudomonas* bacteria	• Oral and injectable forms • Given orally, most absorption occurs in stomach and small intestine • Rapidly distributed • Give 1–2 hours before eating	• penicillin V (V-Cillin K®), penicillin G procaine (Crystacillin®), penicillin G benzathine with penicillin G procaine (Dual Pen®) • amoxicillin (Amoxi-tabs®, Amoxi-drops®, Biomox®, Robamox-V®) • ampicillin (Polyflex®, Omnipen®) • ampicillin with sulbactam (Unasyn®) • amoxicillin with clavulinic acid (Clavamox®) • ticarcillin (Ticar®) • carbenicillin (Pyopen®, Geocillin®) • cloxacillin (Dari-Clox®, Orbenin-DC®) • dicloxicillin (Dynapen®, Pathocil®) • oxacillin (Add-Vantage®) • piperacillin (Pipracil®) • piperacillin with tazobactam (Zosyn®) • ticarcillin with clavulanate (Timentin®) • hetacillin (Hetacin-K®) • nafcillin (Nafcil®)
cephalosporins	• Inhibit cell wall synthesis • Bactericidal • First generation mainly work on gram+ bacteria, second through fourth generation work on gram+ and gram– bacteria with fourth generation having the broadest spectrum (including *Pseudomonas*) • Can cross placenta	• Oral and injectable forms • GI absorption not good; usually administered parenterally • Well distributed to tissues, except CNS • Vomiting and diarrhea may occur when given on empty stomach • If animal is allergic to penicillin, it may be allergic to cephalosporin	• First generation: cephapirin (Cefa-Dri®, Cefa-lak®), cefadroxil (Cefa-drops®, Cefa-tabs®), cefazolin (Kefzol®), cephalexin (Keflex®) • Second generation: cefoxitin (Mefoxin®), cefaclor (Ceclor®), cefotetan (Cefotan®), cefuroxime (Ceftin®) • Third generation: ceftiofur (Naxcel®, Spectramast®, Excenel®), cefovecin (Convenia®), cefoperazone (Cefobid®), cefotaxine (Claforan®), cefpodoxine (Simplicef®), ceftriaxone (Rocephin®) • Fourth generation: cefepime (Maxipime®)
polypeptides	• Inhibit either cell wall or cell membrane synthesis • Bactericidal	• Absorption is poor; used for topical infections or wound lavage	• polymyxcin B (found in Optiprime® ophthalmic ointment) • bacitracin (found in Mycitracin® and Trioptic® ophthalmic ointment)
glycopeptides	• Inhibit cell wall synthesis • Bactericidal	• Treats drug-resistant *Staphylococcus* spp • Injectable form used in veterinary medicine	• vancomycin (Vancocin®)

(Continued)

CLASS OF ANTIBIOTIC	ACTION OF ANTIBIOTIC	CONSIDERATIONS	EXAMPLES
carbapenems	• Inhibit cell wall synthesis • Bactericidal	• Injectable form only • Broadest spectrum beta-lactam antibiotic • Used to treat resistant bacteria	• imipenem-cilastatin (Primaxin®) • meropenem (Merrem®)
monobactams	• Inhibit cell wall synthesis • Bactericidal	• Injectable form only • Gram-negative spectrum of activity	• aztreonam (Azactam®)
aminoglycosides	• Inhibit protein synthesis[en] • Bactericidal (concentration dependent) • Work mainly on gram– bacteria • Can cross placenta	• Injectable form only (except neomycin which is topical) • Not absorbed readily from GI tract; usually given parenterally • Nephrotoxicity and ototoxicity concerns • Do not mix with penicillin in the same syringe (makes penicillin inactive)	• gentamicin (Gentocin®, Garacin®) • neomycin (Biosol®, Mycifradin®) • amikacin (Amiglyde-V®, Amikin®) • tobramycin (Nebcin®) • dihydrostreptomycin (Ethamycin®) • spectinomycin (Adspec®, Spectam®)
tetracyclines	• Inhibit protein synthesis • Bacteriostatic • Work on gram+ and gram– bacteria, as well as rickettsial bacteria, spirochetes, and some protozoa	• Oral and injectable forms • Once given, quickly distributed, sometimes to CNS • Very little metabolism • Bind to calcium, causing side effects (do not give with dairy products or antacids/antidiarrheal drugs) • Can cause yellow discoloration of teeth due to calcium binding	• tetracycline (Panmycin Aquadrops®, Oxy-Tet 100® injectable, Tetracycline HCl® soluble powder) • oxytetracycline (Terramycin®, Liquamycin®) • chlortetracycline (Aureomycin®) • doxycycline (Vibramycin®, Doxirobe® Gel) • minocycline (Minocin®) • demeclocycline (Declomycin®)
chloramphenicol	• Inhibits protein synthesis • Bacteriostatic • Works on gram+ and gram– bacteria as well as rickettsial bacteria	• Oral, injectable, and ointment forms • Readily absorbed into tissues • Side effect of bone marrow suppression makes use not recommended	• chloramphenicol (Chloromycetin®, Viceton®, Amphicol®)
florfenicol	• Inhibits protein synthesis • Bacteriostatic	• Injectable form • Well distributed in body; can achieve therapeutic levels in the CNS	• florfenicol (Nuflor®)
macrolides	• Inhibit protein synthesis • Bactericidal or bacteriostatic	• Well distributed to most body tissues, but not the CNS	• tilmicosin (Micotil®) • tylosin (Tylan®) • erythromycin (Erythro-100®, Erythro-Dry®) • azithromycin (Zithromax®) • tulathromycin (Draxxin®)

(Continued)

Table 14-2 (*Continued*)

CLASS OF ANTIBIOTIC	ACTION OF ANTIBIOTIC	CONSIDERATIONS	EXAMPLES
lincosamides	• Inhibit protein synthesis • Bactericidal or bacteriostatic	• Recommended for abscesses and dental infections	• clindamycin (Antirobe®) • pirlimycin (Pirsue®) • lincomycin (Lincocin®)
aminocoumarins	• Inhibit protein and nucleic acid synthesis; also interferes with cell wall synthesis • Bactericidal • Works mainly on gram+ bacteria	• Used as a combination product in dogs • Used as a mastitis treatment in dry dairy cattle	• novobiocin/tetracycline/prednisolone (Delta Albaplex®) • novobiocin/penicillin G procaine (Albadry Plus®) • novobiocin (Biodry®)
diterpines	• Inhibit protein synthesis • Bacteriostatic • Works on gram+ cocci and limited gram– bacteria	• Premix, solution, and powder formulations • Used to treat pneumonia in swine	• tiamulin (Denagard®)
sulfonamides	• Inhibit folic acid synthesis • Sulfonamides are bacteriostatic • Trimethoprim and ormetroprim are bactericidal and used to potentiate sulfas • Potentiated sulfas are bactericidal	• Can have anti-inflammatory effects • Well distributed through the body, including eye and CNS and synovial fluid • Can cause increased salivation in cats	• sulfadiazine/trimethoprim (Tribrissen®) • sulfadimethoxine (Albon®) • sulfadimethoxine/ormetroprim (Primor®) • sulfasalazine (Azulfidine®) • sulfamethoxazole with trimethoprim (Bactrim®)
fluoroquinolones	• Inhibit DNA function • Bactericidal	• Readily absorbed into tissues and body fluids after oral and parenteral administration	• enrofloxacin (Baytril®) • orbifloxacin (Orbax®) • difloxacin (Dicural®) • marbofloxacin (Zeniquin®) • sarafloxacin (SaraFlox®) • ciprofloxacin (Cipro®) • danofloxacin (Advocin®)
nitrofurans	• Inhibit bacterial enzyme systems • Bactericidal	• Eliminated from body quickly; usually used in urinary tract infections • Banned for use in food-producing animals	• nitrofurazone (Furazone®, NFZ Puffer®) • nitrofurantoin (Macrodantin®) • furazolidone (Topazon®, Furox®)
nitroimidazoles	• Disrupt DNA and nucleic acid synthesis • Bactericidal	• Well absorbed after oral administration • Use with caution in pregnant animals	• metronidazole (Flagyl®)
rifampin	• Disrupts RNA synthesis • Bactericidal or bacterioistatic depending on dose	• Relatively well absorbed from GI tract • Can cause red urine, tears, sweat, and saliva • Usually used in combination with other antibiotics	• rifampin (Rifadin®, Rimactane®)

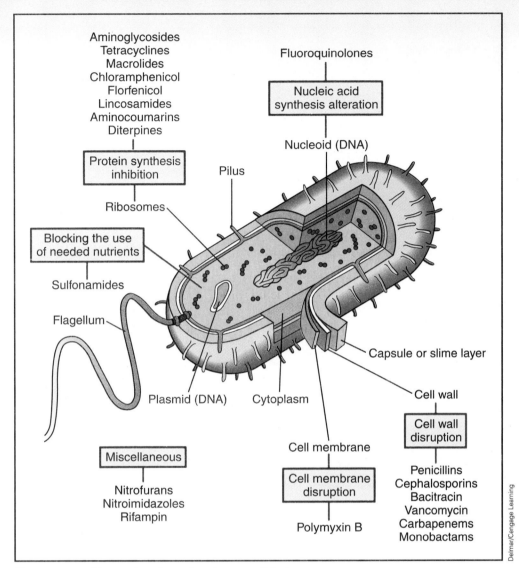

Aminoglycosides
Tetracyclines
Macrolides
Chloramphenicol
Florfenicol
Lincosamides
Aminocoumarins
Diterpines

Protein synthesis inhibition

Ribosomes

Blocking the use of needed nutrients

Sulfonamides

Flagellum

Plasmid (DNA)   Cytoplasm

Miscellaneous

Nitrofurans
Nitroimidazoles
Rifampin

Pilus

Fluoroquinolones

Nucleic acid synthesis alteration

Nucleoid (DNA)

Capsule or slime layer

Cell wall

Cell wall disruption

Penicillins
Cephalosporins
Bacitracin
Vancomycin
Carbapenems
Monobactams

Cell membrane

Cell membrane disruption

Polymyxin B

Delmar/Cengage Learning

Figure 14-7 Sites of action of various antibiotics.

ANTIMICROBIALS FOR FUNGI

Fungi of medical significance are divided into two groups: molds and yeast. Fungal infections in animals are either superficial in nature (e.g., skin infections such as ringworm [Figure 14-8]) or systemic in nature (e.g., internal infections such as blastomycosis). Fungal infections are referred to as *mycoses* and are diagnosed by fungal media or serologic tests (Figure 14-9). Superficial mycoses tend to be diagnosed with dermatophyte test media (in which the test medium changes color if the proper fungi are present) and microscopic identification of fungal structures (fungi are examined under a microscope to observe the hyphae and/or spores). Systemic mycoses are usually diagnosed via serology.

Antifungals are used to treat diseases caused by fungi. Fungi differ from bacteria in that they have a rigid cell wall that is made up of chitin and various polysaccharides and a cell membrane that contains ergosterol. The

Figure 14-8 Ringworm in a Holstein heifer.

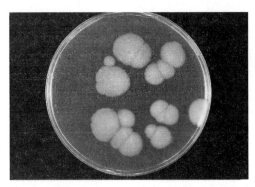

Figure 14-9 (A) Yeast colonies on Sabouraud Dextrose Agar (SDA). (B) Mold colonies on SDA.

composition of the protective layers of the fungal cell makes them resistant to antibiotics; conversely, because of their cellular composition, bacteria are resistant to antifungal drugs. The drugs used to treat fungal infections can be toxic to the host because both fungal and animal cells are similar (both are eukaryotic cells that contain sterols, which accounts for their toxicity) and should not be used indiscriminately. Antifungals may be fungicidal or fungistatic. They work by affecting the fungal cell membrane, interfering with RNA and protein synthesis, and disrupting fungal cell division (Figure 14-10).

Polyene Antifungal Agents

Nystatin and *amphotericin B* are both polyene antifungal agents. Both drugs are poorly absorbed by the gastrointestinal tract and work by binding to the fungal cell membrane. Nystatin is given orally for proliferation of *Candida albicans* fungi in the gastrointestinal tract, a frequent sequela of antibiotic drug therapy. It can also be used in a topical ointment (for skin infections and otitis externa) and an oral suspension applied to the mouth area for treatment of local infections in dogs and cats and crop mycoses in birds. Nystatin is not absorbed from the gastrointestinal tract and passes unchanged in the stool. Side effects are rare and include contact dermatitis with topical use and gastrointestinal upset with oral use.

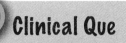

Clinical Que

Fungal infections are difficult to treat, and it takes a long course of drug treatment to resolve these infections.

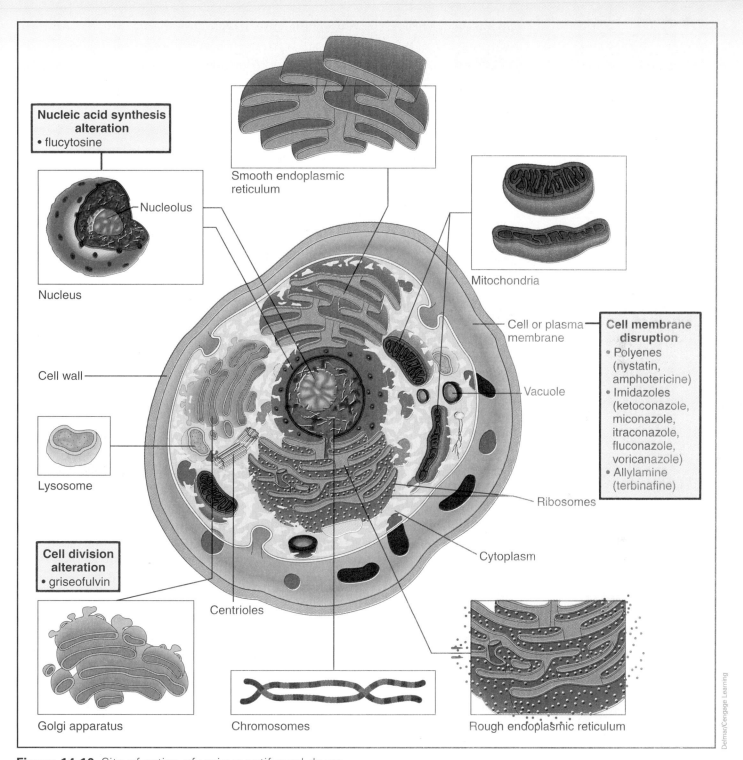

Nucleic acid synthesis alteration
• flucytosine

Nucleolus

Nucleus

Smooth endoplasmic reticulum

Mitochondria

Cell or plasma membrane

Cell membrane disruption
• Polyenes (nystatin, amphotericine)
• Imidazoles (ketoconazole, miconazole, itraconazole, fluconazole, voricanazole)
• Allylamine (terbinafine)

Vacuole

Cell wall

Lysosome

Ribosomes

Cytoplasm

Cell division alteration
• griseofulvin

Centrioles

Golgi apparatus

Chromosomes

Rough endoplasmic reticulum

Figure 14-10 Site of action of various antifungal drugs.

Amphotericin B is given intravenously and is found in creams, lotions, and ointments for topical use. It is used parenterally for the treatment of systemic mycotic infections. Amphotericin B cannot penetrate tissues well, but in tissues it does penetrate well, it remains for a long time; therefore, it is dosed every other day or three times per week. Amphotericin B comes in a vial of powder, and is light and moisture sensitive. It is usually given through a filter system because it can precipitate out of solution. Amphotericin B is used for the systemic infections caused by species of *Blastomyces, Aspergillus, Coccidioides, Histoplasma, Cryptococcus, Mucor,* and *Sporothrix.* Amphotericin B is extremely nephrotoxic (serum urea and creatinine levels and urinalysis should be monitored in animals on this medication) and must be used with care.

Imidazole Antifungal Agents

Ketoconazole and *miconazole* are examples of imidazole antifungal agents that work by causing leakage of the fungal cell membrane. The imidozole (commonly known as azole) antifungals were discovered in the early 1970s with ketoconazole being the first orally active azole antifungal. Imidazoles have a broad spectrum of antifungal activity. These drugs have fewer side effects than amphotericin B, but also have delayed onset of action. Ketoconazole is available in oral and topical forms (shampoos and creams), whereas miconazole comes in parenteral and topical forms. Both treat systemic and some superficial fungal infections, such as those caused by species of *Blastomyces, Coccidioides, Cryptococcus, Candida, Histoplasma, Microsporum, Trichophyton* and *Malassezia* (otitis). Although side effects are less serious with these antifungals, hepatotoxicity and cardiotoxicity have been noted.

Itraconazole is an oral imidazole antifungal. Itraconazole is used to treat systemic mycotic infections such as those caused by species of *Candida, Aspergillus, Cryptococcus, Histoplasma, Blastomyces,* and *Malassezia.* It has also been used to treat dermatophyte infections. Itraconazole has gained favor among small-animal veterinarians because of its increased activity and fewer side effects than ketoconazole, though it can cause gastrointestinal signs.

Fluconazole is another imidazole antifungal agent used to treat systemic mycotic infections such as those caused by species of *Cryptococcus, Blastomyces, Histoplasma,* and *Malassezia.* It can also be used to treat dermatophyte and superficial *Candida* infections. Fluconazole is fungistatic when given orally or intravenously. Fluconazole is especially useful in treating CNS infections. Side effects of fluconazole include vomiting and diarrhea.

Another imidazole antifungal agent, *voriconazole,* is the drug of choice for treating *Aspergillus, Candida, Cryptococcus,* and *Fusarium* organisms that develop resistance to other imidazoles. The activity is comparable to amphotericin B with fewer side effects. Voriconazole is given orally and is able to penetrate the CNS. Side effects of voriconazole include hepatotoxicity, renal toxicity, and anemia. An injectable form of vorizonazole is available, but has increased toxicity associated with its use.

Antimetabolic Antifungal Agents

Flucytosine is an example of an oral antimetabolic antifungal agent that works by interfering with the metabolism of RNA and proteins. Flucytosine is used mainly in combination with other antifungals to treat *Cryptococcus* infections, because it is well absorbed from the gastrointestinal tract. Its main side effects are bone marrow abnormalities.

Superficial Antifungal Agents

Griseofulvin, available only as an oral preparation, treats dermatophyte fungal infections of the skin, hair, and nails. Dermatophytes, which include species of *Microsporum, Trichophyton,* and *Epidermophyton,* are commonly known as *ringworm*. Griseofulvin is fungistatic and works by disrupting fungal cell division. Griseofulvin comes in microsize and ultramicrosize formulas; both forms should be administered with a fatty meal to aid in absorption. The ultramicrosize formulation is better absorbed than the microsize formulation. Griseofulvin can cause gastrointestinal and teratogenic side effects; therefore, it should not be administered to pregnant or breeding animals. Dosing regimens of griseofulvin vary and are usually based on veterinarian preference, ease of owner compliance, and clinical experience.

Terbinafine, an allylamine antifungal agent, is used to treat *Malassezia* and dermatophyte infections in dogs and cats; however, its spectrum of activity also includes species of *Aspergillus, Blastomyces, Cryptococcus, Histoplasma,* and *Sporothrix.* Terbinafine inhibits ergosterol synthesis, a component of the fungal cell membrane. Terbinafine is well tolerated, but side effects may include hepatotoxicity and neutropenia.

Another drug that has been used to treat ringworm infections in cats is lufenuron (Program®). Lyme sulfur is a topical product also used in the treatment of ringworm. These drugs are covered in Chapter 15, which deals with antiparasitics. Topical antifungals used to treat otic infections are covered in Chapter 18, which deals with ophthalmic and otic medications.

Examples of antifungal drugs are listed in Table 14-3.

Table 14-3 Classes of Antifungals and Their Effectiveness

CLASS	MECHANISM OF ACTION	CONSIDERATIONS	EXAMPLES
polyenes	Bind to fungal cell membrane	Not well absorbed; fairly toxic	• amphotericin B (Fungizone®) • nystatin (Panalog®)
imidazoles	Cause leakage of fungal cell membrane	Less toxic; used for systemic mycotic infections and some dermatophyte infections	• ketoconazole (Nizoral®) • miconazole (Monistat,® Conofite®) • itraconazole (Sporanox®) • fluconazole (Diflucan®) • voriconazole (Vfend®)
antimetabolics	Interfere with RNA and protein synthesis	Used mainly in combination with other antifungals to treat *Cryptococcus* infections	• flucytosine (Ancobon®)
superficials	Disrupt fungal cell division	Used for dermatophyte infections	• griseofulvin (Fulvicin-U/F®, Grifulvin V®) • terbinafine (Lamasil®)

ANTIMICROBIALS FOR VIRUSES

Viruses, unlike bacteria, are intracellular invaders that alter the host cell's metabolic pathways. A typical virus particle contains a strand of RNA or DNA surrounded by a lipid coat. The particle penetrates the host cell and causes new RNA or DNA to be produced therein. The newly formed RNA or DNA carries the genetic traits of the virus particle, and effectively paralyzes the host cell's metabolic machinery. **Antiviral** drugs act by preventing viral penetration of the host cell, or by inhibiting the virus's production of RNA or DNA. Antiviral medications are limited in their ability to treat viral infections because viruses are tiny and replicate inside cells, changing how the cell works depending on the type of cell they invade. Antiviral drugs can be toxic to animal cells, and viruses can develop resistance to antiviral drugs. Human antiviral drugs include acyclovir, amantadine, idoxuridine, cytosine arabinoside (ara C), adenine arabinoside (ara A), methisazone, interferon, and interferon inducers; however, only *acyclovir* and *interferon* are discussed here, because of the limited use of the other antivirals in veterinary practice.

Acyclovir (Zovirax® tablets, suspension, and injectable) interferes with the virus' synthesis of DNA (Figure 14-11). Acyclovir is specific for herpes virus infections and is used to treat ocular feline herpes virus. Side effects include blood disorders such as anemia and leukopenia.

Interferon is a protein substance with multiple roles in the body's natural defenses—chief among them is stimulating noninfected cells to produce antiviral proteins. Researchers have demonstrated interferon's ability to

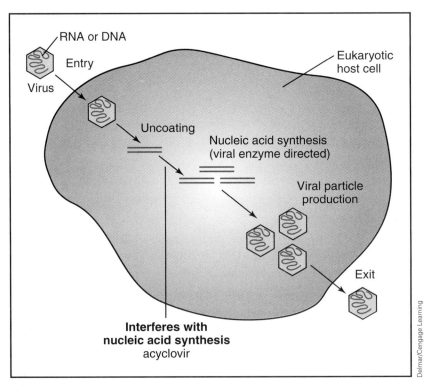

Figure 14-11 Antiviral drugs act by preventing viral penetration of the host cell or by inhibiting the virus's production of RNA or DNA.

protect host cells from a number of different viruses, and they are developing antiviral drugs called *interferon inducers* that stimulate the production and release of interferon. Interferon (Roferon-A®) has been used orally in veterinary practice for the treatment of feline leukemia virus (FeLV) and ocular herpes infections in cats. Side effects are rarely seen in cats.

CONTROLLING GROWTH OF MICROORGANISMS

There are many terms used to describe the way microbial growth can be controlled. In the strictest sense, sterilization is the removal or destruction of all microbes (bacteria, viruses, fungi, parasites, and endospores; this does not include prions, which are not typically killed by standard sterilization techniques). Sterilization is typically achieved by steam under pressure, incineration, or ethylene oxide gas. An example where sterilization is used is the processing of surgical instruments. Asepsis is a term used to describe an environment or procedure that is free of contamination by pathogens. Examples of asepsis include preparation of a surgical field and handwashing. Disinfection is the process of using physical or chemical agents to reduce the number of pathogens on inanimate objects such as surgical equipment or examination tables, and the chemical used in this process is called a **disinfectant**. Disinfectants kill or inhibit the growth of microorganisms on inanimate objects. Disinfection does not guarantee that all pathogens are eliminated. When a chemical is used on skin or other living tissue, the process is called antisepsis, and the chemical is called an **antiseptic**. Antiseptics kill or inhibit the growth of microorganisms on living tissue. Antiseptics and disinfectants may have the same chemicals in them; however, disinfectants are more concentrated or can be left on a surface for a longer period of time. Some disinfectants, such as steam or concentrated bleach, are not suitable for use as antiseptics.

Other terms that further describe disinfectants or antiseptics include the following:

- **germicide**: a chemical that kills microorganisms

- **bactericidal**: a chemical that kills bacteria

- **virucidal**: a chemical that kills viruses

- **fungicidal**: a chemical that kills fungi

- **sporicidal**: a chemical that kills endospores, which are especially resistant to chemicals

- **tuberculocide**: a chemical that kills *Mycobacterium tuberculosis*, the bacterium that causes tuberculosis. Since tuberculosis is only spread by aerosol, it does not matter if a chemical is a tuberculocide in regard to organism transmission. The label *tuberculocide* is important in that it is used as a benchmark for germicidal potency, but not for its effectiveness against tuberculosis spread.

An effective sanitizing and disinfection program is a crucial step in a veterinary hospital, on a farm, or in a research setting. Ideally, a sanitizing and disinfection program should be instituted after animal depopulation and before restocking occurs; however, this does not always happen especially in

Clinical Que

Some bacteria can form endospores when environmental conditions worsen. Bacterial spores are technically called endospores and should not be confused with fungal spores, which are reproductive structures.

veterinary hospitals. The goal of a sanitizing and disinfection program is to reduce the number of pathogens (bacteria, fungi, viruses, and parasites) in the environment, which in turn reduces the potential for diseases to occur in animals. By identifying the type of pathogen and the chemicals that are effective against that type of pathogen, a facility can effectively prevent the spread of infectious agents.

The first step in ridding an area of pathogens is sanitizing. **Sanitizing** is the physical removal of organic material (such as manure, blood, feed, and carcasses) because infectious agents are often protected in these materials and therefore can survive the disinfection process. The sanitizing process includes a dry sanitizing and a wet sanitizing step. Dry sanitizing is the physical removal of organic material, such as the removal of feed, litter, and manure. Wet sanitizing occurs after dry sanitizing and involves the use of water. There are four steps in the wet sanitizing process: soaking, washing, rinsing, and drying. Detergents (categorized as wetting agents) may be used in the soaking and washing steps; however, it is more important to have pressure washers with the proper pressure (500 to 800 psi) to ensure all the organic materials are removed from the facilities. Rinsing with water removes toxic chemicals from the area or instrument to limit destruction of these surfaces. The final step is to dry the area or instrument quickly. Excess moisture can result in bacteria multiplying to higher levels than prior to sanitizing.

The last step in a sanitizing and disinfection program is the disinfection process. This process involves the use of a disinfectant that will reduce or kill the pathogens. There are many different kinds of disinfecting agents. Ideally, these agents should remain stable during storage and use, be easy to apply, not damage or stain the objects they are applied to, be nonirritating to animal tissue, and have the broadest possible spectrum of activity. Low cost is also an advantage because so much disinfecting product is used in a veterinary facility.

When choosing a disinfecting agent, keep in mind the surface it will be applied to and the range of organisms it needs to eliminate. A chemical agent that works well on animal cages may not work well on plastic or rubber equipment. Likewise, a disinfecting agent that works well on viruses may not work well on bacterial endospores.

Some disinfecting agents are less effective in the presence of organic waste (blood, feces, etc.), soap, and hard water. Thorough sanitizing of the surface to be treated is an important step in disinfection.

When purchasing disinfecting agents, make sure to read the package insert in regard to dilution recommendations and special use instructions. Some chemicals work on different microorganisms depending upon how and if they are diluted (Figure 14-12). *Contact time*, the amount of time that the agent spends on the surface, is also critical to the efficacy of the chemical.

Always keep or request Material Safety Data Sheets (MSDSs) on all products used in disinfection. The filing of MSDS and container labeling are important components of each clinic's or facility's hazard communication plan. The Occupational Safety and Health Administration (OSHA), a part of the U.S. Department of Labor, enacted the Hazard Communication Standard in

Clinical Que

If done properly, a good sanitizing can remove 90 percent of the pathogens.

Clinical Que

When diluting a chemical, always start with the quantity of water and then add the chemical concentrate. This avoids splashing of chemical into your eyes. Latex gloves and protective goggles are also recommended during these procedures.

ROCCAL®-D PLUS Veterinary Disinfectant

ROCCAL®-D PLUS

NDC 0009-7308-01, **NDC** 0009-7308-05

VETERINARY AND ANIMAL CARE DISINFECTANT

EFFECTIVE IN 400 PPM HARD WATER AS $(CaCO_1)$

DISINFECTS IN 5% ORGANIC SOIL LOAD

ACTIVE INGREDIENTS:

Didecyl dimethyl ammonium chloride 9.2%

Alkyl (C_{12}, 61%; C_{14}, 23%; C_{16}, 11%; C_{18}, 2.5%; C_8 & C_{10}, 2.5%)

   dimethyl benzyl ammonium chloride 9.2%

Alkyl (C_{12}, 40%; C_{14}, 50%; C_{16}, 10%)

   dimethyl benzyl ammonium chloride 4.6%

bis-n-tributyltin oxide 1.0%

INERT INGREDIENTS: 76.0%

BACTERICIDE, FUNGICIDE, VIRUCIDE

For Veterinary, Laboratory Animal, Kennel and Animal Breeder Facilities.

DANGER

KEEP OUT OF REACH OF CHILDREN

EPA REG. NO. 65020-12-1023   EPA EST. NO. 65020 GA-1

Roccal-D Plus

- is a complete, chemically balanced disinfectant providing clear use solutions even in hard water.

- is a residual bacteriostat and inhibits bacterial growth on moist surfaces.

- contains rust corrosion inhibitors.

- deodorizes by killing most microorganisms that cause offensive odors.

DIRECTIONS FOR USE IN VETERINARY CLINICS, ANIMAL CARE FACILITIES, ANIMAL RESEARCH CENTERS, ANIMAL BREEDING FACILITIES, KENNELS AND ANIMAL QUARANTINE AREAS.

It is a violation of Federal Law to use this product in a manner inconsistent with its labeling. Roccal-D Plus is a one-step germicide, fungicide, soapless cleaner and deodorant effective in the presence of organic soil (5% serum). It is non-selective and when used as directed, will not harm tile, terrazo, resilient flooring, concrete, painted or varnished wood, glass or metals.

- To clean and disinfect hard surfaces, use ½ fluid ounce of Roccal-D Plus per gallon of water. Apply by immersion, flushing solution over treated surfaces with a mop, sponge, cloth or bowl mop to thoroughly wet surfaces. Prepare fresh solutions daily or when solution becomes visibly dirty.

- To clean badly soiled areas, use up to 1½ fluid ounce per gallon of water.

- To disinfect, allow treated surfaces to remain moist for at least 10 minutes before wiping or rinsing.

- To control mold and mildew growth on previously cleaned, hard nonporous surfaces, use ½ fluid ounce per gallon. Allow to dry without wiping. Reapply as new growth appears.

BOOT BATH: Use 1 fluid ounce per gallon in boot baths. Change solution daily and anytime it becomes visibly soiled. Use a nylon bristled brush to clean soils from boots.

DISINFECTING VANS, TRUCKS AND FARM VEHICLES: Clean and rinse vehicles and disinfect with ½ fluid ounce per gallon Roccal-D Plus. If desired, rinse after 10 minutes contact or leave unrinsed.

Do not use Roccal-D Plus on vaccination equipment, needles or diluent bottles as the residual germicide may render the vaccines ineffective.

Roccal-D Plus should not be mixed with other cleaning or disinfecting compounds or products. BROAD SPECTRUM GERMICIDAL ACTION IN HARD WATER AND UNDER SOIL LOAD CONDITIONS: At ½ fluid ounce per gallon (1:256) in official AOAC Use Dilution and Fungicidal Tests, Roccal-D Plus is effective in water up to 400 ppm hardness (as $CaCO_3$) and an organic soil load of 5% serum against the following organisms.

BACTERIA

Pseudomonas aeruginosa ATCC 15442
Salmonella choleraesuis ATCC 10708
Enterobacter aerogenes ATCC 63809
Pasteurella multocida ATCC 7707
Shigella dysenteriae ATCC 13313
Klebsiella pneumoniae ATCC 4352
Enterococcus faecium ATCC 6569
Salmonella gallinarum ATCC 9184
Serratia marcescens ATCC 264
Bordetella avium ATCC 35086
Streptococcus agalactiae ATCC 27916
Mycoplasma gallisepticum ATCC 15302
Mycoplasma gallinarum ATCC 19708
Actinomyces pyogenes ATCC 19411
Actinobacillus pleuropneumoniae ATCC 27088
Corynebacterium pseudotuberculosis ATCC 19410
Rhodococcus equi ATCC 6939
Streptococcus equi var. zooepidemicus ATCC 43079
Staphylococcus aureus ATCC 6538
Salmonella enteriditis ATCC 4931
Streptococcus pyogenes ATCC 9547
Salmonella pullorum ATCC 9120
Escherichia coli ATCC 11229
Alcaligenes faecalis ATCC 8748
Shigella sonnei ATCC 29930
Salmonella typhosa ATCC 6539
Proteus morganii ATCC 25830
Proteus mirabilis ATCC 25933
Mycoplasma iners ATCC 19705
Mycoplasma hypopneumoniae ATCC 25934
Bordetella bronchiseptica ATCC 19395
Streptococcus equi var. equi ATCC 33398

FUNGI

Aspergillus fumigatus ATCC 10894
Trichophyton mentagrophytes var. interdigitale ATCC 9533
Candida albicans ATCC 18804

VIRUSES

Using accepted virus propagation and hard surface test methods, Roccal D-Plus is effective at 1:256 in 400 ppm hard water and 5% serum against the following viruses.

Avian laryngotracheitis ATCC VR-783
Canine parvovirus ATCC VR-953
Infectious bronchitis ATCC VR-22
Transmissible gastroenteritis ATCC VR-763
Mycoplasma gallisepticum ATCC 15302
Infectious bursal disease (Gumboro) ATCC VR-478
Canine distemper onderstepoort strain
Equine herpesvirus ATCC VR-700
Feline infectious peritonitis strain, DF-2
Swine PRRS virus, Reference strain, NVSL
Newcastle's disease ATCC VR-109
Parainfluenza ATCC VR281
Pseudorabies ATCC VR-135
Avian influenza ATCC VR-798
Canine herpesvirus ATCC VR-552
Equine influenza A ATCC VR-297
Porcine parvovirus ATCC VR-742
Vesicular stomatitis virus, Indiana

Figure 14-12 Sample disinfectant package insert. Reprinted with permission from Pharmacia Animal Health, Kalamazoo, MI.

ROCCAL®-D PLUS Veterinary Disinfectant

STORAGE AND DISPOSAL: Store only in tightly closed original container in a secure area inaccessible to children. Do not contaminate water, food or feed by storage and disposal. Disposal: Pesticide wastes are acutely hazardous. Improper disposal of excess pesticide spray mixture, or rinsate is a violation of Federal Law. If these wastes cannot be disposed of by use according to label instructions, contact your State Pesticide or Environmental Control Agency, or the hazardous waste representative at the nearest EPA Regional Office for guidance.

Do not reuse empty container. Rinse container thoroughly before discarding in trash.

PRECAUTIONARY STATEMENTS
HAZARDS TO HUMANS AND DOMESTIC ANIMALS
DANGER — KEEP OUT OF REACH OF CHILDREN

CORROSIVE: Causes severe eye and skin damage. Do not get into eyes, on skin or clothing. Wear goggles or face shield and rubber gloves when handling the concentrate. Harmful or fatal if swallowed. Avoid contamination of food. Wash thoroughly with soap and water after handling.

ENVIRONMENTAL HAZARDS: This product is toxic to fish. Do not discharge effluent containing this product into lakes, streams, ponds, estuaries, oceans or other waters unless in accordance with the requirements of a National Pollutant Discharge Elimination System (NPDES) permit and the permitting authority has been notified in writing prior to discharge. Do not discharge effluent containing this product to sewer systems without previously notifying the local sewage treatment plant authority. For guidance contact your State Water Board or Regional Office of the EPA.

PHYSICAL AND CHEMICAL HAZARDS: Do not use or store near heat or open flame.

STATEMENT OF PRACTICAL TREATMENT: In case of contact, immediately flush eyes or skin with plenty of water for at least 15 minutes. For eyes, call a physician. Remove and wash contaminated clothing before reuse. If ingested, drink promptly a large quantity of raw egg whites, gelatin solution, or if these are not available, drink a large quantity of water. Avoid alcohol. Call a physician immediately.

NOTE TO PHYSICIAN: Probable mucosal damage may contraindicate the use of gastric lavage.

1/IGL

Manufactured Exclusively for

Pharmacia & Upjohn Company
Kalamazoo, MI 49001, USA   **816 538 002**

Figure 14-12 (*Continued*)

Figure 14-13 (A) Hazardous materials must only be disposed of in the containers properly labeled with the biohazard symbol. (B) Any sharp items must be disposed of in a hard plastic "sharps" container.

1988 to educate and protect employees who work with potentially hazardous materials. The hazard communication plan includes the following:

- a written plan that serves as a primary resource for the entire staff. It lists all hazardous materials used in the facility and the name of the person responsible for keeping this list current (Figure 14-13). The written plan also includes where MSDSs are kept, how the MSDSs are obtained, procedures for labeling materials, a detailed description of employee training, and how independent contractors are informed of hazardous materials in the facility.

- an inventory of hazardous materials on the premises.

- current MSDSs for hazardous materials.

- proper labeling of all materials in the facility (including secondary containers, which are diluted, or transferred chemicals that are not in their original containers).

- employee training for every employee working with or exposed to hazardous materials.

  The following must be on all MSDSs:

- product name and chemical identification

- name, address, and telephone number of the manufacturer

- list of all hazardous ingredients

- physical data for the product (liquid, solid, and gas)

- fire and explosion information

- information on potential chemical reactions when the product is mixed with other materials

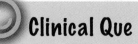

Clinical Que

Many hazardous materials are labeled with key words to watch for, including Caution, Warning, Danger, Toxic, Flammable, Reactive, and Corrosive.

Clinical Que

Quaternary ammonium compounds can be recognized by its chemical name having ammonium in it or having an -nium in its name.

- outline of emergency and sanitizing procedures
- personal protective equipment required when handling the material
- description of any special precautions necessary when using the material

Types of Disinfecting/Antiseptic Agents

The main types of disinfectants that are used in veterinary medicine are phenols, quaternary ammonium compounds, aldehydes, alcohols, halogens, biguanides, ethylene oxide, and others such as hydrogen peroxide.

Phenols and Phenolics

Phenols were the first antiseptics and are found in a variety of products. In 1867, human physician Dr. Joseph Lister began using phenol to reduce infection during surgery. The efficacy of phenol remains one standard to which the actions of other antimicrobial agents are compared. The basic phenol molecule has a 6-carbon ring with an –OH group (Figure 14-14). Phenolics have a phenol molecule that has been chemically modified with the addition of a halogen or organic functional group. This class of chemicals works by destroying the selective permeability of cell membranes, resulting in leakage of cellular material. Phenols and phenolics have intermediate- to low-level disinfection ability. One advantage to this class is that they are effective even in presence of contaminated organic matter (such as vomit, pus, saliva, and feces). Phenols and phenolics are effective against gram-positive bacteria, with some effectiveness against gram-negative bacteria, fungi, and some types of enveloped viruses. Phenols and phenolics are ineffective against endospores and nonenveloped viruses such as parvovirus. This group of chemicals should not be used as antiseptics because they are irritating to skin, can be absorbed systemically, and have been linked to neurotoxicity.

Quaternary Ammonium Compounds

Quaternary ammonium compounds are commonly called the "quats" and have a charged nitrogen with four hydrophobic groups. They work by denaturing protein and disrupting the cell membrane. Most quaternary ammonium compounds are combined with detergents, which allows them to act as surfactants. Surfactants are "surface active" chemicals that reduce surface tension, which makes solvents more effective at dissolving solute (particles). This is accomplished by decreasing the attraction among solvent molecules. Quats are effective against gram-positive and gram-negative bacteria (better on gram-positive than gram-negative), but are not effective against endospores and have limited efficacy on fungi. Quaternary ammonium products work on enveloped viruses, but not nonenveloped viruses such as parvovirus. They act rapidly and are usually not irritating to skin or corrosive to metal. Organic debris, hard water, and soaps inactivate quaternary ammonium compounds; therefore, the site should be free of these materials and dry prior to application.

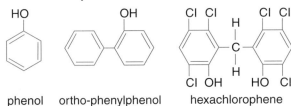

Figure 14-14 Chemical structures of select disinfectants.

Delmar/Cengage Learning

Clinical Que

In general, the decreasing order of resistance of infectious agents is the following:

1. prions
2. endospores (clostrial diseases like tetanus) and acid-fast bacteria (such as *Mycobacterium avium*)
3. protozoal cysts
4. nonenveloped viruses (enteroviruses and adenoviruses)
5. fungi (such as *Candida* and *Aspergillus*)
6. gram-negative bacteria (such as *Pseudomonas, E. coli,* and *Salmonella*)
7. gram-positive bacteria (*Staphylococcus aureus* and *Streptococcus equi*)
8. lipid-enveloped viruses (avian influenza virus).

Aldehydes

Aldehydes are organic compounds that contain a functional group –CHO (carbon-hydrogen-oxygen), known as an aldehyde. Examples of aldehydes are *glutaraldehyde, ortho-phthalaldehyde,* and *formaldehyde.* Glutaraldehyde and ortho-phthalaldehyde work by affecting protein structure. They are rapid, broad-spectrum antimicrobials that kill fungi and bacteria within a few minutes and endospores in about three hours. Glutaraldehyde at an acid pH will act only at a bacterial endospore's surface, whereas at an alkaline pH, it penetrates the endospore. Therefore glutaraldehyde is usually added with an alkalinizing agent (such as a quaternary ammonium compound) to enhance its effectiveness. Viruses are inactivated in a short period of time as well. Sterilization with glutaraldehyde takes 10 hours. Organic debris and hard water do not inactivate glutaraldehydes. Items processed with glutaraldehydes need to be rinsed thoroughly with water in accordance with federal, state, and local regulations. However, they are not used much in veterinary medicine due to high cost, instability, and toxicity issues (fumes are toxic, and proper ventilation is essential).

Formaldehyde can be used in gas or solution form. As a gas, it can disinfect a large area like an incubator; as a liquid, it can disinfect instruments. In the aqueous form, it is referred to as *formalin.* It works by affecting nucleic acids of microbes. Formaldehyde's extreme toxicity and classification as a carcinogen limit its clinical use. Read the MSDS before using formalin. Use formalin only in areas with good ventilation and avoid skin and eye contact and inhalation of vapors. Formalin is most commonly used in fixation of tissue biopsies for pathological examination and in footbaths to treat hairy warts in cattle.

Alcohols

Alcohols, either 70 percent ethyl alcohol or 50 percent or 70 percent isopropyl alcohol, are used alone in aqueous solutions or as solvents in other chemicals. Alcohols provide an intermediate level of disinfection. Alcohols work by coagulating proteins and dissolving membrane lipids. Surprisingly pure alcohol is not as effective as an antimicrobial agent because the denaturation of proteins requires water; therefore, 70 percent alcohol is commonly used on animal skin. In addition, higher concentrations of alcohol can dry the skin and possibly alter the intact barrier that protects animals from pathogen invasion through the skin. Alcohols evaporate quickly, which is an advantage in that they do not leave a residue; however, they may not contact microbes long enough to be effective. Alcohols are nonirritating, nontoxic, and inexpensive. Alcohols work well on both gram-positive and gram-negative bacteria and enveloped viruses, but are ineffective on endospores and nonenveloped viruses. To be antifungal, alcohols must be in contact with fungi for several minutes. Alcohols must be applied in sufficient quantity, at the proper concentration, and for adequate time (several seconds to minutes) to be effective. Alcohol is not recommended as an antiseptic, because of the pain it causes and the denaturing effect it has on proteins. Dirt and organic debris must be removed prior to alcohol application for the alcohol to be effective. It is commonly used to clean skin surfaces prior to injections and when taking blood samples; however, swabbing an animal's skin with alcohol

Clinical Que

The CDC has approved the use of alcohol-based gels as an effective method of handwashing in healthcare facilities. However, alcohol is ineffective against endospores because the endospore coat has little or no lipid in it, and since that is a mechanism by which alcohol works, it would make sense that alcohol is ineffective against this endospore-forming bacteria (such as *Clostridium* sp.).

probably removes more microbes by physical action rather than by chemical action. If the term "tincture" is associated with other antimicrobial agents, this means that the chemical is combined with alcohol. The addition of alcohol makes the chemical more effective than when it is simply dissolved in water.

Halogens

Halogens are very reactive, nonmetalic chemical elements found in the Group VII column of the periodic table. Halogens have an intermediate level of disinfection. Chlorine and iodine agents represent the halogens used in veterinary medicine. They work by interfering with proteins and enzymes of the microbe. Chlorine agents used in veterinary medicine are the hypochlorites, which are made by the reaction of chlorine gas, lye, and water. Hypochlorites typically contain 70 percent available chlorine and kill bacteria, fungi, and enveloped viruses. Chlorine gas can kill endospores (effectiveness depends upon concentration, exposure time, and formulation). Hypochlorite, which is found in bleach, is inexpensive and easy to purchase. The disadvantages of hypochlorite use are its bleaching of fabric, its corrosiveness to metal surfaces, and its vapor that can be irritating to eyes and mucous membranes. Hypochlorites are rapidly inactivated by organic matter, light, and heat. Bleach is routinely used in a 1:10 dilution. It is easily inactivated by organic material and unstable if exposed to light. When diluted, bleach has a shelf life of 24 hours.

Iodine is a black chemical that forms a brown solution when mixed with water or alcohol. Most classes of microbes are killed by iodine if proper concentration and exposure times are used (activity against bacterial endospores is not consistent). Iodine compounds are commonly used as topical antiseptics (used to treat ringworm in cattle) and are the most commonly used product in teat dips. Iodophors are complexes of iodine and a neutral polymer such as polyvinyl alcohol (PVA). This formulation causes slow release of iodine and increases its ability to penetrate skin. Iodophor compounds are marketed as scrubs, solutions, or tinctures. *Scrubs* have soap products added to them, *solutions* are dilutions of iodine and water, and *tinctures* are iodine and dilute alcohol. Iodine products can be corrosive to metals if left in contact for long periods of time, can be irritating to skin in high concentrations, and can stain fabrics and other materials.

Biguanides

Chlorhexidine, a biguanide, is one of the most commonly used disinfects and antiseptics in veterinary practice. Chlorhexidine works by denaturing proteins. It is relatively mild, nontoxic, and fast acting. It is effective against bacteria, some fungi, and enveloped viruses such as feline infectious peritonitis (FIP) and FeLV, but does not work on nonenveloped viruses and endospores. Chlorhexidine is commonly used as a surgical scrub, for cleansing wounds in lower concentrations, and as a teat dip. It can have residual activity of 24 hours because it binds to the outer surface of the skin.

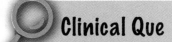

Clinical Que

OSHA recommends bleach more than any other disinfecting solution in medical-based practices. Bleach solutions have a short potency span and must be replaced every two to three days, or mixed up fresh when needed.

Ethylene Oxide

Ethylene oxide, a gas at room temperature, works by destroying DNA and proteins. Ethylene oxide is one of the only gases used for chemical sterilization (often referred to as gas sterilization). This gas is rather penetrating but slow acting. It may require 90 minutes to 3 hours for complete sterilization. Products must then be exposed to air for several hours to get rid of as much residual gas as possible. Ethylene oxide is explosive and rated as a potent carcinogen, so it must be handled and contained carefully. Its benefit is that it can sterilize objects that cannot withstand heat, such as rubber products.

Oxidizing Agents

Oxidizing agents such as hydrogen peroxide and peroxygen compounds are effective against bacteria, endospores, viruses, and fungi. Hydrogen peroxide has been used as a 3 percent solution to kill anaerobic bacteria in deep wounds such as puncture wounds. This chemical causes oxygen to be released when it reacts with cellular products and compromised tissue. Because hydrogen peroxide damages proteins and thus can damage animal tissue, its use on tissue should be limited. It is commonly used for oral infections and surface wound management.

Peroxygen compounds such as peroxyacetic acid or peracetic acid are made by mixing acetic acid with hydrogen peroxide. Most preparations contain peracetic acid and hydrogen peroxide. On degradation, peroxyacetic acid breaks down to acetic acid, water, and oxygen. With both higher temperatures and stronger concentrations, endospores can be destroyed. Only low concentrations of peracetic acid are needed for antimicrobial efficacy. It is corrosive to metals over time if it is not thoroughly rinsed off.

Soaps or detergents have limited bactericidal activity. Their main functions are mechanical removal of microbes. They may contain ingredients effective against some bacteria, but do not work on endospores and have limited use on viruses.

Table 14-4 summarizes the various disinfectants.

Table 14-4 Types of Disinfectants and Antiseptics

DISINFECTANT GROUP	PRODUCT EXAMPLES	USE	ACTION	COMMENTS
phenols and phenolics	• ortho-phenylphenol (Lysol®, Amphyl®, Tek-Trol®, 1 Stroke Environ®) • hexachlorophene (Phisohex®)	Laundry, floors, walls, equipment	Moderately bactericidal, virucidal, and fungicidal	• Action not affected by organic material • Used as a 2–5% solution on contaminated objects
quaternary ammonium compounds	• didecyl dimethyl ammonium chloride (Roccal®-D) • benzalkonium chloride (Zephiran®)	Instruments, rubber, inanimate objects	Moderately bactericidal, virucidal, and fungicidal	• Action not affected by hard water

(Continued)

Disinfectant Group	Product Examples	Use	Action	Comments
aldehydes	• gluteraldehyde (Cidex®, Glutarol®)	Instruments	Highly bactericidal, virucidal, and fungicidal	• Action not affected by organic material or hard water
	• ortho-phthalaldehyde (Cidex-OPA®)	Instruments	Highly bactericidal, virucidal, fungicidal, and tuberculocidal	• All items processed with aldehydes need to be rinsed with water prior to use
	• formaldehyde	Tissue biopsies and footbaths	Highly bactericidal, virucidal, fungicidal, and tuberculocidal	• Use in areas of good ventilation • Avoid skin and eye contact and inhalation of vapors
alcohols	• 70% isopropyl • 50% ethyl alcohol	Instruments Thermometers	Highly bactericidal, some virucidal action, and poor fungicidal action	• 70% solution usually used • Affected by organic material and dirt
halogens	• chlorines: (Chlorox®) • iodophors: (Betadine®, Povidine®)	Chlorines: floors, cages Iodophors: presurgical scrub, thermometers	Moderately to highly bactericidal, highly virucidal, moderately to highly fungicidal, and some sporicidal activity	• Corrosive to surfaces; • Vapors can be irritating • Iodine tincture is about 2% • Iodine tincture is about 2%
biguanide	• chlorhexidine (Nolvasan®, Hibiclens®, Virosan®)	Skin wounds, presurgical scrub, oral cleaning solutions, and cages	Highly bactericidal, moderately virucidal, and poorly fungicidal	• Residual action of about 24 hours due to binding to skin
ethylene oxide		Rubber goods, blankets, and lensed instruments	Highly bactericidal, virucidal, and fungicidal	• "Gas sterilization" for objects that cannot withstand heat • Carefully read MSDS prior to handling • Keep away from flames and sparks
oxidizing agents	• hydrogen peroxide • peracetic acid (Virkon®S)	Cleaning wounds and floors, foot bath	Moderately to highly bactericidal, virucidal, and fungicidal	• Causes oxygen release, which kills anaerobic bacteria • Action not affected by organic matter or hard water

SUMMARY

Antimicrobials, sometimes referred to as anti-infectives, include the drug categories antibiotics, antifungals, and antivirals. Disinfectants and antiseptics also work on these microorganisms.

Antibiotics work on bacteria. Antibiotics may have a narrow spectrum of activity (working only on gram-negative or gram-positive bacteria) or a broad spectrum of activity (working on both [but typically not all] gram-negative and gram-positive bacteria). Antibiotics can also be described as bactericidal (actually kill the bacteria) or bacteriostatic (inhibit the growth of bacteria). Bacteria are either resistant, sensitive, or intermediate to a particular antibiotic. The sensitivity of bacteria to a particular antibiotic is determined by agar diffusion testing (antibiotic sensitivity testing). Concerns regarding antibiotic use include bacterial resistance and antibiotic residues. Antibiotics work by a variety of mechanisms, including inhibition of bacterial cell wall synthesis, damage to the bacterial cell membrane, interference with bacterial metabolism, and impairment of bacterial nucleic acids.

Antifungals work against fungal infections. Fungal infections may be superficial or systemic. Antifungal drugs work in a variety of ways, including binding to the fungal cell membrane, causing leakage of the fungal cell wall, interfering with fungal metabolism, and disrupting fungal cell division.

Antiviral drugs work by preventing viral penetration of the host cell or by inhibiting the virus's production of RNA or DNA. Antiviral drugs have been used in veterinary practice for feline herpes virus and FeLV.

Disinfectants are chemicals that kill or inhibit the growth of microorganisms on inanimate objects. Antiseptics are chemicals that kill or inhibit the growth of microorganisms on animate or living tissue. Each disinfectant and antiseptic has its own spectrum of activity, its own dilution values, and its own application instructions; therefore, labels should be read prior to use.

It's a Wrap

The answers to the questions in this chapter's Setting the Scene case study should be understood after reading this chapter. When a client uses improper terminology or describes a situation incorrectly, how does the veterinary professional educate and not insult a client? When a client asks for the "strongest" antibiotic available, how can the veterinary technician help this client understand that her rationale of antibiotic use is flawed? What is the best way to explain to this client how antibiotics work, what side effects can be expected, and why "stronger" is not a good term to

(continued)

use in describing antibiotics? One way to describe antibiotics to this client is to state that antibiotics work on bacteria and by a variety of different ways. Some antibiotics work on different bacterial structures than others, which makes their effectiveness among different types of bacteria variable. Antibiotics should not be thought of as stronger or weaker than one another. Antibiotics are defined as narrow spectrum (working only on gram-positive or gram-negative bacteria) or as broad spectrum (working on both, but perhaps not all, gram-positive and gram-negative bacteria). The best way to treat bacterial infections is to know exactly what organism is present so that the proper antibiotic is used. Identifying bacteria is done through culture methods. Without culture information, antibiotics may be dispensed to cover the most common organisms that cause the animal's condition (such as pneumonia, urinary bladder infections, etc.). The antibiotic that has the best chance of targeting a particular bacterium and getting to the area where the bacterium is causing disease is the best one to use. It is unclear what the owner is referring to when she wants the strongest antibiotic for her cat. She may mean she wants the broadest-spectrum antibiotic available to make sure that the potential to inhibit the pathogen is maximized or she may be referring to the one that works without causing side effects. Remind this client that antibiotics do not "cure" sneezing. Sneezing and other body defenses like nasal discharge and coughing are ways in which an animal can protect itself from contracting infections or from worsening infection.

What are some consequences of this improper antibiotic use? One consequence of indiscriminate antibiotic use includes the development of antibiotic resistance. Antibiotics should always be prescribed and taken at the proper dose and for the proper time. Recheck examinations allow the veterinary staff to determine whether or not the infection is eliminated. If it is not, additional antibiotics can be prescribed at that time. Another consequence of indiscriminate antibiotic use is alteration of bacterial normal flora, which can lead to this cat developing other health issues such as vomiting and diarrhea. Alteration of bacterial normal flora may also allow overgrowth of yeast in some areas of the body, which may require treatment with antifungal medication.

CHAPTER REVIEW

Matching

Match the drug name with its action.

1. _____ interferon
2. _____ enrofloxacin
3. _____ sulfonamide
4. _____ nystatin
5. _____ erythromycin
6. _____ doxycycline
7. _____ cephalexin
8. _____ gentamicin
9. _____ procaine penicillin G
10. _____ ketoconazole

a. antifungal used to treat *Candida* infections
b. used to treat viral infections
c. fluoroquinolone antimicrobial
d. antimicrobial used widely before the advent of penicillin
e. aminoglycoside antibiotic that may cause nephro- and ototoxicity
f. long-acting penicillin form given only by injection
g. antibiotic known as a cephalosporin
h. tetracycline antibiotic
i. macrolide antibiotic
j. drug used orally and topically to treat fungus infections by causing leakage of the fungal cell membrane

Multiple Choice

Choose the one best answer.

11. Which antibiotics contain the beta-lactam ring?
    a. penicillins and aminoglycosides
    b. quinolones and tetracyclines
    c. macrolides and quinolones
    d. penicillins and cephalosporins

12. Which antibiotics are not recommended for young animals?
    a. penicillins and aminoglycosides
    b. fluoroquinolones and tetracyclines
    c. macrolides and fluoroquinolones
    d. penicillins and cephalosporins

13. Which antibiotic many cause bone marrow suppression if taken systemically or handled improperly?
    a. penicillin
    b. chloramphenicol
    c. enrofloxacin
    d. tetracycline

14. Which antibiotics are used only topically?
    a. penicillins and cephalosporins
    b. polymyxin B and bacitracin
    c. erythromycin and doxycycline
    d. gentamicin and sulfonamides

15. Which antifungal is only given IV?
    a. nystatin
    b. amphotericin B
    c. ketoconazole
    d. flucytosine

16. Griseofulvin is used to treat
    a. gram-positive bacterial infections.
    b. gram-negative bacterial infections.
    c. systemic mycotic infections.
    d. dermatophyte infections.

17. Chlorine and iodine agents represent which group of disinfectants/antiseptics?
    a.   phenols
    b.   aldehydes
    c.   halogens
    d.   alcohols

18. Which antiseptic is used commonly as a surgical scrub, for cleansing wounds, and as a teat dip?
    a.   hydrogen peroxide
    b.   alcohols
    c.   chlorhexidine
    d.   formaldehyde

19. What chemical is used to sterilize rubber goods?
    a.   phenol
    b.   ethylene oxide
    c.   hydrogen peroxide
    d.   alcohol

True/False

Circle a. for true or b. for false.

20. All disinfectants are sporicidal.
    a.   true
    b.   false

Case Studies

21. A 1200-lb. Holstein cow freshened (gave birth) four days ago, which was two weeks earlier than her due date. She refuses to eat her feed today. On PE, the cow shows an elevated temperature of 104.2°F and is lethargic. Vaginal examination reveals that she has retained fetal membranes. The veterinarian decides to treat her with procaine penicillin at a dosage of 15,000 IU/kg/day.
    a.   What is one reason why procaine penicillin is used?
    b.   What would the dose be for this cow in IU?
    c.   The concentration of procaine penicillin is 300,000 IU/mL. How many milliliters will this cow get per dose?

22. A six-month-old M Beagle puppy (25#) has a history of tick infestation from two months ago. The dog is now lame and is running a fever.
    a.   What disease should be considered in this dog?
    b.   What is the usual treatment for this dog?
    c.   Can this drug be used in this dog?

23. An owner of a cattery comes into the clinic and says that her cats have had a positive culture for ringworm. She would like to use a drug on all her cats.
    a.   One drug that works on ringworm is griseofulvin. What do you want to discuss with this owner before the veterinarian prescribes griseofulvin?
    b.   What other drug choice may be used in this situation?

Critical Thinking Questions

24. Antibiotic resistance is frequently in the news. Think about the responsibilities that veterinary professionals have to prevent the development of antibiotic resistance. What considerations are there in administering a course of antibiotic treatment for an animal? What is appropriate in terms of follow-up protocols during antibiotic treatment?

25. Antibiotic use in food-producing animals is a source of concern for veterinary and human health professionals. Investigate research sources to explain why this is a concern and what veterinary professionals and the U.S. government is doing to address the issue.

ADDITIONAL REVIEW MATERIAL

Additional review material can be found on the StudyWARE™ CD included with this text.

REFERENCES

Allen, D. G. (2005). *Handbook of Veterinary Drugs* (3rd ed.). Baltimore, MD: Lippincott Williams & Wilkins.

Brunton, L., Lazo, J., and Parker, K. (2006). *Goodman and Gilman's the Pharmacological Basis of Therapeutics* (11th ed.). NY: McGraw-Hill.

Jeffrey, J. (2008). *Sanitation-Disinfection Basics for Poultry Flocks.* University of California-Davis Web site, http://www.vetmed.ucdavis.edu/.

KuKanich, B. (2008). *A Review of Selected Systemic Antifungal Drugs for Use in Dogs and Cats.* Veterinary Medicine.

Plumb, D. C. (2008). *Plumb's Veterinary Drug Handbook* (6th ed.). Ames, IA: Blackwell Publishing.

Riviere, J., and Papich, M. (2009). *Veterinary Pharmacology and Therapeutics* (9th ed.). Ames, IA: Wiley-Blackwell.

Seymore S. Block. (2001). *Disinfection, Sterilization, and Preservation* (5th ed.) Philadelphia, PA: Lippincott, Williams, & Williams.

Spratto, G. R., and Woods, A. L. (2008). *PDR Nurse's Drug Handbook.* Clifton Park, NY: Delmar Cengage Learning.

Rice, J. (1999). *Principles of Pharmacology for Medical Assisting* (3rd ed.). Clifton Park, NY: Delmar Cengage Learning.

Shimeld, L. A. (1999). *Essentials of Diagnostic Microbiology.* Clifton Park, NY: Delmar Cengage Learning.

CHAPTER 15

ANTIPARASITICS

OBJECTIVES

Upon completion of this chapter, the reader should be able to:

- differentiate between the two main groups of parasites and describe how treatment for each varies.
- differentiate between nematodes, cestodes, and trematodes.
- describe various treatment and prophylactic strategies for nematodes, cestodes, and trematodes.
- describe treatment and prophylactic strategies for coccidia.
- explain why anticoccidial drugs are called coccidiostats and how this affects treatment.
- describe the various types of heartworm preventatives and treatments.
- describe the various forms in which ectoparasite products are available.
- describe the various chemicals used in ectoparasite control.

KEY TERMS

adulticide	antitrematodal drugs
anthelmintics	coccidiostats
anticestodal drugs	ectoparasites
anticoccidial drugs	endoparasites
antinematodal drugs	microfilaricide
antiprotozoal drugs	

Setting the Scene

A farmer has 200 calves that he would like to start on a deworming program. He is interested in giving his calves something that will cover many different parasites and is easy to administer. He would like to know what dewormers are available and whether they come in easy-to-administer forms. How should the veterinary technician explain antiparasitic drugs to clients? Are dewormers safe for cattle of all ages? What side effects does he need to watch for? Understanding the different types of parasites to control or treat, the routes of administration used in parasite control and treatment, and the side effects of these drugs is critical to providing proper care for clients.

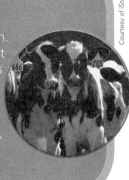

PARASITES AND ANIMAL DISEASE

Parasitism is a relationship between two different organisms in which one of the organisms (the parasite) benefits while the other organism (the host) is harmed. The parasite gains nourishment and a place to live and reproduce from the host. The harm to host animals from a parasite depends on the health of the host and may range from minor illness (failure to thrive or mild lethargy) to generalized impairment (anemia, weakness, weight loss, or even death).

Parasitic infections are common in veterinary medicine and cause a variety of clinical signs in animals depending on the body system they occupy. In addition, parasitic infection is a costly condition to manage in both pets and livestock. Some parasitic infections can be transmitted to people and can create a significant risk to public health. It is important to understand the life cycle of parasites and the mechanisms by which antiparasitic drugs work so that the proper treatment, preventative medication, and sanitation recommendations can be offered to the client.

TYPES OF PARASITES

Parasite control in animals is an important function for the veterinary staff. Clients want to know about the safest and most effective medicines to use in the treatment and prevention of parasitic disease in their pets and livestock. The two major groups of parasites are endoparasites and ectoparasites. **Endoparasites** live within the body of the host and cause internal parasite *infections*; **ectoparasites** live on the body surface of the host and cause external parasite *infestations*.

Parasites can be contracted in a number of different ways, including animal-to-animal contact, ingestion of contaminated food or water, insect transmission, or direct contact with the parasite (walking, lying, or rolling on infected soil). Some parasites that are present on or within the host may not cause clinical signs in the animal, whereas other parasites present on or within the host actively harm the animal. Most intestinal parasites are diagnosed by microscopic fecal examination. Keep in mind that clinical signs may develop with some parasitic infections before eggs are detected in clinical samples, which may make identification and prevention of parasitic disease more difficult.

ENDOPARASITES

Endoparasites live in the host's body. There are two main groups of pathogenic endoparasites in animals: helminths and protozoa. *Helminths* are worms found primarily in the gastrointestinal tract, liver, lungs, and circulatory system. The two major helminth groups are nematodes and platyhelminths (Figure 15-1). *Nematodes* are cylindrical (like a slender tube) and nonsegmented, and may be referred to as *roundworms* because they are round in cross-section. Most

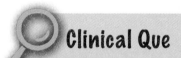

Clinical Que

Ectoparasiticides treat arthropod infestations (insects and arachnids), anthelmintics treat worm infections, antiprotozoals treat protozoan infections, and endectocides treat internal parasitic infections and external parasitic infestations. (Insects include flies, mosquitoes, bots, cuterebra, lice, and fleas. Arachnids include spiders, scorpions, ticks, and mites.)

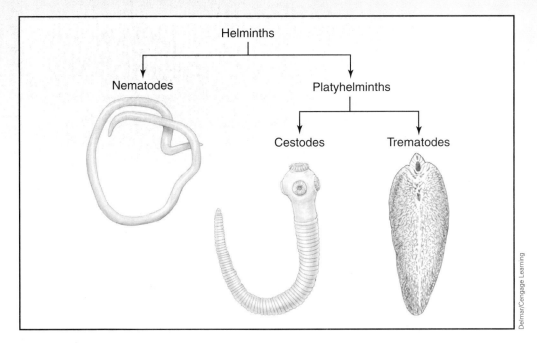

Figure 15-1 Types of helminths

nematodes inhabit the stomach and intestines of domestic animals, wild animals, and birds and are transmitted by the fecal-oral route (an exception are filarial nematodes, which are transmitted by arthropods, have adults that live in blood and lymphatic tissues, and are covered later in this chapter). *Platyhelminths* are flattened and may be referred to as *flatworms*. Flatworms are further divided into *cestodes* (sometimes referred to as *tapeworms*) and *trematodes* (sometimes referred to as *flukes*). There are many types of cestodes and trematodes, and they inhabit various locations in animals. Some cestodes can live in body tissues in the immature form, while other cestodes can live in the intestinal tract in the adult form. Some trematodes live in the bile ducts of ruminants and cause considerable losses to the livestock industry. Table 15-1 summarizes the types of helminths usually seen in veterinary medicine.

Table 15-1 Helminths of Veterinary Significance

MAIN CATEGORY	LOCATION OF PARASITE	SCIENTIFIC AND COMMON NAMES
Nematodes	Abomasal worms in ruminants or stomach worms in monogastric animals	• Barberpole worm *Haemonchus contortus, H. placei* (R) • Brown stomach worm *Ostertagia ostertagi* (R) • Small stomach worm or hairworm *Trichostrongylus axei* (R, H) • *Hyostrongylus rubidus* (Sw) • Large-mouth stomach worm *Habronema muscae* (H)

(Continued)

Table 15-1 *(Continued)*

MAIN CATEGORY	LOCATION OF PARASITE	SCIENTIFIC AND COMMON NAMES
	Intestinal worms	• Small intestinal worms *Cooperia punctata, C. oncophora, C. mcmasteri* (R) • Hookworms *Bunostomum phlebotomum* (R), *Ancylostoma sp* (D, F) • Nodular worms *Oesophagostomum* spp. (R, Sw) • Thread-necked intestinal worm *Nematodirus helvetianus* (R) • Bankrupt worm *Trichostrongylus colubriformis* (R) • Large strongyles *Strongylus vulgaris, S. edentatus, S. equinus, Triodon-tophorus* spp. (R, E) • Small strongyles *Cyathostomum* spp., *Cylicocyclus* spp., *Cylicostephanus* spp., *Cylicodontophorus* spp. (R) • Whipworms *Trichuris suis* (Sw), *Trichuris vulpis* (D) • Threadworms *Strongyloides ransomi* (Sw), *Strongyloides westeri* (E), *Strongyloides stercoralis* (D) • Ascarids *Parascaris equorum* (H), *Toxocara canis* (D), *Toxocara cati* (F), *Toxascaris leonina* (D, F) • Pinworms *Oxyuris equi* (E)
	Circulatory system worms	• Heartworms *Dirofilaria immitis* (D, F)
	Lungworms	• *Dictyocaulus* spp. (R, H) • *Prostostrongylus rufescens* (S, G) • *Muellerius capillaris* (S, G) • *Metastrongylus* spp. (Sw) • *Filaroides* spp. (D)
	Kidney worms	• *Stephanurus dentatus* (Sw)
	Urinary bladder worm	• *Capillaria* spp. (D, F)
	Gastrointestinal worms	• *Habronema* spp. (H) • *Draschia* spp. (H)
	Skin worm	• *Onchocerca* spp. (H)
Cestodes (tapeworms)		• *Moniezia benedeni* (R) • *Taenia* spp. (R, D, F) • *Echinococcus granulosus* (R is intermediate host, D) • *Dipylidium caninum* (D, F) • *Anoplocephala perfoliata, A. magna* (H) • *Paranoplocephala mamillana* (H)
Trematodes (flukes)	Liver fluke Deer liver fluke Lung fluke	• *Fasciola hepatica* (R) • *Fascioloides magna* (R) • *Paragonimus kellicotti* (D, F)

R = ruminants, C = cattle, S = sheep, G = goats, Sw = swine, D = dogs, F = cats, H = horses

Table 15-2 Protozoa of Veterinary Significance

Type of Movement	Location of Parasite	Scientific and Common Names
None	Gastrointestinal tract	• Coccidia *Eimeria* spp. (R, Sw, H)
None	Gastrointestinal tract	• Coccidia *Isospora* spp. (Sw, D, F)
None	Gastrointestinal tract; muscle tissue	• *Toxoplasma gondii* (F)
Flagellum	Gastrointestinal tract	• *Giardia* spp. (R, H, D, F)
None	Gastrointestinal tract	• *Cryptosporidium* spp. (R, H, D, F, Sw)
None	Muscle in ruminants	• *Sarcocystis* spp. (R in muscle)
	Gastrointestinal tract in dogs and cats	• *Sarcocystis* spp. (D, F shed in stool)
	Central nervous system	• *Sarcocystis neurona* (H)
Flagellum	Reproductive tract	• *Tritrichomonas foetus* (C)
None	Circulatory system	• *Babesia* (C, H, D)
None	Circulatory system	• *Cytauxzoon felis* (F)
Flagellum	Circulatory system	• *Trypanosoma* (D, C)
Cilia	Gastrointestinal tract	• *Balantidium coli* (Sw)

R = ruminants, C = cattle, S = sheep, G = goats, Sw = swine, D = dogs, F = cats, H = horses

Protozoa are single-celled parasites found in many species of animals (Table 15-2). Protozoa vary greatly in size, form, and structure with most of them being microscopic. Protozoa are typically categorized by the type of movement they have: cilia, flagella, pseudopodia (amoeba), or no movement.

ANTIPARASITIC DRUGS

Antiparasitic drugs are categorized by the type of parasite they work against. **Anthelmintics** kill or expel worm parasites by acting on metabolic pathways that are present in the invading worm, but absent or significantly different in the animal host. Anthelmintics are further categorized as antinematodal, anticestodal, or antitrematodal. These drugs are given in a variety of ways (Table 15-3). The drug's solubility largely dictates the route of administration. Water-insoluble anthelmintics are usually given orally (suspension, paste, granules), whereas more water soluble compounds can be given orally as a solution, topically as a pour-on, or via injection. The drug particle size also plays a role in the route of administration. Generally, smaller particles are more easily absorbed from the gastrointestinal tract; larger particles (especially those that are insoluble) have minimal absorption from the gastrointestinal tract and therefore may be less toxic to the host.

Clinical Que

Antiparasitic drugs should be used with caution in old, young, pregnant, or debilitated animals. Consult package inserts before applying antiparasitics to these groups of animals.

Clinical Que

Most anthelmintics have withdrawal times; consult package inserts or the Food Animal Residue Avoidance Bank (FARAD) at www.farad.org for these times.

Table 15-3   Administration Routes for Anthelmintics

MAIN ROUTE	EXAMPLE ROUTE	DESCRIPTION	CONCERNS
Oral	Tablets	May be hard tablet, chewable tablet, or bolus (large pill) that animal ingests or is given	• Allows greater control over amount of drug given • Palatability is important (especially with cats and horses)
		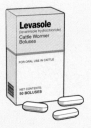	
	Liquid	May be in the form of a solution, suspension paste, paste syringe, or drench	• Drug must be shaken well to ensure adequate mixing of chemical throughout the liquid • Paste syringes contain a precalibrated amount of paste that need not be shaken • Drenches are liquid forms given by mouth that force the animal to drink • Allows greater control over amount of drug given
		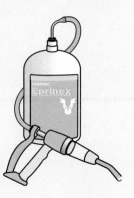	
	Feed additives	May be in feed, added to mineral mixes, added to drinking water, or added to salt blocks	• Treats parasites in large numbers of animals • Allows little control over amount of drug ingested by an individual animal • Is stress free for the animals • Saves expense of rounding up livestock

(Continued)

MAIN ROUTE	EXAMPLE ROUTE	DESCRIPTION	CONCERNS
	Sustained-release	Device is implanted in rumen to allow slow release of the drug over time (especially helpful in treating later stages of parasite larvae)	• Saves time in retreating of animals • Can treat animals over a period (like the complete grazing season)

MAIN ROUTE	EXAMPLE ROUTE	DESCRIPTION	CONCERNS
Injectable	Solution	Given SQ usually (if given IM, may affect the carcass)	• Easy way to administer • Local reaction is possible, but is rare • Allows greater control over amount of drug given • Achieves higher blood levels rapidly • Requires livestock to run through chutes or be crowded into restricted spaces for injection with multidose syringes for treatment • Can control amount of drug given

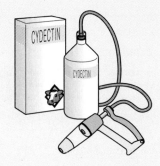

MAIN ROUTE	EXAMPLE ROUTE	DESCRIPTION	CONCERNS
Topical (pour-ons)	Solution	Absorbed through the skin via the sebaceous glands and hair follicles	• Used in large animals • Achieves high blood levels rapidly • Easy to administer • Reduces stress for treated animals • Used for internal and external parasites • Pour-ons are highly concentrated dose forms that may be hazardous

Antinematodal Drugs

Antinematodal drugs work against nematodes. Nematodes are elongated, unsegmented, cylindrical worms that come in a variety of sizes and shapes and infect a variety of organs and organ systems. There are seven groups of antinematodal drugs (Table 15-4). Some drugs are in several of these groups, as they may be effective against more than one type of parasite.

Table 15-4   Types of Antiparasitic Drugs

Benzimidazoles (nematodes)
- thiabendazole
- oxibendazole
- mebendazole (also works on tapeworms)
- fenbendazole (also works on tapeworms, flukes, and *Giardia* protozoa)
- albendazole (also works on tapeworms, flukes, and *Giardia* protozoa)
- oxfendazole (also works on tapeworms and flukes)
- febantel (a probenzimidazole used in combination with other products to broaden its spectrum of activity)

Imidazothiazoles (nematodes)
- levamisole

Tetrahydropyrimidines (nematodes)
- pyrantel pamoate
- pyrantel tartrate
- morantel tartrate

Organophosphates (nematodes; ectoparasites, including bots)
- dichlorvos
- coumaphos

Piperazines (nematodes)
- piperazine

Avermectins (also called macrocyclic lactones) (nematodes; heartworm prevention; ectoparasites, such as bots and grubs)
- ivermectin
- eprinomectin
- selamectin
- moxidectin
- milbemycin oxine
- doramectin

Depsipeptides (nematodes)
- emodepside (used in combination with praziquantel)

Pyrazine derivatives (platyhelminths)
- praziquantel (tapeworms, flukes)
- epsiprantel (tapeworms)

Benzene sulfonamide (flukes)
- clorsulon

Coccidiostats (coccidia)
- sulfadimethoxine
- amprolium
- decoquinate
- nicarbazine
- monensin
- robenidine

Nitroimidazoles (*Giardia*)
- metronidazole

Carbanilide derivative (*Babesia*)
- imidocarb

Aminoquinolone (*Babesia*)
- primaquine

Folic acid antagonist (*Toxoplasma* and *Sarcocystis*)
- pyrimethamine

Triazine (*Sarcocytis*)
- ponazuril

Nitrothiazolyl-salicylamide derivative (*Sarcocystis*)
- nitazoxanide

The *benzimidazole* category of antinematodal drugs has excellent efficacy against nematode infections. They are believed to work by interfering with energy metabolism of the worm. The main benzimidazole is *thiabendazole*, whose effectiveness against ascarids and strongyles and low toxicity (due to limited gastrointestinal absorption) made its introduction a landmark in helminth therapy. Thiabendazole is no longer commonly available in the United States, but because it also has antifungal and anti-inflammatory effects, it may be found in otic preparations (e.g., Tresaderm Otic®) or used to treat ringworm in cattle and aspergillosis in dogs. The other benzimidazoles work at lower dose levels and with broader-spectrum activity. Animals take all the benzimidazoles orally, as a paste, a tablet, a granulated powder, or a suspension (Figure 15-2). Repeated doses of benzimidazoles are advantageous because of their slow killing process. Side effects are rare with benzimidazoles, but may include vomiting, diarrhea, and lethargy. Other benzimidazoles include oxibendazole, mebendazole, fenbendazole, albendazole, and oxfendazole (covered in Table 15-4).

- *Oxibendazole* is used as a horse dewormer and in combination products for dogs. In dogs, liver toxicity has been noted with this product. An example is Anthelcide EQ Equine Wormer Paste®.

- *Mebendazole* is a granular powder used in dogs and horses to treat ascarid, hookworm, and cestode infections. An example is Telmintic®. Liver toxicity has been noted with this product.

- *Fenbendazole* is used in small animals, food animals, horses, birds, and reptiles. It has a wide spectrum of activity, including nematodes, hookworms, whipworms, and the cestode *Taenia pisiformis* in small animals. Fenbendazole may also be used to treat metronidazole-resistant giardiasis. Vomiting and diarrhea have been caused by this drug.

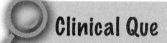

Clinical Que

The most common target for antiparasitic drugs to attack is the parasite's nervous system.

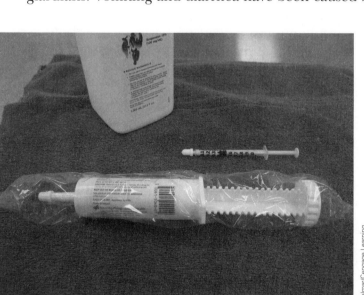

Figure 15-2 Fenbendazole comes in a variety of forms including suspension (which is then aspirated into a syringe to give to the animal) and paste (pastes given to large animals frequently are sold in premeasured syringes)

It has been approved for use in lactating dairy animals. An example is Panacur®, which is found in a variety of forms including granules, suspensions, and pastes.

- *Albendazole* is primarily used in cattle to treat abomasal and intestinal nematodes, flukes, and cestodes. It is not labeled for use in pregnant cows or in dairy cows of breeding age. Albendazole also has a long withdrawal time, which limits it use. An example is Valbazen® suspension.

- *Oxfendazole* is available as an oral paste and solution that is used in horses against strongyles, ascarids, and pinworms. Examples of oxfendazole include Benzelmin® paste and Synanthic®.

A drug in a related group, the probenzimidazoles, is *febantel*. Probenzimidazoles are examples of prodrugs (drugs that must be metabolized in the animal's body to the active form of the drug). Probenzimidazoles are metabolized in the animal's body to true benzimidazoles and then work the same as the benzimidazoles. Febantel is usually added to another antiparasitic. Drontal Plus® tablets is a trade name for a febantel, pyrantel pamoate, and praziquantel combination.

The imidazothiazoles are a second group of antinematodal drugs, of which *levamisole* is an example. Levamisole stimulates the nematode's cholinergic nervous system, leading to paralysis of the parasite. Because levamisole works on the adult nematode's nervous system, it is not ovicidal. Levamisole expels most nematodes within 24 hours, though the worms may be passed alive. Levamisole comes in oral forms, such as pellets, powder to mix into drinking water, suspensions, and pastes. Levamisole is effective against ascarids, strongyles, whipworms, and hookworms and has been used as a **microfilaricide** (drug that eliminates the circulating prelarval stage called microfilaria) in heartworm treatment regimens in dogs (this use is no longer common). Levamisole also has anti-inflammatory and immunostimulant properties. Levamisole toxicity in the host animal is due mainly to its cholinergic effects, which may cause salivation, ataxia, and muscle tremors. Trade name examples of levamisole are Levasole® and Tramisol®.

The tetrahydropyrimidines include *pyrantel pamoate, pyrantel tartrate,* and *morantel tartrate.* These drugs are cholinergic agonists that mimic the action of acetylcholine. This causes initial stimulation and then paralysis of the worm, allowing it to be excreted by the intestine. These drugs come in tablet, paste, suspension (has a somewhat pleasant taste), and top dress forms and are very safe, with few side effects. This group is effective at treating ascarids, pinworms, strongyles, and hookworms. Particles in these suspensions can settle to the bottom of the container; therefore, this product should be shaken thoroughly prior to withdrawing a dose. Trade name examples of this group include Nemex® and Strongid-T® (pyrantel pamoate), Strongid-C® (pyrantel tartrate), and Rumatel® (morantel tartarate).

Organophosphates (OPs) are used against both endoparasites and ectoparasites. OPs have a narrow range of safety and were generally given as a feed supplement to beef and dairy cattle. OPs are one of the few drugs effective against bots in horses. Currently OPs are rarely used in animals due to their side effects; however, they are used in agricultural products, and OP toxicity may be seen in animals. OPs inhibit cholinesterase (are labeled as cholinesterase inhibitors), causing acetylcholine to remain active in the neuromuscular junction of the parasite. They are neurotoxic to parasites and can cause some neurologic side effects in the host such as the development of tremors and hyperexcitability. Other host side effects of OPs result from parasympathetic nervous system stimulation and can be remembered by the acronym SLUDDE (salivation, lacrimation, urination, defecation, dyspnea, and emesis). OPs should not be given to heartworm-positive dogs, because they may cause dyspnea (difficulty breathing) and death due to sudden worm kill-off. OPs should be used with caution in all species; if the OP is not labeled for a particular species, it is better to use an antiparasitic drug that is labeled for that species, rather than using an OP extralabel. OPs are effective against bots, nematodes, strongyles, and pinworms in horses; strongyles in ruminants; hookworms, nematodes, and whipworms in dogs and cats; and ascarids, whipworms, strongyles, and nodule worms in swine. Some generic and trade names of OPs are *dichlorvos* (Task®, Atgard®), and *coumaphos* (Baymix®).

Piperazine compounds consistently eradicate only ascarids in dogs, cats, horses, ruminants, and poultry. They are anticholinergic drugs that block neuromuscular transmission in the parasite. These drugs go in food or drinking water for poultry and swine; dogs and cats usually get tablets. Clients should be made aware that treating animals that have ascarid infections with piperazine will often result in intact worms being vomited or passed in the stool. Piperazines are practically nontoxic and are sold over the counter under such trade names as Pipa-Tabs® and Hartz Advanced Care Once-a-Month Wormer® for Puppies.

The representative drug of the avermectins (also known as macrocyclic lactones) is *ivermectin*. Avermectins bind to certain chloride channels in invertebrate nerve and muscle cells, causing paralysis and death of the parasite. Low doses of ivermectin, given orally and parenterally, work well against a wide range of internal and external parasites such as nematodes, bots, lice, mange mites, and grubs. The low dose also allows it to be injected subcutaneously. Occasionally ivermectin can cause rapid death of parasites, resulting in allergic reactions, inflammation, and swelling in the area of the body where the parasite dies. Ivermectin has also produced adverse reactions in collies and collie crosses. Trade names for ivermectin products are Heartgard®, Heartgard Plus®, and Ivomec®. The "Plus" products contain other antiparasitic agents to broaden their spectrum of activity. Another avermectin is *moxidectin*. This is the chemical used in ProHeart-6®, the six-month injectable heartworm preventative, Cydectin® in cattle, and Quest 2 percent Equine Oral Gel®. Side effects of this group of antiparasitics include stupor and convulsions. (This group of drugs is also covered under heartworm treatments later in this chapter.)

Depsipeptides are antiparasitic agents effective against nematodes (ascarids and hookworms) that work at the neuromuscular junction by stimulating

Clinical Que

Ivermectin and related compounds are not effective against cestodes or trematodes.

Clinical Que

Most parasite infections may be diagnosed via microscopic identification of eggs or oocysts (antigen or antibody tests are also available for some parasite infections). Egg identification is important so that the most appropriate antiparasitic agent can be used to treat the affected animal. Repeating the fecal examination after treatment (fecal recheck times vary with the parasite) is equally important to ensure that the treatment was effective.

presynaptic receptors in the secretin receptor family causing paralysis and death of the parasite. *Emodepside* is the representative drug in this category, which is effective against *Toxocara cati, Toxascaris leonine,* and *Ancyclostoma tubaeforme.* Emodepside is found in a combination topical product (Profender® Spot On) that also contains praziquantel (which also covers the cestodes *Dipylidium caninum, Echinococcus granulosus,* and *Taenia taeniaeformis*). It is not used in kittens less than eight weeks of age. Side effects include lethargy, salivation, vomiting, and neurological signs (tremors) usually following the animal's licking of the application site. Excessive grooming of the application site may also be seen.

Anticestodal Drugs

Anticestodal drugs are used for the treatment of cestode infections in animals (Table 15-4). Cestodes are segmented flatworms that consist of a scolex (head), neck, and a variable number of proglottids (segments). Cestodes increase in length by producing new proglottids from the neck area; therefore, the oldest proglottids are at the distal end of the worm. A cestode will attach itself to the intestinal wall by its scolex, so dislodging the scolex is an important part of cestode treatment. Some anticestodal drugs are used in combination with the antiparasitic drugs already discussed.

Praziquantel works by increasing the cestode's cell membrane permeability, resulting in disintegration of the worm's outer tissue covering. This disintegration includes the scolex, so the cestode cannot relodge itself into the bowel wall and begin producing new proglottids. Because the cestode disintegrates, proglottid segments are not observed in the stool. Praziquantel works on all of the cestode genera (*Taenia, Echinococcus,* and *Dipylidium*), but flea control is needed to eliminate the cestode *Dipylidium caninum* because fleas are involved in this tapeworm's life cycle. Praziquantel is effective against juvenile and adult cestodes but not eggs; therefore, proper hygiene such as hand washing after cleaning a litterbox or disposing of feces is essential. Side effects include anorexia, vomiting, diarrhea, and lethargy. Droncit® is a trade name of praziquantel that is available for treating tapeworms of dogs and cats (*Dipylidium caninum* and *Echinococcus granulosus*) as an oral tablet or IM injection (injection stings, so the oral route is preferred). Profender® (emodepside plus praziquantel) is a combination product used topically in cats greater than 1 kilogram and older than eight weeks of age for treatment and control of hookworm (*Ancyclostoma tubaeforme*), roundworm (*Toxocara cati*), and tapeworm (*Dipylidium caninum* and *Taenia taeniaeformis*) infections. Additional side effects for Profender® were listed under emodespide. Quest® Plus is an equine oral gel that contains moxidectin and praziquantel and is labeled for treatment of *Anoplocephala perfoliata, Anoplocephala magna,* and *Paranoplocephala mamillana* tapeworms in horses (as well as those parasites covered by moxidectin).

Epsiprantel, known by its trade name Cestex®, is approved for dogs and cats greater than seven weeks of age and is effective against juvenile and adult *Taenia* and *Dipylidium* cestodes, but not *Echinococcus* species. Epsiprantel causes disintegration of the cestode, so proglottid segments are not observed in the stool. Like praziquantel, proper flea control is needed to control *Dipylidium* tapeworms.

Fenbendazole, a benzimidazole covered earlier, is effective against *Taenia* spp. of cestodes, but not *Dipylidium caninum*. Unless other parasites that can be treated with fenbendazole are present in the animal, praziquantel or epsiprantel are recommended.

Antitrematodal Drugs

Antitrematodal drugs are used in the treatment of *trematodes*, the flat, leaf-shaped helminths whose bodies lack segmentation (Table 15-4). Trematodes are commonly referred to by the area they infect (e.g., the liver trematode or liver fluke). Trematodes have intermediate hosts, the first of which is almost always a snail. Snails are found in wet, rainy environments; therefore, managing trematode infections requires more than just drug treatment. Clorsulon, albendazole, and praziquantel are all antitrematodal drugs.

Clorsulon, a benzene sulfonamide, inhibits the trematode's enzyme systems used in energy production and robs the trematode of its energy. Clorsulon is a drench and is effective against the adult and immature form of the liver trematode, *Fasciola hepatica*, in cattle. Curatrem® is an over-the-counter product of clorsulon. Slaughter withdrawal time is eight days when used as labeled. Withdrawal times in milk have not been established; therefore, use of clorsulon is not recommended in female dairy animals of breeding age. Ivomec® Plus is a combination product containing clorsulon and ivermectin. The addition of ivermectin expands the spectrum of activity; however, it also lengthens the withdrawal time to 49 days prior to slaughter and is not used in female dairy cattle of breeding age.

Albendazole is effective against a variety of parasites, including *Fasciola hepatica* adults in cattle. It is a benzimidazole, and it works by interfering with the energy metabolism of the worm. Albendazole is also effective against some nematodes. Valbazen® is a trade name albendazole product that is a broad-spectrum anthelmintic (it also works on cestodes and nematodes). Both albendazole and clorsulon have withdrawal times associated with their use. Milk withdrawal times for albendazole have not been established; therefore, it is not approved for use in lactating animals.

Praziquantel was covered earlier in the anticestode section. It may also be used to treat lung trematodes in dogs and cats.

Anticoccidials and Other Antiprotozoals

Coccidiosis is a protozoan (*Coccidia*) infection that causes various intestinal disorders—some serious and even fatal—in various species. Coccidiosis (always prevalent in the poultry industry) can be managed by adding **coccidiostats (anticoccidial drugs)** to the daily feed ration. Coccidia have a complex life cycle that makes them difficult to treat. Coccidia infections tend to involve management problems with sanitation procedures. Coccidiostats do not kill the coccidia parasite, so cleaning of contaminated

Clinical Que

Ruminant cestodes rarely cause problems and can be treated with fenbendazole. Cestodes in horses (*Anoplocephala* spp.) can be pathogenic and should be treated with praziquantel or a double dose of pyrantel pamoate.

Clinical Que

Coccidia infections in carnivores are caused by *Isospora* spp., and coccidia infections in herbivores are caused by *Eimeria* spp. If dogs have *Eimeria* infections, it is due to their ingestion of herbivore feces.

Clinical Que

"Controlled infection" is the goal for treating coccidial infections. Drugs and sanitation play prominent roles in this control.

waste from housing facilities and proper disinfection are key in removing coccidia from an environment. Providing clean food and water and avoiding overcrowded conditions also help to keep coccidia organisms in check. Sulfa drugs produced in the 1940s were the first effective coccidiostats; however, the protozoans developed sulfa-resistant strains. One widely used sulfa drug, *sulfadimethoxine* (Albon®), is still used in poultry, dogs, and cats for treatment of coccidia. Sulfadimethoxine does not eliminate the infection that the animal already has, but it reduces the spread of disease by reducing the number of oocysts shed. This reduced shedding reduces the spread of coccidia to other animals. More effective drugs introduced through the 1970s included *nicarbazine* (Maxiban 72® for chickens), *amprolium* (Corid® for calves), *monensin* (Coban 60® for poultry, Rumensin® for cattle and goats), *decoquinate* (Deccox® for cattle and goats), and *robenidine* (Robenz Type A Medicated Article® for chickens). These drugs mainly work by affecting some aspect of the protozoan's metabolism (amprolium competes with the protozoa for thiamine, monensin binds with ions to affect mitochondrial function). These drugs work at various stages in the coccidian life cycle. Most of these drugs, when given to animals that are actively shedding oocyts, will not cure those animals, but may reduce the number of oocysts shed into the environment. Most of these drugs are given orally in feed, water, or by liquid or tablets. Some coccidiostats such as monensin need to be handled carefully, which includes wearing protective clothing buttoned to the neck and wrist and a washable hat, impervious elbow-length gloves, and a half face-piece respirator with dust cartridge. Hands should be thoroughly washed after handling monensin. Monensin should not be used in equine species as it has caused death in horses.

In addition to coccidia, other parasitic protozoa cause disease in animals. *Giardia* is another protozoan parasite seen in animals for which antiprotozoal drugs are used. For the organism *Giardia lamblia*, there is a preventative vaccine called GiardiaVax®, as well as drugs for treating the organism. GiardiaVax® is an annual vaccine labeled for use in dogs to prevent infection by *Giardia* protozoa; it also reduces the shedding of cysts. Initially, two injections are given SQ about two to four weeks apart. Treatment of giardiosis includes *metronidazole* (Flagyl®), *fenbendazole* (Panacur®), and *albendazole* (Valbazen®). Metronidazole is a nitroimidazole drug believed to work by entering the protozoal cell and interfering with its ability to function and replicate. Fenbendazole and albendazole were covered earlier in this chapter. References should be consulted prior to using these drugs in pregnant or lactating animals, to assess the drug's safety in these animals. Metronidazole has been experimentally identifed as causing mutations.

Another protozoan, *Babesia* sp., is a blood parasite transmitted by ticks. It can be treated subcutaneously or intramuscularly with *imidocarb* (Imizol®). Imidocarb, a carbanilide derivative, appears to have a cholinergic effect on the protozoan. It is slowly metabolized and therefore is not used in food-producing animals because of increased withdrawal times. Side effects of imidocarb include salivation, dyspnea, and restlessness. *Primaquine* (generic) is an aminoquinoline drug used to treat *Babesia felis* in cats. It has a very narrow therapeutic range, and monitoring of the patient's CBC helps identify if myelosuppression is occurring. Side effects include vomiting (giving with food may alleviate vomiting) and myelosuppression. Using tick prevention can reduce infections with *Babesia* sp.

Equine protozoal myeloencephalitis is a neurological disease in horses that is caused by the protozoan *Sarcocystis neurona*. The immature forms of this protozoan are ingested by the horse (dead end host), and then they enter the bloodstream, replicate, and migrate to the central nervous system. *Pyrimethamine* (Daraprim®) is an antiprotozoal tablet used to treat toxoplasmosis in small animals and equine protozoal myeloencephalitis (EPM) in horses. Pyrimethamine works by inhibiting an enzyme that converts one form of folic acid used for metabolism in parasites to an inactive form of folic acid. Side effects include anorexia, vomiting, and myelosuppression. Pyrimethamine may be combined with *sulfadiazine* (ReBalance®) for enhanced efficiency of treating EPM.

Other drugs approved for use in treating EPM are ponazuril and nitazoxanide. *Ponazuril* (Marquis®) is a triazine antiprotozoal drug that targets the plasmid body found in some protozoa. Ponazuril may also be used in dogs and cats to treat toxoplasmosis and coccidiosis. Side effects include the development of blisters on the mouth or skin rashes. *Nitazoxanide* (Navigator®) is a nitrothiazolyl-salicylamide derivative available as an oral paste used to treat EPM. Side effects include fever, anorexia, and lethargy.

HEARTWORM MEDICATIONS

Heartworm disease is a parasitic disease of dogs and other canines, but has also been identified in cats, ferrets, and occasionally humans. This disease is caused by the filarial nematode *Dirofilaria immitis*, transmitted by mosquitoes. Adult heartworms usually live in the pulmonary artery, but can also be found in the right ventricle and right atrium and, rarely, in the vena cava. Female adult heartworms produce microfilariae. Microfilariae circulate in the bloodstream and are taken up by mosquitoes when they feed on an infected animal's blood. In the mosquito, the microfilariae undergo changes until they become infective third-stage larvae. When these mosquitoes bite other animals, the third-stage larvae are transmitted to the new host. These larvae develop into adults and ultimately reach the aforementioned locations. The life cycle of heartworm disease is illustrated in Figure 15-3.

Medications to prevent and treat this disease, as well as guidelines issued by the American Heartworm Society, have been improved and monitored to find the best methods of curtailing this disease. Managing this disease involves preventing third-stage larvae from reaching maturity, eliminating adult heartworms, and eliminating circulating microfilariae.

Third-stage larvae can be kept from reaching maturity by the use of preventative medication. When given to a heartworm-negative dog, such a drug will prevent any larvae transmitted by a mosquito vector from reaching the adult stage. When given at higher doses, these medications also kill circulating

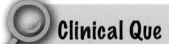
Clinical Que

Microfilariae are prelarval stages of *Dirofilaria immitis*. The term "microfilaria" is not synonymous with larva.

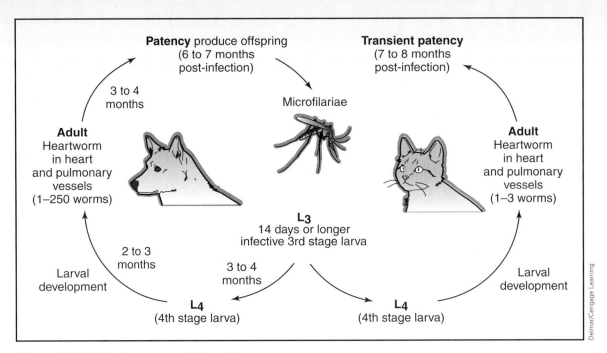

Figure 15-3 The heartworm life cycle

heartworm microfilariae in heartworm-positive dogs. Examples of preventative drugs include the following:

- *ivermectin* (Heartgard®, Heartgard Plus®, Heartgard® for Cats, Iverhart™ Plus, Tri-Heart™ Plus). This monthly preventative is given orally to eliminate the tissue stage of heartworm microfilaria. Ivermectin revolutionized heartworm prevention in dogs, allowing dosing to go from once daily to once monthly (there are now other drugs that work monthly or semiannually). Heartgard Plus®, Tri-Heart™ Plus, and Iverhart™ Plus also contain pyrantel pamoate, which is effective against ascarids and hookworms. At higher dosages, ivermectin is used after **adulticide** treatment in heartworm-positive dogs. Ivermectin is not recommended for puppies less than six weeks of age, and it should be used with caution in collies or collie mixes. Collies and collie mixes should be observed for at least eight hours after administration of ivermectin for development of side effects. Side effects for all dogs include neurologic signs such as salivation, ataxia, and depression.

    Ivermectin effectiveness varies with dosage. At low dosages (6 mcg/kg), it is used as a heartworm preventative in dogs and cats that acts by blocking larval development. At moderate dosages (50 mcg/kg), it is used to kill microfilariae in dogs. At high dosages (200 mcg/kg), it will kill most internal parasites. Ivermectin doses must be calculated accurately. It is given in dosages of *mcg/kg*, which is a thousandfold lower than most drugs. Most calculation errors occur because the mcg unit is overlooked.

- *milbemycin* (Interceptor®, Sentinel®). This is a monthly preventative given orally that eliminates the tissue stage of the heartworm microfilariae by interfering with invertebrate neurotransmission.

Both products are also effective against hookworms, ascarids, whipworms, and *Demodex* mites in dogs. Milbemycin has been used as a heartworm preventative in cats. Sentinel® also contains lufenuron for flea control. Side effects from either product are uncommon.

- *selamectin* (Revolution®). This monthly preventative is applied topically and absorbed systemically for the prevention of heartworm disease, and the prevention and control of fleas, ear mites, and sarcoptic mange in dogs. It has some activity against the American dog tick (*Dermacentor variabilis*), but not the deer tick (*Ixodes scapularis*), which is the carrier of Lyme disease. It is effective for treatment of ascarids and hookworms in cats. Side effects include alopecia at the application site, vomiting, and diarrhea.

- *moxidectin* (ProHeart-6®). This six-month injection interrupts early larval development and thus prevents heartworm disease. Its side effects include neurologic and gastrointestinal signs. If used to replace another heartworm preventative, the first dosage of moxidectin must be given within one month of stopping the original preventative.

Adult heartworms may develop in the right ventricle because of improper or no use of preventatives. Adulticides are used to kill the adult heartworm in heartworm-positive dogs. *Melarsomine* (Immiticide®) is an arsenical drug given by deep IM injection once daily for two days. The drug is given in the epaxial muscles on either side of the vertebral column, in the area between the L3 and L5 vertebrae. Melarsomine is administered with a longer needle (1.5 inches) to place it deep in the muscle and with a small-gauge needle (22- or 23-gauge needle depending on the dog's weight) to prevent leakage of drug from the injection site. Melarsomine appears to be less toxic than thiacetarsamide (Caparsolate®, the former adulticide treatment) because it contains a different form of the arsenic compound. It is not recommended for animals with large numbers of heartworms in the right ventricle, right atrium, or vena cava (referred to as *caval syndrome* when the vena cavae are involved). Coughing, gagging, and lethargy are side effects of this drug. Dogs should be kept quiet after receiving adulticide treatment. Nephrotoxicity and hepatotoxicity may occur with melarsomine, but less so than with thiacetarsamide.

Any circulating microfilariae that appear after infection with adult heartworms must be eliminated using a microfilaricide drug. Microfilaricide drugs are given six weeks after administration of the adulticide and include the following:

- *ivermectin* (discussed previously). As a microfilaricide, it is given at a higher dose orally and monitored for neurologic side effects. The dog then returns in three weeks for a blood test; if positive, the dog is reevaluated for adult heartworms. If negative, the dog is started on a monthly preventative.

- *milbemycin* (discussed previously).

- *levamisole* (Levasole®, Tramisol®). This drug is given orally for a week or longer depending on the dosage. It is now used infrequently, due to the increasing use of ivermectin and milbemycin.

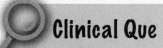

Clinical Que

All manufacturers of heartworm preventative recommend that dogs be microfilaria-negative before they are started on a preventative.

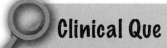

Clinical Que

Glucocorticoids and aspirin are sometimes given after adulticide treatment to prevent emboli; however, the use of these drugs is currently discouraged because of their paradoxical protective effect on adult female worms.

ECTOPARASITES

Ectoparasites live on the outside of the animal's body, but can still cause disease in livestock and pets and loss of revenue in farm animals. Accurate identification of the ectoparasite is necessary to select the most appropriate drug to use for control. Ectoparasites include flies (bots and maggots), grubs, lice, fleas, mites, ticks, and mosquitoes.

ECTOPARASITE DRUGS

Ectoparasites are controlled in many different ways. Ectoparasiticides are externally applied through the use of sprays (prediluted and concentrated), dips, pour-ons, ear tags, collars, spot-ons, shampoos, dusts, and foggers. Ectoparasiticides can also be delivered as oral medications. Table 15-5

Table 15-5 Application Methods for Ectoparasiticides

TYPE OF PRODUCT	ADVANTAGES	DISADVANTAGES
Prediluted sprays (include sprays for animals and premise sprays)	• Convenient and easy to use (apply from head to tail, avoiding eyes, mouth, and nose) • Usually has quick kill • May have residual effects • Available for animal and environment	• Water-based sprays do not penetrate oily coats or fabrics well • Alcohol-based sprays may be drying and irritating to skin
Concentrated Sprays	• Concentrated form may offer cost savings • Can be diluted at different concentrations for different ectoparasites	• Error in dilution may occur • Diluted product may not have long shelf life
Yard spray/Kennel spray	• Offer residual effects	• Can only be used on environment • Efficacy varies
Dips	• Offer residual effects	• Must be diluted properly • Animal should be shampooed first • Animal must dry with dip product on—cannot rinse product off
Pour-ons	• Can ensure that an individual animal is treated • May treat many animals at a time with proper application devices	• Activity of drug may be limited if applied to unclean animal (e.g., animal with caked mud or manure on its hide) • May be applied incorrectly, resulting in limited value of the treatment or development of toxicity (application varies; may be along the backline from shoulders to the pelvis or in single spot)
Shampoos	• Rinse well • May contain medication effective against parasites	• May only contain products for cleaning the coat • No residual effect even if medication present • May have to be diluted before use • Must leave on animal for a specified time prior to rinsing

(Continued)

TYPE OF PRODUCT	ADVANTAGES	DISADVANTAGES
Dusts or powders	• Can be used in animals that do not tolerate sprays	• Do not provide quick kill • May irritate and dry skin
Foggers	• Work well in large, open rooms • Quick method for environmental control	• Product does not get everywhere needed (in corners or furniture); however, coverage can be improved if premise spray is used with the fogger • Can be toxic to fish; must cover food products when applying
Oral products	• No mess • Works for a period of time	• May not kill all stages of the ectoparasite • May have systemic effects • Ectoparasite may have to take a blood meal for the medication to be effective
Topical long-acting (Spot-ons)	• Long-lasting • May work for multiple parasites • May work for different stages of parasite development • Work by providing area of repellent near application site	• May cause skin problem at site of application • Causes oiliness at site of application • Animal should avoid bathing or swimming with some products • May not be usable on young, old, or sick animals
Injectables	• Long-lasting in some cases • Easier for owners who do not want to administer medication to the animals themselves	• May cause adverse reaction at injection site • Should be given by veterinary staff

summarizes various forms of ectoparasiticides, and Figure 15-4 illustrates application of spot-on formulas.

Always read product labels to determine if there is a need to follow any safety procedures when using these products. The veterinary technician may need to

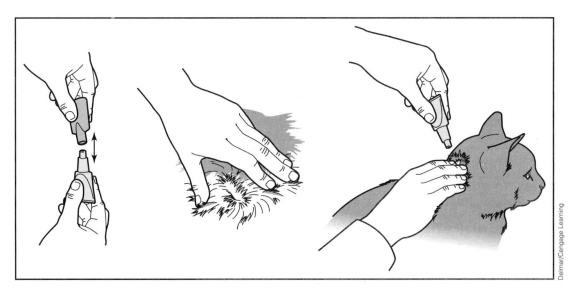

Figure 15-4 Spot-on application. (A) Applicator is held upright, and snap applicator tip is bent or cap is twisted off (depending on the manufacturer). (B) Animal's hair is parted to expose skin. (C) Applicator is squeezed and contents are applied to skin.

wear aprons and waterproof gloves when applying some products. Proper disposal of excess or unused product needs to be done in accordance with state pesticide and federal (EPA) guidelines to avoid groundwater or wildlife contamination. Animals treated with residual products may have to have their activity limited to avoid environmental contamination (e.g., cattle may not be permitted to enter lakes, ponds, or streams for a designated time, to avoid wildlife contamination). Proper ventilation and/or disposal of bedding material in the area in which the product is applied may also be recommended. Always consult MSDS information, package inserts, or drug handbooks prior to use.

Table 15-6 lists chemical agents for ectoparasite treatment in animals. Table 15-7 lists drugs used for monthly or longer prevention of fleas, ticks, and heartworm disease.

Table 15-6   Chemical Products for Ectoparasite Control

PRODUCT CATEGORY	GENERIC AND TRADE NAME EXAMPLES	EFFICACY
Pyrethrins and pyrethroids: • Names end in *-rin* or *-thrin* • Pyrethrins are natural plant products • Pyrethroids are synthetic pyrethrins	• *pyrethrin* (Mycodex Shampoo®, Bio Spot Shampoo®) • *d-trans allethrin* (Duocide Spray®) • *permethrin* (ProTICall®)	• Very safe • Quick kill • Adulticide, insecticide/miticide • Often manufactured with other products such as imidacloprid and pyriproxyfen • Acts on the parasite's nerve cell membrane to disrupt the sodium channel current, which delays repolarization and paralyzes the parsite • May have limited residual effects • Form labeled for dogs may be too high a concentration for cats • In small animals used primarily for fleas and ticks in dogs • In large animals used primarily for flies, lice, mites, mosquitoes, ticks, and keds • Commonly used in sprays, dips, foggers, pour-ons, insecticidal ear tags, and premise sprays
Insect growth regulators (IGR): • Include insect development inhibitors and juvenile hormone mimics	• *methoprene* (Ovitrol® and Siphotrol®) • *pyriproxyfen* (Nylar®, Vectra™)	• Products with IGR provide the flea with high levels of IGR, which mimics the insect's juvenile hormone (JH). Fleas need low levels of JH to molt to the next stage; high levels interrupt normal molting, so the insect stays in the larval stage and eventually dies • When combined with an adulticide (permethrin, fipronil, phenothrin, etc.), it kills all stages of the parasite, making reinfestation unlikely • Do not have adulticide activity by themselves • Found in sprays and flea collars

(Continued)

Product Category	Generic and Trade Name Examples	Efficacy
Chitin synthesis inhibitor	• *lufenuron* (Program®) • *lufenuron and milbemycin* (Sentinel®) • *lufenuron and nitenpyram* (Capstar® Flea Management System, Program® Flavor Tabs)	• Chitin is an insect protein that gives strength and stiffness to its body; chitin synthesis inhibitors prevent proper formation of this protein • Lufenuron is an oral tablet given to dogs and a suspension or tablet given orally or an injection given SQ to cats to protect against fleas for 1 month • Fleas that feed on blood containing lufenuron continue to lay eggs, but the eggs fail to develop normally • Does not kill adult fleas
Neonicotinoid	• *nitenpyram* (Capstar®) • *nitenpyram and lufenuron* (Capstar® Flea Management System, Program® Flavor Tabs)	• Neonicotinoid compound that binds and inhibits nicotinic acetylcholine receptors • Tablet that kills adult fleas within 30 minutes (nitenpyram not effective against eggs or immature forms if used alone) • Can safely give a dose as often as one per day • Can use on puppies and kittens older than four weeks and weighing more than 2 pounds
	• *dinotefuran and pyriproxyfen* (Vectra™ for Cats, Vectra™ for Cats and Kittens)	• Third-generation neonicotinoid (dinotefuran) and IGR (pyriproxyfen) that can be used on kittens older than eight weeks of age • Once-monthly topical spot-on that kills adult fleas within six hours and controls the development of all flea stages
Synergists	• *piperonyl butoxide* (Ecto-foam®, Adams Flea & Tick Dust II®) • *N-octyl bicycloheptene dicarboximide* (Ectokyl 3X Flea & Tick Shampoo®, Pyrethrin Plus Shampoo®)	• Have limited activity against arthropods; (some inhibit insect metabolic enzymes); however, they increase the efficacy of pyrethrins and pyrethroids • Found in sprays, shampoos, dusts, and dips
Imidacloprid	• *imidacloprid* (Advantage®)	• Acts as an insect neurotoxin by binding to the niconyl receptor (inhibits cholinergic activity of the parasite) • Marketed for dogs and cats • Applied topically at the back of the neck, but is not absorbed into the blood • Kills adult fleas on contact • Has four-week residual effect
	• *imidacloprid and permethrin* (K9 Advantix®)	• Works synergistically to rapidly paralyze and kill parasites • Kills fleas (adult and larval stages) and kills and repels mosquitoes, and ticks (deer, American dog, Brown dog, lone star) • For use on dogs and puppies seven weeks of age or older • Not for use on cats

(Continued)

Table 15-6   *(Continued)*

PRODUCT CATEGORY	GENERIC AND TRADE NAME EXAMPLES	EFFICACY
	• *imidacloprid and moxidectin* (Advantage Multi® for Dogs, Advantage Multi® for Cats)	• Kills adult fleas, adult and immature hookworms, adult roundworms, adult whipworms, and prevents heartworm disease in dogs over seven weeks of age • Kills adult fleas, ear mites, adult and immature hookworms, adult roundworms, and prevents heartworm disease in cats over nine weeks of age • Once-monthly topical solution
Phenylpyrazole	• *fipronil* (Frontline®) • *fipronil with methoprene* (Frontline Plus®)	• Fipronil interferes with chloride channels of insects, which overstimulates their nervous system causing death of adult fleas, ticks, and chewing lice (Figure 15-5) • Methoprene is an insect growth regulator, which makes the combination product effective against flea eggs and flea larvae • Is applied topically, but is not absorbed into the blood • Collects in the oils of the skin and hair follicles and continues to be released over a period of time resulting in residual activity (spreads over the body in 24 hours) • Kills newly emerged adult fleas before they can lay eggs • Residual activity even after bathing • Labeled for treatment of fleas, ticks, and chewing lice when used monthly and control of sarcoptic mange with repeated treatments • Not for use in puppies or kittens less than eight weeks of age
Semicarbazone	• *metaflumizone* (ProMeris® for Cats)	• Blocks sodium influx, which is needed to propagate a nerve impulse in fleas that causes a reduction in feeding, paralysis, and death of adult fleas • Once-monthly topical spot-on labeled for use in cats greater than eight weeks of age for treatment and prevention of fleas • Hypersalivation may be seen if cat licks application site
	• *metaflumizone plus amitraz* (ProMeris® Duo for Dogs)	• Combination product with amitraz is a topical spot-on for control and prevention of adult fleas and ticks (*Ixodes ricinus, Ixodes hexagonus, Dermacentor reticulates, Dermacentor variabilis,* and *Rhipicephalus sanguineus*) in dogs over eight weeks of age; extra-label use for demodectic and sarcoptic mange mites • Hypersalivation may be seen if dog licks application site

(Continued)

PRODUCT CATEGORY	GENERIC AND TRADE NAME EXAMPLES	EFFICACY
Avermectin	• *selamectin* (Revolution®)	• Interferes with postsynaptic stimulation of the muscle fiber in arthropods or neurons in nematodes causing paralysis and death • In cats works on adult fleas and eggs, heartworms, ear mites, hookworm, and roundworms • In dogs works on adult fleas and eggs, heartworms, ear mites, sarcoptic mange, and American dog tick
	• *ivermectin* (Ivomec®, Equell® paste, Equimax®, Eqvalan®)	• Injectable or oral solution used for some ectoparasites and endoparasites • Potentiates insect's GABA neural and neuromuscular transmission • Blood-feeding ectoparasites (such as fleas, mites, and lice) are killed much better with ivermectin than superficial, nonblood feeders (like *Cheyletiella* spp.) • Effective against the following ectoparasites: bots, grubs, lice, and mites (including demodex)
	• *doramectin* (Dectomax®)	• Injectable and topical solution used for some ecto- and endoparasites. • Used as a pour-on in cattle for biting lice and mites; used in dogs and cats for generalized demodicosis
	• *eprinomectin* (Ivomec®, Eprinex®)	• Topical pour-on for beef and dairy cattle • Used to treat and control GI nematodes and ectoparasites (cattle grubs, lice, mange mites, and horn flies), ear mites (*Psorpotes cuniculi*) in rabbits • Increases permeability of the cell parasite's membrane to chloride ions resulting in paralysis and death of the parasite. • No milk or meat withdrawal
Acetylcholine receptor agonist	• *spinosad* (Comfortis®)	• One-month oral chewable flea protection for treatment and prevention of fleas in dogs 14 weeks of age or older • Causes involuntary muscle contractions in fleas that leads to seizures, paralysis, and death • Fast acting and long acting • Flea death begins in 30 minutes and is complete in four hours • May cause systemic side effects such as vomiting, anorexia, lethargy, or diarrhea • Administration with food increases its effectiveness

(Continued)

Table 15-6 (*Continued*)

PRODUCT CATEGORY	GENERIC AND TRADE NAME EXAMPLES	EFFICACY
Formamidines	• amitraz (Mıtaban® Dip, PrevenTIC® Collar, ProMeris® for dogs, Taktic®EC)	• Used for treatment of demodectic mange and scabies in dogs; ticks, mange mites, and lice in beef and dairy cattle and swine (there is a withdrawal time in swine) • Alpha-2 agonist and monoamine oxidase inhibitor that causes excess adrenergic activity in the parasite's nervous system • Animals may show sedation for 24–72 hours following treatment • Toxic to cats and rabbits (although has been used on cats in diluted form) • Use gloves and protective clothing when applying to animals; wash hands and arms after application to animal • Use in well-ventilated area • Flammable until diluted with water • Yohimbine and atipamezole are antagonists of amitraz
Sulfurated lime solution	• *lime sulfur* (Lym Dyp®, LimePlus Dip®)	• Provides antimicrobial and antiparasitic activity through the formation of pentathionic acid and hydrogen sulfide after application • Used in the treatment of sarcoptic and notoedric mange, cheyletiellosis, chiggers, fur mites, and lice; also demodicosis in cats • Also effective for the treatment of ringworm • May stain light-colored animals • Used as a rinse or dip following dilution
Repellents	• *DEET* (Blockade®) • *butoxypolypropylene glycol* (VIP® Fly Repellent Ointment)	• Used to repel mosquitoes, flies, and gnats • May be used in combination with pyrethrins and pyrethroids • Include sprays, ear tags, and topicals for ear tips
Rotenone	• *rotenone* (generic)	• Uncouples oxidative phosphorylation • Used in dips and pour-on liquids • Toxic to fish; consider pesticide runoff possibilities if rotenone is used to treat insects on plants • Not commonly used due to the availability of safer chemicals
D-limonene	• *D-limonene* (VIP Flea Dip and Shampoo®)	• Extract of citrus peel that has some insecticidal activity • Provides quick kill • No residual • Pleasant smell • Used with other products

Table 15-7   Drugs for Monthly (or Longer) Prevention of Fleas, Ticks, and Heartworm Disease

EFFECTIVE FOR FLEAS	EFFECTIVE FOR TICKS	EFFECTIVE FOR PREVENTION OF HEARTWORM DISEASE
• *dinotefuran and pyriproxyfen* (Vectra™ for Cats, Vectra™ for Cats and Kittens) • *dinotefuran, pyriproxyfen, and permethrin* (Vectra™ 3D) • *fipronil* (Frontline®) • *fipronil with methoprene* (Frontline Plus®) • *imidacloprid* (Advantage®) • *imidacloprid and moxidectin* (Advantage® Multi for dogs) • *imidacloprid and permethrin* (K9 Advantix®, Proticall®) • *lufenuron* (Program®) • *metaflumizone* (ProMeris® for Cats) • *metaflumizone plus amitraz* (ProMeris® Duo for Dogs) • *nitenpyram* (Capstar®) • *nitenpyram and lufenuron* (Capstar® Flea Management System, Program® Flavor Tabs) • *permethrin* (ProTICall®) • *selamectin* (Revolution®) • *spinosad* (Comfortis®)	• *amitraz*(PrevenTIC® collar) • *fipronil* (Frontline®) • *fipronil with methoprene* (Frontline Plus®) • *permethrin* (ProTICall®) • *selamectin* (Revolution®) (labeled for American dog tick only)	• *ivermectin* (Heartgard®, Heartgard Plus®, Tri-Heart Plus®, and Iverhart™ Plus) (Plus products also work on some intestinal parasites) • *milbemycin* (Interceptor®, Sentinel® [also works on some intestinal parasites]) • *moxidectin* (ProHeart®-6 works for six months) • *selamectin* (Revolution®) (also works on ear mites)

Chemicals used to treat ectoparasites may also have adverse effects on the animal being treated. Animals should always be observed for the development of local reactions to the chemicals being applied. Some chemicals, such as OPs, may cause additional problems such as excessive salivation, vomiting, diarrhea, and muscle tremors in treated animals. Animals may need to be a certain age before a particular chemical can safely be used on them. Animals (and humans) may be able to ingest certain products, such as collars, that may result in toxicity. Concurrent use of other medications may also cause toxicity in animals treated with ectoparasiticides. All of these factors must be considered before any product is dispensed. Always consult the animal's medical record and MSDS information, package inserts, or drug handbooks prior to use and/or dispensing of these products.

SUMMARY

Antiparasitic drugs work on endoparasites and/or ectoparasites. Endoparasites are found in the body and include parasitic worms such as nematodes, cestodes, and trematodes. Coccidia, a type of protozoan, are also endoparasites treated in veterinary medicine.

Clinical Que

Spot-on describes medication applied topically in small amounts to a localized area. Some spot-ons, such as older OP flea prevention products, may be absorbed systemically. Others work topically and are absorbed systemically (flea, tick, and heartworm prevention products), and still others only work topically (flea and tick prevention products). Spot-ons that are absorbed systemically have the potential to cause more side effects in animals (Figure 15-5).

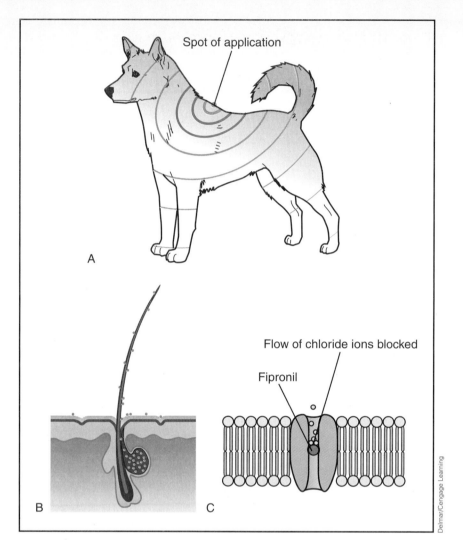

Figure 15-5 (A) Spot-on flea prevention is applied to the skin and is quickly spread across the body. (B) Insecticide is stored in a hair's sebaceous gland, which can secrete the compound for a month or more. (C) Fipronil is a flea prevention that blocks the flow of chloride ions that otherwise interrupt nerve signals of the flea. This blockage of chloride ions hyperexcites a flea's central nervous system and causes the flea to seizure and die.

Antinematodal drugs include benzimidazoles, imidazothiazoles, tetrahydro-pyrimidines, and OPs. Anticestodal drugs include praziquantel and epsiprantel. Antitrematodal drugs include clorsulon, albendazole, and praziquantel. Anticoccidial drugs, also known as coccidiostats because they inhibit but do not kill coccidia, include sulfadimethoxine, nicarbazine, amprolium, monensin, decoquinate, and robenidine. Antiprotozoal drugs include metronidazole, febendazole, albendazole, and imidocarb. These drugs are summarized in Table 15-8.

Heartworm disease is caused by the parasite *Dirofilaria immitis*. Multiple strategies exist for controlling heartworm disease, including preventing third-stage larvae from reaching maturity (prophylaxis), eliminating adult heartworms

Table 15-8 Internal Parasite Drugs

PARASITE CATEGORY	DRUG CATEGORY AND EXAMPLES
Antinematodal	**Benzimidazoles** oxibendazole (Anthelcide EQ Equine Wormer Paste®) mebendazole (Telmintic®) fenbendazole (Panacur®) albendazole (Valbazen®) oxfendazole (Benzelmin® Paste and Synanthic®) • febantel (Drontal Plus® also contains pyrantel pamoate and praziquantel combination) **Imidazothiazoles** • levamisole (Levasole® and Tramisol®) **Tetrahydropyrimidines** • pyrantel pamoate (Nemex® and Strongid-T®) • pyrantel tartrate (Strongid-C®) • morantel tartrate (Rumatel®) **Piperazines** • piperazine (Pipa-Tabs® and Hartz Advanced Care Once-a-Month Wormer® for Puppies) **Avermectins or macrocyclic lactones** • ivermectin (Heartgard®, Heartgard Plus®, Ivomec®) • eprinomectin (Ivomec®, Eprinex®) • selamectin (Revolution®) • moxidectin (ProHeart-6®, Cydectin®, Quest 2% Equine Oral Gel) • milbemycin oxine (Interceptor®, Sentinel®) • doramectin (Dectomax®) **Depsipeptides** • emodepside (Profender®, which also contains praziquantel)
Anticestodals	**Pyrazine derivatives** • praziquantel (Droncit®, Profender® [which also contains emodepside], and Quest® Plus [which also contains moxidectin]) • epsiprantel (Cestex®) **Benzimidazoles** • albendazole (Valbazen®) • mebendazole (Telmintic®) • fenbendazole (Panacur®) • oxfendazole (Benzelmin® Paste and Synanthic®)
Antitrematodals	**Benzene sulfonamide** • clorsulon (Curatrem®, Ivomec® Plus, which also contains ivermectin) **Benzimidazoles** • albendazole (Valbazen®) • fenbendazole (Panacur®) • oxfendazole (Benzelmin® Paste and Synanthic®) **Pyrazine derivative** • praziquantel (Droncit®, Profender® [which also contains emodepside], and Quest® Plus [which also contains moxidectin])

(Continued)

Table 15-8 (Continued)

PARASITE CATEGORY	DRUG CATEGORY AND EXAMPLES
Antiprotozoals	**Coccidiostats** • sulfadimethoxine (Albon®) • amprolium (Corid®) • decoquinate (Deccox®) • nicarbazine (Maxiban 72®) • monensin (Coban 60®, Rumensin®) • robenidine (Robenz Type A Medicated Article®) **Nitroimidazoles** • metronidazole (Flagyl®) **Benzimidazoles** • fenbendazole (Panacur®) • albendazole (Valbazen®) **Carbanilide derivative** • imidocarb (Imizol®) **Aminoquinoline** • primaquine (generic) **Folic acid antagonist** • pyrimethamine (Daraprim®) • pyrimethamine plus sulfadiazine (ReBalance®) **Triazine** • ponazuril (Marquis®) **Nitrothiazolyl-salicylamide derivative** • nitazoxanide (Navigator®)

(adulticide therapy), and eliminating circulating microfilariae (microfilaricide therapy). These drugs are summarized in Table 15-9.

Ectoparasites live on the outside of animals and include flies, grubs, lice, fleas, mites, and ticks. There are a variety of ways to apply ectoparasiticidic chemicals and a wide variety of products used to control ectoparasites.

Table 15-9 Heartworm Medication

DRUG CATEGORY	EXAMPLES
Adulticide	• Melarsomine (Immiticide®)
Microfilaricides	• ivermectin (Heartgard®, Heartgard Plus®, Heartgard® for Cats, Iverhart™ Plus, Tri-Heart™ Plus) • milbemycin (Interceptor®, Sentinel®) • levamisole (Levasole®, Tramisol®)
Preventatives	• ivermectin (Heartgard®, Heartgard Plus®, Heartgard® for Cats, Iverhart™ Plus, Tri-Heart™ Plus) • milbemycin (Interceptor®, Sentinel®) • selamectin (Revolution®) • moxidectin (ProHeart-6®)

It's a Wrap

The answers to the questions in this chapter's Setting the Scene case study should be understood after reading this chapter. This client would like to know what dewormers are available for calves and whether they come in easy-to-administer forms. How should the veterinary technician explain antiparasitic drugs to clients? Are dewormers safe for cattle of all ages? Dewormers used in livestock are typically chosen based on where the animals live (regionally and type of environment), age of the animal, and withdrawal times, which can differ based on whether animals produce meat or milk. Ivermectin-type products are often used topically. The product used in young animals may differ from the older animals because of withdrawal times. For example, doramectin may be used in calves because it has the longest-lasting effect on parasites, but it also has the longest slaughter hold, making it undesirable in some older animals. Doramectin cannot be used in lactating cows, so this group of animals will need a different deworming product. Dairy cows may be treated with an eprinomectin product because it is labeled for use in lactating animals.

What side effects does he need to watch for? Side effects to watch for in animals treated with topical dewormers include reaction at the application site and neurologic signs if given at extremely high doses. Neurologic signs include ataxia, listlessness, and occasionally death. The most common side effect observed is ataxia. Side effects of orally administered dewormers include vomiting, diarrhea, and neurologic signs (tremors and ataxia).

CHAPTER REVIEW

Matching

Match the drug name with its action.

1. _____ pyrethroids
2. _____ fenbendazole
3. _____ avermectin
4. _____ organophosphate
5. _____ clorsulon
6. _____ sulfadimethoxine
7. _____ melarsomine
8. _____ IGR
9. _____ fipronil
10. _____ pyrantel pamoate

a. safe antinematodal drug that should be shaken prior to use
b. chemical that mimics juvenile hormone of insects
c. antitrematodal
d. group of quick-kill chemicals found in flea products
e. coccidiostat
f. commonly used antinematodal that includes the brand name product Panacur®
g. topical flea and tick treatment with residual activity
h. cholinesterase inhibitor that is used against endo- and ecto-parasites and has a narrow safety margin
i. category of antiparasitic that works against endo- and ecto-parasites and includes the drug ivermectin
j. adulticide used to treat heartworm disease

Multiple Choice

Choose the one best answer.

11. What chemical is used to treat demodectic mange in dogs?
    a. amitraz
    b. fipronil
    c. imidacloprid
    d. methoprene

12. What is a disadvantage of using oral forms of flea prevention?
    a. These products work for a specified period of time.
    b. The flea needs to take a blood meal to get the medication.
    c. They provide quick knockdown effects.
    d. These products must be diluted before use.

13. Cleaning of contaminated waste from the environment is important when treating an animal with
    a. antinematodals.
    b. anticestodals.
    c. antitrematodals.
    d. coccidiostats.

14. Ivermectin can be used in treating all stages of heartworm disease except
    a. the third-stage larvae.
    b. the adults.
    c. the microfilariae.
    d. the juveniles.

15. Antitrematodal drugs include all of the following *except*
    a. clorsulon.
    b. albendazole.
    c. epsiprantel.
    d. praziquantel.

16. Ectoparasiticidic chemicals used to quickly treat the indoor environment are known as
    a. yard sprays.
    b. dusts.
    c. foggers.
    d. topical long-acting.

17. Treatment of *Giardia lamblia* infections should include
    a. metronidazole.
    b. praziquantel.
    c. epsiprantel.
    d. ivermectin.

18. Heartworm preventative can be given at the following intervals:
    a. hourly
    b. monthly
    c. six months
    d. both b and c are correct

True/False

Circle a. for true or b. for false.

19. All heartworm preventatives also control intestinal parasites.
    a. true
    b. false

20. All flea products also control ticks.
    a. true
    b. false

Case Studies

21. A four-month-old F Springer Spaniel (20#) is brought to the clinic for a puppy check. The owner is concerned about the dog having worms, because she also has young children. She would like to get a dewormer for this dog.
    a. What advice should be given to this owner before a dewormer is prescribed?
    b. The owner brings in a stool sample and cestode eggs are found. What medication should be given to this client?

22. An owner calls the clinic and wants to know what is available for prevention of heartworm disease in his dogs. What types of heartworm prevention are there?

Critical Thinking Questions

23. What are the signs of organophosphate toxicity? Knowing these signs, what is a possible antidote?

24. Why is it important to do a fecal examination before dispensing a dewormer for a puppy or a kitten?

ADDITIONAL REVIEW MATERIAL

Additional review material can be found on the StudyWARE™ CD included with this text.

BIBLIOGRAPHY

Allen, D. G. (2005). *Handbook of Veterinary Drugs* (3rd ed.). Baltimore, MD: Lippincott Williams & Wilkins.

American Heartworm Society: www.heartwormsociety.org.

Broyles, B., Reiss, B., and Evans, M. (2007). *Pharmacological Aspects of Nursing Care* (7th ed.). Clifton Park, NY: Delmar Cengage Learning.

Companion Animal Parasite Council: http://www.capcvet.org/index.html

Dryden, M. (2007). Total parasite control-piecing the puzzle together, Proceedings of the North American Veterinary Conference, Orlando, FL.

Plumb, D. C. (2008). *Plumb's Veterinary Drug Handbook* (6th ed.). Ames, IA: Blackwell Publishing.

Spratto, G. R., and Woods, A. L. (2008). *PDR Nurse's Drug Handbook*. Clifton Park, NY: Cengage Learning.

Veterinary Pharmaceuticals and Biologicals (12th ed.). (2001). Lenexa, KS: Veterinary Healthcare Communications.

CHAPTER 16

ANTI-INFLAMMATORY AND PAIN-REDUCING DRUGS

OBJECTIVES

Upon completion of this chapter, the reader should be able to:

- explain the role of inflammation in protecting the body from damage.
- describe the inflammatory pathway involving arachidonic acid.
- describe how steroidal and nonsteroidal anti-inflammatory drugs affect the inflammatory pathway.
- contrast the advantages and disadvantages of using glucocorticoid drugs.
- outline how the negative feedback pathway affects glucocorticoid production and how this affects use of glucocorticoid drugs.
- describe the use of glucocorticoid drugs in animals.
- describe the use of NSAIDs in animals.
- differentiate between the types of NSAIDs and list the advantages and disadvantages of each.
- explain the role of antihistamines in controlling inflammation.

KEY TERMS

analgesics

antihistamines

anti-inflammatory drugs

antipyretic

corticosteroids

glucocorticoids

iatrogenic

inflammation

mineralocorticoids

nonsteroidal anti-inflammatory drugs (NSAIDs)

steroidal anti-inflammatory drugs

Setting the Scene

A horse owner calls the clinic, requesting an exam on his five-year-old quarter horse gelding, which has been showing signs of lameness for two days. On PE, the veterinarian noted that the horse is pointing his left front foot (heel is off the ground). A nerve block is done to determine the location of the injury. Radiographs are also taken, and it is determined that the horse has degenerative changes in the navicular bone. The horse is put on phenylbutazone and referred to a farrier for proper shoeing. The owner asks the veterinary technician if the phenylbutazone will fix his horse's lameness. Will it? Do analgesics and anti-inflammatory drugs cure disease? Can they cause additional problems for the animal?

INFLAMMATION

Inflammation is a natural response of living tissue to injury and infection. It is a useful and normal process that occurs in the body as a nonspecific protective mechanism. A series of events is triggered when inflammation occurs, including vascular changes and release of chemicals that help the body destroy harmful agents at the injury site and repair damaged tissue. Inflammation may be caused by chemical, physical, or biological agents; however, the response is fairly uniform regardless of the cause. Beneficial effects, such as vasodilation and increased permeability of blood vessels, allow blood substances and fluid to leave the plasma and go to the injured site in the *early* or *vascular phase* of inflammation. Following the brief vascular phase is the *delayed* or *cellular phase*, which can last hours to days after the injury occurs. Accumulation of leukocytes (white blood cells) in inflamed tissue, reduced blood flow, and widespread tissue damage occur during the cellular phase. Chemicals such as histamine, prostaglandin, and bradykinin are released, which can result in bronchoconstriction, vasodilation, shock, and pain in the cellular phase. The activity of neutrophils, monocytes, and lymphocytes also increases during this phase.

Clinically, inflammation presents as redness, heat, swelling, pain, and decreased range of motion. These signs of inflammation are summarized in Table 16-1 and Figure 16-1.

The inflammatory pathway is a response by tissues to initiate the healing process. When tissue is damaged, phospholipids in the cell membrane are broken down by the enzyme phospholipase. This breakdown of phospholipids by phospholipase produces arachidonic acid. The enzyme cyclooxygenase breaks down arachidonic

Table 16-1 Signs of Inflammation

Sign	Explanation
Pain	Pain is due to tissue swelling and release of chemicals such as prostaglandin.
Heat	Heat is due to increased blood accumulation and pyrogens (fever-producing substances) that interfere with temperature regulation.
Redness	Redness occurs in the early phase of inflammation due to blood accumulation in the area of tissue injury from chemical release (such as prostaglandins and histamine).
Swelling	Swelling occurs in the delayed phase of inflammation because kinins dilate arterioles and increase capillary permeability. This increased capillary permeability allows plasma to leak into the interstitial tissue at the injury site.
Decreased range of motion	Function is lost due to fluid accumulation at the injury site. Pain also decreases mobility to an area.

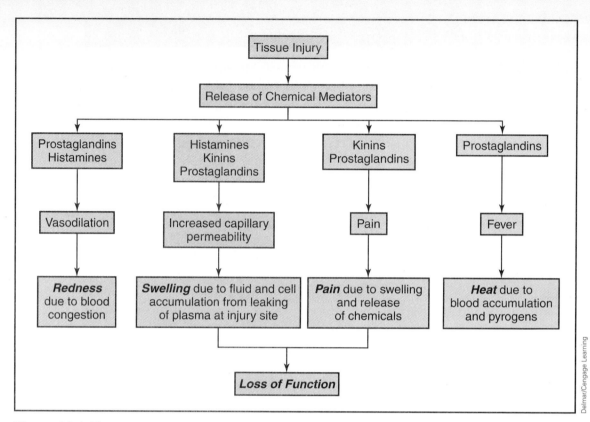

Figure 16-1 Tissue response to injury.

acid to prostaglandins and thromboxanes. Another enzyme, lipoxygenase, can also break down arachidonic acid, producing leukotrienes. Prostaglandins and leukotrienes are chemicals that induce inflammation. Thromboxanes cause aggregation of platelets. Figure 16-2A shows the inflammatory pathway.

INFLAMMATION REDUCERS

Although inflammation is a useful protective response to injury, there are times when inflammation should be reduced. Drugs that reduce inflammation are known as **anti-inflammatory drugs**. In an attempt to control inflammation when necessary, two main groups of drugs are used: steroidal anti-inflammatory drugs and nonsteroidal anti-inflammatory drugs. **Steroidal anti-inflammatory drugs** work by blocking the action of the enzyme phospholipase. **Nonsteroidal anti-inflammatory drugs (NSAIDs)** work by blocking the action of the enzyme cyclooxygenase (Figure 16-2B). A third drug category used in controlling inflammation is the antihistamines, which counteract the action of histamine, a chemical produced in inflammation.

Steroidal Inflammation Reducers

Corticosteroids are hormones produced by the adrenal cortex, the outer part of the adrenal gland. A *hormone* is a chemical substance produced in one part of the body that is transported to another part of the body where it influences and regulates cellular and organ activity. Two groups of corticosteroids synthesized

Clinical Que

Steroidal and nonsteroidal anti-inflammatory drugs both have anti-inflammatory effects, but they are not chemically related.

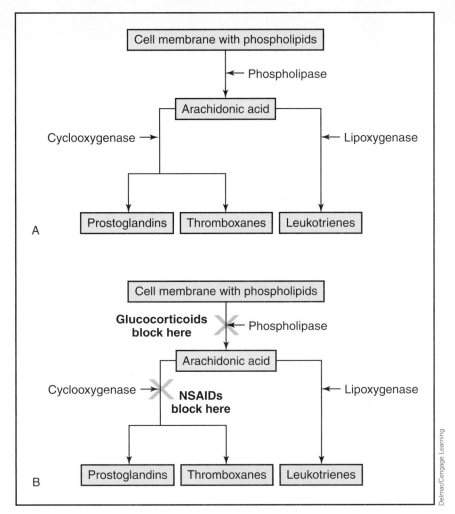

Figure 16-2 Tissue Response to injury (A) The inflammatory pathway (B) Drug therapy and the inflammatory pathway.

by the adrenal cortex have important veterinary applications: **glucocorticoids** (cortisol is the main form in animals) and **mineralocorticoids** (aldosterone is the main form in animals). The mineralocorticoids primarily help the body retain sodium and water. Mineralocorticoids also help maintain the fluid and electrolyte balance crucial for body functions. Glucocorticoids have anti-inflammatory effects because they inhibit phospholipase, an enzyme that damages cell membranes (thus, glucocorticoids help stabilize cell membranes). Glucocorticoids raise the concentration of liver glycogen and increase blood glucose levels. Glucocorticoids also affect carbohydrate, protein, and fat metabolism, as well as muscle and blood cell activity. This section examines the anti-inflammatory applications of glucocorticoids.

Glucocorticoids are produced by the adrenal cortex in response to adrenocorticotropic hormone (ACTH) secretion by the anterior pituitary gland. The anterior pituitary gland secretion of ACTH is controlled by a releasing factor from the hypothalamus. These substances work in conjunction with each other through a process called *negative feedback* (see Chapter 10). When blood levels of glucocorticoid are low, the hypothalamus secretes releasing factor.

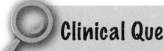

Clinical Que

The terms "glucocorticoid" and "corticosteroid" may be used interchangeably; however, corticosteroids include both glucocorticoids and mineralocorticoids. When describing drugs, remember that some have only glucocorticoid function while others have both glucocorticoid and mineralocorticoid function (and could be called corticosteroids).

Clinical Que

Glucocorticoids are so named because they stimulate the increase of glucose levels.

This releasing factor then signals the anterior pituitary gland to secrete ACTH. ACTH in turn signals the adrenal cortex to produce glucocorticoid. Likewise, if blood levels of glucocorticoid are high, the hypothalamus stops secreting releasing factor. This decreased secretion of releasing factor causes the anterior pituitary to produce less ACTH, resulting in less glucocorticoid production by the adrenal cortex. This check-and-balance system (Figure 16-3) keeps some hormones, including glucocorticoid levels, at the proper blood concentration.

Glucocorticoid drugs are used to treat many conditions, including inflammatory conditions, allergic responses, and systemic diseases (Figure 16-4).

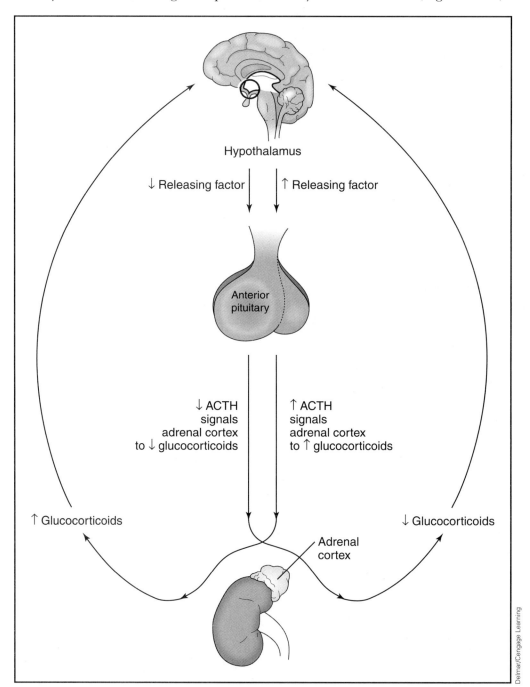

Figure 16-3 Negative feedback mechanism for glucocorticoid production

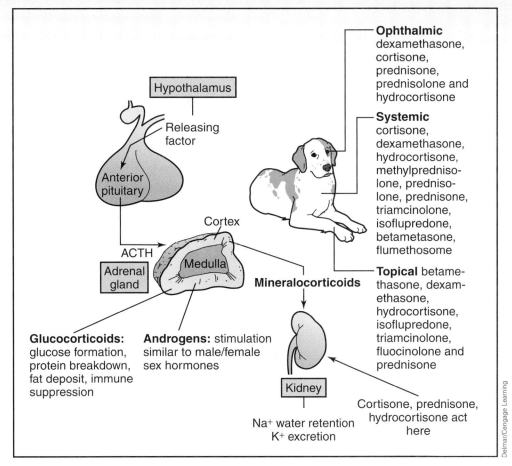

Figure 16-4 The adrenal cortex makes glucocorticoids, mineralocorticoids, and androgens. Glucocorticoids are used clinically to reduce inflammation and may be administered to the eyes (ophthalmic), systemically, and topically. The type of glucocorticoid and its route of administration are illustrated in the right column.

Inflammatory conditions that may require glucocorticoid drugs are autoimmune disorders, shock, ocular inflammation, musculoskeletal inflammation, spinal cord inflammation due to intervertebral disc disease, and lameness in horses. Allergic responses such as drug reactions, contact dermatitis, pruritus, and anaphylaxis may require glucocorticoid drugs. Systemic diseases such as Addison's disease (adrenocortical insufficiency) and treatment of some forms of cancer may benefit from glucocorticoid drug use. Glucocorticoid drugs can also be used to terminate late-stage pregnancies.

The benefits of glucocorticoid drugs include the following:

- They reduce inflammation, therefore they reduce pain (they are the "feel-good" drugs).
- They relieve pruritus (itching).
- They reduce scarring by delaying healing.
- They reduce tissue damage.

The drawbacks to glucocorticoid drugs include the following:

- They delay wound healing.

- They increase the risk of infection by decreasing attraction of leukocytes to the damaged site.

- They may cause gastrointestinal ulceration and bleeding.

- They increase the risk of corneal ulceration if corneal damage already exists.

- They can induce abortion in some species.

A wide variety of glucocorticoid drugs, often referred to simply as *steroids* or *cortisone drugs*, are synthetically produced (Figure 16-5). The glucocorticoid drugs manufactured in the laboratory have significantly greater anti-inflammatory effects than adrenally produced glucocorticoids. Glucocorticoid drugs can be short-acting (duration of action less than 12 hours), intermediate-acting (duration of action between 12 and 36 hours), or long-acting (duration of action more than 48 hours). Glucocorticoid drugs come in many formulations, including aqueous solutions, alcohol solutions, suspensions, and tablets. These drugs have several routes of administration: oral, parenteral, and topical. The water-soluble formulations can be injected IV, IM, or SQ. Repository or long-acting depot products are not given IV.

Glucocorticoid drugs may also be found in preparations containing antibiotics and antifungals. Some of the glucocorticoids used in veterinary medicine are listed in Table 16-2.

Even though glucocorticoid drugs are widely used in veterinary medicine, their use should be monitored for side effects. Some side effects of systemic glucocorticoid use include polyuria, polydipsia, polyphagia, suppressed healing, gastric ulcers, thinning of skin, and muscle wasting. Systemic glucocorticoid use can also affect the blood cell values of the CBC. Lymphopenia (decreased number of lymphocytes), eosinopenia (decreased number of eosinophils), monocytopenia (decreased number of monocytes), neutrophilia, and thrombocytopenia can be seen in peripheral blood with systemic glucocorticoid use. These decreases are due to movement of those cells to the tissue, and the increases are due to movement of marginalized neutrophils from blood vessel walls to the circulation. Another important consequence of systemic glucocorticoid use is the development of iatrogenic Cushing's disease (hyperadrenocorticism). Cushing's disease is a condition caused by overproduction of glucocorticoids (due to an adrenal tumor or

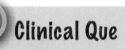

Clinical Que

A glucocorticoid can usually be recognized by the *-sone* ending in the generic name.

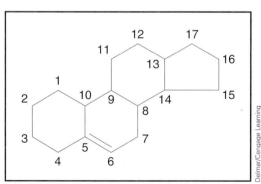

Figure 16-5 Steroid molecule

Delmar/Cengage Learning

Table 16-2 Glucocorticoids Used in Veterinary Medicine

GENERIC NAME	EXAMPLE(S) OF TRADE NAMES	PREPARATION	DURATION OF ACTION
cortisone	Cortone®	Tablets, injection, ophthalmic ointment/suspension	Short
hydrocortisone	Cortef®, Solu-Cortef®	Tablets, oral suspension, topical cream, lotion, ointment, injection	Short
isoflupredone	Predef 2x®, Tritop®	Injection, cream, ointment	Intermediate
methylprednisolone	Medrol®	Tablets	Intermediate
methylprednisolone acetate	Depo-Medrol®	Suspension for injection	Intermediate
prednisone/ prednisolone	Deltasone®, Meticorten®	Tablets, injection, ophthalmic ointment	Intermediate
prednisolone sodium succinate	Solu-Delta-Cortef®	Injection	Intermediate
triamcinolone	Vetalog®	Tablets, suspension for injection, topical cream	Intermediate
betamethasone	Betasone®	Injection, topical, tablets, syrup	Long
dexamethasone	Azium®	Tablets, injection, elixir/ solution, ophthalmic suspension	Long
flumethasone	Flucort®	Injection	Long
fluocinolone	Synalar®	Topical cream and solution	Long

pituitary disease). Animals with Cushing's disease are PU, PD, and lethargic; have bilaterally symmetrical alopecia (hair loss); and have a pendulous abdomen caused by the catabolic (breakdown) effects of glucocorticoids. **Iatrogenic** means caused by the treatment ordered by physicians/veterinarians. Iatrogenic hyperadrenocorticism occurs when an animal has been on glucocorticoids for a period of time or on excessive doses of glucocorticoids. Animals with iatrogenic Cushing's disease appear clinically similar to animals with Cushing's disease. Side effects of topical glucocorticoid use include localized irritation and skin atropy.

In healthy animals, when glucocorticoid levels rise, the hypothalamus signals the anterior pituitary gland to decrease ACTH production. This in turn slows glucocorticoid production. When glucocorticoid supplements are given, the animal's body responds the same way: by signaling the hypothalamus to release inhibitory factor that tells the anterior pituitary gland to decrease ACTH

Clinical Que

Hydrocortisone, cortisone, and prednisone have both glucocorticoid and mineralocorticoid activity; therefore, they can also be used as replacement therapy for patients with adrenal insufficiency.

Clinical Que

Prednisone or prednisolone? Prednisone must be converted to prednisolone by the liver; therefore, animals with liver dysfunction should be prescribed prednisolone.

production. As long as the treatments are being given, the adrenal cortex is no longer stimulated to produce glucocorticoid, and in time the adrenal cortex can shrink or atrophy. The atrophied adrenal cortex cannot immediately start making its own glucocorticoids if glucocorticoid treatment is suddenly stopped. The animal's decreased blood levels of glucocorticoid will decrease further when the treatments are stopped and the body fails to make its own glucocorticoids. Decreased blood levels of glucocorticoid result in weakness, lethargy, vomiting, and diarrhea. This shortage of glucocorticoid results in iatrogenic Addison's disease (hypoadrenocorticism), a condition caused by insufficient glucocorticoid in the animal's system. To prevent iatrogenic Addison's disease, glucocorticoids are given at the lowest dosage possible and are withdrawn gradually to allow the adrenal cortex time to resume production of natural glucocorticoids.

Some key points about glucocorticoid therapy are as follows:

- The anti-inflammatory effects of glucocorticoid drugs do not cure any specific disease.

- The anti-inflammatory effects of glucocorticoid drugs may help disseminate some infectious agents that may be present in the body. When given at therapeutic dosages, glucocorticoids primarily alter cell-mediated immunity, which affects fungal agents to the greatest degree. When glucocorticoids are given at higher dosages, they impair both cell-mediated and humoral immunity, thus compromising viral and bacterial agents in addition to fungal agents. Adequate antifungal drugs or antibiotics with antifungal properties may prevent this problem.

- The anti-inflammatory effects of glucocorticoid drugs help chronic diseases that flare up periodically and in the absence of a known cause: arthritis, tendonitis, bursitis, conjunctivitis, and various forms of dermatitis.

- Caution should be used when giving high dosages of glucocorticoid drugs to pregnant animals (one of their uses is to terminate late-stage pregnancies).

- Whenever possible, glucocorticoid drugs should be applied locally to avoid disturbing the hormonal balance of adrenally produced steroids. Even though some absorption may occur, the low systemic glucocorticoid level maintains normal adrenal cortex function while also maintaining a high local level of the drug.

- Use alternate-day dosing at the lowest possible dose whenever possible, to prevent iatrogenic Cushing's disease (hyperadrenocorticism) (Figure 16-6).

- Avoid continuous glucocorticoid use.

- Taper animals off glucocorticoid drugs to prevent iatrogenic Addison's disease (hypoadrenocorticism).

- Glucocorticoid drugs should not be used in animals with corneal ulcers.

Figure 16-6 Hyperadrenocorticism (Cushing's Disease) may be seen with inappropiate use of glucocorticoids.

Nonsteroidal Inflammation Reducers

In addition to steroidal anti-inflammatory agents, veterinarians also use anti-inflammatory drugs known as *NSAIDs*. Veterinarians use many kinds of NSAIDs: the phenylbutazone family (pyrazolone derivatives), the ibuprofen family (propionic acid derivatives), the COX-2 inhibitors, flunixin meglumine, the indol acetic acid derivates, the fenamates, and the classical salicylates such as aspirin. Table 16-3 lists some NSAIDs used in animals.

NSAIDs work by either completely or selectively inhibiting cyclooxygenase, which has two forms: cyclooxygenase-1 (commonly called COX-1) and cyclooxygenase-2 (commonly called COX-2) (Figure 16-7). COX-2 is believed to be more involved in inflammation, whereas COX-1 is believed to be more

Table 16-3 NSAIDs Used in Animals

SALICYLATES	PYRAZOLONE DERIVATIVES	PROPIONIC ACID	COX-2 INHIBITORS	MISCELLANEOUS NSAIDS	INDOL ACETIC ACID DERIVATIVE	FENAMATES	TEPOXALIN
• aspirin	• phenylbutazone (Buazolidin®, Equipalazone®, Butatron®)	• ibuprofen (Motrin®, Advil®) • ketoprofen (Ketofen®, Orudis®) • carprofen (Rimadyl®) • naproxen (Equiproxen®, Aleve®)	• deracoxib (Deramaxx®) • meloxicam (Metacam®) • firocoxib (Equioxx®, Previcox®)	• flunixin meglumine (Banamine®, Finadyne®) • diclofenac sodium (Surpass®)	• etodolac (EtoGesic®)	• meclofenamic acid (Arquel®) • tolfenamic acid (Tolfedine®)	• tepoxalin (Zubrin®)

Courtesy of Dr. Mark Jackson, DVM, PhD, Glascow University

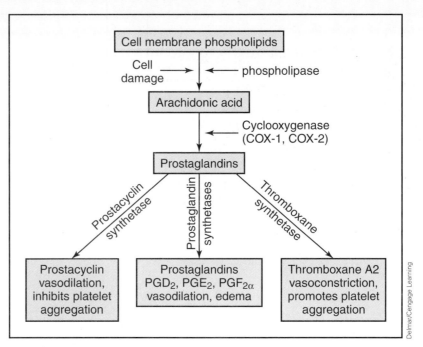

Figure 16-7 Cyclooxygenase activity and its effect on prostaglandin production

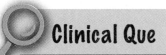

Clinical Que

NSAIDs are also referred to as prostaglandin inhibitors because they inhibit cyclooxygenase, the enzyme that converts arachidonic acid to prostaglandin.

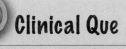

Clinical Que

Buffered or coated aspirin still carries a risk of causing gastrointestinal ulceration and bleeding.

involved with the stomach (hence the gastrointestinal side effects often associated with NSAIDs). NSAIDs tend to have fewer side effects than glucocorticoid drugs and produce analgesia, inhibition of platelet aggregation, and fever reduction (which is not true of glucocorticoid drugs). Some side effects of NSAID use include gastric ulceration and bleeding, bone marrow suppression, and tendency to bleed. Renal papillary necrosis is a significant side effect of NSAID use in animals with any hypotensive condition such as dehydration, blood loss, or prolonged anesthesia. Prolonged hypotension results in cells not receiving enough oxygen and in cell death due to tissue hypoxia. This scenario is usually prevented by the release of prostaglandin E2 by the kidney; however, NSAIDs block beneficial prostaglandin from being released. The following are the most common NSAID types.

Salicylates

The salicylates are the oldest anti-inflammatory agents and include *aspirin*, also known as *acetylsalicylic acid* or *asa*. Aspirin is a potent inhibitor of prostaglandin synthesis (by inhibition of cyclooxygenase). Prostaglandins accumulate at tissue injury sites, causing pain and inflammation. Aspirin is an analgesic, **antipyretic** (fever-reducing), and anti-inflammatory, and a reducer of platelet aggregation. Aspirin is used for the treatment of degenerative joint disease (commonly known as osteoarthritis), as an adjunct to postadulticide treatment of heartworm parasites, to give pain relief, to reduce fever, and in the treatment of some forms of shock. Cats cannot metabolize aspirin as rapidly as other species because the enzyme needed to metabolize aspirin (glucuronyl transferase) is less effective in cats; therefore, aspirin must be used with caution in cats (Figure 16-8). If aspirin is used to treat a cat's condition, it is given at lower dosages and less frequently (e.g., every 48 to 72 hours). Aspirin comes in many forms, including adult and baby strengths, buffered forms, enteric-coated forms, and combination forms. Care must be taken so as not to dispense a product that contains other drugs (such as a glucocorticoid, calcium, or other pain medication) if the other drug is not also prescribed. Some gastric side effects can be decreased if aspirin is given with food.

Aspirin is commonly given as a home remedy by animal owners who may not associate the administration of this drug to signs of toxicity, which arise over time. It is important for the veterinary technician to be aware of the signs of salicylate toxicity. Signs of salicylate toxicity include gastrointestinal problems (anorexia, abdominal pain, gastrointestinal upset, vomiting, diarrhea, melena, and lethargy), respiratory problems (panting in most animal species; however, cats with CNS depression may also have decreased respirations from salicylate toxicity), neurological problems (restlessness, anxiety, incoordination, and seizures), bleeding problems (salicylates alter platelet function resulting in bleeding disorders), and kidney failure (PU, PD, vomiting, diarrhea, anorexia, and production of dilute urine).

Clinical Que

Aspirin is an analgesic and anti-inflammatory agent (via prostaglandin inhibition), an antipyretic (via blocking prostaglandin mediators of pyrogens), and reducer of platelet aggregation (via thromboxane inhibition).

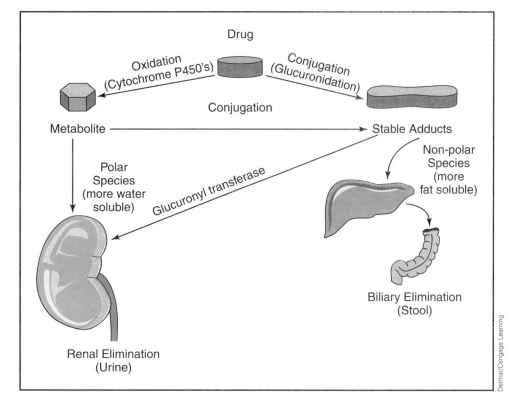

Figure 16-8 Role of glycuronyl transferase in drug metabolism. Metabolized drugs are mainly cleared from the body through chemical modification (biotransformation) in the liver. The goal of metabolism is to make drugs either more water soluble (for excretion in the urine) or more fat soluble (for excretion in the bile, and then into the feces). The liver does this either by cytochrome P450 enzymes, which chemically change drugs, or through conjugation enzymes such as glucuronyl transferases that link one chemical to another. Cats have decreased activity of glucuronyl transferase and therefore cannot metabolize certain drugs such as aspirin as effectively as other animal species

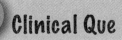

Pyrazolone Derivatives

Phenylbutazone, an inhibitor of prostaglandin synthesis by the inhibition of cyclooxygenase, is the representative drug in this category. Phenylbutazone (or "bute" as it is commonly referred to by horse owners) is frequently used in equine medicine for pain associated with the musculoskeletal system such as laminitis and bone pain. The liver metabolizes phenylbutazone, and it is highly protein bound, so care must be taken when bute is given to an animal with low albumin (blood protein) levels. Phenylbutazone is a mild to moderate analgesic, antipyretic, and anti-inflammatory agent. It comes in paste, tablet, powder, and bolus forms for oral use, as well as gel and injection forms. Side effects include oral and gastrointestinal erosions and ulcers, diarrhea, and renal necrosis. In 2003, the FDA-DVM instituted a ban on the use of phenylbutazone in dairy cattle because this drug is known to cause blood disorders in humans.

Propionic Acid Derivatives

Propionic acid derivatives contain the *-fen* group of NSAIDs, which includes *ibuprofen* (Motrin®, Advil®), *ketoprofen* (Ketofen®, Orudis®), *carprofen* (Rimadyl®), and *naproxen* (Equiproxen®, Aleve®). Carprofen and naproxen block cyclooxygenase, while ketoprofen blocks both cyclooxygenase and lipoxygenase. These products are used for their analgesic properties; however, some of these products are also antipyretic and anti-inflammatory. Ketoprofen and naproxen are approved for use in horses, and carprofen is approved for use in dogs, so the use of human over-the-counter products is no longer recommended. Side effects include gastrointestinal ulceration, vomiting (in species that can vomit), and anorexia. Carprofen is believed to target COX-2, which acts in only inflammation and not on gastrointestinal receptors, so it has limited gastrointestinal side effects. However, rare liver toxicities to carprofen have been noted, especially in Labrador retrievers, which were over-represented in the population of dogs reported by the public as having an adverse reaction to the drug. Before beginning carprofen therapy, it is recommended that blood work be done to assess liver enzymes. Carprofen is approved for food animals in some European and Asian countries, but its use in these animals is extralabel in the United States. Since flunixin is approved for use in cattle in the United States and has similar indication, carprofen use in cattle is not legal unless it can be justified why flunixin is ineffective in a particular animal. Naproxen is not recommended for use in dogs because even a single dose may cause perforated gastric ulcers.

Selective COX-2 Inhibitors

Newer NSAIDs now on the market were developed to be more selective in their inhibition of prostaglandins by targeting COX-2 to a significantly greater degree than to the protective COX-1 when administered at their recommended dosages. One drug in this group is *deracoxib* (Deramaxx®), which is a member of the coxib class of NSAIDs that works by selectively inhibiting COX-2. COX-2 is responsible for the synthesis of inflammatory mediators. Deracoxib has limited inhibition of COX-1, which is an enzyme responsible for physiological processes such as platelet aggregation and protection of the gastric mucosa. Deracoxib is indicated for the control of postoperative pain and inflammation associated with orthopedic surgery in dogs weighing more than four pounds. It is available in a chewable tablet that should be given after eating to increase its bioavailability. Side effects include vomiting, anorexia, diarrhea, and blood abnormalities such as elevated kidney and liver values. Dogs should have pre-drug blood testing prior to the use of deracoxib.

Meloxicam (Metacam®) is an NSAID of the oxicam group that acts by selective inhibition of COX-2, which in turn inhibits prostaglandin synthesis. It also inhibits leukocyte infiltration into the inflamed area. Meloxicam has anti-inflammatory, analgesic, and antipyretic effects and is used mainly in small animals. In dogs, it is available as an oral suspension that is applied to the animal's food and is used for the alleviation of inflammation and pain in both acute and chronic musculoskeletal disorders. In cats, meloxicam is approved as a one-time SQ injection prior to surgery for control of postoperative pain and inflammation associated with surgery. Meloxicam is used infrequently in large animals; however, it may be used in cattle in combination with appropriate antibiotic therapy as an injection to treat acute respiratory infection to reduce clinical signs in affected animals. In dairy cattle, it is also used to treat mastitis.

Meloxicam should not be used in pregnant or lactating animals, or in animals with liver, cardiac, kidney, or gastrointestinal problems. Meloxicam should not be used in combination with other NSAIDs or glucocorticoid drugs. Side effects include anorexia, vomiting, diarrhea, and lethargy.

Firocoxib (Equioxx® and Previcox®) is an oral selective COX-2 inhibitor in the coxib class labeled for use in dogs and horses for control of pain and inflammation associated with osteoarthritis. It is used to treat fever, pain, and/or inflammation. Firocoxib is available as a chewable tablet for dogs and an oral paste for horses. It should be used with caution in animals with preexisting renal, hepatic, or cardiovascular dysfunction. Careful patient monitoring should be done when used in dehydrated animals. Side effects include vomiting and anorexia.

Carprofen (Rimadyl®) is an oral and injectable drug believed to selectively target COX-2 that was previously covered in the propionic acid derivative section.

Flunixin Meglumine

This NSAID is a potent inhibitor of cyclooxygenase and is labeled for horses and cattle as Banamine® and Finadyne®. It is used extra-label in other species with extreme caution and typically only for a single dose. *Flunixin* is an injectable (IV or IM) analgesic, antipyretic, and anti-inflammatory drug commonly used for musculoskeletal and colic pain. In cattle, IM use is recommended due to prolonged withdrawal times when given IV. It is also used for the treatment of shock, intervertebral disc disease, and pain secondary to surgery and parvovirus infection. Flunixin is a potent analgesic: The onset of analgesia can begin in about 15 minutes.

Indol Acetic Acid Derivative

The representative drug in this group is *etodolac* (EtoGesic®). It is labeled as an anti-inflammatory agent and analgesic for osteoarthritis in dogs. It works by inhibiting cyclooxygenase and is believed to be a more selective inhibitor of COX-2 than COX-1. It also has the benefit of once-daily oral dosing. Side effects include vomiting, diarrhea, lethargy, and hypoproteinemia. Liver enzymes should be assessed one week after initiation of therapy.

Fenamates

Meclofenamic acid (Arquel®) is the representative drug of the fenamate group and is an inhibitor of cyclooxygenase. It is used as an analgesic and anti-inflammatory in equine medicine for osteoarthritis. It comes in granule form that is mixed in the feed.

Tolfenamic acid (Tolfedine®) is another fenamate approved for use in dogs and cats in Canada and Europe for reduction of pain and inflammation and for the treatment of acute mastitis and respiratory tract disease in cattle. Tolfenamic acid has significant antithromboxane activity and is not recommended for use presurgically because of its effect on platelet function. Side effects include vomiting and diarrhea.

Dual-Pathway NSAID

Tepoxalin (Zubrin®) is a rapid-disintegrating tablet used for the control of pain and inflammation associated with osteoarthritis in dogs. Tepoxalin tablets dissolve in the dog's mouth within seconds, which enhances owner compliance and helps ensure complete dosing. Tepoxalin is an NSAID that blocks both parts of the arachidonic acid cycle (the cyclooxygenase and lipoxygenase pathways). Tepoxalin decreases prostaglandin production at the site of inflammation (COX-1 inhibition), decreases edema and pain (COX-2 inhibition), and decreases gastric irritation and ulceration (lipoxygenase inhibition). Side effects of tepoxalin include gastrointestinal and renal problems. Tepoxalin is distributed in blister packs to avoid any contact with moisture, which would cause rapid disintegration of the drug.

Diclofenac Sodium

Diclofenac sodium is an NSAID that is a nonspecific cyclooxygenase inhibitor (both COX-1 and COX-2). It may also have some lipooxygenase inhibition activity. Diclofenac sodium reduces prostaglandin production, thus reducing pain, fever, and inflammation. Diclofenac sodium is available as a topical cream (Surpass®) and is approved for topical use in horses for local control of joint pain and inflammation. It is labeled for use in hock, knee, fetlock, and pastern joints of horses for up to 10 days. Reported side effects with topical use are minimal. Rubber gloves should be worn when applying diclofenac sodium to horses.

Clinical Que

As a class, cyclooxygenase inhibitory NSAIDs produce gastrointestinal, kidney, and liver side effects.

Miscellaneous Anti-Inflammatory Drugs

There are many anti-inflammatory drugs used in veterinary medicine that do not fall into one of the categories already described. The following drugs have anti-inflammatory properties, but may also have other actions such as providing analgesia or reducing cellular damage.

Dimethyl Sulfoxide

Dimethyl sulfoxide (DMSO) is an anti-inflammatory drug, but is also well known for its ability to penetrate skin and serve as a carrier of other drugs. DMSO also causes vasodilation. DMSO works by inactivating the superoxide radicals produced by inflammation. This mechanism traps oxygen radicals, reducing cellular damage. DMSO is labeled for reduction of swelling via topical application, but has been used extra-label for a variety of things, such as IV for treatment of CNS swelling and topically following perivascular injection of irritating solutions. This drug is available as a gel and solution and is also found in a combination ear product containing glucocorticoid (Synotic®). Caution should be used when applying DMSO, as it can cause skin irritation, can leave a garlic taste in the mouths of both patient and medical personnel, and can cause birth defects in some species. Always wear rubber gloves when applying DMSO. A bandage may be applied over the area of application to prevent owner and animal contact with DMSO; however, this may cause burning of the treated area.

Orgotein

This superoxide dismutase drug inactivates superoxide radicals (similar in action to DMSO). Palosein® is an example of *orgotein* and is used IM in horses for joint disease. It can also be given intra-articularly to horses and SQ to dogs.

Osteoarthritis Treatments

The treatment of osteoarthritis is either surgical or nonsurgical. Nonsurgical treatment is typically attempted before surgical treatment and is successful in managing many cases of osteoarthritis. Medical management, in addition to dietary management and managed physical activity, is achieved with controlling inflammation. Anti-inflammatory drugs are used frequently in the treatment of inflammation secondary to osteoarthritis and were discussed previously. Additional pharmacological treatments for osteoarthritis include the following.

Glycosaminoglycans

Glycoproteins, known as *proteoglycans*, form part of the extracellular matrix of connective tissue like cartilage. Proteoglycans are high-molecular-weight substances and contain many different polysaccharide side chains. These

Clinical Que

In the United States, flunixin meglumine is the only NSAID labeled for use in beef and dairy cattle (IV only). Extravascular (IM or SQ) injections are considered illegal by AMDUCA.

polysaccharides make up most of the proteoglycan structure; hence, proteoglycans resemble polysaccharides more than they do proteins. The polysaccharide groups in proteoglycans are called *glycosaminoglycans* or *GAGs*. GAGs include hyaluronic acid, chondroitin sulfate, dermatan sulfate, keratan sulfate, and heparin sulfate. All of the GAGs contain derivatives of glucosamine or galactosamine. Glucosamine derivatives are found in hyaluronic acid and heparin sulfate. Galactosamine derivatives are found in chondroitin sulfate.

- *Hyaluronic acid* is normally part of the joint fluid in animals. When hyaluronic acid is given intra-articularly (it is FDA approved via this route), it is believed to help cushion degenerating joints. This cushioning helps relieve joint pain and improve mobility. Some animals may develop local reactions to hyaluronic acid, but such reactions usually subside within 24 to 48 hours. A trade name drug is Hyalovet®.

- *Polysulfated glycosaminoglycans* (PSGAGs) appear in a drug that is a semisynthetic mix of glycosaminoglycans from bovine cartilage. It is known as Adequan® and can be given intra-articularly or IM in horses and dogs. It is reported to promote the production of joint fluid from the synovial membrane and has some anti-inflammatory action.

- *Glucosamine* and *chondroitin sulfate* are believed to play a role in the maintenance of cartilage structure and function. Glucosamine may be found in products alone (Maxi GS®) or in combination with chondroitin sulfate (Cosequin®). Glucosamine/chondrotin sulfate is a neutraceutical that is approved by the USDA for its analgesic and anti-inflammatory activity. The combination product comes in tablet and food supplement forms. Chondroitin sulfate in combination with hyaluronic acid is an FDA-approved drug.

Antihistamines

Antihistamines are drugs that counteract the effect of histamine. Histamine causes bronchoconstriction and inflammatory changes when it is released from mast cells. When histamine combines with tissue receptors, it causes dilation of blood vessels, increased capillary permeability, smooth muscle spasming, and increased glandular secretion (Figure 16-9). Antihistamines, or H_1 blockers, compete with histamine for receptor sites, thus preventing a histamine response. Antihistamines do not have an effect after histamine attaches to its receptor site. H_1 blockers are also called *histamine antagonists*. There are two types of histamine receptors: H_1, which constrict extravascular smooth muscles when stimulated, and H_2, which increase gastric secretions when stimulated (see Chapter 11). Side effects of antihistamines include drowsiness, dry mucous membranes, and stimulation of the central nervous system in high doses.

H_1 blockers are used to treat pruritus, laminitis in large animals, motion sickness, anaphylactic shock, and some upper respiratory conditions. Examples of H_1 blockers are listed in Table 16-4.

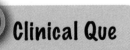

Clinical Que

Antihistamines do not affect the release of histamine, but act instead by competing with histamine for its receptor sites.

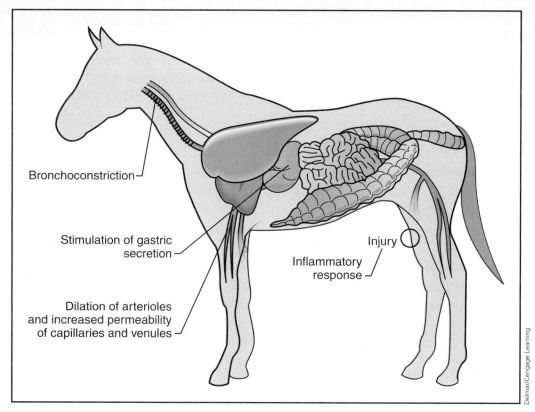

Figure 16-9 Effects of histamine in the body

Bronchoconstriction

Stimulation of gastric secretion

Dilation of arterioles and increased permeability of capillaries and venules

Injury

Inflammatory response

Delmar/Cengage Learning

Table 16-4 Antihistamines Used in Veterinary Medicine

GENERIC NAME	TRADE NAME EXAMPLES
cetirizine	• Zyrtec®
clemastine	• Tavist®
chlorpheniramine	• Chlor-Trimeton®
cyproheptadine	• Periactin®
dimenhydrinate	• Dramamine®
diphenhydramine	• Benadryl® • Histacalm Shampoo®
hydroxyzine	• Atarax®
meclizine	• Bonine® • Antivert® • D-Vert®
pyrilamine maleate	• Histavet-P®
triplelennamine	• PBZ® • PBZ-SR® • Re-Covr®
trimeprazine	• Temaril-P® (combination product with prednisolone)

IMMUNOMODULATORS

Immunomodulation is the adjustment of the immune response to a desired level. This adjustment could include immunopotentiation (enhancement of the immune response), immunosuppression (reducing the immune response), and induction of the immunologic tolerance (nonreactivity to an antigen).

Cyclosporine is a selective immunomodulator that suppresses T lymphocyte activity by inhibiting the production of cytokines such as interleukin-2, interleukin-4, and alpha-interferon by helper T lymphocytes, thereby inhibiting T lymphocyte activation and proliferation. Cyclosporine is considered one of the most effective immunosuppressant agents available. It has anti-inflammatory and antipruritic properties. Atopica® is a cyclosporine developed for the treatment of atopic dermatitis in dogs. Initially, it is given orally daily until clinical improvement is seen (usually four to eight weeks) and then reduced to treatment every second day. Once clinical signs of atopic dermatitis are controlled, it is given every three to four days. Side effects include gastrointestinal disturbances such as vomiting and diarrhea that occur within the first few weeks of administration and are generally transient. It is recommended that bacterial and fungal infections be treated before using cyclosporine because the inhibition of T lymphocytes may decrease the immune response. Cyclosporine is also covered in Chapter 20.

Piroxicam is an NSAID (nonselective COX inhibitor) that has anti-inflammatory, analgesic, and antipyretic properties. Its anti-inflammatory properties are due to inhibition of prostaglandin synthesis; however, its narrow therapeutic index and side effects such as gastrointestinal ulceration, renal necrosis, and peritonitis have limited its use (there are also many other NSAIDs on the market that are safer). Piroxicam (Feldene®) has antitumor effects by limiting production of prostaglandin-E2 (PGE_2), which usually stimulates growth of some tumor cell lines and inhibits T lymphocyte production. By limiting PGE_2 production, piroxicam is believed to help T lymphocyte production and inhibit growth of some tumor cells. Piroxicam is used as an adjunct treatment of urinary bladder transitional cell carcinoma and may be beneficial in treatment of squamous cell carcinoma, mammary adenocarcinoma, and transmissible venereal tumor. Piroxicam has a very long half-life and may have teratogenic effects.

Gold salts such as auranofin (Ridaura®) and gold sodium thiomalate (Aurolate®) have anti-inflammatory and immunomodulating effects. Gold is engulfed by macrophages where it inhibits phagocytosis and lysosomal enzyme activity. Gold also inhibits release of histamine and production of prostaglandins. Auroanfin also suppresses helper T cells, comes in capsule form, and is used to treat pemphigus and idiopathic arthritis in dogs. Gold sodium thiomalate (and aurothioglucose, another gold salt that may be discontinued) comes in injectable form, is given IM, and is used to treat pemphigus in small animals.

PAIN RELIEVERS

Pain is an unpleasant sensory event of the peripheral and central nervous systems and an emotional and cognitive experience associated with actual or potential tissue damage (see Chapter 7). Pain is classified as physiologic (the body's protective mechanism to avoid tissue injury) or pathologic (that which arises from tissue injury and inflammation or from damage to a portion of the nervous system). Pathologic pain can be further divided into categories such as nociceptive (peripheral tissue injury) or neuropathic (damage to the peripheral nerves or central nervous system). Nociceptive pain occurs as a result of the activation of the peripheral nociceptive nervous system by noxious stimuli such as heat, cold, intense mechanical force, and chemical irritants (Figure 16-10). Neuropathic pain results from sensory axons in the peripheral or central nervous system without peripheral nociceptive stimulation. Inflammation, modulated by leukotrienes, nerve growth factor, bradykinin, prostaglandins, histamine, hydrogen ions, and neurosignaling chemicals such as norepinehrine and neuropeptides, can also produce pain.

Analgesics are drugs that relieve pain without causing loss of consciousness. The proper analgesic is valuable in treating pain. Selection of an analgesic is based on many things, including the following:

- effectiveness of the agent. Mild to moderate pain can be controlled by NSAIDs, but severe pain may have to be controlled by potent analgesics such as opioid products.

- duration of action. An analgesic with a short duration of action may be sufficient for brief postsurgical pain; moderate to severe chronic pain may require use of an analgesic with a longer duration of action.

- duration of therapy. Some analgesics are highly effective but harmful with prolonged use. Long-term safety must be considered when long-term therapy is needed.

- available routes of drug administration. For long-term pain relief, oral therapy is the most convenient form for owners. Injectable routes that can be used in the clinic may be time consuming and expensive for the client.

Analgesics fall into two categories: narcotics (opioid) and nonnarcotics. Narcotic analgesics were covered in Chapter 7. Nonnarcotic analgesics include aspirin, phenylbutazone, ibuprofen, ketoprofen, carprofen, naproxen, deracoxib, meloxicam, firocoxib, flunixin meglumine, etodolac, meclofenamic acid, tolfenamic acid, and tepoxalin. These drugs and their actions were covered in the anti-inflammatory section of this chapter.

Another nonnarcotic analgesic is acetaminophen. This drug class includes the drug Tylenol®, which has analgesic effects but limited, if any, antipyretic and anti-inflammatory effects. Most consider acetaminophen to have no anti-inflammatory effects, and it is rarely used in veterinary

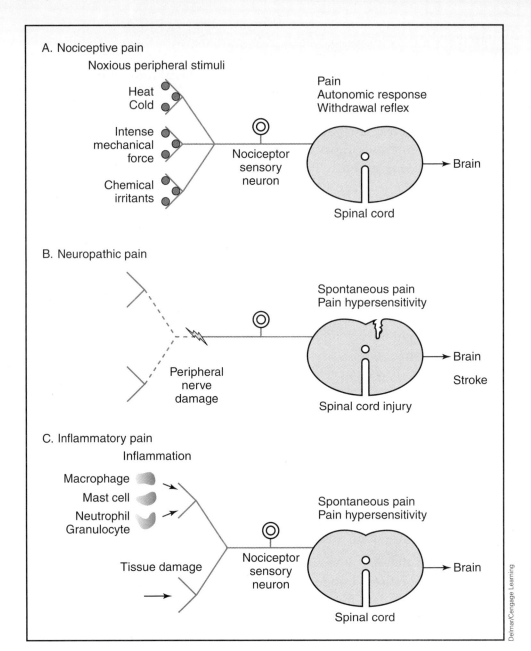

Figure 16-10 Types of pain

medicine unless it is in the combination product codeine-acetaminophen. Acetaminophen reduces fever by its effect on the hypothalamus of the brain, and likely reduces the perception of pain by effects in the brain that are not yet clearly identified. Gastrointestinal side effects of acetaminophen are rare due to its mechanism of action, though it can cause liver and kidney dysfunction in all animals. Acetaminophen toxicity is treated with acetylcysteine. Acetaminophen is contraindicated in cats at any dosage. Cats lack the enzyme glucuronyl transferase needed to process acetaminophen, which leads to a buildup of toxic metabolites that cause blood disorders in cats.

Table 16-5 summarizes the drugs covered in this chapter.

Table 16-5 Drugs Covered in This Chapter.

DRUG CATEGORY	EXAMPLE
Glucocorticoids	• betamethasone • cortisone • dexamethasone • flucinolone • hydrocortisone • isoflupredone • methylprednisolone • methylprednisolone acetate • prednisolone • prednisone • prednisolone sodium succinate • triamcinolone
NSAIDs	• aspirin • deracoxib, maloxicam, firocoxib • diclofenac sodium • etodolac • flunixin meglumine • ibuprofen, ketoprofen, carprofen, naproxen • meclofenamic acid, tolfenamic acid • phenylbutazone • tepoxalin
Free Radical Scavengers	• DMSO • orgotein
Osteoarthritis treatments	• chondroitin sulfate • diclofenac sodium • glucosamine • hyaluronic acid • PSGAGs
Antihistamines	• cetirizine • clemastine • chlorpheniramine • cyproheptadine • dimenhydrinate • diphenhydramine • hydroxyzine • meclizine • pyrilamine maleate • tripelennamine • trimeprazine
Immunomodulators	• cyclosporine • piroxicam • gold salts (auranofin, gold sodium thiomalate)
Analgesics	• narcotics (see Chapter 7) • nonnarcotic (NSAIDs; see anti-inflammatories section)

SUMMARY

There are two main categories of anti-inflammatory drugs: glucocorticoids (glucocorticoid is a type of corticosteroid) and nonsteroidal anti-inflammatories (NSAIDs) (Figure 16-11). Glucocorticoid drugs work by inhibiting phospholipase, and most NSAIDs work by inhibiting cyclooxygenase. Side effects are more severe with glucocorticoid drugs and include PU/PD, polyphagia, suppressed healing, gastric ulcers, thinning of skin, and muscle wasting. Iatrogenic Cushing's disease may be seen in animals given glucocorticoid drugs for long periods of time. Care must be taken when stopping glucocorticoid drug use so that the negative feedback mechanism of the hypothalamus-anterior pituitary-adrenal gland is given time to reestablish itself thus preventing the development of iatrogenic Addison's disease.

Glucocorticoid drugs can be short-, intermediate-, or long-acting. They can usually be recognized by the *-sone* ending of their generic names.

Most NSAIDs are good analgesic, antipyretic, and anti-inflammatory drugs. The main side effect of their use is gastrointestinal ulceration and

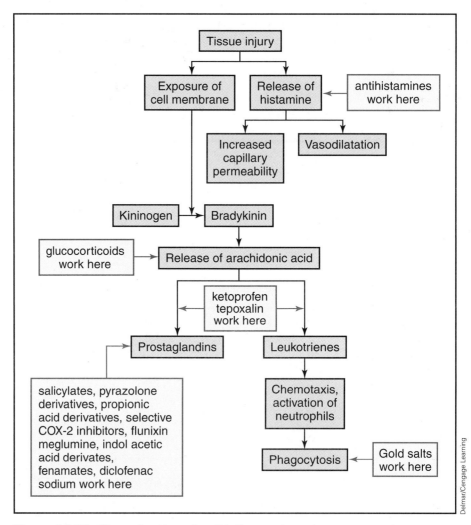

Figure 16-11 Sites of action of anti-inflammatory drugs

bleeding, which may be diminished with the use of selective COX-2 inhibitors (such as deracoxib, meloxicam, and firocoxib). Other examples of NSAIDs include aspirin, ibuprofenlike products, phenylbutazone, flunixin meglumine, etodolac, meclofenamic acid, tolfenamic acid, tepoxalin, and diclofenac sodium. Miscellaneous anti-inflammatory drugs are DMSO and orgotein.

Antihistamines work by blocking histamine attachment to H_1 receptors. Antihistamines are used as anti-inflammatory drugs for many diseases, including pruritus and laminitis.

Analgesics relieve pain without causing loss of consciousness. Analgesics fall into two categories: narcotic (opioid) and nonnarcotic (NSAIDs). Acetaminophen reduces the perception of pain through its effect on the hypothalamus of the brain. Acetaminophen has limited, if any, antipyretic and anti-inflammatory effects and is not commonly used in veterinary practice unless it is in the combination product codeine-acetaminophen.

Osteoarthritis treatments include treating pain associated with the disease, but may also include a variety of drugs such as glycosaminoglycans (hyaluronic acid, PSGAGs, glucosamine, and chondroitin sulfate), and orgotein. Immunomodulators are substances that adjust the immune response to a desired level and may include immunopotentiation, immunosuppression, and induction of immunologic tolerance. Immunomodulators include cyclosporine, piroxicam, and gold salts.

It's a Wrap

The answers to the questions in this chapter's Setting the Scene case study should be understood after reading this chapter. In this case study, the owner asks the veterinary technician if phenylbutazone will fix his horse's lameness. Will it? Do analgesics and anti-inflammatory drugs cure disease? Can they cause additional problems for the animal?

Anti-inflammatory drugs like phenylbutazone work by inhibiting cyclooxygenase, an enzyme that promotes the formation of prostaglandin from arachidonic acid in the cell membrane. Some anti-inflammatory drugs also have analgesic and antipyrexic properties.

In this horse, degenerative changes are observed in the articular cartilage of the navicular bone. Anti-inflammatory drugs will not repair cartilage (cure the disease); they simply reduce inflammation, which effects swelling, pain, and decreased motion. Analgesics do not repair cartilage either; they simply reduce pain. Phenylbutazone will make the horse feel better but will do nothing to repair the injury. One danger in administering anti-inflammatory and analgesic drugs to animals with

(continued)

injuries is that when they feel less pain, they may overexert themselves, further damaging the injured tissue. Masking the injury may make the animal feel better than it actually is.

Other dangers associated with anti-inflammatory drug use is the development of gastrointestinal problems (ulceration and bleeding), nephrotoxicity, bone marrow suppression, and bleeding tendencies (from decreased platelet aggregation).

CHAPTER REVIEW

Matching

Match the drug name with its action. Answers may be used more than once.

1. _____ dexamethasone
2. _____ prednisone
3. _____ dimethyl sulfoxide (DMSO)
4. _____ carprofen
5. _____ phenylbutazone
6. _____ aspirin
7. _____ flunixin meglumine
8. _____ triamcinolone
9. _____ methylprednisolone
10. _____ diphenhydramine

a. nonsteroidal anti-inflammatory drug
b. glucocorticoid drug or "steroid"
c. antihistamine
d. free radical scavenger

Multiple Choice

Choose the one best answer.

11. Which type of anti-inflammatory drug has the greatest amount of anti-inflammatory effects?
    a. Glucocorticoid drugs, because they block phospholipase, which occurs first in the inflammatory pathway.
    b. NSAIDs, because they block cyclooxygenase, which occurs first in the inflammatory pathway.
    c. Antihistamines, because they block histamine, which is the only inflammatory mediator.
    d. All of the categories work equally.

12. Iatrogenic disease is caused by
    a. endocrine disease.
    b. the treatment.
    c. steroids.
    d. glucocorticoid drugs.

13. Propionic acid derivatives are NSAIDs that
    a. can be identified by the *-fen* ending of their generic names.
    b. block phospholipase.
    c. are analgesic, anesthetic, and anti-inflammatory.
    d. all of the above are correct.

14. Which NSAID works by inhibiting cyclooxygenase-2 and not cyclooxygenase-1?
   a. flunixin meglumine
   b. DMSO
   c. deracoxib
   d. aspirin

15. Indol acetic acid derivatives like etodolac (EtoGesic®) have the benefit of
   a. being labeled for use in all species.
   b. being available in chewable and injectable forms.
   c. being available in gel and solution forms.
   d. once-daily oral dosing.

16. What ending on the generic name of some antihistamines may indicate that they are antihistamines?
   a. *-hist*
   b. *-ane*
   c. *-amine*
   d. *-ate*

17. What type of anti-inflammatory drug is sometimes used for motion sickness in animals?
   a. glucocorticoid drugs
   b. NSAIDs
   c. PSGAGs
   d. antihistamines

18. What NSAID is commonly used for pain relief of lameness and colic in horses?
   a. aspirin
   b. phenylbutazone
   c. ibuprofen
   d. etodolac

19. What anti-inflammatory drug readily penetrates skin and must be handled cautiously to avoid absorption of the drug by the person giving the treatment?
   a. flunixin meglumine
   b. cortisone
   c. DMSO
   d. PSGAG

20. What NSAID is commonly used and labeled for use in large animals, and should be used with caution (if at all) in small animals?
   a. flunixin meglumine
   b. etodolac
   c. DMSO
   d. carprofen

Case Studies

21. A 10-year-old M/N German Shepherd (95#) is brought to the clinic for ongoing lameness. His PE is normal except for pain on palpation of the hips. The veterinarian requests an X-ray of the dog's hips. The X-rays reveal that this dog has hip dysplasia. The owner would like some pain medication for this dog.
   a. What are some choices for pain and inflammation relief in this dog?
   b. The owner would like to try aspirin in this dog. What side effects should the owner be told about?
   c. The dog develops diarrhea on aspirin. The owner would now like to try carprofen. What test would you recommend that she have done on her dog prior to starting carprofen?
   d. The owner would like to give the dog medication only once daily. What are some possible choices?

22. The owner of a 10-year-old M/N Dachshund calls the clinic to ask for advice about her dog. Another veterinarian had put the dog on a 10-day course of prednisone for intervertebral disc disease. The owner wants to stop giving the dog prednisone immediately, because the dog is urinating in the house.
   a. What could happen as a result of suddenly withdrawing high doses of glucocorticoids from this dog after two weeks of treatment with the drug?
   b. How can this consequence be prevented from occurring?

Critical Thinking Questions

23. What considerations must be taken into account when deciding on the best approach to long-term use of anti-inflammatory agents?

24. What are some problems associated with glucocorticoid use?

ADDITIONAL REVIEW MATERIAL

Additional review material can be found on the StudyWARE™ CD included with this text.

REFERENCES

Allen, D. G. (2005). *Handbook of Veterinary Drugs* (3rd ed.). Baltimore, MD: Lippincott Williams & Wilkins.

Conzemius, M. (2008). *Pain Management Methods for Treating Osteoarthritis.* NAVC Proceedings, 2008, Orlando, FL.

Diesel, A., and Moriello, K. (May 2008). A busy clinician's review of cyclosporine. In *Veterinary Medicine,* Vol. 103, No. 5. pp. 266–273.

Medical economics. *PDR for Nutritional Supplements* (2001). Montvale, NJ: Thomson Healthcare.

Perkowski, S. Z. NSAIDs for the maintenance of acute pain in dogs and cats. NAVC Veterinary Proceedings, 2007, Orlando, FL.

Plumb, D. C. (2008). *Plumb's Veterinary Drug Handbook* (6th ed.). Ames, IA: Blackwell Publishing.

Reiss, B., and Evans, M. (2007). *Pharmacological Aspects of Nursing Care* (7th ed., revised by Bonita E. Broyles. Clifton Park, NY: Delmar Cengage Learning.

Smith, G., Davis, J., Tell, L., Webb, A., and Riviere J. (March 2008). Extralabel use of nonsteroidal anti-inflammatory drugs in cattle. *JAVMA,* Vol. 232, No. 5, March 1, 2008, pp. 697–701.

Spratto, G. R., and Woods, A. L. (2008). *PDR Nurse's Drug Handbook.* Clifton Park, NY: Cengage Learning.

Veterinary Pharmaceuticals and Biologicals (12th ed.). (2001). Lenexa, KS: Veterinary Healthcare Communications.

Veterinary Values (5th ed.). (1998). Lenexa, KS: Veterinary Healthcare Communications.

Wynn, R. (1999). *Veterinary Pharmacology.* Cambridge, MA: Harcourt Learning Direct.

CHAPTER 17
DRUGS FOR SKIN CONDITIONS

KEY TERMS

antipruritics	pruritus
antiseborrheics	seborrhea
antiseptics	seborrhea oleosa
astringents	seborrhea sicca
caustics	soaks
counterirritants	topical
dressings	
keratolytics	
immunomodulators	

Setting the Scene

A farmer tells the veterinary technician that one of his children's fair calves has a skin infection and that he has been treating it with the ointment once provided for his other cattle. He says the skin infection looked exactly like the bacterial infection he saw in his cattle last year, so he thought this ointment would be an easy way to get rid of the infection. The fair is rapidly approaching, and he wants to clear this infection before the calf is judged. He does not want to treat the calf with oral antibiotics because there is a meat withdrawal time for these drugs. Are there additional tests the veterinarian may want to perform on this animal or are there additional questions to ask this farmer?

Courtesy of iStockPhoto

473

BASIC SKIN ANATOMY AND PHYSIOLOGY

The integumentary system consists of the skin and its appendages, such as glands, hair, fur, wool, feathers, scales, claws, nails, and hooves. Considered one of the largest organs in the body, the integumentary system is involved in many processes.

Skin plays a role in the immune system (acting as a physical barrier to infection and containing special antigen-processing cells), waterproofs the body, prevents fluid loss, and provides a site for vitamin D synthesis. Sebaceous glands lubricate the skin and discourage bacterial growth on the surface. Sweat glands regulate body temperature and excrete wastes through sweat. Hair helps control body heat loss and is a sense receptor. Nails, hooves, and claws protect the surface of the distal phalanx.

The skin is made up of three layers: the epidermis, dermis, and subcutaneous layers (Figure 17-1). The outermost or most superficial layer is the *epidermis*. The epidermis is several cell layers thick and does not contain blood vessels. The epidermis is dependent on the deeper layers for nourishment. The thickness of the epidermis varies greatly from region to

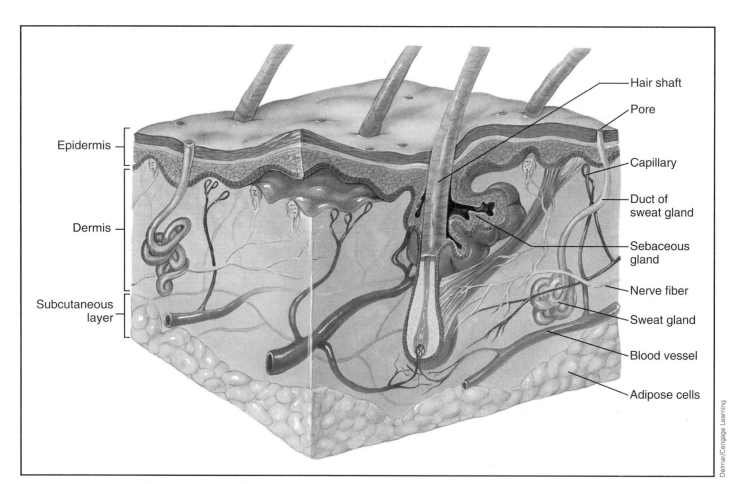

Figure 17-1 Skin layers and structures

region in any animal and from species to species. The epidermis is made up of stratified squamous epithelium. The base of the epidermal layer is known as the *basal layer*. Cells multiply and push upward in an orderly fashion from the basal layer. As the epidermal cells are pushed further to the surface from the underlying blood supply, they die, and their cytoplasm is converted to keratin. The basal layer also contains melanocytes, which produce and contain the black pigment, *melanin*, that gives color to the skin and hairs.

The *dermis* (or *corium*) is the layer directly deep to the epidermis. The dermis is composed of blood vessels, lymph vessels, nerve fibers, and the accessory organs of the skin (glands and hair follicles), all situated in a tight mesh of collagen strands.

The *subcutaneous layer*, or *hypodermis*, is located deep to or under the dermis and is composed of connective tissue. The subcutaneous (SQ) layer contains a large amount of fat.

DRUGS USED IN THE TREATMENT OF SKIN DISORDERS

Many drugs used in the treatment of skin disorders have been discussed in previous chapters. This section addresses the use of **topical** treatments (agents applied to a surface that affect the area to which they are applied) and their role in skin disease management. *Alopecia*, or hair loss resulting in bald patches, has been addressed in preceding sections on the endocrine system (Chapter 10), antimicrobial drugs (Chapter 14), and antiparasitic drugs (Chapter 15). The role of anti-inflammatory agents was discussed in Chapter 16.

On Top of Bacteria and Fungi

Topical antibacterial agents are used to prevent infection associated with minor skin abrasions and to treat superficial skin infections caused by susceptible bacteria (Figure 17-2). Topical use of antibiotics can lead to the development of sensitivity to the drug being used, so any reactions to topical medication must be documented in the animal's medical record. Several antibiotic agents are often combined in a single product to take advantage of the different antibacterial spectrum of each drug.

Antifungal drugs are most commonly employed in the treatment of two types of fungal infections of the skin: dermatophytes and yeast. *Dermatophyte* infections are commonly referred to as *ringworm* infections (see Figure 14-8) and include the genera *Microsporum, Trichophyton,* and *Epidermophyton*. Dermatophyte organisms can live only on dead keratin tissue and can be successfully eliminated only if the infected area is free of fungus. Therapy for dermatophytes typically involves oral and/or topical treatments for several weeks.

Topical antibacterials and antifungals are listed in Table 17-1. This is only a partial list of products available, and their spectrum of activity should be

Clinical Que

An animal's hair or fur may limit the use of creams and ointments, but may be improved by clipping or shaving the area. Animals may also remove topically applied drugs by licking which removes drug before it penetrates the skin and may produce systemic side effects if large amounts of drug are ingested.

Clinical Que

Use caution when applying antibiotics to extensively damaged skin, as appreciable amounts of drug may be absorbed systemically.

Clinical Que

Ringworm is a contagious and zoonotic disease. Ideally, affected animals should not have contact with any other animals.

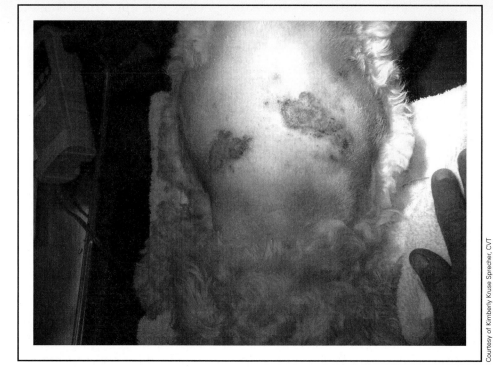

Courtesy of Kimberly Kruse Sprecher, CVT

Figure 17-2 Pyoderma in a dog.

reviewed in Chapter 14 on antimicrobials. Topical antiparasitic agents are covered in Chapter 15. Other topical treatments are covered in Chapter 18 on ophthalmic and otic medications.

Table 17-1 Topical Antibacterials and Antifungals

CATEGORY	EXAMPLES
Topical antibacterials	• bacitracin and polymyxin: Mycitracin®, Vetro-Biotic®
	• clindamycin: Cleocin T®
	• gentamicin: found in topical sprays like gentamicin sulfate with betamethasone and Gentacin Spray®
	• mupirocin (broad-spectrum antibiotic that inhibits bacterial protein synthesis): Muricin®
	• nitrofurazone: Fura-zone®, Furacin® ointment, cream, and solution, NFZ® Puffer (powder)
	• neomycin: Panalog®, Animax® (both contain nystatin/neomycin/thiostrepton and triamcinolone ointment)
	• silver sulfadiazine: works against *Pseudomonas* sp. (Silvadene®)
	• thiostrepton (gram + spectrum of activity): Panalog®, Animax® (both contain nystatin/neomycin/thiostrepton and triamcinolone ointment)

(Continued)

CATEGORY	EXAMPLES
Topical antifungals	• clotrimazole: Clotrimazole Solution 1%®
	• copper naphthenate: Kopertox® (used to treat thrush in horses)
	• ketoconazole: Nizoral®, Ketochlor®
	• lime sulfur: LymDyp®
	• miconazole: Conofite®, Micaved Spray 1%®, Resizole Leave-in Conditioner®
	• nystatin: Panalog®
	• terbinafine: Lamisil®

On Top of Itching

Pruritus, or itching, may be associated with many skin and systemic diseases. When skin is damaged, an influx of inflammatory cells and release of inflammatory mediators and cytokines occur in the epidermis and dermis. This influx of cells and chemicals leads to proliferation of keratinocytes which leads to scaling. Pruritus is often seen with inflammation. Topical **antipruritics** provide slight to moderate relief of itching, and are usually used in conjunction with oral medications such as antihistamines and glucocorticoids. Some products listed in other categories may have antipruritic effects, but control of itching is not their main function. Topical antipruritics include the following.

Nonsteroidal Topical Antipruritics

These products give temporary relief of itching and are usually used in combination with oral products.

Topical Anesthetics

Local anesthetics inhibit the conduction of nerve impulses from sensory nerves, thereby reducing pain and pruritus. They are generally used topically to minimize discomfort associated with allergies, insect bites, and burns. Local anesthetics are poorly absorbed from intact skin, but can be absorbed through damaged skin. Most topical anesthetic drugs can be recognized by the *-caine* ending in the name, and include such agents as *lidocaine, tetracaine, benzocaine,* and *pramoxine.* Trade name examples include DermaCool® with Lidocaine, Xylocaine® (lidocaine), Dermoplast® (benzocaine), Relief® Spray and ResiPROX® (pramoxine with oatmeal), and Pontocaine® (tetracaine).

Topical Soothing Agents

Colloidal oatmeal (oats ground into an extremely fine powder) is believed to have soothing and anti-inflammatory effects when applied topically. Colloidal oatmeal also coats, moisturizes, and protects skin. It is found in colloidal oatmeal shampoos such as Epi-Soothe® and Oatderm Soothing Shampoo®.

Topical Antihistamines

Antihistamines are antipruritics that provide temporary relief of pain and itching associated with allergic reactions and sensitive skin. Products contain *diphenhydramine* (an antihistamine to control itching) and oatmeal (to cleanse and soothe irritated skin). Examples of this category are Histacalm Shampoo® and Histacalm Spray®.

Topical Glucocorticoids

Topically applied glucocorticoids are very effective in alleviating inflammatory signs. Glucocorticoids have anti-inflammatory and antipruritic action. Glucocorticoids also decrease swelling by maintaining the cellular integrity of the capillary, thus preventing leakage of large proteins from the blood and loss of excessive fluid into the tissue space. When applied to the skin, they interfere with normal immune responses and reduce redness, itching, and edema. They also slow the rate of skin cell production.

The effectiveness of a topical glucocorticoid depends on the potency of the drug used, the vehicle used to carry the glucocorticoid to the skin, the thickness and integrity of the skin at the application site, and the amount of moisture present in the skin. Damaged skin at the application site may increase the amount of drug absorbed into the bloodstream and result in systemic side effects.

The least potent topical glucocorticoid is *hydrocortisone*. It is suitable for long-term topical use. Topical glucocorticoids containing a fluorine atom in their structure are among the most potent products (e.g., *fluocinolone*). These products should be used sparingly.

Another difference between glucocorticoid drugs is the duration of action. Short-acting glucocorticoids, such as hydrocortisone and *cortisone*, have a duration of action of less than 12 hours. Intermediate-acting glucocorticoids, including *prednisone, prednisolone, triamcinolone,* and *methylprednisolone,* have a duration of action between 12 and 36 hours. Long-acting glucocorticoids, such as *betamethasone, flumethasone,* and *dexamethasone,* have a duration of action greater than 48 hours.

Topical glucocorticoid products may be combined with other ingredients such as antibiotics to broaden their action. Examples of products containing topical glucocorticoids include Gentocin Topical Spray® (betamethazone and gentamicin), Vetalog Cream® and Genesis Topical Spray® (triamcinolone), Panolog Cream® (triamcinolone and nystatin, neomycin and thiostrepton), DermaCool-HC®, Corticalm Lotion ® (hydrocortisone), Synalar Cream® (fluocinolone), Neo-Synalar® (fluocinolone/neomycin), and Tritop® (isoflupredone acetate).

On Top of Seborrhea

Seborrhea dermatitis is a skin condition characterized by abnormal flaking or scaling of the outermost layer of the epidermis. Seborrhea accompanied by increased production of sebum (oil) is referred to as **seborrhea oleosa**.

Seborrhea without increased production of sebum is referred to as **seborrhea sicca**. A variety of products are used to treat seborrhea (**antiseborrheics**). One important group is the **keratolytics**, which remove excess keratin and promote loosening of the outer layers of the epidermis. Keratolytics break down the protein structure of the keratin layer, permitting easier removal of this compacted material. Keratolytics are found in medicated shampoos to help in the treatment of seborrhea. Topical antiseborrheics include the following:

- *sulfur*, which is keratolytic, antipruritic, antibacterial, antifungal, and antiparasitic. Sulfur slows down epidermal cell proliferation and tends to be nonirritating and nonstaining to the animal. Sulfur is used to treat seborrhea sicca. It is not a good degreasing agent and is not as drying as other antiseborrheic agents. Products include SebaLyt Shampoo®, Sebolux Shampoo®, Allerseb-T Shampoo®, and NuSal-T Shampoo®.

- *salicylic acid*, which is keratolytic, antipruritic, and antibacterial. Salicylic acid lowers skin pH resulting in increased hydration of keratinocytes and is used to treat seborrhea sicca and hyperkeratotic skin disorders. Products include SebaLyt Shampoo® and KeraSolv Gel®.

- *coal tar*, which is keratolytic and degreasing. Tar suppresses epidermal growth and DNA synthesis. It is irritating and may stain light-colored haircoats. Coal tar is used to treat seborrhea oleosa and may be irritating in cats. Products include Mycodex Tar and Sulfur Shampoo®, NuSol-T®, and LyTar Shampoo®.

- *benzoyl peroxide*, which is keratolytic, antipruritic, antibacterial, and degreasing. It lowers skin pH and disrupts microbe cell membranes and is used to treat seborrhea dermatitis in which overproduction of oil occurs. It is also used to treat moist dermatitis (hot spots), pyoderma, stud tail, and a variety of skin lesions that are moist and/or are contaminated with bacteria. Products include Pyoben Gel®, Sulf/OxyDex Shampoo®, and OxyDex Gel®.

- *selenium sulfide*, which is keratolytic, degreasing, and antifungal. It interferes with hydrogen-bond formation in keratin and is used to treat seborrhea and eczema. Use of selenium sulfide may result in subsequent irritation. Products include Seleen Plus Medicated Shampoo® and Selsun Blue Shampoo®.

- *phytosphingosine*, which is a ceramide (waxy material meant to mimic the normal lipid composition of the epidermis). Ceramides are part of the intracellular cement of skin cells and are key molecules in the natural defense of skin. Phytosphingosine also helps maintain moisture balance in the skin, which helps control dry, scaly skin. It may also have anti-inflammatory and antimicrobial properties. Products include Douxo® Seborrhea Shampoo and Douxo® Seborrhea Spot-on.

Clinical Que

Medicated shampoos are used to treat skin conditions and should not be used for routine bathing. Types of medicated shampoos include antipruritic, antibacterial, antifungal, and antiseborrheic. For maximum effectiveness they typically should be left on the skin for 5 to 15 minutes before rinsing.

Clinical Que

"Hot spots" is a common term for acute moist dermatitis. Hot spots are treated with multiple drugs, including glucocorticoids, antibiotics, and benzoyl peroxide, in combination with shaving of the affected area.

Miscellaneous Topical Drugs

Table 17-2 lists other categories of drugs used topically to treat skin disorders. Table 17-3 reviews the drugs covered in this chapter.

Table 17-2 Miscellaneous Topical Drugs

CATEGORY	ACTION	EXAMPLES
Astringents: agents that constrict tissues	• Stop discharge by precipitating protein • Have some antibacterial properties • Used to treat moist dermatitis and other moist skin lesions	• Stanisol® (contains salicylic acid, tannic acid, and boric acid) • Tanni-Gel® (contains isopropyl alcohol, salicylic acid, tannic acid, and benzocaine)
Antiseptics: substances that kill or inhibit the growth of microorganisms on living tissue	• Acetic acid is effective against *Pseudomonas* sp.	• Acetic acid is usually found in ear preparations; products include Fresh-Ear®, Oti-Clens®, and ClearEar Cleansing Solution®
	• Alcohols are bactericidal, astringent, and cooling	• 70% or 90% isopropylalcohol, 70% ethyl alcohol
	• Benzalkonium chloride is antibacterial and antifungal	• Benzalkonium chloride is found in Myosan Cream® and Dermacide®
	• Chlorhexidine is bactericidal, fungicidal, and partially virucidal • Chlorhexidine solution has a 24-hour residual effect	• Chlorhexidine products include Nolvasan® Solution, Scrub, and Ointment; Chlorhexi-Derm Shampoo®
	• Iodine is bactericidal fungicidal, virucidal, and sporicidal (as a 1% solution) • Povidone liberates free iodine and is preferred over iodine solution or tincture (containing alcohol) because it is less irritating to skin	• Iodine is found in Betadine® Solution and Scrub, Tincture of Iodine, Lugol's® Solution, and Xenodine® Spray
	• Propylene glycol is antibacterial and antifungal	• Propylene glycol products are typically used as solvents or vehicles for other drugs (goes by generic name)
	• Triclosan is antibacterial	• Triclosan products include Sebalyt Shampoo® and Triclosan Shampoo
Soaks and **dressings:** substances applied to areas to draw out fluid or relieve itching	• Aluminum acetate (Burow's solution) is drying and mildly antiseptic. It is used as a soak to relieve itching and inflammatory discharge	• Aluminum acetate is found in Domeboro® powder and tablets
	• Magnesium sulfate is used in wound dressings to draw fluid out of tissues	• Magnesium sulfate is found in Epsom salts
Caustics: substances that destroy tissue	• Destroy tissue • Used to treat warts and excessive granulation tissue (proud flesh in horses)	• Silver Nitrate Stick Applicators®, Stypt-Stix® (silver nitrate). Note: silver nitrate products can stain • Equine HoofPro®, Copper Suspension®, HoofPro+® (copper sulfate)

(Continued)

CATEGORY	ACTION	EXAMPLES
Counterirritants: substances that produce irritation and inflammation in areas of chronic inflammation	• Thought to increase blood supply to the area, which in turn brings WBC, antibodies, and so on to the area to stimulate healing; carry away kinins to relieve pain	• Lin-O-Gel® (alcohol, camphor, menthol, and iodine) • Equi-Gel® (camphor, menthol, thymol, witch hazel, and isopropyl alcohol)
Immunomodulators: substances that have an effect on the immune system (either immunostimulatory or immunosuppressive)	• *Imiquimod* stimulates patient's monocytes and macrophages to induce regression of viral protein production. Used to treat sarcoid in horses, squamous cell carcinoma, feline herpes viral dermatitis, and localized solar dermatitis	• Aldara® (imiquimod)
	• *Tacrolimus* inhibits T lymphocyte activation. Used to treat atopic dermatitis, discoid lupus erythematosus, pemphigus erythematosus or foliaceous, and perianal fistulas	• Protopic® (tacrolimus)
	• *Pimecrolimus* inhibits T lymphocyte activation. Used to treat atopic dermatitis, discoid lupus erythematosus, pemphigus erythematosus or foliaceous, and perianal fistulas	• Elidel® (pimecrolimus)
Retinoids	• Stimulates cellular mitotic activity. Used to treat chin acne, callous pyoderma, and footpad hyperkeratosis	• Retin-A® (tretinoin)

Table 17-3 Drugs Covered in This Chapter

DRUG CATEGORY	EXAMPLES
Topical antibiotics	bacitracin, clindamycin, gentamicin, mupirocin, polymyxin, nitrofurazone, neomycin, silver sulfadiazine, thiostrepton
Topical antifungals	clotrimazole, copper naphthenate, ketoconazole, lime sulfur, miconazole, nystatin, terbinafine
Nonsteroidal topical antipruritics	topical anesthetics: lidocaine, benzocaine, tetracaine, pramoxine topical soothing agent: oatmeal topical antihistamines: diphenhydramine
Topical glucocorticoids	hydrocortisone, betamethazone, triamcinolone, fluocinolone, isoflupredone acetate
Antiseborrhea agents	sulfur, salicylic acid, coal tar, benzoyl peroxide, selenium sulfide, phytosphingosine
Astringents	products containing: salicylic acid, tannic acid, boric acid, isopropyl alcohol
Antiseptics	alcohols, chlorhexidine, propylene glycol, acetic acid, iodine, benzalkonium chloride, triclosan
Soaks and dressings	aluminum acetate (Burow's solution), magnesium sulfate
Caustics	silver nitrate, copper sulfate
Counterirritants	products containing alcohol, camphor, menthol, iodine, thymol, witch hazel
Immunomodulators	imiquimod, tacrolimus, pimecrolimus
Retinoids	tretinoin

SUMMARY

Topical medications are used to treat a variety of skin conditions, including pruritus, scaling, and infection. Types of topical medications are antibiotics, antifungals, antiparasitics, antipruritics, antiseborrheals, and miscellaneous. Topical antibiotics and antifungals are commonly used in combination products to widen their spectrum of activity. The spectrum of activity of antibiotics and antifungals is covered in Chapter 14; examples of topical antibiotics are found in Table 17-1. Topical antiparasitic products are covered in Chapter 15. Topical antipruritic drugs tend to be used in conjunction with systemic antipruritic medications; this class includes glucocorticoids and nonsteroidals (topical anesthetics, colloidal oatmeal products, and antihistamines). Ingredients in topical antiseborrheals include sulfur, salicylic acid, coal tar, benzoyl peroxide, selenium sulfide, and phytosphingosine. Miscellaneous topical medications include astringents, antiseptics, soaks and dressings, caustics, and counterirritants.

It's a Wrap

The answers to the questions in this chapter's Setting the Scene case study should be understood after reading this chapter. Are there additional tests the veterinarian may want to perform on this animal, or are there additional questions to ask this farmer whose children's fair calf has a skin infection?

The veterinary technician needs to tactfully tell this farmer it is not a good practice to hold onto medications used in other animals. Without a physical examination, it is difficult to know whether or not this infection is the same one his other cattle had last year. Since this is a calf he is intending to take to the fair, he must make sure that he is treating the infection with the proper medication. One condition that should be tested for and examined in this calf is ringworm. Ringworm is usually diagnosed visually, but it is better to diagnose this disease with a Wood's light or fungal assay, or both. Antibiotics would be ineffective against the organism that causes ringworm, which is a superficial fungal infection.

Another condition this calf may have is parasitic infestation, which can be diagnosed visually or with a skin scraping (for mites). Since parasitic infections are treated with different medications than bacterial infections and can be transmitted to other animals, it is important to give the farmer the proper medication for his animals.

CHAPTER REVIEW

Matching

Match the drug name with its action.

1. _____ triamcinolone

2. _____ sulfur

3. _____ benzoyl peroxide

4. _____ iodine

5. _____ acetic acid

6. _____ magnesium sulfate

7. _____ chlorhexidine

8. _____ coal tar

9. _____ selenium sulfide

10. _____ salicylic acid

a. keratolytic, antipruritic, antibacterial shampoo and gel ingredient used to treat hot spots, pyoderma, stud tail, and other moist lesions

b. keratolytic, degreasing shampoo ingredient that may be irritating and stain light-colored haircoats; used to treat seborrhea sicca

c. glucocorticoid used to treat pruritus

d. keratolytic, antipruritic, antimicrobial shampoo ingredient that is nonirritating and nonstaining

e. keratolytic, antipruritic, antibacterial shampoo ingredient that is used to treat hyperkeratotic skin disorders

f. keratolytic, degreasing, antifungal shampoo ingredient used to treat seborrhea and eczema

g. antiseptic that is bactericidal, fungicidal, and is effective against some viruses

h. antiseptic that is bactericidal, fungicidal, virucidal, and sporicidal (as a 1% solution)

i. chemical that is effective against *Pseudomonas* sp. and is found in ear preparations

j. component of Epsom salts

Multiple Choice

Choose the one best answer.

11. Which of the following is an example of a nonsteroidal topical antipruritic?
- a. miconazole
- b. polymyxin
- c. fluocinolone
- d. oatmeal

12. Which of the following shampoo ingredients is used as a keratolytic to treat seborrhea?
- a. sulfur
- b. oatmeal
- c. chlorhexidine
- d. propylene glycol

13. What group of chemicals works by destroying tissue at the site of application?
- a. keratolytics
- b. antifungals
- c. antiseborrheals
- d. caustics

14. What chemical is found in products like Betadine® and Lugol's® Solution?
- a. chlorhexidine
- b. iodine
- c. propylene glycol
- d. acetic acid

15. What group of drugs works by drawing fluid out of tissues?
    a. caustics
    b. counterirritants
    c. dressings
    d. antipruritics

16. Which group of skin products works by precipitating protein, thus stopping discharge?
    a. antiseptics
    b. caustics
    c. astringents
    d. soaks

17. Which of the following products has an antibacterial, antifungal, and glucocorticoid drug in it?
    a. Nitrofurazone®
    b. Conofite®
    c. Kopertox®
    d. Panalog®

True/False

Circle a. for true or b. for false.

18. The epidermis does not contain blood vessels.
    a. true
    b. false

19. Keratolytics promote loosening of the outer layers of the epidermis.
    a. true
    b. false

20. Counterirritants work by calming the inflammatory response or counteracting inflammation.
    a. true
    b. false

Case Studies

21. A five-year-old M/N Dachshund (15#) presents to the clinic with oily and scaly ear margins. On PE the dog appears normal and has normal TPR. The veterinarian recommends a biopsy of the affected area, which is performed. The biopsy report reveals that this dog has a form of seborrhea.
    a. Describe types of shampoo ingredients and their actions.
    b. The owner wants to know how long to leave these products on her dog's skin. What should this client be told?

22. A five-month-old F Labrador retriever (40#) presents to the clinic with excessive licking of the ventral abdomen. On PE, the TPR are normal, the dog is excited, and is in good flesh. Examination of the ventral abdomen reveals multiple bumps, some of which appear to have pus in them, and pink skin. The veterinarian feels that this dog has puppy pyoderma.
    a. What does the veterinary technician want to check for in this dog?
    b. After the technician reexamined the dog's skin, there were no other abnormalities observed. The veterinarian decides to treat this dog with a topical antibiotic. Is there another ingredient that would be helpful to have in this product?
    c. What is a disadvantage to using a topical product in this case?
    d. What is an advantage to using a topical product in this case?

Critical Thinking Questions

23. One action of glucocorticoids is to suppress the immune system. Why would an animal benefit from glucocorticoids if it has a bacterial infection?

24. What are some nonmedication therapies that can be used in an animal that has a skin condition?

ADDITIONAL REVIEW MATERIAL

Additional review material can be found on the StudyWARE™ CD included with this text.

REFERENCES

Allen, D. G. (2005). *Handbook of Veterinary Drugs* (3rd ed.). Baltimore, MD: Lippincott Williams & Wilkins.

Amundson Romich, J. (2008). *An Illustrated Guide to Veterinary Medical Terminology* (3rd ed.). Clifton Park, NY: Thomson Delmar Learning.

Bloom, P. B. (2007). *New Drugs in Veterinary Dermatology*. In NAVC Veterinary Proceedings, Orlando, FL.

Plumb, D. C. (2008). *Plumb's Veterinary Drug Handbook* (6th ed.). Ames, IA: Blackwell Publishing.

Reiss, B., and Evans, M. (2007). *Pharmacological Aspects of Nursing Care* (7th ed., revised by Bonita E. Broyles). Clifton Park, NY: Delmar Cengage Learning.

Spratto, G. R., and Woods, A. L. (2008). *PDR Nurse's Drug Handbook*. Clifton Park, NY: Delmar Cengage Learning.

Veterinary Pharmaceuticals and Biologicals (12th ed.). (2001). Lenexa, KS: Veterinary Healthcare Communications.

Veterinary Values (5th ed.). (1998). Lenexa, KS: Veterinary Healthcare Communications.

Wynn, R. (1999). *Veterinary Pharmacology*. Cambridge, MA: Harcourt Learning Direct.

CHAPTER 18

OPHTHALMIC AND OTIC MEDICATIONS

OBJECTIVES

Upon completion of this chapter, the reader should be able to:

- describe the anatomy and physiology of the eye.
- describe the basic pathophysiology of the various types of ocular disease.
- explain the indications for ophthalmic medication, including diagnostic drugs, miotics, mydriatics, intraocular pressure reducers, and other topical forms of medications.
- describe the anatomy and physiology of the ear.
- describe the basic pathophysiology of various otic conditions.
- explain the indications for otic medications, including topical antibiotics, antifungals, antiparasitics, drying agents, cleansing agents, dewaxing agents, anti-inflammatories, and local anesthetics.

KEY TERMS

alpha-adrenergic agonists

beta-adrenergic blockers

carbonic anhydrase inhibitors

cycloplegics

dewaxing agents

drying agents

glaucoma

immunomodulating agents

keratoconjunctivitis sicca (KCS)

lacrimogenics

miotics

mydriatics

osmotic diuretics

otitis externa

otitis interna

otitis media

Setting the Scene

A farmer comes into the clinic and says that some of his calves have developed pinkeye (a bacterial eye infection known medically as infectious keratoconjunctivitis). He remembers that he has ointment at home that he used to treat his cat's eye infection when the cat got a scratch on the cornea during a fight. He wonders if he can use the same ointment for his calves. Are there any questions that may be necessary to ask him? Can antibiotics in ophthalmic ointments be absorbed systemically, thereby raising concerns about withdrawal times with these calves?

BASIC OCULAR SYSTEM ANATOMY AND PHYSIOLOGY

The ocular system is responsible for vision and is comprised of the eyes and *adnexa* (surrounding structures). The eyes are the receptor organs for sight. The eye consists primarily of a multilayered sphere called the *globe*. The fibrous outer layer of the globe is the *sclera*. The sclera maintains the shape of the eye. The anterior portion of the sclera is transparent and is called the *cornea*. The cornea provides some of the focusing power of the eye.

The *choroid* is the opaque middle layer of the globe. The choroid contains blood vessels and supplies blood to the entire eye. The choroid consists of the *iris* (the pigmented, muscular diaphragm that helps regulate the amount of light entering the pupil), the *pupil* (a circular opening in the center of the iris), the *lens* (the clear, flexible, and curved capsule located behind the iris and pupil that is responsible for most of the focusing power of the eye), and the *ciliary body* (the thickened extension of the choroid that assists in accommodation of the lens). The term *uvea* refers to the iris, ciliary body, and choroid.

The inner layer of the globe is the *retina*. The retina is the nervous tissue layer of the eye that receives images. The retina contains specialized cells called *rods* and *cones* that convert visual images to nerve impulses that travel from the

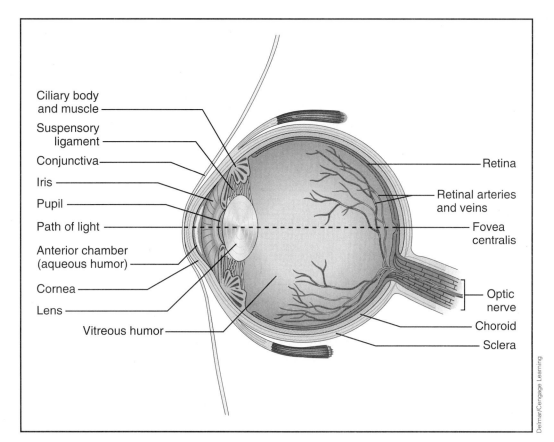

Figure 18-1 Cross-section of the globe of the eye.

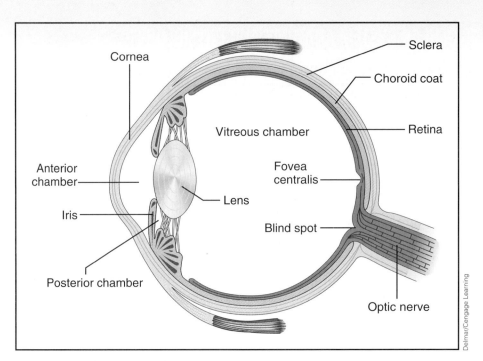

Figure 18-2 Chambers of the eye.

eye to the brain via the optic nerve. The *optic disk* is the region of the eye where nerve endings of the retina gather to form the optic nerve. Figure 18-1 shows the structures of the globe.

The eye is divided into compartments as well: the *anterior compartment* (also known as the *aqueous chamber*) is the anterior one-third of the globe in front of the lens and is divided into the anterior and posterior chambers by the iris. The *vitreous compartment* (also known as the vitreous chamber) is the posterior two-thirds of the eye behind the lens. The anterior compartment contains watery aqueous humor, and the vitreous compartment contains the gelatinous vitreous body. Figure 18-2 shows the compartments of the eye.

The adnexa, or accessory organs, of the eye include the following:

- the orbit (the bony cavity of the skull that contains the globe).
- the eye muscles (seven muscles that control eye movement).
- the eyelids (upper and lower lids that protect the eye from injury, foreign material, and excessive light and the nictitating membrane or third eyelid).
- the eyelashes (hairlike structures that protect the eye).
- the conjunctiva (the mucous membranes that line the underside of each eyelid and cover the front surface of the eyeball).
- the lacrimal apparatus (structures that produce, store, and remove tears).

Figure 18-3 shows the adnexa of the eye.

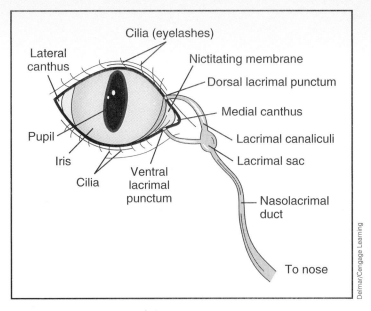

Delmar/Cengage Learning

Figure 18-3 Adnexa of the eye.

OPHTHALMIC DRUGS

The ophthalmic drugs covered in this chapter are mainly topical medications. Topical ophthalmic drugs take the form of either eye drops (solutions and suspensions) or ointments. Solutions are the least irritating, but provide the shortest contact time (they therefore must be applied frequently). Suspensions are similar to solutions but need to be shaken well prior to administration and may block tubing if given through a lavage system. When using topical ophthalmic drugs, veterinarians take into account drug penetration, frequency of drug application, and ability of the owner to apply the drug formulation. Other routes of administration are listed in Table 18-1.

Drug Penetration

Topical ophthalmic drugs penetrate the cornea by diffusion and tend to be absorbed into the anterior chamber having little effect in the posterior or vitreous chambers. Water-soluble drugs (such as the salts atropine sulfate and pilocarpine hydrochloride) penetrate the cornea well, while highly polar drugs (such as hydrophilic antibiotics such as beta-lactam, sulfa, and aminoglycoside antibiotics) penetrate the cornea poorly. Atropine and pilocarpine are polar (penetrate the cornea poorly) and water soluble (penetrate the cornea well), which together affect their rate of corneal penetration. Hydrophilic drugs often reach target tissues by blood vessel uptake into the sclera and accumulation in the ciliary body. More lipophilic drugs, such as fluoroquinolone antibiotics, penetrate the cornea more readily and diffuse through the pupil against aqueous humor flow to the posterior chamber. Water-soluble drugs penetrate the

Clinical Que

Higher concentrations of topical drugs can result in higher concentration of drug in the anterior chamber; however, higher concentrations of drug may be irritating, resulting in more tearing and loss of drug through lacrimation.

Table 18-1 Routes of Administration for Ophthalmic Drugs

ROUTE OF ADMINISTRATION	TISSUES REACHED	FORMULATION	ADVANTAGES	DISADVANTAGES	COMMENTS

Topical	• Cornea
	• Conjunctiva
	• Anterior uvea
	• Eyelids
	• Nasolacrimal system
		• Drops
		  • Solutions
		  • Suspensions
			• Easier administration in small animals
			• Minimal visual impairment
			• Lower incidence of contact dermatitis
				• More difficult administration in large animals
				• Less contact time with cornea
				• More frequent application than ointments
				• Diluted by tear production
				• More systemic absorption possible
					• More than 1 drop rarely indicated per dose
					• Allow 5 minutes between drops
					• Instill in order of least viscous to most viscous
					• Instill water-based products prior to oil-based products

		• Ointments	• Longer contact time • Less frequent administration • Protects cornea from drying • Not diluted by tear production • Less expensive than drops	• Temporary blurring of vision • More difficult administration • More contact dermatitis • Do not use with penetrating corneal wounds • Difficult to determine exact dose	• Ensure that client understands that the tube should not contact the eye
Subconjunctival injection	• Cornea • Anterior uvea	• Sterile solutions and suspensions	• Longer duration of action • Higher anterior chamber concentrations versus topical	• May create scar tissue • Temporary pain	• Improved compliance • Used for drugs with poor corneal penetration
Retrobulbar injection	• Vitreous chamber • Optic nerve	• Sterile solutions and suspensions	• Used to instill local anesthetic for removal of bovine eye • Antibiotics may be given via this route to treat intraocular infections or retrobulbar cellulitis	• Infrequently used except for treatment of orbit or posterior half of the globe	• Primarily used for local anesthetic
Systemic drugs	• Eyelids • Vitreous chamber • Optic nerve • Anterior uvea • Nasolacrimal system	• Oral • Intramuscular • Subcutaneous • Intravenous	• Allows drug penetration to areas where topical therapy is inadequate	• Systemic toxicity • Does not reach cornea	• See information under specific drug section

corneal stroma layer and lipid-soluble drugs penetrate the corneal epithelium making a combination of water- and lipid-soluble properties in a topical ophthalmic preparation desirable (drugs that have both water- and lipid-soluble properties are called differentially soluble).

Frequency of Drug Application

Ointments are usually administered less frequently than eye drops. Owners may prefer to apply eye medication less frequently, because of schedule issues and apprehension about putting medication in the animal's eye; therefore, owners should be questioned about their preference before ophthalmic medication is dispensed. Ointments tend to blur an animal's vision and should be administered at the appropriate time to avoid excessive vision problems.

Ease of Application

Some clients have better success at administering ointments, and other clients have better success with drops. The client's preference may be taken into account in these cases. Figures 18-4A and B demonstrate administration of ophthalmic ointment and drops into the eye.

Ophthalmic drug preparations contain active drug and vehicles to extend the contact time of the drug. Vehicles that provide lubrication and increased contact time include methylcellulose, hydroxyethylcellulose, and polyvinyl alcohol. Polyvinylpyrrolidone is a viscous vehicle that produces an artificial mucus when used with ethylene glycol polymers. Aqueous bases mix readily with tears and are washed away with tears while oily bases (such as lanolin, mineral oil, and petrolatum) provide lubrication and moisturizing of the cornea and conjunctiva (especially in patients who cannot blink).

Clinical Que

Ophthalmic ointments tend to last longer than ophthalmic drops; therefore, they can be administered less frequently.

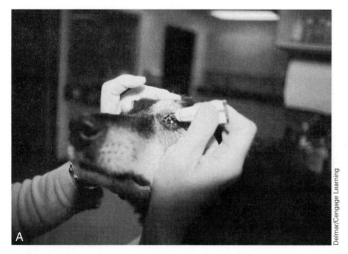

Figure 18-4 (A) Administering ophthalmic ointment in the eye. (B) Administering ophthalmic drops in the eye. When giving ophthalmic medications, it is important not to touch the tip of the applicator to the eye to avoid contamination of the drug.

Diagnostic Drugs

Diagnostic drugs are used to locate lesions or foreign objects in the eye. Diagnostic drugs include topical anesthetics and fluorescein sodium.

Topical anesthetics are used during removal of foreign bodies from the eye and performance of comprehensive eye examinations in which instruments are applied to the cornea to measure intraocular pressure (Figure 18-5). Corneal anesthesia is accomplished in about one minute and lasts for about ten minutes. Examples include *proparacaine hydrochloride* (Ophthaine®, Ophthetic®) and *tetracaine hydrochloride* (Pontocaine®, Alcaine®). Opened bottles should be stored in the refrigerator, used within the expiration dates, protected from light, and any discolored solutions should be discarded.

Fluorescein sodium is used to detect corneal scratches (the stain is orange until it adheres to a corneal defect, where it appears green), foreign bodies (which are surrounded by orange), and patency of the nasolacrimal duct (dye appears in the nasal secretions in animals with an intact nasolacrimal duct). Fluorescein impregnates an individually wrapped sterile strip to which sterile saline is added to allow drops to fall onto the cornea for application to the eye (Figure 18-6). After a few seconds, excess fluorescein is flushed from the eye with sterile saline. Stain will be retained in areas of full-thickness corneal epithelial loss because fluorescein is water soluble. Because the outer layer of the intact cornea is fat soluble, the drug cannot penetrate or adhere to an intact cornea. The stroma layer beneath the outer layer is water soluble. If the outer layer is damaged due to corneal injury or corneal ulceration, the fluorescein adheres to the water-soluble stroma layer when the dye is rinsed with sterile saline.

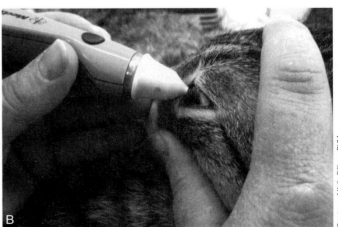

Figure 18-5 Detection of intraocular pressure by tonometry. (A) A Shiotz tonometer, an example of an applanation tonometer, measures corneal resistance to an applied force. Before intraocular pressure readings are taken, topical anesthetic is applied to the cornea, to allow the tonometer to be placed directly on the cornea. (B) A pneumatic tonometer detects intraocular pressure by measuring corneal resistance to a puff of air.

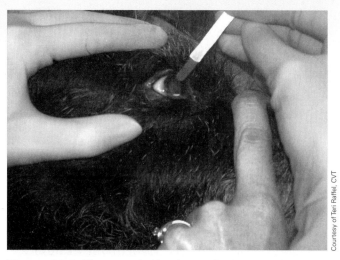

Courtesy of Teri Raffel, CVT

Figure 18-6 Fluorescein dye is applied to the cornea to allow visualization of corneal defects.

Pupil Closing

Miotics are cholinergic drugs that constrict the pupil. They are used to treat open-angle glaucoma because they lower the intraocular pressure by increasing outflow of aqueous humor. Miotics generally promote the outflow of aqueous humor. *Pilocarpine*, a topical cholinergic drug, is the most commonly used miotic. Systemic absorption of pilocarpine is possible, but uncommon. The onset of action is usually 10 to 30 minutes and duration of action is from 4 to 8 hours. Side effects of pilocarpine include local irritation and redness. Trade names of pilocarpine include Piloptic® and Isopto-Carpine®.

Pupil Opening

Mydriatics are drugs that dilate pupils. **Cycloplegics** are drugs that paralyze the ciliary muscles (ciliary muscles function in controlling the shape of the lens) and may minimize pain due to ciliary spasm. These drugs may be used in combination or alone to achieve desired outcomes. Examples include the following:

- *atropine*, an anticholinergic drug used for the treatment of acute inflammation of the anterior uvea and as an aid in examination of the retina (by dilating the pupil to allow visualization of the retina). Atropine produces mydriasis and cycloplegia by blocking the effect of acetylcholine on the sphincter muscle of the iris and the muscles of the ciliary body. Peak effect for mydriasis is 30 to 40 minutes and 1 to 3 hours for cycloplegia. The main side effect of atropine is dry mouth (decreased saliva production). Some animals will salivate following topical atropine administration due to gastric upset caused by ingestion of atropine following drainage of the drug through the nasolacrimal duct. Its use is contraindicated in animals with glaucoma, because it increases intraocular pressure, and in animals with keratoconjunctivitis sicca (KCS or dry eye), because it decreases tear production. Atropine

Clinical Que

Drugs that rapidly penetrate mucous membranes will rapidly enter the bloodstream through the cornea.

Clinical Que

Healing of incisions or corneal ulcers is slowed when anything is placed in the eye. Ointments affect these processes more than other preparations; therefore, ophthalmic drops are typically used following eye surgery.

(Atrophate® or generic) is usually obtained generically and is available as an ophthalmic solution and ointment.

- *homatropine*, an anticholinergic drug used for eye examination and treatment of uveitis. It has a faster onset and shorter duration of action than atropine, making it a more desirable diagnostic agent. It produces mydriasis and cycloplegia, but less so than atropine. The side effects are the same as for atropine. Trade names include Isopto Homatropine® and Homatrocel® Ophthalmic.

- *phenylephrine*, a sympathomimetic drug used to evaluate eye diseases such as uveitis and Horner's syndrome. It may be used prior to conjunctival surgery to decrease hemorrhage. It produces mydriasis, but not cycloplegia. It also produces vasoconstriction. Side effects include ocular discomfort, tearing, and rebound miosis. Trade name examples are Mydfrin 2.5 percent® and Neo-Synephrine®.

- *tropicamide*, an acetylcholine receptor blocker used for fundic examination. It is a rapid-acting mydriatic and has slight cycloplegic effect. It has a more rapid onset and shorter duration of action than atropine, making it a more desirable diagnostic agent. Its side effects include local discomfort and dry mouth. It is contraindicated in animals with glaucoma or KCS. Mydriacyl® and Opticyl® are trade names of tropicamide.

- *epinephrine*, a sympathomimetic drug that reduces intraocular pressure, produces mydriasis, and aids in the diagnosis of Horner's syndrome (Figure 18-7). It is used to prevent glaucoma in the unaffected eye, but should not be used in cases of closed-angle glaucoma. Epinephrine may cause ocular discomfort when applied to the eye. Epifrin® is a trade name of epinephrine.

Courtesy of Kimberly Kruse Sprecher, CVT

Figure 18-7 Horner's syndrome in a cat. Horner's syndrome is caused by injury of the cervical sympathetic innervation to the eye and produces clinical signs that include sinking of the eyeball (enophthalmus), ptosis of the upper eyelid, pupil constriction, and prolapse of the third eyelid.

Pressure Reducing

Glaucoma is a group of diseases characterized by increased intraocular pressure (Figure 18-8). The disorder can be caused by an acquired structural defect within the eye (primary glaucoma); it may be the consequence of another ocular disease or trauma (secondary glaucoma); or it may be the result of a genetic defect (congenital glaucoma). If left untreated, the increase in intraocular pressure can damage the nervous tissue of the eye, mainly the retina and optic nerve, resulting in blindness.

In healthy eyes, aqueous humor is constantly being produced by the ciliary process located behind the iris. Its production is controlled by enzyme systems, mainly *carbonic anhydrase*. Once the aqueous humor enters the eye, it passes from the posterior chamber through the pupil and into the anterior chamber. From there it is drained from the eye through a spongelike substance called the *trabecular meshwork*. Figure 18-9A demonstrates aqueous humor flow through a healthy eye.

When intraocular pressure increases, the outflow mechanism for aqueous humor is blocked. If the iris occludes the trabecular meshwork, normal outflow of aqueous humor is prevented, and the animal is said to have *narrow-angle glaucoma* (Figure 18-9B). If there is no change in the chamber angle of the eye, but aqueous humor outflow is impeded because of degenerative changes, the animal is said to have *open-angle glaucoma.*

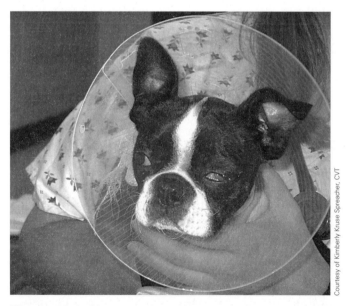

Figure 18-8 Glaucoma in a dog.

Drugs used to decrease intraocular pressure as a result of glaucoma include the following:

- miotics, such as *pilocarpine*, which cause constriction of the pupil (discussed previously in this chapter). Miotics may be used less commonly in the treatment of glaucoma with the advent of topical prostaglandins; however, their use is still common in veterinary patients due to their low cost.

Clinical Que

Mydriatics are used to aid in eye exams, to relieve inflammation associated with uveitis (inflammation of the iris, ciliary body, and choroid) and keratitis (inflammation of the cornea), to break up or prevent adhesions between the iris and the lens, and to prepare an animal for ocular surgery.

Clinical Que

Mydriatics are contraindicated in an animal with glaucoma because they relax iris muscles and hinder aqueous humor outflow from the eye.

- prostaglandins, such as *latanaprost* (Xalantan®), *bimatoprost* (Lumigan®), and *travoprost* (Travatan®), used topically can reduce intraocular pressure by increasing outflow of aqueous humor. Profound miosis is seen with the use of latanaprost. Prostaglandins can be identified by the "prost" in the generic name.

- **carbonic anhydrase inhibitors (CAIs)**, which interfere with the production of carbonic acid and thus lead to decreased aqueous humor formation. The decrease in aqueous humor decreases intraocular pressure. Carbonic anhydrase inhibitors were originally developed as diuretics and are given orally and parenterally. Side effects include vomiting, diarrhea, and weakness. Examples of systemic CAIs are *acetazolamide* (Diamox®), *dichlorphenamide* (Daranide®), and *methazolamide* (Neptazane®). Side effects of these drugs include gastrointestinal disturbances (such as anorexia and diarrhea), dehydration, and urinary crystal development. Topical CAIs such as *brinzolamide HCl* (Azopt®) and *dorzolamide HCl* (Trusopt®) may be used to replace systemic CAIs. Local irritation may been seen with the use of topical CAIs.

- **beta-adrenergic blockers**, sympatholytic drugs which decrease the production of aqueous humor and thus decrease intraocular pressure. Heart and respiratory monitoring is indicated because sympatholytic drugs produce the systemic side effects of bradycardia, hypotension, and bronchospasms. This category of drug tends to be used with primary glaucoma to prevent development of disease in both eyes. They are used topically and may cause blurred vision, so owners should be made aware of this possibility. Examples in this group include *timolol maleate* (Timoptic®), *betaxolol hydrochloride* (Betoptic®), *carteolol* (Ocupress®), *levobunolol* (Betagan®), and *metipranolol* (OptiPranolol®).

- **alpha-adrenergic agonists** are sympathomimetic drugs that reduce aqueous humor secretion and thus decrease intraocular pressure. This category of drug is typically combined with other drugs to aid in decreasing intraocular pressure. Examples in this group include *aproclonidine* (Iopidine®) and *brimonidine* (Alphagan®) and are applied topically. Aproclonidine may cause diarrhea and vomiting in dogs and cats; brimonidine is better tolerated in animals than aproclonidine.

- **osmotic diuretics** increase the volume of urine excreted by the kidneys, which promotes the release of water from the tissues and are used prior to surgery or as an emergency treatment of glaucoma. They are given IV to decrease vitreous humor volume and rapidly decrease intraocular pressure. Side effects of osmotics include electrolyte imbalances, cardiovascular problems, and gastrointestinal problems such as vomiting. Examples include *mannitol* (Osmitrol®) and *glycerin* (Glyrol®, Osmoglyn®).

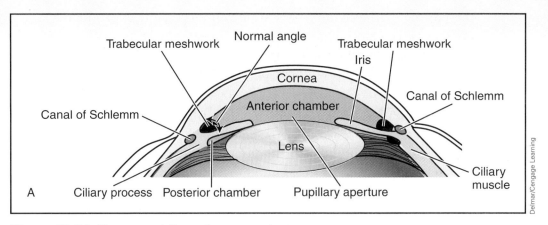

Figure 18-9A The normal flow of aqueous humor.

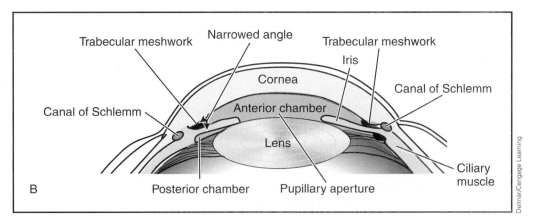

Figure 18-9B In narrow-angle glaucoma, the outflow of aqueous humor is impeded, resulting in an increase in intraocular pressure

Dry Eye Repairers

Keratoconjunctivitis sicca (KCS or dry eye) is a disease in which tear production is decreased, resulting in persistent mucopurulent conjunctivitis and corneal scarring and ulceration. KCS is common in dogs and is thought to be immune mediated (Figure 18-10). Treatment for KCS includes the following:

- *artificial tears*, which are isotonic solutions that are pH buffered to serve as a lubricant for dry eyes and to alleviate eye irritation from KCS. There are a large variety of over-the-counter artificial eye products (Table 18-1).

- antibiotic-glucocorticoid preparations. These combination products may be used if corneal ulcers do not exist. These products are also listed in Table 18-1.

- **lacrimogenics**, which increase tear production. The lacrimal glands are under parasympathetic control, so giving drugs that stimulate the parasympathetic nervous system will increase tear production. *Pilocarpine* is a lacrimogenic that was formerly used topically and orally in food to increase tear production in animals with KCS. Side effects, such as vomiting and diarrhea, have limited its use for KCS treatment.

- **immunomodulating agents**, which adjust the immune response to a desired level. *Cyclosporine* is an immunomodulating agent that interferes with interleukin production by T lymphocytes. This interference stops local inflammation, resulting in improved tear production after several weeks of treatment. Commercial products containing cyclosporine are mixed with oil to enable topical application. Systemic absorption is not seen following topical treatment. A trade name of cyclosporine is Optimmune®. *Tacrolimus* is another drug equally as effective as cyclosporine and may be used in cyclosporine-resistant cases of KCS. Tacrolimus is a drug that is compounded at this time.

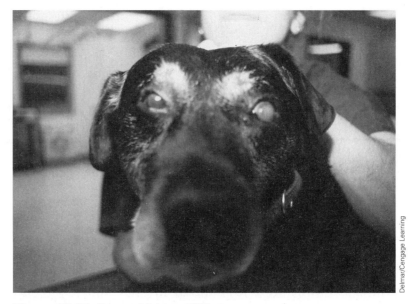

Figure 18-10 Dog exhibiting KCS.

All the Rest

Ophthalmic drugs are used to treat bacterial and fungal eye infections, inflammatory and allergic conditions, and pruritus. Small animals receive most of the ophthalmic drugs veterinarians use, because ophthalmic preparations used on food-producing animals may accumulate in tissues. Because they are most often used to treat multiple infections, most ophthalmic preparations are mixtures of more than one drug. A typical ophthalmic preparation may contain an antibacterial such as neomycin, a second antibacterial such as polymyxin B, and an anti-inflammatory such as prednisolone. Table 18-2 summarizes those drugs.

Table 18-2 Ophthalmic Anti-Infectives, Anti-Inflammatories, and Tear Supplements

CATEGORY	EXAMPLE	ACTION
Topical antibacterial drugs (read labels to determine if glucocorticoids are used in these preparations)	bacitracin (Mycitracin®, Trioptic-P®, Vetropolycin®, Neosporin® Ophthalmic)	Works against gram-positive organisms; usually found in combination with neomycin and polymyxin B
	polymyxin B (Mycitracin®, Trioptic-P®, Vetropolycin®, Neosporin® Ophthalmic)	Works against gram-negative organisms; usually used in combination with bacitracin and neomycin
	oxytetracycline (Terramycin®)	Broad-spectrum antibiotic that is effective against *Chlamydia sp.*; may be used in combination with other drugs
	tetracycline (Achromycin®, Aureomycin®)	Broad-spectrum antibiotic that is effective against *Chlamydia sp.*
	aminoglycosides • gentamicin (Gentocin®, Garamycin®, Genoptic®, OptVet®) • tobramycin (Tobrex®) • neomycin (Mycitracin®, Trioptic-P®, Vetropolycin®, Neosporin® Ophthalmic)	Work against *Staphylococcus sp.* and gram-negative organisms, including *Pseudomonas sp.*; may be formulated alone or with corticosteroids; neomycin is a broad-spectrum antibiotic usually used in bacitracin and in combination with polymyxin B
	erythromycin (Ilotycine Ophthalmic®, Ak-Mycin®)	Broad-spectrum antibiotic; usually used for gram-positive infections
	fluoroquinolones • ciprofloxacin (Ciloxan®) • norfloxacin (Chibroxin®) • ofloxacin (Ocuflox®) • moxifloxacin (Vigamox®)	Broad-spectrum antibiotic
	chloramphenicol (Bemacol®, Chlorbiotic®, Chloricol®, Vetrachloracin®, Chlorasol®)	Broad-spectrum antibiotics; handle with care due to human side effects; cannot use in food-producing animals
	sulfonamides • sulfacetamide (Bleph-10®)	Broad-spectrum antibiotic
Topical antifungal drugs*	natamycin (Natacyn-Ophthalmic®)	Works in treating mycotic keratitis (mainly *Fusarium sp., Candida sp., Aspergillus sp.*)
Topical antiviral drugs	• idoxuridine (Stoxil®) • trifluridine (Viroptic®) • vidarabine (Vir-A Ophthalmic®) • acyclovir (compounded)	Used to treat viral infections of the eye, mainly in cats (ocular herpes). These drugs interrupt viral replication and are virostatic, not virucidal; treatment must be continued past clinical resolution
Topical glucocorticoid drugs	• prednisolone acetate drops (Pred Mild®, Econopred®, PredForte®) • prednisolone sodium phosphate drops (generic brands) • dexamethasone drops and ointment (Decadron Phosphate®, Maxidex®) • loteprednol etabonate (Alrex®)	Glucocorticoids are used to treat inflammation of the conjunctiva, sclera, cornea, and anterior chamber. Penetration to the vitreous chamber and eyelids is poor. Glucocorticoids delay healing and should not be used in patients with corneal ulcers, fungal infections, or viral infections

(Continued)

Table 18-2 *(Continued)*

CATEGORY	EXAMPLE	ACTION
	• triple antibiotic with hydrocortisone (Neobacimyx H®, Trioptic-S®, Vetropolycin HC®) • triple antibiotic with dexamethasone (Maxitrol®) • neomycin with isoflupredone acetate (Neo-Predef®) • gentamicin with betamethasone (Gentocin Durafilm®) • chloramphenicol with prednisolone (Chlorasone®) • sulfacetamide with prednisolone (Blephamide Liquifilm®)	
Topical nonsteroidal anti-inflammatory drugs	• flurbiprofen sodium (Ocufen®) • ketorolac tromethamine (Acular®) • diclofenac sodium (Voltaren®) • suprofen sodium (Profenol®) • bromfenac (Xibrom®) • nepafenac (Nevanac®)	Topical NSAIDs are used to treat inflammation, usually after surgery
Tear supplements	• artificial tears (Bion Tears®, Liquifilm Tears®, Hypotears®, Adsorbotear®) • lubricants (Lacri-Lube S.O.P.®, Akwa Tears®, DuraTear® Naturale)	Artificial tears are isotonic, pH-buffered solutions that lubricate dry eyes and provide eye irrigation. Lubricants are petrolatum-based products that lubricate and protect eyes (mainly used during anesthesia, in which the eyes may remain open while tear production is reduced)

*Other topical antifungal drugs used to treat ocular infections include amphotericin B, povidone iodine, miconazole, and itraconazole. Topical forms of these antifungal drugs are compounded.

BASIC EAR ANATOMY AND PHYSIOLOGY

The ear is the sensory organ that allows hearing and helps maintain balance. The ear is divided into three parts: outer, middle, and inner. The outer ear consists of the *pinna* (also known as the *auricle*) and the external auditory canal. The pinna catches sound waves and transmits them into the external auditory canal. The external auditory canal transmits the sound from the pinna to the tympanic membrane.

The middle ear begins with the *tympanic membrane* or *eardrum*. The middle ear contains the tympanic membrane, auditory ossicles, eustachian tube, oval window, and round window. The sound waves are transmitted from the tympanic membrane to the auditory ossicles, which are three small bones in the middle ear. The auditory ossicles transmit sound waves past the eustachian tube (involved in air-pressure equilibrium) to the oval window. From the oval window, the sound waves enter the inner ear.

Clinical Que

All dogs and cats can get otitis, but those dogs with long pendulous ears tend to have more chronic ear problems.

The inner ear is a series of labyrinths or canals. The vestibule, cochlea, and semicircular canals make up the bony labyrinth. The vestibule and semicircular canals are responsible for sensing balance and equilibrium and the cochlea is responsible for hearing. The structures of the inner ear receive the sound waves that are then relayed to the brain. Figure 18-11 shows the structures of the ear.

A common ear problem seen in veterinary medicine is inflammation. **Otitis externa** is inflammation of the pinna and external auditory canal. It is commonly seen in dogs and cats. Many things, including bacteria, parasites, yeast, allergies, systemic disease, and neoplasia, can cause otitis externa. Clinical signs include head shaking, ear scratching, and discharge from the ear. Factors that predispose animals to developing otitis externa include anatomical and conformational factors (long pendulous ears, narrowed ear canals, and excessive hair in the external ear canal), excessive moisture (from frequent swimming or bathing), iatrogenic factors (from inappropriate treatment by veterinarians),

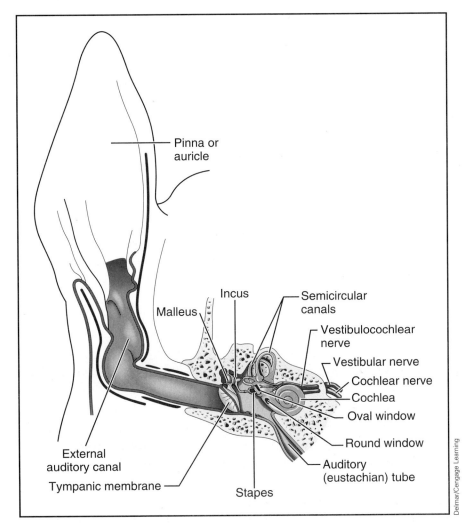

Figure 18-11 Cross-section of ear structures.

and obstructive ear disease (polyps or tumors). Factors that perpetuate otitis externa (those not responsible for initiation of otitis externa, but may cause the condition to continue even after the primary problems are resolved) include bacteria, yeast, underlying disease (such as endocrine diseases like hypothyroidism and hyperadrenocorticism, atopy, autoimmune disease, and keratinization disorders), and otitis media. **Otitis media** is inflammation of the middle ear. Otitis media may be difficult to diagnose because it may be clinically silent (have no signs) or have signs of purulent discharge and head shaking. Otitis media may develop secondary to extension of otitis externa through a ruptured tympanic membrane or extension up the eustachian tube following an upper respiratory infection. Other causes of otitis media include polyps, trauma, and infection. **Otitis interna** is an inner ear infection. Signs of otitis interna include head tilt toward the affected side, ataxia, and possibly nausea and vomiting.

The primary goal of controlling otitis is to determine and treat its primary cause whether it involves removal of foreign bodies, treatment of underlying disease, or management of atopy. Elimination of predisposing factors is an additional goal of treatment and prevention of otitis reoccurrence. It is important to identify the underlying cause of otitis before treatment, so that therapy can be designed to achieve optimal results. Examination usually includes otoscopic examination, ear cytology, and culture-and-sensitivity testing if indicated by the suspected presence of infection. Infectious agents, excessive exudates, and inflammation of the ear can be reduced through various ear cleaning protocols, topical therapies, and systemic therapies or a combination of these. The tympanic membrane should be examined to help choose appropriate cleaning agents (nonotoxic solutions such as isotonic saline are used for flushing debris from the ear canal when the tympanic membrane is ruptured). Cleaning and flushing the ear canal is an important step in treating otitis because it removes exudates that cause inflammation (exudate promotes more inflammation and tissue destruction) and may improve the efficacy of topical medications because the cleaning process removes the barrier between the medication and the target tissue/infectious agents. The use of systemic drugs may be needed if flushing of the ear canal is limited by narrowing of the ear canal caused by severe inflammation and edema.

Clinical Que

The only liquid that is safe to use for flushing ears with ruptured tympanic membranes is sterile isotonic saline solution. All antiseptics should be avoided. Otic drugs that cannot be used with ruptured tympanic membranes include aminoglycosides, chlorhexidine, chloramphenicol, and iodine compounds.

Clinical Que

It is important to wash your hands after applying topical antifungal agents, to avoid the spread of fungal infections.

OTIC MEDICATIONS

Among the many drug combinations veterinarians use to alleviate ear diseases are antifungal agents to eradicate fungi, glucocorticoids to reduce inflammation and pruritus, antibiotics to treat bacterial infections, antiparasitics to treat ear mites, and local anesthetics to reduce pain. Other types of otic preparations include cleansers, drying agents, and cerumen (earwax) dissolvers that are used in combination with otic medications to treat or prevent disease. Many otic preparations are a combination of these ingredients. Table 18-3 lists the various otic drugs.

Table 18-3 Otic Drugs

CATEGORY	EXAMPLE	ACTION

Topical antibiotic otic drugs	• aminoglycosides:
	  • gentamicin (Otomax®, Getocin Otic Solution®, Tri-Otic®, GentaVed Otic®);
	  • neomycin sulfate (Tresaderm®, Panalog®, Tritop®)
		Broad-spectrum antibiotics that are usually combined with a glucocorticoid, antifungal agent, and/or antiparasitic agent. Neomycin sulfate products also contain thiostreptin, a gram-positive antibiotic; tympanic membrane should be intact when using this medication; aminoglycosides are ototoxic.

	• thiostreptin (Tresaderm®, Panalog®, Tritop®)	Gram-positive antibiotic that is usually combined with a glucocorticoid, antifungal agent, and/or antiparasitic agent.
	• chloramphenicol (Liquichlor®, Chlora-Otic®)	Broad-spectrum antibiotic usually combined with a glucocorticoid and/or topical anesthetic. Do not use in food-producing animals. Handle this drug with caution.
	• fluoroquinolones: • enrofloxacin (Baytril Otic®)	Broad-spectrum antibiotic that is combined with *silver sulfadiazine* (has both antifungal and broad-spectrum antibacterial properties).
Topical antiparasitic otic drugs	• thiabendazole (Tresaderm®)	Used to treat ear mites in dogs and cats. Thiabendazole has antiparasitic and antifungal properties. Preparation may contain an antibiotic (such as neomycin) and a glucocorticoid (such as dexamethasone).
	• *pyrethrins* (Mita-Clear®, Cerumite®, Aurimite®)	Used to clear mite infestations of the ear; treatment should continue for at least three weeks.
	• *milbemycin oxime* (MilbeMite®)	Used to treat ear mite infestations in cats and kittens four weeks of age and older. Treatment consists of administering solution from one tube per ear as a single treatment. MilbeMite® is available in a foil pouch that contains two tubes of solution (one tube for each ear). Milbemycin oxime is also available orally.
	• *ivermectin*: • injectable (Ivomec®)	Injectable treatment that is given SQ extra-label to treat ear mites in dogs and cats.
	• topical (Acarexx 0.01 percent suspension®)	Otic solution is labeled for use in cats and kittens (more than four weeks of age) and is packaged in ampules.
	• *selamectin* (Revolution®)	Once-monthly treatment for ear mites in cats; applied to dorsal cervical skin; (also protects against heartworms, roundworms, hookworms, mange mites, the American dog tick, and fleas).
Topical otic antifungal agents	• *clotrimazole* (Otibiotic®, Otomax®)	Works against *Malassezia, Microsporum, Trichophyton, Epidermophyton,* and *Candida* fungi; antifungal agent is combined with gentamicin and betamethasone.

(Continued)

Table 18-3 *(Continued)*

CATEGORY	EXAMPLE	ACTION
	• *nystatin* (Dermagen®, Panalog®, Derma-Vet®, Dermalone®)	Works against *Candida* fungi; antifungal agent is combined with neomycin sulfate, thiostreptin, and triamcinolone.
	• *miconazole* (Conofite®)	Works against *Microsporum* and *Trichophyton*; local irritation may be seen with miconazole treatment.
	• *thiabendazole* (Tresaderm®)	Works against *Microsporum* and *Trichophyton*; also contains antiparasitic, antibiotic, and glucocorticoid.
Topical otic drying agents	Various products that contain *salicylic acid, acetic acid, boric acid,* or *tannic acid.* Examples include Dermal Dry®, VetMark Ear Powder®, OtiRinse Cleansing/ Drying Ear Solution®, Oti-Care-B®	Reduces moisture in the ear to help prevent or treat certain infections of the ear. Ears should be cleaned prior to putting in drying agents.
Topical otic cleansing agents/local anesthetics	Various products contain antibiotics, antiseptics (such as *chlorhexidine* and *povidone iodine*), and/or anesthetics/ soothing agents (such as *lidocaine*). Examples include Solvaprep®, Epi-Otic®, Oti-Clens®, Fresh-Ear®	Used to clean ears and control odor. Also used for gentle flushing of the ear using a bulb syringe or tubing. Ears should be dried thoroughly after use.
Topical otic dewaxing agents	Various products that contain cerumen softeners or drying agents such as *benzyl alcohol, cerumene,* and similar chemicals. Examples include Cerulytic® and Cerumene®.	Used to remove debris and wax before treatment with topical medications, and to aid in wax removal by flushing of the ear with a bulb syringe or tubing.
Topical anti-inflammatories	• *fluocinolone plus DMSO* (Synoptic®)	DMSO enhances percutaneous absorption of glucocorticoids. Avoid contact with human skin to reduce risk of absorbing the drug.
	• *hydrocortisone* (Buro-Otic®, Clearx® Ear Treatment)	Glucocorticoids are used to reduce inflammation association with local irritation.

SUMMARY

Ophthalmic drugs are used to treat conditions of the eyes; they include miotics, mydriatics, and drugs to decrease intraocular pressure and treat KCS. Miotics cause pupillary constriction and are used for treatment of glaucoma. Mydriatics cause pupillary dilation and are used in the treatment of inflammatory disorders, as well as an aid in ocular exams. Drugs that decrease intraocular pressure are used to treat glaucoma, and include miotics, CAIs, beta-adrenergic blockers (sympatholytics), and osmotic diuretics. Keratoconjunctivitis sicca or dry eye is thought to be immune mediated and is treated with artificial tears, antibiotic-glucocorticoid preparations, and immunomodulators. Many ophthalmic drugs are topical preparations, such as antibiotics, antifungals, antivirals, anti-inflammatories (both steroidal and nonsteroidal), and artificial tears and lubricants.

Otic drugs are used to treat conditions of the ear—mainly bacterial, fungal, or parasitic infections. Drying agents, cleansing solutions, dewaxing agents, and local anesthetics can aid in the treatment and prevention of ear infections.

Tables 18-4 and 18-5 summarize the drugs covered in this chapter.

Table 18-4 Ophthalmic Drugs Covered in This Chapter

CATEGORY	EXAMPLE
Topical ophthalmic anesthetics	• proparacaine hydrochloride • tetracaine hydrochloride
Diagnostic stain	• fluorescein
Miotic	• pilocarpine
Mydriatic	• phenylephrine
Mydriatics/cycloplegics	• atropine • homatropine • tropicamide • epinephrine
Prostaglandins	• latanaprost • bimatoprost • travoprost
Carbonic anhydrase inhibitors	• acetazolamide (systemic) • dichlorphenamide (systemic) • methazolamide (systemic) • brinzolamide HCl (topical) • dorzolamide HCl (topical)
Beta-adrenergic blockers	• timolol maleate • betaxolol • carteolol • levobunolol • metipranolol
Alpha-adrenergic agonists	• aproclonidine • brimonidine
Osmotic diuretics	• mannitol • glycerin
Lacrimogenic	• pilocarpine

(Continued)

Table 18-4 *(Continued)*

CATEGORY	EXAMPLE
Immunomodulators	• cyclosporine • tacrolimus
Topical antibacterial ophthalmic drugs	• bacitracin • neomycin • polymyxin B • oxytetracycline • tetracycline • gentamicin • tobramycin • erythromycin • ciprofloxacin • norfloxacin • ofloxacin • moxifloxacin • chloramphenicol • sulfacetamide
Topical antifungal ophthalmic drugs	• natamycin • amphotericin B • povidone iodine • miconazole • itraconazole
Topical antiviral ophthalmic drugs	• idoxuridine • trifluridine • vidarabine • acyclovir
Topical glucocorticoid ophthalmic drugs	• prednisolone acetate • prednisolone sodium phosphate • dexamethasone • triple antibiotic with hydrocortisone • neomycin with isoflupredone acetate • gentamicin with betamethasone • chloramphenicol with prednisolone • loteprednol etabonate • triple antibiotic with dexamethasone • sulfacetamide with prednisolone
Topical nonsteroidal anti-inflammatory ophthalmic drug	• flurbiprofen sodium • ketorolac tromethamine • diclofenac sodium • suprofen sodium • bromfenac • nepafenac
Tear supplement	• artificial tears • lubricants

Table 18-5   Otic Drugs Covered in This Chapter

Category	Example
Topical antibiotic otic drugs	• gentamicin • neomycin sulfate • thiostreptin • chloramphenicol • enrofloxacin
Topical antiparasitic otic drugs	• thiabendazole (neomycin/thiabendazole/ dexamethasone solution) • pyrethrins • milbemycin oxime • ivermectin • selamectin
Topical otic antifungal agents	• clotrimazole • nystatin • miconazole • thiabendazole
Topical otic drying agents	• salicylic acid • acetic acid • boric acid • tannic acid
Topical otic cleansing agents/local anesthetics	antibiotics, antiseptics (such as chlorhexidine and povidone iodine), and/or anesthetics/ soothing agents (such as lidocaine) used to clean the ear
Topical otic dewaxing agents	cerumen softeners or drying agents containing chemicals such as benzyl alcohol, cerumene, and similar chemicals
Topical anti-inflammatories	• fluocinolone plus DMSO • hydrocortisone

It's a Wrap

The answers to the questions in this chapter's Setting the Scene case studies should be understood after reading this chapter. The first case study about calves with pinkeye asked if there are any questions that may be necessary to ask the client prior to him applying an ophthalmic ointment that was prescribed for another animal in these calves' eyes. Can antibiotics in ophthalmic ointments be absorbed systemically, thereby raising concerns about withdrawal times with these calves?

Antibiotics used in small animals may or may not be the same as those used in large animals. It is important to know which medication was used for the cat and if this antibiotic ointment is effective for the treatment of pinkeye in cattle. In other words, is the antibiotic effective against the organism that causes pinkeye?

Cats with ocular infections are often treated with ophthalmic ointments or drops. Pinkeye in cattle is often not treated topically due to the labor involved in treating cattle on a daily or twice-daily basis. Antibiotics such as penicillin mixed with an anti-inflammatory drug like dexamethasone are typically injected into the dorsal eyelid to treat pinkeye in cattle. Oxytetracycline is an antibiotic labeled specifically for use in cattle to treat pinkeye. If antibiotics are given systemically, withdrawal times must be taken into consideration.

The second case study is about a cat with ear mites and asks what should be done prior to treating this cat's ear? Before treating animals with ear disease, it is important to determine what type of organism is causing the disease. It is also important to know if the tympanic membrane is intact because some cleaning solutions and topical medications cannot be used in animals with ruptured tympanic membranes. Otoscopic examination of the tympanic membrane is recommended before any type of intervention is done in ear care, including ear cleaning.

What questions should the owner be asked regarding other animals in the household? Knowing how contagious diseases are transferred among animals is important for both owners and veterinary professionals. Multipet households face additional challenges in regard to contagious diseases and their control. In the case of ear mites, it is important to know if there are other animals in the household. If so, they must also be treated at the same time as this cat.

CHAPTER REVIEW

Matching

Match the drug name with its action.

1. _____ fluorescein
2. _____ mannitol
3. _____ cyclosporine
4. _____ pilocarpine
5. _____ topical otic drying agents
6. _____ pyrethrin
7. _____ atropine
8. _____ chloramphenicol
9. _____ acetazolamide
10. _____ ivermectin

a. mydriatic used to treat acute inflammation of the anterior uvea
b. antiparasitic otic solution used to treat ear mites
c. extra-label ear mite medicine given SQ to dogs and cats
d. dye used to detect corneal defects such as scratches and ulcers
e. broad-spectrum antibiotic that should not be used in food-producing animals
f. carbonic anhydrase inhibitor used to treat glaucoma
g. products containing salicylic acid, acetic acid, boric acid, or tannic acid
h. immunomodulating agent that stimulates tear production and is used to treat KCS
i. miotic used to treat open-angle glaucoma
j. osmotic diuretic used as an emergency treatment for glaucoma

Multiple Choice

Choose the one best answer.

11. Otomax® is the trade name of a topical otic solution containing a glucocorticoid, antifungal, and
    a. chloramphenicol.
    b. neomycin sulfate.
    c. gentamicin.
    d. enrofloxacin.

12. When treating patients with topical antibiotic otic drugs, it is important to
    a. refrigerate all medications.
    b. determine whether the tympanic membrane is intact.
    c. handle the drugs with caution and not use them in food-producing animals.
    d. treat for parasites as well.

13. Proparacaine and tetracaine are examples of
    a. miotic drugs.
    b. topical anesthetics.
    c. mydriatic drugs.
    d. osmotic drugs.

14. Which group of ophthalmic drugs promotes the outflow of aqueous humor by lifting the iris away from the filtration angle area?
    a. miotic drugs
    b. topical anesthetics
    c. mydriatic drugs
    d. osmotic drugs

15. Which group of ophthalmic drugs causes pupillary dilation, allowing examination of the retina?
    a. miotic drugs
    b. topical anesthetics
    c. mydriatic drugs
    d. osmotic drugs

16. Trifluridine works on what category of microorganisms?
    a. bacteria
    b. fungi
    c. parasites
    d. viruses

17. Which drugs should not be used in patients with corneal ulcers or scratches?
    a. cyclosporine
    b. flurbiprofen
    c. glucocorticoids
    d. lubricants

18. What is one advantage of using ophthalmic ointment rather than ophthalmic drops?
    a. You can get more doses from a tube of ointment.
    b. Ointments always contain glucocorticoids to reduce inflammation.
    c. Ointments last longer and therefore the client has to treat the animal less frequently.
    d. Ointments make the animal's vision blurry, making it rest more while it is sick.

True/False

Circle a. for true or b. for false.

19. All mydriatics are also cycloplegics.
    a. true
    b. false

20. Most otic treatments for ear mites are effective with a single application.
    a. true
    b. false

Case Studies

21. A three-year-old F/S Poodle (8#) comes into the clinic with a history of rubbing her eyes. On PE her TPR is normal; she is timid, but otherwise normal according to her owner. The veterinarian wants to perform an eye exam. The veterinarian already has Schirmer tear test strips (to check for tear production). All of the ophthalmic equipment is already in the clinic.
    a. What products should be gathered so that the veterinarian can perform this dog's eye exam?
    b. It is determined that this dog has a corneal scratch from playing with the owner's cat. What product helped diagnose the corneal scratch?
    c. What product is contraindicated in this poodle?
    d. The veterinarian wants to prescribe an antibiotic for the owner to apply to the dog's eye. The owner works all day. What product form should be recommended for this client to use on his pet?

22. A three-month-old M DSH kitten presents to the clinic with excessive ear scratching. A PE shows that the kitten is thin, but has normal TPR and does not appear to have signs of a respiratory infection. The kitten is not current on his vaccinations.
    a. Knowing the age of the kitten, what would be a good thing to check him for?
    b. If the kitten is positive for the observation performed in part a, what are some treatment options for this patient?

Critical Thinking Questions

23. Veterinary ophthalmology has become a big business. Research eye diseases of canines and how they can be prevented. Consider the use of genetic screening for inherited eye diseases, DNA testing (Optigen®) and Canine Eye Registration Foundation (CERF) examination. What is the role of a veterinary professional in working with clients who maintain purebred canines and felines as breeding stock?

24. Hearing loss can occur in animals. Some forms of hearing loss are iatrogenic. Name some nontopical medications that can cause ototoxicity.

ADDITIONAL REVIEW MATERIAL

Additional review material can be found on the StudyWARE™ CD included with this text.

BIBLIOGRAPHY

Allen, D. G. (2005). *Handbook of Veterinary Drugs* (3rd ed.). Baltimore, MD: Lippincott Williams & Wilkins.

Amundson Romich, J. (2008). *An Illustrated Guide to Veterinary Medical Terminology* (3rd ed.). Clifton Park, NY: Delmar Cengage Learning.

Bertone, J., and Horspool, L. (2004). *Equine Clinical Pharmacology*. Elsevier Health Systems.

Engler, K. S. (2007). *The Good, the Bad, and the Smelly:* Otitis externa *Reviewed*, in NAVC Conference Veterinary Proceedings, pp. 938–940.

Reiss, B., and Evans, M. (2007). *Pharmacological Aspects of Nursing Care* (7th ed., revised by Bonita E. Broyles). Clifton Park, NY: Delmar Cengage Learning.

Plumb, D. C. (2008). *Plumb's Veterinary Drug Handbook* (6th ed.). Ames, IA: Blackwell Publishing.

Spratto, G. R., and Woods, A. L. (2008). *PDR Nurse's Drug Handbook*. Clifton Park, NY: Delmar Cengage Learning.

Veterinary Pharmaceuticals and Biologicals (12th ed.). (2001). Lenexa, KS: Veterinary Healthcare Communications.

Veterinary Values (5th ed.). (1998). Lenexa, KS: Veterinary Healthcare Communications.

Wynn, R. (1999). *Veterinary Pharmacology*. Cambridge, MA: Harcourt Learning Direct.

CHAPTER 19

FLUID THERAPY AND EMERGENCY DRUGS

OBJECTIVES

Upon completion of this chapter, the reader should be able to:

- describe body fluid composition, location, and percentages in animals.
- describe how estimation of the level of dehydration is determined based on physical exam findings.
- describe various routes of fluid administration.
- differentiate between crystalloids and colloids.
- give examples of isotonic, hypotonic, and hypertonic crystalloid solutions and describe their uses.
- give examples of colloid solutions and describe their uses.
- list various fluid additives and describe their uses.
- practice calculating fluid therapy volumes based on maintenance values, rehydration values, and ongoing fluid loss values.
- determine the rate of fluid administration based on adult and pediatric administration sets.
- describe various administration sets and fluid delivery systems.
- detail the importance of monitoring fluid therapy and the parameters to be monitored.
- describe basic emergency protocol for an animal in respiratory or cardiac arrest.
- list various emergency drugs and their uses in emergency procedures.

KEY TERMS

adult administration
  sets

colloids

crystalloids

extracellular fluid
  (ECF)

fluid overload

hypertonic

hypotonic

intracellular fluid
  (ICF)

isotonic

maintenance fluid

oncotic pressure

ongoing fluid losses

osmolality

osmotic pressure

pediatric
  administration sets

rehydration fluid

solutes

tonicity

Setting the Scene

New clients bring in a recently adopted six-week-old puppy for examination because he has developed diarrhea. The owners say that the dog has seemed lethargic, but assumed it was because he was in new surroundings. The dog has not been vaccinated nor has had any fecal parasite exams performed on his stool. On PE, the veterinarian notes that the puppy is moderately dehydrated, quiet, and slightly resentful of

(continued)

abdominal palpation. During the PE, the puppy defecates a moderate amount of very watery, blood-tinged diarrhea. The veterinarian orders a fecal examination and blood work to be done on this dog. A fecal antigen test for parvovirus is also performed. Pending the results of the tests, the dog is hospitalized, isolated from other animals, and fluid and antibiotic therapy are initiated. The owners ask why fluids have to be given to this puppy, because he is so small and does not drink a lot of water anyway. Explain to the owners why this puppy needs fluids, where he is losing fluids, and what type of fluids he will get. Is this puppy more at risk of developing dehydration than an older animal? What is the quickest way to get fluids into this puppy? Explain the basics of fluid therapy to this owner.

BASICS OF BODY FLUID

Water is the primary body fluid and is vital for normal cellular function. Body water is distributed among three types of "compartments": cells, blood vessels, and tissue spaces. These tissue spaces exist between blood vessels and cells; therefore, body water may be described as *intracellular* (within the cell), *intravascular* (within the blood vessels), or *interstitial* (in the tissue spaces between blood vessels and cells) depending on its location. Fluid within the cell is further classified as **intracellular fluid (ICF)**, whereas intravascular fluid and interstitial fluid together are classified as **extracellular fluid (ECF)**. About two-thirds of body water is intracellular and is found mainly in skeletal muscle, blood cells, bone cells, and adipose cells. The remaining one-third of body water is extracellular and is found in plasma (about 25 percent) and in the interstitial fluid between cells (about 75 percent) (Figure 19-1).

In healthy animals, a state of equilibrium exists between the amount of water taken in and the amount of water lost in normal physiologic processes. Fluid loss occurs from urinary, gastrointestinal, respiratory, and skin sources (evaporation); fluid intake comes from ingestion of liquids and food as well as from metabolism. The kidneys are the primary regulators of the volume of water within the body. When body water is insufficient, urine volume diminishes. Conversely, when animals drink an excessive amount of water, their urinary output increases.

Body water contains *solutes*, substances such as sodium that dissolve in a *solvent* (which is water in biological systems). *Electrolytes* are substances that split into ions (charged particles) when placed in water. *Cations* are positively charged ions and *anions* are negatively charged ions. The electrolytes found in ICF and ECF are essentially the same; however, their concentrations vary between these two compartments. In an effort to establish equilibrium, body water moves from a less concentrated solution (one with fewer solute particles per unit of solvent) to a more concentrated solution (more solute particles per unit of solvent). This is known as *movement along a concentration gradient* or *osmosis*.

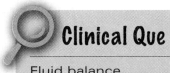

Clinical Que

Fluid balance depends on electrolyte balance.

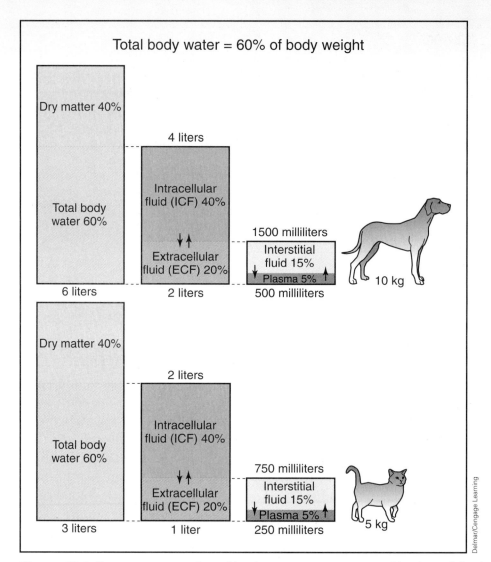

Figure 19-1 Compartments of total body water as a percent of body weight. In a 10 kg dog, 60% of the body weight is total body water (6 L); 40% of the total body water is intracellular fluid (4 L) and 20% is extracellular fluid (2 L); 15% of the extracellular fluid is interstitial fluid (1.5 L) and 5% is plasma (0.5 L). In a 5 kg cat, 60% of the body weight is total body water (3 L); 40% of the total body water is intracellular fluid (2 L) and 20% is extracellular fluid (1 L); 15% of the extracellular fluid is interstitial fluid (0.75 L) and 5% is plasma (0.25 L).

The primary ions in the body are sodium (Na^+), potassium (K^+), chloride (Cl^-), phosphate (PO_4^{-3}), and bicarbonate (HCO_3^-). Sodium is the primary extracellular cation, and potassium is the primary intracellular cation. Chloride is the primary extracellular anion, and phosphate is the primary intracellular anion. Bicarbonate ions are also extracellular (Figure 19-2).

Electrolytes are at equilibrium when the concentration of each electrolyte in the body fluid compartments is held within a limited normal range. Because electrolytes are dissolved in body fluids (mainly water), electrolyte and fluid balance are closely related. Therefore, when fluid volume changes, the electrolyte concentration changes.

Clinical Que

Remember that diffusion is the movement of molecules or solutes from an area of high concentration to an area of low concentration. Osmosis is the movement of water through a selectively permeable membrane from an area of lesser *solute* concentration to an area of greater solute concentration.

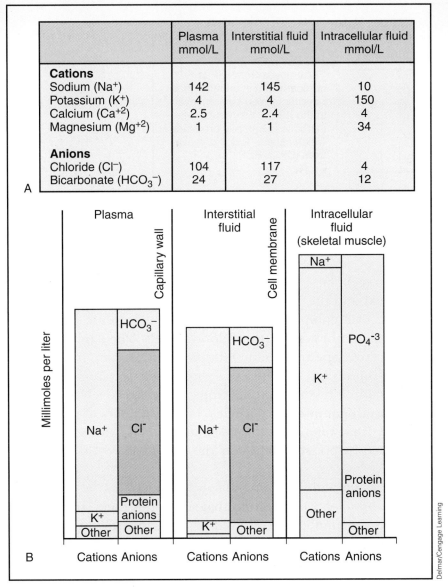

	Plasma mmol/L	Interstitial fluid mmol/L	Intracellular fluid mmol/L
Cations			
Sodium (Na$^+$)	142	145	10
Potassium (K$^+$)	4	4	150
Calcium (Ca^{+2})	2.5	2.4	4
Magnesium (Mg^{+2})	1	1	34
Anions			
Chloride (Cl$^-$)	104	117	4
Bicarbonate (HCO$_3^-$)	24	27	12

Figure 19-2 (A) Values of select cations and anions in plasma, interstitial fluid, and intracellular fluid. (B) Electrolyte balance in plasma, interstitial fluid, and intracellular fluid.

The primary regulation of electrolyte balance is through reabsorption of cations. Anions follow the cations; water follows the ions. Because sodium and potassium are the predominant cations, they are the most important in regulating electrolyte balance. If an animal has an elevated sodium concentration in the ECF, osmotic pressure will increase, causing water to move from the intracellular compartment to the extracellular compartment. **Osmotic pressure** is the pressure or force that develops when two solutions of different concentrations are separated by a selectively permeable membrane. Think of osmotic pressure as the force that draws water across a selectively permeable membrane. To establish osmotic equilibrium, water moves from the less concentrated solution into the more concentrated solution (Figure 19-3).

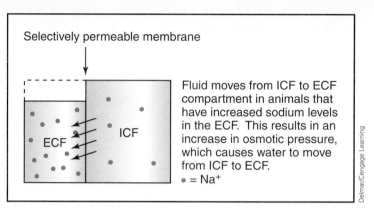

Figure 19-3 Alteration in sodium concentration in ECF causes osmotic pressure to increase, which in turn causes water to move from the intracellular compartment to the extracellular component.

Clinical Que

If the serum osmolality is not within the normal range, a fluid imbalance should be suspected. Serum osmolality is obtained through a laboratory by analyzing serum or by calculation using sodium, glucose, chloride, and potassium concentrations; serum osmolality reflects plasma osmolality.

Clinical Que

Plasma is the liquid portion of blood, containing water, glucose, electrolytes, fats, gases, proteins, bile pigment, and clotting factors. Serum is the fluid portion of blood obtained after separating whole blood into its solid and liquid components after it has clotted.

Within the extracellular space, fluid shifts between the intravascular space (blood vessels) and the interstitial space (tissues) to maintain a fluid balance within the ECF compartment. Fluid exchange occurs only across the walls of capillaries, not across other blood vessels. The capillary membrane acts as a selectively permeable membrane by permitting free passage of water and small solutes but is relatively impermeable to plasma proteins. The amount of osmotic pressure that develops at the selectively permeable membrane is due to the number of particles. Large numbers of particles in the vascular space draw water into the vasculature. Fluid flows only when there is a difference in pressure between the intravascular fluid and the interstitial fluid.

Plasma osmolality is the ratio of solute to solvent (water) in the body. Cells are affected by the osmolality of the fluid that surrounds them. In the body, **osmolality** measures the number of dissolved particles regardless of their size per kilogram of water. Sodium is the largest contributor of particles to osmolality. Plasma osmolality is expressed in units called *osmoles* or milliosmoles per kilogram of water. A solution that has a higher concentration of solute will have a greater plasma osmolality.

FLUID THERAPY

In normal, healthy animals, fluid and electrolytes are balanced. In sick animals, this balance of fluid and electrolytes may be disrupted, making fluid therapy necessary.

In healthy adult animals, approximately 60 percent of the body weight is water; in healthy neonates, the percentage is closer to 80 percent of the body weight. Since neonates have a proportionally larger percent of body weight as water, dehydration can be more severe more quickly in neonates.

The basis for fluid therapy rests on the animal's hydration status. Hydration status can be determined by assessing the patient's history, physical exam status (checking body weight, skin turgor, pulse rate and quality, capillary refill time (CRT), moistness of mucous membranes, and appearance of the eyes), and

Table 19-1 Estimating Level of Dehydration

DEHYDRATION PERCENTAGE	PHYSICAL EXAM FINDINGS
<5%	History of vomiting or diarrhea, but no abnormalities noted on PE History of not eating or drinking well or at all prior to presentation
5%	Tacky mucous membranes; Abdomen feels "doughy" (free peritoneal fluid has been reabsorbed)
6–8%	Mild to moderate decrease in skin turgor; Dry mucous membranes; Slight tachycardia; Corneas appear dull (less tear production)
10–12%	Marked decrease of skin turgor; Dry mucous membranes Weak and rapid pulse; Tachycardia; Slow capillary refill time; Sunken eyes; Mild CNS depression
>12%	Severe dehydration can lead to death

urine production. Table 19-1 provides guidelines for estimating level of dehydration. Keep in mind that as patients become dehydrated, they steal fluid from one area to replace fluid in the area that lost fluid. The first place dehydrated patients steal fluid is from the interstitial space, then from the intracellular space, and finally from the intravascular space. The reason the clinical signs of tachycardia and weak pulses do not appear until the patient is more severely dehydrated is because these signs are associated with decreased intravascular fluid volume only. Since the intravascular space is the last to compensate for dehydration in an animal, it makes sense that signs associated with intravascular fluid space are observed in only severely dehydrated patients. Elevations of laboratory values (such as TP [total protein], BUN [blood urea nitrogen], creatinine, PCV [packed cell volume], and urine specific gravity) may also be seen in dehydrated animals.

Fluid therapy can replace water and electrolytes such as sodium, potassium, and chloride, and restore hydrogen ion balance disrupted by ill health, disease, or trauma. Dehydration from water loss can result from systemic diseases (especially those causing polyuria, diarrhea, or excessive vomiting). Sodium deficiency (*hyponatremia*) results from reduced sodium intake or excessive sodium loss through urination. Potassium deficiency (*hypokalemia*) results from prolonged reduced intake or excessive loss of potassium in gastrointestinal fluids or urine. Chloride loss results from increased secretion and subsequent loss of gastric juice or intestinal fluids. Excessive loss of these electrolytes disturbs the balance of hydrogen ions leading to acid-base abnormalities. Fluid therapy, in addition to maintaining hydration and electrolytes, can also be used to deliver medications.

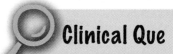 **Clinical Que**

Approximately 60 percent of a healthy adult animal's body weight is water (young and lean animals have a greater percentage of their body weight as water, while geriatric and obese animals have a lesser percentage of their body weight as water).

 **Clinical Que**

One liter of water weighs one kilogram, therefore, a 100-lb adult dog (about 45 kg) has 60 percent or 27 kg of water in its body.

HOW DO WE GET IT THERE?

Fluids can be administered by a variety of routes. Each route has its advantages and disadvantages. The main routes of fluid administration are listed here and demonstrated in Figure 19-4.

Oral (PO) fluids are used for short-term illness and in small animals and neonates. The oral route is the safest route of fluid administration in conscious animals. Oral fluids can be administered by stomach tube, dosing syringe, bottle, nasogastric tube, or gastrostomy tube. Disadvantages of oral fluid administration include possible aspiration of the fluid, the inability to use it in vomiting animals, and less rapid absorption when compared with other methods. Oral fluids are indicated for animals that are anorexic or have diarrhea without vomiting, and for neonatal dehydration. Oral fluids can be water alone or preferably electrolyte solutions. Electrolyte preparations are usually in powder form that are mixed with water. Examples include Hydrolyte®, Oralite®, and Re-Sorb®.

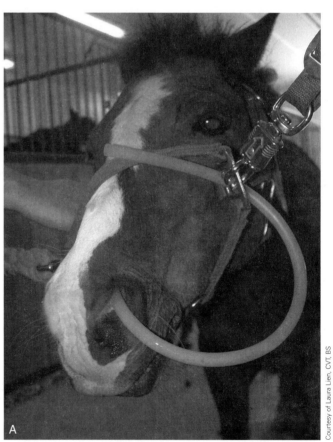

Courtesy of Teri Raffel, CVT

Courtesy of Laura Lien, CVT, BS

Figure 19-4 Routes of fluid administration: (a) oral administration; ways to give oral fluids include with a syringe or via a nasogastric tube. (b) SQ administration; SQ fluids are isotonic fluids given to correct mild to moderate dehydration in noncritically ill patients. (c) IV administration; IV is the preferred route of fluid administration for moderately to severely dehydrated and hypovolemic animals. (d) IO administration; IO administration of fluids is rapid and is in small animals, birds, and pocket pets especially when access to a vein is compromised. IO fluids are given through a long needle with internal stylet attached to a handle, which allows cannulation of the bone marrow for intraosseous infusion

(Continued)

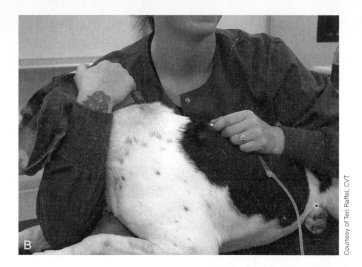

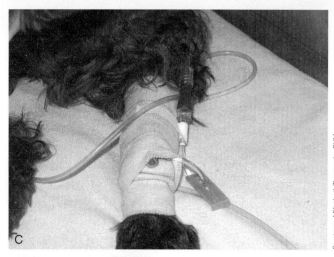

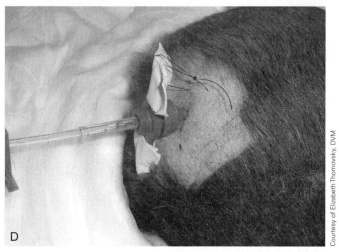

Figure 19-4 *(Continued)*

Subcutaneous (SQ) fluids are used to correct mild to moderate dehydration in noncritically ill patients. In severely dehydrated animals or animals in shock, peripheral vasoconstriction limits the distribution of fluids from the subcutaneous space to where they are needed. Additionally, in animals requiring large volumes of fluids, the subcutaneous route does not allow for administration of enough volume. Finally, subcutaneous fluids cannot include additives (which may cause skin sloughing) and are simply isotonic fluids.

Subcutaneous fluids are usually administered by gravity flow through an 18- or 20-gauge needle. SQ fluids are given in a variety of locations, including the flank region and dorsally along the back between the scapulae. Many prefer to give SQ fluids in the flank region to allow more efficient drainage of fluid in case of infection. The volume of fluid that can be administered SQ is limited by the animal's skin elasticity and may range from 10 to 150 mL per site depending on the animal species being treated. Typically, 5 to 10 mL of fluid per pound of body weight is given per injection site in small animals. Animals differ in their ability to tolerate the infused load comfortably, and multiple sites may be needed to administer the total amount of fluids required. All SQ fluids tend to be absorbed in six to eight hours. Disadvantages of SQ fluid administration include the possibility of infection (especially when given dorsally along the back), subcutaneous edema formation, a slower absorption rate than other routes, and the inability to use hypertonic, hypotonic, or irritating solutions.

Intravenous (IV) fluids are the preferred route for moderately to severely dehydrated and hypovolemic animals. IV routes are best for correcting hypotension because they allow for rapid delivery at a precise dosage. Various tonicities of fluids can be used. An injection site is prepared using aseptic technique and a sterile IV catheter is seated in the vein. Sterile technique is then used in the administration of fluids or drugs through the IV catheter. Disadvantages of IV fluid therapy include an increased possibility of **fluid overload** (more fluid going into the animal than is coming out of the animal), injury to and inflammation of the catheterized blood vessels, and potential extravascular placement of fluids. IV fluid administration requires close monitoring, asepsis, and catheter care, which requires that animals must stay and be monitored in a hospital or clinic.

Intraosseous (IO) fluids are particularly useful in small animals, birds, and pocket pets because they provide direct access to the vascular space. IO administration of fluids is also useful when access to a vein is compromised and should be used if an IV catheter cannot be placed. IO fluids are given via the bone marrow, using sterile technique, through a needle proportionally sized to bone size. A bone marrow needle and stylet are used for mature animals, and a spinal needle and stylet are used for younger animals. IO fluids are most commonly given in the femur (through a site prepared over the trochanteric fossa) or in the humerus (through a site prepared over the greater tubercle). Fluids, whole blood, plasma, and/or drugs (other than drugs that suppress the bone marrow) can be given IO. IO fluids are absorbed rapidly. Many veterinarians do not practice this route of administration often, so they do not possess confidence in their skills and may be reluctant to try it. Other disadvantages include the possibility of bone infection and the need to avoid growth plates (although growth plate fractures are not common because the bone is so soft in animals in which IO catheters are placed).

Rectally (pr) administered fluids may be a good route of fluid administration in young animals, unless diarrhea is present. Electrolyte absorption is good and absorption is rapid, but this route is not commonly utilized because it is difficult to keep fluid from running out of the rectum.

Intraperitoneal (IP) fluids are given when IV access is not available. Isotonic fluids are administered with 16- to 20-gauge needles. The injection site must be as aseptic as possible and is usually located just lateral to ventral midline, between the umbilicus and pelvis. Disadvantages of IP fluid administration include the possibility of sepsis, the inability to use IP routes in animals awaiting abdominal surgery, and the inability to use hypertonic solutions. Due to the disadvantages associated with IP fluids and because they are not absorbed much more quickly than SQ fluids, they are not recommended for use in animals.

WHAT CAN WE GIVE?

Crystalloids and colloids are the two categories of fluids used in fluid therapy. **Crystalloids** are sodium-based electrolyte solutions or solutions of glucose in water that are commonly used to replace lost fluid and electrolytes. The composition is similar to plasma fluid. Crystalloids are further described by their tonicity and are categorized as isotonic, hypotonic, or hypertonic.

Tonicity is based on a measurement called osmolality. Recall that osmolality is the osmotic pressure of a solution based on the number of particles per kilogram of solution. Osmotic pressure is the ability of **solutes** (particles) to attract water (causing osmosis). Not all particles contribute to osmolality. Sodium and glucose provide most of the particles to determine osmotic pressure. Normal osmolality of blood and ECF is 290 to 310 mOsm/kg. The osmolality of **isotonic** solutions is the same as the fluid component of blood and extracellular water; thus isotonic solutions produce no significant changes in the osmolality of blood. **Hypotonic** solutions have osmolality lower than that of the fluid component of blood and may cause the red blood cells to swell if given in extremely large quantities due to lowering of the blood osmolality. Red blood cell swelling happens following hypotonic fluid administration because the osmolality inside the red blood cell is greater than that in the blood, causing fluid to diffuse into the red blood cell and cause them to swell. The osmolality of **hypertonic** solutions is greater than that of the fluid component of blood and may cause the red blood cells to shrink. However, in proper concentrations hypertonic solutions can cause fluid to shift into the intravascular space without causing changes in the red blood cells.

Red blood cell swelling due to use of hypotonic solutions and red blood cell shrinkage due to use of hypertonic solutions is rare because the body quickly restores normal osmolality in most cases by having fluid move into or out of the bloodstream readily. Figure 19-5 summarizes solution tonicity.

Clinical Que

Sodium is important in maintaining acid-base balance and normal heart function, and in the regulation of osmotic pressure in body cells. Hyponatremia can result from severe vomiting or diarrhea, excessive diuresis or diuretic use, renal failure, and wound drainage.

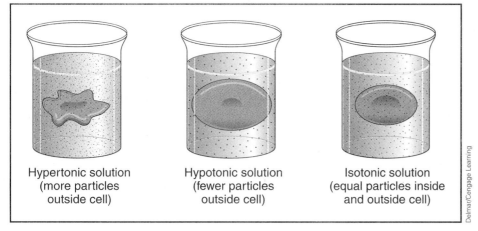

Hypertonic solution
(more particles
outside cell)

Hypotonic solution
(fewer particles
outside cell)

Isotonic solution
(equal particles inside
and outside cell)

Delmar/Cengage Learning

Figure 19-5 Solution tonicity. Blood cells placed in hypertonic solution lose fluid in an attempt to equalize the osmolality in the cell to the solution. Blood cells placed in hypertonic solution gain fluid in an attempt to equalize the osmolality in the cell to the solution. Blood cells placed in isotonic solution have neither a net gain nor loss of fluid.

Isotonic crystalloids have the same sodium concentration as the ECF. About one-quarter to one-third of the total fluid crystalloid volume infused will remain in the vascular space when isotonic crystalloids are administered with the remainder distributing to the extracellular space. Examples of isotonic fluid include *0.9 percent sodium chloride* (also known as *isotonic saline, normal saline* [NS], or *physiologic saline solution* [PSS]), *lactated Ringer's solution (LRS), Normosol®*, and *Plasmalyte®* (Figure 19-6A).

Isotonic saline (0.9 percent sodium chloride) contains only sodium and chloride ions and is used to expand plasma volume and to correct hyponatremia (decreased sodium levels) or metabolic alkalosis. Because of its sodium content, saline should not be used in patients with heart failure or those that have sodium retention due to liver disease. Isotonic saline causes dilutional effects (reduced potassium, total protein, and packed cell volume) and in large volumes can cause acidifying effects. Large amounts of isotonic solution can potentiate or precipitate hypokalemic, hyperchloremic metabolic (nonrespiratory) acidosis.

Lactated Ringer's solution (LRS) is the fluid of choice in many disease situations. LRS is a saline and lactate solution with electrolytes added. LRS contains

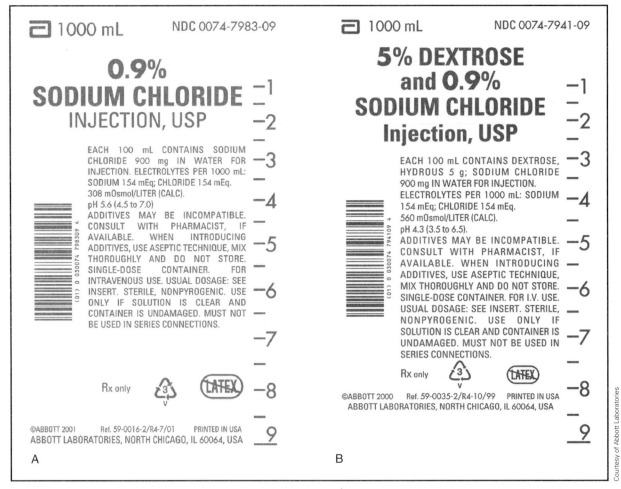

Figure 19-6 Examples of different types of fluids. (A) 0.9% sodium chloride is an example of an isotonic solution. (B) 0.9% normal saline with 5% dextrose is an example of a hypertonic solution.

sodium, potassium, chloride, calcium, and lactate ions. It does not contain magnesium. It has a reduced sodium content as compared with 0.9 percent sodium chloride. The lactate molecule found in LRS is converted in the liver to bicarbonate and is helpful in the treatment of acidosis. LRS is not recommended for use in animals with liver failure because the lactate conversion cannot occur. LRS cannot be given with blood products because the calcium will be chelated by the anticoagulants.

Normosol® is a solution with less sodium, more potassium, more magnesium, less chloride, and no calcium as compared with LRS. It is an all-purpose replacement fluid and has acetate as a buffer. Plasmalyte® is a solution that has less chloride, more magnesium, and no calcium as compared with LRS. Plasmalyte® uses gluconate as a base that is converted in skeletal muscle to bicarbonate. Plasmalyte® cannot be given SQ because it is irritating to tissues due to its low pH.

Hypotonic crystalloid solutions are fluids with less sodium concentration than ECF; thus, they dilute the extracellular sodium and cause a portion of the fluid to move out of the intervascular space into the extracellular space (basically they deliver water). An example of a hypotonic solution is *5 percent dextrose in water (D₅W)*, which is given IV and immediately diffuses out of the intravascular space dividing itself between the ICF and ECF compartments. Two-thirds of the fluid goes to ICF, and one-third goes to the ECF (other crystalloids retain about 1/3 to 1/2 in the intravascular space). D₅W is used to treat hypernatremia and as a fluid supplement in patients that cannot tolerate sodium. Other examples of hypotonic solutions include *1/4 NS (0.25 percent normal saline)* and *1/2 NS (0.45 percent normal saline)*.

Hypertonic fluids have high sodium concentrations that will increase the sodium concentration of the ECF and cause water to move out of the cells into the extracellular space. Examples of hypertonic solutions are *0.9 percent normal saline with 5 percent dextrose* and *3 percent normal saline* (Figure 19-6B). Hypertonic solutions can draw fluid into the intravascular space, thus replacing fluid volume rapidly within blood vessels. Hypertonic solutions can aid in the treatment of shock and edema reduction because fluid is drawn from the interstitial component into the vascular space, which reduces the potential for development of peripheral or pulmonary edema. One problem associated with all hypertonic solutions is that what causes them to be hypertonic only stays in the vasculature for a short time. Once the solute, dextrose or sodium, diffuses out of the intravascular space, all the fluid drawn into the intravascular space diffuses out again. Therefore, the effects of hypertonic solutions are only transient effects.

Isotonic, hypotonic, and hypertonic solutions and their properties are summarized in Table 19-2. Common fluid component abbreviations are listed in Table 19-3.

One adverse reaction associated with all types of fluids administered via the parenteral route is fluid overload. Fluid overload (also known as circulatory overload) is a condition in which the body's fluid requirements are met and the administration of fluid occurs at a rate that is greater than the rate at which the body can use or eliminate the fluid. In other words, fluid overload is the

Table 19-2 Types of Crystalloid Solutions

CRYSTALLOID SOLUTION	[NA+] (mEq/L)	OSMOLALITY (mOml/L)	[CL-] (mEq/L)	[K+] (mEq/L)	[MG+2] (mEq/L)	[CA+2] (mEq/L)	LACTATE (mEq/L)	ACETATE (mEq/L)	GLUCONATE (mEq/L)	DEXTROSE (g/L)
*Isotonic Saline (0.9% NaCl)	155	310	155	0	0	0	0	0	0	
*Lactated Ringer's	130	273	109	4	0	4	28	0	0	
*Normosol®	140	295	98	5	3	0	0	27	23	
#D5W		253								50
#1/4 Normal Saline	34	69	34	0	0	0	0	0	0	
#1/2 Normal Saline	77	155	77	0	0	0	0	0	0	
+0.9% Normal Saline with 5% Dextrose	77	405	77	0	0	0	0	0	0	50
+3% Normal Saline	513	1030	513	0	0	0	0	0	0	

*isotonic solutions
#hypotonic solutions
+hypertonic solutions

Table 19-3 Common Fluid Component Abbreviations

ABBREVIATION	SOLUTION COMPONENT
D	Dextrose
W	Water
S	Saline
NS	Normal Saline (0.9% NaCl)
NaCl	Sodium Chloride
RL	Ringer's Lactate
LRS	Lactated Ringer's Solution

administration of more fluid than the body can handle. The amount of fluid and the rate of fluid administration that will cause fluid overload depend on several factors including the patient's cardiac and renal function. Signs of fluid overload include respiratory changes (dyspea or tachypnea), fluid deposition in the subcutaneous space, or weight gain in patients.

Colloids are fluids with large molecules that enhance the oncotic pressure of blood, causing fluid to move from the interstitial and intracellular spaces into the vascular space. **Oncotic pressure** is the osmotic pressure exerted by colloids in a solution (basically it is the pressure exerted by the plasma proteins). The plasma proteins tend to pull water into the circulatory system. Colloids do not readily diffuse across cell membranes, because of their large size. Colloids are used for vascular space expansion in treating hypovolemic shock and for treating severe chronic disease in which hypoproteinemia is seen. Natural colloids include *plasma, albumin,* and *whole blood.* Synthetic colloids include *dextrans, hydroxyethyl starch,* and *hemoglobin glutamer-200.* The advantage of using colloids versus hypertonic solutions is that colloids will stay in the vascular space, while sodium will diffuse out of the blood vasculature with the use of hypertonic solutions.

Whole blood is commonly used to treat severe anemia and cases of severe blood loss. Whole blood contains all the cellular (RBC, WBC, and platelets) and plasma (proteins and clotting factors) components of blood. It provides RBCs for carrying oxygen to tissues. If used for platelet replacement, whole blood should be used within six hours of collection; if using to supply coagulation factors, whole blood is good for up to 30 days.

The *hematocrit,* also known as the *packed cell volume,* is the volume of red blood cells in proportion to the intravascular fluid volume, expressed as a percentage. It is one indication of fluid loss or gain. A decreased hematocrit reading may be due to hemodilution, blood loss, increased RBC destruction, or failure to produce RBCs. Causes of decreased hematocrit readings include anemia of hemorrhage, nonregenerative anemia, autoimmune hemolytic anemia, and anemia of chronic inflammation (formerly known as anemia of chronic disease). These causes of low hematocrit readings may be indications

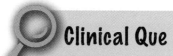

Clinical Que

Crystalloid solutions can be isotonic (have about the same sodium concentration as blood), hypotonic (have less sodium concentration than blood), or hypertonic (have greater sodium concentration than blood).

for whole blood use (transfusions). An increased hematocrit reading can indicate intravascular fluid loss because the total number of RBCs does not change, yet the proportion of RBCs to intravascular fluid increases. Therefore, the amount of intravascular fluid must be decreasing. In these cases, use of whole blood is not indicated.

Whole blood is collected and stored in any of several anticoagulants, including citrate phosphate dextrose adenine (CPSA-1), acid citrate dextrose (ACD), storage medium for blood (SMB), and heparin. Whole blood can be collected in glass vacuum bottles, plastic bags, and plastic syringes. Glass is inert to stored blood, yet is breakable, more expensive, and requires more storage space than plastic bags. It is usually recommended that blood be stored for 28 to 30 days. Blood group typing and testing of donors for infectious diseases transmissible via blood are recommended. Blood is usually collected via the jugular vein and is typically given using a 20- or 23-gauge indwelling catheter. As a guide, 2.2 mL/kg of whole blood raises the packed cell volume by 1 percent when the packed cell volume of the transfused blood is 40 percent.

Plasma is a natural colloid solution that is easy to collect and can be stored frozen for long periods of time (for years if at −40 to −70°C). Plasma contains albumin and globulins (important plasma proteins) that can aid in the treatment of liver disease for coagulation factor deficiency. Fresh frozen plasma is prepared by separating blood into plasma and cells and freezing the plasma component within six hours after whole blood collection. Fresh plasma and fresh frozen plasma, when used as a source of coagulation factors, can be given at a rate of 10 to 20 mL/kg IV once daily for control of bleeding caused by coagulopathies. Frozen plasma should be used within six hours of thawing to maintain the clotting proteins. Adverse reactions to plasma are rare but include fever, urticaria, and hypotension.

Albumin is the main protein in the blood and constitutes about 50 percent of the total protein. The main function of albumin is to maintain the colloid osmotic pressure of blood. Without albumin (and its role in maintaining colloid osmotic pressure), fluid would readily leave the vascular space and accumulate in the tissues, leading to swelling. Transfusions of albumin can be helpful in restoring body protein and in expanding the plasma volume. Too much albumin or albumin administered too rapidly can cause fluid to be retained in the vessels of the lung. Adverse reactions to albumin include fever, anorexia, sudden death, and a one to two week delayed allergic reaction that can be fatal. Currently, the only albumin products available are human albumin products that are available in a variety of percents and if used should be used with caution.

Dextran, in saline or dextrose, is a synthetic colloid solution with large polysaccharides derived from sugar beets. Dextran comes in two concentrations, 40 (Rheomacrodex®) and 70 (Dextran 70®, Gentran 70®). Dextran can be used to treat cases of shock, but its use is limited by allergic reactions and clotting problems seen in animals. To restore blood volume, dextran 40 is given at a rate of 0.7 g/lb/day, and dextran 70 is given at a rate of 0.9 g/lb/day. Adverse reactions to dextran include an increase in bleeding time, anaphylactic reactions,

Clinical Que

Colloids are frequently called volume expanders, because they function like plasma proteins in blood and help maintain oncotic pressure.

development of acute renal failure (dextran 40), and gastrointestinal effects such as abdominal pain and vomiting (dextran 70). Dextrans are currently unavailable in the United States.

Hydroxyethyl starch or *hetastarch* (Hespan®) is a synthetic colloid that expands plasma volume with fewer side effects than dextran. Hetastarch is a combination of hydroxyethyl starch in normal saline and is used to treat hypovolemic shock and hypoproteinemia. The rate of hetastarch administration is 10 to 20 mL/kg/day. Possible side effects include allergic reactions (urticaria, periorbital edema, and dyspnea) and coagulopathies. Anaphylactic reactions are seen less commonly with hetastarch than with dextran.

Hemoglobin glutamer-200 (Oxyglobin®) is a hemoglobin-based, oxygen-carrying fluid that increases plasma and total hemoglobin (Hb) concentration, resulting in increases in arterial oxygen delivery. The Hb in this product is a purified, polymerized bovine Hb diluted in a modified lactated ringer's solution and has an osmolality of approximately 300 mOsm/kg. This Hb distributes oxygen via the plasma instead of the red blood cells. Following infusion of hemoglobin glutamer-200, the plasma and total Hb concentrations increase, but the hematocrit may decrease because of the hemodilution resulting from this product's colloidal properties. Therefore, special considerations are taken into account when assessing oxygen-carrying capacity. Oxyglobin® also causes vasoconstriction and increases blood pressure. Hemoglobin concentration is measured to monitor animals given hemoglobin glutamer-200, and this value is multiplied by 3 to estimate the animal's hematocrit.

Oxyglobin® is used IV for the treatment of anemia in dogs and cats because it increases the oxygen-carrying capacity of blood while reducing blood viscosity. It can also be administered to dogs and cats that have become hemodiluted (PCV < 15 to 18 percent) from crystalloids and that are suffering from chronic anemia (PCV < 10 percent). Oxyglobin® use does not require cross-matching of blood type prior to administration. This product can be administered through any IV line and should not be shaken prior to use (shaking results in foaming of the product). It is compatible with any other IV fluid, but it should not be combined with other products in its bag. Oxyglobin® is stable for up to three years when stored at 2° to 30°C, but should be used within 24 hours after its foil overwrap is opened. Oxyglobin®, like other colloids, helps keep other fluids in the vascular space, so it works well to give oxyglobin concurrently with other crystalloids, especially in dehydrated animals. Side effects include mucous membrane and urine discoloration, vomiting, circulatory overload (signs include coughing, dyspnea [difficulty breathing], and pulmonary edema), and fever. Oxyglobin® is contraindicated in dogs with congestive heart failure or other impairment of cardiac or renal output.

WHAT CAN WE ADD?

Special additives in the crystalloid fluids described earlier may help improve patient recovery and reverse of signs of disease (Figure 19-7A–D). Some additives may precipitate with some types of fluids, so read the product inserts before adding supplements to fluid bags. The following additives are commonly added to fluids.

Fifty percent dextrose

Dextrose is a carbohydrate used in IV fluids to correct hypoglycemia. When a patient is prone to hypoglycemia, dextrose may be added to the fluids. Dextrose supplementation may be needed by patients with increased metabolic needs, such as anorexic patients with sepsis. Dextrose is not added to the fluids as a calorie source, but serves as an energy source for the brain. Dextrose

$$V_1 \times C_1 = V_2 \times C_2$$
$$(1000 \text{ mL})(2.5\%) = V_2 \times 50\%$$
$$\frac{2500 \text{ (mL)(\%)}}{50\%} = V_2$$
$$50 \text{mL} = V_2$$

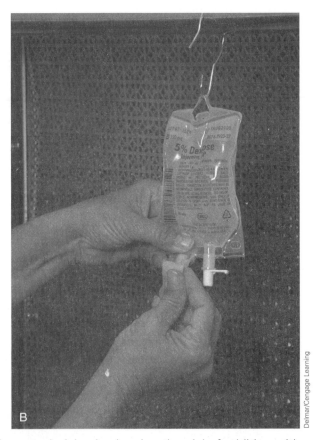

Figure 19-7 Compounding a solution. (A) The veterinary technician is cleaning the vial of additive with alcohol. (B) The bag of solution is also cleaned with alcohol. (C) Medication is added to the bag of solution for compounding. (D) The compounded solution is labeled.   *(continued)*

Delmar/Cengage Learning

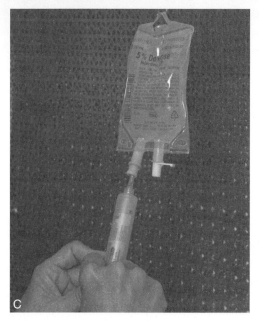

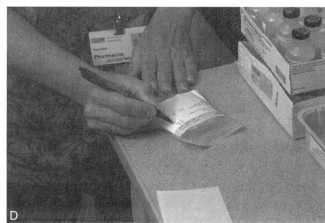

Figure 19-7 *(Continued)*

is typically added to fluids to make a 2.5 to 5 percent solution. If adding 50 percent dextrose to a 1000 mL bag of fluids to make a 2.5 percent solution, 50 mL of 50 percent dextrose must be added.

$$V_1 \times C_1 = V_2 \times C_2$$
$$(1000 \text{ mL})(5\%) = V_2 \times 50\%$$
$$\frac{5000 \text{ (mL)}(\%)}{50\%} = V_2$$
$$100 \text{ mL} = V_2$$

If adding 50 percent dextrose to a 1000 mL bag of fluids to make a 5 percent solution, 100 mL of 50 percent dextrose must be added.

Potassium

Potassium is necessary for the contraction of smooth, cardiac, and skeletal muscles and other physiologic processes. Potassium must be consumed daily because it cannot be stored in the body and is easily lost through the kidneys in urine. Potassium is usually supplemented to anorexic and/or patients with significant fluid losses from diuresis. Hypokalemia may result in lethargy, muscle weakness, and vomiting. Potassium is given IV via a slow drip to avoid cardiac problems. The normal serum potassium level is 3.5 to 5.5 mEq/L. The term "milliequivalent" (mEq) is used to express the number of ionic charges of each electrolyte on an equal basis. It measures the chemical activity of ions. A mEq is one one-thousandth of an equivalent. An equivalent is defined as the weight (g) of an element that will combine with 1 g of H⁺. Milliequivalents are based on the molecular weight, valence number, and milligrams of element. When calculating the amount of potassium to give, use mEq/mL as a concentration to calculate the dose. The supplement used to add potassium is KCl and is available in a concentration of 2 mEq/mL (Figure 19-8). Table 19-4 can be used as a guide to potassium supplementation.

Sodium bicarbonate

Sodium bicarbonate plays a vital role in the acid-base balance of the body. Sodium bicarbonate is added to fluids to correct metabolic acidosis since sodium bicarbonate has basic properties. Metabolic acidosis may be seen with conditions such as severe shock, diabetic acidosis, severe renal disease, and cardiac arrest. Sodium bicarbonate is available in 8 percent (1 mEq/mL) and 5 percent (0.6 mEq/mL) concentrations. The amount of sodium bicarbonate added is based on the bicarbonate deficit. The bicarbonate deficit is determined by the following equation:

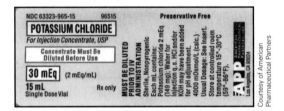

Figure 19-8 Potassium chloride is an additive that can be added to solution and is measured in milliequivalents (mEq) per mL.

	AMOUNT OF POTASSIUM TO ADD TO A 250 mL FLUID BAG	AMOUNT OF POTASSIUM TO ADD TO A 1000 mL FLUID BAG
Table 19-4 Guide to Potassium Supplementation in Animals		
SERUM POTASSIUM LEVELS		
3.5–5.5 (normal)	5 mEq	20 mEq
3.0–3.4	7 mEq	28 mEq
2.5–2.9	10 mEq	40 mEq
2.0–2.4	15 mEq	60 mEq
<2.0	20 mEq	100 mEq

> Bicarbonate deficit = patient's serum bicarbonate value – normal serum bicarbonate value (24 mEq/L)

To determine the number of mEq of sodium bicarbonate to administer, the following equation is used:

> mEq sodium bicarbonate supplement = bicarbonate deficit × 0.3 × animal's weight (kg)

The following is an example of calculating sodium bicarbonate supplementation. If a 30 kg dog has a bicarbonate level of 12 mEq/L, how many mEq of bicarbonate should be administered to this animal?

Calculate the bicarbonate deficit

> 24 mEq/L – 12 mEq/L = 12 mEq/L

Use the equation for bicarbonate supplementation:

> Bicarbonate deficit × 0.3 × animal's weight in kg = bicarbonate supplement (mEq)
>
> 12 mEq/L × 0.3 × 30 kg = 108 mEq

Side effects with sodium bicarbonate administration include development of alkalosis and other electrolyte abnormalities. Replacement must be given slowly over several hours to avoid these side effects. Sodium bicarbonate is incompatible with several solutions and should be mixed only after reading package inserts or references. Supplementation of fluids with sodium bicarbonate is rare.

Calcium

Calcium is necessary for the functioning of nerves and muscles, the clotting of blood, the building of bones and teeth, and other physiologic processes. Calcium is given to patients with hypocalcemia due to diseases such as milk fever, eclampsia, and endocrine disorders. Calcium may be in the form of calcium gluconate or calcium chloride, or in combination with other electrolytes such as magnesium, potassium, phosphorus, and dextrose. Calcium is available as *calcium gluconate 10 percent for injection, calcium gluconate 23 percent* (for large animals), *calcium chloride 10 percent for injection,* and *combination products* like Cal-Dextro® and Norcalciphos® (which also contain phosphorus, potassium, and/or dextrose). Calcium supplementation is calculated using specific dosages and usually given to effect. Cardiac and respiratory rate and rhythm should be monitored when giving calcium. Hypotension, cardiac arrhythmias, and cardiac arrest may be seen when IV calcium is given too rapidly.

Vitamins

Water-soluble vitamins may be added to fluids because they are lost rapidly by anorexic or debilitated animals. Vitamin B complex is frequently added to fluids. In general, 2 to 3 mL of B complex is given per 1000 mL of fluid. B vitamins can cause pain at the injection site if given SQ, so proper restraint is important when administering fluids containing B vitamins. Vitamin B is sensitive to light and is stored in dark vials away from direct light.

HOW MUCH DO WE GIVE?

When calculating fluid replacement therapy, keep in mind that animals require fluids for the following:

- **rehydration** (to correct body water loss due to dehydration).

- **maintenance** (to replace body water lost daily via normal body functions).

- replacement of **ongoing fluid losses** (to replace body water lost through vomiting and diarrhea).

Rehydration Volumes

Dehydration in animals can cause many health problems. Things to remember with dehydration include the following:

- Dehydration affects younger animals much more rapidly than older animals.

- Older patients with chronic disease require more fluids than other animals.

- Animals need more fluids if they are active or if the weather is hot or humid.

- Drugs such as glucocorticoids and diuretics will alter fluid and electrolyte requirements.

Clinically, the amount of fluid needed to correct dehydration deficits can be determined from the degree of skin turgor, CRT, abdominal palpation (doughy or not), dull corneal appearance, sunken eyes, and clinical history (vomiting, diarrhea, and anorexia). These parameters were listed in Table 19-1.

The amount of fluid needed to rehydrate an animal is based on the estimated percent of dehydration (the estimated amount that the animal is dehydrated). To calculate this value, take the estimated percent dehydration and multiply it times the weight of the animal in kilograms. This is the fluid deficit in *liters*. To calculate the value in milliliters, take the estimated percent dehydration, multiply it times the animal's weight in kilograms, and then multiply it by 1000. When performing this calculation, the percent dehydration should be in decimal form (e.g., 10 percent = 0.10, 5 percent = 0.05).

Maintenance Fluid Volumes

Maintenance fluid is the volume of fluid needed by the animal on a daily basis to maintain body function. Maintenance fluid volumes can be determined based on the amount of fluid lost from sensible and insensible losses. Sensible losses, which are body water lost in urine and feces, can be measured. Insensible body water losses are the result of normal metabolic processes but not easily measured; such losses occur through sweating, ventilation, and mucous membrane evaporation.

Many different values are used for the determination of maintenance fluid volumes; therefore, consulting with the veterinarian about the preferred volume is necessary. One value typically used is 50 mL/kg/day in adult animals and 110 mL/kg/day in young animals. These values will replace both sensible and insensible body water losses. A value typically used for calculation of insensible losses only is 20 mL/kg/day. These values are multiplied by the animal's weight in kg to determine the volume needed in mL.

Ongoing Fluid Loss Volumes

Animals that are losing additional fluid amounts due to vomiting or diarrhea need to have this fluid loss replaced. This number is determined by estimation. If there is a vomiting dog in the clinic, the amount of fluid loss can be estimated by monitoring the quantity and frequency of vomiting episodes. This volume is then added to the other volumes described earlier.

EXAMPLES OF FLUID CALCULATIONS

The following are some fluid calculation examples to illustrate the preceding descriptions of fluid replacement therapy.

Example 1: An adult dog weighing 50 lb needs maintenance fluids. Calculate the amount of fluids this dog needs per 24 hours.

Step 1: Convert 50 lb to kg.

$$50 \text{ lb} \times \text{kg}/2.2 \text{ lb} = 22.7 \text{ kg}$$

Step 2: Multiply the weight in kg by the maintenance fluid value (in this case, use 50 mL/kg/day).

$$22.7 \text{ kg} \times 50 \text{ mL/kg/day} = 1136 \text{ mL per day}$$

Example 2: An adult 14 lb cat with 3 percent dehydration (mild to no evidence of clinical dehydration, but a history of fluid loss) comes into the clinic. The cat is to be kept npo (nothing by mouth). Calculate a fluid dose for this cat.

Step 1: Convert 14 lb to kg.

$$14\ lb \times kg/2.2\ lb = 6.4\ kg$$

Step 2: Multiply the weight in kg by the maintenance fluid value (in this case, use 50 mL/kg/day).

$$6.4\ kg \times 50\ mL/kg/day = 318\ mL\ per\ day$$

Step 3: Calculate replacement for dehydration.

$$3\% = 0.03$$
$$0.03 \times 6.4\ kg = 0.192\ L$$
$$0.192\ L \times 1000\ mL/L = 192\ mL$$

Step 4: Add all values together to determine total daily fluids for this cat.

$$318\ mL + 192\ mL = 510\ mL\ of\ fluid\ for\ day\ 1$$

Example 3: The cat in example 2 stays at the clinic overnight and begins vomiting. It is estimated that the cat vomited about 100 mL over the evening. Recalculate this cat's fluid needs for the next day, assuming that the cat is still 3 percent dehydrated on day 2.

Step 1: Take the amount of fluid calculated in example 2, and add the volume lost through vomiting.

$$510\ mL + 100\ mL = 610\ mL\ of\ fluid\ for\ day\ 2$$

Example 4: A mature, 450 kg horse with severe obstructive colic is estimated to be 8 percent dehydrated when presented to the clinic. This horse needs fluids stat (immediately). Calculate a daily fluid replacement volume for this horse.

Step 1: The animal's weight is already in kilograms, so no conversion is needed.

Step 2: Multiply the animal's weight in kilograms by the maintenance fluid value (in this case, use 50 mL/kg/day).

450 kg × 50 mL/kg/day = 22,500 mL

Step 3: Calculate replacement for dehydration.

8% = 0.08
0.08 × 450 kg = 36 L
36 L × 1000 mL/L = 36,000 mL

Step 4: Add all calculated values to determine total daily fluids for this animal.

22,500 mL + 36,000 mL = 58,500 mL or 58.5 L

HOW FAST DO WE GIVE IT?

The rate of fluid replacement parallels the severity of dehydration. In a clinical setting, the fluid replacement rate tends to be based on clinical judgment, but some general rules apply.

- Fluids are usually replaced rapidly at first, especially in cases of shock, and then tapered to a maintenance dose. Shock fluid values are always higher than maintenance fluid values (consult with shock fluid values in the Emergency Drug section of this chapter).

- Rate of replacement of the deficit depends on the cardiovascular and renal status of the patient. If these two systems are not functioning properly, the rate of replacement may have to be decreased so as to not overload these systems.

- Fluid input often does not necessarily equal output initially; transient imbalance is frequently the optimal course of therapy. Hour-by-hour evaluation is necessary to adjust fluid rates until the patient is stable.

Ideally, fluids should be given over a 24-hour period. Fluids are administered via fluid administration sets that deliver a constant number of drops/mL of fluid. With **adult administration sets**, this value is typically 15 gtt/mL. With **pediatric administration sets**, this value is typically 60 gtt/mL. Always check the administration set you are using, because the number of drops per mL can vary with the manufacturer (some are 10 gtt/mL, while others are 20 gtt/mL) (Figure 19-9A and B). Using the preceding examples, calculate the drip rate.

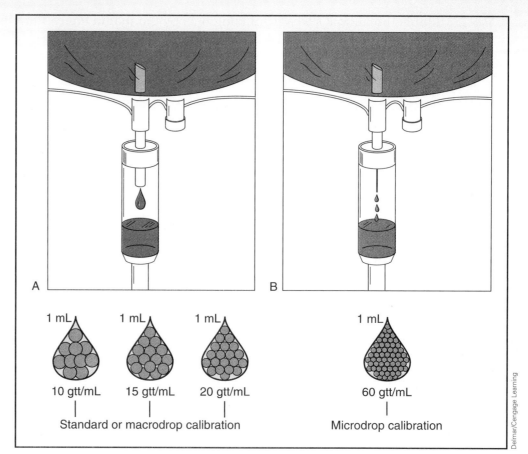

Figure 19-9 Administration sets deliver a constant number of drops per mL with the amount of drops differing based on whether the set contains a macrodrop or microdrop drip chamber. (A) Macrodrop drip chambers typically deliver 15 gtt/mL, but may also deliver 10 gtt/mL and 20 gtt/mL depending on the manufacturer. (B) Microdrop drip chambers typically deliver 60 gtt/mL.

Example 1: It was previously determined that an adult dog weighing 50 lb needs 1136 mL of fluid per day. Use an adult administration set that delivers 15 gtt/mL to calculate the fluid drip rate.

Step 1: Using an adult administration set that delivers 15 gtt/mL, take this value, and multiply it by the fluids needed per day.

$$\frac{1136\,mL}{24\,hours} \times 15\,gtt/mL = \frac{17{,}040\,gtt}{24\,hours}$$

It is difficult to count drops for 24 hours; therefore, hours should be converted to minutes

Step 2: Convert the value for 24 hours to minutes.

$$\frac{17{,}040\,gtt}{24\,hours} \times \frac{1\,hour}{60\,minutes} = \frac{17{,}040\,gtt}{1440\,minutes}$$

Step 3: Take the two values and divide to get drops per minute.

$$\frac{17{,}040 \text{ gtt}}{1440 \text{ minutes}} = 118 \text{ gtt/minute or (rounded up) 12 gtt/minute}$$

Since in most cases, veterinary technicians do not want to stand and count drops for a full minute; therefore, this value can be further reduced to shorter time intervals to facilitate monitoring. For example, 12 gtt/minute = 6 gtt/ 30 seconds = 3 gtt/15 seconds

Example 2: This cat needs 472 mL of fluid per day. Use a pediatric administration set that delivers 60 gtt/mL to calculate the fluid drip rate.

> *Step 1:* Using a pediatric administration set that delivers 60 gtt/mL, take this value, and multiply it by the fluids needed per day.

$$\frac{472 \text{ mL}}{24 \text{ hours}} \times \frac{60 \text{ gtt}}{\text{mL}} = \frac{28{,}320 \text{ gtt}}{24 \text{ hours}}$$

Step 2: Convert the value for 24 hours to minutes.

$$\frac{30{,}600 \text{ gtt}}{24 \text{ hours}} \times \frac{1 \text{ hour}}{60 \text{ minutes}} = \frac{28{,}320 \text{ gtt}}{1440 \text{ minutes}}$$

Step 3: Take the two values and divide to get drops per minute.

$$\frac{28{,}320 \text{ gtt}}{1440 \text{ minutes}} = 19.7 \text{ gtt/minute or (rounded up) 20 gtt/minute}$$

The drops in the pediatric administration set are smaller; therefore, more drops are needed to deliver 1 mL.

Example 3: This cat needs 572 mL per day. Use a pediatric administration set that delivers 60 gtt/mL to calculate the fluid drip rate.

> *Step 1:* Take this value, and multiply it by the fluids needed per day.

$$\frac{572 \text{ mL}}{24 \text{ hours}} \times \frac{60 \text{ gtt}}{\text{mL}} = \frac{34{,}320 \text{ gtt}}{24 \text{ hours}}$$

Step 2: Convert the value for 24 hours to minutes.

$$\frac{34{,}320 \text{ gtt}}{24 \text{ hours}} \times \frac{1 \text{ hour}}{60 \text{ minutes}} = \frac{34{,}320 \text{ gtt}}{1440 \text{ minutes}}$$

Step 3: Take the two values and divide to get drops per minute.

$$\frac{34,320 \text{ gtt}}{1440 \text{ minutes}} = 23.8 = \text{gtt/minute or (rounded up) 24 gtt/minute}$$

Example 4: This horse needs 51,300 mL per day. Use an adult administration set that delivers 15 gtt/mL to calculate the fluid drip rate.

Step 1: Take this value, and multiply it by the fluids needed per day.

$$\frac{51,300 \text{ mL}}{24 \text{ hours}} \times 15 \text{ gtt/mL} = \frac{769,500 \text{ gtt}}{24 \text{ hours}}$$

Step 2: Convert the value for 24 hours to minutes.

$$\frac{769,500 \text{ gtt}}{24 \text{ hours}} \times \frac{1 \text{ hour}}{60 \text{ minutes}} = \frac{769,500 \text{ gtt}}{1440 \text{ minutes}}$$

Step 3: Take the two values and divide to get drops per minute.

$$\frac{769,500 \text{ gtt}}{1440 \text{ minutes}} = 534.4 \text{ gtt/minute or (rounded down) 534 gtt/minute}$$

Step 4: Take the value per minute and divide by 60 to get drops per second.

$$\frac{534.4 \text{ gtt}}{\text{minutes}} \times \frac{1 \text{ minute}}{60 \text{ seconds}} = 8.9 \text{ gtt/second or (rounded up) 9 gtt/second}$$

This is still pretty difficult to measure, but is more manageable than 534 gtt/minutes. Estimation will be needed in this case.

Example 5: A dog is receiving 50 gtt/minute of LRS administered through an adult drip set that delivers 15 gtt/mL. The dog needs to receive a total of 350 mL of LRS for its treatment. How many hours will it take for this dog to get its total dose?

Step 1: The value for the amount of fluid this dog needs is already provided. Take this value and determine how many drops of it are delivered per unit of time desired.

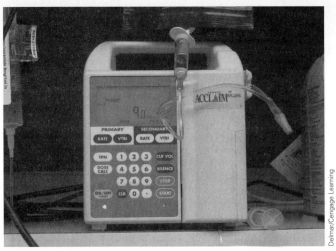

Figure 19-10 An infusion pump applies a set amount of pressure so that a set volume is infused over a set period of time. Manufacturers supply special volumetric tubing to use with their infusion devices to ensure accurate and consistent IV infusions. The flow rate is programmed into the device in milliliters per hour.

$$350 \text{ mL} \times \frac{15 \text{ gtt}}{\text{mL}} \times \frac{1 \text{ minute}}{50 \text{ gtt}} \times \frac{1 \text{ hour}}{60 \text{ minutes}} = 1.75 \text{ hours}$$

WHAT DO WE USE TO GIVE IT?

Fluids are administered by fluid bags attached to administration sets or by fluid infusion pumps. Fluid bags deliver fluids by gravity, and the rate can be adjusted by the diameter of the administration line delivering the fluids. Administration sets may have roller clamps that can be adjusted to increase or decrease the amount of fluid delivered. Administration sets may also have screw clamps or slide clamps to control the diameter of the administration line. *Infusion pumps* are machines on which flow rates are set, and the total amount to be given is entered (Figure 19-10). These pumps then give the desired amount of fluid at the desired rate. They can be readjusted when fluid delivery is either too fast or too slow.

KEEPING WATCH

Fluid administration should be monitored to make sure the animal is not getting too much or too little fluid for its needs. Additionally, fluid needs change with the changing health status of the animal. Physical findings such as nasal secretions (increased serous secretions indicate too much fluid), lung sounds (harsh lung sounds indicate edema from too much fluid), tachypnea (rapid respiratory rate), and demeanor of the animal should be monitored regularly. Laboratory values such as hematocrit and total protein can be measured in the clinic to make sure the animal's hydration status is satisfactory. Measuring urine specific gravity can also help to ensure appropriate hydration.

The amount of fluid given can be monitored in a variety of ways. Fluid bags have milliliter increments on them that should be monitored to make sure the animal is receiving the correct amount of fluid over time. White tape may be applied to the bags to keep track of fluid volume delivered per hour, which is extremely helpful if more than one person is monitoring the fluids. Some administration sets have volume control chambers (burettes) to allow easier measurement of fluid volumes. These chambers can be filled with the exact amount of fluid to be delivered for that treatment; when the chamber is empty, the animal has received that volume. Fluid pumps will also keep track of the number of milliliters of fluid administered for

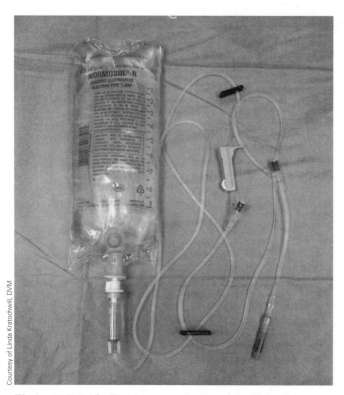

Figure 19-11A Fluid bag and administration set.

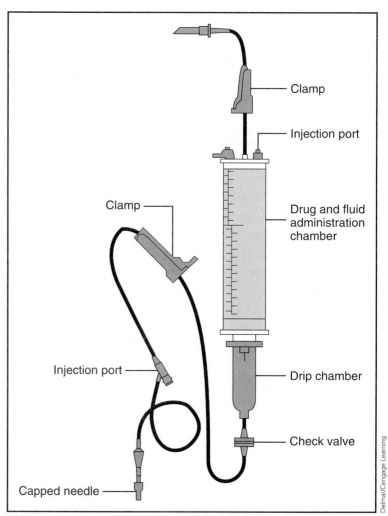

Figure 19-11B A volume control set.

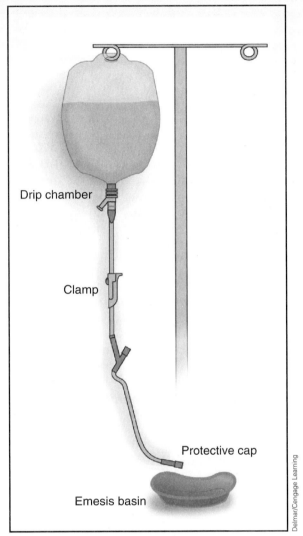

Figure 19-11C Priming the intravenous infusion equipment. To prepare the administration set for IV infusion, close all clamps, fill half of the drip chamber, remove the protective cap, release the clamps, and allow the solution to clear all air from the tubing. Be certain that all air has been removed from the tubing. Then reclamp the tubing and replace the cap.

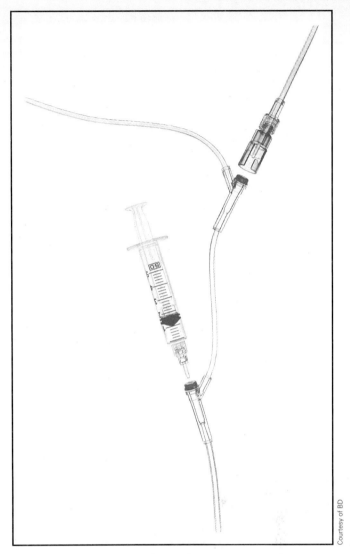

Figure 19-11D A needleless syringe system is designed to prevent accidental needlesticks during parenteral administration.

each patient. Examples of fluid bags and administration sets are shown in Figures 19-11A–D.

After the patient has been rehydrated and seems to be improving, the fluid plan is readjusted depending on the animal's situation. Eventually all fluid administration will be stopped, but in some situations abruptly stopping fluids may lead to the redevelopment of dehydration due to increased urine production by the kidneys. Tapering the fluids allows the animal's body to adjust to the decrease in fluids provided for it and helps to alleviate this issue.

ADMINISTRATION THROUGH A SPECIAL ADMINISTRATION CHAMBER (E.G., SOLUSET® OR BURETROL®)

1. Wash your hands.

2. Follow the manufacturer's directions for priming the setup. After priming the setup, clamp the administration tubing below the drip chamber.

3. Allow 10–15 mL of the fluid being administered intravenously to flow into the drug administration chamber.

4. Close the clamp between the bag and administration chamber.

5. Cleanse the injection site on the administration chamber with alcohol.

6. Inject the medication to be administered into the chamber.

7. Open the clamp between the bag and drug administration chamber, and add the appropriate amount of fluid to the administration chamber.

8. Clamp the tubing above the administration chamber.

9. Gently agitate the drug administration chamber to mix the fluids.

10. Open the clamp below the chamber.

11. Establish the flow rate appropriate to permit administration of the required amount of medication within the specified time period.

12. Once the medication has been administered, open the clamp above the administration chamber to resume administration of the fluid as ordered.

13. Chart the procedure including the date, time, medication, dosage, amount of fluid infused, and patient's reaction to the procedure.

EMERGENCY DRUGS

Animals with life-threatening disease or injury have limited physiologic reserves in order to survive, making timely therapy life saving and delayed therapy futile. A variety of conditions are considered emergencies, and they cannot all be covered in this chapter. This section attempts to address dogs and cats brought into a clinic for primary respiratory or cardiac arrest. Prompt identification of a patient's problems is critical to their survival. Time is important, and seconds can make a difference, so a plan and emergency work area should be part of all clinic setups.

The goal of emergency treatment is to maintain adequate oxygenation of vital organs. Oxygenation of vital organs is the goal of both the respiratory and cardiovascular systems. The primary goal of the respiratory system is to promote gas exchange. Air enters the nose or mouth, is warmed and humidified, passes down the respiratory tree, is filtered, and enters the smaller airways such as the bronchioles and alveoli. The primary goal of the cardiovascular system is perfusion of organs and tissues. Blood reaches the alveoli through the pulmonary vasculature as it leaves the heart via the right ventricle to the pulmonary arteries. The pulmonary arteries branch into smaller capillaries that allow red blood cells to move through them in a single file, allowing contact with the alveoli where oxygen diffuses out of the alveoli and binds to Hb as carbon dioxide diffuses off the Hb molecule into the alveoli for expiration. Flow of air and delivery of blood to the level of the alveoli are both needed for proper gas exchange.

Restoring ventilation and correcting tissue hypoxia (a life-threatening condition in which oxygen delivery is inadequate to meet metabolic demands) and acidosis (abnormally low blood pH) are key factors in helping animals survive emergency situations. Oxygen delivery is the product of blood flow and oxygen content; thus, hypoxia may result from alterations in tissue perfusion, decreased oxygen partial pressure in the blood, or decreased oxygen-carrying capacity. Hypoxia may also result from reduced oxygen transport from the vasculature to the cells or impaired utilization within the cells. Acidosis may be either metabolic or respiratory. Metabolic acidosis typically is the result of conditions such as vomiting, diarrhea, or diseases such as renal disease or diabetic ketoacidosis. Respiratory acidosis results from a failure to exhale carbon dioxide from the lungs as quickly as it forms in respiring tissues. Carbon dioxide accumulates in the blood and tissues where it forms carbonic acids. An inadequate supply of oxygen ultimately hinders aerobic metabolism and depletes molecular energy.

When presented with an animal in respiratory or cardiac arrest, keep in mind the basic life support ABCs:

A = establish *airway*

B = *breathe* for the animal

C = maintain *circulation* with thoracic compressions and IV fluids

A = Airway

The patient should be observed for signs of breathing (the chest wall and abdomen should be moving). If the animal is not trying to breathe, ventilatory support should be intiated. Tissue hypoxia and acidosis are minimized if ventilatory support is begun immediately. Establishing an airway may include passing an endotracheal tube, suctioning to clear an airway, or performing a tracheostomy. The airway must be free of blood, vomit, dirt, and mucus.

B = Breathing

It is best to deliver 100 percent oxygen by positive-pressure ventilation via an endotracheal tube with an inflated cuff. Manual artificial resuscitators (such as Ambu® bags not attached to an oxygen source) deliver room air (Figure 19-12), which is only about 21 percent oxygen. Mouth-to-tube delivery is only about 17 percent oxygen. Initially, the rate of oxygen delivery should include two breaths of one to two seconds' duration to see if spontaneous respiration will begin. If spontaneous respiration does not occur, artificial

Clinical Que

When making the first contact with the owner of an animal that needs emergency care, obtain the following information:

- nature of the illness/injury
- condition of the animal
- time injury/event occurred or was noticed
- any preexisting illness and medications
- age, breed, sex, and weight of animal, if available

Clinical Que

Always wear gloves when handling an animal that is bleeding. It is difficult to assess whether the blood is from the animal or from a person who may have been bitten or scratched while helping the animal.

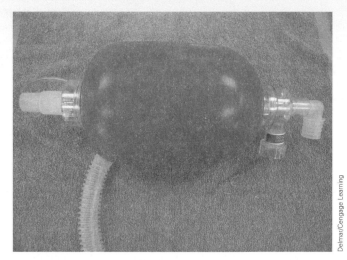

Delmar/Cengage Learning

Figure 19-12 An AMBU® bag is an example of a manual artificial resuscitator that delivers room air, which is only about 21% oxygen. Positive-pressure ventilation via an endotracheal tube with an inflated cuff delivers 100% oxygen and is the preferred method to deliver oxygen in an emergency situation.

respiration is initiated. Ventilation is usually performed at a rate of 10 to 12 breaths/minute, which is a hyperventilatory rate. The ventilation is continued by giving breaths, looking for normal chest expansion, and allowing normal exhalation to occur. The volume of oxygen delivered should produce a normal chest expansion.

C = Circulation

If there is no pulse, external (closed chest) cardiac compression is begun. Small animals weighing less than 15 kg are placed in right lateral recumbency, and compressions are done over the heart (at about the fourth to fifth intercostal space at the costochondral junction). Large dogs of more than 15 kg are placed in dorsal recumbency, and compressions are done over the sternum at the highest point. The rate of compressions is usually about 80 to 100 compression per minute, with enough force to compress the chest wall by about 30 percent. Chest compressions should be continuous, with no pauses during administration of ventilatory breaths, placement of IV catheters and fluids, endotracheal intubation, ECG assessment, palpation of pulses, administration of medications, and/or application of abdominal/pelvic wraps. If closed chest CPR fails to produce a peripheral pulse, blood flow can be augmented by internal cardiac compression.

Heart rhythm analysis is important in cardiopulmonary resuscitation and, along with ECG placement, should occur early in cases of cardiopulmonary arrest (Figure 19-13). The arrhythmias of concern are asystole, ventricular tachycardia, ventricular fibrillation, pulseless electrical activity, and sinus bradycardia. The use of emergency drugs may be initiated during external compressions. These drugs should be given intravenously, intratracheally, or intraosseously. ECG monitoring and end tidal CO_2 monitoring (to see if ventilation is adequate) should be performed to monitor drug therapy.

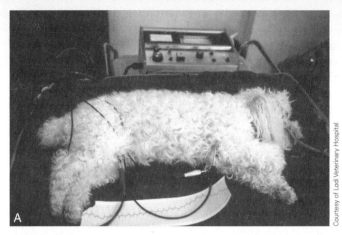
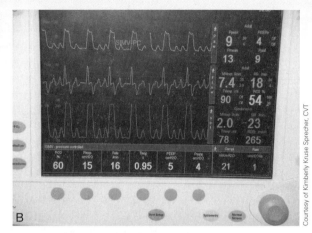

Figure 19-13 ECG placement and heart rhythm analysis are important to access early in cases of cardiopulmonary arrest. A) dog having an ECG; B) heart tracings are seen on the ECG monitor.

A central line is the preferred route of emergency drug administration during CPR; however, rarely is a central line in place before cardiopulmonary arrest and the time needed to place one during an emergency situation is too time consuming. A peripheral IV catheter is the next most preferred route of emergency drug administration, followed by intraosseous administration (in the tibial crest, femoral trochanteric fossa, and proximal humerus), and finally by intratracheal administration. Medication given via a peripheral catheter should be given as a bolus injection, followed by 0.9 percent NaCl IV and raising of the extremity for 10 to 20 seconds. Chest compression should be given for two minutes after drug administration via a peripheral vein before checking the ECG. Medication given via the IT route should be diluted with 5 to 10 mL of sterile water and the drug dosage increased by 3 to 10 times the IV dosage. Emergency drugs include, but are not limited to, those listed in Table 19-5.

Table 19-5 Emergency Drugs

DRUG	WHEN USED	CANINE AND FELINE DOSAGE	SHORTHAND DOSAGE	CONCENTRATION
Atropine	Bradycardia; repeat every three to five minutes for a maximum of three doses	0.04 mg/kg IV, IO; 0.08 mg/kg IT	0.8 mL per 10 kg; 1.6 mL per 10 kg	0.5 mg/mL
Crystalloid solution	To correct shock; give in aliquots of fluid (1/4 to 1/3 shock dose as a bolus first, reassess patient, then repeat)	90 mL/kg/hr (dog); 60 mL/kg/hr (cat)		
Dexamethasone	Use only for severe allergic reaction	4 mg/kg IV	20 mL per 10 kg	2 mg/mL

(Continued)

Drug	When Used	Canine and Feline Dosage	Shorthand Dosage	Concentration
diazepam	To control status epilepticus	0.5 mg/kg IV	1.0 mL per 10 kg	5 mg/mL
epinephrine	Asystole (no heartbeat); repeat doses should be given every three to five minutes	0.1 mg/kg IV, IO; 0.2 mg/kg IT	1 mL per 10 kg; 2 mL per 10 kg	1:1000
hemoglobin glutamer-200 (Oxyglobin®)	If animal is anemic, this drug is used to increase systemic oxygen content. This product is a hemoglobin-based, oxygen-carrying fluid that increases plasma and total hemoglobin concentration. Also given as colloid and for pressor support (to maintain adequate mean arterial pressure)	10–30 mL/kg IV at a rate of up to 10 mL/kg/hr (dogs); 2.5–5 mL/kg IV (cats). If given too rapidly or to an animal with congestive heart failure, it may result in circulatory overload		
lidocaine	Ventricular arrhythmias	2.0–4.0 mg/kg IV IO (dog); 4.0–10 mg/kg IT (dog); 0.2 mg/kg IV, IO, IT (cats use with caution)	1–2 mL per 10 kg; 2–5 mL per 10 kg; 1 mL per 5 kg	20 mg/mL
mannitol	Use only for cerebral edema	250 mg/kg IV	50 mL per 10 kg	5% = 50 mg/mL
naloxone	Opioid reversal	0.02–0.04 mg/kg IV; 0.04–0.1 mg/kg IT	0.5–1.0 mL per 10 kg; 1.0–2.5 mL per 10 kg	0.4 mg/mL
prednisone sodium	Use only for severe allergic reaction	30 mg/kg IV	30 mL per 10 kg	100 mg/10 mL or 500 mg/10 mL

IV fluid therapy should be administered if the patient is hypovolemic. IV fluids should not be administered at shock dosages (90 mL/kg for dogs and 45 mL/kg for cats) unless the patient was hypovolemic prior to cardiopulmonary arrest. To correct shock, the fluid should be given in aliquots (1/4 to 1/3 of the shock dose as a bolus first). The patient should then be reassessed, and the fluid dose repeated.

Cardiovascular and pulmonary function must be closely monitored for several hours/days following a successful resuscitation to make sure that cardiac arrest does not recur. Complications of cardiopulmonary resuscitation can cause the patient additional health concerns and must be addressed as they arise. More than 50 percent of animals that undergo CPR once will arrest a second time and require CPR again.

SUMMARY

Body fluids make up about 60 percent of an adult animal's body weight (in neonates, body fluids make up closer to 80 percent of body weight). Body water is divided into intracellular (within cells) and extracellular (in plasma and interstitial fluid). Fluid therapy is needed when an animal cannot maintain fluid and electrolyte balances. Fluid therapy can be administered orally, SQ, IP, IV, and IO.

The types of fluids used in fluid therapy are categorized as either crystalloid (sodium-based electrolyte solutions or solutions of glucose in water) or colloid (fluids with large molecules that enhance the oncotic force of blood, causing fluid to move into the vascular space). The ability of a solution to cause water movement is referred to as its tonicity. Isotonic fluids have solute concentrations similar to that of blood; therefore, fluid has no net movement either into or out of the blood vessel. Hypotonic fluids have less solute concentration than blood, potentially causing red blood cells to take in fluid and swell; the net movement of fluid is into the cell. Hypertonic fluids have more solute concentration than blood, causing cells to give up fluid and shrink; the net movement of fluid is out of the cell. Additives can be administered with fluids in certain disease states. Additives include 50 percent dextrose, potassium, sodium bicarbonate, calcium, and water-soluble vitamins.

The amount of fluid to be given to an animal is based on rehydration values, maintenance values (sensible and insensible fluid values), and ongoing fluid loss values. The rate of fluid administration parallels the severity of dehydration and is usually determined for a 24-hour period. Rate of fluid administration is determined based on the use of either an adult or a pediatric administration set. Fluid administration should be monitored and adjusted based on the health and recovery status of the patient.

In emergency situations, airway patency, assessment of breathing, and maintenance of circulation are key in an animal's treatment. The goal of emergency treatment is to maintain adequate oxygenation of vital organs. The use of emergency drugs may be initiated early in the course of cardiopulmonary resuscitation. These drugs should be given intravenously, intratracheally, or intraosseously. ECG monitoring and end tidal CO_2 monitoring (to see if ventilation is adequate) should be performed to monitor drug therapy. IV fluid therapy should be administered if the patient is hypovolemic. Various protocols for emergency treatments are available.

It's a Wrap

The answers to the questions in this chapter's Setting the Scene case study should be understood after reading this chapter. Why does this puppy need fluids, where is he losing fluids, and what type of fluids will he get?

Dehydration occurs when the body loses more fluid than it takes in. In this case, the puppy is losing fluid through its stool (diarrhea). When the puppy's fluid supply is decreased, a variety of complications may arise. These complications include electrolyte imbalance, which can affect muscle and nerve function, low blood pressure, increased heart rate, and increased respiratory rate. To avoid these physiological responses, fluids are given to counteract the effects of dehydration.

Is this puppy more at risk of developing dehydration than an older animal? Young animals are more at risk for developing dehydration since a larger percentage of their body weight is water.

What is the quickest way to get fluids into this puppy? The most commonly used method for rapid administration of fluids into animals are the intravenous and intraosseous routes.

The concern with this puppy is that he may have parvovirus, a highly contagious gastrointestinal virus in dogs. Parvovirus infection in puppies causes acute enteritis with vomiting, hemorrhagic diarrhea, and elevated temperature. The enteric form of parvovirus can cause severe dehydration from damage to the intestinal epithelium. Animals are usually infected by ingestion of fecal material from infected animals. Infected animals can shed virus in feces for about two weeks. Another form of parvovirus infection affects the heart and causes sudden death in puppies that are between four and eight weeks of age.

Parvovirus is diagnosed via clinical signs and a fecal antigen test. Treatment consists of prompt, intensive care, including fluid therapy and supportive care.

CHAPTER REVIEW

Matching

Match the fluid name with its action.

1. _____ 0.9 percent sodium chloride
2. _____ D5W
3. _____ whole blood
4. _____ LRS
5. _____ 0.9 percent normal saline with 5 percent dextrose
6. _____ dextran
7. _____ plasma protein solutions
8. _____ Normosol®
9. _____ hetastarch
10. _____ albumin

a. natural colloid
b. synthetic colloid
c. isotonic crystalloid
d. hypotonic crystalloid
e. hypertonic crystalloid

Multiple Choice

Choose the one best answer.

11. What may happen to red blood cells placed in a hypotonic solution?
    a. They may shrink.
    b. They may swell.
    c. They remain the same.
    d. They become nucleated.

12. What may happen to red blood cells placed in a hypertonic solution?
    a. They may shrink.
    b. They may swell.
    c. They remain the same.
    d. They become nucleated.

13. What happens to red blood cells when placed in an isotonic solution?
    a. They may shrink.
    b. They may swell.
    c. They remain the same.
    d. They become nucleated.

14. The best way to get oxygen to a patient is via
    a. positive-pressure ventilation from a manual artificial resuscitator.
    b. positive-pressure ventilation from an endotracheal tube with an inflated cuff.
    c. positive-pressure ventilation from mouth-to-tube delivery.
    d. all of the above are about equal.

15. Which emergency drug is given if there is no heartbeat?
    a. epinephrine
    b. sodium bicarbonate
    c. mannitol
    d. dexamethasone

16. What emergency drug is given when the patient has a slow heartbeat?
    a. mannitol
    b. atropine
    c. prednisone sodium succinate
    d. sodium bicarbonate

True/False

Circle a. for true or b. for false.

17. When dextrose is added to fluids, it is meant to serve as a calorie source.
    a.   true
    b.   false

18. When calculating rehydration fluid values, the percent dehydration should be in decimal form.
    a.   true
    b.   false

19. Adult administration sets always deliver fluid at 15 gtt/mL, and pediatric administration sets always deliver fluid at 60 gtt/mL.
    a.   true
    b.   false

20. CPR methods for small animals (<15 kg) and larger animals (>15 kg) are the same.
    a.   true
    b.   false

Case Studies

21. A mature, 500-kg horse with severe obstructive colic and 8 percent clinical dehydration presents to the clinic. This horse needs fluids stat (immediately).

    a.   Calculate fluid therapy for this animal. Assume a maintenance value of 50 mL/kg/day.

        This horse will be given its daily fluid replacement with the following:

- 2.5 percent dextrose. This comes in 50 percent dextrose in 500-mL bottles.
- 0.9 percent NaCl. This comes in granular form.
- 20 mEq/L KCl. This comes in 30-mL bottles that contain 2 mEq/mL.
- Using the volume concentration method or dimensional analysis method (unit cancellation), calculate these additives. Remember that percents equal g/100 mL (e.g., 5 percent = 5 g/100 mL).

    b.   How much NaCl in grams (assume that sodium levels are normal in this patient) need to be given to this horse?

    c.   How much dextrose in grams and milliliters need to be given to this horse? How many bottles is this?

    d.   How much KCl in milliliters (assume that potassium levels are normal in this patient) need to be given to this horse? How many bottles is this?

    e.   How much sterile water does this horse need?

22. A dog in your clinic is in asystole and the veterinarian needs a dose of IV epinephrine calculated for a 25# dog (dosage is 0.1 mg/kg). What is the dose of epinephrine (in mg and mL) that needs to be given to this dog based on the fact that the epinephrine available in the clinic is at a concentration of 1:1000?

Critical Thinking Question

23. Why is the concept of hydration important? Explain what you know about this physiological concept.

ADDITIONAL REVIEW MATERIAL

Additional review material can be found on the StudyWARE™ CD included with this text.

CHAPTER REFERENCES

Aldrich, J. (March 2005). Global Assessment of the Emergency Patient. In *Veterinary Clinics of North America: Emergency Medicine*, Vol. 35, No. 2, pp. 281–306. Philadelphia: Saunders.

Allen, D. G. (2005). *Handbook of Veterinary Drugs* (3rd ed.). Baltimore, MD: Lippincott Williams & Wilkins.

Chan, D. L. (May 2008). Colloids: Current Recommendations. In *Veterinary Clinics of North America: Advances in Fluid, Electrolyte, and Acid-Base Disorders*, Vol. 38, No. 3, pp. 587–593, Philadelphia: Saunders.

Hoskins, J. D. (May 2008). Fluid therapy: Finding the best options for perfusion, oxygen supply. *DVM Magazine*, pp. 4S–8S.

Mazzaferro, E. M. (May 2008). Complications of Fluid Therapy. In *Veterinary Clinics of North America: Advances in Fluid, Electrolyte, and Acid-Base Disorders*, Vol. 38, No. 3, pp. 607–619. Philadelphia: Saunders.

Muir III, W. W. (January 2007). Intra- and perioperative fluid therapy: What's really needed. In NAVC Conference Veterinary Proceedings, pp. 201–204.

Pachtinger, G. E. (May 2008). Assessment and Treatment of Hypovolemic States. In *Veterinary Clinics of North America: Advances in Fluid, Electrolyte, and Acid-Base Disorders*, Vol. 38, No. 3, pp. 629–643. Philadelphia: Saunders.

Plumb, D. C. (2008). *Plumb's Veterinary Drug Handbook* (6th ed.). Ames, IA: Blackwell Publishing.

Plunkett, S. J., and McMichael, M. (February 2008). Cardiopulmonary Resuscitation in Small Animal Medicine: An Update. *Journal Veterinary Internal Medicine*, 22(1): 9–25.

Rudloff, E., and Kirby, R. (May 2008). Fluid Resuscitation and the Trauma Patient. In *Veterinary Clinics of North America: Advances in Fluid, Electrolyte, and Acid-Base Disorders*, Vol. 38, No. 3, pp. 645–652. Philadelphia: Saunders.

Spratto, G. R., and Woods, A. L. (2008). *PDR Nurse's Drug Handbook*. Clifton Park, NY: Delmar Cengage Learning.

CHAPTER 20

ANTINEOPLASTIC AND IMMUNOSUPPRESSIVE DRUGS

OBJECTIVES

Upon completion of this chapter, the reader should be able to:

- describe how cancer develops.
- explain the theory behind antineoplastic drug use.
- describe the five stages of the cell cycle.
- differentiate between CCNS and CCS antineoplastic drugs.
- give examples of CCNS and CCS antineoplastic drugs.
- explain growth fraction and doubling time as they relate to cancer cells.
- describe the use of biologic response modifiers.
- list various biologic response modifiers.
- describe the role of immunosuppressive drugs on the cell cycle.

KEY TERMS

anticancer agents

antineoplastic drugs

biologic response
 modifiers (BRMs)

cell-cycle nonspecific
 (CCNS)

cell-cycle specific
 (CCS)

chemotherapeutic
 agents

doubling time

growth fraction

immunosuppressive
 drugs

pulse dosing

Setting the Scene

A client's boxer has been diagnosed with lymphosarcoma. Owner and dog are sent to a referral clinic for a complete workup and discussion of treatment options. The client calls his pet's primary clinic after taking his dog to the veterinary oncologist, who offered him a treatment plan consisting of two antineoplastic agents. The owner would like to have the treatments given at his pet's primary clinic so that he will not have to drive 100 miles to the referral clinic. The oncologist is willing to prescribe the medication and let the client have the treatments done at the primary clinic. How are antineoplastic agents handled? How are these agents administered? What needs to be discussed with this client?

Courtesy of iStockPhoto

NEOPLASMS

Cancer is a disease process that can affect an animal at any age. All cancers start as a single cell that is genetically different from the other cells in the surrounding tissues. This genetically different cell divides, passing along its abnormalities to daughter cells, eventually producing a tumor or neoplasm that has characteristics quite different from the original cell and tissue. Cancerous cells lose cellular differentiation and organization (exhibit anaplasia), which eventually leads to their inability to function normally. Cancerous cells also grow without the normal homeostatic control that regulates normal cell growth and control (they exhibit autonomy). Autonomy allows the cells to grow without control, resulting in the formation of a tumor.

Over time as neoplastic cells grow uncontrollably, they invade and damage healthy tissue. Neoplasms can spread (or metastasize) from the site of origin to other areas of the body that are favorable for cell growth. The abnormal cells release enzymes that generate new blood vessels in an area to supply nutrients and oxygen to the cells, which aid their growth. Over time the cancerous cells rob the host cells of nutrients and energy and block normal lymph and blood vessels, leading to a loss of function of the normal cells.

An animal's immune system can damage some neoplastic cells by production of T lymphocytes, antibodies, interferons, and tissue necrosis factor (TNF). All of the mechanisms try to eliminate the cancerous cells from the body before they become uncontrollable and threaten the animal's life. Once a neoplasm has grown and enlarged, it may overwhelm the animal's immune system, which is then unable to control further neoplastic growth.

CANCER-FIGHTING DRUGS

Antineoplastic drugs, also called **anticancer agents** and **chemotherapeutic agents**, stop the cancerous activity of malignant cells. Clinically useful antineoplastic drugs act against characteristics unique to malignant cells, including their rapid cell division and growth, their different rate of cellular drug uptake, and their increased cellular response to selected anticancer drugs. Unfortunately, some factors present in malignant cells also occur in other parts of the body. For example, rapid cell division and growth occur in the bone marrow, gastrointestinal tract, reproductive organs, and hair follicles, making these body systems vulnerable to the effects of antineoplastic agents.

Some antineoplastics act at certain phases of the cell cycle (Figure 20-1). The five phases of the cell cycle are the following:

1. G_1 phase: enzymes that are needed for DNA synthesis are produced.
2. S phase: DNA synthesis and replication.
3. G_2 phase: RNA and protein synthesis.
4. M phase: mitosis phase involving cell division.
5. G_0 phase: resting phase.

 Clinical Que

The treatment of cancer is generally most successful when the cancerous cells are localized and have not been disseminated throughout the body. When cancerous cells are localized and accessible (such as some forms of skin cancer), the optimal treatment may be surgical removal of the affected tissue, with or without chemotherapy.

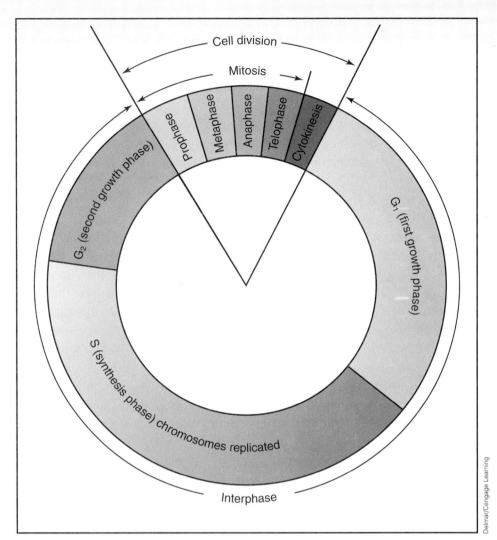

Figure 20-1 Stages of the cell cycle.

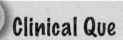

Clinical Que

All antineoplastic agents are cytotoxic (poisonous to cells) and therefore interfere with normal as well as neoplastic cells.

Clinical Que

Antineoplastic drugs alter animal cells in a variety of ways, and they tend to have greater impact on abnormal cells that make up the neoplasm (tumor) than on normal cells.

Nonspecific versus Specific

Cancer cells move more quickly through the phases of the cell cycle than do normal cells. Some antineoplastic agents work during any phase of the cell cycle (known as **cell-cycle nonspecific** or **CCNS**); others act during a specific phase of the cell cycle (known as **cell-cycle specific** or **CCS**). CCNS drugs kill the cell during the dividing and resting phases. CCS drugs are effective against rapidly growing cancer cells. Table 20-1 and Figure 20-2 summarize antineoplastic agents.

Growth fraction and doubling time are two factors that play a role in cancer cell response to antineoplastics. **Growth fraction** is the percentage of the cancer cells that are actively dividing. A high growth fraction is seen when the cells are dividing rapidly, and a low growth fraction occurs when the cells are dividing slowly. In general, antineoplastics are more effective against cancer cells that have a high growth fraction. Leukemias and some lymphomas have high growth fractions and tend to respond better to antineoplastic treatment. Mammary carcinomas tend to have a low growth fraction and thus tend

Table 20-1   Antineoplastic Agents Used for Chemotherapy in Animals

DRUG TYPE	DRUG CATEGORY	EXAMPLES	CELL-CYCLE EFFECT
CCNS	*Alkylating agents:* Cross-link DNA to inhibit its replication	• *busulfan* (Myleran®, Busulfex®) • *carboplatin* (Paraplatin®) • *chlorambucil* (Leukeran®) • *cisplatin* (Platinol®) • *cyclophosphamide* (Cytoxan®, Neosar®) • *dacarbazine* (DTIC-Dome®) • *lomustine* (CeeNu®) • *mechlorethamine* (Mustargen®) • *melphalan* (Alkeran®) • *thiotepa* (Ledertepa®, Tespamin®)	• Work on all phases of the cell cycle, but are more effective in the G_1 and S phases • Tend to be used for lymphoproliferative diseases, osteosarcoma, mast cell tumors, and carcinomas
	Antitumor antibiotics: Inhibit DNA, RNA, and (in some cases) protein synthesis	• *bleomycin* (Blenoxane®) • *dactinomycin,* also known as *actinomycin D* (Cosmegen®) • *doxorubicin* (Adriamycin®) • *mitoxantrone* (Novantrone®) • *streptozocin* (Zanosar®)	• Work on all phases of the cell cycle, but doxorubicin is more effective at the S phase • Tend to be used for lymphoproliferative diseases, sarcomas, and carcinomas
	Steroid drugs: Action may include anti-inflammatory effects, suppression of bone marrow cells, reduction of edema, and suppression of tumor growth	• androgens (danazol [Danocrine®]) • estrogens (ECP®, Synovex C®) • glucocorticoids (prednisone, dexamethasone) • progestins (Component E-C®)	• Work on all phases of the cell cycle, but more effective at the S and M phases • Tend to be used for lymphoproliferative diseases, reproductive cancers, mast cell tumors, and CNS tumors
CCS	*Antimetabolites:* Cell-cycle-specific drugs that affect the S phase (involving DNA synthesis)	• *azathioprine* (Imuran®): antagonizes purine base metabolism • *5-fluorouracil* (Adrucil®, 5-FU®): pyrimidine base analogs • *cytarabine* (Cytosar-U®): pyrimidine base analog • *hydroxyurea* (Hydrea®, Droxia®): inhibits thymidine incorporation into DNA • *mercaptopurine* (Purinethol®): antagonizes purine base metabolism • *methotrexate* (Mexate®): inhibits folic acid synthesis • *thioguanine* (Tabloid®): blocks purine nucleotides	• Work on S phase of the cell cycle. They either inhibit folic acid, which is needed for the synthesis of proteins and DNA, or by being an analog of pyrimidine or purine (which are bases occurring in DNA and RNA) and incorporating into the DNA/RNA molecule • Tend to be used for lymphoproliferative diseases and carcinomas
	Alkaloids: Chemicals derived from plants; stop cancer cell division (mitosis)	• *vinblastine* (Alkaban-AQ®, Velbane®) • *vincristine* (Oncovin®, Vincasar®)	• Work on the M phase of the cell cycle, inhibiting mitosis and causing cell death • Tend to be used for lymphoproliferative diseases, mast cell tumors, sarcomas, and carcinomas

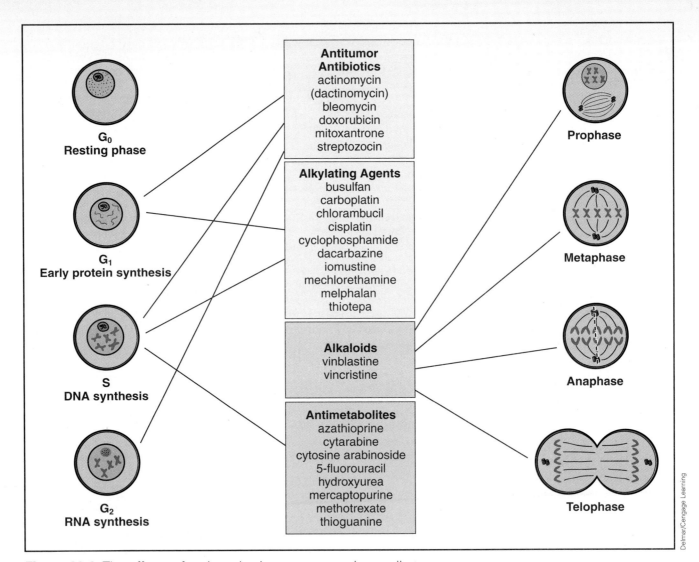

Figure 20-2 The effects of antineoplastic agents on various cell stages.

to respond unfavorably to antineoplastic treatment. In general, small, early forming, and fast-growing tumors respond better to antineoplastic drugs. Solid tumors have a large percentage of cells in the G_0 phases, so these tumors have a low growth fraction and tend to be less sensitive to antineoplastic drugs. **Doubling time** is the time required for the number of cancer cells to double. When tumors age and enlarge, their growth fraction decreases and their doubling time increases.

Administering Antineoplastics

Veterinarians often administer antineoplastic drugs in various protocols or therapeutic combinations—a procedure called *combination therapy*. The advantage to combination therapy is that a tumor is more effectively destroyed when bombarded with drugs that target different sites or act in different ways.

Pulse dosing is a method of delivering some types of chemotherapeutic agents that produce escalating levels of drug early in the dosing interval followed by a prolonged dose-free period. This type of drug delivery system offers therapeutic advantages such as reduced dose frequency and greater compliance. In comparison to intermittent dosing, pulse dosing front loads the chemotherapeutic agent, allowing an extended dose-free period during which the concentration of chemotherapeutic agent falls close to zero. Pulse dosing is the preferred way to administer some chemotherapeutic agents because continual administration of low doses tends to produce tumor resistance to the drug.

Calculation of antineoplastic drug doses usually is based on body surface area (BSA) in square meters. Body surface area is determined from body weight in kilograms using prepared charts (see Appendix H). A dosage for cisplatin is 20 mg/m²/day. A dog's weight of 10 kg (22 lb) would translate into 0.46 m². To calculate the dose for this dog, take 20 mg/m²/day × 0.46 m² = 9.2 mg/day. The concentration of cisplatin is 1 mg/mL, so this dog would need 9.2 mg/day × 1 mL/1 mg = 9.2 mL of cisplatin per day.

When preparing and administering antineoplastics, great care must be taken to ensure safe handling. Table 20-2 lists proper handling techniques for antineoplastic agents.

When administering antineoplastic agents IV, it is recommended to infuse unmedicated IV solution first through the same administration set/catheter, and to check for blood return, pain, redness, or edema before starting to administer the solution containing the antineoplastic agent. After giving the antineoplastic agent, infuse unmedicated solution IV through the same administration set/catheter before withdrawing needles or IV sets, to ensure that residual antineoplastic agents do not remain in or on this equipment.

Immune Enhancers

Biologic response modifiers (BRMs) are agents used to enhance the body's immune system. These drugs are used in conjunction with antineoplastic protocols to help the animal mount an immune response against the tumor.

Interferons are a group of naturally occurring proteins that have antitumor and antiviral effects. Interferons regulate lymphocytes and block replication of viral cells. Three major types of interferon have been identified: alpha, beta, and gamma. *Alpha interferons* are the group used to treat tumors and have also been used to treat viral infections in cats, such as FeLV and FIP. Commercially available products include Roferon-A® injection, Intron A® injection, Alferon N® injection, and Actimmune® injection. These products must be stored refrigerated or frozen. They can also be diluted and frozen in the appropriate dose size. They are typically diluted in a solution of 3 million IU/mL in 1 L of sterile saline. Side effects include fever and lethargy.

Colony stimulating factors (CSFs) stimulate the growth, maturation, and differentiation of bone marrow stem cells. They have been used to treat neutropenia in dogs and cats. Types include G-CSF, which stands for

Clinical Que

Some antineoplastic drugs need to be refrigerated before reconstitution; others do not. Always read labels to ensure proper handling and storage of these drugs.

Clinical Que

Advise clients to report pain, redness, or edema, near the injection site after treatment.

Table 20-2   Guidelines for Handling Antineoplastic Agents

- Antineoplastics should be handled by trained personnel.

- Handling of antineoplastic agents by pregnant women is generally contraindicated.

- OSHA recommends that antineoplastic agents should be prepared under a vertical laminar flow hood (Figure 20-3). When pharmaceutical practice calls for the use of aseptic techniques and a sterile environment, a horizontal laminar flow hood is sufficient. However, while this type of unit provides product protection, it may expose the operator and the other room occupants to aerosols generated during drug preparation procedures. A vertical laminar flow hood provides both product and operator protection by filtering incoming and exhaust air through a high-efficiency particulate air (HEPA) filter. It is unlikely that the majority of practices will have vertical laminar flow hoods. Preparation of drugs without such equipment is a compromise on safety. In this situation, drugs should be prepared in an area away from people and animals, without doors, windows, or draughts, and the maximum possible personal protection used. This should include gloves, disposable arm sleeves and a gown, goggles, and a mask.

- Use latex gloves to protect the skin. Gloves made of polyvinyl chloride may be permeable to some cytotoxic agents. Wearing two pairs of latex gloves is recommended because all gloves are porous to some degree.

- Good handwashing before and after handling drugs is essential.

- Prevent contact of antineoplastic agents with skin or mucous membranes. If this occurs, wash the area immediately with large volumes of water, document the contact, and seek medical assistance.

- Before drug preparation, don a disposable, nonpermeable surgical gown with closed front and knit cuffs that completely cover the wrists.

- Goggles (or eye shield) and particulate filtration masks are recommended to avoid ocular and respiratory contact.

- If reconstituting the drug, vent the vial at the beginning of the procedure. Special vent needles are available and should be used. This lowers the internal pressure in the vial and reduces the risk of spilling or spraying the solution when the needle is withdrawn from the vial.

- Use leur-lock syringes and IV equipment that have screw-on attachments to prevent spillage of antineoplastic agents.

- Wipe all external surfaces of syringes and bottles with alcohol. All disposable equipment, gloves, and bottles should be placed in a separate disposable leakproof puncture-resistant plastic bag specifically marked for chemotherapy and collected for incineration.

- Wear latex gloves when disposing of vomit, urine, or feces from animals receiving antineoplastics.

- Maintain a record of all exposure during preparation, administration, cleanup, and spills.

granulocyte colony stimulating factor; and GM-CSF, which stands for granulocyte macrophage colony stimulating factor. G-CSF is marketed as *filgrastim* (Neupogen®), and GM-CSF is marketed as *sargramostim* (Leukine®). Side effects of filgrastim include bone pain, diarrhea, and anorexia. Use with caution during pregnancy, lactation, or with myeloid cancers, as it acts as a growth factor for any tumor type. Side effects of sargramostim include hypotension.

Interleukins are a group of chemicals that play various roles in the immune system. *Interleukin-2* (Proleukin®) promotes replication of antigen-specific T cells and has been used to treat some cancers in dogs and cats. Side effects include fever, allergic reaction, and malaise.

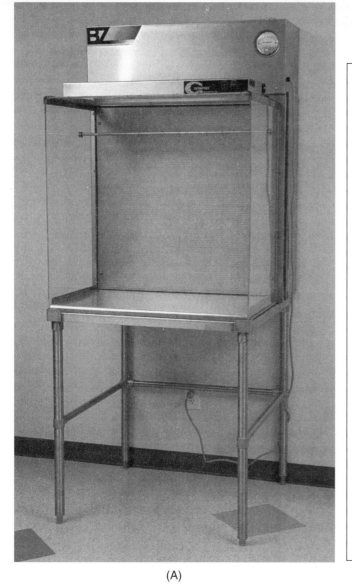

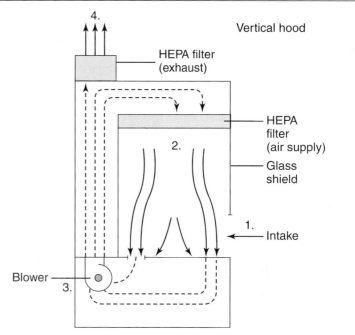

1. Room air enters the laminar airflow. This makes up about 30% of the air in the hood.

2. HEPA-filtered air enters and makes up 70% of the air in the hood.

3. Air from the work area is drawn down into the base and pulled back through the unit.

4. Air is exhausted after being filtered through carbon or HEPA filters.

(A)   (B)

Figure 20-3 (A) Vertical laminar flow hoods are used for mixing chemotherapeutic agents. (B) The principle of air flow in a vertical laminar flow hood.

Acemannan (Acemannan Immunostimulant®) is a chemical used in the treatment of fibrosarcomas and mast cell tumors in dogs and cats. It is a potent stimulator of macrophage activity. It is administered as a prelude to surgery and by concurrent intraperitoneal and intralesional injections weekly for a minimum of six treatments. Prior to use, it must be reconstituted with sterile diluent, and it is only good for four hours after rehydration. Acemannan is a USDA-labeled biologic and is not an FDA-approved product.

Monoclonal antibodies are identical immunoglobulin molecules that have cytotoxic effects on tumor cells. Monoclonal antibody therapy uses antibodies made in the laboratory (instead of the animal), and once they are given to the patient, they recruit other immune cells to the tumor to destroy cancer cells.

They may be conjugated with antineoplastics and/or other agents to deliver them directly to the tumor cells.

Immune Dehancers

Immunosuppressive drugs are drugs that work mainly by interfering with one of the stages of the cell cycle or by affecting cell messengers. They are used primarily in treating immune-mediated disorders in which the immune system is overactive. Examples include the following:

- *cyclosporine* (Optimmune® Ophthalmic Ointment, Sandimmune® oral and for injection, Atopica®), which inhibits the proliferation of T lymphocytes. It is used in veterinary medicine for managing keratoconjunctivitis sicca (KCS) in dogs and for immune-mediated skin disorders. Side effects include nephrotoxicity and vomiting.

- *azathioprine* (Imuran® tablets and injectable), which affects cells in the S phase of the cell cycle (Table 20-1). It also inhibits T and B lymphocytes and is used mainly in dogs for immune-mediated diseases such as systemic lupus and immune-mediated hemolytic anemias. The main side effect of azathioprine is bone marrow suppression. This drug is not recommended for use in cats or pregnant animals.

- *cyclophosphamide* (Cytoxan®), which interferes with DNA and RNA replication and thus disrupts nucleic acid function. It is used in conjunction with antineoplastic agents and for immune-mediated diseases such as autoimmune hemolytic anemia. Side effects include bone marrow suppression and gastrointestinal signs (diarrhea and vomiting).

- *mercaptopurine* (Purinethol®), which is converted into a ribonucleotide that acts as a purine antagonist that inhibits RNA and DNA synthesis. It is used as an adjunctive therapy for lymphosarcoma, leukemia, and severe rheumatoid arthritis. Side effects include diarrhea, vomiting, and bone marrow suppression.

Enzymes

The use of a therapeutic enzyme *L-asparaginase* (Elspar®, Oncaspar®, Erwinase®), offers a different approach to treating tumor cells. L-asparaginase works by hydrolyzing asparagines into aspartic acid and ammonia. Malignant cells are dependent on an exogenous source of asparagine for survival. Normal cells are able to synthesize asparagine and thus are less affected by rapid depletion of this substance. By giving the enzyme L-asparaginase, asparagine is quickly converting to aspartic acid and ammonia, making it unavailable to the tumor cells. L-asparaginase is used for the treatment of lymphomas, mast cell tumors, and thrombocytopenia. Side effects include pain at the injection site, hypotension, and diarrhea.

Enzyme Inhibitors

Enzyme-inhibiting drugs offer another approach to treating tumor cells by controlling their growth. *Toceranib* (Palladia®) is a tyrosine kinase inhibitor that is approved for treating grade II or III, recurrent, cutaneous mast cell tumors in dogs. Tyrosine kinase is an enzyme that is overexpressed in some types of tumors such as mast cell tumors; therefore, inhibiting this enzyme interferes with cell communication and cell growth resulting in tumor cell death and diminished blood supply to the tumor. Toceranib is an oral tablet that is administered at an initial dosage with dose reductions and dose interruptions used to manage adverse reactions such as neutropenia, diarrhea, anorexia, lethargy, and vomiting. Toceranib does not produce adverse reproductive effects on bitches because it is not known to cause fetal abnormalities. Clients should be advised to wash their hands after handling the tablets, to wear gloves when cleaning up body fluids from their dogs receiving toceranib (stool and vomit), and to launder towels and soiled items contaminated with stool and vomit from treated dogs separately from other laundry (Figure 20-4).

Client Information Sheet

Palladia
(toceranib phosphate) tablets

This summary contains important information about PALLADIA. You should read this information before you start giving your dog PALLADIA and review it each time the prescription is refilled as there may be new information. This sheet is provided only as a summary and does not take the place of instructions from your veterinarian. Talk with your veterinarian if you do not understand any of this information or if you want to know more about PALLADIA.

What is PALLADIA?
⊙ PALLADIA, a tyrosine kinase inhibitor, is a drug used to treat mast cell tumors, a common form of cancer that affects dogs.
⊙ PALLADIA works in two ways:
  • By killing tumor cells.
  • By cutting off the blood supply to the tumor.
⊙ Your veterinarian has decided to include PALLADIA as a part of your dog's treatment plan for mast cell tumor. Other types of treatment, such as surgery, drug treatment and/or radiation may be included in the plan. Be sure to speak with your veterinarian about all parts of your dog's treatment plan.

What do I need to tell my veterinarian about my dog before administering PALLADIA?
⊙ Tell your veterinarian about all other medications your pet is taking, including: prescription drugs; over the counter drugs; heartworm, flea & tick medications; vitamins and supplements, including herbal medications.
⊙ Tell your veterinarian if your dog is pregnant, nursing puppies, or is intended for breeding purposes.

How do I give PALLADIA to my dog?
⊙ PALLADIA should be given to your dog by mouth (orally).
⊙ PALLADIA may be hidden inside a treat; be certain your dog swallows the entire tablet(s).
⊙ Follow your veterinarian's instructions for how much and how often to give PALLADIA.
⊙ See the **Handling Instructions** section below in order to administer PALLADIA safely to your dog.

How will PALLADIA affect my dog?
⊙ PALLADIA may help shrink your dog's tumor. Like other cancer treatments, it can be difficult to predict whether your dog's tumor will respond to PALLADIA, and if it does respond, how long it will remain responsive to PALLADIA. Regular check ups by your veterinarian are necessary to determine whether your dog is responding as expected, and to decide whether your dog should continue to receive PALLADIA.

What are some possible side effects of PALLADIA?
⊙ Like all drugs, PALLADIA may cause side effects, even at the prescribed dose. Serious side effects can occur, with or without warning, and may in some situations result in death.
  • The most common side effects which may occur with PALLADIA include diarrhea, decreased/loss of appetite, lameness, weight loss and blood in the stool.
⊙ **Stop PALLADIA immediately and contact your veterinarian if you notice any of the following changes in your dog:**
  • Refusal to eat
  • Vomiting or watery stools (diarrhea), especially if more frequent than twice in 24 hours
  • Black tarry stools
  • Bright red blood in vomit or stools
  • Unexplained bruising or bleeding
  • Or if your dog experiences other changes that concern you
⊙ There are other side effects which may occur. For a more complete list, ask your veterinarian.

Handling Instructions
What do I need to know to handle PALLADIA safely?
Because PALLADIA is an anti-cancer drug, extra care must be taken when handling the tablets, giving the drug to your dog, and cleaning up after your dog.
⊙ PALLADIA is not for use in humans.
⊙ You should keep PALLADIA in a secure storage area out of the reach of children.
⊙ Children should not come in contact with PALLADIA. Keep children away from feces, urine, or vomit of treated dogs.
⊙ If you are pregnant, a nursing mother, or may become pregnant and you choose to administer PALLADIA to your dog, you should be particularly careful and follow the handling procedures described below.
⊙ PALLADIA prevents the formation of new blood vessels in tumors. In a similar manner, PALLADIA may affect blood vessel formation in the developing fetus and may harm an unborn baby (cause birth defects). For pregnant women, accidental ingestion of PALLADIA may have adverse effects on pregnancy.
⊙ If PALLADIA is accidentally ingested by you or a family member, seek medical advice immediately. It is important to show the treating physician a copy of the package insert or label. In cases of accidental human ingestion of PALLADIA, you may experience gastrointestinal discomfort, including vomiting or diarrhea.

The following handling procedures will help to minimize exposure to the active ingredient in PALLADIA for you and other members of your household:
⊙ Anyone who administers PALLADIA to your dog should wash their hands after handling tablets.
⊙ When you or others are handling the tablets:
  • Do not split or break the tablets to avoid disrupting the protective film coating.
  • PALLADIA tablets should be administered to your dog immediately after they are removed from the bottle.
  • Protective gloves should be worn if handling broken or moistened tablets. If your dog spits out the PALLADIA tablet, the tablet will be moistened and should be handled with protective gloves.
  • If the PALLADIA tablet is "hidden" in food, make sure that your dog has eaten the entire dose. This will minimize the potential for exposure to children or other household members to PALLADIA.
⊙ Cleaning up after your dog:
  • Because PALLADIA is present in the stool, urine and vomit of dogs under treatment, you must wear protective gloves to clean up after your treated dog.
  • While your dog receives PALLADIA, place the stool, feces or vomit, and any disposable towels used to clean up in a plastic bag which should be sealed for general household disposal. This will minimize the potential for exposure to people in contact with the trash.
  • You should not wash any items soiled with stool, urine or vomit from your dog with other laundry.

This client information sheet gives the most important information about PALLADIA. For more information about PALLADIA, talk with your veterinarian.

To report a suspected adverse reaction call Pfizer Animal Health at 1-800-366-5288.

Made in Italy
Distributed by Pharmacia & Upjohn Company
Division of Pfizer Inc, New York, NY 10017

Issued March 31, 2009

Delmar/Cengage Learning

Figure 20-4 A client information sheet. Client information sheets should be given to clients whose pets are receiving chemotherapeutic drugs. Special handling procedures are clearly explained in the client information sheet and are critical in achieving client compliance with safety and handling of chemotherapeutic drugs.

Adverse Reactions

Antineoplastic agents cause adverse reactions on rapidly dividing normal cells, such as those in the gastrointestinal tract and blood. Table 20-3 lists the general adverse reactions caused by antineoplastic agents.

SUMMARY

Antineoplastic drugs work by blocking phases of the cell cycle. The groups of antineoplastic drugs are alkylating agents, antitumor antibiotics, antimetabolites, steroids, and alkaloids. Antineoplastic drugs are classified as either CCNS or CCS depending on whether or not they work on a specific part of the cell cycle. They can be used singly or in combination with other antineoplastic drugs.

Table 20-3 General Adverse Reactions to Antineoplastic Agents

ADVERSE REACTIONS	PATIENT CONSIDERATIONS
Bone marrow suppression	• Low white blood cell counts increase the risk of infection • Low platelet counts may result in bleeding • Low red blood cell counts leading to anemia may cause lethargy
Gastrointestinal effects	• Anorexia may be attributed to foul taste left in the mouth from antineoplastic agents and/or gastrointestinal upset • Gastrointestinal upset and vomiting may be secondary to antineoplastic use • Diarrhea may be secondary to antineoplastic use and could lead to dehydration
Alopecia	• Hair loss varies; the animal should be kept out of the cold and should avoid excessive sunlight
Infertility	• Infertility secondary to antineoplastic use may be permanent
Cardiotoxicity	• Damage to the heart from certain antineoplastic drugs such as doxorubicin may be permanent
Nephrotoxicity	• Kidney failure from certain antineoplastic drugs such as cisplatin may be seen especially in animals receiving pulse dosing of medication

Biologic response modifiers are used to enhance the body's immune system. BRMs include interferon, CSFs, interleukin, acemannan, and monoclonal antibodies.

Immunosuppressive drugs are used to combat overactive immune responses that cause immune-mediated disease. Examples of immunosuppressive drugs are cyclosporine, azathioprine, cyclophosphamide, and mercaptopurine.

Enzymes work by speeding up reactions. In the treatment of tumors, L-asparaginase works by speeding up the conversion of asparagine, a product needed by malignant cells, to aspartic acid and ammonia, by-products that the malignant cells cannot use or metabolize.

The general side effects of antineoplastic agents include bone marrow suppression, gastrointestinal effects, alopecia, infertility, cardiotoxicity, and nephrotoxicity.

Table 20-4 lists the drugs covered in this chapter.

Table 20-4 Drugs Covered in This Chapter

CATEGORY	GENERIC NAME	TRADE NAME(S)
Alkylating agents	busulfan	Myleran®, Busulfex®
	carboplatin	Paraplatin®
	chlorambucil	Leukeran®
	cisplatin	Platinol®
	cyclophosphamide	Cytoxan®, Neosar®
	dacarbazine	DTIC-Dome®
	lomustine	CeeNu®
	mechlorethamine	Mustargen®
	melphalan	Alkeran®
	thiotepa	Ledertepa®, Tespamin®
Antitumor antibiotics	bleomycin	Blenoxane®
	dactinomycin (actinomycin D)	Cosmegen®
	doxorubicin	Adriamycin®
	mitoxantrone	Novantrone®
	streptozocin	Zanosar®
Steroid drugs	androgens (danazol)	Danocrine®
	estrogens	ECP®, Synovex C®

(Continued)

Table 20-4   *(Continued)*

Category	Generic Name	Trade Name(s)
	glucocorticoids (prednisone, dexamethasone)	generic, Azium®
	progestins	Component E-C®
Antimetabolites	azathioprine	Imuran®
	5-fluorouracil	Adrucil®, 5-FU®
	cytarabine	Cytosar-U®
	hydroxyurea	Hydrea®, Droxia®
	mercaptopurine	Purinethol®
	methotrexate	Mexate®
	thioguanine	Tabloid®
Alkaloids	vinblastine	Alkaban-AQ®
	vincristine	Oncovin®, Vincasar®
Biologic response modifiers	interferon	Roferon-A® injection, Intron A® injection, Alferon N® injection, Actimmune® injection
	filgrastim	Neupogen®
	sargramostim	Leukine®
	interleukin-2	Proleukin®
	acemannan	Acemannan® Immunostimulant
	monoclonal antibodies	Canine Lymphoma Monoclonal Antibody 231®
Immunosuppressives	cyclosporine	Optimmune® Ophthalmic Ointment, Sandimmune®, Atopica®
	azathioprine	Imuran®
	cyclophosphamide	Cytoxan®
	mercaptopurine	Purinethol®
Enzyme	L-asparaginase	Elspar®, Oncaspar®, Erwinase®
Enzyme inhibitor	Toceranib	Palladia®

It's a Wrap

The answers to the questions in this chapter's Setting the Scene case study should be understood after reading this chapter. How are antineoplastic agents handled? How are these agents administered? What needs to be discussed with this client?

The risk of exposure to antineoplastic agents is greatest during drug preparation and administration, with the primary routes being inhalation of aerosols, direct contact, and ingestion of spilled or improperly handled drugs. Two other routes of exposure important to veterinarians and their clients include handling of discarded items that have come in contact with chemotherapy (e.g., syringes, catheters, and gloves) and contact with excreta from patients treated with antineoplastic agents. Animal care workers in the veterinary clinic are at particular risk for exposure because they clean the animal cages, and they should be adequately trained in special handling procedures for these patients. Pregnant workers should avoid any exposure to antineoplastics and other drugs that may affect the fetus (see Appendix O). It is also important for veterinarians to give veterinary clients information on appropriate handling of drug and animal waste.

Most antineoplastics are expensive and should be prepared in a biological safety cabinet, which most veterinary clinics do not have. For these reasons, pharmacists are frequently asked to prepare chemotherapy for administration in the veterinary clinic and to compound and/or dispense prescriptions for at-home administration by the client. When a pharmacist prepares an antineoplastic agent for administration by a veterinarian in his/her clinic, the veterinarian should ask that a Material Safety Data Sheet (MSDS) and the drug package insert accompany the medication. The pharmacist is also a valuable resource for information on emergency procedures in the event of accidental exposure or leakage of the drug, as well as potential side effects, drug interactions, precautions, and contraindications associated with a particular drug.

Chemotherapy safety should be discussed with clients before the animal is discharged. While it is important to point out potential hazards associated with human exposure to these drugs, it is also important to avoid frightening clients. When antineoplastics are dispensed for at-home administration, medication vials, syringes, and bags must always be identified clearly with chemotherapy labels. The prescription label should also contain clear directions for use and disposal instructions. A procedure for

(continued)

safe disposal should be established between the veterinarian and pharmacist before dispensing. Medication should always be dispensed in a child-resistant container. Follow-up with clients is very important to make sure that the medicine is being given correctly and handled safely.

Antineoplastic drugs commonly used in pets are eliminated primarily in the urine and/or feces. In general, a 24-hour posttreatment period for special handling of chemotherapy patients has been recommended. This is the time frame when urine or feces are most likely to contain excreted drug and metabolites. Some drugs, however, are metabolized and eliminated more slowly than others. Individual animals may be slower to metabolize some drugs due to other factors, such as concurrent drug therapy interactions or altered physiological state. Therefore, a 72-hour special handling for patients receiving chemotherapy may be more appropriate. An even longer period may be needed for some drugs, such as carboplatin. Labeling of the medication should include a specified time period for special handling of animal waste.

Clients should be given written educational materials. Questions and answers should be reviewed with the client face to face (either by the veterinarian or pharmacist) to make sure there is a clear understanding of hazards and precautions.

CHAPTER REVIEW

Matching

Match the drug name with its action.

1. _____ cisplatin

2. _____ doxorubicin

3. _____ interferon

4. _____ interleukins

5. _____ vincristine

6. _____ acemannan

7. _____ cyclophosphamide

8. _____ azathioprine

9. _____ cyclosporine

10. _____ methotrexate

a. CCS antineoplastic that is an antimetabolite

b. CCNS antineoplastic that is an antitumor antibiotic

c. BRM that is used to treat tumors and viral infections

d. BRM that is used to treat fibrosarcomas and mast cell tumors

e. immunosuppressive that inhibits T cell proliferation

f. CCNS antineoplastic that has immunosuppressive activity

g. immunosuppressive that inhibits B and T cells

h. CCNS antineoplastic that is an alkylating agent

i. group of BRMs that promotes replication of T cells

j. CCS antineoplastic that is an alkaloid

Multiple Choice

Choose the one best answer.

11. Which antineoplastic drug inhibits folic acid synthesis?
- a. methotrexate
- b. 5-fluorouracil
- c. cytarabine
- d. vincristine

12. The percentage of cancer cells that are actively dividing is called the
- a. doubling time.
- b. synthesis phase.
- c. malignant phase.
- d. growth fraction.

13. Drugs used to enhance the body's immune system are
- a. CCNS drugs.
- b. CCS drugs.
- c. BRMs.
- d. immunosuppressive drugs.

14. Which type of interferon is used to treat tumors and viral infections in cats?
- a. alpha
- b. beta
- c. gamma
- d. omega

15. Which immunosuppressive drug is used orally, as an injectable, and as an ophthalmic ointment?
- a. cyclophosphamide
- b. azathioprine
- c. cyclosporine
- d. monoclonal antibodies

16. Which category of CCNS antineoplastic drug inhibits DNA replication and is most effective in the G_1 and S phases of the cell cycle?
- a. alkylating agents
- b. antitumor antibiotics
- c. steroid drugs
- d. antimetabolites

17. Which category of antineoplastic drug consists of analogs of bases occurring in DNA and incorporates into the DNA molecule?
- a. antitumor antibiotics
- b. antimetabolites
- c. alkaloids
- d. steroid drugs

18. Calculation of antineoplastic drugs is based on
- a. conversion of weight in kg to body surface area in square centimeters.
- b. conversion of weight in kg to body surface area in square meters.
- c. conversion of weight in pounds to body surface area in square feet.
- d. conversion of weight in pounds to body surface area in square inches.

True/False

Circle a. for true or b. for false.

19. CCNS antineoplastics work on any phase of the cell cycle.
- a. true
- b. false

20. Antineoplastics are safe to handle using usual methods.
- a. true
- b. false

Case Study

21. A 15-year-old F/S Beagle (25#) is presented to the clinic after being diagnosed with malignant lymphoma at a referral clinic. Part of the multidrug treatment protocol includes methotrexate 2.5 mg/m² po sid.
    a. Using the chart provided in Appendix H, calculate an oral dose of methotrexate for this beagle.
    b. Methotrexate comes in 2.5 mg scored tablets. How many pills will this dog be given per dose?
    c. What advise should be given to the client when administering the pills to the dog?

Critical Thinking Questions

22. Since many antineoplastic drugs can cause diarrhea, what problems might arise in patients who develop diarrhea secondary to cancer chemotherapy?

23. What are some other treatment options for animals with cancer?

ADDITIONAL REVIEW MATERIAL

Additional review material can be found on the StudyWARE™ CD included with this text.

BIBLIOGRAPHY

Allen, D. G. (2005). *Handbook of Veterinary Drugs* (3rd ed.). Baltimore, MD: Lippincott Williams & Wilkins.

Moini, J. (2005). *The Pharmacy Technician: A Comprehensive Approach.* Clifton Park, NY: Delmar Cengage Learning.

Plumb, D. C. (2008). *Plumb's Veterinary Drug Handbook* (6th ed.). Ames, IA: Blackwell Publishing.

Spratto, G. R., and Woods, A. L. (2008). *PDR Nurse's Drug Handbook.* Clifton Park, NY: Cengage Learning.

Veterinary Pharmaceuticals and Biologicals (12th ed.). (2001). Lenexa, KS: Veterinary Healthcare Communications.

Veterinary Values (5th ed.). (1998). Lenexa, KS: Veterinary Healthcare Communications.

Wynn, R. (1999). *Veterinary Pharmacology.* Cambridge, MA: Harcourt Learning Direct.

CHAPTER 21
VACCINES

OBJECTIVES

Upon completion of this chapter, the reader should be able to:

- differentiate between specific and nonspecific immunity.
- differentiate between active and passive immunity.
- differentiate between natural and artificial immunity.
- explain what a vaccine is and why it is used.
- differentiate between inactivated, modified live, live, and recombinant vaccines.
- describe the use of toxoids, antitoxins, and antisera.
- differentiate between autogenous, polyvalent, and monovalent vaccines.
- describe routine care and handling of vaccines.
- list various vaccines given to animals.

KEY TERMS

active immunity
adjuvant
antibody
antibody titer
antigen
antiserum
antitoxins
artificial immunity
attenuated (modified-live) vaccine
autogenous vaccines
bacterin
core vaccine
inactivated (killed) vaccine

live vaccines
monovalent vaccine
natural immunity
noncore vaccine
nonspecific immunity
passive immunity
polynucleotide (DNA) vaccine
polyvalent (multiple-antigen) vaccine
recombinant vaccines
specific immunity
subunit vaccine
toxoids
vaccines

Setting the Scene

A cat owner receives a reminder card from the clinic, noting that her cats are due to receive their annual vaccinations. She makes an appointment and brings her cats to the clinic at the appointed time. She asks about the safety of vaccines because she had read in her cat magazine that some vaccines might cause cancer. Is the veterinary technician able to tell her how safe (or unsafe) vaccines are? Does one type of vaccine cause more adverse effects than other vaccines? Should she vaccinate her cats at all? Should she vaccinate them annually? Explain the importance of vaccination protocols and how they protect animals from disease.

Courtesy of iStockPhoto

PROTECTION AGAINST DISEASE

An animal can protect itself against disease in a variety of ways. Some ways, classified as **nonspecific immunity**, include things like physical barriers (intact skin), mucus production, inflammation, fever, and phagocytosis of foreign material. Nonspecific defense mechanisms are directed equally against all pathogens and are the initial defense against invading agents. Other ways animals protect themselves are specific for a particular antigen. **Specific immunity** takes over when the nonspecific mechanisms fail. Some advantages to specific immunity are that it is targeted for a specific antigen and it has memory. Specific immunity arises from the T and B lymphocytes. Both types of lymphocytes originate in the bone marrow; however, lymphocytes that mature in the thymus are called T lymphocytes, whereas those that mature in the bone marrow are called B lymphocytes. T lymphocytes are responsible for *cell-mediated immunity*, in which T lymphocytes directly attack the invading antigen. Cell-mediated immunity is important in protecting against intracellular bacterial or viral infections, fungal diseases, and protozoal diseases. After the first exposure to an antigen, the cell-mediated response takes about six days to yield optimal protection. B lymphocytes are responsible for *antibody-mediated immunity*, in which activated B lymphocytes produce antibodies that react with antigen. Antibody-mediated immunity (also known as *humoral immunity*) is important in protecting against extracellular phases of systemic viral and bacterial infections and for protection against endotoxin- and exotoxin-induced diseases. After the first exposure to an antigen, the humoral response takes about 14 days to reach optimal protection.

There are four ways to acquire specific immunity: actively, passively, naturally, or artificially. *Active* and *passive* refer to whose immune system reacts to the antigen; *natural* and *artificial* refer to how the immunity is obtained. Table 21-1 and Figure 21-1 give examples of the various types of immunity.

- **Active immunity** arises when an animal receives an **antigen** (anything that stimulates the immune response) that activates the B and T lymphocytes and causes that animal to produce **antibodies** (proteins made by activated B lymphocytes called plasma cells to counteract antigens). The animal actively makes antibodies. A disadvantage to active

Clinical Que

Cell-mediated and humoral immunity are interdependent processes.

Table 21-1 Examples of Immunity

EXAMPLE	TYPE OF IMMUNITY
Dog getting parvovirus from another dog	Naturally acquired active immunity
Dog gets vaccinated for parvovirus	Artificially acquired active immunity
Horse gets tetanus antitoxin	Artificially acquired passive immunity
Foal gets maternal antibodies through colostrum	Naturally acquired passive immunity

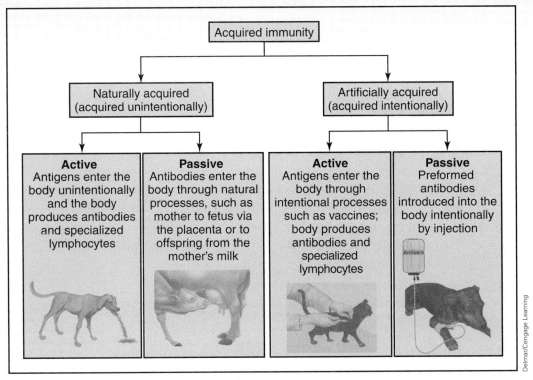

Figure 21-1 Types of acquired immunity.

immunity is that it takes time to develop, but its benefit is that it creates memory. *Memory* in this context is the capacity of the immune system to respond more rapidly and effectively to subsequent antigenic challenge from the same antigen, because of the formation of memory cells. Active immunity lasts a relatively long time and is capable of restimulation. Active immunity can be natural or artificial.

- **Passive immunity** arises when an animal receives antibodies from another animal. Antibodies (or antitoxins) are produced in a donor animal by active immunization. Once the antibody level is high enough, blood is taken from the donor animal, and the immunoglobulin portion is precipitated and purified. Ideally, the immunoglobulin can then be given to similar species because it is not recognized as foreign; however, cross-species use does occur in many cases where we use artificial passive immunity. Since there is introduction of foreign protein in cross-species use of immunoglobulin, the animal needs to be monitored for signs of anaphylaxis when these products are used. The benefit of passive immunity is that it provides immediate onset of protection; however, the animal is only temporarily protected for a short time, and since passive immunity does not create memory, the animal becomes susceptible to reinfection. Passive immunity can be natural or artificial.

- **Natural immunity** is any immunity acquired during normal biological experiences.

- **Artificial immunity** is any immunity acquired through medical procedures.

Clinical Que

Antibody-producing
tissues cannot
distinguish between
an antigen that
is capable of
causing disease
(a live antigen), an
attenuated antigen,
or a killed antigen.

Clinical Que

Vaccines are
biological frauds;
they mimic infections
by pathogens to
trigger the immune
response into
mounting a reaction.

Clinical Que

In general, it is easier
to induce protection
against viruses than
against bacteria.

Why Vaccinate?

In order to speak to someone about the need for or effectiveness of vaccines, the veterinary technician should know why vaccinates are given, and what the vaccine consists of that keeps an animal from contracting a particular illness. Several vaccines, against several different diseases, are given over the course of an animal's life to stimulate a response specific to that particular antigen. Vaccines trigger specific immune responses to help fight future infections from a specific agent, so animals receiving a vaccine may develop immunity to the inducing antigen only.

The body must be exposed to the antigen to activate a defense against the next exposure to the antigen. The point of a vaccine is to produce immunity without producing the disease. Vaccines are not administered after contracting an illness or when first coming down with an illness because by then it is too late to prevent an animal from getting sick.

WHAT IS A VACCINE?

The term *vaccine* comes from the Latin word *vacca* (cow), because the cowpox virus was used in the first preparation for active immunization against smallpox. A **vaccine** is a suspension of weakened, live, or killed microorganisms or selected proteins normally associated with these organisms administered to prevent, improve, or treat an infectious disease (Figure 21-2). Exposure to a weakened, live, or killed version of a harmful microorganism stimulates the immune system to develop immunity. The goal of vaccination is to cause an immune response without having the patient suffer the full course of the disease.

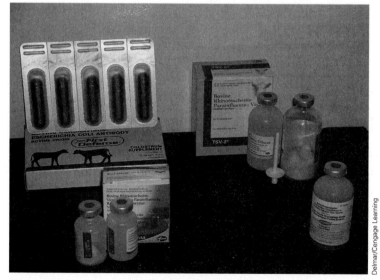

Figure 21-2 Examples of vaccine selection (clockwise from lower left): (1) Modified live vaccine. Freeze-dried powder must be reconstituted with liquid before usage. (2) An oral vaccine (bolus) used to provide passive immunity to newborn calves. (3) A modified-live vaccine that is to be administered intranasally. The white nasal cannula shown in front of the bottles is applied to the end of a syringe to administer the vaccine into the nasal passage. (4) A killed vaccine that is ready to be administered.

Types of Vaccines

A vaccine should be considered from the standpoints of antigen selection, effectiveness, ease of administration, safety, and cost. In natural immunity, an infectious agent stimulates B and T lymphocytes to create memory cells. With vaccination, the objective is to obtain the same result with a modified version of the microorganism or its parts.

Vaccines can be categorized in a variety of ways. Some ways of categorizing vaccines are by the properties of the antigen (such as killed, modified live, live, recombinant, and DNA), type of vaccine (such as toxin, antitoxin, antisera, and autogenous), and the number of antigens in the vaccine (such as monovalent and polyvalent). Types of vaccine and their descriptions are as follows:

Inactivated (Killed)

Inactivated or killed vaccines are made from microorganisms, microorganism parts, or microorganism by-products that have been chemically treated, heated, or exposed to gamma radiation to kill the microorganism. In this process, the antigenic structure of the microorganism is kept intact so that it can stimulate an immune response. An example of a killed vaccine is the rabies vaccine. The advantages to inactivated or killed vaccines are that they are usually safe, stable, and unlikely to cause disease (they do not replicate in the host). Disadvantages to inactivated or killed vaccines include the need for repeated doses to ensure protection, resulting in increased client cost. Also, most killed vaccines contain an adjuvant and/or preservatives that may cause reactions.

- **Adjuvants** are substances that enhance the immune response. There are four types of adjuvants: depot, particulate, immunostimulatory, and mixed. *Depot adjuvants* protect antigens from rapid degradation, thus contributing to a prolonged immune response. Examples of depot adjuvants are aluminum phosphate, aluminum hydroxide, alum, and Freund's incomplete adjuvant (water-in-oil emulsion). *Particulate adjuvants* deliver antigen in such a way that both cell-mediated and humoral immunity are enhanced by stimulation of antigen processing. Examples of particulate adjuvants are liposomes and microparticles. *Immunostimulatory adjuvants* promote cytokine (proteins that help mediate cellular interaction and regulate cell growth and section) production. Immunostimulatory adjuvants stimulate macrophages, lymphocytes, or antigen processing. Examples of immunostimulatory adjuvants are lipopolysaccharide, glucans, dextran sulfate, and complex microbial products such as anaerobic *Corynebacterium* sp. *Mixed adjuvants* combine a particulate or depot adjuvant with an immunostimulatory agent. An example of a mixed adjuvant is Freund's complete adjuvant (killed *Mycobacterium tuberculosis* incorporated into water-in-oil emulsion).
- If bacteria are the microbe inactivated (killed), the "vaccine" is known as a bacterin. Bacteria are usually killed with formaldehyde and incorporated with alum or aluminum hydroxide adjuvants. Because the bacteria in bacterins are dead, the immunity produced is relatively short, usually

lasting no longer than 1 year and sometimes less. Examples of diseases in which bacterins are used include swine erysipelas (*Erysipelothrix rhusiopathiae*) and strangles (*Streptococcus equi*).

- **Toxoids** are a special type of vaccine used to stimulate an active immune response against toxins instead of microorganisms. The toxin is inactivated by heat or chemicals (mainly formaldehyde) but is still able to stimulate antibody production. Toxoids have shorter durations of effectiveness than other vaccines and usually contain adjuvants and preservatives. Toxoids are usually incorporated with an alum adjuvant such as aluminum hydroxide and are available for most clostridial diseases and infections caused by toxigenic *Staphylococcus* spp. An example of a toxoid is tetanus toxoid used in the immunization against tetanus caused by *Clostridium tetani.*

- If only part of the microorganism is used, it is called a **subunit vaccine**. Subunit vaccines were developed to isolate the most important part of the microorganism—the part needed to produce the desired immune response—and eliminate the parts that can cause adverse reactions or interfere with the immune response. Many diseases require more than one protein for the vaccine to be completely effective, so many times subunit vaccines do not provide cross protections.

Clinical Que

Safe and effective vaccines should mimic the natural protective response, not cause disease, have long-lasting effects, and be easy to administer.

Attenuated (Modified-Live)

Attenuated or **modified-live vaccines** (MLVs) contain microorganisms that go through a process of losing their virulence (called *attenuation*).The microorganisms are altered so that their virulence is reduced; however, they must be able to replicate within the patient to provide immunity. The level of attenuation is critical to the success of MLVs; underattenuation will result in residual virulence and disease, while overattenuation will result in an ineffective vaccine. The body will mount an immune response to modified-live vaccines, but they do not usually produce disease. MLVs stimulate both cell-mediated and humoral immunity better than killed vaccines; therefore, the immunity produced by MLV vaccines usually lasts longer than immunity produced by killed vaccines. MLVs also have better efficacy and quicker stimulation of cell-mediated immunity than killed vaccines. Disadvantages of MLVs include possible abortion when given to pregnant animals, and some MLVs can produce mild forms of the disease and can be shed into the environment. Handling and storage of MLVs are critical and are discussed later in this chapter. An example of a MLV is canine parvovirus.

Microorganisms are made avirulent by genetic manipulation, growth on media or cells to which they are not naturally adapted, through prolonged tissue culture, or through the use of chemicals. Modified-live microorganisms may be grown in media containing adjusted levels of chemicals that trigger and enhance mutations of that microorganism. These mutations change the microorganism's metabolism in a way that alters its ability to cause disease. Temperature-sensitive vaccine microorganisms develop this way. Temperature-sensitive microorganisms lose their ability to grow at the animal's normal body temperature, though they can grow at other temperatures (like those of the ocular or nasal mucosa). Chemically altered microorganisms are considered safer than modified-live microorganisms. They produce the same level of immunity, but the duration of the immunity is shorter. MLVs are typically freeze-dried and need to be reconstituted prior to administration.

Live

Live vaccines are made from live microorganisms that may be fully virulent (able to cause disease) or avirulent. *Brucella abortus* RB51 is a live, attenuated vaccine used in cattle and must be handled with great care due to the possibility of disease initiation in humans accidentally exposed while handling the vaccine. The advantages to live vaccines are that fewer doses are needed to achieve active immunity, they provide longer immunity, adjuvants are not needed (thus eliminating possible reactions), and they are inexpensive. Disadvantages include cost of vaccine production to avoid killing the organisms and possible residual virulence, not only for the animal for which the vaccine is produced but also for other animals (hence the requirement of careful handling). Live vaccines may revert to a fully virulent type or spread to unvaccinated animals, causing persistent infection and disease within a group of animals.

Recombinant

Recombinant vaccines are made in a variety of ways. The general principle of recombinant vaccines is that a gene or part of a microorganism is isolated and removed from one organism (usually the pathogen) and inserted into another microorganism. The microorganisms are "recombined" to make something new. The advantages to recombinant vaccines include fewer side effects, effective immunity, and varied routes of administration. Increased cost is a disadvantage. The USDA classifies the different types of recombinant techniques as follows:

- Antigens generated by gene cloning (category I). Recombinant genes (DNA) coding for a surface or other molecule are isolated from the pathogen. This DNA is then inserted into a nonpathogenic cloning vector (bacterium, yeast, or other cell), and the recombinant antigen is expressed. This technique yields large amounts of purified antigen that can be produced and used in a vaccine. The first successful use of gene cloning to prepare an antigen in this way involved the foot-and-mouth disease virus.

- Genetically attenuated organisms (category II). Genes may be deleted from a pathogenic microorganism modifying its genes so that the bacterium becomes irreversibly attenuated. Gene deletions can also result in the microorganism's inability to replicate so that it cannot cause disease. These genetically altered microorganisms are then used to produce a vaccine. This vaccine stimulates the immune response, yet has a low risk of producing the disease. An example of a vaccine developed from gene deletion is the pseudorabies virus vaccine for swine.

Clinical Que

MLVs offer quicker protection in the face of an outbreak than killed vaccines.

- Live recombinant organisms (category III). Microorganisms may have select parts that stimulate the immune response. Genes that code for those select parts can be inserted into a recombinant organism to produce a vaccine containing only the antigenic part or the recombinant microorganism. The first category III vaccine approved by the USDA was against the Newcastle disease virus.

Polynucleotide

Polynucleotide or **DNA vaccines** directly inject DNA that encodes for foreign antigens into bacterial plasmids (circular pieces of DNA) that act as vectors. When the genetically engineered plasmid is injected intramuscularly into an animal, it may be taken up by host cells where the DNA is transcribed into mRNA and translated into endogenous vaccine protein. Transfected host cells express the vaccine protein allowing the animal to develop neutralizing antibodies and cytotoxic T lymphocytes. Polynucleotide vaccines are ideal for those microorganisms that are difficult or dangerous to grow in the laboratory. An advantage to polynucleotide vaccines is that it is possible to select only the genes for the antigen of interest. DNA vaccine examples are feline immunodeficiency virus and canine melanoma vaccine.

Antitoxins

Antitoxins are substances that contain antibodies (immunoglobulins) obtained from an animal that has been hypersensitized to neutralize toxins. The immunity produced is short lived because the immunity rendered is passive. Antitoxins may also contain preservatives that can cause adverse reactions. An example of an antitoxin is tetanus antitoxin. Tetanus antitoxin is obtained by injecting *Clostridium tetani* toxins that have been denatured and made nontoxic by treatment with formaldehyde into donor horses, allowing time for antibody levels to rise in these horses, collecting blood, separating plasma from the blood cells, and using the protein that contains the antibodies to the toxin for passive immunization. Tetanus antitoxin is given to animals when they are at risk of developing tetanus secondary to a contaminated deep puncture wound.

Antiserum

Antiserum is antibody-rich serum obtained from a hypersensitized or actually infected animal. Antiserum is produced in a similar fashion as antitoxin except that antibodies are collected from the plasma. The immunity produced is immediate but shortlived because the immunity rendered is passive. Antiserum may also contain preservatives that can cause adverse reactions. Examples of antisera that have been developed are canine distemper, feline panleukopenia, and bovine anthrax.

Autogenous

Autogenous vaccines are produced for a specific disease problem in a specific area from a sick animal. For example, an organism may be isolated

Clinical Que

Toxoid vs. Antitoxin. Toxoids are inactivated toxins used like a vaccine, while antitoxins are antibodies produced in response to a toxin and are used when exposure to a toxin is likely. Toxoids provide active immunity, while antitoxins provide passive immunity.

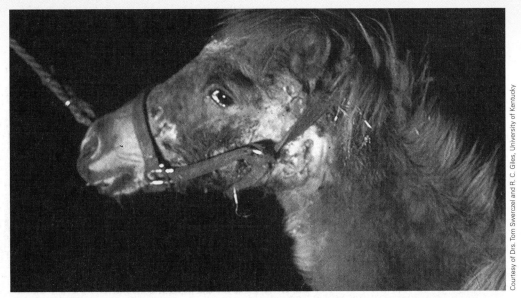

Figure 21-3 A horse with strangles caused by *Streptococcus equi*. An autogenous vaccine for *Streptococcus equi* can be developed from horses on farms experiencing outbreaks of strangles.

from a farm where an infectious disease outbreak is occurring and made into a vaccine for that specific farm. The microorganisms are grown in culture, killed, and mixed with an adjuvant. These vaccines may contain endotoxin and other by-products found in the culture; therefore, they should be used with caution. An example of an autogenous vaccine is *Streptococcus equi* developed from horses on farms experiencing outbreaks of strangles (Figure 21-3).

Multiple-Antigen and Single-Antigen Vaccines

Multiple-antigen vaccines contain more than one antigen and are referred to as **polyvalent**. Polyvalent vaccines contain a mixture of different antigens and can be convenient to administer because fewer injections are needed. For a polyvalent vaccine to be approved, the manufacturer must show that each part of the vaccine induces the same level of immunity as does the single-antigen (referred to as **monovalent**) vaccine. Table 21-2 compares polyvalent and monovalent vaccines.

Other Vaccine Possibilities

Some other possible mechanisms for vaccine production include use of peptides (using chemically synthesized protein fragments that antibodies recognize and bind to), anti-idiotypes (an antigen-binding region of a given antibody can be antigenic in a different animal, causing that animal to produce antibodies to the original antibody), and synthetic peptides (chemical synthesis of a protein component of a microorganism). Some of these methods have been utilized to a degree; however, adverse effects and cost have limited their use to date.

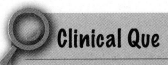

Clinical Que

The effectiveness of a polyvalent vaccine should be clinically indistinguishable from that of monovalent vaccines.

Table 21-2 Comparison of Polyvalent and Monovalent Vaccines

VACCINE TYPE	ADVANTAGES	DISADVANTAGES
Polyvalent	• Convenient • Fewer injections to administer • Less expensive when giving multiple vaccine agents	• Rate of adverse reactions increases as the number of antigens increases
Monovalent	• Can select only the desired antigens to administer • If an animal has had an adverse reaction in the past, can give different vaccines on different days or exclude certain ones	• Using several monovalent vaccines exposes patients to higher amounts of proteins and possibly adjuvants • Adverse reactions increase when giving many monovalent vaccines at a time (although this can be decreased by giving at different times)

Table 21-3 summarizes the type of response stimulated by each vaccine type. Table 21-4 compares the advantages and disadvantages of various types of vaccines.

Maternally Derived Antibodies

Vaccines are not effective when given in the presence of maternal antibodies that come primarily from colostrum (maternal antibodies may also be transmitted via the placenta). Young animals receive a small amount of naturally acquired passive immunity from their mother's milk, in the form of colostrum, ingested during the first few days of nursing. Maternally derived antibodies from colostrum provide disease resistance for a few weeks. Maternally derived antibodies also hinder successful vaccination of animals very early in life. This

Table 21-3 Immune Response Stimulated by Particular Vaccines

RESPONSE	VACCINE TYPE
Stimulates antibody response	• Killed (inactivated) • Subunit • Toxoid • Modified-live (attenuated) • Polynucleotide
Stimulates cellular response	• Live • Modified-live (attenuated) • Recombinant • Polynucleotide

Table 21-4 Advantages and Disadvantages of Vaccine Types

Killed Vaccines, Subunit Vaccines, and Toxoids

ADVANTAGES

- No risk of reverting to virulent form
- No risk of vaccine microorganism spreading between animals because microorganism not shed in environment
- Decreased abortion risk
- More stable storage
- No mixing, which decreases the risk of contamination (does not need reconstitution)

DISADVANTAGES

- More likely to cause allergic reactions and postvaccination lumps
- Two initial doses needed at least ten days apart
- Slower onset of immunity
- Immune response may not be as strong or as long as MLV
- Generally stimulates humoral immunity
- Tend to be more expensive than MLV
- Full protection may not develop until two to three weeks after last immunization

Modified-live (attenuated) and live vaccines

ADVANTAGES

- One initial dose is usually sufficient for protection (additional boosters may be required)
- More rapid protection
- Less likely to cause allergic reactions or postvaccination lumps
- Prolonged protection
- Better at stimulating cell-mediated immunity

DISADVANTAGES

- Could revert to the virulent form
- Could produce disease in immunosuppressed animals
- Could produce an excessive immune response
- Some risk of abortion
- Must be handled and mixed with additional care
- Risk of contamination during mixing because they need to be reconstituted
- Require multiplication in the host

temporary maternal protection provides variable antibody levels among animals and decreases by six to nine weeks. To continue and enhance this protection, young animals should receive vaccinations and booster vaccinations to ensure that an appropriate immune response has occurred. Booster vaccinations are recommended in young animals, because effective vaccination varies among individuals, due to variable levels of maternally derived antibodies. Since it is impossible to predict the exact time of loss of material immunity, the initial vaccination series will generally consist of some vaccines that must be administered multiple times to achieve a satisfactory level of immunity.

Vaccine Allergies and Adverse Consequences

Vaccines are the only safe, reliable, and effective way of protecting animals against the major infectious diseases of animals. Nevertheless, vaccine use is not risk free. Residual virulence and toxicity, allergic reactions, disease in immunodeficient hosts, neurological complications, and undesirable fetal affects are some of the significant risks associated with vaccine use. Only

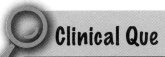

Clinical Que

Once an animal is born, successful active immunization is effective only after passive immunity has waned.

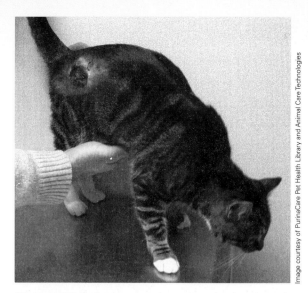

Image courtesy of PurinaCare Pet Health Library and Animal Care Technologies

Figure 21-4 A postvaccine sarcoma in a cat.

licensed vaccines should be used, and the manufacturer's recommendations should be followed. The likelihood of an adverse reaction should be weighed against the benefits to the animal prior to vaccine administration.

Always ask clients about allergies in their animals prior to immunization. Previous vaccine reactions should be reported to the veterinary medical staff prior to vaccination and recorded in the medical record. Advise clients of what to look for in regard to vaccine reactions and what they should do if they observe these reactions. Typical vaccine reactions include pain at the injection site (sting produced by some inactivating agents in the vaccine), local reactions at the injection site (localized swelling may be firm and warm to the touch, typically appearing one day postvaccination and lasting for about one week), and systemic reactions (allergic reactions ranging from mild reactions such as hives, wheezing, swelling, difficulty in breathing, fever, lethargy, vomiting, and salivation). More severe vaccine reactions include hypotension and shock due to anaphylaxis (type I hypersensitivity reactions that occur immediately or within a few minutes or hours after exposure to an antigen), intense local inflammation (type III hypersensitivity reactions), and immune diseases such as vaccine-associated sarcoma (*sarcomas* are tumors of connective-tissue origin) in cats (Figure 21-4) (type IV hypersensitivity reactions) and autoimmune hemolytic anemia in dogs. All forms of hypersensitivity are more commonly seen with multiple injections of antigens and therefore tend to be seen more with the use of killed vaccines. Some reactions are associated with a particular vaccine such as the modified-live canine distemper vaccine, which can cause some dogs to develop neurological signs (such as seizures) after being vaccinated.

ISSUES OF VACCINE USE

Vaccines play an important role in controlling and preventing infectious diseases. Veterinarians have greatly reduced the incidence of various infectious diseases by establishing vaccination protocols and educating clients about the importance of vaccinating their animals. To ensure that vaccination protocols are successful, the following items should be addressed.

Care and Handling

Inappropriate handling of vaccines may lead to their inactivation. Vaccines (especially MLVs) are sensitive to sunlight, excessive heat, and freezing. Vaccines should be ordered from a reputable company that delivers them in shipments with cold packs. Upon delivery, vaccines should be unpacked and stored in the refrigerator. Frozen vaccines may have experienced cell death, and overheated vaccines will have been rendered inactive.

Some vaccines must be reconstituted before administration. Use only the diluent provided by the manufacturer, as these diluents are pH sensitive for each vaccine product. Vaccines that have to be reconstituted should be mixed as near to the time of administration as possible. Vigorous mixing of vaccines may inactivate the proteins in these products. Vaccine can be purchased in multiple- or single-dose vials. Multiple-dose vials carry the additional risk of contamination with each use. Live vaccines may be rendered ineffective by the use of chemicals to sterilize the syringe or by excessive use of alcohol when cleaning the skin. Table 21-5 lists vaccine handling guidelines.

Route of Administration

Vaccines should be given by the route identified by the manufacturer. The level of immune protection can be ensured only when given by the proper route; some vaccines deliver local and perhaps systemic immunity (as with intranasal vaccines) and some deliver systemic immunity (as with intramuscular and

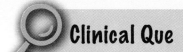

Clinical Que

The production of veterinary vaccines is controlled by the Animal and Plant Health Inspection Service of the USDA.

Table 21-5 Vaccine Handling Guidelines

- Protect vaccines from temperature extremes.
- Protect vaccines from ultraviolet light.
- Use only diluents supplied by the manufacturer for a specific product. These diluents are pH sensitive for a particular product.
- Vigorous mixing (the kind that causes foaming) of vaccines may inactivate the proteins in these products.
- Do not mix vaccines in the same syringe unless recommended by the manufacturer.
- Vaccines should be used as soon as possible after reconstitution.
- Do not use chemically sterilized syringes when administering vaccines.
- Use the entire recommended dose of vaccine.
- Using multiple-dose vials may result in inadequate mixing and therefore unequal distribution of antigens and adjuvant.
- Multiple-dose vials have an increased risk of contamination.
- Use proper animal restraint when administering vaccines.
- Use the route of administration recommended by the manufacturer.
- Clean the injection site of any dirt or debris.
- Document vaccine administration in the animal's medical record. Documentation should include vaccine type, name, manufacturer, serial number, expiration date of vaccine, date of administration, route of administration, and administration site.

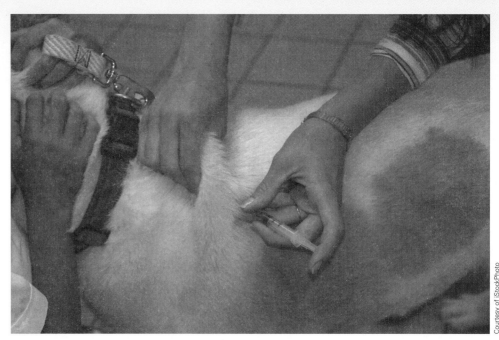

Courtesy of iStockPhoto

Figure 21-5 Dog receiving a subcutaneous vaccine.

subcutaneous vaccines) (Figure 21-5). Route of vaccine administration in food-producing animals is also important, as IM vaccines can cause more tissue damage (and hence damage to muscles used in meat processing) than SQ vaccines.

Using Vaccines

Mixing vaccines together to reduce the number of injections an animal receives is not recommended. The mixing of vaccines can result in one vaccine component interfering with another vaccine component, meaning that the animal does not receive an adequate dose. Vaccines should be given in different sites and those sites recorded in the medical records in case of vaccine reaction. For food-producing animals, most vaccines have warnings regarding slaughter.

The use of partial doses of vaccines is not recommended. The dose of vaccine required to induce an immune response is species specific, not based on the weight of the animal.

Causes of vaccine failure is categorized in Figure 21-6.

Patient Considerations

Routine physical examination prior to vaccine administration is recommended. Animals in good health will respond well to vaccination. Patient factors to consider when vaccinating animals include the following:

- animal age at vaccination. Age recommendations for vaccination are provided by the manufacturer to avoid age-specific risks of infection and complications, to ensure that the animal has the ability to mount an appropriate immune response, and to account for potential interference by maternal antibodies.

Clinical Que

Any animal that receives less than the standard dose of vaccine, or receives a vaccine through a nonstandard route or at a nonstandard site, should be revaccinated.

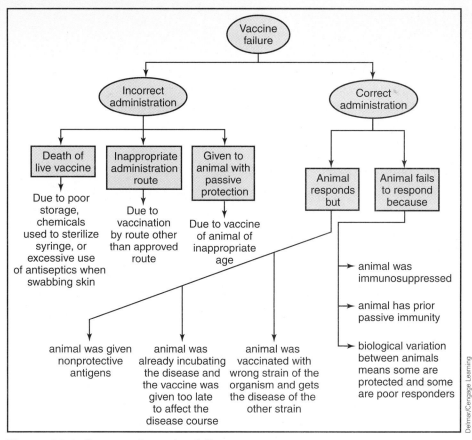

Figure 21-6 Causes of vaccine failure.

- illness. Animals that are debilitated, malnourished, ill, or have a high fever should not be vaccinated in most circumstances. Animals that are not healthy cannot mount the proper immune response to achieve protection.

- use of concurrent medication. Animals that are receiving immuno-suppressive drugs pose an interesting problem regarding vaccination. Animals on high doses of glucocorticoids should not receive MLVs. Using killed vaccines in animals that are receiving high levels of glu-cocorticoids may also be a problem, as the animal may not be able to mount an adequate immune response. Each case should be evaluated individually as to the dose of immunosuppressive drug, the duration of treatment, and the risk of not vaccinating the animal. The same theory applies to animals that are immunocompromised by infection, like FeLV in cats and various cancers in all animals.

- pregnancy. The use of MLVs in pregnant and nursing animals is not recommended, as microorganism particles can be shed and affect the offspring. Some MLVs can pass the microorganism to the fetus, result-ing in disease in the offspring. There may be circumstances in which the benefits of vaccination outweigh these risks, so each case should be evaluated on its own.

- patient's environment. Animals that have the opportunity to be exposed to other animals, are living in clusters, and are living in an area where

Clinical Que

Vaccination should be avoided in conjunction with other procedures that may induce immune system responses.

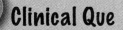

the infectious agent exists are at higher risk for acquiring that disease than animals that are not. Animals in these situations should be vaccinated; others should be assessed on a case-by-case basis. For example, an inside cat in a single-cat household should not receive every vaccine manufactured for cats. The benefit of vaccinating for a particular disease should be weighed against the potential for exposure to an infectious agent versus the risks associated with vaccination.

VACCINE PROTOCOLS

The practice of annual vaccination has been a hallmark of veterinary practice for many years. The concept of annual booster vaccinations is universally credited to the manufacturers of veterinary vaccines. The manufacturers included the "recommendation" to revaccinate annually on the product information sheets accompanying the vaccines. It is important to note that this recommendation was not based entirely on scientific research. The original source of the recommendation is not known, but may have been extrapolated from early experimental evidence on duration of immunity for rabies vaccines. Only recently have studies ever been undertaken by veterinary biologics manufacturers to determine the duration of immunity bestowed by their products, excluding rabies vaccines, because this was not required until recently for product approval. To receive a license to produce a specific vaccine for sale, the manufacturer is required only to produce evidence that the product is safe and effective in providing immunity against a specific invader. With the exception of rabies vaccines, manufacturers were under no obligation to prove that this annual revaccination interval is necessary to provide immunity. This means there is really no data regarding how long immunity lasts in a vaccinated animal (with the exception of the rabies vaccine).

One way to assess when revaccination is necessary is the **antibody titer**. When an animal is exposed to a foreign protein (such as the surface of a virus or bacteria), a specific immune response against that particular protein (the antigen) occurs. This specific immune response results in the production of antibodies. It takes time for the animal to mount this specific immune response on its first exposure to the antigen. During this first exposure to the antigen, the animal produces memory cells. Memory cells have the ability to make antibody against that specific antigen in a much shorter time frame upon subsequent exposures. Immunity against a particular antigen due to memory cell development can last a very long time, even a lifetime.

Antibody titers are serum tests that express the level of antibody to a particular antigen in a particular individual. Antibody titers are expressed as 1:2, 1:4, and so on, which represent the dilution at which an immune response is still adequate. An antibody titer of 1:4 is better than an antibody titer of 1:2 since it indicates that the animal's serum could be diluted further and still have adequate levels of antibodies present. Antibody titers that fall below the accepted dilution are generally considered inadequate. Although serum antibodies can be monitored in vaccinated animals, tests have not been standardized, and there is no consensus regarding the interpretation of these antibody titers. The correlation between antibody titer and protection against challenge exposure to the disease has not been thoroughly investigated in veterinary medicine. A low titer may not necessarily mean that the animal lacks protection to subsequent exposure to the antigen. Until further research on antibody titers is done, positive antibody titers should be interpreted as the animal having developed memory cells to that particular antigen.

Core versus Noncore

Vaccines for dogs and cats are now being described as belonging to one of two categories: core vaccines and noncore vaccines. **Core vaccines** are recommended for all individual animals because the consequences of infection are severe, infection poses a substantial zoonotic potential, disease prevalence is high, the organism is easily transmitted to others of its species, and/or the vaccine is safe and efficacious.

Noncore vaccines are recommended only for individual animals deemed to be at high risk for contact with the organism. Vaccination with noncore vaccines is based on evaluation of all risk factors, including vaccine safety and efficacy.

VACCINE EXAMPLES

The following is a list of some of the more common animal vaccines used in the United States for disease prevention.

Vaccines Available for Dogs

- canine distemper, adenovirus type 2 [infectious canine hepatitis], leptospira, parainfluenza, and parvovirus (DA$_2$LPP)
- coronavirus (CV)
- lyme
- rabies
- *Bordetella bronchiseptica* (kennel cough)
- canine herpes virus (CAV-1)
- *Giardia lamblia*
- canine melanoma
- canine influenza (H3N8)

Vaccines Available for Cats

- feline viral rhinotracheitis, feline calicivirus, feline panleukopenia (FVRCP)
- *Chlamydia psittaci* (C); this may be found in an FVRCP-C vaccine
- feline leukemia (FeLV)

- rabies
- feline infectious peritonitis (FIP)
- *Giardia lamblia*
- dermatophyte (ringworm)
- feline immunodeficiency virus (FIV)

Vaccines Available for Cattle

- rotavirus
- coronavirus
- *E. coli*
- *Brucella abortus* RB51
- clostridial disease (*C. perfringens* types B, C, and D; *C. chauvoei, C. novyi, C. septicum,* and *C. sordelli*); may also contain *C. hemolyticum*
- infectious bovine rhinotracheitis (IBR)
- parainfluenza-3 (PI-3)
- bovine viral diarrhea (BVD)
- bovine respiratory syncytial virus (BRSV)
- leptospirosis
- tetanus
- *Campylobacter*
- *Mannheimia hemolytica*
- *Pasteurella multocida*
- *Moraxella bovis*
- *Hemophilus somnus*

Vaccines Available for Sheep

- *Clostridium perfringens* types C and D
- tetanus
- soremouth (contagious ecthyma)
- foot rot
- caseous lymphadentitis
- *Chlamydia* sp. (enzootic abortion)
- *Campylobacter fetus* (vibriosis)
- *E. coli*
- parainfluenza
- rabies

Vaccines Available for Goats

- *Clostridium perfringens* types B, C, and D
- soremouth (contagious ecthyma)
- foot rot
- caseous lymphadentitis
- *E. coli*
- parainfluenza

Vaccines Available for Swine

- *Clostridium perfringens* types C
- *E. coli*
- *Salmonella*
- *Streptococcus suis*
- Porcine reproductive and respiratory syndrome virus
- *Bordetella bronchiseptica*
- *Leptospira* sp.
- porcine circovirus
- rotavirus
- transmissible gastroenteritis virus
- *Mycoplasma hyopneumoniae*
- parvovirus
- *Hemophilus pleuropneumonia*
- *Erysipelothrix rhusiopathiae*
- *Hemophilus parasuis*
- *Pasteurella multocida*
- swine influenza
- *Lawsonia intracellulausis*
- pseudorabies

Vaccines Available for Horses

- Eastern, Western, and Venezuelan equine encephalitis (EEE/WEE/VEE)
- equine viral rhinopneumonitis
- equine influenza
- strangles (*Streptococcus equi*)
- equine viral arteritis

- equine monocytic ehrlichiosis (Potomac horse fever)
- rabies
- tetanus toxoid (tetanus antitoxin is used in horses with wounds and no history of or expired tetanus toxoid vaccine)
- anthrax
- equine protozoal myelitis
- West Nile virus

Vaccines Available for Ferrets

- canine distemper (make sure it is labeled for ferrets)
- rabies

SUMMARY

Vaccines are an important part of veterinary practice to prevent infectious diseases in animals. Vaccines confer artificially acquired active immunity. A vaccine is a suspension of weakened, live, or killed microorganisms or selected proteins normally associated with these organisms administered to prevent, improve, or treat an infectious disease. Vaccines can be categorized in a variety of ways. Some ways of categorizing vaccines are by the properties of the antigen (such as killed, modified live, live, recombinant, and DNA), type of vaccine (such as toxin, antitoxin, antisera, and autogenous), and the number of antigens in the vaccine (such as monovalent and polyvalent). Types of vaccines include inactivated (killed), modified-live (attenuated), live, recombinant, polynucleoid, toxoids, antitoxins, and antisera. Vaccines are also described as autogenous, polyvalent, or monovalent.

Vaccines are the only safe, reliable, and effective way of protecting animals against the major infectious diseases of animals; however, vaccine use is not risk free. Residual virulence and toxicity, allergic reactions, disease in immunodeficient hosts, neurological complications, and undesirable fetal affects are some of the significant risks associated with vaccine use.

Vaccines must be handled properly to ensure their efficacy. Vaccines are sensitive to sunlight, excessive heat, and freezing. Mixing vaccines together to reduce the number of injections may result in one vaccine component interfering with another vaccine component. Vaccines should be given by the route identified by the manufacturer. The level of immune protection can be ensured only when given by the proper route; some vaccines deliver local and perhaps systemic immunity (as with intranasal vaccines) and some deliver systemic immunity (as with intramuscular and subcutaneous vaccines). Vaccines are given by a variety of routes, including IM, SQ, and IN.

Many patient factors should be considered when developing a vaccination plan for animals. These include potential for exposure to the infectious agent, age of the animal, health status of the animal, and environmental conditions.

One way to access a vaccination protocol is by antibody titers (serum tests that express the level of antibody to a particular antigen in a particular individual). Antibody titers that fall below the accepted dilution are generally considered inadequate. Vaccines are described as core and noncore. Core vaccines are recommended for all individual animals because the consequences of infection are severe, infection poses a substantial zoonotic potential, disease prevalence is high, the organism is easily transmitted to others of its species, and/or the vaccine is safe and efficacious. Noncore vaccines are recommended only for individual animals deemed to be at high risk for contact with the organism. Vaccination with noncore vaccines is based on evaluation of all risk factors, including vaccine safety and efficacy.

It's a Wrap

The answers to the questions in this chapter's Setting the Scene case study should be understood after reading this chapter. Is the veterinary technician able to tell this client how safe (or unsafe) vaccines are? Does one type of vaccine cause more adverse effects than other vaccines? Should she vaccinate her cats at all? Should she vaccinate them annually? Explain the importance of vaccination protocols and how they protect animals from disease.

It is important for animals to have protective antibody levels against infectious diseases to which they may be exposed. Titer tests may be performed to determine if an animal has a specific level of antibodies. A titer test is a procedure that measures the antibody levels in blood. If by chance one of the infectious agents has a low antibody level, the animal can be independently vaccinated for that disease alone rather than having the animal's immune system bombarded with agents he or she does not need.

A common practice in veterinary medicine is annual revaccination. There may be no immunologic basis for annual revaccination because immunity to viruses may persist for years or for the life of the animal. Successful vaccination against most bacterial pathogens produces immunologic memory that remains for years, allowing an animal to develop a protective anamnestic (secondary) response when exposed to virulent organisms. Some veterinarians feel that only the immune response to toxins requires boosters (e.g., tetanus toxin booster in humans is recommended once every 7 to 10 years). The practice of annual vaccination should be examined with regard to its efficacy unless it is required by law (i.e., certain states require annual revaccination for rabies). Some vaccines are only needed every three years, and some clinics recommend titer testing to determine whether a vaccine is needed.

(continued)

One concern, however, is that titer levels may not accurately indicate the immune status. A titer is a reflection of the quantity of circulating antibodies (immunoglobulins) to a given antigen. Cells in the body produce the antibody. These cells retain the ability to produce antibodies toward a given antigen for quite a long time, usually for life. On reexposure, they can produce antibody within 48 hours. A low or absent titer, therefore, does not mean the body is unprotected.

Vaccine success or failure can also be affected by the age of the animal. With young animals, maternal antibodies may be passed from the mother to the kittens via the umbilical cord and via colostrum (the first milk). Maternal antibodies may protect the young animal, but they may also interfere with vaccination. For this reason, young animals are often vaccinated several times in the hope that a vaccination will be given shortly after the maternal antibody levels diminish to a level that will not interfere with vaccination.

Most vaccines are administered as "cocktails," which are a combination of many vaccines in one shot. Vaccines can also be given individually to minimize stress on an animal's immune system. If an animal has an immune system imbalance or chronic disease, veterinarians may postpone vaccination until the animal is in an optimal state of health. Vaccinating animals when their immune system is compromised may lead to a worsening of their present problems or a weakening of their immune system. Limiting vaccination to appropriately indicated vaccines could greatly reduce complications from vaccination reactions.

There are a variety of concerns regarding vaccines, including the following:

- Vaccines may or may not provide any protection. This may result from poor vaccine performance (as with feline leukemia virus, feline infectious peritonitis virus, and ringworm vaccines).

- There may be a lack of exposure risk to a particular infectious disease (therefore, not all vaccines need to be given to every animal at all times). Each animal and its risk of contracting a particular disease need to be considered on a case-by-case basis.

- Some vaccines may cause side effects that cause a different illness than the diseases they are designed to prevent.

- Polyvalent vaccines may cause more adverse reactions than monovalent vaccines.

- Vaccines with adjuvants may cause more adverse reactions than those without adjuvants.

CHAPTER REVIEW

Matching

Match the vaccine type with its properties.

1. _____ toxoid	a. vaccine with only one antigen
2. _____ live	b. vaccine with a mixture of antigens
3. _____ polyvalent	c. vaccine in which microorganisms have been chemically treated or heated to kill the microorganism
4. _____ inactivated	d. vaccine in which genes or parts of a microorganism are moved from one organism to another
5. _____ autogenous	e. vaccine made from live microorganisms that may be virulent or avirulent
6. _____ recombinant	f. vaccine against toxins
7. _____ modified-live	g. vaccine in which microorganisms go through a process of losing their virulence (are attenuated)
8. _____ antitoxin	h. vaccine that contains antibodies to neutralize toxins
9. _____ monovalent	i. vaccine that is antibody-rich serum from a hypersensitized animal
10. _____ antisera	j. vaccine produced for a specific disease in a specific area

Multiple Choice

Choose the one best answer.

11. A substance added to a vaccine that enhances the immune response is a/an
    a.   bacterin.
    b.   recombinant.
    c.   adjuvant.
    d.   attenuation.

12. The process of a microorganism losing its virulence is termed
    a.   recombinant.
    b.   attenuation.
    c.   avirulence.
    d.   hypopathogenicity.

13. A farmer has an outbreak of salmonellosis on his farm, and a specific vaccine is made for those bacteria. This type of vaccine is a/an
    a.   antitoxin.
    b.   antisera.
    c.   autogenous.
    d.   autovalent.

14. A vaccine made from bacteria is called a/an
    a.   bacterin.
    b.   prokaryotin.
    c.   attenuated.
    d.   modified.

15. People who get sick from hepatitis A food poisoning are given immunoglobulins to provide immediate protection. This is an example of
    a.   naturally acquired active immunity.
    b.   artificially acquired active immunity.
    c.   artificially acquired passive immunity.
    d.   naturally acquired passive immunity.

True/False

Circle a. for true or b. for false.

16. FVRCP is an example of a polyvalent vaccine.
   a.   true
   b.   false

17. Vaccines are made to protect against viruses and viral diseases.
   a.   true
   b.   false

18. A horse with a puncture wound and no history of tetanus prophylaxis would receive tetanus toxoid.
   a.   true
   b.   false

19. One advantage to live vaccines is that they do not need adjuvants.
   a.   true
   b.   false

20. Rabies vaccine is an example of a monovalent vaccine.
   a.   true
   b.   false

Critical Thinking Questions

21. Some practitioners recommend monovalent vaccines to their clients because they feel that giving one agent at a time is optimal. Other practitioners recommend polyvalent vaccines in the hope of avoiding some vaccine reactions by giving one injection with multiple agents in it. Provide arguments for why some people prefer monovalent vaccines while others prefer polyvalent vaccines.

22. Arboviruses are a group of viruses that are transmitted by arthropods (arbo is short for arthropod-borne). There are vaccines for some types of arboviruses. What are some other ways to prevent infection by arboviruses?

ADDITIONAL REVIEW MATERIAL

Additional review material can be found on the StudyWARE™ CD included with this text.

BIBLIOGRAPHY

Blakeslee, D. (1996). *Vaccines and the immune system,* JAMA HIV/AIDS Resource Center, http://www.amc_assn.org/special/hiv/newsline/briefing/vacc.htm.

Canine Vaccination Protocols (December 2008). Supplement to Compendium: Continuing Education for Veterinarians, Vol. 30, No. 12(A), pp. 2–11.

Cooper, G. (2000). *Considering the Innocuous Vaccine.* Madison Area Technical College, Madison, WI.

Elston, T. (1998, January 15). AAFP/AFM vaccination guidelines. *Journal of the American Veterinary Medical Association,* 212(2): 227–241.

Feline Vaccination Protocols (August 2008). Supplement to Compendium: Continuing Education for Veterinarians, Vol. 30, No. 8(B), pp. 2–11.

Immunostimulation. (2005). *The Merck Veterinary Manual* (9th ed.). Rahway, NJ: Merck & Co, Inc.

Tizard, I. R. (2004). *Veterinary Immunology an Introduction* (7th ed.). Philadelphia, PA: Saunders.

Understanding vaccines. (2002, January). *NebGuide Nebraska Coorperative Extension G02-1445,* http://www.ianr.unl.edu/pubs/animaldisease/g1445.htm

CHAPTER 22
BEHAVIOR-MODIFYING DRUGS

OBJECTIVES

Upon completion of this chapter, the reader should be able to:

- outline the various forms of behavior modification.
- discuss why behavior-modifying drugs are not a simple solution to behavior modification.
- list the various categories of behavior-modifying drugs and describe their mechanisms of action.
- describe the factors considered when selecting which category of drug to use to treat a behavior disorder.

KEY TERMS

antianxiety drugs

antidepressant drugs

gamma-aminobutyric acid (GABA)

monoamine oxidase inhibitors (MAOIs)

pheromones

selective serotonin reuptake inhibitors (SSRIs)

tricyclic antidepressants (TCAs)

Setting the Scene

The owner of a male German Shepherd dog comes into the clinic for an opinion about his dog, which has been very protective of him lately and will not let other people or dogs near him. His dog is particularly aggressive toward other male dogs. This owner is very concerned and wants to know what he can do to prevent his dog from injuring others or their animal. He suggests filing the dog's teeth or extracting them altogether, building an enclosure for the dog, or surrendering the dog to the animal shelter. He thinks that if he could get some medication to calm his dog down, the dog would not be so aggressive. What suggestions should be given to this owner? Is medication the best answer for management of this dog's behavior? What are some ways to avoid behavior problems? What can be done in behavior cases where rapid resolution of the animal's problem behavior is needed?

Courtesy of iStockPhoto

BEHAVIOR PROBLEMS IN ANIMALS

Behavior problems and solutions are a rapidly expanding area of interest and knowledge in veterinary medicine. The use of drugs to treat problem behaviors is only a small part of treating animal behavior problems. Correctly diagnosing the condition, examining social conditions that affect the situation, and altering external stimuli are all parts of behavior treatment.

Although drug therapy to treat behavior problems may seem simple and convenient, it is not the first or only choice the veterinary community has for modifying an animal's behavior. Owners need to know about the potential problems that behavior modification drugs may cause. Owners of animals with diagnosed behavior problems, who are considering the use of behavioral drugs, need to be made aware of potential side effects of long-term drug use—the most notable are the development of liver, kidney, and cardiovascular problems. Periodic monitoring of blood values (CBC, serum chemistries) and ECGs are needed, because animals on behavior-modifying drugs may not show signs of disease until they have progressed to a dangerously unhealthy state.

Clients should also be aware that many behavior-modifying drugs are used extra-labelly. Extra-label drug use requires a veterinarian/client/patient relationship and compliance with the Animal Medicinal Drug Use Clarification Act of 1994 (see Chapter 1). Clear treatment plans for using behavioral drugs should be formulated. Continual monitoring of the animal's health and assessment of whether the behavior-modifying drug is helping the animal should also be done.

It may seem easy for the owner of an animal with diagnosed behavior problems to reach for medication to solve the animal's behavior. However, it should be emphasized that behavior modification includes retraining the animal and eliminating external stressors for the animal.

Clinical Que

With many behavior-modifying drugs, it takes six to eight weeks of treatment before any change is seen.

Clinical Que

Antianxiety drugs are also referred to as anxiolytics because they prevent feelings of tension or fear.

CLASSES OF BEHAVIOR-MODIFYING DRUGS

The types of drugs used for behavior modification include antianxiety agents, antidepressants, hormones, and pheromones.

Antianxiety Drugs

Anxiety manifests itself in multiple forms in animals. Some examples of problems associated with anxiety include separation anxiety, excessive vocalization, whining, whimpering, and inappropriate urination. **Antianxiety drugs** attempt to decrease or eliminate these behaviors.

Antihistamines

Antihistamines inhibit the H_1 receptor site from binding histamine, a natural substance in the body that is released in response to tissue damage. Most antihistamines also produce some degree of sedation because they suppress the central nervous system (Figure 22-1). The use of antihistamines as behavior-modifying

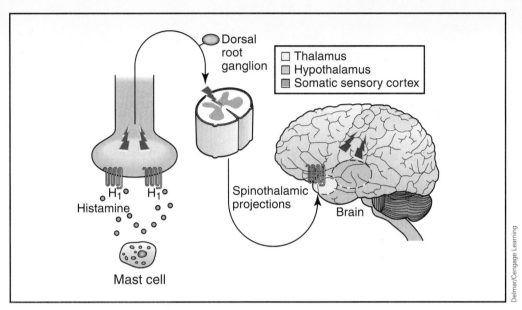

Figure 22-1 Mast cells are the main source of histamine, which may act on
H₁ receptors to transmit itch sensation via spinothalamic projections to the brain.
H₁ receptors present in the wakefulness center of the hypothalamus are modulated
by centrally acting H₁ receptor antagonists, which also produces sedation

drugs centers on their side effect of CNS depression. Antihistamines have been used in the treatment of anxiety and the behaviors associated with anxiety (e.g., inappropriate urination, pacing with vocalization, and travel-related anxiety). Antihistamines are also used to control pruritus (itching). Animals may develop anxiety associated with pruritus, and the antipruritic effects of antihistamines appear to lessen this anxiety. Antipruritic effects of antihistamines are achieved at higher doses than are the antianxiety effects. Many classes of antihistamines have anticholinergic effects, so side effects may include urinary retention, increased heart rate, and increased respiratory rate. Antihistamines used in behavior modification include the following:

- *hydroxyzine* (Atarax®), which produces greater sedative effects than other antihistamines and can last four to six hours. Hydroxyzine has minimal anticholinergic effects.

- *diphenhydramine* (Benadryl®) has lesser sedative effects and may be used with some success for pruritus. It is also used in cases in which mild sedation may be helpful, such as travel-related anxiety.

Benzodiazepines

Benzodiazepines are chemically related compounds that are used to relieve anxiety. Benzodiazepines are anxiolytics because they prevent feelings of tension or fear. Benzodiazepines appear to work on the limbic system of the brain by potentiating the inhibitory action of **gamma-aminobutyric acid (GABA)**, an amino acid that helps mediate nerve impulse transmission in the CNS (Figure 22-2). Benzodiazepines bind to specific sites in the brain, and this

Clinical Que

Many behavior-modifying drugs found in different categories have similar functions and mechanisms of action. Therefore, their specific effect(s) may depend on the dosage used.

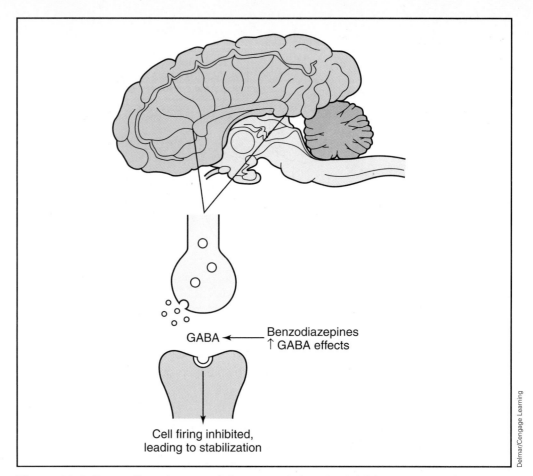

Figure 22-2 Benzodiazepines work by potentiating the inhibitory action of gamma-aminobutyric acid (GABA), an amino acid that helps mediate nerve impulse transmission in the CNS.

binding appears to produce sedation and relieve anxiety. This group of drugs causes little drowsiness at normal therapeutic doses, does not readily cause the development of tolerance to their antianxiety effect, and does not interfere significantly with metabolism of other drugs. One of the major advantages of benzodiazepines is their rapid onset of action producing relief from anxiety in 20 to 60 minutes with the appropriate drug and dosage. Benzodiazepines cause muscle relaxation that is independent of its sedative effect. At low dosages, benzodiazepines have a sedative effect; at moderate dosages, they have antianxiety effects; and at high dosages, they facilitate sleep. Benzodiazepines have been used to treat some forms of aggression (especially in cats), urine spraying, and noise phobias. Side effects include tolerance to the sedative effects of the drug and increased muscle spasticity. Hepatotoxicity is seen with benzodiazepine use in cats. *Flumazenil* (Romazicon®) is a benzodiazepine antagonist used to reverse the benzodiazepine effects after either therapeutic use or overdose. Examples in this group are C-IV controlled substances and include the following:

- *diazepam* (Valium®),
- *chlordiazepoxide* (Librium®),

Clinical Que

Benzodiazepines are not dependent on their ability to cause sedation in order to modify behavior.

- *lorazepam* (Ativan®),
- *flurazepam* (Dalmane®),
- *alprazolam* (Xanax®),
- *clorazepate* (Tranxene® T-tab). Clorazepate may also be combined with clidinium (Librax®) to treat irritable bowel syndrome.

Phenothiazines

Phenothiazine drugs have similar chemical structures and work by antagonism of dopamine. The neurotransmitter dopamine is unequally distributed in the brain, but is found in high concentrations in the limbic system. The limbic system is involved in control of body metabolism, body temperature regulation, wakefulness, vomiting, and hormonal balance. Increased dopamine levels are associated with some psychotic diseases in humans such as schizophrenia.

When phenothiazines are used as antianxiety drugs in animals, they suppress both normal and abnormal behavior. They have been used to treat aggression, but this treatment is now under question. Increased aggression may be exhibited in an animal given phenothiazines due to drug suppression of learned behavior by the animal that normally controls its natural aggressive behavior. Animals that are naturally aggressive can learn through conditioning to suppress an aggressive response; however, phenothiazines can cause such learned behavior to be ignored in some animals, allowing the natural aggression to manifest itself. This phenomenon makes an animal's reactions more unpredictable. Side effects of phenothiazines include CNS (sedation and increased likelihood of seizure development), anticholinergic (dry mouth and constipation), and cardiovascular (hypotension) problems. Examples of phenothiazines include the following:

- *chlorpromazine* (Thorazine®)
- *acepromazine* (PromAce®)
- *promazine* (Sparine®)
- *perphenazine* (Trilafon®)
- *prochlorperazine* (Compazine®)

Azapirones

Buspirone (BuSpar®) is an azapirone antianxiety drug that is chemically different from the other antianxiety drugs, and does not cause sedation like most of the other antianxiety drugs. Buspirone is believed to work by blocking serotonin. Buspirone creates no convulsant or withdrawal problems. It is used to treat urine spraying in cats and anxiety-associated aggression. It usually takes weeks of treatment before a clinical response is seen. Side effects include gastrointestinal problems.

Anticonvulsants

Carbamazepine (Tegretol®) is a carboxamide anticonvulsant drug that is used to treat seizurelike anxiety in dogs. Carbamazepine reduces the propagation

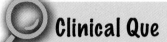

Clinical Que

When discontinuing the use of behavior-modifying drugs, withdrawal should generally be gradual, to avoid potential return of behavior problems or the development of side effects.

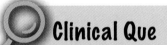

Clinical Que

Acepromazine may cause protrusion of the penis in male large animals that may last two hours.

of abnormal impulses in the brain by blocking sodium channels, thereby inhibiting the generation of repetitive action potentials in the epileptic focus. Carbamazepine is absorbed slowly from the gastrointestinal tract, is distributed erratically following oral administration, and is primarily metabolized by the liver by the cytochrome P450 system. Carbamazepine induces the hepatic cytochrome P450 system, and its half-life decreases with chronic administration, which may enhance the metabolism of other drugs. The main side effect of carbamazepine use is hepatotoxicity.

Amino Acid Derivative

L-theanine is an amino acid derivative that is found in the leaves of *Camellia sinensis*, an evergreen scrub used to make tea, and in the enzymatically manufactured product suntheanine. *L-theanine* increases the levels of the inhibitor neurotransmitter GABA, as well as serotonin and dopamine in the brain. In veterinary medicine, it is available as a chewable tablet (Anxitane®) to help alleviate fear and anxiety in dogs and cats. L-theanine is a nutraceutical product that produces minimal side effects and does not cause drowsiness. It is available in two sizes: S for cats and dogs weighing less than 22 pounds, and M and L for dogs weighing 22 pounds or more.

Barbiturates

Clinical Que

Long-term treatment with phenobarbital requires blood monitoring to avoid hepatotoxicity.

The antianxiety action of barbiturates was once attributed to their ability to cause CNS depression (sedation), but it is currently thought to be due to their effects on the GABA receptor. Barbiturates can cause liver problems (as they induce liver enzymes) and heavy tranquilization. These problems have led to decreased use. Long-term use of barbiturates for behavior modification is questionable because of variable and unpredictable results. An example of a barbiturate used to modify behavior is *phenobarbital* (Luminal®), which has been used to control vocalization in cats in small doses.

Antidepressants

Antidepressant drugs are used in veterinary medicine to treat various mood changes, including aggression, and cognitive dysfunction in animals. It is theorized that depression results from a deficiency of norepinephrine, dopamine, or serotonin, which are all neurotransmitters in key areas of the brain. Normally, the transmission of nerve impulses between two nerves or between a nerve and tissue takes place via the release of neurotransmitters from their storage sites at the nerve terminal (Figure 22-3). After the neurotransmitter combines with the appropriate receptors, there are several mechanisms that can reduce the concentration of neurotransmitter in the synaptic space. One mechanism involves reuptake of the neurotransmitter by the nerve terminal from which it was released. Another mechanism involves destruction of the neurotransmitter by the enzyme monoamine oxidase (MAO). The main neurotransmitters involved in behavioral disorders are summarized in Table 22-1.

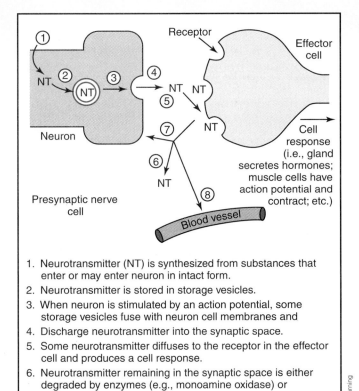

1. Neurotransmitter (NT) is synthesized from substances that enter or may enter neuron in intact form.
2. Neurotransmitter is stored in storage vesicles.
3. When neuron is stimulated by an action potential, some storage vesicles fuse with neuron cell membranes and
4. Discharge neurotransmitter into the synaptic space.
5. Some neurotransmitter diffuses to the receptor in the effector cell and produces a cell response.
6. Neurotransmitter remaining in the synaptic space is either degraded by enzymes (e.g., monoamine oxidase) or
7. It reenters the neuron (reuptake into the presynaptic nerve) and is again stored by storage vesicles or
8. It diffuses out of the synaptic cleft and is removed by the vascular system.

Figure 22-3 Nerve impulse transmission.

Table 22-1   Neurotransmitters Involved in Behavior Disorders

• acetylcholine	Neurotransmitter that communicates between nerves and muscles that is also important as the preganglionic neurotransmitter throughout the autonomic nervous system and as the postganglionic neurotransmitter in the parasympathetic nervous system and in several pathways in the brain.
• norepinephrine	Catecholamine that is released by nerves in the sympathetic branch of the autonomic nervous system and is high in some areas of the brain such as the limbic system.
• dopamine	Neurotransmitter found in high concentrations in certain parts of the brain and is involved in the coordination of impulses and responses.
• serotonin	Neurotransmitter found in the limbic system that is important in arousal and sleep as well as in preventing depression.
• gamma-aminobutyric acid (GABA)	Neurotransmitter found in the brain that inhibits nerve activity and is important in preventing overexcitability or stimulation such as seizure activity.

Tricyclic Antidepressants

Tricyclic antidepressants (TCAs) have a three-ring (tricyclic) structure and work by interfering with the reuptake of neurotransmitter by the presynaptic nerve cell (Figure 22-4). This interference increases the concentration of neurotransmitter at postsynaptic receptors in the CNS. Tertiary amine TCAs inhibit serotonin and norepinephrine reuptake, while secondary amine TCAs inhibit norepinephrine reuptake. TCAs have been used to treat separation anxiety, pruritic conditions, and compulsive disorders in animals. If side effects occur with TCAs, the use of TCA metabolites may give favorable results. Side effects of TCAs include anticholinergic effects (dry mouth, constipation, urinary retention, and increased heart rate), liver problems, and thyroid effects (sick euthyroid condition and interference with thyroid medication). Examples of TCAs are *amitriptyline* (Elavil®), *imipramine* (Tofranil®), *clomipramine* (Clomicalm®, Anafranil®), and *doxepin* (Sinequan®).

Monoamine Oxidase Inhibitors

Monoamine oxidase inhibitors (MAOIs) work by inhibiting the enzyme MAO, thus reducing the destruction of neurotransmitters such as dopamine, norepinephrine, epinephrine, and serotonin and increasing their free level in the CNS. Monoamine oxidase enzymes consist of two types: monoamine oxidase

Clinical Que

Cats are more sensitive to TCAs than dogs.

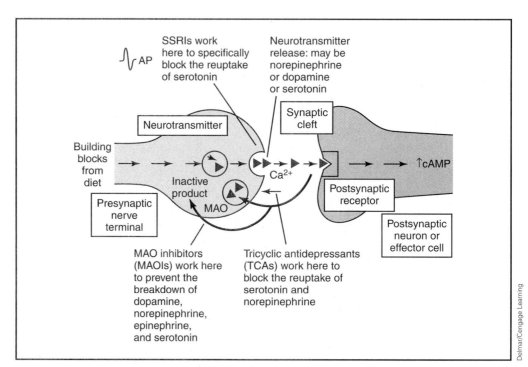

Figure 22-4 Sites of action for the antidepressants: monoamine oxidase inhibitors (MAOI), selective serotonin reuptake inhibitors (SSRI), and tricyclic antidepressants (TCA).

A is an enzyme involved in the metabolism of serotonin, norepinephrine, and epinephrine, while monoamine oxidase B is an enzyme involved in the metabolism of dopamine. MAOIs irreversibly inhibit MAO. It may take weeks to months to see the antidepressant effect of MAOIs. The only MAOI approved for use in dogs is *selegiline* (Anipryl®). Selegiline blocks the reuptake of dopamine and is used to treat cognitive dysfunction in aging dogs. Side effects include hypotension, drowsiness, and anticholinergic effects (dry mouth, increased heart rate, and constipation). MAOIs that may be used extra-label in animals include *phenelzine* (Nardil®), *isocarboxazid* (Marplan®), and *tranylcypromine sulfate* (Parnate®).

Selective Serotonin Reuptake Inhibitors

Selective serotonin reuptake inhibitors (SSRIs) are as effective as TCAs, but have fewer side effects and tolerance problems. SSRIs selectively inhibit serotonin reuptake, resulting in increased serotonin neurotransmission. This group is structurally diverse; uses include treating depression, aggression, anxiety, phobias, and compulsive disorders (Figure 22-5). Examples of SSRIs are *fluoxetine* (Reconcile®, Prozac®), *sertraline* (Zoloft®), *paroxetine* (Paxil®), *fluvoxamine* (Luvox®), *escitalopram* (Lexapro®), and *citalopram* (Celexa®). Side effects of SSRIs include lethargy, reduced appetite, vomiting, shaking, diarrhea, and restlessness.

Hormones

The hormones used or altered in treatment of behavior problems are the reproductive hormones progestins, estrogens, and testosterone.

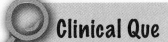

Clinical Que

Hypertensive problems can occur in animals given both MAOIs and food rich in tyramine (aged cheese, yogurt, and chicken liver).

Courtesy of iStockPhoto

Figure 22-5 Psychogenic hair loss in a cat due to excessive grooming.

Progestins and estrogens are used as behavior-modifying drugs mainly because of their calming effects, which are created by their suppression of the excitatory effects of glutamine (an amino acid) and their suppression of malelike behaviors. Side effects of these drugs include mammary gland hyperplasia, endometrial hyperplasia, pyometra, bone marrow suppression, and endocrine disorders. Drugs in this category include the following:

- *diethylstilbestrol* (DES), a synthetic estrogen that has a different chemical structure than natural estrogens but similar pharmacological effects. It is a second-line drug used to treat urinary incontinence. Other drugs, such as bethanecol and phenylpropanolamine, are the preferred, first-line treatment. DES is not commercially available in veterinary or human products in the United States; however, it can be obtained from compounding pharmacies. The limited availability of DES in the United States has caused a reduction in its use in veterinary medicine. Side effects include myelosuppression (resulting in nonregenerative anemia, thrombocytopenia, and neutropenia) and the development of pyometra and endometrial hyperplasia in intact female animals.

- *medroxyprogesterone acetate* (Depo-Provera®, Provera®), a synthetic progestin used to treat aggression, because of its calming and/or feminizing actions. It is also used to treat malelike behaviors of mounting, territorial marking (urination), intermale aggression, and roaming. Side effects include increased appetite and/or thirst, lethargy, personality changes, mammary changes (neoplasms, milk production, and enlargement), diabetes mellitus, and pyometra.

- *megestrol acetate* (Ovaban®, Megace®), a synthetic progestin approved by the FDA to postpone estrus and alleviate false pregnancies in dogs. It is used extra-label to treat urine spraying, anxiety, and aggression in cats. It is used to treat eosinophilic ulcers and dermatitis conditions. Megestrol acetate may cause endocrine problems (iatrogenic Addison's disease and diabetes mellitus) that are usually resolved when treatment is discontinued. Megestrol acetate is rarely used as a behavior-modification drug due to the development of other drugs with fewer side effects.

Testosterone is an androgenic steroid responsible for secondary sex characteristics of the male, and it produces malelike behaviors. Drugs that inhibit testosterone production or block enzymes that convert testosterone to dihydrotestosterone (its potent form) have been used to treat aggressive behavior in male dogs. An androgen antagonist, *delmadinone* (Terdak®), has also been used to treat aggression in male dogs. Side effects of delmadinone include decreased libido and fertility. Finasteride (Proscar®) is an androgen hormone inhibitor that inhibits the enzyme that converts testosterone to its potent form.

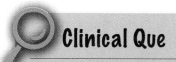

Clinical Que

Because of the number and seriousness of side effects associated with hormonal drugs, they should be the last drugs tried in treating animal behavioral problems.

Finasteride has been used in humans to treat benign prostatic hyperplasia, and may play a future role in veterinary medicine.

Pheromones

Pheromones are chemicals that trigger a natural behavioral response usually within or between members of the same species. There are many different types of pheromones such as sex pheromones and territorial pheromones. The first pheromone was identified in 1956 in insects. In insects, behavior (and pheromone response) is fairly predictable; however, this is not always true in mammals making pheromone use unpredictable in its effectiveness. In most animals, pheromones have a general calming effect. Examples of pheromones used in veterinary medicine include the following:

> **Clinical Que**
>
> Hormonal drugs that are reproductive in nature should not be used in pregnant animals.

- *dog appeasing hormone (DAP)* is a substance produced by lactating females to provide a feeling of comfort and safety to the young. A synthetic version of this pheromone is used in a variety of products to make the dog feel safe and calm. DAP diffuser (D. A. P.® Diffuser) comes in a plug-in diffuser that continuously releases the pheromone into the environment; DAP spray (D. A. P.® Spray) is sprayed into the environment 15 minutes prior to a stressful situation like kenneling or a car ride; DAP collar (D. A. P.® Collar) is used for introduction of a dog to a new environment. DAP is used to treat separation anxiety, destruction, excessive barking, house soiling, phobias, and excessive licking. The diffuser should not be touched with wet hands or metal objects when it is plugged in nor should it be touched with uncovered hands or immediately after use.

- *feline facial pheromone (FFP)* is a substance made by cat cheek gland secretions that is used as a plug-in diffuser (Feliway® Diffuser) or spray (Feliway® Spray) that is used to treat urine marking or spraying, avoidance of social contact, stressful situations, intercat aggression, vertical scratching, or appetite loss. The spray is applied to objects one to two times daily for 30 days and then every two to three days for maintenance.

- *equine appeasing pheromone (EAP)* is a maternal pheromone found in the "wax area" close to the mammae of a nursing mare. EAP is available as a spray (Modipher EQ® Spray), is administered in two sprays into each naris 30 minutes before a stressful event, and is used in horses to alleviate stressful situations such as transport, shoeing, clipping, new environments, and training.

No significant side effects are noted for the use of pheromones. The use of diffusers may take up to 72 hours for the chemical to saturate the area, so effects of the product may not be immediate.

Table 22-2 summarizes behavior disorders in which drug therapies are used, and Table 22-3 lists some behavior-modifying drugs.

Table 22-2   Behavior Disorders and Some of Their Drug Therapies

BEHAVIORAL DISORDER	DRUGS USED FOR THIS DISORDER
Urine spraying/marking	<ul><li>benzodiazepines such as alprazolam, diazepam, and chlordiazepoxide</li><li>buspirone</li><li>TCAs such as clomipramine</li><li>SSRIs such as paroxetine and fluoxetine</li><li>pheromones</li></ul>
Aggression	<ul><li>TCAs such as amitriptyline</li><li>benzodiazepines such as diazepam</li><li>SSRIs such as fluoxetine</li><li>hormones such as medroxyprogesterone acetate and megestrol acetate</li></ul>
Obsessive-compulsive disorders	<ul><li>TCAs such as amitriptyline</li><li>antihistamines (especially when due to pruritus)</li><li>benzodiazepines such as diazepam</li><li>pheromones</li></ul>
Anxiety	<ul><li>phenothiazines such as acepromazine</li><li>TCAs such as amitriptyline</li><li>buspirone</li><li>antihistamines (especially when due to pruritus)</li><li>benzodiazepines such as diazepam and clorazepate</li><li>pheromones</li></ul>
Noise phobias	<ul><li>benzodiazepines such as alprazolam and clorazepate</li><li>buspirone</li><li>TCAs such as clomipramine</li><li>pheromones</li></ul>
Fear aggression	<ul><li>TCAs such as amitriptyline and clomipramine</li><li>buspirone</li><li>benzodiazepines such as diazepam</li></ul>

Table 22-3   Behavior-Modifying Drugs

DRUG CATEGORY	DRUG	FUNCTION	EXAMPLES
Antianxiety agents	benzodiazepines	• Promote the inhibitory neurotransmitter GABA in the brain • Used to treat separation anxiety, phobias, aggression, and urine spraying	• diazepam (Valium®) • lorazepam (Ativan®) • alprazolam (Xanax®) • chlordiazepoxide (Librium®) • flurazepam (Dalmane®) • clorazepate (Tranxene® T tab)
	antihistamines	• Cause CNS depression, resulting in sedation	• hydroxyzine (Atarax®) • diphenhydramine (Benadryl®)
	phenothiazines	• Work as dopamine antagonists	• chlorpromazine (Thorazine®) • acepromazine (PromAce®) • promazine (Sparine®) • perphenazine (Trilafon®) • prochlorperazine (Compazine®)
	azapirones	• Serotonin blockers • Used to treat urine spraying	• buspirone (BuSpar®)
	anticonvulsants	• Blocks sodium channels thus inhibiting the generation of repetitive action potentials • Used to treat seizurelike anxiety	• carbamazepine (Tegretol®)
	amino acid derivative	• Increases GABA concentrations in the brain • Increases serotonin and dopamine in the brain • Nutraceutical product	• L-theanine (Anxitane®)
Antidepressants	tricyclics	• Prevent uptake of neurotransmitters • Used to treat separation anxiety, obsessive licking, hypervocalization, and urine spraying	• amitriptyline (Elavil®) • clomipramine (Clomicalm®, Anafranil®) • imipramine (Tofranil®) • doxepin (Sinequan®)
	selective serotonin reuptake inhibitors	• Inhibit serotonin uptake from the synapse • Used to treat obsessive licking, phobias, separation anxiety, and aggression	• fluoxetine (Prozac®) • sertraline (Zoloft®) • paroxetine (Paxil®) • fluvoxamine (Luvox®) • citalopram (Celexa®) • escitalopram (Lexapro®)
	monoamine oxidase inhibitors	• Block dopamine uptake from the synapse • Used to treat canine cognitive dysfunction or dementia	• selegiline (Anipryl®) • phenelzine (Nardil®) • isocarboxazid (Marplan®) • tranylcypromine sulfate (Parnate®)
Hormones	synthetic progestins	• Believed to correct hormonal imbalance	• megestrol acetate (Megace®, Ovaban®) • medroxyprogesterone (Depo-Provera®)

(Continued)

Table 22-3   *(Continued)*

Drug Category	Drug	Function	Examples
	synthetic estrogens	• Believed to correct hormonal imbalance	• diethylstilbestrol (DES) (Stilphostrol®)
	testosterone inhibitors	• Inhibit testosterone production or conversion to its potent form	• delmadinone (Terdak®) • finasteride (Proscar®)
Pheromones	DAP	Believed to provide chemical in environment that provides feeling of comfort and safety used to treat separation anxiety, destruction, excessive barking, house soiling, phobias, and excessive licking	• dog appeasing pheromone (D. A. P.® Spray, D. A. P.® Diffuser, D. A. P.® Collar)
	FFP	Believed to provide substance that marks territory so that cat does not mark territory by urine marking used to treat urine marking/spraying, avoidance of social contact, stressful situations, intercat aggression, vertical scratching, or appetite loss	• feline facial pheromone (Feliway® Diffuser, Feliway® Spray)
	EAP	Believed to provide chemical found in mammae of nursing mare used to alleviate stressful situations such as transport, shoeing, clipping, new environments, and training	• equine appeasing hormone (Modipher EQ® Spray)

SUMMARY

The use of drugs to treat problem behaviors is an expanding area of veterinary medicine. Remember that only a small part of treating animals with behavior problems involves drug therapy. Correctly diagnosing the condition, examining social conditions that affect the situation, and altering external stimuli are all parts of behavior modification.

When behavior modification drugs are used, owners need to be made aware of the potential problems these drugs may cause. Clients should also be told that many behavior-modifying drugs are used extra-label. It should be stressed to owners of animals with diagnosed behavior problems that behavior modification includes retraining the animal and eliminating external stressors for the animal, in addition to the seemingly easy drug treatment.

Classes of drugs used in the treatment of behavior disorders include anti-anxiety drugs, antidepressant drugs, hormonal drugs, and pheromones. Many behavior-modifying drugs fall into multiple categories, especially when used at varying dosages.

It's a Wrap

The answers to the questions in this chapter's Setting the Scene case study should be understood after reading this chapter. What suggestions should be given to the owner of the aggressive male German Shepherd dog? Is medication the best answer for management of this dog's behavior? What are some ways to avoid behavior problems? What can be done in behavior cases where rapid resolution of the animal's problem behavior is needed?

This owner should be advised that there is no "magic pill" that will alleviate this dog's aggression. Behavior modification must also be instituted along with medication to ensure that this dog's aggression is dealt with effectively. It would be beneficial for the dog to have a PE and some baseline blood work done to determine if he has some systemic disease that may be attributing to his aggression. Another thing the owner could do is to have the dog neutered. Neutering may help curb some of his aggressive nature.

The next step would be to take the dog to a trainer who can help him with the dog's aggression. Training may include the use of special halters, leash command training, and limiting physical punishment for the dog since it may escalate the dog's degree of aggression. Until the dog's behavior has been modified, the owner has to take special precautions to make sure the dog is well supervised and not allowed to roam. Any discussion of the dog's aggression and the plans to alleviate it should be documented in his medical record.

Ideally, behavior counseling begins when an owner brings in a puppy for his/her first PE, so that behavior problems can be avoided. Spending time with a client when an animal is first introduced into a household and continuing behavior discussions during the pet's first years of life may prevent some behavioral issues from developing.

CHAPTER REVIEW

Matching

Match the behavior-modifying drug with its classification.

1. _____ megestrol acetate
2. _____ DAP
3. _____ acepromazine
4. _____ buspirone
5. _____ amitriptyline
6. _____ carbamazepine
7. _____ fluoxetine
8. _____ hydroxyzine
9. _____ chlordiazepoxide
10. _____ selegiline

a. MAOI
b. SSRI
c. benzodiazepine
d. antihistamine
e. phenothiazine
f. progestin hormone
g. azapirone
h. TCA
i. pheromone
j. anticonvulsant

Multiple Choice

Choose the one best answer.

11. Which class of behavior-modifying drugs works by potentiating the inhibitory action of GABA?
    a. antihistamines
    b. benzodiazepines
    c. phenothiazines
    d. azapirones

12. Which drug inhibits the reuptake of serotonin?
    a. fluoxetine
    b. selegiline
    c. phenobarbital
    d. amitriptyline

13. Which group of drugs works by inhibiting the enzyme that destroys neurotransmitters?
    a. SSRI
    b. TCA
    c. MAOI
    d. GABA

14. Which drug can cause diabetes mellitus during its use?
    a. diethylstilbestrol
    b. medroxyprogesterone acetate
    c. megestrol acetate
    d. delmadinone

15. Which group of drugs works by dopamine antagonism?
    a. azapirones
    b. phenothiazines
    c. progestins
    d. TCAs

16. Which group of drugs is both antianxiolytic and antipruritic?
    a. antihistamines
    b. benzodiazepines
    c. phenothiazines
    d. SSRIs

True/False

Circle a. for true or b. for false.

17. The use of behavior-modifying drugs is the most effective way to treat behavior problems in animals.
   a.   true
   b.   false

18. Most behavior-modifying drugs are approved by the FDA for animal use.
   a.   true
   b.   false

19. Most behavior-modifying drugs cause sedation.
   a.   true
   b.   false

20. Most behavior-modifying drugs start affecting an animal's behavior immediately.
   a.   true
   b.   false

Case Study

21. A four-year-old M/N Pug is presented for aggression toward its owner for the past year. The dog's history includes neutering at six months of age and routine vaccination, deworming, and heartworm testing. The owner states that the aggression started about a year ago: The dog bit her on the hand while she was brushing him. Since then, the owner has taken the dog to a groomer. The dog then began biting if he was not fed or petted on demand. The owner can no longer pick up the dog and is afraid to move the dog off the bed or chair or out of her lap. On PE, the dog is difficult to restrain and has to be muzzled. The results of the PE and baseline laboratory work (CBC, chemistry panel, and UA) are normal. The dog is diagnosed with dominance aggression.
   a.   Is the first treatment option for this dog a behavior-modifying drug?
   b.   What are some reasons why this dog feels dominant over his owner?
   c.   What suggestions can be made regarding this case?

Critical Thinking Questions

22. Describe reasons why sedative drugs, like acepromazine, would not be a good choice for treating animals with aggression.

23. SSRIs have become popular drugs for treating obsessive-compulsive disorders in animals. Give some examples of obsessive-compulsive disorders in animals.

ADDITIONAL REVIEW MATERIAL

Additional review material can be found on the StudyWARE™ CD included with this text.

BIBLIOGRAPHY

Allen, D. G. (2005). *Handbook of Veterinary Drugs* (3rd ed.). Baltimore, MD: Lippincott Williams & Wilkins.

American Hospital Formulary Service (AHFS). (2003). Bethesda, MD, American Society of Health-System Pharmacists, pp. 2155–2251.

Curtis, T. M. (2007a). *Comprehensive Approaches to Separation Anxiety in Dogs.* In Veterinary Proceedings of the North American Veterinary Conference, Orlando, FL.

Curtis, T. M. (2007b). *Feline Elimination Concerns.* In Veterinary Proceedings of the North American Veterinary Conference, Orlando, FL.

Plumb, D. C. (2008). *Plumb's Veterinary Drug Handbook* (6th ed.). Ames, IA: Blackwell Publishing.

Reiss, B., and Evans, M. (2007). *Pharmacological Aspects of Nursing Care* (7th ed.). Clifton Park, NY: Thomson Delmar Learning.

Spratto, G. R., and Woods, A. L. (2008). *PDR Nurse's Drug Handbook.* Clifton Park, NY: Cengage Learning.

Veterinary Pharmaceuticals and Biologicals (12th ed.). (2001). Lenexa, KS: Veterinary Healthcare Communications.

Veterinary Values (5th ed.). (1998). Lenexa, KS: Veterinary Healthcare Communications.

CHAPTER 23
HERBAL THERAPEUTICS

OBJECTIVES

Upon completion of this chapter, the reader should be able to:

- outline the various forms of alternative medicine.
- differentiate between Western and Chinese herbal medicine.
- describe quality control concerns regarding herbal supplements.
- describe the different forms and routes of administration of herbs.
- describe properties of commonly used herbs.
- explain ways to ensure proper client education regarding herb use in animals.

KEY TERMS

acupuncture

alternative medicine

botanical medicine

Chinese traditional herbal medicine

chiropractic

complementary medicine

holistic veterinary medicine

homeopathy

nutraceutical medicine

physical therapy

poultices

Western herbal medicine

Setting the Scene

A client comes into the clinic and says she is interested in maintaining her cat's health by using herbal supplements. She uses a variety of herbal supplements herself and wants to know if she can use some of these on her cat. The veterinary technician checks her file and sees that her cat has had a heart murmur auscultated on PE for the past few years, but has not had a cardiac workup yet. The technician is aware that people can buy herbal supplements without a prescription and without their doctor's advice. Can herbal supplements really cause any damage? Is it possible to know for sure? What herbal supplements could the technician recommend (or discourage) based on this cat's history?

Courtesy of iStockPhoto

Clinical Que

Some veterinarian's concerns about the use of alternative and complementary therapies center around their belief in the lack of objective documentation for the safety and efficacy regarding these therapies.

Clinical Que

Some herbs that are classified as alternative treatments have been used for many years. Consider the use of psyllium (bulk laxative), witch hazel (astringent), and aloe vera (burn treatment).

ALTERNATIVES

The use of alternative and complementary therapies has been well received by some members of the veterinary community and caused great concern to other members. Regardless of the veterinary viewpoint on these therapies, the numbers of animal owners using alternative therapies have increased over the years, many times without the knowledge of the veterinarian treating the animal. The use of the Internet, especially Web site message boards on which pet owners praise the benefits of these therapies, has been a contributing factor in the growth of these therapies.

WHAT'S IN A NAME?

Alternative and complementary medicine consists of multiple diverse disciplines. The term alternative medicine applies to treatments or therapies that are outside accepted conventional medicine. Complementary medicine implies that these therapies can be used with or in addition to conventional treatment. The use of both terms gives a more inclusive definition to this group of therapies.

The American Veterinary Medical Association (AVMA) has set up guidelines for veterinary alternative and complementary medicine. Included in this field are the following:

- veterinary acupuncture and acutherapy, which consists of the examination and stimulation of body points by use of acupuncture needles, injections, and other techniques for the diagnosis and treatment of numerous conditions in animals.

- veterinary chiropractic, which is the examination, diagnosis, and treatment of animals through manipulation and adjustments of specific joints and cranial structures.

- veterinary physical therapy, which is the use of noninvasive techniques for the rehabilitation of animal injuries. This includes stretching, range-of-motion exercise, hydrotherapy, massage therapy, and application of heat and cold (Figure 23-1).

- veterinary homeopathy, a medical discipline in which animal conditions are treated by administration of substances that are capable of producing clinical signs in healthy animals. These substances are used therapeutically in very small doses. The theory behind homeopathy is that the signs caused by a substance (such as increased mucus production or coughing) are needed clinically to resolve the condition. If these substances are given in small doses, the signs will be produced and help treat the condition.

- veterinary botanical medicine, which uses plants and plant derivatives as therapeutic agents.

- nutraceutical medicine, which uses micronutrients, macronutrients, and other nutritional supplements as therapeutic agents. Examples of

Courtesy of Kris Jensen, MS, PT, SCS

Figure 23-1 Dog receiving physical therapy.

nutraceutical substances include glucosamine, chondroitin sulfate, and coenzyme Q-10.

- **holistic veterinary medicine** is a comprehensive approach to health care using both alternative and conventional diagnostic techniques and therapeutic approaches. Holistic veterinarians may use a combination of methods (including those previously listed) to treat a patient's condition. Holistic medicine is a philosophy of practice that looks at the patient as a whole: body, mind, and spirit.

The rest of this chapter covers the use of herbs in veterinary medicine.

WEST VERSUS EAST

The use of herbs in human and veterinary medicine can be divided into two main groups: Western and Chinese (also known as Eastern). The central tenet of Western herbal medicine is that individuals have an inner force that works to maintain physical, emotional, and mental health. Herbalists who practice Western herbal medicine believe that many diseases occur because an individual's inner force or natural immune system is out of balance.

Western herbal remedies have been used as far back as 3000 B.C. in ancient Egypt. Ancient Greece, Rome, and the Middle East also used herbal remedies. When Columbus arrived in America, New World plants became available to Europeans, and blended plant use (New World and European plants) increased. For centuries, most Western medicine consisted of herbal medicine.

Only since World War II has medicine relied less on plants and more on synthetic drugs.

Chinese traditional herbal medicine is based on a holistic philosophy of life that emphasizes the relationship among the mental, emotional, and physical components of each individual, as well as the importance of harmony between individuals, their social groups, and the greater population. Chinese traditional herbal medicine attempts to restore health through corrections of imbalances within a patient's body or between the patient and natural order.

Chinese traditional herbal medicine, which includes the use of herbs and medicinal plants in combination, can be traced to three Chinese emperors: Fu Si (2852 B.C.), Shen Nong (3494 B.C.), and Huang Di (2697 B.C.). The principles are based on Taoism, a philosophy that emphasizes following the right path (Tao) in order to find one's place within the universe. This philosophy emphasizes herbal medicine and daily dietary habits. Prescriptions for herbal medicines were formulated to correct excesses or deficiencies of *yin* (cold, moisture, dimness, inward movement, quietness, and slowing) and *yang* (heat, dryness, brightness, outward movement, forceful action, and speed), blockages in the flow of *qi* (the universal life force), localized organ disorders, and emotional problems associated with physical illness. Besides the balance of Yin and Yang, the Chinese viewed organ function differently than did people in the West and used Chinese traditional herbal medicine to address the flow of energy (qi) in the body. Eastern herbal medicine is used within the framework of TCM. Many of the same herbs are used in Eastern and Western herbal medicine; however, Chinese "herbs" sometime include animal parts or minerals. Chinese herbal formulas involve the use of multiple herbs in one remedy, whereas in Western herbal medicine typically one herb is used at a time.

Why Herbs?

Herbal supplements are one of the fastest growing segments of the dietary supplement market in the United States. It is not clear what is driving this trend, but there appear to be some consistent patterns. One reason may be the desire for a more holistic approach to health care, especially by older people who have more chronic ailments. This rationale extends the holistic approach to animals, especially as they age.

A second reason for the increased use of herbal supplements in animals may be that people believe conventional treatments have real or perceived limitations. They may perceive a higher incidence of adverse effects with chemically based products than with "natural" products. Another reason for this trend is that use of the Internet, combined with the advertising practices of the companies that produce herbal supplements, have influenced people's buying behavior. If people use herbal supplements for themselves, they tend to want to use them on their animals as well.

Another factor that may affect people's desire to use herbs in their animals is that herbs have been used for a long time. Human ailments have been treated for thousands of years with herbs. In addition, many conventional drugs are derived from herbs. The assumption is that herbs are safe and do not have the side effects seen in some conventional drugs.

Clinical Que

Some of the most potent and toxic chemicals come from plants. Herbs are considered "natural," but that does not mean that they never produce serious side effects. Side effects are reported for all types of drugs (including those approved by the FDA), whether they are natural or not.

However, keep in mind that herbal supplements can cause problems in some patients. Herbs can interact with conventional medications in many ways, including enhancing the effect of the medicine. Herbs can also increase or reduce the bioavailability of prescription drugs. Herbs that are safe for humans may not be safe for animals. For example, garlic is used in humans as an antimicrobial and cholesterol-lowering agent. In animals, garlic may cause anemia if given for long periods of time.

Quality Control

Herbs cannot be sold and promoted in the United States as drugs because herbal supplements do not require FDA approval. Under the 1994 Dietary Supplement Health and Education Act (DSHEA), herbs, vitamins, minerals, amino acids, and other natural substances are considered nutritional or food supplements. The DSHEA permits general health claims such as "improves heart health" and "promotes regularity" as long as the label also has a disclaimer stating that supplements are not approved by the FDA and are not intended to diagnose, treat, cure, or prevent any disease. The claims must be truthful and not misleading and supported by scientific evidence. Safety and efficacy are not guaranteed, and no outside monitoring programs are currently available to identify and judge the potency of these herbs. Because herbal products are unregulated, the concentrations of active ingredients can vary between dose forms (ginseng products have been found to vary 200-fold between capsule and liquid forms). Currently the FDA is working with several trade organizations to develop guidelines for herbal supplements. Some manufacturers have devised their own parameters for quality control. Table 23-1 gives information on factors affecting herb quality.

Table 23-1 Factors Affecting Herb Quality

FACTOR	CONSIDERATIONS
Environment	• Is the herb wild-grown or commercially grown? • Were fertilizers used for growing the herb? • What was the climate and rainfall during the growing season of the herb?
Plant	• Was the correct part of the plant used (seeds, stalks, or roots)? • Was the plant part harvested at the correct time? • Were the plants mature enough or young enough to yield the desired active ingredients of the herb? • Is the correct genus and species of the herb being used?
Handling	• After harvest, was the herb properly handled? • After harvest, were temperature, sunlight, and humidity maintained in the proper range?

Herb Forms and Administration

Herbs can be given to patients in a variety of forms: capsules and tablets, ointments, extracts, **poultices**, compresses, teas, and in bulk. Table 23-2 summarizes forms of herbs.

Herbal treatments are administered in many different ways. They can be taken orally as a drink; applied to the skin as a cream, ointment, or poultice; taken internally in tablet or capsule form; and/or added to water for bathing.

Sometimes herbs are given in combination with each other. The belief is that a combination of herbs creates an additive effect of the advantages of the different herbs, while diluting the toxic effects of one herb with the benefits of other herbs in the combination.

Table 23-2 Herb Forms

HERB FORM	PROPERTIES
Capsules and tablets	• Effective and convenient to give • Variety of sizes available
Ointments	• Topical application for specific areas • Should be applied to intact skin (because pets often lick their wounds it is important to be sure the ingredients are nontoxic)
Extracts	• Liquids come in water, alcohol tincture, and glycerin forms (alcohol tinctures are not well tolerated by pets. The dose can be diluted in an equal amount of hot water to evaporate off the alcohol) • Powders come in capsules and tablets • Standardized commercially to provide a certain number of milligrams of active ingredient per dose • Efficacy is more consistent; however, patient acceptance is poor
Poultices and compresses	• A poultice is made by boiling fresh or dried herbs, squeezing out excess liquid, cooling the herb, and applying it to the skin. The herb is then wrapped with gauze to hold the material in place. This is left on for a few minutes to hours. • A compress is made with an herbal extract (not the plant part) by soaking a cloth in the extract. The cloth is then applied to the skin.
Teas	• Best-known form for human patients • Most active ingredients of herbs are water soluble and therefore available to the patient in tea form • Commercial teas usually do not have adequate amount of herb to be effective
Bulk	• Herb purchased in loose form • Form may be chopped, powdered, granular extract (most common form for Chinese herbs), or fresh

HERBAL COUNTDOWN

The following are some of the more popular herbal supplements used by people. Clients may be interested in using these same herbs for their animals.

Ginkgo

Ginkgo, *Ginkgo biloba,* is the earth's oldest living tree species; many individual trees live for more than 1000 years (Figure 23-2). Gingko is the only tree to survive the atomic blasts of Hiroshima and Nagasaki. It has been used for centuries in China to help improve memory and to treat respiratory problems in humans.

The active ingredients in ginkgo are ginkgo flavone glycosides and the terpene lactones. It has been used in cases of vascular insufficiency, to enhance the utilization of oxygen and glucose by the brain, and to reestablish perfusion in areas of ischemia (deficiency of blood flow). It has also been used to treat depression. In animals, gingko has been used to reduce aging effects on the nervous system, to reduce hypertension, and as a general tonic in geriatric patients.

An ingredient in gingko is a potent inhibitor of platelet-aggregating factor. Animals using aspirin, NSAIDs, and heparin should be given gingko with care. Gingko should be stopped at least one week prior to surgery. Gingko may also inhibit cytochrome P450 enzyme and may induce hypoglycemia. Animals with diabetes mellitus should not take gingko.

St. John's Wort

St. John's Wort, *Hypericum perforatum* (Figure 23-3), got its common name because it is believed that the flowers of St. John's Wort have their brightest appearance on June 24th, the birthday of John the Baptist. In Europe, people use St. John's Wort as a food coloring and flavor additive and topically for the treatment of burns and wounds.

Figure 23-2 Ginko (*Ginkgo biloba*).

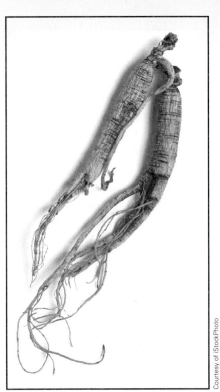

Figure 23-3 St. John's Wort (*Hypericum perforatum*).

Figure 23-4 Ginseng (*Panax ginseng*).

About 10 groups of components are active ingredients in St. John's Wort. One component, hyperforin, regulates the effects of serotonin. In humans, St. John's Wort is used to treat mild to moderate depression. It has been extensively studied in humans and appears to work well in some patients. In animals, St. John's Wort has been used to treat behavior disorders such as lick granulomas, aggression, separation anxiety, and obsessive-compulsive disorders.

St. John's Wort can induce cytochrome P450 and alter the pharmacodynamics of other drugs. It should not be used with other antidepressant medications. It can also affect blood pressure and cause photosensitivity (especially in livestock) in higher doses.

Ginseng

Ginseng, *Panax ginseng* (Figure 23-4), has been part of traditional Chinese medicine for more than 2000 years. It has been used to treat almost every ailment, but is popular as a general tonic and as a medication to help cope with stress.

The main active ingredients are ginsenosides (13 have been identified) that are thought to be responsible for increasing energy, countering stress, and enhancing physical performance. Ginseng also seems to stimulate natural killer cell activity, resulting in improved ability to fight infection. Ginsenosides also interfere with coagulation and platelet function. In animals, ginseng has been used to treat extreme weight loss (such as that secondary to cancer), anorexia, systemic infections, and feline leukemia.

The dosage of ginseng used in animals should be standardized to the ginsenoside content of the product being used. One problem with dosing ginseng is the great variability in commercial ginseng products and variation of the ingredient concentration from what is represented on the label. Ginseng seems to be well tolerated, but it is recommended not to give more to an animal in an attempt to get better results. The inhibition of platelet aggregation has raised some safety concerns with increased dosages or increased frequency of use. Ginseng use should be discontinued at least one week prior to elective surgery. Ginseng can also affect blood sugar levels in nondiabetic and diabetic animals, increase blood pressure and heart rate, increase gastrointestinal motility, and stimulate the brain to the point of seizures (when large doses are used).

Garlic

Garlic, *Allium sativum* (Figure 23-5), has been used for more than 5000 years in humans, for a variety of conditions including parasitic infections, respiratory problems, and poor digestion. Louis Pasteur confirmed garlic's antimicrobial action, and Albert Schweitzer used it in the treatment of amoebic dysentery.

Garlic's activity is due to its allicin content. Allicin is an odorless amino acid that, when exposed to air (when the garlic is broken), will produce allicin. Allicin is an unstable compound that gives garlic its characteristic odor. Therefore, any garlic preparation that is odorless is probably medically worthless. Allicin is converted to other sulfur compounds, such as ajoene, which are the active ingredients of garlic.

Garlic is used for numerous purposes, including reduction of cholesterol and triglyceride levels and hypertension. It is believed that garlic has anticarcinogenic properties, especially against colon, stomach, and prostate cancers. It also has an inhibitory effect on *Heliobacter pylori*, which has been linked to the development of gastric ulcers. Garlic has also been used to treat parasitic (roundworm, tapeworm, and hookworm infections) and fungal infections (ringworm). In animals, it has been used for flea prevention (this use has not

Courtesy of iStockPhoto

Figure 23-5 Garlic (*Allium sativum*).

been clinically proven), for respiratory problems, and to acidify urine (testing has not proven this latter claim). Garlic is high in potassium and should be used with caution in animals with electrolyte imbalances.

Dosages of garlic vary with the preparation and desired effect. Precautions concerning garlic use revolve around inhibition of platelet aggregation and prolongation of bleeding times. Garlic should be discontinued at least one week prior to any surgical procedure. In high dosages, garlic can cause gastrointestinal upset, clotting problems, and Heinz body anemia in cats. Because garlic enters the milk, these gastrointestinal problems can be passed to suckling young. Cats are especially susceptible to garlic toxicity.

Echinacea

Echinacea purpurea (Figure 23-6) is a wildflower native to North America, commonly known as the purple coneflower. Historically, it has been used for a variety of conditions such as infections, joint pain, abscesses, and burns. Echinacea was part of the *National Formulary of the United States* until 1950, when antibiotic use increased and interest in echinacea waned. With the increasing problems of antibiotic resistance, echinacea's benefits have been reexamined.

The actions of echinacea are due to polysaccharides called fructofuranosides. These compounds contribute to tissue regeneration, regulation of the inflammatory response, and a mild cortisonelike effect. Echinacea stimulates phagocytosis and natural killer cell activity. Most evidence shows that echinacea is effective in limiting the severity of and shortening the duration of some infections.

Courtesy of iStockPhoto

Figure 23-6 Echinacea (*Echinacea purpurea*).

Dosage of echinacea varies with the preparation of the herb. It is usually given at higher dosages early in the course of disease for better results. Minimal side effects are seen with this herb.

Saw Palmetto

Saw palmetto, *Serenoa repens* (Figure 23-7), was used by Native Americans in the treatment of urinary tract infections. By the early twentieth century, it was being used to treat benign prostatic hypertrophy in humans.

The active ingredients of saw palmetto are fatty acids from the berries, which produce an enzyme to prevent the conversion of testosterone to dihydrotestosterone (DHT). Saw palmetto does not alter the overall size of the prostate, but shrinks the inner prostatic epithelium. Saw palmetto has been used to treat benign prostatic hyperplasia, to stimulate appetite, to treat interstitial cystitis in cats, and as a mild diuretic. Saw palmetto extract is well tolerated, and few adverse side effects are noted other than mild gastrointestinal problems.

Evening Primrose

Evening primrose, *Oenothera biennis* (Figure 23-8), contains gamma-linolenic acid, which is a fatty acid in the omega-6 family. Evening primrose oil has been used in humans for treating premenstrual syndrome, rheumatoid arthritis, diabetic neuropathy, and eczema. In cats and dogs it is used to maintain healthy skin, normal growth in young, and lactation in females.

Dosage of evening primrose oil varies with the condition being treated. It is recommended that evening primrose oil be taken with food to increase

Figure 23-7 Saw palmetto (*Serenoa repens*).

Figure 23-8 Evening primrose (*Oenothera biennis*).

absorption. Benefits may not be seen for several months in treating chronic conditions. Because this is an oil supplement, higher dosages can lead to loose stools and abdominal cramps. Evening primrose oil may lower the seizure threshold in some people on phenothiazine medication, so use caution when giving it to animals with a history of seizures. Evening primrose oil is also used in shampoos to relieve dry, pruritic skin.

Goldenseal

Goldenseal, *Hydrastis canadensis* (Figure 23-9), gets its name from the fact that when its stem is broken near the base, the scar resembles the gold wax used to seal letters. Native Americans used goldenseal as a dye and medicinally for skin and eye conditions and diarrhea. Because of overharvesting in the 1980s, goldenseal is considered an endangered species.

The active ingredients of goldenseal are the isoquinoline alkaloids— especially a substance called berberine. Berberine is found in the stem bark of the plant as well as in the roots. Goldenseal is used mainly as an antibacterial and antiparasitic drug. Its antiparasitic activity is primarily against amoebae, such as *Entamoeba* sp., and the protozoan *Giardia*.

Dosage for oral use of goldenseal varies, but it should be taken only for a limited time (usually two to three weeks). High doses of goldenseal for prolonged periods of time can cause cardiac problems and central nervous stimulation. Goldenseal should not be used in pregnant animals because it stimulates uterine contractions.

Cranberry

Cranberry, *Vaccinium macrocarpon* (Figure 23-10), is a North American heritage herb that has been used for food and as medicine mainly for the treatment of urinary tract infections. Its name comes from the Pilgrims, who called it *crane berry* because the plant's stem looks like the neck and head of a crane.

Courtesy of the Pennsylvania Department of Conservation and Natural Resources

Figure 23-9 Goldenseal (*Hydrastic canadensis*).

Figure 23-10 Cranberry (*Vaccinicum macrocarpon*).

It was believed that cranberry acidified urine, but that is now known to be false. The primary effect of cranberry is interference with the attachment of urinary pathogens (such as *E. coli*) to the urinary bladder wall. Cranberry has been used to treat urinary tract infections in cats. Cranberry compounds may also protect the stomach from adhesion of *H. pylori* and prevent the development of gastric ulcers. High levels of cranberry consumption can lead to diarrhea and stomach problems. Prolonged use can lead to the formation of kidney stones.

Valerian

Valerian, *Valeriana officinalis* (Figure 23-11), is a plant native to Europe and North America. Valerian produces small, rose-colored flowers that bloom on a four-foot-high stem. Historically, it has been used as a sedative and antianxiety drug.

Figure 23-11 Valerian (*Valeriana officinalis*).

Valerian's effects have been attributed to several volatile oils, including valeric acid, valerenic acid, and valepotriates. Valerenic acid inhibits the breakdown of gamma-aminobutyric acid (GABA). Because GABA is inhibitory, hindering its breakdown results in higher levels of GABA, which decreases central nervous system activity. Valerian is primarily used as a sleep aid, to calm hyperactivity in dogs, and for mild tranquilization. Adverse reactions to valerian are mild. Care should be taken if giving valerian with other depressant drugs.

Hawthorn Berry

Hawthorn, *Crataegus oxyacantha* (Figure 23-12), is a small tree or shrub that is a member of the rose family. The hawthorn produces berries that were used by the ancient Chinese to make a fermented beverage. The New York Medical Journal first published a report about the use of hawthorn berries in the treatment of heart disease in October 1896. The hawthorn was hung over the doorway in the Middle Ages to prevent the entry of evil spirits.

Hawthorn berries are used as a heart and valvular tonic typically in animals in the early stages of congestive heart failure. There does not appear to be one active ingredient in Hawthorn; however, flavonoid and proanthocyanidins have been described as active compounds in this herb. Hawthorn and its extracts act as mild positive inotropes and angiotensin-converting enzyme (ACE) inhibitors that strengthen heart contractility and stabilize the heart against arrhythmias by prolonging the refractory period (the period after a heartbeat when the heart cannot beat again). The flavonoids also decrease capillary leakiness and dilate coronary arteries, thus increasing blood flow to the heart. Hawthorn berry may be used in animals with kidney disease due to its ability to enhance blood flow and in animals being treated with cardiotoxic chemotherapeutic drugs. Hawthorn berry should be used

Courtesy of iStockPhoto

Figure 23-12 Hawthorn (*Crataegus oxyacantha*).

with caution in animals with hypotension and hypertrophic cardiomyopathy; however, it appears to have fewer side effects than other inotropic drugs such as digitalis.

Ginger

Ginger, *Zingiber officinale* (Figure 23-13), has been used as a medicine in Asian, Indian, and Arabic herbal traditions since ancient times. In China, ginger has been used to aid digestion and treat stomach upset, diarrhea, and nausea for more than 2000 years. Ginger is native to Asia, where it has been used as a culinary spice for at least 4400 years. Ginger grows in fertile, moist, tropical soil, and its underground stem (rhizome) is used as an herb. The stem of the ginger plant extends about 12 inches above ground and has long, narrow, and green leaves, and white or yellowish-green flowers.

The active ingredient in ginger is gingerol, which converts to shogaol and zingerone (pungent phenol compounds) when dried and stored. Ginger is used to treat arthritis because it is a vasodilator that may increase blood flow to joints and is an anti-inflammatory herb that inhibits prostaglandin. It relieves indigestion by promoting aboral gastrointestinal motility and reducing gastric upset and vomiting. Ginger has been used in animals to treat intestinal disorders, motion sickness, and gastric upset secondary to the use of nonsteroidal anti-inflammatory drugs (NSAIDs). Ginger may also inhibit platelet clumping; therefore, animals should not take ginger for one week prior to surgery. It may also raise body temperature and should not be used in animals with a fever. Ginger should not be used in pregnant or hypoglycemic animals or those with clotting disorders.

Courtesy of iStockPhoto

Figure 23-13 Ginger (*Zingiber officinale*).

Milk Thistle

Milk thistle, *Silybum marianum* (Figure 23-14), has been used since Greek and Roman times as an herbal remedy for a variety of ailments especially liver problems. In the late 19th and early 20th centuries physicians in the United States used milk thistle seeds to relieve congestion of the liver, spleen, and kidneys. Milk thistle is native to the Mediterranean, but can now be found widespread throughout the world. This plant grows in dry, sunny areas; has stem branches at the top; and reaches a height of 4 to 10 feet. Its flowers are red-purple, and its fruit is small, hard skinned, brown, spotted, and shiny.

The active ingredient of milk thistle is silymarin, which consists of a group of compounds called flavonolignans. Flavonolignans are protective and restorative to the liver. These substances are hepatoregenerative (help repair liver cells damaged by toxic substances) and hepatoprotective (keep new liver cells from being destroyed by these same substances), reduces inflammation, and has potent antioxidant effects. Side effects are rare and may include loose stools when animals are given large doses.

Most milk thistle products are standardized preparations extracted from the fruits (seeds) of the plant; however, the shells, seeds, and leaves are also used. Most preparations are standardized to contain 70 to 80 percent of flavonolignans (silibinin, silychristin, and silydianin), collectively known as silymarin.

Figure 23-14 Milk thistle (*Silybum marianum*).

ADVICE TO CLIENTS

Given the increasing public awareness and use of herbs and other complementary and alternative treatment methods, the veterinary profession needs to provide reliable information to clients so that responsible choices can be made. One way to provide reliable information is by developing regulation of herbal supplements. The National Animal Supplement Council has developed a Compliance Plus program, a nonregulatory program that applies to non-food-producing animals and the reporting of adverse effects of complementary and alternative treatments. Members of the National Animal Supplement Council would also be required to implement labeling, manufacturing, and quality control standards for their industry. This may help with the development of standards for the herbal supplement and product industry.

Another way to provide reliable information is by educating clients about herb use and possible side effects or interactions. A list of complementary and alternative medicine veterinary associations is found in Table 23-3. The following are some guidelines for dealing with patients in a clinical setting.

- Ask each client whether they give herbs or other supplements to their animals.

- Inform clients that herb–drug interactions exist.

- Encourage the use of standardized products from respected manufacturers.

- Use herbal therapies in recommended doses; more is not better.

- Avoid herbs with known toxicities.

- Herb use in pregnant or nursing animals, the very young, or the very old may not be wise.

- Accurate diagnosis of the animal's condition is essential to evaluate all therapeutic options.

- Document all herb or supplement use in the animal's medical record.

Table 23-3 Complementary and Alternative Medicine Veterinary Associations

American Holistic Veterinary Medical Association	http://ahvma.org
Academy of Veterinary Homeopathy	http://theavh.org
American Veterinary Chiropractic Association	www.animalchiropractic.org
American Academy of Veterinary Acupuncture	www.aava.org
International Veterinary Acupuncture Society	www.ivas.org
Veterinary Botanical Medicine Association	www.vbma.org

SUMMARY

The use of complementary and alternative medicine and therapies has increased in recent years. Multiple disciplines of complementary and alternative medicine are recognized by the AVMA.

The central tenet of Western herbal medicine is that individuals have an inner force that works to maintain physical, emotional, and mental health. Herbalists who practice Western herbal medicine believe that many diseases occur because an individual's inner force or natural immune system is out of balance. Chinese traditional herbal medicine is based on a holistic philosophy of life that emphasizes the relationship among the mental, emotional, and physical components of each individual, as well as the importance of harmony between individuals, their social groups, and the greater population. Chinese traditional herbal medicine attempts to restore health through corrections of imbalances within a patient's body or between the patient and natural order.

Herbal supplements have been used for thousands of years for a variety of conditions. Examples of herbs used in veterinary medicine include gingko, St. John's Wort, garlic, and ginseng. It is important to use a preparation with the correct part of the plant (root, stem, or seed) when using herbs, and to be aware of possible herb–drug interactions. Some herbs have side effects and may have to be discontinued prior to surgery. Always question clients about whether they use herbal supplements for their animals.

Table 23-4 lists the herbs discussed in this chapter.

Table 23-4 Herbs Covered in This Chapter

HERB	BODY SYSTEM CATEGORY	EFFECT	CAUTIONS	FORM USED
Cranberry	• Urogenital	• Prevent and treat urinary infections • Reduces urine calcium to help prevent stones	• None known	Berries and dried powder are used for juice and capsules
Echinacea	• Immune	• Immunostimulant • Anti-inflammatory	• None known	Roots are used for capsules, tinctures, and compresses
Evening primrose	• Skin • Mucous membranes	• Atopic eczema	• Avoid in animals with blood-clotting disorders	Extracted oil is used in ointments and compresses
Garlic	• Cardiovascular	• Antibiotic activity to some species of gram+ and gram− bacteria • Cholesterol and blood pressure reduction • Antiparasitic • Antifungal against ringworm	• Clotting problems may develop • Larger doses can cause gastrointestinal problems • Heinz body anemia may be seen with long-term use	Fresh pulp and capsules are given internally

(Continued)

HERB	BODY SYSTEM CATEGORY	EFFECT	CAUTIONS	FORM USED
Ginger	• Gastrointestinal • Skeletal	• Digestive aid • Anti-inflammatory • Vasodilation	• May inhibit platelet clumping, so should be discontinued before surgery • May increase body temperature and should not be used in animals with fever • Do not use in pregnant or hypoglycemic animals or those with clotting disorders	Roots used in powder or capsule form; also found in teas, tablets, and candied form
Ginkgo	• Nervous • Cardiovascular	• Reduces aging-associated brain problems • Anticoagulant • Neutralizes free radicals that damage cells	• Do not use long term in animals with clotting problems	Tinctures, tea infusions, capsules, oil used internally and externally
Ginseng	• Immune	• Immunostimulant that prevents disease • Stimulant of nervous system • Antifatigue drug	• Gastrointestinal problems if used long term	Roots used for internal administration with capsules and teas
Goldenseal	• Gastrointestinal • Urinary	• Antibiotic activity • Antiparasitic against amoeba and *Giardia* • Less commonly used for urinary infections and gastroenteritis	• Do not use in pregnant animals (stimulates the uterus) • Diarrhea in large doses • Nervous system effects	Dried roots are used in tincture, capsule, and tea forms
Hawthorn berry	• Cardiovascular • Renal	• Positive inotrope • ACE inhibitor • Increases renal blood flow	• Do not use if animal is hypotensive or has hypertrophic cardiomyopathy	Leaves and flowers are for tinctures and infusions
Milk thistle	• Hepatic • Splenic • Renal	• Hepatoregenerative • Hepatoprotective • Anti-inflammatory • Antioxidant	• None known	Liquid extract from fruit used for capsule or injections; shells, seeds, and leaves are also used
St. John's wort	• Nervous	• Antianxiety • Antidepressant • Anti-inflammatory	• Photosensitization	Dried leaves and flowering tops used for tea infusion and tincture (internally), oil (externally)
Saw palmetto	• Urogenital	• Anti-inflammatory • Appetite stimulant	• Diarrhea in large amounts	Berries are used in an extract given internally
Valerian	• Nervous	• Tranquilizer • Less commonly used as a diuretic and antispasmotic	• None known	Root is used dried or fresh in tincture form

It's a Wrap

The answers to the questions in this chapter's Setting the Scene case study should be understood after reading this chapter. Can herbal supplements really cause any damage? Is it possible to know for sure? What herbal supplements could the technician recommend (or discourage) based on this cat's history? Staying current on the use of herbal supplements is key to maintaining safety and efficacy of these agents in animals. Continuing education (meetings, journals, books, and monitored Web sites) plays a major role in the success or failure of herbal supplement use in animals.

Pet owners try to cure their pet's ailments with many home remedies and supplements recommended by their friends. They may not seek out veterinary advice about treatments they want to use on their pets.

Herbal supplements are really medicinal plants and hence may have side effects associated with their use. They can interact with prescribed medications and may have side effects. This cat should have a through cardiac workup to assess the status of any heart condition before supplementation is considered.

Human herbal supplements are regulated as dietary supplements under the Dietary Supplemental Health and Education Act of 1994 (DSHEA). Herbal supplements are not "prescribed" for a particular condition but are recommended for structural and functional claims ("good for the immune system"). They do not carry specific health claims, such as preventing joint disease, and they are not regulated for quality or purity (only the name of the plant and the plant part used are provided on the label).

Herbs to avoid with cardiac disease include, but are not limited to, gingko (can cause heart palpitations and interfere with aspirin, coumadin, and heparin) and ginseng, which can increase heart rate and blood pressure. Problems often occur when high dosages of herbs are used.

CHAPTER REVIEW

Matching

Match the herb with its action or body system it affects.

1. _____ echinacea
2. _____ milk thistle
3. _____ valerian
4. _____ cranberry
5. _____ garlic
6. _____ evening primrose
7. _____ St. John's wort
8. _____ ginger
9. _____ hawthorne berry
10. _____ ginkgo

a. skin
b. nervous system tranquilizer
c. digestive aid
d. immune system
e. reduces aging-associated brain problems
f. urogenital
g. liver
h. ACE inhibitor
i. antidepressant activity
j. antimicrobial activity

Multiple Choice

Choose the one best answer.

11. Which type of alternative medicine is based on the belief that individuals have an inner force that works to maintain physical, emotional, and mental health?
    a. Western herbal medicine
    b. veterinary botanical medicine
    c. nutraceutical medicine
    d. Chinese traditional herbal medicine

12. Which type of alternative medicine is based on a holistic philosophy of life that emphasizes the relationship among the mental, emotional, and physical components of each individual, as well as the importance of harmony between individuals, their social groups, and the population at large?
    a. Western herbal medicine
    b. veterinary botanical medicine
    c. nutraceutical medicine
    d. Chinese traditional herbal medicine

13. What herb form involves boiling fresh or dried herbs, squeezing out excess liquid, cooling the herb, and applying it to the skin?
    a. dried root
    b. poultice
    c. compress
    d. dried stem

14. Which herb has been used to treat urinary tract infections in cats because it inhibits binding of some bacteria to the urinary bladder wall?
    a. catnip
    b. valerian
    c. cranberry
    d. saw palmetto

15. Which herbal supplement may cause photosensitivity?
    a. gingko
    b. ginseng
    c. cranberry
    d. St. John's Wort

16. Which herbal supplement has been used as a mild tranquilizer in dogs?
    a. valerian
    b. saw palmetto
    c. ginseng
    d. St. John's Wort

17. What is the active ingredient in catnip?
    a. nepatalactone
    b. ginsenoside
    c. allicin
    d. fructofuranoside

True/False

Circle a. for true or b. for false.

18. Because herbal supplements are used in humans and animals, they are regulated by the FDA.
    a. true
    b. false

19. Herbs are natural; therefore they can be used safely in pregnant and nursing animals.
    a. true
    b. false

20. All parts of herbs have equal potency and effectiveness.
    a. true
    b. false

Case Study

21. A 10-year-old DSH cat is brought into the clinic for lethargy. Her owner states that the cat has been acting "tired" and does not seem to want to eat. On PE, the cat's TPR are normal, a large amount of gas is palpated in the intestines, and some loose feces are seen in the perineal region. Blood is collected from this cat for a CBC, chemistry screen, and thyroid level. The cat's CBC results show that this cat is anemic, which is probably the cause of her lethargy. The cat's fecal exam reveals that the cat also has roundworms and hookworms. When told of the test results, the cat's owner says they cannot be correct, because she has been treating the cat with garlic supplements to get rid of the parasitic infection.
    a. What should this owner be told about garlic supplementation?
    b. What additional information should be obtained from this client regarding her cat?
    c. Could the garlic be causing any of the cat's health problems? Why?

Critical Thinking Questions

22. What are some problems that may be encountered when seeking information on the use of herbal supplements in animals?

23. Herbal toxicities may be seen in animals with owners who are overzealous about the use of herbs in their pets. Overzealous owners may give their animals "too much of a good thing" and cause more harm than good. What are antidotes for herbal overdoses?

ADDITIONAL REVIEW MATERIAL

Additional review materials can be found on the StudyWARE™ CD included with this text.

BIBLIOGRAPHY

AltVetMed (complementary and alternative veterinary medicine), http://www.altvetmed.com.

Bearman, J. (2009). The AnShen News, http://www.anshenvet.com.

Guidelines on Alternative and Complementary Therapies. (1996). Schaumburg, IL: American Veterinary Medical Association.

Libster, M. (2002). *Delmar's Integrative Herb Guide for Nurses.* Clifton Park, NY: Thomson Delmar Learning.

Messonnier, S. (2001). *Natural Health Bible for Dogs and Cats: Your A-Z Guide to Over 200 Conditions, Herbs, Vitamins, and Supplements.* New York: Three Rivers Press.

Nolen, R. (2002, August 15). Facing crackdown, dietary supplement companies promise changes. *Journal of the American Veterinary Medical Association, 221*(4): 479–483.

PDR for Herbal Medicines (2nd ed.). (2000). Montvale, NJ: Thomson Medical Economics.

PDR for Nutritional Supplements. (2001). Montvale, NJ: Thomson Medical Economics.

Pet Education.com, http://www.peteducation.com.

Ramey, D. (2003, June 15). Regulatory aspects of complementary and alternative veterinary medicine. *Journal of the American Veterinary Medical Association, 222*(12): 1679–1682.

Rosenthal, M. (January 2009). What if your clients want CAM for their pets? *Veterinary Forum, 26*(1): 24–35.

Schoen, A., and Wynn, S. (1998). *Complementary and Alternative Veterinary Medicine Principles and Practice.* St. Louis, Moseby.

Watkins, R. (2002, March). Herbal therapeutics: The top 12 remedies—Part I. *Emergency Medicine Journal, 34*(3): 12–19.

Watkins, R. (2002, March). Herbal therapeutics: The top 12 remedies—Part II. *Emergency Medicine Journal, 34*(4): 43–54.

Watkins, R. (2002, March). Herbal therapeutics: The top 12 remedies—Part III. *Emergency Medicine Journal, 34*(5): 33–39.

Wynn, S., and Fougere, B. (2006). *Veterinary Herbal Medicine.* St. Louis: Moseby.

APPENDIX A
PROPER USE OF NEEDLES AND SYRINGES

- Syringes are calibrated to allow precise measurement of medication. Determine if the calibrations represent metric units, apothecary units, or international units.

- For the greatest accuracy use the syringe volume that is closest to the volume of medicine to be delivered. For example, do not use a 60 cc syringe to measure 2 cc of fluid volume. The best choice in this case would be a 3 cc syringe.

- Disposable syringes are sterilized, prepackaged, nontoxic, nonpyrogenic, and ready for use.

- Needle size is inversely proportional to its number. For example, 16-gauge needles have large lumens, and 25-gauge needles have small lumens.

- Selection of needle size is based on many factors including the following:

  - The amount and viscosity of the medication. Thick, oily medications are given through needles with larger lumens; thinner medications are given through needles with smaller lumens.

  - The route of administration. Intradermal injections are given with small-lumen needles (like 25 gauge) and subcutaneous injections in large animals are given with large-lumen needles (like 16 or 18 gauge).

  - The species, size, and age of the animal.

- Selection of needle length is based on route of administration. To give a deep muscle injection, a needle of appropriate length must be used.

- Needles must be disposed of in a rigid, puncture-resistant container (Figure A-1).

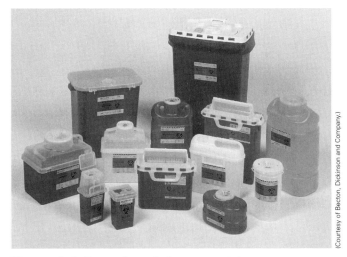

(Courtesy of Becton, Dickinson and Company.)

Figure A-1 Examples of sharps containers.

APPENDIX B
WITHDRAWING (ASPIRATING) MEDICATION FROM A VIAL

STANDARD PRECAUTIONS

Purpose

Medication is supplied in a variety of packaging.
Medication from a vial must be aspirated into a syringe for parenteral injection.

Equipment/Supplies

Vial of medication
Alcohol wipes
Appropriate sterile syringe and needle unit
Disposable gloves
Sharps container

Procedure Steps

1. Check the veterinarian's order and assemble equipment.
2. Wash hands. Apply gloves if needed.
3. Select the proper-sized sterile needle and syringe for the medication and administration route. If necessary, attach the needle to the syringe.
4. Obtain the vial of medication and check the label against the veterinarian's order. Check for correct medication, dose, route, and time. Check medication expiration dates.
5. Remove the metal or plastic cap from the vial. If the vial has been opened previously, clean the rubber stopper by applying an alcohol wipe in a circular motion (Figure B-1A).
6. Remove the needle cover—pull it straight off.
7. Inject air into the vial as follows:
   a. Hold the syringe pointed upward at eye level. Pull back the plunger to take in a quantity of air equal to the ordered dose of medication.
   b. Hold the vial upright according to personal preference. Take care not to touch the rubber stopper.
   c. Insert the needle through the rubber stopper of the vial. Inject the air by pushing in the plunger (Figure B-1B).
8. Withdraw the medication: Hold the vial and the syringe steady. Pull back on the plunger to withdraw the measured dose of medication. Measure accurately. Keep the tip of the needle below the surface of the liquid; otherwise, air will enter the syringe. Keep the syringe at eye level (Figure B-1C).

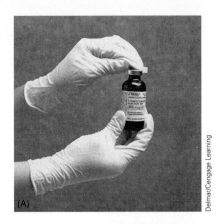
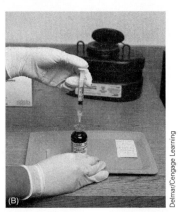

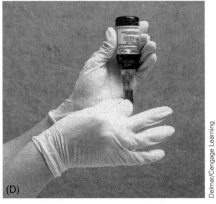
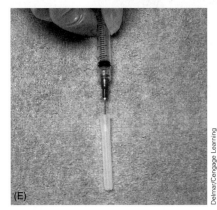

Figure B-1 (A) Disinfect the rubber stopper on the medication vial with an alcohol swab. (B) Keeping the bevel of the needle above the fluid level, inject an amount of air equal to the medication quantity to be withdrawn. (C) Hold the syringe pointed upward at eye level and keep the bevel of the needle in the medication. Pull back the plunger and aspirate the quantity of medication ordered. (D) Tap the syringe to eliminate air bubbles. Your hand should hold the syringe while you tap it. (E) After the correct dose has been withdrawn, recover the sterile needle. To safely recap needles, the "one-hand" technique can be used. First, place the cap on a flat surface and remove your hand from the cap. With one hand, hold the syringe and use the needle to "scoop up" the cap. When the cap covers the needle completely, use the other hand to secure the cap with the needle hub. Be careful to handle the cap at the bottom only (near the hub).

9. Check the syringe for air bubbles. Remove them by tapping sharply on the syringe (Figure B-1D).
10. Remove the needle from the vial. Replace the sterile needle cover using the "one-hand" technique described in Figure B-1E or by using a safety syringe that shields the point of the needle to prevent accidental needle sticks (refer to Figure 3-6).
11. Check the vial label against the veterinarian's order.
12. The dose is now ready for injection.
13. Return a multiple-dose vial to the proper storage area (cabinet or refrigerator). Dispose of unused medication in a single-dose vial according to facility procedure. (Remember, disposal of a controlled substance must be witnessed and the proper forms signed.)
14. Discard used syringe and needle unit immediately after use in a sharps container (Figure B-2).

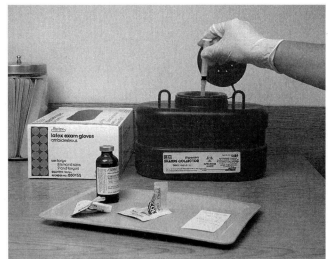

Figure B-2 Dispose of the used syringe-needle unit in a sharps container.

APPENDIX C
WITHDRAWING (ASPIRATING) MEDICATION FROM AN AMPULE

Purpose

Medication is supplied in a variety of packaging.

An ampule is a sterile, glass, single-dose container of liquid medication. It is aspirated into a syringe for parenteral injection.

Equipment/Supplies

Ampule of medication
Alcohol wipes
Sterile gauze sponges
Appropriate sterile syringe and needle unit
Disposable gloves
Sharps container

Procedure Steps

1. Check the veterinarian's order and assemble equipment.

2. Wash hands. Apply gloves if needed.

3. Obtain the ampule of medicine and check the label against the veterinarian's orders. Check for correct medication, dose, route, and time. Check medication expiration date.

4. Flick ampule of medication (medication will often get "trapped" above the neck of the ampule). A sharp flick of the wrist will help force all the medication down below the neck of the ampule into the body of the ampule (Figure C-1A). This is important to ensure that all medication is available in the body of the ampule. If some of the medication remains trapped above the neck in the top of the ampule, some medication will not be available for use and it is possible to give an incorrect dose, especially if the patient is to receive the entire contents of the ampule.

5. Thoroughly disinfect the ampule neck with an alcohol wipe. Since the needle will enter the opening of the ampule, wiping the neck of the ampule prior to removal of the top ensures disinfection of the neck or opening of the ampule. Check label (second time).

6. With a sterile gauze, wipe dry the neck of the ampule. Completely surround the ampule with the gauze and forcefully snap off the top of the ampule by pushing the top away from you (Figure C-1B). This ensures safety from possible injury from broken glass. Discard top in sharps container.

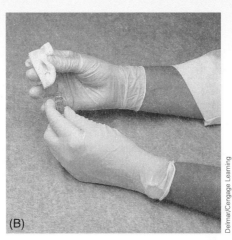
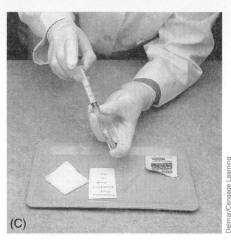

Figure C-1 (A) Hold ampule by the top and force all the medication into the bottom of the ampule by a snap of the arm and wrist. (B) Remove the top from the ampule with a sterile gauze wrapped around the neck of the ampule. Turn your hand up and out simultaneously. (C) Aspirate the required dose into the syringe. Flick away any air bubbles that cling to the side of the syringe prior to administration of the drug to the patient.

7. Place opened ampule down on medicine tray.

8. Select the proper-sized sterile syringe and needle unit for the medication and administration route. Aspirate the required dose into the syringe (Figure C-1C).

9. Discard used syringe and needle unit immediately after use in a sharps container.

APPENDIX D
RECONSTITUTING A POWDER MEDICATION FOR ADMINISTRATION

STANDARD PRECAUTIONS

Purpose

Drugs for injection may be supplied in a powdered (dry) form and must be reconstituted to a liquid for injection. A diluent (usually but not always sterile saline) is added to the powder, mixed well, and the appropriate dose drawn up to be administered.

Equipment/Supplies

Vial of medication
Diluent
Alcohol wipes
Two appropriately sized syringe and needle units
Disposable gloves
Sharps container

Procedure Steps

1. Check the veterinarian's order and assemble equipment.

2. Wash hands. Apply gloves if needed.

3. Select the proper-sized sterile syringe and needle unit for the medication and administration route before reconstituting powder medication (Figure D-1A).

4. Obtain vial of medicine and check the label against the veterinarian's orders. Check for correct medication, dose, route, and time. Check medication expiration date.

5. Remove tops from diluent and powder medication containers and wipe with alcohol wipes (Figure D-1B).

6. Insert the needle of a sterile syringe and needle unit through the rubber stopper on the vial of diluent that has been cleansed with an alcohol wipe. The syringe and needle unit should have an amount of air in it equal to the amount of diluent to be withdrawn (Figures D-1C and D).

7. Withdraw the appropriate amount of diluent to be added to the powder medication (Figures D-1E and F). Cover the sterile needle on the syringe containing the appropriate amount of diluent.

8. Add this liquid to the powder medication container that has been cleansed with an alcohol wipe (Figure D-1G).

9. Remove needle and syringe from vial with powder medication and diluent and discard them into a sharps container (Figure D-1H).

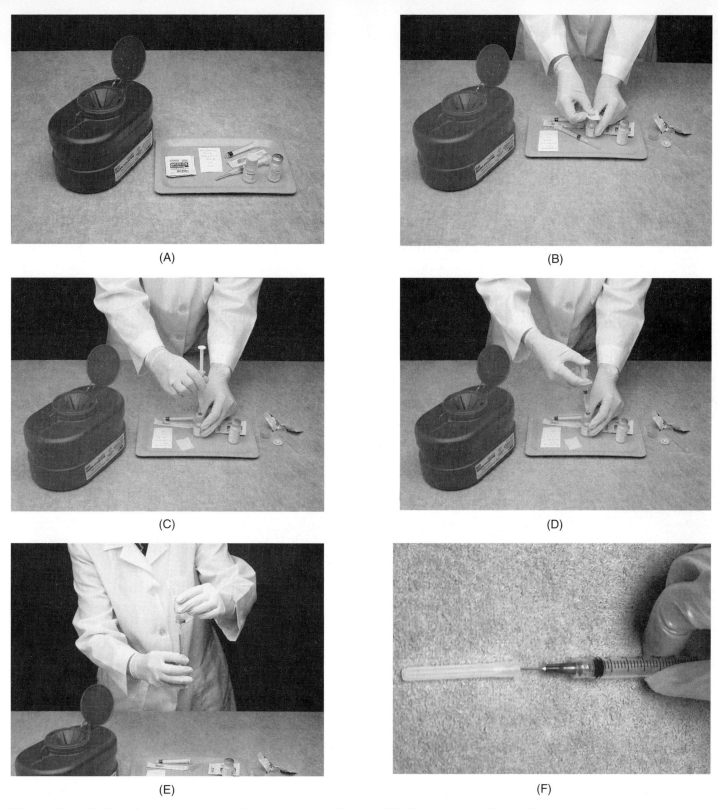

(A)

(B)

(C)

(D)

(E)

(F)

Figure D-1 (A) Supplies for reconstituting powder medication. (B) Remove tops from diluent and powdered medication containers. Wipe top of each with an alcohol wipe. (C) Prepare to inject air in an amount equal to the amount of diluent being removed from the diluent vial. (D) Inject air into the diluent vial. (E) Prepare to separate the vial from the syringe and needle unit after withdrawing diluent. (F) Cover the sterile needle on the syringe containing diluent, using the "one-hand" technique described in Appendix B.

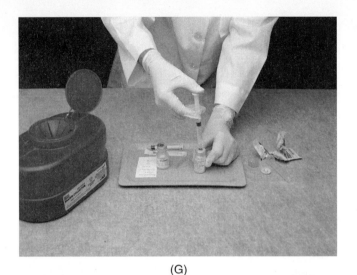

(G)

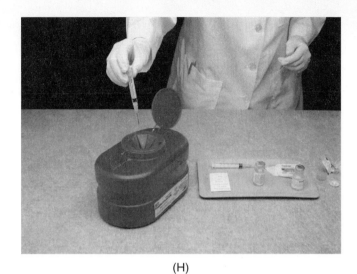

(H)

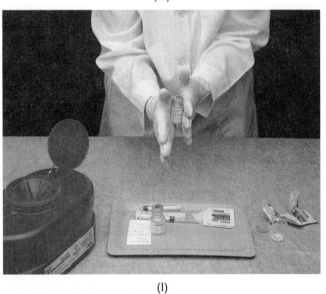

(I)

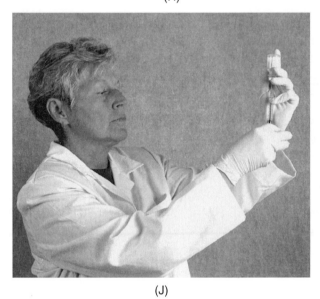

(J)

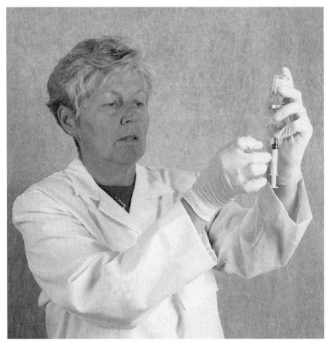

(K)

Figure D-1 (G) Inject diluent into the vial containing powdered medication. Before injecting, clean the top of the vial again with an alcohol wipe. (H) Discard the syringe needle unit used for diluent in a sharps container. (I) Roll the vial of powdered medication with diluent between the palms of your hands to mix it well. Label the vial with the date and time mixed, amount of diluent added, strength of dilution, your initials, and the expiration date. (J) Use a second sterile syringe and needle unit to draw up the prescribed dose of medication ordered by the veterinarian. (K) Flick away any air bubbles that cling to the side of the syringe prior to administration of the reconstituted drug to the patient.

10. Roll the vial between the palms of the hands to completely mix together the powder and diluent (Figure D-1I). Label the multiple-dose vial with the dilution or strength of the medication prepared, the date and time, your initials, and the expiration date.

11. With a second sterile syringe and needle, withdraw the desired amount of medication (Figure D-1J).

12. Flick away any air bubbles that cling to the side of the syringe prior to administering the reconstituted drug to the patient (Figure D-1K).

13. Return the multidose vial to the proper storage area (cabinet or refrigerator). Dispose of unused medication in a single-dose vial according to facility procedure. (Remember, disposal of a controlled substance must be witnessed and the proper forms signed.)

14. Discard used syringe and needle unit immediately after use in a sharps container.

APPENDIX E
The Dos and Don'ts of Drug Administration

Preparation

- Wash hands before preparing medication.
- Check for drug reactions in patient's record.
- Check medication order with veterinarian.
- Check label on drug container to make sure proper drug is being used.
- Check expiration date on drug label; use only if date is current.
- Verify drug dose.
- Recheck drug calculation with another veterinary professional.

Administration

- Check patient's record or hospital identification to ensure drug is being dispensed/administered to the proper patient.
- Dispense/administer only those drugs that you have prepared or have rechecked with another veterinary professional.
- Explain proper administration technique if dispensing drug to a client.
- Have animal in appropriate position depending on the route of drug administration.
- Use universal precautions when handling needles.
- Stay with the patient until all of the drug is administered or have a monitoring system in place for drugs given over a period of time.
- Discard needles and syringes in appropriate container.
- Discard unused drugs in appropriate manner; controlled substances must be recorded if they are discarded.
- Discard unused solutions or store unused stable solutions appropriately until their next use (refrigerator). Write date and time opened and your initials on the label.
- Keep controlled substances locked up and store key appropriately.

Recording

- Report drug error immediately to veterinarian and document error in the medical record.
- Record in the patient's chart the drug given, dosage, dose, time, route, and your initials or signature. Also record any drug reactions in the patients chart.
- Record controlled substance use in the controlled substance log.

THE DON'TS OF DRUG ADMINISTRATION

- Do not be distracted when preparing medications.

- Do not give drugs prepared by others without substantial communication with that veterinary professional.

- Do not use drugs from containers with labels that are difficult to read or whose labels are partially removed or have fallen off.

- Do not transfer drugs from one container to another.

- Do not pour drugs into your hands.

- Do not give drugs for which the expiration date has passed.

- Do not guess about drugs and drug dosages. Ask the veterinarian when in doubt.

- Do not leave unlabeled medication by the patient.

- Do not use drugs that have sediment, are discolored, or are cloudy (and should not be).

- Do not give drugs to patients who have had an allergic reaction to them.

- Do not give a drug to a patient if the client states that the drug looks different from the drug their animal has been receiving. Always check the order and the drug to make sure you are dispensing/administering the proper medication.

APPENDIX F
INVENTORY MANAGEMENT

WHY KEEP A DRUG INVENTORY?

- It determines what is physically available for use and what is on order.
- It is a source of manufacturers/distributors with addresses, telephone numbers, and representatives' names.
- It involves receiving, unpacking, listing, and pricing of drugs and supplies, which help veterinary technicians to be aware of drug use and cost trends.

TYPES OF INVENTORY MANAGEMENT

- Alphabetical: all drugs and supplies arranged from A to Z
- Classification: drugs are categorized by use
- Categoric: capsules/tablets in one area, injectables in another, and so on
- Numeric: each drug is assigned a number and stored accordingly

MONITORING CURRENT INVENTORY

- Should be done at least annually
- Computer programs provide instant monitoring of inventory
- Hard-copy methods include ledgers and inventory cards
- Should include the following:
  - item
  - supplier
  - expiration dates
  - cost per unit
  - quantity purchased
  - quantity on hand
  - date of current inventory

CONTROLLED SUBSTANCE INVENTORY

- Purchase records for schedule II–V drugs (schedule I drugs are not utilized in veterinary clinics) must be kept for at least two years.

- Schedule II drugs must be logged separately.

- Inventory follow-up must be done every two years.

- Storage areas must be locked (preferably double-locked) or in a safe attached to a concrete floor.

- Records must be easily retrievable.

RECEIVING SHIPMENTS

- Check packing slip to make sure that the correct items, quantity of items, size and strength of medication, and so on, were received. Make sure the package was sent to the correct address.

- Call the distributor immediately if there is a problem.

- Refuse shipment of damaged items. Remember to retain the shipping slip and a copy.

- Make sure the shipment was handled correctly (was not frozen, left in heat, etc.)

- Once the shipment has been checked, file the packing slip for back orders, individual companies, and so on.

- When unpacking, make sure to rotate stock so that items that will expire first are used first and any outdated items can be returned to the supplier for a possible credit.

- Check dating and make sure items can be used within the expiration dates (you may need to request longer dating if an item is not used frequently).

- Wait for back orders or cancel them and order through another company.

APPENDIX G
UNIT CONVERSIONS AND MATH REFERENCES

PREFIXES FOR METRIC UNITS
1,000,000 = mega- = M
1000 = kilo- = k
100 = hecto- = h
10 = deka- = dk
0.1 = deci- = d
0.01 = centi- = c
0.001 = milli- = m
0.000001 = micro- = μ or mc
0.000000001 = nano- = n
0.000000000001 = pico- = p

METRIC-TO-METRIC CONVERSIONS
Linear measure: base unit is meters (m)
  1 m = 100 centimeters (cm)
  1 m = 1000 millimeters (mm)
  1 m = 1,000,000 micrometers or microns
    (mcm or μ)
  1000 m = 1 kilometer (km) or 0.001 km = 1 m
Volume measure: base unit is liter (L)
  1 L = 100 centiliters (cL)
  1 L = 1000 milliliters (mL)
  1000 L = 1 kiloliter (kL)
Weight measure: base unit is gram (g)
  1 g = 100 centigrams (cg)
  1 g = 1000 milligrams (mg)
  1 g = 1,000,000 micrograms or 0.000001 g = 1
    microgram (mcg or μg)
  0.001 mg = 1 mcg or 1 μg (a gram is 1000
    times a milligram; a milligram is 1000
    times a microgram); 1 mg = 1000 mcg
    or μu
  1000 g = 1 kg

**METRIC, HOUSEHOLD, AND APOTHECARY
CONVERSIONS**
Length
  Metric base unit = meter
  1 meter = 1.0936 yards
  1 centimeter = 0.39370 inch
  1 inch = 2.54 centimeters
  1 kilometer = 0.62137 mile
  1 mile = 5280 feet or 1.6093 kilometers
  1 foot = 0.3048 meter

Mass
  Metric base unit = gram
  1 kilogram = 2.2 pounds
  1 pound = 453.59 grams
  1 pound = 16 ounces
  1 grain = 65 milligrams
  15 grains = 1 gram
  1 dram = 3.888 grams
  1 ounce = 28.35 grams
  1 ton = 2000 pounds
  1 gram = 0.035274 ounces

Volume
  Metric base unit = liter
  1 liter = 1.0567 quarts
  1 gallon = 4 quarts
  1 gallon = 8 pints
  1 pint = 2 cups = 16 fluid ounces
  1 cup = 8 ounces
  1 gallon = 3.7854 liters
  1 quart = 32 fluid ounces
  1 quart = 0.94633 liter
  1 minim = 0.06 milliliter
  1 fluid dram = 3.7 milliliters
  1 fluid ounce = approximately 30 milliliters
  1 milliliter = 1 cubic centimeter
  1 teaspoon = 5 milliliters

Temperature
  $°F = 1.8°C + 32$
  $°C = \dfrac{°F - 32}{1.8}$

DOSE CALCULATIONS
To calculate a dose:
1. Convert the animal's weight from pounds
   to kilograms (unless the dosage is given in
   pounds).
2. Multiply the animal's weight in kg by the
   dosage to get the animal's drug dose.
3. Divide the dose by the drug concentration,
   with the unit of drug to administer in
   the numerator (set up units so that they
   cancel).

APPENDIX H

CONVERSION OF BODY WEIGHT IN KILOGRAMS TO BODY SURFACE AREA IN METERS (FOR DOGS AND CATS)

DOGS

kg	m²	kg	m²
0.5	0.06	26.0	0.88
1.0	0.10	27.0	0.90
2.0	0.15	28.0	0.92
3.0	0.20	29.0	0.94
4.0	0.25	30.0	0.96
5.0	0.29	31.0	0.99
6.0	0.33	32.0	1.01
7.0	0.36	33.0	1.03
8.0	0.40	34.0	1.05
9.0	0.43	35.0	1.07
10.0	0.46	36.0	1.09
11.0	0.49	37.0	1.11
12.0	0.52	38.0	1.13
13.0	0.55	39.0	1.15
14.0	0.58	40.0	1.17
15.0	0.60	41.0	1.19S
16.0	0.63	42.0	1.21
17.0	0.66	43.0	1.23
18.0	0.69	44.0	1.25
19.0	0.71	45.0	1.26
20.0	0.74	46.0	1.28
21.0	0.76	47.0	1.30
22.0	0.78	48.0	1.32
23.0	0.81	49.0	1.34
24.0	0.83	50.0	1.36
25.0	0.85		

Although the above chart is compiled for dogs, it can also be used for cats. A formula for more precise values is

$$BSA \text{ in } m^2 = \frac{K \times W^{2/3}}{10^4}$$

BSA = Body surface area
m² = square meters
W = weight in grams
K = 10.1 (dogs)
        10.0 (cats)

Adapted from *Veterinary Values* (1998). (5th ed.). Lenexa, KS: Veterinary Healthcare Communications.

APPENDIX I
METRIC CONVERSIONS GUIDE

When doing metric conversions, use this table to make it easier.

3		2		1	
kg	g	cg	mg
kL	L	cL	mL
km	m	cm	mm

Each time you need to convert something to a larger or smaller measurement, move left or right on the table. When you cross a line with a number, move the decimal point that many places to the left or right, depending on your conversion. If you pass through more than one line, add up the numbers to know the number of places to move the decimal. For example, how many cm are there in 5 m? You move from m to cm crossing only the line with the number 2; hence, you add 2 decimal places or zeros in this case (in the direction of your move). How many km are in 20 mm? You move from mm to km crossing 3 lines (a line with the number 1, a line with the number 2, and a line with the number 3). Adding $1 + 2 + 3 = 6$. You move the decimal point 6 places to the left (the direction of your move), and your answer is 0.000020 km.

Be sure to validate your answer by making sure the larger metric unit has the smaller numerical value.

APPENDIX J
MATHEMATICS REVIEW

FRACTIONS AND DECIMALS

A fraction indicates a portion of a whole number. There are two types of fractions: common fractions (such as ½) and decimal fractions (such as 0.5). Common fractions are usually referred to as fractions, and decimal fractions are usually referred to as decimals.

Fractions

A fraction is an expression of division, with the denominator (bottom number) indicating the total number of equal-sized parts into which the whole is divided and the numerator (top number) indicating how many of those parts are being considered.

$\frac{1}{4}$ numerator
denominator

The whole is divided into four equal parts (denominator) and one part (numerator) is considered.

$\frac{1}{4}$ = 1 part of 4 parts, or $\frac{1}{4}$ of the whole.

The fraction $\frac{1}{4}$ may also be read as 1 divided by 4.

There are four types of fractions: proper, improper, mixed, and complex. Proper fractions have a numerator less than the value of the denominator giving it a value less than 1. Improper fractions have numerators greater than or equal to the value of the denominator giving it a value greater than or equal to 1. Mixed number fractions have a whole number and a proper fraction that are combined and always have a value greater than 1. Complex fractions have a numerator, denominator, or both that contain a decimal, fraction, or mixed number. Examples of the types of fractions are shown below.

Proper fraction	$\frac{1}{2}$
Improper fraction	$\frac{5}{4}$
Mixed numbers	$1\frac{1}{4}$
Complex fraction	$\frac{\frac{1}{2}}{\frac{2}{4}}$

To perform dose calculations, it may be necessary to convert among different types of fractions and reduce them to lowest terms. Simple rules of adding, subtracting, multiplying, and dividing fractions need to be understood so that working with fractions becomes automatic.

The value of a fraction can be expressed in many ways. Equivalent fractions have the same value; however, their numerator and denominator may be different because if both terms of the fraction (numerator and denominator) are multiplied or divided by the same nonzero number the form of the fraction is changed, but the value of the fraction remains the same. For example, ½ and ¾ and ⅜ all have a value of half of the whole; however, the numerators and denominators are different.

When calculating doses, it is easier to work with fractions using the smallest possible numbers by reducing the fraction to its lowest terms or simplifying the fraction. To reduce a fraction to lowest terms, divide both the numerator and denominator by the largest nonzero whole number that will go evenly into both the numerator and denominator. If both the numerator and denominator cannot be divided evenly by a nonzero whole number other than one, the fraction is already in lowest terms.

Reduce $\dfrac{6}{8}$ to lowest terms.

2 is the largest number that will divide evenly into both 6 (numerator) and 8 (denominator).

$$\frac{6}{8} \;=\; \frac{6 \div 2}{8 \div 2} \;=\; \frac{3}{4} \quad \text{in lowest terms.}$$

Sometimes this reduction takes several steps. Always check a fraction to see if it can be reduced further.

$$\frac{1,000}{2,000} \;=\; \frac{1,000 \div 500}{2,000 \div 500} \;=\; \frac{2}{4} \quad \text{(not in lowest terms)}$$

$$\frac{2}{4} \;=\; \frac{2 \div 2}{4 \div 2} \;=\; \frac{1}{2} \quad \text{(in lowest terms)}$$

To find an equivalent fraction in which both the numerator and denominator are larger, multiply both the numerator and denominator by the same nonzero number.

Enlarge $\dfrac{1}{5}$ to the equivalent fraction in tenths.

$$\frac{1}{5} \;=\; \frac{1 \times 2}{5 \times 2} \;=\; \frac{2}{10}$$

Converting Fractions

Converting fractions allows calculations to be performed with greater ease and permits the expression of answers in simplest terms. To convert a mixed number to an improper fraction with the same denominator, multiply the whole number by the denominator, and add the numerator. Place that value in the numerator, and use the denominator of the fraction part of the mixed number.

$$3\frac{1}{2} \;=\; \frac{(3 \times 2) + 1}{2} \;=\; \frac{6 + 1}{2} \;=\; \frac{7}{2}$$

To convert an improper fraction to an equivalent mixed number or whole number, divide the numerator by the denominator. Any remainder becomes the numerator of a proper fraction that should be reduced to lowest terms.

$$\frac{10}{4} \;=\; 10 \div 4 \;=\; 2\frac{2}{4} \;=\; 2\frac{1}{2}$$

Comparing Fractions

When calculating drug doses, it helps to know when the value of one fraction is greater or less than another. The relative sizes of fractions can be determined by comparing the numerators when the denominators are the same or comparing the denominators if the numerators are the same. If the denominators are the same, the fraction with the larger numerator has the greater value.

Compare $\frac{3}{5}$ and $\frac{2}{5}$

Denominators are both 5
Numerators: 3 is greater than 2

$\frac{3}{5}$ has a greater value

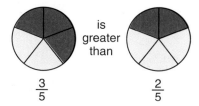

$\frac{3}{5}$ is greater than $\frac{2}{5}$

If the numerators are the same, the fraction with the smaller denominator has the greatest value. A smaller denominator means it has been divided into fewer pieces, so each one is larger.

Compare $\frac{1}{2}$ and $\frac{1}{4}$

Numerators are both 1
Denominators: 2 is less than 4

$\frac{1}{2}$ has a greater value

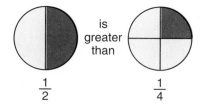

$\frac{1}{2}$ is greater than $\frac{1}{4}$

Addition and Subtraction of Fractions

To add or subtract fractions, all the denominators must be the same. To determine the least common denominator, find the smallest whole number into which all denominators will divide evenly. Once the least common denominator is determined, convert the fractions to equivalent fractions with the least common denominator.

Find the equivalent fractions with the least common denominator for $\frac{3}{8}$ and $\frac{1}{3}$

Find the smallest whole number into which the denominators 8 and 3 will divide evenly. In this case, the least common denominator is 24. Convert the fractions to equivalent fractions with 24 as the denominator.

$$\frac{3}{8} = \frac{3 \times 3}{8 \times 3} = \frac{9}{24} \qquad \frac{1}{3} = \frac{1 \times 8}{3 \times 8} = \frac{8}{24}$$

The fraction $\frac{3}{8}$ was enlarged to $\frac{9}{24}$ and the fraction $\frac{1}{3}$ was enlarged to $\frac{8}{24}$. Both fractions have the same denominator.

Add or subtract the numerators, place that value in the numerator, and use the least common denominator as the denominator. Convert to a mixed number and/or reduce the fraction to the lowest term, if possible.

$$\frac{1}{4} + \frac{3}{4} + \frac{2}{4}$$

Find the common denominator. This fraction already has a common denominator.

Add the numerators and use the common denominator:

$$\frac{1+3+2}{4} = \frac{6}{4}$$

Convert to a mixed number and reduce to lowest terms:

$$\frac{6}{4} = 1\frac{2}{4} = 1\frac{1}{2}$$

$$1\frac{1}{10} - \frac{3}{5}$$

Find the least common denominator: 10. The number 10 is the smallest number that both 10 and 5 will equally divide into.

Convert to equivalent fractions in tenths.

$$1\frac{1}{10} = \frac{11}{10}$$

$$\frac{3}{5} = \frac{3 \times 2}{5 \times 2} = \frac{6}{10}$$

Subtract the numerators, and use the common denominator:

$$\frac{11-6}{10} = \frac{5}{10}$$

Reduce to lowest terms:

$$\frac{5}{10} = \frac{1}{2}$$

Multiplying and Dividing Fractions

To multiply fractions, multiply numerators and denominators to arrive at the product (answer). Cancellation of terms simplifies and shortens the process of multiplying fractions. Cancellation is based on the fact that the division of both the numerator and denominator by the same nonzero whole number does not change the value of the resulting number.

$$\frac{1}{8} \times \frac{8}{9} = \frac{1}{\cancel{8}_{1}} \times \frac{\cancel{8}^{1}}{9} = \frac{1}{1} \times \frac{1}{9} = \frac{1}{9}$$

When multiplying a fraction by a nonzero whole number, first convert the whole number to a fraction with a denominator of 1; the value of the number remains the same.

$$\frac{2}{3} \times 4$$

No terms to cancel.
Convert the whole number to a fraction.

$$\frac{2}{3} \times 4 = \frac{2}{3} \times \frac{4}{1}$$

Multiply numerators and denominators:

$$\frac{2}{3} \times \frac{4}{1} = \frac{8}{3}$$

Convert to a mixed number:

$$\frac{8}{3} = 8 \div 3 = 2\frac{2}{3}$$

To multiply mixed numbers, first convert them to improper fractions, and then multiply.

$$3\frac{1}{2} \times 4\frac{1}{3}$$

Convert:

$$3\frac{1}{2} = \frac{7}{2}$$

$$4\frac{1}{3} = \frac{13}{3}$$

$$\frac{7}{2} \times \frac{13}{3}$$

Cancel if possible. In this case, no numbers can be cancelled.

Multiply: $\frac{7}{2} \times \frac{13}{3} = \frac{91}{6}$

Convert to a mixed number:

$$\frac{91}{6} = 15\frac{1}{6}$$

When dividing fractions, the divisor (number to the right of the division sign) is inverted, and the operation is changed to multiplication. Once inverted, cancellation is done (if possible), the dividend (fraction being divided or the first number) is multiplied by the inverted divisor, and the quotient (answer) is obtained.

$$\frac{1}{4} \div \frac{2}{7} = \frac{1}{4} \times \frac{7}{2} = \frac{7}{8}$$

Dividend   Divisor   Changed to multiplication   Inverted divisor   Quotient

To divide mixed numbers, first convert them to improper fractions.

$$1\frac{1}{2} \div \frac{3}{4}$$

Convert: $\frac{3}{2} \div \frac{3}{4}$

Invert divisor and change ÷ to x: $\frac{3}{2} \times \frac{4}{3}$

Cancel: $\frac{3}{2} \times \frac{4}{3} = \frac{1}{1} \times \frac{2}{1}$

Multiply: $\frac{1}{1} \times \frac{2}{1} = \frac{2}{1}$

Simplify: $\frac{2}{1} = 2$

Decimals

Decimal fractions are fractions with a denominator of 10, 100, 1,000, or any power of 10. Decimals appear to be whole numbers because of the way they are written, but they always have a numeric value of less than 1. The words for all decimal fractions end in th(s) such as tenths, hundredths, and thousandths. When decimals are read, they may or may not be read with the starting zero (the zero to the left of the decimal point), but is written with the zero to emphasize the decimal point.

$$0.1 = \frac{1}{10}$$

$$0.01 = \frac{1}{100}$$

$$0.001 = \frac{1}{1,000}$$

Decimal numbers are numeric values that include a whole number, a decimal point, and a decimal fraction. Examples of decimal numbers are 5.24 and 43.762. When working with decimals, always consider the decimal point as the center that separates whole and fractional amounts. The position of the numbers in relation to the decimal point indicates the place value of the numbers.

Decimal numbers are read by stating the whole number first, the decimal point as and, and the decimal fraction by naming the value of the last decimal place. For example, 5.227 is read as five and two hundred twenty-seven thousandths.

Comparing Decimals

To compare decimal fractions, the value is determined by its position to the right of the decimal point. Zeros added after the last digit of a decimal fraction do not change its value, but demonstrate an increased level of precision of the value being reported (therefore, should only be used to indicate increased precision levels).

17.3 = seventeen and three tenths.

17.03 = seventeen and three hundredths.

When comparing decimal fractions, consider the value in the tenths position first, then the value in the hundredths position second, etc.

In this case, 17.3 is greater than 17.03.

A common error when comparing decimals is to overlook the decimal place values and misinterpret higher numbers for greater amounts and lower numbers for lesser amounts. To accurately compare decimal amounts, align the decimal points and add zeros so that the numbers being compared have the same number of decimal places. Adding zeros to the end of a decimal fraction for purposes of comparison does not change the original value.

A dog is prescribed 0.5 mg of a drug.
The recommended maximum drug dose
is 0.25 mg and the minimum recommended
dose is 0.125 mg. Comparing decimals, it
is clear that the drug dose ordered (0.5 mg)
is not within the recommended range.

0.125 mg (recommended minimum dose)
0.250 mg (recommended maximum dose)
0.500 mg (ordered dose)

The dose ordered is actually twice the
recommended maximum dose.

Adding and Subtracting Decimals

Adding and subtracting decimals is similar to adding and subtracting whole numbers. It is important to line up decimal points when adding or subtracting decimals. Zeros may need to be added to the end of decimal fractions to make all decimal numbers of equal length.

$$1.25 \ + \ 1.75 \ = \ \begin{array}{r} 1.25 \\ + \ 1.75 \\ \hline 3.00 = 3 \end{array}$$

Add 0.9, 0.65, 0.27, 4.712

$$\begin{array}{r} 0.900 \\ 0.650 \\ 0.270 \\ + \ 4.712 \\ \hline 6.532 \end{array}$$

$$0.725 \ - \ 0.5 \ = \ \begin{array}{r} 0.725 \\ - \ 0.500 \\ \hline 0.225 \end{array}$$

$$12.5 \ - \ 1.5 \ = \ \begin{array}{r} 12.5 \\ - \ 1.5 \\ \hline 11.0 = 11 \end{array}$$

Multiplying and Dividing Decimals

Multiplying and dividing decimals is similar to multiplying and dividing whole numbers. It is important to line up the decimal points when multiplying or dividing decimals. To multiply decimals, multiply without concern for decimal point placement. Count off the total number of decimal places in both of the decimals multiplied. Move the decimal point in the product (answer) by moving it to the left of the number of places counted.

$$1.72 \ x \ 0.9 \ = \ \begin{array}{r} 1.72 \\ x \quad 0.9 \\ \hline 1.548 \end{array}$$
(2 decimal places)
(1 decimal place)
(The decimal point is located
3 places to the left because
a total of 3 decimal places
are counted)

When multiplying by a decimal power of 10, move the decimal point as many places to the right as there are zeros in the multiplier.

0.001 x 1,000

The multiplier 1,000 has 3 zeros; move
the decimal point 3 places to the right.

0.001 x 1,000 = 0.001. = 1

When dividing decimals, the problem should be set up the same way as for division of whole numbers. To divide decimals, move the decimal point in the divisor (number divided by) and the dividend (number divided) the number of places needed to make the divisor a whole number. Then place the decimal point in the quotient (answer) above the new decimal point place in the dividend.

$$
\begin{array}{r}
40.3 \text{ (quotient)} \\
2.5\overline{)100.75} \\
\underline{100} \\
07 \\
\underline{00} \\
75 \\
\underline{75}
\end{array}
$$

100.75 ÷ 2.5 =
(dividend)   (divisor)

When dividing a decimal by a power of 10, move the decimal point to the left as many places as there are zeros in the divisor.

0.65 ÷ 10

The divisor 10 has 1 zero; move the decimal point 1 place to the left.

0.65 ÷ 10 = .0.65 = 0.065

The zero placed to the left of the decimal to avoid confusion and to emphasize that this is a decimal.

Rounding Decimals

For dose calculations, it will be necessary to compute decimal calculations to thousandths (3 decimal places) and round to hundredths (2 decimal places) or tenths (1 decimal place) for the final answer. To round a decimal to hundredths, drop the number in the thousandths place and do not change the number in the hundredths place if the number in thousandths place is 4 or less. Increase the number in the hundredths place by 1 if the number in the thousandths place is 5 or more. To round a decimal to tenths, drop the number in the hundredths place and do not change the number in the tenths place if the number in hundredths place is 4 or less. Increase the number in the tenths place by 1 if the number in the hundredths place is 5 or more.

Tenths
Hundredths
Thousandths

\ | /

All rounded to hundredths (2 decimal places)

0.123 = 0.12
5.325 = 5.33
0.30 = 0.3

Tenths
Hundredths

| /

All rounded to tenths (1 decimal place)

0.13 = 0.1
0.75 = 0.8
0.95 = 1.0

Conversion Between Fractions and Decimals

For doing dose calculations, fractions may need to be converted to decimals and decimals may need to be converted to fractions. To convert a fraction to a decimal, divide the numerator by the denominator.

$$\frac{1}{4} = 4\overline{)1.00} = 0.25$$
$$\frac{8}{}$$
$$\frac{20}{20}$$

To convert a decimal to a fraction, express the decimal number as a whole number in the numerator of the fraction. Then express the denominator of the fraction as the number 1 followed by as many zeros as there are places to the right of the decimal point. Reduce the resulting fraction to lowest terms.

Convert 0.125 to a fraction.

Numerator: 125

Denominator: 1 followed by 3 zeros = 1,000

Reduce: $\frac{125}{1,000} = \frac{1}{8}$

FRACTION	DECIMAL FRACTIONS	DESCRIPTION
Table Appendix O:	Fractions and Their Related Decimal Fractions	
6/10	0.6	Because 10 has 1 zero, the decimal point of 6 is moved to the left **once.**
24/100	0.24	Because 100 has 2 zeros, the decimal point of 24 is moved to the left **twice.**
445/1000	0.445	Because 1000 has 3 zeros, the decimal point of 445 is moved to the left **three** places.

RATIOS AND PERCENTS

Veterinary professionals need to understand ratios and percents to be able to accurately interpret, prepare, and administer a variety of medications and treatments.

Ratios

Like a fraction, a ratio is used to indicate the relationship of one part of a quantity to the whole. The two quantities are written as a fraction or separated by a colon. The terms of a ratio are the numerator (always to the left of the colon) and the denominator (always to the right of the colon) of a fraction. Like fractions, ratios should be stated in lowest terms.

In a boarding facility, 5 of the boarders are dogs.
If there are 35 total animals in the boarding facility,
what is the ratio of dogs to all boarders?

5 dogs to 35 boarders = 5 dogs per 35 boarders = $\frac{5}{35} = \frac{1}{7}$

This is the same as a ratio of 5:35 or 1:7

Percents

A type of ratio is a percent. A percent is per hundred parts.

$$5\% = 5 \text{ percent} = 5/100 = \frac{5}{100} = 0.05$$

Converting among Ratios, Percents, Fractions, and Decimals

Once ratios, percents, fractions, and decimals are understood, converting from one to the other can be done. To convert a percent to a fraction, delete the % sign. Then write the remaining number as the numerator. Write 100 as the denominator and reduce the answer to lowest terms.

$$10\% = \frac{10}{100} = \frac{1}{10}$$

To convert a percent to a ratio, delete the % sign. Then write the remaining number as the numerator. Write 100 as the denominator and reduce the answer to lowest terms. Then express the fraction as a ratio.

$$25\% = \frac{25}{100} = \frac{1}{4} = 1{:}4$$

To convert a percent to a decimal, delete the % sign. Then divide the remaining number by 100, which is the same as moving the decimal point two places to the left.

$$25\% = \frac{25}{100} = 25 \div 100 = .25. = 0.25$$

To convert a decimal to a percent, multiply the decimal number by 100, which is the same as moving the decimal point two places to the right. Then add the % sign. When converting a decimal to a percent, always move the decimal point two places so that the resulting percent is the larger number.

$$0.25 \times 100 = 0.25. = 25\%$$

To convert a ratio to a percent, convert the ratio to a fraction. Then convert the fraction to a decimal. Lastly, convert the decimal to a percent.

Convert 1:1,000 epinephrine solution to the equivalent concentration expressed as a percent.

$$1{:}1{,}000 = \frac{1}{1{,}000} \quad \text{(ratio converted to fraction)}$$

$$\frac{1}{1{,}000} = .001. = 0.001 \quad \text{(fraction converted to decimal)}$$

$$0.001 = 0.00.1 = 0.1\% \quad \text{(decimal converted to percent)}$$

Thus 1:1,000 epinephrine solution = 0.1% epinephrine solution

Comparing Percents and Ratios

Veterinary professionals frequently administer solutions with the concentration expressed as a percent or ratio. If there are two solutions to be given intravenously, it is important for the veterinary technician to know that the 0.9% solution is less than a 5% solution. A 0.9% solution means that there are 0.9 parts of solid per 100 total parts (0.9 parts is less than one whole part, so it is less than 1%). A 5% solution has 5 parts of solid per 100 total parts (or more than five times 0.9 parts). When comparing these two solution, it is determined that 5% solution is much more concentrated than the 0.9% solution.

To compare a solution concentration expressed as a fraction percent to a solution expressed as a decimal percent, one of the solutions must be converted to the other form to clarify values and to allow easier comparison of the concentrations.

Compare a $\frac{1}{3}$ % solution to a 0.45% solution.

$$\frac{1}{3}\% = \frac{\frac{1}{3}}{100} = \frac{1}{3} \div \frac{100}{1} = \frac{1}{3} \times \frac{1}{100} = \frac{1}{300} = 0.003\bar{3}$$

$$0.45\% = \frac{0.45}{100} = 0.0045$$

0.0045 is greater than $0.003\bar{3}$

To compare solutions expressed as ratios, they can also be converted to decimal form to clarify values and to allow easier comparison of the concentrations.

Compare a 1:1,000 solution to a 1:100 solution.

$$1:1,000 = \frac{1}{1,000} = 0.001$$

$$1:100 = \frac{1}{100} = 0.010 \text{ (zero added for comparison)}$$

0.010 is greater than 0.001

SOLVING SIMPLE EQUATIONS FOR X

One way to set up dose calculations is in the simple equation form. Typically answers should be expressed in decimal form because those are used most frequently in dose calculations and administration. Keep in mind when setting up simple equations that the unknown quantity is represented by X, the number 1 can be dropped in front of the X because any number multiplied by 1 is the same number, and dividing a number by 1 does not change the value of X.

$\frac{100}{200} \times 1 = X$ is the same as $\frac{100}{200} = X$

Reducing to lowest terms: $\frac{100}{200} = X = \frac{1}{2} = X$

Convert to decimal form: $\frac{1}{2} = 0.5$

Answer is $0.5 = X$

$\frac{3}{5} \times 2 = X$

Convert: express 2 as a fraction: $\frac{3}{5} \times \frac{2}{1} = X$

Multiply fractions: $\frac{3}{5} \times \frac{2}{1} = \frac{6}{5} = X$

Convert to a mixed number: $\frac{6}{5} = 1\frac{1}{5} = X$

Convert to decimal form: $1\frac{1}{5} = 1.2$

Answer is $1.2 = X$

RATIO-PROPORTION: CROSS MULTIPLYING TO SOLVE FOR X

A proportion is two ratios that are equal or an equation between two equal ratios. A proportion is written as two ratios separated by an equal sign, such as 5:10 = 10:20. Some calculations have the unknown X as a different term in the equation. To determine the value of the unknown X, it may be necessary to cross multiply in a proportion. In a proportion, the product of the means (the two inside numbers) equals the product of the extremes (the two outside numbers). Finding the product of the means and the extremes is cross multiplying.

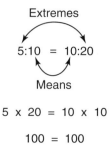

Extremes

5:10 = 10:20

Means

$$5 \times 20 = 10 \times 10$$

$$100 = 100$$

Since ratios are the same as fractions, the same proportion can be expressed as 5/10 = 10/20. The fractions are equivalent or equal. The numerator of the first fraction and the denominator of the second fraction are the extremes, and the denominator of the first fraction and the numerator of the second fraction are the means. If two fractions are equivalent, their cross products are also equal.

$$\text{Extreme } \frac{5}{10} \text{ Mean} \qquad \frac{10}{20} \text{ Extreme}$$

Mean 10   Extreme 20

Cross-multiply to find the equal products
of the means and extremes.

When one of the quantities in a proportion is unknown, a letter may be substituted for this unknown quantity. Solving the equation will find the value of the letter. Dividing or multiplying each side of an equation by the same nonzero number produces an equivalent equation. Dividing each side of an equation by the same nonzero number is the same as reducing or simplifying the equation. Multiplying each side by the same nonzero number enlarges the equation.

$$25X = 100$$

Simplify the equation to find X.

Divide both sides by 25, the number before X.

$$\frac{\overset{1}{\cancel{25}}X}{\underset{1}{\cancel{25}}} = \frac{\overset{4}{\cancel{100}}}{\underset{1}{\cancel{25}}}$$

$$\frac{1X}{1} = \frac{4}{1}$$ (Dividing or multiplying a number by 1 does not change its value. 1X is understood to be X)

$$X = 4$$

Replace X with 4 in the same equation to check your work.

$$25 \times 4 = 100$$

$$\frac{X}{160} = \frac{2.5}{80}$$

Cross-multiply: $\dfrac{X}{160} \diagup\diagdown \dfrac{2.5}{80}$

$$80 \times X = 2.5 \times 160$$

$$80X = 400$$

Simplify: $$\frac{\overset{1}{\cancel{80}}X}{\underset{1}{\cancel{80}}} = \frac{\overset{5}{\cancel{400}}}{\underset{1}{\cancel{80}}}$$

$$X = 5$$

FINDING THE PERCENTAGE OF A QUANTITY

Veterinary professionals also need to be able to find a given percentage or part of a quantity. Percentage describes a part of a whole quantity. In other words, the percentage is equal to some known percent multiplied by the whole quantity. To find a percentage of a whole quantity, change the percent to a decimal. Then multiply the decimal by the whole quantity. Remember that the word times means multiply.

An adult dog weighing 25 lb (11.37 kg) needs approximately 570 mL of daily maintenance fluids. The veterinarian asks to be told when the dog has received 75% of its daily maintenance fluids. Determine 75% of this dog's maintenance fluids.

Percentage = Percent x whole quantity

Let X represent the unknown.

Change 75% to a decimal: $75\% = \dfrac{75}{100} = .75. = 0.75$

Multiply 0.75 x 570 mL: X = 0.75 x 570 mL = 427.5 mL

Therefore, 75% of 570 mL is 427.5 mL

APPENDIX K
EUTHANASIA PROCEDURE

Once the decision has been made by the client to euthanize his/her animal, additional decisions need to be made regarding the euthanasia procedure itself. To make the decisions easier for clients, veterinary professionals need to help with the following:

- Where and when should the euthanasia take place?

- Who will be present during the euthanasia?

- How will the body be disposed of after euthanasia, or will ashes be returned to the client?

- Will a necropsy of the animal be performed?

- How will payment for services be satisfied?

- How will the client get home following the procedure?

Once these decisions have been made, the euthanasia procedure should be done professionally and with compassion for the animal and client. The following list is a set of recommended guidelines to help staff provide the type of environment needed for clients during this time:

- Schedule appointment times to allow additional time before and after the euthanasia procedure to ensure the client does not feel rushed during this difficult time. A more private examination room should be used to make sure that the owner does not feel other clients are intruding on his or her space and feels there is enough time and space for adequate grieving. Facial tissues and brochures on grieving should be available in these rooms.

- Make sure that the owner can tell that the animal is comfortable and cared for. One way to do this is by covering the examination table with a towel or blanket or having the owner bring the animal's favorite blanket to use during the procedure. Remember animals may lose control of their bladder or bowel during euthanasia, so this should be taken into account when choosing the covering and explained to the client.

- Determine whether the animal needs to be tranquilized prior to euthanasia to counteract aggression or extreme apprehension.

- Catheterize a peripheral vein to ensure smooth delivery of the euthanasia solution. To avoid undue client stress, this can be done by taking the animal out of the examination room and returning to the room when the catheter is in place. The catheter may be placed in an easily accessible vein or in a vein in the hind leg to allow the owner to hold the animal or pet its head during the procedure. Once the catheter is in place, the owner should be allowed to spend private time with the animal prior to administration of euthanasia solution.

- Prior to administration of euthanasia solution, some veterinary professionals will inject saline to ensure the catheter's patency. Injection of an anesthetic, such as an ultrashort-acting barbiturate, can minimize the animal's apprehension or excitement prior to injection of the euthanasia solution.

- Advise the owner of how euthanasia drugs work by informing them that barbiturate drugs are central nervous system depressors. When euthanasia solutions that contain large amounts of barbiturates are given, unconsciousness occurs first, then breathing stops due to respiratory depression, and finally the heart stops due to cardiac arrest. The animal may display some emotional behaviors such as fear (struggling and vocalization) due to cerebral cortex involvement. Occasionally, the animal may take a deep breath prior to respiratory failure. Clients should be advised that these things may happen so that they may prepare themselves in case these events happen.

- Assure the client that the patient is dead by ausculting the thorax with a stethoscope and shining a pen light into the animal's eyes before pronouncing that the patient is dead. If the client chose not to be present in the room, allow them to view the animal's body after expiration. Make the animal's body as presentable as possible prior to client viewing by cleaning any blood from the fur, placing the tongue in the mouth, closing the eyes, and removing any bandages or tape. Remember, this is the last (and lasting) impression the client has of his/her animal.

- Offer the client a list of grieving resources so that they can contact these associations when they feel the need to do so (see the following list). Tell the client that they can call and talk to the veterinary staff regarding the euthanasia of their animal if needed.

GRIEF SUPPORT WEB SITES

American Veterinary Medical Association at www.avma.org (under care for pets)

Pet Bereavement Counseling at www.petloss.org

Pet Loss Support Page at www.pet-loss.net

Doctors Foster and Smith Pet Education: Grief and the Loss of a Pet at
    www.peteducation.com/article.cfm?c=0+1278+1494&aid=635

Pet Loss-A Reference to References at www.superdog.com/coping-.html

APPENDIX L
VITAMIN SUPPLEMENTS

WATER-SOLUBLE VITAMINS

VITAMIN	FUNCTION	EXAMPLE	ADMINISTRATION ROUTE
vitamin C	• antioxidant • needed for collagen formation • maintains capillaries, bones, and teeth	• ascorbic acid parenteral injection • OTC brands	IM, SQ, IV, oral
thiamine (vitamin B$_1$)	• part of coenzyme needed for cellular respiration • helps support nervous system	• Vita-Jec® Thiamine • generic • part of B complex • OTC brands	IM, SQ, IV, oral
riboflavin (vitamin B$_2$)	• part of coenzymes needed for cellular respiration • involved in oxidation of fat and protein	• part of B complex product • OTC brands	IM, SQ, IV, oral
niacin (nicotinic acid)	• part of coenzymes needed for cellular respiration • involved in oxidation of fat and protein	• part of B complex product • OTC brands	IM, SQ, IV, oral
folacin (folic acid)	• coenzyme needed for production of hemoglobin • helps in formation of DNA	• part of B complex product • OTC brands	IM, SQ, IV, oral
vitamin B$_6$	• coenzyme used for synthesis of hormones and hemoglobin • helps with CNS control	• part of B complex product • OTC brands	IM, SQ, IV, oral
pantothenic acid (d-panthenol)	• part of coenzyme needed for oxidation of carbohydrates and fats • aids in formation of hormones and some neurotransmitters	• part of B complex product • OTC brands	IM, SQ, IV, oral
vitamin B$_{12}$ (cyanocobalamin)	• part of coenzyme needed for synthesis of nucleic acids and myelin • involved in RBC production	• vitamin B$_{12}$ injection • part of B complex product • OTC brands	IM, SQ, IV, oral
biotin	• coenzyme needed for amino acid and fatty acid metabolism	• part of B complex product • OTC brands	IM, SQ, IV, oral
choline	• part of coenzyme needed for metabolism	• part of B complex product • OTC brands	IM, SQ, IV, oral

FAT-SOLUBLE VITAMINS

VITAMIN	FUNCTION	EXAMPLE	ADMINISTRATION ROUTE
vitamin A	• antioxidant synthesized from beta-carotene • needed for healthy eyes, skin, hair, mucous membranes, and bone growth	• A-D Injection® • Vitamin AD Injection® • OTC brands	IM, SQ, oral
vitamin D	• a group of steroid vitamins needed for development and maintenance of bone and teeth	• A-D Injection® • Vitamin AD Injection® • OTC brands	IM, SQ, oral
vitamin E	• antioxidant that prevents oxidation of vitamin A and polyunsaturated fatty acids • prevents white muscle disease in sheep and calves	• vitamin E/selenium combinations (Bo-Se®, L-Se®, Mu-Se®, E-Se®, Seletoc®) • Vitamin E injectable • OTC brands	IM, SQ, oral
vitamin K	• needed for synthesis of blood clotting • also used in the treatment of rodenticide poisoning	• AquaMEPHYTON®, Mephyton®, Konakion® (all of these products are vitamin K_1)	IM, SQ, IV, oral

APPENDIX M
MANAGING ANIMAL TOXICITIES

I: INITIATE LIFE SUPPORT MEASURES

Life support measures are aimed toward maintaining normal respiratory, cardiovascular, and neurologic function.

CLINICAL SIGN	DRUG
bradycardia	atropine
CNS depression	naloxone
hyperactivity	diazepam
metabolic acidosis	sodium bicarbonate
seizures	diazepam, phenobarbital

II: DECONTAMINATION

The goal of any decontamination process is to reduce further exposure of the animal to the toxicant. Animals must be continually monitored even after decontamination to make sure they do not develop signs of toxicity from other routes of exposure. The following table lists ways to decontaminate animals.

APPLICATION	DRUG	COMMENTS
Adsorbents	• activated charcoal	Used with recent ingestions to adsorb the toxins from the body. It is used at a 10:1 activated charcoal-to-toxicant ratio. It is contraindicated with caustic agents.
Bathing	• detergent or shampoo	Removes most dermal exposures.
Cathartics	• sorbitol • sodium sulfate • magnesium sulfate	Causes diarrhea to help move toxicants out of the body. May be combined with activated charcoal.
Emetics	• syrup of ipecac • 3% hydrogen peroxide • apomorphine • xylazine • salt and water	Good for recent exposures. Contraindicated with corrosive agents, in animals with altered mental status, and in rabbits/rodents.

III: CHELATION THERAPY

Chelators have the ability to form chelator-metal complexes that enhance the elimination of metal. Chelators work best soon after metal exposure. Any possible metal source should be removed to avoid further contamination.

CHELATOR	METAL(S) CHELATED	COMMENTS
EDTA	• lead • zinc	Given in 5% dextrose; dosage may have to be adjusted based on renal function. Multiple treatments may be needed, but should not exceed 5 days.

CHELATOR	METAL(S) CHELATED	COMMENTS
deferoxamine mesylate	• iron	Usually supplemented by gastric lavage with magnesium hydroxide.
DMSA (2,3-dimercaptosuccinic acid)	• arsenic • lead • mercury	May reduce gastrointestinal absorption of lead, with limited side effects.
diphenylthiocarbazone	• thallium	Renal function should be monitored and dose/treatment adjusted if needed.
penicillamine	• lead • copper • mercury	Side effects include proteinuria and hematuria; may see vomiting, anorexia, or CNS depression with this drug.
Prussian blue	• thallium	Stains feces blue (ferric-cyanoferrate).

IV: ANTIDOTAL THERAPIES

Antidotal therapies are used to counteract the effect of toxicants in various ways. Some stabilize vital signs, decrease exposure, facilitate toxicant removal, or antagonize the toxicant at its site of action. Supportive therapy should also be considered in these cases, as the patient will benefit from a combined approach to toxin treatment.

IF POISONED BY	USE DRUG
acetaminophen	• acetylcysteine
anticoagulant rodenticides	• vitamin K_1
atropine	• physostigmine
carbon monoxide	• oxygen
cholecalciferol (vitamin D)	• calcitonin
cyanide	• sodium nitrite, sodium thiosulfate
digoxin, digitoxin	• sheep-origin antibody fragments with digoxin
ethylene glycol	• alcohol dehydrogenase inhibitors such as 4-methylpyrazole and ethanol
fluoride	• calcium borogluconate
gallamine	• atropine
heparin	• protamine sulfate
metocurine	• atropine
NSAIDs	• sucralfate, metoclopramide
opiates	• naloxone
organophosphates	• atropine
organophosphate insecticides	• pralidoxime chloride (2-PAM)
snakebite (rattlesnake, copperhead, water moccasin)	• antivenin
vecuronium	• atropine

APPENDIX N
VETERINARY COMPANY WEB SITES

3M Animal Care Products: http://www.mmm.com
Abbott Labs: http://abbott.com
Activon Products: http://www.activon.com
Addison Biological Laboratory: http://www.addisonlabs.com
Agri Laboratories: http://www.agrilabs.com
Airex Laboratories: http://www.bullenairx.com
Air-Tite Products Co.: http://www.air-tite.com
Alex C. Fergusson Inc.: http://www.afco.net
Allergan: http://www.allergan.com
Allied Monitor: http://www.alliedmonitor.com
Alpharma Inc.: http://www.alpharma.com
Amgen: http://wwwext.Amgen.com
Anchor Division of Boehringer Ingelheim Vetmedica, Inc.: http://www.bi-vetmedica.com
Antec International: http://www.antecint.com
Argent Chemical Laboratories: http://www.argent-labs.com
Aristavet Pharmaceuticals LLC: http://www.aristavet.com
Balchem Corporation: http://www.balchem.com
Bayer: http://www.bayer.co.uk/products/animalhealth.html and http://www.bayerus.com/ah
Bimeda, Inc: http://www.bimeda.com
Bio-ceutic Division of Boehringer Ingelheim Vetmedica: http://www.bi-vetmedica.com
Biocor Animal Health: http://www.biocorah.com
Bio-Derm Laboratories, Inc.: http://www.biogroom.com
Biopure Corporation: http://www.biopure.com
Bio-Tek Industries, Inc.: http://www.bio-tekdisinfectant.com
Biovance Technologies: http://www.biovance.com
Bioyme Incorporated: http://www.biozymeinc.com
Blue Ridge Pharmaceuticals, Inc: http://www.brpharma.com
Boehringer Ingelheim Animal Health: http://www.boehringer-ingelheim.com/biahi/
Bou-Matic: http://www.boumatic.com
Bristol-Myers Squibb: http://www.bms.com
Butler Company: http://www.wabutler.com
Champion Protector/Performer Brand: http://www.agrilabs.com
Colorado Serum Company: http://www.colorado-serum.com
Combe, Inc.: http://www.combe.com
Conklin Company, Inc.: http://www.conklin.com
Dermapet: http://www.dermapet.com
DMS Laboratories: http://www.rapidvet.com
Doctors Foster and Smith: http://www.drsfostersmith.com
Durvet, Inc.: http://www.durvet.com
DVM Formula: http://www.dvmformula.com
DVM Pharmaceuticals, Inc.: http://www.DVMPharmaceuticals.com
Elanco Animal Health: http://www.elanco.com/
Eli Lilly: http://www.lilly.com
Equicare Products: http://www.farnam.com

EVSCO Pharmaceuticals: http://www.evscopharm.com
Farnam Companies: http://www.farnam.com
First Priority, Inc.: http://www.prioritycare.com
Fort Dodge Animal Health: http://www.wyeth.com/divisions/fort_dodge.asp
GlaxoSmithKlein: http://www.gsk.com
Grand Laboratories, Inc.: http://www.grandlab.com
Halocarbon Laboratories: http://www.halocarbon.com
Happy Jack, Inc.: http://www.happyjackinc.com
Hawthorne Products, Inc.: http://www.hawthorne-products.com
Heska Corporation: http://www.heska.com/
IDEXX Laboratories: http://www.idexx.com
Immucell Corporation: http://www.immucell.com
Immvac, Inc: http://www.immvac.com
IVAX: http://www.ivax.com
Johnson & Johnson: http://www.johnsonandjohnson.com/home.html
Jorgensen Labs, Inc.: http://www.jorvet.com
JPI Jones Animal Health: http://www.soloxine.com
Luitpold Pharmaceuticals, Inc.: http://www.luitpold.com
Maine Biological Laboratories: http://www.mainebiolab.com
Med-Pharmex, Inc.: http://www.med-pharmex.com
Merial: http://www.merial.com/main.html
Monsanto Company: http://www.monsanto.com/dairy/
Moorman's Inc.: http://www.moormans.com
Neogen Corporation: http://www.neogen.com
Novartis Animal Health: http://www.novartis.com/
Pet-Ag, Inc.: http://www.petag.com
Pfizer Animal Health: http://www.pfizer.com
Pharmacia & Upjohn Company: http://www.pnuanimalhealth.com
Pharmaderm: http://www.pharmaderm.com
Phoenix Pharmaceutical, Inc.: http://www.phoenixpharmaceutical.com
Professional Biological Company: http://www.colorado-serum.com
Prolabs Ltd.: http://www.agrilabs.com
Rhone-Poulenc: http://www.rpan.com/
Schering-Plough: http://sp-animalhealth.com
Sergeant's Pet Products: http://www.sergeants.com
Synbiotics Corporation: http://www.synbiotics.com
Techmix, Inc.: http://www.techmixinc.com
Vedco, Inc.: http://www.vedcom.com/
Vet-A-Mix: http://www.lloydinc.com
Veterinary Dynamics, Inc.: http://www.thegrid.net/vdi
Veterinary Products Laboratories: http://www.vpl.com
Vetlife: http://www.vetlife.com
Vets Plus, Inc.: http://www.vets-plus.com
Vet Tek, Inc.: http://www.durvet.com
Vetus Animal Health c/o Burns Veterinary Supply: http://www.burnsvet.com
Vineland Laboratories: http://www.vinelandlabs.com
Warner Lambert: http://www.warner-lambert.com
Wildlife Pharmaceuticals, Inc.: http://www.wildpharm.com
Zinpro Corporation: http://www.zinpro.com

APPENDIX O
DRUGS PEOPLE SHOULD AVOID DURING PREGNANCY

DRUG CATEGORIES

The FDA established five pregnancy categories of drugs to demonstrate the potential for systemically absorbed drugs to cause birth defects. The differentiation between categories is based on the degree of documentation (reliability) and the risk–benefit ratio of the drugs. Regardless of the designated Pregnancy Category or presumed safety, no drug should be administered during pregnancy unless it is clearly needed.

Category A: Adequate studies in pregnant women have not demonstrated a risk to the fetus in the first trimester of pregnancy, and there is no evidence of risk in later trimesters.

Category B: Animal studies have not demonstrated a risk to the fetus, but there are no adequate studies in pregnant women, or animal studies have shown an adverse effect, but adequate studies in pregnant women have not demonstrated a risk to the fetus during the first trimester of pregnancy, and there is no evidence of risk in later trimesters.

Category C: Animal studies have shown an adverse effect on the fetus, but there are no adequate studies in humans; the benefits from the use of the drug in pregnant women may be acceptable despite its potential risks, or there are no animal reproduction studies and no adequate studies in humans.

Category D: There is evidence of human fetal risk, but the potential benefits from the use of the drug in pregnant women may be acceptable despite its potential risks.

Category X: Studies in animals or humans demonstrate fetal abnormalities or adverse reaction; reports indicate evidence of fetal risk. The risk of use in a pregnant woman clearly outweighs any possible benefit.

DRUGS TO AVOID TAKING WHILE PREGNANT*

The following is a list of drugs women should avoid taking or using while pregnant or contemplating pregnancy:

- ACE (angiotensin converting enzyme) inhibitors such as enalapril, benazepril, and captopril
- Aminopterin
- Androgens
- Anticonvulsants such as phenytoin, valproic acid, trimethadione, paramethadione, and carbamazepine
- Antineoplastic drugs
- Carbimazole/methimazole
- DES (diethylstilbestrol)
- Doxycycline
- Isotretinoin (acne medications)
- Lithium
- Methotrexate
- Penicillamine

- Streptomycin
- Tetracycline
- Thalidomide
- Thiouracil/propylthiouracil
- Warfarin

ROUTES OF HUMAN EXPOSURE TO VETERINARY DRUGS

People are exposed to veterinary drugs by five main routes: dermal, ocular, injection, inhalation, and ingestion.

- Dermal exposure is contamination of the skin with the product. The amount of drug absorbed and the speed at which it is absorbed depends upon whether or not the skin is intact, the body part contaminated, and the drug itself. An example of dermal exposure is when injectable prostaglandins are spilled onto the skin.

- Ocular exposure is contamination of the eyes with the product. Common ways of contaminating eyes include wiping your eyes with contaminated hands or blow-back (using the wrong size needle, which allows the drug to flow slowly, resulting in needle obstruction. When pressure is placed on the syringe, the medication sprays back into the face). Ocular exposure can occur with any injectable drug, but especially those that are viscous.

- Accidental injection usually occurs as a result of inappropriate restraint of the animal, carelessness, and improper disposal of sharps (needles, syringes, and blades). Any injectable drug or vaccine may enter the body accidentally.

- Inhalation exposure usually results from inhaling powders, mists, or gases. Inhalation of powders and mists can occur when using fly or flea sprays or chemical disinfectants. Inhalation of gases can occur with improper scavenging units on anesthesia machines or when gases are produced while mixing chemicals. Protective clothing and proper masks are needed to avoid inhalation exposure.

- Ingestion exposure may occur from inappropriate handling of the product. Inappropriate handling is typically due to removing or holding the plastic case in your teeth or improper handwashing after working with drugs.

WAYS TO PREVENT CONTAMINATION WITH VETERINARY DRUGS

- Always wear impermeable gloves when working with drugs to protect your skin from contact.
- If the drug accidentally comes into contact with your skin, immediately wash the affected area with soap and water.
- Use correct injection techniques to minimize the possibility of a needle stick.
- Keep needles properly covered until used.
- Ensure proper restraint of the animal prior to injection.
- Ensure proper syringe adjustment.
- Ensure proper needle placement onto the syringe.
- Use properly cleaned needles, and change needles often.

- Never attempt to straighten a bent needle.
- Never carry a loaded syringe in your pocket.
- Know where all MSDS are for drugs and chemicals used in the practice and read them prior to handling veterinary drugs.

DRUGS TO AVOID HANDLING WHILE PREGNANT*

The following is a list of veterinary drugs that women should avoid handling while pregnant or contemplating pregnancy

- Chemotherapy/cytoxic drugs (used in treatment of neoplasia)
- Dimethyl sulfoxide (DMSO) (anti-inflammatory, drug carrier)
- Glucocorticoids such as dexamethasone (anti-inflammatory, immune suppressant)
- Live vaccines like brucellosis vaccine
- Mitotane (hyperadrenocorticism treatment)
- Organophosphate products such as fly tags (parasite control)
- Oxytocin (promotes uterine contraction and induces labor in animals at term)
- Progesterones such as altrenogest (estrus synchronization, estrus prevention, behavioral drug)
- Prostaglandins (estrus synchronization, abortion drug, pyometra treatment)
- Tilmicosin (antibiotic that can cause cardiovascular problems and death in humans)

*Always consult a physician regarding use of drugs during pregnancy and other possible workplace hazards in veterinary practice.

APPENDIX P
DRUGS LISTED ALPHABETICALLY AND CLASSIFIED BY CATEGORY

DRUG	DRUG CATEGORY/ TREATMENT	DRUG	DRUG CATEGORY/ TREATMENT
Acarbose	Oral hypoglycemic agent	Amiodarone	Antiarrhythmic
Acemannan	Immunomodulator	Amitraz	Demodex and sarcoptic
Acepromazine	Antiemetic		mange
	Sedative		Ear mites
	Preanesthetic	Amitriptyline	Anxiety separation
	Restraint		Behavior (urine spraying)
Acetaminophen	Analgesic	Amlodipine	Calcium channel blocker
Acetazolamide	Carbonic anhydrase		(vasodilator)
	inhibitor (glaucoma	Ammonium chloride	Urinary acidifier
	treatment)	Amoxicillin	Antibiotic
	Metabolic acidosis	Amoxicillin/clavulanic	Antibiotic
Acetic acid	Antiseptic	acid	
Acetylcysteine	Mucolytic	Amphotericin B	Antifungal
	Acetaminophen toxicity	Ampicillin	Antibiotic
Acetylsalicylic acid	NSAID	Amprolium	Anticoccidial
(Aspirin)	Antiplatelet	Apomorphine	Emetic
	Antipyretic	Aproclonidine	Glaucoma
Acid citrate dextrose	Anticoagulant	Aspirin (acetylsalicyclic	NSAID
ACTH	Diagnostic agent for	acid)	Antiplatelet
	adrenal cortex disease		Antipyretic
Actinomycin-D	Antineoplastic Antibiotic	Atenolol	Hypertension reducer
Activated charcoal	GI adsorbant		Antiarrhythmic
Acyclovir	Antiviral	Atipamezole	Reversal agent for
Adrenocorticotropic	Diagnostic agent for		medetomidine
hormone (ACTH)	adrenal cortex disease	Atracurium	Neuromuscular blocker
Albendazole	Antinematodal	Atropine	Increases heart rate
Albuterol	Bronchodilator		Agent for cholinergic toxins
Alcohol	Antiseptic		Antidiarrheal
Alcuronium	Neuromuscular blocker		Antiemetic
Alfentanil	Induction agent		Mydriatic (ocular)
Allopurinol	Urate urolithiasis		Cycloplegic
Alprazolam	Behavioral disorders	Auranofin	Anti-inflammatory
Altrenogest	Pregnancy maintenance		Immunomodulator
	Estrus suppression	Aurothioglucose	Anti-inflammatory
Aluminum acetate	Antipruritic		Immunomodulator
	Soak	Azaperone	Tranquilizer
Aluminum hydroxide	Antacid		Sedative
	Phosphate binder	Azathioprine	Antineoplastic
Aluminum carbonate	Antacid		antimetabolite
Amikacin	Antibiotic		Immune mediated
Amiloride	Diuretic		diseases
Aminopentamide	Antidiarrheal	Azithromycin	Antibiotic
	Antiemetic	Aztreonam	Antibiotic
Aminophylline	Bronchodilator	Bacitracin	Antibiotic
Aminopropazine	Smooth muscle relaxant		Growth promotant

Drug	Drug Category/ Treatment	Drug	Drug Category/ Treatment
Baclofen	Urethral obstruction	Carnitine (l-carnitine)	Nutritional supplement for cardiomyopathy
BAL (dimercaprol)	Chelating agent		
Barium sulfate	Diagnostic agent	Carprofen	NSAID
Beclomethasone dipropionate	Anti-inflammatory	Carteolol	Glaucoma
		Castor oil	Stimulant laxative
Benazepril	Vasodilator	Catecholamines	Cardiac contractility enhancers
Bendroflumethiazide	Diuretic		
Benzalkonium chloride	Antiseptic	Cefaclor	Antibiotic
Benzocaine	Local anesthetic	Cefadroxil	Antibiotic
Benzoyl peroxide	Antiseborrheic	Cefamandole	Antibiotic
Betamethasone	Anti-inflammatory	Cefazolin sodium	Antibiotic
Betaxolol hydrochloride	Glaucoma	Cefataxime	Antibiotic
Bethanechol chloride	Cholinergic for urinary bladder atony	Cefepime	Antibiotic
		Cefixime	Antibiotic
Bimatoprost	Prostaglandin (ocular)	Cefonicid	Antibiotic
Bisacodyl	Stool softener	Cefoperazone	Antibiotic
	Cathartic	Cefotaxime	Antibiotic
Bismuth subsalicylate	GI protectant	Cefotetan	Antibiotic
Bleomycin	Antineoplastic antibiotic	Cefovecin	Antibiotic
Boldenone undecylenate	Anabolic steroid	Cefoxitin sodium	Antibiotic
Boric acid	Drying solution	Cefpodoxine	Antibiotic
Bran	Bulk-forming laxative	Ceftazidime	Antibiotic
Bretylium	Ventricular fibrillation	Ceftizoxime	Antibiotic
Brinzolamide	Glaucoma	Ceftriaxone	Antibiotic
Bromfenac	NSAID (topical)	Ceftiofur	Antibiotic
Bromocriptine mesylate	Hyperadrenocorticism	Cefuroxime	Antibiotic
	Pseudopregnancy	Cephalexin	Antibiotic
Buparvaquone	Antiprotozoal	Cephalosporin	Antibiotic
Bupivacaine	Local anesthetic	Cephalothin	Antibiotic
Buprenorphine	Analgesic	Cephapirin	Antibiotic
Buspirone	Anxiety relieving	Cephradine	Antibiotic
Busulfan	Antineoplastic antibiotic	Cerumene	Dewaxing solution
Butorphanol tartrate	Antitussive	Cetirizine	Antihistamine
	Analgesic	Chloral hydrate	Restraint
	Preanesthetic		Sedative
	Sedative		Preanesthetic
Butoxypolypropylene glycol	Ectoparasite repellent	Chlorambucil	Antineoplastic alkylinizing agent
Butyl hyoscine	Cholinergic for urinary bladder atony	Chloramphenicol	Antibiotic
Calcitonin	Hypercalcemia treatment	Chlordiazepoxide	Irritable bowel syndrome
	Cholecalciferol toxicosis treatment		Antianxiety
		Chlorhexidine	Antiseptic
Calcitrol	Parathyroid	Chlorinated hydrocarbons	Ectoparasitic
Calcium carbonate	Antiulcer	Chlorothiazide	Diuretic
Calcium chloride	Calcium supplement		Udder edema
	Ventricular systole	Chlorpheniramine	Antihistine
Calcium gluconate	Calcium supplement		Antipruritic
	Ventricular systole	Chlorpromazine	Antiemetic
Captopril	Vasodilator		Tranquilizer
Carbachol	Miotic (ocular)	Chlorpyrifos	Ectoparasitic (OP)
Carbadox	Antibiotic	Chlortetracycline	Antibiotic
	Growth promotant		Growth promotant
Carbamazepine	Anticonvulsant	Choline chloride	Hepatic lipidosis
Carbaryl	Ectoparasitic	Chondroitin sulfate	Cartilage supplement
Carbenicillin	Antibiotic	Chorionic gonadotropin (hCG)	Luteinizer of follicular cysts
Carbimazole	Hyperthyroidism		Ovulation inducer
Carboplatin	Antineoplastic alkalinizing agent		Induces descent of inguinal testis

Drug	Drug Category/Treatment	Drug	Drug Category/Treatment
Cimetidine	Antiulcer	Dacarbazine	Antineoplastic aklylating agent
Ciprofloxacin	Antibiotic		
Cisapride	Antiemetic	Dactinomycin	Antineoplastic antibiotic
	Gastrointestinal ileus	Danazol	Synthetic hormone
Cisplatin	Antineoplastic alkylinizing agent		Thrombocytopenia
		Danofloxacin	Antibiotic
Citalopram	Behavior modifier	Dantrolene	Functional urethral obstruction
Clemastine	Antihistamine		
Clenbuterol	Brochodilator		Malignant hyperthermia
Clavulanic acid/amoxicillin	Antibiotic	Decoquinate	Anticoccidal
Clindamycin	Antibiotic	DEET	Ectoparasiticide
Clomipramine	Antianxiety	Delmadinone	Testosterone inhibitor
	Lick granuloma prevention	Demecarium	Glaucoma
		Demeclocycline	Antibiotic
Clonazepam	Antianxiety	Deracoxib	NSAID
Clopidogrel bisulfate	Anticoagulant	Desmopressine acetate	Central diabetes insipidus
Cloprostenol sodium	Prostaglandin (synthetic)		Von Willebrand's disease
	Induction of parturition	Desoxycorticosterone acetate (DOCA, DOCP)	Adrenal cortex drug (mineralocorticoid)
Clorazepate	Phobia prevention		
	Anticonvulsant	Detomidine	Sedative
Clorsulon	Antitrematodal		Analgesia
Closantel	Antiparasitic	Dexamethasone	Anti-inflammatory
Clotrimazole	Antifungal		CNS trauma
Cloxacillin	Antibiotic		Adrenocortical collapse
Coal tar	Antiseborrheic	Dexmedetomidine	Preanesthetic
Codeine	Antitussive		Sedative
	Analgesic		Analgesic
Copper naphthenate	Antifungal	Dexpanthenol	Intestinal atony
Corticosteroids	Anti-inflammatory		Prokinetic agent
Corticotropin (ACTH)	Diagnostic agent for adrenocortical insufficiency	Dextran 40 and dextran 70	Plasma expanders for shock
		Dextromethorphan	Antitussive
		Dextrose	Corrects hypoglycemia
Cortisone	Anti-inflammatory (glucocorticoid)	Diazepam	Sedative
			Restraint
Cosyntropin	Diagnostic agent for adrenocortical insufficiency		Anticonvulsant
			Appetite stimulant
			Muscle relaxant
Coumaphos	Ectoparasitic	Diazinon	Ectoparasitic (OP)
Coumarin derivatives	Anticoagulants	Diazoxide	Oral hypoglycemic agent
Cyanocobalamine	Vitamin B_{12}	Dichlorphenamide	Carbonic anhydrase inhibitor
	Aids in RBC production		
Cyclophosphamide	Antineoplastic alkylinizing agent	Dichlorvos	Hook-, whip-, and roundworm antiparasitic
	Immune mediated disease	Diclofenac	NSAID
		Dicloxacillin	Antibiotic
Cyclosporine	Immunosuppressant	Dicyclomine	Cholinergic for urinary bladder atony
	Ophthalmic form for keratoconjunctivitis sicca	Difloxacin	Antibiotic
		Diethylcarbamazine	Heartworm prophylaxis
Cypionate	Androgen (testosterone)	Diethylstilbestrol	Estrogen used for hormone-responsive urinary incontinence
Cyproheptadine	Appetite stimulant		
	Bronchospasm control		
	Hyperadrenocorticism		Mismating
Cytarabine	Antimetabolite antineoplastic	Digitoxin	Cardiac contractility enhancer (heart failure)
			Tachyarrhythmias
Cythioate	Ectoparasitic (OP)		
Cytosine arabinoside	Antineoplastic	Diphenoxylate	Antidiarrheal

Drug	Drug Category/ Treatment	Drug	Drug Category/ Treatment
Digoxin	Cardiac contractility enhancer (heart failure)		Anaphylaxis treatment
	Tachyarrhythmias		Bronchodilator
Dihydrostreptomycin	Antibiotic		Adrenergic agonist (mydriatic)
Diltiazem	Antiarrhythmic	Eprinomectin	Ectoparasitic
Dimenhydrinate	Motion sickness	Epsiprantel	Anticestodal
Dimercaprol	Chelating agent for arsenic	Equine appeasing pheromone (EAP)	Pheromone (behavior modifier)
Dimethyl sulfoxide (DMSO)	Anti-inflammatory	Ergonovine maleate	Uterine involution
	Vehicle for drug transport	Erythromycin	Antibiotic
Dinoprost	Prostaglandin	Erythropoietin	Stimulates RBC production
Dinotefuran	Ectoparasitic		
Diphenhydramine	Antihistamine	Escitalopram	Behavior modifier
	Antipruritic	Estradiol cypionate	Mismating
Diphenoxylate	Antidiarrheal for irritable bowel		Urinary incontinence
			Induction of estrus
			Pyometra
Dirlotapide	Appetite depressant		Growth promotant
D-limonene	Ectoparasitic	Ethacrynic acid	Diuretic
DMSO (dimethyl sulfoxide)	Anti-inflammatory	Ethanol 20%	Ethylene glycol toxicity
	Vehicle for drug transport	Etodolac	NSAID
Dobutamine hydrochloride	Cardiac contractility enhancer	Etorphine	Narcotic analgesic
Docusate calcium	Stool softener	Famotidine	Antiulcer
Docusate potassium	Stool softener	Febantel	Antiparasitic
Docusate sodium succinate	Stool softener	Felbamate	Anticonvulsant
		Feline appeasing pheromone (FAP)	Pheromone (behavior modifier)
Dog appeasing pheromone (DAP)	Pheromone (behavior modifier)	Fenbendazole	Antiparasitic (nematodes, cestodes, trematodes)
Dolasetron	Antiemetic		
Domperidone	Prokinetic agent	Fenoterol	Bronchodilator
Dopamine	Cardiac contractility enhancer	Fentanyl	Narcotic analgesic
			Analgesic
	Renal vasodilator		Preanesthetic
Doramectin	Ectoparasitic	Fenthion	Ectoparasitic
Dorzolamide	Glaucoma	Ferrous sulfate	Antianemia (iron supplement)
Doxapram	Respiratory stimulant		
Doxepin	Behavior modifier		Hemostatic drug
Doxorubicin	Antineoplastic antibiotic	Filgrastim	Biologic response modifier
Doxycycline	Antibiotic		
Doxylamine	Antihistamine	Finasteride	Testosterone inhibitor
Droperidol	Sedative	Fipronil	Ectoparasitic
d-trans allethrin	Ectoparasitic	Firocoxib	NSAID
Edrophonium chloride	Cholinergic (used to diagnose myasthenia gravis)	Florfenicol	Antibiotic
		Flucinolone	Anti-inflammatory
			Immunosuppressant
EDTA	Anticoagulant	Fluconazole	Antifungal
	Chelating agent	Flucytosine	Antifungal
Emodepside	Antinematodal	Fludrocortisone	Adrenal cortex drug (corticosteroid)
Enalapril	Vasodilator		
Enanthate	Androgen (testosterone)	Flumazenil	Benzodiazepine antagonist
Enrofloxacin	Antibiotic		
Ephedrine	Adrenergic (used to treat urinary sphincter incontinence)		Antianxiety
		Flumequine	Antibiotic
		Flumethasone	Anti-inflammatory
	Bronchodilator	Flunixin meglumine	NSAID
Epinephrine	Cardiac contractility enhancer (cardiac arrest)		Endotoxic shock
		Fluocinolone	Glucocorticoid (topical)

Drug	Drug Category/ Treatment	Drug	Drug Category/ Treatment
Fluorescein	Ocular stain	Hydrocodone bitartrate	Cough suppressant
Fluoroquinolones	Antibiotics	Hydrocortisone	Anti-inflammatory
Fluorouracil	Antineoplastic		Replacement therapy
	antimetabolite	Hydrocortisone sodium	Shock succinate
Fluoxetine	Behavior modifier		Hypoadrenocortical crisis
Fluprostenol	Prostaglandin (synthetic)	Hydroflumethiazide	Diuretic
Flurazepam	Sedative	Hydrogen peroxide	Emetic
	Behavior modifier	Hydromorphone	Narcotic analgesic
Flurbiprofen sodium	NSAID	Hydroxyurea	Antineoplastic
Fluticasone	Anti-inflammatory		antimetabolite
Fluvoxamine	Behavior modifier	Hydroxyzine	Antihistamine
Folic acid	Dietary supplement	Hypertonic saline	Shock (plasma volume
Follicle stimulating	Induction of estrus		expander)
hormone (FSH)	Male hypogonadism	Ibuprofen	NSAID
	treatment	Idoxuridine	Herpesviral keratitis
Fomepizole	Alcohol dehydrogenase	Imidacloprid	Ectoparasitic
	inhibitor for ethylene	Imidazole	Antifungal
	glycol toxicity	Imidocarb	Babesiosis
Furazolidone	Antibiotic/antiprotozoal	Imipenem-cilastatin	Antibiotic
Furosemide	Diuretic	Imipramine	Separation anxiety
	Udder edema		Urethral sphincter
	Antihypertensive		incontinence
Gabapentin	Anticonvulsant		Narcolepsy
Gallamine	Neuromuscular blocker	Imiquimod	Immunomodulator
Gentamicin sulfate	Antibiotic	Insulin	Diabetes mellitus (lowers
Glimepiride	Oral hypoglycemic agent		blood glucose)
Glipizide	Oral hypoglycemic agent	Interferon	Immunomodulating
Glucosamine	Cartilage supplement		Antiviral
Glyburide	Oral hypoglycemic agent	Interleukin-2	Immunomodulator
Glycerin	Glaucoma	Ipecac syrup	Emetic
	Osmotic diuretic	Iron dextran	Iron deficiency anemia
Glycopyrrolate	Anticholinergic	Isocarboxazid	Behavior modifier (MAOI)
	(for bradycardia/	Isoetharine	Bronchodilator
	bronchodilation)	Isoflupredone	Anti-inflammatory
Gold sodium thiomalate	Anti-inflammatory		Ketosis
	Immunomodulator	Isoflurophate	Glaucoma
Gonadorelin (GnRH)	Gonadotropic drug	Isoflurane	Inhalant anesthetic
	Ovarian cysts	Isoniazid	Antibiotic
	Induction of ovulation	Isopropamide	Sinus bradycardia
Granisetron	Antiemetic		Antidiarrheal
Griseofulvin	Antifungal	Isoproterenol	Cardiac contractility
Growth hormone	Growth promotant		enhancer
Guaifenesin	Skeletal muscle relaxant		Bradycardia
	Expectorant	Isoxsuprine	Laminitis Navicular
Halothane	Inhalant anesthetic	hydrochloride	disease
hCG	Gonadotropin	Itraconazole	Antifungal
Hemoglobin glutamer-200	Hemoglobin replacer	Ivermectin	Antinematodal
	Colloid		Ectoparasitic
Heparin	Anticoagulant	Kanamycin	Antibiotic
Hetacillin	Antibiotic	Kaolin/pectin	GI protectant
Hetastarch	Plasma volume expander	Ketamine hydrochloride	Chemical restraint
Homatropine	Mydriatic		Anesthesia
Hyaluronic acid	Synovitis	Ketoconazole	Antifungal
	Joint fluid replacer		Hyperadrenocorticism
Hydralazine	Vasodilator	Ketoprofen	NSAID
Hydrochlorothiazide	Diuretic	Ketorolac tromethoamine	NSAID
	Udder edema	L-asparaginase	Antineoplastic enzyme
	Antihypertensive		Immune mediated
	Calcium oxalate uroliths		thrombocytopenia

Drug	Drug Category/ Treatment	Drug	Drug Category/ Treatment
Lactulose	Laxative	Melarsomine	Antinematodal (for
	Hepatic encephalopathy		Dirofilariasis)
Lansoprazole	Antiulcer	Melengesterol acetate	Progesterone
	Esophageal reflux	(MGA)	Growth promotant
Lasalocid	Growth promotant	Meloxicam	Anti-inflammatory
	Anticoccidial	Melphalan	Antineoplastic alkylating
Latanaprost	Prostaglandin (ocular)		agent
Levamisole	Antiparasitic	Meperidine	Analgesic
	Immune stimulation	hydrochloride	Sedative
Levetiracetam	Anticonvulsant	Mepivacaine	Local anesthetic
Levobunolol	Glaucoma	Mercaptopurine	Immunosuppressant
Levothyroxine sodium (T_4)	Thyroid supplement	Meropenem	Antibiotic
Lidocaine	Local anesthetic	Metaflumizone	Ectoparasitic
	Epidural anesthetic	Metaproterenol sulfate	Bronchodilation
	Antiarrhythmic	Methadone	Analgesia
	Gastrointestinal ileus	hydrochloride	
Lime sulfur	Ectoparasitic	Methazolamide	Carbonic anhydrase
Lincomycin	Antibiotic		inhibitor
	Growth promotant	Methicillin	Antibiotic
Liothyronine (T_3)	Thyroid supplement	Methimazole	Thyroid drug
Lisinopril	Vasodilator		(hyperthyroid
Lomustine	Antineoplastic alkylating		treatment)
	agent	Methionine	Urinary acidifier
Loperamide	Antidiarrheal		Laminitis
Lorazepam	Sedative	Methocarbamol	Muscle relaxant
Loteprednol	Glucocorticoid (topical)		Controls toxicosis
L-theanine	Antianxiety		Anticonvulsant
Lufenuron	Ectoparasitic	Methohexital	Anesthesia induction
Luteinizing hormone (LH)	Ovulation stimulator	Methoprene	Ectoparasitic (IGR)
Magnesium gluconate	Hypomagnesemia	Methotrexate	Antineoplastic
Magnesium hydroxide	Antacid		antimetabolite
	Laxative	Methoxamine	Adrenergic
Magnesium oxide	Hypomagnesemia	Methoxyflurane	Inhalant anesthetic
Magnesium sulfate	Osmotic laxative	Methscopolamine	Antidiarrheal
	Ventricular tachyarrhythmias	Methylcellulose	Laxative
	Hypomagnesemia	Methylene blue	Nitrate/nitrite toxicities
	Wound dressing		Cyanide toxicities
Mannitol	Diuretic	Methylprednisolone	Anti-inflammatory
	Glaucoma treatment	Methylprednisolone	Shock treatment
	Oliguric renal failure	sodium	succinate
Marbofloxacin	Antibiotic		Spinal cord trauma
Maropitant citrate	Antiemetic	4-methylpyrazole	Alcohol dehydrogenase
Mebendazole	Antinematodal		inhibitor (ethylene
Mechlorethamine	Antineoplastic alkylating		glycol toxicity)
	agent	Methyltestosterone	Androgen
Meclizine	Antihistamine	Methylxanthines	Bronchodilators
	Motion sickness	Metipranolol	Glaucoma
Meclofenamic acid	Analgesic	Metoclopramide	Antiemetic
Medetomidine	Chemical restraint		Gastrointestinal ileus
hydrochloride	Sedative		Prokinetic agent
Medroxyprogesterone	Progestin	Metocurine	Neuromuscular blocker
	Prevents abortion	Metoprolol	Antihypertensive
	Aggressive behavior	Metronidazole	Antibiotic
	Skin conditions		Antiprotozoal
Megestrol acetate	Progestin	Mexiletine	Antiarrhythmic
	Prevents estrus	Mibolerone	Androgen that prevents
	Treats inflammatory		estrus
	conditions	Miconazole	Antifungal
	(eosinophilic placques)		(dermatophyte)

Drug	Drug Category/Treatment	Drug	Drug Category/Treatment
Midazolam	Sedative	N-octyl bicycloheptene	Ectoparasitic
	Preanesthetic	dicarboximide	
Milbemycin	Prophylactic for heart-,	Norepinephrine	Resuscitation
	hook-, round-, and	Norfloxacin	Antibiotic
	whipworms	Norgestomet	Synchronized breeding
Mineral oil	Laxative	Nortriptyline	Behavior modification
Minocycline	Antibiotic	Novobiocin	Antibiotic
Minoxidil	Vasodilator	Nystatin	Antifungal
Mirtazapine	Appetite stimulant	Oatmeal	Topical soothing agent
	Antiemetic	Ofloxacin	Antibiotic
Misoprostol	Antiulcer	Omeprazole	Antiulcer
Mitotane (o,p'-DDD)	Adrenal cortex drug		Esophageal reflux
Mitoxantrone	Antineoplastic antibiotic	Ondansetron	Antiemetic
Monensin	Growth promotant	Opium	Antidiarrheal
	Anticoccidial		Narcotic analgesic
Monoclonal antibody	Immunomodulator	Orbifloxacin	Antibiotic
Morantel tartrate	Antinematodal	Organophosphate	Ectoparasitic
Morphine sulfate	Analgesia	Orgotein	Analgesic (osteoarthritis)
	Preanesthetic	Oxacillin	Antibiotic
Moxidectin	Antiparasitic	Oxazepam	Appetite stimulant
Moxifloxacin	Antibiotic	Oxfendazole	Antiparasitic
Mupirocin	Antibiotic	Oxibendazole	Antiparasitic
Mycobacterium cell wall	Immunostimulant	Oxymorphone	Sedative
	fraction		Restraint
	Sarcoid treatment		Preanesthetic
Nafcillin	Antibiotic		Analgesic
Naloxone	Respiratory stimulant	Oxytetracycline	Antibiotic
	Reversal agent for		Growth promotant
	morphine	Oxytocin	Hormone (enhances
Naltrexone	Behavior modification		uterine contractions)
	Reversal agent for		Stimulates milk let-down
	morphine		Mastitis, metritis
Nandrolone deconate	Anabolic steroid; bone	2-PAM (pralidoxime chloride)	Organophosphate toxicity
	marrow stimulant	Pancrealipase	Enzyme
Naproxen	Anti-inflammatory	Pancuronium bromide	Paralytic agent
Natamycin	Antifungal	Paramethasone	Anti-inflammatory
N-butylscopolammonium	Antidiarrheal	Paregoric	Antidiarrheal
bromide		Paromomycin	Antiprotozoal
Neomycin	Antibiotic	Paroxetine	Behavior modifier
Neostigmine	Cholinergic (myasthenia	Penicillamine	Lead poisoning (chelating
	gravis)		agent)
	Gastrointestinal atony		Copper hepatopathy
Nepafenac	NSAID (topical)		Cystine uroliths
Nicarbazine	Coccidiostat	Penicillin	Antibiotic
Nicergoline	Adrenergic for urinary		Growth promotant
	bladder atony	Pentazocine	Analgesia
Nicotinamide	Ketosis	Pentobarbital	Anticonvulsant
Nifedipine	Calcium channel blocker		Anesthetic
	(vasodilator)		Euthanasia solution
Nitazoxanide	Antiprotozoal	Pergolide mesylate	Hyperadrenocorticism
Nitenpyram	Ectoparasitic	Permethrin	Ectoparasitic
Nitrofurantoin	Antibiotic	Perphenazine	Fescue toxicity
Nitrofurazone	Antibiotic		Antiemetic
Nitroglycerin	Vasodilator		Antianxiety
	Laminitis	Petroleum products	Laxative
Nitroprusside	Vasodilator		Hairball preventative
Nitrous oxide	Analgesic (inhalant)	Phenelzine	Behavior modifier (MAOI)
Nizatidine	Antacid	Phenobarbital	Anticonvulsant

Drug	Drug Category/ Treatment	Drug	Drug Category/ Treatment
Phenoxybenzamine	Adrenergic (urinary bladder atony, laminitis, antidiarrheal)	Pregnant mare serum gonadotropin (PMSG)	Estrus and ovulation promoter
		Primaquine	Babesiosis
Phenylbutazone	NSAID	Primidone	Anticonvulsant
Phenylephrine	Vasodilator	Probiotics	Reestablishment of normal flora
	Mydriatic		
Phenylpropanolamine	Adrenergic (urinary sphincter enhancer)	Procainamide	Antiarrhythmic
		Prochlorperzine/ isopropamide	Antiemetic
Phenytoin	Anticonvulsant		
Phosmet	Ectoparasitic (OP)		Tranquilizer
Physostigmine	Cholinergic	Progesterone	Suppresses estrus
	Atropine toxicity		Maintains pregnancy
Phytonadione	Hemostatic drug	Promazine hydrochloride	Sedative
Phytosphingosine	Antiseborrheic		Preanesthetic
Pilocarpine	Glaucoma	Propantheline	Antidiarrheal
	Miotic		Sinus bradycardia
	Lacrimogenic		Rectal relaxation
Pimecrolimus	Immunomodulator	Proparacaine	Topical anesthetic (ocular)
Pimobendan	Vasodilator	Propionate	Androgen (testosterone)
	Cardiac contractility enhancer	Propofol	Anesthetic
		Propranolol	Vasodilator
Piperacillin	Antibiotic		Treats tachycardia
Piperazine	Antinematodal (roundworms)	Propylene glycol	Ketosis
			Solvent
Piperonyl butoxide	Ectoparasitic		Drug vehicle
Pirlimycin	Antibiotic	Propylthiouracil	Hyperthyroidism
Piroxicam	NSAID	Prostaglandins	Muscle stimulant
Poloxalene	Frothy bloat		Abortion
Polycarbophil	Laxative	Protamine sulfate	Heparin antagonist
Polyethylene glycol	Laxative	Pseudoephedrine	Nasal decongestant
Polymerized methyl silicone	Antifoaming agent	Psyllium hydrophilic muccilloid	Laxative
Polymyxin B	Antibiotic	Pyrantel pamoate	Antinematodal (round- and hookworms)
Polysulfated	Cartilage protectant glycosaminoglycans	Pyrantel tartrate	Antinematodal (round- and hookworms)
Ponazuril	Antiprotozoal	Pyrethrins	Ectoparasitics
Potassium bromide	Anticonvulsant	Pyridostigmine	Cholinergic (myasthenia gravis)
Potassium chloride	Potassium supplement		
Potassium citrate	Calcium oxide urolithiasis	Pyrilamine maleate	Antihistamine
	Urinary alkalinizer	Pyrimethamine	Toxoplasmosis
Potassium iodide	Antifungal	Pyriproxyfen	Ectoparasitic (IGR)
Povidone iodine	Antiseptic	Quinidine	Antiarrhythmic
Pralidoxime chloride (2-PAM)	Organophosphate toxicity	Ramipril	Vasodilator
		Ranitidine	Antiulcer
Pramoxine	Local anesthetic	Reserpine	Fescue toxicity
Praziquantel	Anticestodal	Rifampin	Antibiotic
Prazosin	Vasodilator	Robenidine	Anticoccidial
	Adrenergic for urinary bladder atony	Rotenone	Ectoparasitic
		Roxarsone	Growth promotant
Prednisolone	Anti-inflammatory		Swine dysentery
	Immunosuppressant	Salicylic acid	Antiseborrheic
	Shock	Salmeterol	Bronchodilator
	Hypoadrenocorticism	Sarafloxacin	Antibiotic
Prednisone	Anti-inflammatory	Sargramostim	Colony stimulating factor
	Immunosuppressant	Selamectin	Ectoparastic
	Shock	Selegiline	Adrenal cortex drug (pituitary dependent)
	Hypoadrenocorticism		
Prednisone sodium succinate	Allergic reactions		Behavior modifier

Drug	Drug Category/ Treatment	Drug	Drug Category/ Treatment
Selenium	Antiseborrheic	Terfenadine	Antihistamine
Selenium and alpha tocopherol	Selenium/Vitamin E deficiency	Testosterone	Androgen
			Produce teaser animal
Sertraline	Behavior modifier		Growth promotant
Sevoflurane	Inhalant anesthetic	Tetanus antitoxin	Tetanus treatment
Silver nitrate	Hemostatic drug	Tetracaine	Local anesthetic
	Caustic	Tetracycline	Antibiotic
Silver sulfadiazine	Antibiotic	Theophylline	Bronchodilator
	Antifungal	Thiabendazole	Antiparasitic
Sodium bicarbonate	Hyperkalemia	Thiacetarsamide	Heartworm adulticide
	Renal failure	Thiamine hydrochloride	Lead poisoning
	Acidosis		Thiamine deficiency
Sodium iodide	Antibiotic	Thiamylal sodium	Anesthetic induction
	Antifungal		General anesthesia (large animals)
Sodium nitrite	Cyanide toxicity		Tranquilization
Sodium phosphate with sodium biphosphate	Osmotic laxative	Thiopental sodium	Anesthetic induction (thiopentone)
Sodium sulfate	Osmotic cathartic		General anesthesia (large animals)
Sodium thiosulfate	Cyanide toxicity		
	Arsenic toxicity		
Somatotropin (bovine) (BST)	Growth promotant	Thioguanine	Antineoplastic antimetabolite
	Increases milk production	Thiostrepton	Antibiotic
Spectinomycin	Antibiotic	Thiotepa	Antineoplastic alkylating agent
Spinosad	Ectoparasitic		
Spironolactone	Diuretic	Thrombin	Hemostatic drug
Stanozolol	Anabolic steroid	Thyroid releasing hormone	TRH stimulation test
	Anemia	Thyroid stimulating hormone	TSH stimulation test
Staphage lysate	Immune stimulant		
Streptomycin	Antibiotic	Thyroxine (T$_4$)	Hypothyroidism (thyroid supplement)
Streptozocin	Antineoplastic antibiotic		
Succinylcholine	Paralytic	Tiamulin	Antibiotic
Sucralfate	Antiulcer	Ticarcillin	Antibiotic
Sulbactam	Antibiotic	Ticarcillin/clavulanic acid	Antibiotic
Sulfacetamide	Antibiotic	Tiletamine-zolazepam	Restraint
Sulfadiazine	Antibiotic		Anesthetic
Sulfasozine	Antibiotic	Tilmicosin	Antibiotic
Sulfadimethoxine	Anticoccidial	Timolol maleate	Glaucoma
Sulfadimethoxine/ ormetoprim	Antibiotic	Tiopronin	Cysteine urinary calculi
		Tobramycin	Antibiotic
Sulfamethoxazole/ trimethoprim	Antibiotic	Tocainide	Antiarrhythmic
		Toceranib	Antineoplastic enzyme inhibitor
Sulfaquinoxaline	Anticoccidial		
Sulfasalazine	Antibiotic	Tolazoline	Reversal for xylazine
	Inflammatory bowel disease	Tolfenamic acid	NSAID
		Tramadol	Analgesic
Sulfathiazole	Antibiotic		Antitussive
	Antiprotozoal	Tranexamic acid	Hemorrhage
Sulfisoxazole	Antibiotic	Tranylcypromine sulfate	Behavior modifier (MAOI)
Sulfonamides	Antibiotic	Travoprost	Prostaglandin (ocular)
Sulfur	Antiseborrheic	Trenbolone acetate	Growth promotant
Suprofen sodium	NSAID (topical)	Tretinoin	Retinoid
Syrup of ipecac	Emetic	Triamcinolone	Anti-inflammatory
Tacrolimus	Immunomodulator	Triamcinolone acetate	Anti-inflammatory
Tannic acid	Drying solution	Trichlormethiazide	Diuretic
Tazobactam	Antibiotic	Triclosan	Antiseptic
Tepoxalin	NSAID	Trifluridine	Herpesviral keratitis
Terbinafine	Antifungal	Triiodothyronine (T$_3$)	Hypothyroidism (thyroid supplement)
Terbutaline	Bronchodilator		

Drug	Drug Category/ Treatment	Drug	Drug Category/ Treatment
Trilostane	Hyperadrenocorticism	Vidarabine	Antiviral
Trimeprazine	Antihistamine	Vinblastine	Antineoplastic
Trimethobenzamide	Antihistamine	Vincristine	Antineoplastic alkaloid
Triamterene	Diuretic	Virginiamycin	Growth promotant
Tripelennamine	Antihistamine		Antibiotic
	CNS stimulant	Vitamin K$_1$	Anticoagulant
Tropicamide	Mydriatic	Voriconazole	Antifungal
Tulathromycin	Antibiotic	Warfarin	Anticoagulant
Tylosin	Antidiarrheal	Xylazine	Emetic
	Antibiotic		Anesthetic
	Growth promotant	Yohimbine	Neurologic stimulant
Valproic acid	Seizures		Reversal agent for
Vancomycin	Antibiotic		xylazine
Vasopressin	Diabetes insipidus	Zeranol	Growth promotant
	drug	Zinc gluconate	Neutering drug
Vecuronium	Neuromuscular blocker	neutralized by arginine	
Verapamil	Calcium channel blocker	Zolazepam	Antianxiety
	(antiarrhythmic)	Zonisamide	Anticonvulsant

APPENDIX Q
ANSWER KEY TO TEXT REVIEW EXERCISES

Chapter Review: 1

2. b
4. g
6. c
8. h
10. i
12. c
14. c
16. a
18. b
20. b

Chapter Review: 2

2. d
4. e
6. a
8. i
10. h
12. d
14. a
16. c
18. b
20. a

Chapter Review: 3

2. b
4. c
6. a
8. d
10. f
12. c
14. b
16. d
18. b
20. a

Chapter Review: 4

2. d
4. a
6. b
8. e
10. i
12. a

14. a
16. b
18. a
20. d

Chapter Review: 5

2. d
4. i
6. f
8. c
10. e
12. b
14. c
16. d
18. d
20. c
22. nothing orally
24. three times daily
26. orally
28. tablet or tablespoon
30. without

Chapter Review: 6

2. h
4. i
6. d
8. b
10. f
12. b
14. b
16. d
18. c
20. d
22. b
24. b
26. a
28. 1550 mg
30. 2550 mL
32. 0.015 g
34. 0.400 L
36. 15 mcg
38. 1 kg
40. 20.08 in
42. 4.54 kg

44. 3125 mL
46. split it into quarters
48. 0.98 in
50. 10 cm
52. 38.9°C
54. 77°C
56. 2 mL
58. 0.8 mL
60. 1.0 mL
62. 2.1 mL diluent for an initial concentration of 350 mg/mL; Give 2.1 mL; 1 full dose; If reconstituting with sterile water for injection, it is good for 24 hours at room temperature (at 350 mg/mL concentration) and 3 days at refrigerated temperatures
64. single dose, Abraxis, 401701G, 20 mL

Chapter Review: 7

2. i
4. d
6. a
8. f
10. g
12. b
14. a
16. b
18. c
20. b

Chapter Review: 8

2. d
4. j
6. h
8. i
10. c
12. d
14. b
16. a
18. b
20. d

Chapter Review: 9

2. b
4. e
6. f
8. h
10. j
12. d
14. c
16. c
18. b
20. c

Chapter Review: 10

2. d
4. g
6. b
8. f
10. j
12. a
14. d
16. a
18. b
20. b
22. Novolin N, NPH, intermediate-acting
24. Humalog, regular, short-acting
26. Lantus, glargine, ultralong-acting
28. U-40; 15 units
30. U-100; 57 units

Chapter Review: 11

2. a
4. j
6. d
8. b
10. i
12. c
14. a
16. c
18. b
20. b

Chapter Review: 12

2. i
4. b
6. c
8. f
10. d
12. a
14. a
16. a
18. d
20. d

Chapter Review: 13

2. d
4. a
6. c
8. d
10. c
12. b
14. d

Chapter Review: 14

2. c
4. a
6. h
8. e
10. j
12. b
14. b
16. d
18. c
20. b

Chapter Review: 15

2. f
4. h
6. e
8. b
10. a
12. b
14. b
16. c
18. d
20. b

Chapter Review: 17

2. d
4. h
6. j
8. b
10. e
12. a
14. b
16. c
18. a
20. b

Chapter Review: 19

2. d
4. c
6. b
8. c
10. a
12. a
14. b
16. b
18. a
20. b

Chapter Review: 21

2. e
4. c
6. d
8. h
10. i
12. b
14. a
16. a
18. b
20. a

Chapter Review: 23

2. g
4. f
6. a
8. c
10. e
12. d
14. c
16. a
18. b
20. b

Chapter Review: 16

2. b
4. a
6. a
8. b
10. c
12. b
14. c
16. c
18. b
20. a

Chapter Review: 18

2. j
4. i
6. b
8. e
10. c
12. b
14. a
16. d
18. c
20. b

Chapter Review: 20

2. b
4. i
6. d
8. g
10. a
12. d
14. a
16. a
18. b
20. b

Chapter Review: 22

2. i
4. g
6. j
8. d
10. a
12. a
14. c
16. a
18. b
20. b

A

absorption the movement of drug from the site of administration into the fluids of the body that will carry it to the site(s) of drug action

acetylcholine neurotransmitter that transmits nerve impulses

acetylcholinesterase an enzyme that destroys acetylcholine

active immunity a reaction that occurs when an animal receives an antigen that activates the B and T lymphocytes and produces memory

active transport movement of particles against the concentration gradient, from a region of low concentration to a region of high concentration

acupuncture examination and stimulation of body points by use of acupuncture needles, injections, and other techniques

adjuvant a substance that enhances the immune response by increasing the stability of a vaccine in the body

adrenal cortex the outer part of the adrenal gland

adrenergic antagonist a drug that inhibits the action of the adrenergic receptors; also called sympatholytic drug. Adrenergic antagonists are divided into alpha and beta; beta-adrenergic antagonists are used in the treatment of hypertension and alpha-adrenergic antagonists

are used to decrease the tone of internal urethral sphincters and to decrease blood pressure.

adrenergic blocking agents drugs that block the effects of the adrenergic neurotransmitters; also called *sympatholytics*

adrenergic drugs drugs that simulate the sympathetic nervous system; also called *sympathomimetics*

adrenocortical insufficiency a progressive condition associated with adrenal atrophy, usually caused by immune-mediated inflammation; also known as *Addison's disease.*

adrenocorticotropic stimulating to the adrenal cortex

adsorbents drugs that bind substances such as bacteria, digestive enzymes, and/or toxins to protect the intestinal mucosa from damage

adult administration set device consisting of a set of tubing, drip chamber, and clamp used to administer fluid that typically delivers 15 drops per milliliter

adulticides drugs that kill adult parasites, especially heartworms

affinity strength of binding between a drug and its receptor

afterload the force needed to push blood out of the ventricles; the impedance to ventricular emptying presented by aortic pressure

agonists drugs that bind to a cell receptor and cause action

alpha-adrenergic agonists drugs that selectively stimulate alpha-adrenergic receptors. The alpha-adrenergic receptor has two subclasses α_1 and α_2.

alternative treatments and/ or therapies that are outside accepted conventional medicine

aminoglycosides a group of broad-spectrum antibiotics that affect protein synthesis of bacteria (primarily gram-negative bacteria)

anabolic tissue building

anabolic steroids drugs with the steroid structure that aid in tissue building

analgesics pain relievers

androgen any steroid hormone that promotes male characteristics

anesthetics drugs that interfere with the conduction of nerve impulses and are used to produce loss of sensation

anestrus period of the estrous cycle when the animal is sexually quiet

angiotensin a vasoconstrictive substance found in blood; its inactive form (**angiotensin I**) is converted to the active form (**angiotensin II**) by the enzyme renin

antacid drug that neutralizes HCl and reduces pepsin activity

antagonists drugs that inhibit or block the response of a cell when bound to its receptor

anterior toward the front (of a body part) (term is not used extensively in veterinary medicine other than to describe organs such as the pituitary gland or eye chambers)

anthelmintics drugs that kill worm parasites in the gastrointestinal tract, liver, lungs, and circulatory system

anthrax a fast-killing bacterial infection caused by *Bacillus anthracis*

antianxiety drugs agents that lessen anxiety but do not make the animal drowsy

antiarrhythmics drugs used to treat arrhythmias or abnormal variations in heart rhythm

antibiotic a chemical substance produced by a microorganism that has the capacity, in dilute solutions, to kill (bactericidal activity) or inhibit the growth (bacteriostatic activity) of bacteria

antibiotic resistance condition in which bacteria continue to multiply despite administration of a particular antibiotic

antibiotic residue presence of an antibiotic or its metabolites in animal tissues or food products

antibody protein made by activated B-lymphocytes (plasma cells) to counteract a specfic antigen

antibody titer a serum test that expresses the level of antibody to a particular antigen in a particular individual

anticancer drugs drugs that stop the cancerous activity of malignant cells without disturbing the chemistry of healthy cells; also called *antineoplastic agents* and *chemotherapeutic agents*

anticestodal drugs drugs that kill or inhibit cestode worms (tapeworms)

anticholinergic drugs drugs that treat vomiting, diarrhea, and excess gastric secretion by blocking parasympathetic nerve impulses; also called *parasympatholytics*

anticoagulants agents that prevent blood clotting

anticoccidial drugs drugs that inhibit coccidia organisms; also called *coccidiostats*

anticonvulsants drugs that inhibit seizures (convulsions)

antidepressant drugs drugs used to counteract depression in humans; may be used to treat various mood changes and behavior problems such as aggression and cognitive dysfunction in animals

antidiarrheals drugs that decrease peristalsis to reverse diarrhea

antiemetics substances that prevent vomiting

antifoaming agents drugs that reduce or prevent the formation of foam

antifungals drugs that prevent or treat disease spread by fungi

antigen anything that stimulates immune response

antihistamines drugs that block the effects of histamine

antihypertensives drugs used to decrease high blood pressure

anti-inflammatory drugs drugs that reduce inflammation

anti-infectives drugs that counteract infectious agents

antimicrobials substances made in the chemical laboratory that kill microorganisms or suppress their multiplication or growth

antinematodal drugs agents that kill or inhibit nematodal worms (roundworms)

antineoplastic agents see *anticancer drugs*

antiprotozoal drugs drugs that kill or inhibit protozoal organisms

antipruritic substance that controls itching

antipyretic fever reducing

antiseborrheic substance that counteracts abnormal flaking and/or scaling of the outermost layer of the epidermis

antiseptics substances that kill or inhibit the growth of microorganisms on living tissue

antiserum antibody-rich serum obtained from a hypersensitized or actually infected animal

antisialogues drugs that decrease salivary flow

antitoxins substances that neutralize toxins by use of antibodies obtained from a hypersensitized animal

antitrematodal drugs drugs that kill or inhibit trematodes (flukes)

antitussives drugs that suppress coughing

antiulcer drugs drugs that prevent erosions of the gastrointestinal mucosa

antivirals drugs that prevent viral penetration of the host cell, or inhibit the virus's production of DNA or RNA

apothecary system a system of liquid units of measure used chiefly by pharmacists; also called the "common system"

aqueous prepared in water

arrhythmia a variation from the normal beating of the heart; often causes reduced cardiac output due to the abnormal pumping activity

artificial immunity immunity acquired through medical procedures

astringent   agent that constricts tissues

atrophy   wasting; decreasing in size

attenuated vaccine   see *modified-live vaccine*

autogenous vaccine   vaccine produced for a specific disease in a specific area from a sick animal

B

bacilli   rod-shaped bacteria

bacitracin   a polypeptide antibiotic used primarily as a topical medication for skin, mucous membranes, and eyes

bactericidal   agent that kills bacteria

bacterin   inactivated bacteria that serve as a vaccine against bacteria

bacteriostatic   agent that inhibits or slows the growth of bacteria

benzimidazole   category of antinematodal drug that has excellent efficacy against nematode infections. They are believed to work by interfering with energy metabolism of the worm.

benzimidazole-pyridazinones   inodilators that increase force of contraction (positive inotrope) and cause widening of the blood vessels (vasodilation)

beta-adrenergic blocker   drug that inhibits beta-adrenergic receptors

beta-2-adrenergic agonists   drugs that stimulate beta receptors

beta-lactamase   enzyme produced by some bacteria that destroys the beta-lactam ring of some antibiotics (such as penicillin and cephalosporin)

bioavailability   the degree to which a drug is absorbed and reaches the circulation

bioequivalency   similar ability to be absorbed and to produce similar blood levels after absorption

biologic   a medicinal preparation made from living organisms or their products; examples include vaccines and serums

biologic response modifiers (BRMs)   agents used to enhance the body's immune system

biotransformation   the chemical alteration of drug molecules into metabolites by the body cells of animals; also called *metabolism*

blood-enhancing drugs   drugs that affect red blood cells and their production

bolus   a concentrated mass of pharmaceutical preparation or mass of food

bolus administration   injection of a drug in a minute amount of fluid, with a syringe and needle only

botanical medicine   discipline that uses plants and plant derivatives as therapeutic agents

brand name   the name of a drug, usually capitalized, which establishes proprietary recognition of a particular manufacturer's product; also called *proprietary name, trade name*

BRMs   see *biologic response modifiers*

broad-spectrum antibiotics   drugs that act on both gram-positive and most gram-negative bacteria

bronchi   air passages located between the trachea and bronchioles

bronchodilators   drugs that widen the airway for better breathing

bulk laxatives   drugs that absorb water and swell to form an emollient gel that stimulates peristalsis and fecal elimination

C

calcium channel blockers   drugs that block the influx of calcium ions into the myocardial cells

carbonic anhydrase inhibitor   category of drugs that interfere with the production of carbonic acid, which leads to decreased aqueous humor formation

carcinogenicity   the ability or tendency to produce cancer

cardiac glycosides   drugs used to increase the strength of cardiac contractions

catecholamines   a group of chemically related compounds having sympathomimetic action (e.g., epinephrine and norepinephrine)

cathartics   harsh laxatives that result in a soft to watery stool and abdominal cramping

cc   see *cubic centimeter*

cell-cycle nonspecific (CCNS)   anticancer drug that works on any phase of the cell cycle

cell-cycle specific (CCS)   anticancer drug that works on a specific phase of the cell cycle

cell mediated   affected by cellular rather than chemical elements

centi-   prefix meaning one hundredth of a unit

central nervous system (CNS)   the portion of the nervous system consisting of the brain and spinal cord

cephalosporins   broad-spectrum antibiotics effective against most gram-positive and many gram-negative organisms

chemotherapeutic agents   see *anticancer drugs*

Chinese traditional herbal medicine   discipline based on a holistic philosophy of life that emphasizes the relationship among the mental, emotional, and physical components of the individual, as well as the importance of harmony among individuals, their social groups, and the general population

chiropractic   examination, diagnosis, and treatment of animals through manipulation and adjustments of specific joints and cranial structures

chloramphenicol   a broad-spectrum antibiotic not considered first-line because of its toxicity

cholinergic agonists   drugs that function to enhance the effects mediated by acetylcholine in the central nervous system, the peripheral nervous system, or both; also known as *parasympathomimetics*

cholinergic blocking agent   see *anticholinergic drugs*

cholinergic drugs   agents that mimic the action of the parasympathetic nervous system; also called *parasympathomimetics*

cholinergic receptors   sites stimulated by the nerve terminal's release of acetylcholine

chronic study   see *long-term toxicity test*

clindamycin   an antibiotic that is a derivative of lincomycin

clinical trials   tests done in the target species to determine if the drug is safe and effective in that species

CNS   see *central nervous system*

CNS stimulant   drug that reverses CNS depression

coccidiosis   a protozoan infection that causes various intestinal disorders, some serious and even fatal, in various species

coccidiostats   see *anticoccidial drugs*

colloid solution   fluids with large molecules that enhance the oncotic force of blood, causing fluid to move from the interstitial and intracellular spaces into the vascular spaces; also referred to as colloids

combination therapy   the administration of various combinations of anticancer drugs

***Compendium of Veterinary Products* (*CVP*)**   a compendium of United States veterinary standards

competitive nondepolarizers   drugs that block activity at the neuromuscular junction; also called *curarizing agents*

complementary medicine   therapies that can be used with or in addition to conventional treatment

compounding   the preparation, mixing, assembling, packaging, and/or labeling of a drug based on a prescription drug order from a licensed practitioner for an individual patient

controlled substances   drugs considered dangerous because of their potential for abuse or misuse and whose use is monitored by the Drug Enforcement Administration

conversion factor   number used with either multiplication or division to change a measurement from one unit of measurement to its equivalent in another unit of measure. A conversion factor always has a value of 1.

core vaccine   vaccine recommended for all individual animals because the consequences of infection are severe, infection poses a substantial zoonotic risk, disease prevalence is high, the organism is easily transmitted to others of its species, and or the vaccine is safe and efficacious

corpus luteum   a "yellow body" that produces progesterone in pregnant animals

corticosteroids   hormones produced by the adrenal cortex

crystalloid solution   sodium-based electrolyte solution or solution of glucose in water that has a composition similar to that of plasma fluid; also referred to as crystalloids

cubic centimeter (cc)   the liquid volume that would fill a one-centimeter-square cube

curare alkaloids   toxic botanical extracts originally used as arrow poisons in South America

curarizing agents   see *competitive nondepolarizers*

cycloplegic   drug that paralyzes the ciliary muscles of the eye and minimizes pain due to ciliary spasm

D

DEA   see *United States Drug Enforcement Administration*

decongestant   drug that decreases congestion of the nasal passages by reducing swelling

depot preparation   see *repository medication*

dewaxing agent   substance used to remove debris and wax (usually from the ear)

diabetes insipidus   a disease characterized by increased thirst and urination, caused by failure of the posterior pituitary to release sufficient antidiuretic hormone

diabetes mellitus a broad term denoting a group of diseases characterized by insulin deficiency

diestrus period of the estrous cycle after metestrus; a short phase of inactivity in polyestrous animals

diffusion movement of atoms, ions, or molecules from an area of high concentration to an area of low concentration

digestive enzyme enzyme supplement used in animals in which the pancreas is not producing sufficient enzymes for digestion

digitalis a derivative of the foxglove plant used to treat cardiac failure

digitalis drugs see *cardiac glycosides*

direct marketing purchase of a drug directly from the company that makes it

disinfectant chemical that kills or inhibits the growth of microorganisms on inanimate objects

distribution physiological movement of drugs from the systemic circulation to the tissues

distributors agencies that purchase drugs from the manufacturing companies and resell the drugs to veterinarians; also called *wholesalers*

diuresis increased excretion of urine

diuretics drugs that increase the volume of urine excreted by the kidneys and thus promote release of water from the tissues

dosage amount of drug per animal species' body weight or measure (e.g., 100 mg/kg)

dosage interval how frequently the dosage is given (e.g., bid)

dosage regimen the dosage and the dosage interval together (e.g., 100 mg/kg bid)

dose the amount of a drug that is administered at one time to achieve the desired effect

doubling time time required for the number of cancer cells to double in mass

dram a fluid volume measure in the apothecary system; consists of 60 minims

dressing substance applied to an area to draw out fluid or relieve itching

drug abuse the illicit use of an illegal drug or the improper use of a legal prescription drug

Drug Enforcement Administration see *United States Drug Enforcement Administration*

drying agent product that reduces moisture (usually used for the ear)

E

ECF see *extracellular fluid*

ectoparasite a parasite living on the surface of the host's body

edema abnormal intercellular fluid accumulation

effective dose-50 (ED$_{50}$) the dose of a test drug that causes a defined effect in 50 percent of the animals that take it

elimination removal of a drug from the body; also called *excretion*

emetic drug that induces vomiting

emulsion a mixture of two immiscible liquids, one being dispersed throughout the other in small droplets

endoparasite a parasite that lives within the body of the host

epidural within the epidural space (also known as the "subarachnoid space")

ergot derivative of rye fungus used to induce labor and treat migraines

erythromycin A broad-spectrum antibiotic produced by a strain of the bacterium *Streptomyces erythreus*

erythropoietin protein made in the kidneys that stimulates red blood cell formation from bone marrow stem cells

estrogen a hormone that promotes female sex characteristics

estrous cycle the entire sexual cycle in the female

estrus heat (in a sexual or reproductive sense); the period of the estrous cycle in which the female is receptive to the male

euthanasia solution chemicals used to humanely end an animal's life

excretion see *elimination*

expectorants drugs that enable or increase the coughing-up and swallowing of material from the lungs

expiration date the date before which a drug meets all specifications and after which the drug can no longer be used

extracellular fluid (ECF) fluid found in the intravascular and interstitial spaces

extra-label drug drug used in a manner not specifically described on the FDA-approved label

F

facilitated diffusion diffusion that utilizes a carrier molecule

FDA see *United States Food and Drug Administration*

feed conversion efficiency rate at which food is converted to tissue

first-line drug   the veterinarian's first drug of choice for a particular use

fluid dram   60 minims

fluid ounce   a common standard unit used to measure liquid volume

fluid overload   a condition in which more fluid is going into the animal than is coming out of the animal

follicular phase   stage of the estrous cycle in which the graafian follicle is present and produces estrogen

Food and Drug Administration   see *United States Food and Drug Administration*

fungicidal   chemical that kills fungi

G

gamma-aminobutyric acid   an amino acid that helps mediate nerve impulse transmission in the CNS; abbreviated GABA

gastroenteritis   inflammation of the lining of the stomach and intestines

generic company   business that sells, under its own company name, drugs that are no longer under patent protection

generic equivalent   drug determined to be the therapeutic equal of a brand-name drug for which the patent has expired

generic name   the official identifying name of a drug agreed upon by the FDA and the company of origin; also called *nonproprietary name*

germicide   chemical that kills microorganisms

glaucoma   group of diseases resulting in increased intraocular pressure

glucagon   a hormone that increases blood glucose levels by promoting the breakdown of liver glycogen into glucose, which exits the liver and enters the bloodstream

glucocorticoids   a type of corticosteroid that raises the concentration of liver glycogen and blood sugar in the body; also used as anti-inflammatory agents and to terminate late-stage cattle pregnancies

glycogen   the chief carbohydrate storage material in animals

gonadotropins   hormones that stimulate the gonads (testes in males and ovaries in females)

grain   the basic unit of weight measurement in the apothecary system

gram (g)   the metric standard unit of weight measurement

gram-negative bacteria   strains of bacteria that are decolorized by the Gram's stain process; they have cell walls more complex than those of gram-positive bacteria

gram-positive bacteria   strains of bacteria that resist decolorization by the Gram's stain process; they have cell walls less complex than those of gram-negative bacteria

Gram stain   a staining and decolorizing procedure that divides bacteria into gram-positive and gram-negative strains

growth fraction   the percentage of cancer cells that are actively dividing

growth promotants   chemical compounds that can improve growth; also called "anabolic agents"

H

half-life   the time required for the amount of drug in the body to be reduced by half of its original level

hemostatic drug   agent that helps promote clotting

heparin   anticoagulant drug

holistic medicine   comprehensive approach to health care that uses both alternative and conventional diagnostic techniques and therapeutic approaches

homeopathy   medical discipline in which animal conditions are treated by administration of very small doses of substances that are capable of producing clinical signs in healthy animals

hormone   a chemical substance produced in one part of the body that is transported to another part of the body, where it influences and regulates cellular and organ activity

household system   system of measure in which an approximate dose is acceptable; based on 1 pound (lb) containing 16 ounces (oz)

hydrophilic   water loving

hyperadrenocorticism   disease caused by hyperactivity of the adrenal cortex; may be caused by tumors of the pituitary gland, tumors of the adrenal cortex, or excessive treatment with corticosteriods

hyperthyroidism   excessive functional activity of the thyroid gland

hypertonic   having osmolality greater than that of blood and extracellular water

hypogonadism   the decreased function and retarded growth and sexual development of the gonads (testes in males and ovaries in females)

hypokalemia   potassium deficiency

hypothyroidism the underproduction of thyroxine and the effects of that underproduction

hypotonic having osmolality lower than that of blood and extracellular water

I

iatrogenic caused by treatment

IC see *intracardiac*

ICF see *intracellular fluid*

ID see *intradermal*

IM see *intramuscular*

immiscible incapable of mixing

immunomodulating agent drug that adjusts the immune response to the desired level

immunosuppressive drug drug that decreases the immune response by interfering with one of the stages of the cell cycle or by affecting cell messengers to the immune system

inactivated vaccine vaccine made from microorganisms, microorganism parts, or microorganism by-products that have been chemically treated or heated to kill the microorganisms; also known as *killed vaccine*

indwelling catheter a catheter specially designed to remain in place

induce to increase the rate of metabolism of an enzyme system

inflammation a natural response of living tissue to injury and infection; involves vascular changes and release of chemicals that help the body destroy harmful agents at the injury site

infusion of fluid administration of large volumes of fluid continuously over extended periods of time

inhalation taking in by breathing

inotropy the force of contraction

insulin a hormone formed in the pancreas that responds primarily to a rise in blood glucose and promotes storage of glucose in the liver as glycogen

interferon a protein substance with multiple roles in the body's natural defenses, chief among them stimulating noninfected cells to produce antiviral proteins

intermediate describes the ability of bacteria to sometimes be affected and sometimes not be affected by a particular antibiotic

intermittent therapy treatment done by diluting a drug dose in a small volume of fluid and administering it via an indwelling catheter for a period of 30 to 60 minutes

intra-arterial (IA) within an artery

intra-articular within the joint

intracardiac (IC) within the heart

intracellular fluid (ICF) fluid within the cell

intradermal (ID) within the dermis

intramammary within or into the mammary gland

intramedullary within the marrow cavity of a bone

intramuscular (IM) within the substance of a muscle

intraosseous (IO) within a bone

intraperitoneal (IP) within the peritoneal or abdominal cavity

intrathecal within the subarachnoid space of the meninges

intravenous (IV) within a vein

ionized charged

ion trapping a condition in which a drug molecule changes from its ionized (charged) form to the nonionized (uncharged) form as it moves from one body compartment to another body compartment

islets of Langerhans the pancreatic cells that secrete insulin

isotonic having osmolality the same as blood and extracellular water

IV see *intravenous*

K

keratoconjunctivitis sicca (KCS) disease in which tear production is decreased, resulting in persistent mucopurulent conjunctivitis and corneal scarring and ulceration; also known as "dry eye"

keratolytic substance that removes excess keratin and promotes loosening of the outer layers of the epidermis

killed vaccine see *inactivated vaccine*

kilo- metric prefix meaning one thousand times the unit

kinetics the scientific study of motion

L

lacrimogenic drug that increases tear production

laxative medicine that loosens the bowel contents and encourages evacuation

lethal dose (LD$_{50}$) the dose of a test drug that kills 50 percent of the animals that take it

lincomycin an antibiotic produced by a strain of *Streptomyces* that works by inhibiting protein synthesis

lipophilic fat loving

liter (L) the metric standard unit used to measure liquid volume

live vaccine vaccine made from live microorganisms, which may be fully virulent or avirulent

loading dose   initial dose of a drug given to get the drug concentration up to the therapeutic range in a very short period of time

long-term toxicity test   a toxicity test of a drug that lasts from three months to two years; also called *chronic study*

loop diuretics   diuretics that block sodium reabsorption in Henle's loop

loop of Henle   a U-shaped renal tubule that is part of the nephron

luteal phase   stage of the estrous cycle in which the corpus luteum is present and produces progesterone

M

macrolides   antibiotics with large molecular structure and many-membered rings

maintenance dose   dose of drug that maintains or keeps the drug in the therapeutic range

maintenance fluid   fluid volume needed by the animal on a daily basis to maintain body function

margin of safety   the difference in magnitude between a drug's effective dose and its lethal dose; LD_{50}/ED_{50}

materia medica   the study of the physical and chemical characteristics of materials used as medicines

MIC   see *minimum inhibitory concentration*

measurement   the use of standard units to determine the final weight or volume of substances

metabolism   the chemical alteration of drug molecules into metabolites by the body cells of animals; also known as *biotransformation*

meter   the metric standard unit used to measure length

metestrus   period of the estrous cycle after sexual receptivity

metric system   system of measure developed by the French that is based on factors of 10

micro-   prefix meaning one millionth of a unit

microfilaricide   drug that kills microfilaria of *Dirofilaria immitis*

microgram (mcg)   a unit of metric weight equivalent to one-thousandth of a milligram or one-millionth of a gram

microsomal enzymes   enzymes found primarily in the liver that are involved in the biotransformation (metabolism) of many drugs

milli-   a Latin prefix meaning one thousandth of a unit

milliliter (mL)   a metric value of liquid volume equal to one thousandth of a liter

mineralocorticoids   corticosteroids that function primarily to help the body retain sodium and the kidneys to secrete potassium

minim   the liquid volume of a drop of water from a standard medicine dropper

minimum inhibitory concentration (MIC)   the lowest concentration of a particular antibiotic that visually inhibits the growth of bacteria

miotic   drug that constricts the pupil

miscible   able to be mixed

modified-live vaccine (MLV)   vaccine made with microorganisms that have undergone a process to nullify or reduce their virulence; also known as *attenuated vaccine*

monograph   a written account about a single drug

monoamine oxidase inhibitors (MAOIs)   drugs that work by inhibiting the enzyme MAO, thus reducing the destruction of neurotransmitters such as dopamine, norepinephrine, epinephrine, and serotonin and increasing their free level in the CNS

monovalent vaccine   single-antigen vaccine

mucolytics   drugs designed to liquify thick mucous secretions for better removal from the lungs

mucosal protective drug   drug that inhibits pepsin; may or may not cover an ulcer to protect it

mydriatic   drug that dilates the pupil

N

narcotic antagonist   see *opioid antagonist*

narrow-spectrum antibiotics   drugs that act specifically on the gram-positive or the gram-negative families of bacteria

natural immunity   immunity acquired during normal biological experiences

nebulize   administer as a fine spray for inhalation

neuroleptanalgesics   agents that are a combination of an opioid and a tranquilizer or sedative

neuromuscular blockers   drugs that produce paralysis

neuromuscular junction   the point where a nerve fiber meets the muscle it affects

noncore vaccine   vaccine recommended only for individual animals deemed to be at high risk for contact with the organism

nonionized   uncharged

nonparenteral   administered through the gastrointestinal tract (orally)

nonproprietary name   see *generic name*

nonspecific immunity   defense mechanisms that are directed against all pathogens and are the initial defense against invading agents

nonsteroidal anti-inflammatory drug (NSAID)   drug that reduces inflammation by inhibiting prostaglandin production (e.g., aspirin and phenylbutazone)

nutraceutical medicine   discipline that uses micronutrients, macronutrients, and other nutritional supplements as therapeutic agents

O

ongoing fluid loss   fluid loss due to vomiting, diarrhea, or other pathology

ophthalmic drugs   drugs formulated as eye drops and eye ointments to be instilled into the eye or applied directly to the eyelid

opioid antagonist   drugs that block the binding of opioids to their receptors; also known as narcotic antagonists

oral   by mouth

osmolality   concentration of a solution in terms of osmols of solutes per kilogram of solvent

osmotic diuretics   drugs that consist of large molecules that have limited capability of being reabsorbed into the blood from the kidney tubule, resulting in large amounts of fluid lost through urine

osmotic pressure   the ability of particles to attract water

oncotic pressure   osmotic pressure exerted by colloids in a solution (basically it is the pressure exerted by the plasma proteins)

OTC   see *over-the-counter drug*

otitis externa   outer ear infection

otitis interna   inner ear infection

otitis media   middle ear infection

ovarian follicle   the ovum and its encasing cells

over-the-counter (OTC) drug   drug for which no prescription is needed

oxytocic   labor producing

P

package insert   information provided with the drug bottle or vial by the manufacturer to clarify the drug's properties and uses

parameter   the intensity of effect measured on a subjective scale of either 1 through 3 or 1 through 10

parasympatholytics   see *anticholinergic drugs*

parasympathomimetics   see *cholinergic drugs*

parenteral   administered by routes other than the gastrointestinal tract

particulates   composed of several particles

passive diffusion   movement of atoms, ions, or molecules from an area of high concentration to an area of low concentration

passive immunity   protection conferred when an animal receives antibody from another animal

pediatric administration set   device consisting of a set of tubing, drip chamber, and clamp used to administer fluid that typically delivers 60 drops per milliliter

peristalsis   a wavelike contraction of longitudinal and circular muscle fibers that moves food through the gut

phagocytosis   engulfing solids into the cell by surrounding the particle and forming a vesicle

pharmacodynamics   the study of a drug's mechanisms of action and its biological and physiological effects

pharmacokinetics   the study of the absorption, distribution, biotransformation, and excretion of drugs

pharmacology   study of the history, sources, and physical and chemical properties of drugs; also includes study of how drugs affect living systems

pharmacotherapeutics   field that examines the treatment of disease with medicines

pharmacotherapy   treatment of disease with medicine

pheromones   chemicals that trigger a natural behavioral response usually within or between members of the same species

physical therapy   use of noninvasive techniques for the rehabilitation of animal injuries

Physician's Desk Reference (PDR)   a compendium of United States human drug standards

pinocytosis   engulfing fluids into the cell by surrounding the particles and forming a vesicle

polymyxin B   a polypeptide antibiotic effective against gram-negative bacteria only

polypeptide   a chain of amino acids

polypeptide antibiotics   agents composed of polypeptides that are effective against bacteria

polyuria   the formation and excretion of a large volume of urine

polyvalent vaccine   multiple-antigen vaccine

posterior   toward the rear (of a body part): term is not used frequently in veterinary medicine other than to describe organs such as pituitary gland and organ parts such as eye chambers

potentiated   chemically combined with another drug to enhance the effects of both

poultice   a form of herbal treatment made by boiling fresh or dried herbs, squeezing out excess liquid, cooling the herb, and applying it to the skin

powder   a dry, granulated version of a drug mixed with inert bulking and flavoring agents to allow dilution of the drug

preclinical studies   series of tests performed on laboratory animals to determine the safety and effectiveness of a drug

preliminary studies   tests performed to determine if a product will perform as expected

preload   the volume of blood entering the right side of the heart; the ventricular end-diastolic volume

prescription   an order, written by a licensed veterinarian, to a pharmacist to prepare the prescribed medicine, to affix the directions, and to sell the preparation to the client

prescription drug   drug that is limited to use under the supervision of a veterinarian or physician

probiotic   substance that seeds the gastrointestinal tract with beneficial bacteria

proctoscopy   examining the rectum with a proctoscope

proestrus   period of the estrous cycle before sexual receptivity

progesterone   a female sex hormone produced and secreted by the corpus luteum

prokinetic agents   gastrointestinal drugs that increase the motility of parts of the GI tract to enhance movement of material through it

proprietary name   see *brand name*

prostaglandins   group of naturally occurring, chemically related chemicals that stimulate uterine contractions and other smooth muscle

protectant   drug that coats the gastrointestinal mucosa

protocols   therapeutic combinations

pruritus   itching

psychoactive drug   a drug that affects the mind or behavior

pulse dosing   method of delivering some types of chemotherapeutic agents that produce escalating levels of drug early in the dosing interval followed by a prolonged dose-free period

purgatives   harsh cathartics that cause watery stool and abdominal cramping

Q

quinolones   synthetic, broad-spectrum antibiotics

R

receptor   specialized cell binding component that is a three-dimensional protein or glycoprotein located on the cell surface, in the cytoplasm, or within the nucleus of the cell

recombinant vaccine   vaccine containing a gene or part of a microorganism that was removed from one organism and inserted into another organism, thus "recombining" the genetics of both to make something new

rehydration volume   fluid volume used to correct dehydration

renal tubules   tubes within the nephron of the kidney

renin   an enzyme released by the kidney that activates angiotensin (a potent vasoconstrictor)

repository preparation   a drug injected with a substance that delays absorption; also called *depot preparation*

resistant   the ability of a microorganism to withstand doses of a drug that will kill or inhibit the growth of most members of its species

respiratory stimulants   drugs that stimulate breathing

ruminant   an animal with one glandular stomach and three esophageal outpouches that regurgitates and chews undigested food (e.g., cattle, sheep, goat)

S

SC   see *subcutaneous*

seborrhea   a skin condition characterized by abnormal flaking and/or scaling of the outermost layer of the epidermis

seborrhea oleosa   seborrhea with increased production of sebum (oil)

seborrhea sicca   seborrhea with decreased production of sebum (oil)

second-line drugs   drugs that may have toxic side effects or are used only when the first-line choice cannot be used or is ineffective

sedatives   drugs that allay irritability and excitement

selective serotonin reuptake inhibitors (SSRIs)   drugs that selectively inhibit serotonin reuptake, resulting in increased serotonin neurotransmission.

This group is used in treating depression, aggression, anxiety, phobias, and compulsive disorders

sensitive   the ability of a microorganism to be killed or its growth inhibited by an antibiotic

shelf life   the time during which a drug remains stable and effective for use

short-term tests   tests that occur in the hours following a drug test dose; checks for adverse reactions

short-term toxicity tests   toxicity test of a drug that lasts from the start of the test to about three months; used to assess adverse drug reactions that occur during short-term use of the drug

soak   substance applied to an area to draw out fluid or relieve itching

solute   the dissolved substance of a solution

solution   water that holds a dissolved substance; also called the *vehicle*

solvent   the dissolving substance of a solution

spasmolytic   drug that relieves muscle spasticity

special tests   tests done to determine specific outcomes, such as carcinogenicity or teratogenicity

specific immunity   defense mechanisms against a particular antigen that arise from the B and T lymphocytes

spectrum of action   the range of bacteria that an antibiotic or antimicrobial will eradicate

sporicidal   chemical that kills spores, which are especially resistant to chemicals

SQ   see *subcutaneous*

status epilepticus   ongoing seizures

steady state   point at which drug accumulation and elimination are balanced

steroidal anti-inflammatory drugs   group of anti-inflammatory drugs that have the steroid configuration of carbon atoms in four interlocking rings (e.g., glucocorticoids such as cortisol)

stool softeners   drugs that soften and lubricate the fecal mass

strangles   an equine streptococcal infection of the nose and throat

Streptomyces erythreus   a bacterium that produces erythromycin

subcutaneous   beneath the layers of the skin; abbreviated SC, SQ, or subQ

subdural   between the dura mater and the arachnoid membrane of the meninges; below the dura mater

sulfonamides   a group of drugs that inhibit the synthesis of folic acid to block the growth of a wide variety of bacteria

suspension   a finely divided, undissolved substance dispersed in water

subQ   see *subcutaneous*

subunit vaccine   vaccine that contains the part of the antigen needed to produce the desired immune response and eliminates the parts that can cause adverse reactions or interfere with the immune response

sympathetic drugs   drugs that function to enhance the effects mediated by epinephrine or norepinephrine in the central nervous system, the peripheral nervous system, or both; these drugs decrease intestinal

motility, decrease intestinal secretions, and inhibit the action of sphincters

sympatholytics   see *adrenergic blocking agents*

sympathomimetics   see *adrenergic drugs*

systems-oriented screen   a test of a drug's effect on a particular physiological system

T

tablet   medicine mixed with an inert binder and molded or compressed into a hard mass

teratogenicity   the capacity to cause birth defects

testosterone   the primary male sex hormone

tetracyclines   broad-spectrum antibiotics effective against many gram-negative and gram-positive bacteria

therapeutic index   an expression synonymous with margin of safety

therapeutic range   the drug concentration in the body that produces the desired effect in the animal with minimal or no signs of toxicity

therapy   treatment of disease

thiazides   diuretics that act directly on the renal tubules to block sodium reabsorption and promote chloride ion excretion

thyroid hormone   a collective term for two active hormones found in the thyroid gland: *thyroxine* and *tri-iodothyronine*

thyroid replacement therapy   the use of thyroid hormone to treat hypothyroidism

thyroxine (T_4)   a thyroid hormone that acts as a catalyst in the body and influences many metabolic, growth, reproductive, and immune effects

tolerance   reduced effect from the use of a substance resulting from repeated use

tonicity   a property based on an osmolality; describes the amount of solutes in a solution

topical   a medication that goes on the skin or mucous membranes

topical otic dewaxing agent   see *dewaxing agent*

topical otic drying agent   see *drying agent*

total daily dose   the entire amount of drug delivered in 24 hours

total dose   the total amount of a drug given to an animal in one administration

toxicity evaluation   an evaluation to see if a drug causes convulsions, seizures, or changes in blood pressure, heart rate, respiration rate, sleep, or muscle tension

toxoid   vaccine against a toxin

trade name   see *brand name*

tranquilizers   drugs that calm anxiety

tricyclic antidepressants (TCAs)   drugs that have a three-ring (tricyclic) structure and work by interfering with the reuptake of neurotransmitter by the presynaptic nerve cell

tri-iodothyronine (T$_3$)   a thyroid hormone that acts as a catalyst in the body and influences many metabolic, growth, reproductive, and immune effects

tuberculocide   chemical that kills *Mycobacterium tuberculosis*

U

United States Drug Enforcement Administration (DEA)   the federal agency (in the United States) that enforces the controlled substances laws and regulations of the United States and brings to the criminal and civil justice system those involved in violating the laws regarding the use, manufacture, or distribution of controlled substances

United States Food and Drug Administration (FDA)   the federal agency (in the United States) formed in 1906 that approves or disapproves drugs to ensure the safety, purity, and effectiveness of drugs produced and used in the United States.

United States Pharmacopeia (USP)   a publication that is the recognized legal drug standard of and for the United States

urinaryacidifers   drugs that lower the pH of urine

urinary alkalinizers   drugs that raise the pH of urine

urinary incontinence   loss of voluntary control of micturition

urolith   a stone in the urinary system

V

vaccine   a suspension of weakened, live, or killed microorganisms administered to prevent, improve, or treat an infectious disease

vasoconstriction   narrowing of the blood vessels

vasodilators   drugs used to widen arteries and/or veins, which alleviates vessel constriction and improves cardiac output

vehicle   see *solution*

veterinarian/client/patient relationship   a relationship between a veterinarian and client for which the veterinarian assumes the responsibility for making clinical judgments regarding the health of the animal(s) (patients) and the need for medical treatment, and the client agrees to follow the veterinarian's instructions

Veterinary Pharmaceuticals and Biologicals (VPB)   a compendium of United States veterinary drug standards

veterinary pharmacology   the study and use of drugs in animal health care

virucidal   chemical that kills viruses

volatilize   rapidly evaporate

W

Western herbal medicine   discipline based on belief that individuals have an inner force that works to maintain physical, emotional, and mental health

wholesalers   see *distributors*

withdrawal time   period of time after drug administration during which the animal cannot be sent to market for slaughter and the eggs or milk must be discarded, because of the potential for drug residues to persist in these products

INDEX

Note: Page numbers in **bold** refer to definition of the term

System Requirements

- Operating systems: Microsoft Windows XP w/SP 2, Windows Vista w/ SP 1, Windows 7
- Processor: Minimum required by Operating System
- Memory: Minimum required by Operating System
- Hard Drive Space: 15MB
- Screen resolution: 1024 × 768 pixels
- CD-ROM drive
- Sound card & listening device required for audio features
- Flash Player 10. The Adobe Flash Player is free, and can be downloaded from http://www.adobe.com/products/flashplayer/

Setup Instructions

1. Insert disc into CD-ROM drive. The StudyWare™ installation program should start automatically. If it does not, go to step 2.
2. From My Computer, double-click the icon for the CD drive.
3. Double-click the *setup.exe* file to start the program.

Technical Support

Telephone: 1-800-648-7450
8:30 A.M.-6:30 P.M. Eastern Time
E-mail: delmar.help@cengage.com

StudyWare™ is a trademark used herein under license.
Microsoft® and Windows® are registered trademarks of the Microsoft Corporation.
Pentium® is a registered trademark of the Intel Corporation.